Coral Reefs of the World

Volume 19

Coral Reefs of the World is a series presenting the status of knowledge of the world's coral reefs authored by leading scientists. The volumes are organized according to political or regional oceanographic boundaries. Emphasis is put on providing authoritative overviews of biology and geology, explaining the origins and peculiarities of coral reefs in each region. The information is so organized that it is up to date and can be used as a general reference and entry-point for further study. The series will cover all recent and many of the fossil coral reefs of the world.

Prospective authors and/or editors should consult the Series Editors B.M. Riegl and R.E. Dodge for more details. Any comments or suggestions for future volumes are welcomed:

Dr. Bernhard M. Riegl/Dr. Richard E. Dodge
Nova Southeastern University
Dania Beach, FL 33004
USA
e-mail: rieglb@nova.edu and dodge@nova.edu

Erik Cordes • Furu Mienis
Editors

Cold-Water Coral Reefs of the World

Springer

Editors
Erik Cordes
Department of Biology
Temple University
Philadelphia, PA, USA

Furu Mienis
Department of Ocean Systems
NIOZ Royal Netherlands Institute for Sea Research
Den Burg, The Netherlands

ISSN 2213-719X ISSN 2213-7203 (electronic)
Coral Reefs of the World
ISBN 978-3-031-40896-0 ISBN 978-3-031-40897-7 (eBook)
https://doi.org/10.1007/978-3-031-40897-7

This Springer imprint is published by the registered company Springer Nature Switzerland AG
The registered company address is: Gewerbestrasse 11, 6330 Cham, Switzerland

Paper in this product is recyclable.

The editors would like to dedicate this book to the next generation of ocean explorers who will use these thoughts and words to improve our understanding of the distribution and function of the cold water coral reefs of the world.

Foreword

Cold-Water Coral Reefs of the World is being published 150 years after HMS *Challenger* set sail on the epic global expedition that laid the foundations of modern oceanography and deep-sea biology. Charles Wyville Thomson from the University of Edinburgh and his team pushed the technologies of the day to dredge life from 5.5 km depth—an astonishing achievement. They collected a huge diversity of species from the deep seafloor and laid to rest Edward Forbes' azoic theory that the deep sea would be a lifeless desert much like the deep Aegean Forbes had sampled in the 1840s. Henry Mosely was the naturalist on board *Challenger* who took charge of working on the corals they sampled. Of the four thousand or so new species they discovered, Moseley described 48 new scleractinian corals.

Challenger was a stunning scientific success for many reasons, but perhaps the most important is that the guiding principles of the expedition weren't just to explore the deep sea for its own sake but to test big ideas. The azoic theory was a powerful motivator. Would the deep seabed across the world be largely lifeless or would there be weird, stalked crinoids, corals, and other animals as Michael Sars had reported from the coasts of Norway and Finland, and Wyville Thomson had himself seen in the NE Atlantic during the HMS *Porcupine* and *Lightning* expeditions (1868–1870)? Fuelled by Darwin's theory of evolution by natural selection published in 1859, scientists also speculated that the deep sea could be a refuge for living fossils like the stalked crinoids Michael Sars had discovered. It took 23 years to finish the 50 volumes of the *Challenger* report where the results were beautifully presented—the deep sea was rich with species, although not the storehouse of living fossils some had predicted.

If we look at the development of cold-water coral reef research since the pioneering days of the eighteenth and nineteenth centuries, we can see how it has evolved through discrete eras, and how big ideas have shaped this evolution. When an area is first explored, like the North Atlantic in the days of Sars and Wyville Thomson, people were discovering for the first time that cold-water corals formed extensive reef-like habitats. In his book *The Depths of the Sea*, Wyville Thomson was so surprised by the amount of *Lophelia pertusa* in his dredge hauls west of the Shetland Islands that he wrote "... the handsome branching *Lophohelia prolifera* [*Lophelia pertusa*] ... forms stony copses covering the bottom for many miles, the clefts of its branches affording fully appreciated shelter to multitudes of *Arca nodulosa*, *Psolus squamatus*, *Ophiopholis aculeata*, and other indolent 'commensals'" (Fig. 1).

Thus, the first research era laid the foundations discovering where cold-water corals occur and what lives with them. Thanks to quantum leaps in acoustic mapping technologies and autonomous survey platforms, we now have the tools to map vast areas of the deep seafloor in unprecedented detail. While much remains to be done, we can now say a lot more about where cold-water coral reefs are found and have a far better understanding of how important they are as habitats for other species.

Once we know an area supports cold-water corals, it's natural to ask what makes the area suitable for the corals to grow and proliferate. This book provides a state-of-the-art review of what we know about cold-water coral reefs and the habitats they form around the world. Much of this research relies on another big idea that catalysed cold-water coral reef research in the 1990s and early 2000s. Like the azoic theory over a century earlier, the hydraulic theory that

Fig. 1 Illustration of the "handsome branching *Lophohelia profilera*" (=*Lophelia pertusa*) from a specimen collected by dredge "at depths of 300 to 600 fathoms" by the H.M.S. *Porcupine*. From "The Depths of the Sea" by C. Wyville Thomson, MacMillan and Co. 1873

cold-water corals relied on a chemosynthetic food chain fuelled by hydrocarbon seepage was a massive spur to researchers wanting to test this idea. Would corals show tell-tale evidence that they were fuelled by chemosynthesis, or would they be linked to food chains fuelled by primary productivity at the ocean's surface?

The last two decades have seen fantastic studies showing that cold-water coral reefs are indeed coupled to surface productivity and not reliant on a methane seep food chain—no small feat given how remote and inaccessible these habitats are. In some settings, like the giant coral carbonate mounds discovered in the NE Atlantic, it even seems the corals are such effective ecosystem engineers that the mounds they form preferentially focus food input compared to adjacent areas.

As we've unravelled these food supply stories, we've also seen a quantum leap in our understanding of cold-water coral reef habitats with studies now beginning to combine classical habitat mapping with ecosystem function assessments so we can start to extrapolate spatial models of biomass, respiration, and, perhaps soon, the blue-carbon storage potential of these remarkable places.

But in parallel with the renaissance of scientific interest in cold-water coral reefs over the last two decades has come very clear evidence that human activities have had serious impacts. I vividly remember my first cold-water coral research expedition in 1997. We were working west of Shetland in the areas Wyville Thomson described as supporting stony copses of coral covering many miles. After many days trying we only recovered clumps of dead coral and one or two tiny fragments of living *Lophelia pertusa*—but this was an area that had been heavily trawled for years beforehand.

By the early 2000s, cold-water corals reefs became a poster child for deep-sea conservation and since then many have been closed to bottom trawling. Important legally binding measures, including the United Nations resolutions on vulnerable marine ecosystems, have also helped bring a far more precautionary approach to deep-sea fisheries management, including in areas beyond national jurisdiction.

However, area-based management tools, including marine protected areas, offer little to no defence against the greatest threat to life on Earth—anthropogenic climate change. The oceans of the world have already absorbed over 90% of anthropogenic warming and around 25% of our carbon dioxide emissions. The sea is warming, becoming more acidic and seeing declines in the oxygen vital to life. The scleractinian corals that build cold-water reef frameworks are canaries in the coal mine of deep ocean climate change. As this book eloquently shows, reef framework-forming cold-water corals only occupy relatively restricted environmental niches. The palaeo-record shows us how quickly cold-water corals vanish from the fossil record, with recent work pointing to how important altered food supply can be in controlling cold-water coral reef histories. Cold-water scleractinian corals also grow in seawater less supersaturated with calcium carbonate than enjoyed by their counterparts in the tropics. This means that many of the cold-water coral reefs that grow today in seawater above the aragonite saturation horizon will be exposed to undersaturated water corrosive to their skeletons within the next 50–100 years. It seems that seawater warming, and increased coral bleaching, may be the most urgent threat to tropical warm water coral reefs with ocean acidification the more insidious threat to cold-water coral reefs.

Given the scale of the challenges facing ocean ecosystems in general, and cold-water coral reefs in particular, what can we do? While climate change impacts will transcend reserve boundaries, we must do all we can to understand and limit other pressures. At the time of writing, the potential for commercial deep-sea mining is closer than at any point in history. Decisions to mine, or not to mine, need to be made in full understanding of the implications on deep-sea ecosystems—including the implications of mining plumes on the seabed and the water column through which cold-water coral larvae disperse.

Huge strides in the policy and governance world are now also being made. In March 2023, delegates at the United Nations finalised negotiations for a legally binding treaty text to conserve and sustainably use marine biodiversity of areas beyond national jurisdiction. Now subject to ratification and implementation, this new "BBNJ High Seas Treaty" is the single best opportunity to align the complex and often contradictory legal regime through which marine biodiversity is managed beyond national jurisdiction—an important development for how humanity will sustainably manage activities across the deep and open ocean.

On top of progress in the regulatory and governance regime, the results in this book show that people, from scientists to policymakers, are ready for a new era during which we will move from understanding the ecological importance of cold-water coral reefs and the threats they face to actively restoring them. To achieve this, we need to be radical and apply the scientific knowledge outlined in this book on the environmental niches of cold-water coral reefs to identifying climate change-resilient refuges suitable for restoration. We need to upscale our efforts to understand the population genetics of cold-water coral reefs so we preserve hidden genetic diversity. We need to understand the role of the coral microbiome and the importance of local adaptations. These are hugely challenging goals, but alongside maximum efforts to reduce carbon dioxide emissions, it is time to enter a new era of cold-water coral reef restoration coupled with climate-resilient and ecologically linked networks of protected areas.

We simply cannot lose cold-water coral reefs as we begin to understand just how important they are.

University of Edinburgh
Edinburgh, UK
11 April 2023

J. Murray Roberts

Contents

Editors and Contributors

About the Editors

Erik Cordes is Full Professor and Vice Chair of the Department of Biology at Temple University in Philadelphia, PA, USA. He received his M.S. from Moss Landing Marine Labs and his Ph.D. from Penn State University and was a postdoctoral fellow at Harvard University. He has worked on the ecology of the deep sea for over 30 years, spent almost 2 years at sea on over 30 research expeditions and has made 47 dives in manned submersibles (including one to 6000 m depth). He is a self-described ocean explorer and ecological oceanographer whose research is focused on deep-sea coral reefs, natural hydrocarbon seeps and hydrothermal vents, as well as their conservation. He has organised and led expeditions to the east coast of the USA, the Gulf of Mexico, the Caribbean Sea, the Pacific coast of Costa Rica and the Phoenix Islands Protected Area in the central Pacific. He actively collaborates with professional communicators, visual artists, filmmakers and musicians to bring the deep sea to the widest audience possible.

Furu Mienis is a senior scientist in the Department of Ocean Systems at the NIOZ Royal Netherlands Institute for Sea Research. She received her M.Sc. from the Vrije Universiteit Amsterdam and did her Ph.D. at NIOZ on near-bed environmental conditions influencing cold-water coral growth and reef and mound development, followed by a postdoctoral fellowship at MARUM (Bremen, Germany). Her research focuses on the interplay between geochemistry, ecology and physical oceanography. She focuses on understanding the environmental conditions and dynamics, including transport and (food) particle supply mechanisms that influence the functioning of deep-sea ecological hotspots, including cold-water coral reefs and sponge grounds. Over the last few years, she has also studied sedimentological and physical processes in canyons to define if canyons are pathways or sinks for organic carbon and relate organic matter distributions to benthic and pelagic faunal activity and abundance. She led and participated in over 35 scientific research cruises in the North Atlantic, studying the deep sea from the Caribbean to the Arctic. In parallel to her more fundamental research, she dedicates time to create awareness about the hidden treasures in the deep sea, as well as works on the technical development of deep-sea bottom observatories, which resulted in several new observational and autonomous sampling tools.

Contributors

Steven Auscavitch Department of Biology, Boston University, Boston, MA, USA

Amy R. Baco Department of Earth, Ocean and Atmospheric Sciences, Florida State University, Tallahassee, FL, USA

Angelo F. Bernardino Departamento de Oceanografia, CCHN, Universidade Federal do Espirito Santo, Goiabeiras, Vitória, ES, Brazil

Virginia Biede Department of Earth, Ocean and Atmospheric Sciences, Florida State University, Tallahassee, FL, USA

W. Ben Brantley Department of Earth, Ocean and Atmospheric Sciences, Florida State University, Tallahassee, FL, USA

Sandra Brooke Coastal and Marine Lab, Florida State University, St. Teresa, FL, USA

Pål Buhl-Mortensen Institute of Marine Research, Bergen, Norway

Stephen Cairns Department of Invertebrate Zoology, National Museum of Natural History, Smithsonian Institution, Washington, DC, USA

Malcolm R. Clark National Institute of Water and Atmospheric Research (NIWA), Wellington, New Zealand

Erik E. Cordes Department of Biology, Temple University, Philadelphia, PA, USA

Andrew Davies Department of Biological Sciences, University of Rhode Island, Kingston, RI, USA

Amanda Demopoulos US Geological Survey, Wetland and Aquatic Research Center, Gainesville, FL, USA

Norbert Frank Institut für Umweltphysik (IUP) & Institut für Geowissenschaften, Universität Heidelberg, Heidelberg, Germany

André Freiwald Marine Research Department, Senckenberg am Meer, Wilhelmshaven, Germany

Ryan Gasbarro Department of Biology, Temple University, Philadelphia, PA, USA

Daniela Y. Gaurisas Departamento de Oceanografia, CCHN, Universidade Federal do Espirito Santo, Goiabeiras, Vitória, ES, Brazil

Andrea Gori Departament de Biologia Evolutiva, Ecologia i Ciències Ambientals, BarcelonaUniversitat de Barcelona, Barcelona, Spain

Institut de Recerca de la Biodiversitat (IRBio), Universitat de Barcelona, Barcelona, Spain

Jordi Grinyó Department of Ocean Systems, NIOZ Royal Netherlands Institute for Sea Research, Den Burg, The Netherlands

Dierk Hebbeln Center for Marine Environmental Sciences (MARUM), University of Bremen, Bremen, Germany

Sebastian J. Hennige School of GeoSciences, Kings Buildings, University of Edinburgh, Edinburgh, UK

Veerle A. I. Huvenne National Oceanography Centre, University of Southampton, Southampton, UK

Claudio Lo Iacono Institut de Ciències del Mar, Consejo Superior de Investigaciones Científicas (ICM-CSIC), Barcelona, Spain

Jay J. Lunden Department of Biology, Temple University, Philadelphia, PA, USA

Beatriz Mejia-Mercado Department of Earth, Ocean and Atmospheric Sciences, Florida State University, Tallahassee, FL, USA

Furu Mienis Department of Ocean Systems, NIOZ Royal Netherlands Institute for Sea Research, Den Burg, The Netherlands

Nicole Morgan Department of Earth, Ocean and Atmospheric Sciences, Florida State University, Tallahassee, FL, USA

Cheryl L. Morrison US Geological Survey, Eastern Ecological Science Center, Kearneysville, WV, USA

Martha S. Nizinski National Systematics Laboratory, NOAA Fisheries, Department of Invertebrate Zoology, Smithsonian National Museum of Natural History, Washington, DC, USA

Covadonga Orejas Centro Oceanográfico de Gijón, Instituto Español de Oceanografía, Gijón, Spain

Hanse-Wissenschaftskolleg, Institute for Advanced Study, Delmenhorst, Germany

Frank A. Parrish Pacific Islands Fisheries Science Center, National Oceanic and Atmospheric Administration, Honolulu, Hawaii, USA

Andrea M. Quattrini Department of Invertebrate Zoology, Smithsonian Institution, National Museum of Natural History, Washington, DC, USA

John Reed Cooperative Institute of Ocean Research, Exploration and Technology, Harbor Branch Oceanographic Institute, Florida Atlantic University, Fort Pierce, FL, USA

E. Brendan Roark Department of Geography, Texas A&M University, College Station, TX, USA

Harry Roberts Louisiana State University, Baton Rouge, LA, USA

Murray Roberts University of Edinburgh, Edinburgh, UK

Ashley A. Rowden National Institute of Water and Atmospheric Research (NIWA), Wellington, New Zealand

Leslie Sautter Department of Geology and Environmental Geosciences, College of Charleston, Charleston, SC, USA

Harvey Seim Department of Earth, Marine and Environmental Sciences, University of North Carolina at Chapel Hill, Chapel Hill, NC, USA

Paulo Y. G. Sumida Departamento de Oceanografia Biologica, Instituto Oceanográfico, Universidade de São Paulo, São Paulo, Brazil

Tracey Sutton Oceanic Ecology Lab, Nova Southeastern University, Dania Beach, FL, USA

Marco Taviani Istituto di Scienze Marine, Consiglio Nazionale delle Ricerche (ISMAR-CNR), Bologna, Italy

Stazione Zoologica Anton Dohrn, Villa Comunale, Napoli, Italy

Jürgen Titschack Center for Marine Environmental Sciences (MARUM), University of Bremen, Bremen, Germany

Marine Research Department, Senckenberg am Meer, Wilhelmshaven, Germany

Robert E. Todd Department of Physical Oceanography, Woods Hole Oceanographic Institution, Woods Hole, MA, USA

Dianne M. Tracey National Institute of Water and Atmospheric Research (NIWA), Wellington, New Zealand

Andrew Wheeler School of Biological, Earth and Environmental Science, Environmental Research Institute, University College Cork, Cork, Ireland

Claudia Wienberg Center for Marine Environmental Sciences (MARUM), University of Bremen, Bremen, Germany

A Global View of the Cold-Water Coral Reefs of the World

1

Erik E. Cordes, Furu Mienis, Ryan Gasbarro, Andrew Davies, Amy R. Baco, Angelo F. Bernardino, Malcolm R. Clark, André Freiwald, Sebastian J. Hennige, Veerle A. I. Huvenne, Pål Buhl-Mortensen, Covadonga Orejas, Andrea M. Quattrini, Dianne M. Tracey, Andrew J. Wheeler, and Claudia Wienberg

Abstract

Cold-water corals (CWC) form reef structures in continental margin and seamount settings at tropical, temperate, and even some polar latitudes. This global distribution makes them more widespread than shallow-water reefs, while their role in these regions is no less important than the influence that shallow-water coral reefs have on shallow, tropical systems. They create habitat structure, host endemic species, enhance elemental cycling, alter current flow, sequester carbon, and provide many other ecosystem services that we are just beginning to understand. This introductory chapter to The Cold-Water Coral Reefs of the World reviews historical and recent information, reveals new findings from reefs that have been discovered only recently, and presents key avenues for future research. Global distribution and environmental data are synthesized into an ensemble model that described the niche of key species of framework-forming corals. Using an algorithm to distinguish coral colony occurrence from coral reef and mound occurrence, we further describe the subset of conditions under which CWC form reefs. This effort reveals new areas that are highly likely to host undiscovered CWC reef habitats and provides a framework for future ocean exploration. We are on the cusp of understanding the critical role that CWC reefs play in the world ocean, and this chapter and this book helps to set the

Supplementary Information: The online version contains supplementary material available at https://doi.org/10.1007/978-3-031-40897-7_1.

E. E. Cordes (✉)
Department of Biology, Temple University, Philadelphia, PA, USA
e-mail: erik.cordes@temple.edu

F. Mienis
Department of Ocean Systems, NIOZ Royal Netherlands Institute for Sea Research, Den Burg, The Netherlands

R. Gasbarro
Department of Biology, Temple University, Philadelphia, PA, USA

Southwest Fisheries Science Center, NOAA, Santa Cruz, California, USA

A. Davies
Department of Biological Sciences, University of Rhode Island, Kingston, RI, USA

Graduate School of Oceanography, University of Rhode Island, Narragansett, Rhode Island, USA

A. R. Baco
Department of Earth, Ocean and Atmospheric Sciences, Florida State University, Tallahassee, FL, USA

A. F. Bernardino
Departamento de Oceanografia, CCHN, Universidade Federal do Espirito Santo, Goiabeiras, Vitória, ES, Brazil

M. R. Clark
National Institute of Water and Atmospheric Research (NIWA), Wellington, New Zealand

A. Freiwald
Marine Research Department, Senckenberg am Meer, Wilhelmshaven, Germany

S. J. Hennige
School of GeoSciences, Kings Buildings, University of Edinburgh, Edinburgh, UK

V. A. I. Huvenne
National Oceanography Centre, University of Southampton, Waterfront Campus, Southampton, UK

P. Buhl-Mortensen
Institute of Marine Research, Bergen, Norway

C. Orejas
Centro Oceanográfico de Gijón, Spanish Institute of Oceanography (IEO-CSIC), Gijón, Spain

A. M. Quattrini
Department of Invertebrate Zoology, Smithsonian Institution, National Museum of Natural History, Washington, DC, USA

C. Wienberg
Center for Marine Environmental Sciences (MARUM), University of Bremen, Bremen, Germany

C. Wienberg
Center for Marine Environmental Sciences (MARUM), University of Bremen, Bremen, Germany

C. Wienberg
Center for Marine Environmental Sciences (MARUM), University of Bremen, Bremen, Germany

E. Cordes, F. Mienis (eds.), *Cold-Water Coral Reefs of the World*, Coral Reefs of the World 19,
https://doi.org/10.1007/978-3-031-40897-7_1

stage for future efforts to determine their global impact and potential threats to the ecosystem services they provide.

Keywords

Deep sea · History of oceanography · Coral reefs · Coral physiology · Environmental controls · Predictive habitat modeling

1.1 Introduction

Cold-water corals, often referred to as deep-water corals, include scleractinian framework-forming corals that create reef structures (Fig. 1.1) on continental margins, ridges, and seamounts worldwide, from the equator to beyond the polar circles. Recent advances in our exploration of the ocean floor have brought about the realization of the ubiquity of coral reefs, not just in well-lit shallow waters, but also deeper in the light-limited mesophotic regions (50–200 m), and extending to the dark, deep ocean from 200 to over 1000 m depth (Cairns 2007; Roberts et al. 2009). The role that corals play in these deep ecosystems is no less important than the influence that shallow-water coral reefs have on tropical systems, including habitat creation that increases local species richness and abundance (Cordes et al. 2008; Buhl-Mortensen et al. 2010; Linley et al. 2017; Kazanidis et al. 2021), enhanced biogeochemical cycling (Cathalot et al. 2015; de Froe et al. 2019; Maier et al. 2021), altered current flow and particle trapping (Titschack et al. 2009; Corbera et al. 2022a), and other important ecological processes that we are just beginning to quantify (Cordes et al. 2021). We are on the cusp of understanding the critical role that cold-water coral reefs play in the world ocean, and this book helps to lay the foundation on which this knowledge will be built in the future.

The rapidly evolving state of knowledge of CWC at the global scale has not been synthesized since the comprehensive book on cold-water coral biology, geology, and ecology by Roberts et al. (2009), although information has been updated in the two United Nations (UN) World Ocean Assessments (Cordes et al. 2016a, 2021), and more topic-specific (Rossi et al. 2017) or regional-scale volumes (Orejas and Jimenez 2019). Rather than repeat the fundamental information contained in those publications, this volume focuses specifically on CWC *reef* ecosystems in different regions of the world. This volume of the *Coral Reefs of the World* series synthesizes recent and historical information, reveals new data from reefs that have been discovered only recently, and presents key avenues for future research. In this introductory chapter, we endeavor to synthesize some of the work presented in the subsequent chapters of the book in both a qualitative assessment of what we have learned from the cutting-edge science presented in the subsequent chapters and also by synthesizing the observations from these disparate regions of the world and creating predictive habitat models using state-of-the-art oceanographic and bathymetric data products to examine the global distribution and habitat characteristics of CWC reefs and the main scleractinian taxa that form them.

C. Wienberg
Center for Marine Environmental Sciences (MARUM), University of Bremen, Bremen, Germany

C. Wienberg
Center for Marine Environmental Sciences (MARUM), University of Bremen, Bremen, Germany

1.1.1 History of Cold-Water Coral Research

Cold-water corals were first recognized in the 1700s in Norway, with the original description of *Madrepora pertusa* (=*Lophelia pertusa*, =*Desmophyllum pertusum*, see note below) by Carl von Linné (Linnaeus) published in the book *Systema Naturae* in 1758. Subsequent studies by dredge and trawl then discovered the first known CWC reefs off the coast of Norway in the late 1700s (Pontoppidan 1753; Gunnerus 1768) (Fig. 1.2). Intensive oceanographic research in the North Atlantic beginning in the 1800s with the expeditions of the H.M.S. *Lightning* and *Porcupine* and the U.S. Coast Guard Steamers *Hassler* and *Blake* (among others) revealed more of the diversity and some of the distribution of corals in the region, but they remained poorly understood until relatively recently.

The debate over the use of the term "reef" to describe structures generated by some cold-water scleractinians dates to this early era of exploration and Sars (1865) descriptions of the Norwegian reefs (Sars 1865). The term "reef" originates from the Old Norse "rif," which is a nautical term meaning a hazardous "rib" of rock, sand, or biological material that lies close to the sea surface. In his report of the material from the H.M.S. *Porcupine*, Jeffreys (1869) states "The presence of corals at great depth will also materially alter the views generally received of the depth at which reef-builders may work, and modify to a certain extent Darwin's theory of the reefs and their mode of growth" (Jeffreys 1869). This statement elicited a strong response from Duncan (1870) who inspected the dredged material and came to Darwin's defense, concluding "I have no hesitation in asserting that there is not one species found in these deep seas which is 'reef-building' in its habit or whose structures resemble those of the true reef forms. Mr. Darwin's theory is therefore as yet as strong as ever." (Duncan 1870). While (most of) Mr. Darwin's theories remain strong, we would argue that the good Mr. Duncan's conclusion is unfortunately incorrect.

Fig. 1.1 Cold-water coral (CWC) reefs of the world. (**a**) Two different color morphs of *Lophelia pertusa* from a reef at 870 m water depth in the Explorer Canyon, Celtic Margin, NE Atlantic, a part of The Canyons Marine Conservation Zone (image courtesy of National Oceanography Centre, UK). (**b**) An alfonsino fish (*Beryx* sp.) swimming above a mound of *L. pertusa* on the central Blake Plateau at approximately 800 m depth (image courtesy of the NOAA Ocean Exploration, Windows to the Deep 2019). (**c**) Living *L. pertusa* reef with octocorals and a galatheid squat lobster on top of an Angolan cold-water coral mound in 350 m water depth (image courtesy of MARUM ROV SQUID, Bremen, Germany). (**d**) *Madrepora oculata* reef formation with abundant chirostylid crabs from approximately 800 m depth in the northern Gulf of Mexico (Image courtesy of the ECOGIG consortium and the Ocean Exploration Trust). (**e**) *Enallopsammia profunda* reef structure with Brisingidae sea-stars, Crinoidea feather stars, and an Oreostomatid fish on pinnacles at 575 m depth on Orona Atoll, Phoenix Islands, Kiribati (image courtesy of Erik Cordes and the Schmidt Ocean Institute). (**f**) A solitary orange roughy (*Hoplostethus atlanticus*) swimming over colonies of *Solenosmilia variabilis* with brisingids, crinoids, and various Hexactinellida sponges at 1200 m depth on the flanks of the Valerie Seamount, Louisville Ridge, New Zealand (image collected using NIWA's deep towed imaging system deployed off RV Tangaroa, 2014)

The classical definition of a biological "reef" says that it is a "rigid, wave-resistant framework constructed by large skeletal organisms" (Lowenstam 1950), which precludes its use for deep-water structures. To avoid this issue, various authors have applied different terms for accumulations of coral colonies and skeletons over the years, including "massifs,"

Det første Forsøg
paa
Norges Naturlige Historie,
forestillende
Dette Kongeriges Luft, Grund, Fielde, Vande, Væxter,
Metaller, Mineralier, Steen-Arter, Dyr, Fugle,
Fiske og omsider Indbyggernes Naturel, samt
Sædvaner og Levemaade.
Oplyst med Kobberstykker.
Den viise og almægtige Skaber til Ære, saavel som hans
fornuftige Creature til videre Eftertankes Anledning,
af
ERICH PONTOPPIDAN Dr.
Episc. Bergens. Membr. Reg. Societ. Scient. Hafniens.
Kiøbenhavn, 1752.
Trykt i de Berlingske Arvingers Bogtrykkerie, ved Ludolph Henrich Lillie.

Fig. 1.2 Drawing of *Lophelia pertusa* in Pontoppidan (1755), The natural history of Norway. Stone-trees [Scleractinia], pp. 159–160. *"The fisherman often sell coral bushes to the apothecaries at Bergen, and, upon asked, what is their opinion about the origin and growth of this marine vegetable, they answer, that sometimes a white drop is observed to fall from the branches of the old coral, as well as from the sea-trees, as if it were milk or seed, and where this falls a vegetable is produced according to its species. As to its medical uses it has the character of being absorbent, refrigerative, emollient, astringent, and strengthening, which may be true, when the tincture of it, consisting of the extracted salts and oil, is administered inwardly. The little beads, made of the coral [Corallium rubrum] are endued with any such singular virtue that when applied externally, or hung about the neck, they are a preservative against the apoplexy, the plague, and other contagions. All over the East they wear necklaces and bracelets of coral beads brought from Marseilles. Possible could white coral be brought into fashion, a diligent search might procure as great a quantity in our seas"*

"banks," "mounds," and "bioherms" (Le Danois 1948; Teichert 1958; Wilson 1979; Mullins et al. 1981; Frederiksen et al. 1992; Hovland et al. 1994; Mortensen et al. 1995; Freiwald et al. 1997). However, the parallel use of different terms for the same coral structures can be confusing and misleading. The great similarities with respect to morphology and diversity of associated animals between deep-water and shallow-water reefs lend credence to the term "coral reef" as the best one to explain what kind of biogenic structures we are talking about. Wood (1998) defined the term "reef" as "... a discrete carbonate structure, formed *in situ*, of bound organic compounds that develop topographic relief upon the sea floor." Applying this term to the variety of structures we have discovered, both shallow and deep, since Darwin's time, we now know that reefs are widespread throughout the ocean to bathyal depths.

The global distribution of CWCs was first revealed by the great oceanographic expeditions including that of the HMS *Challenger*, which sailed the world between 1872 and 1876, led by Charles Wyville Thomson. The *Challenger* expedition visited the northeast, southeast, and southwest Atlantic, and the equatorial and southwest Pacific, bringing back the first specimens of numerous CWC species (Quelch 1886). Additional global deep-sea expeditions further revealed corals to be present across a wide range of latitudes and large depth ranges in the World Ocean (Roberts et al. 2009). Technological advances in oceanographic sampling gear and the use of soundings to determine depth led to the first maps of the ocean floor at the resolution necessary to see distinct seafloor features. It was Marie Tharp's compilation of oceanographic data that provided the first ocean basin-scale views of the ocean floor in the 1950s, and a global map in 1977, which resolved features on the scale of tens of kilometers (Heezen and Tharp 1977).

Modern ecological and geological studies of CWC reefs were also being conducted in the mid-1900s, and discoveries of large bioherms or lithoherms along with detailed studies of the biology of living corals came from the eastern North Atlantic (Dons 1944; Le Danois 1948) and Mediterranean (Pérès and Picard 1964) but also off of the coasts of the southeastern US (Stetson et al. 1962), Gulf of Mexico (Moore and Bullis Jr 1960), Africa (Squires 1959), and New Zealand (Squires 1964). The early sampling gear to reveal these corals included lifting of telegraph cables (Ralph and Squires 1962) or the use of more traditional grabs, sleds, and trawls (McKnight and Estcourt 1978). Many bycatch samples of corals, both live and skeleton, have also come from the use of bottom trawls, long-lines, and midwater trawls used on commercial vessels and fisheries research vessels (Mortensen et al. 2008; Dias et al. 2020). With the advent of seafloor camera systems deployed from vessels followed by submersible technologies in the 1960s and the proliferation of their use in the 1970s and 1980s came the first, and continually improving, observations of CWC habitats. Once this imagery was available, these features could truly be confirmed to be "reefs," with the beauty and rich biodiversity associated with these habitats beginning to be realized (Squires 1959; Mackay et al. 2014). Global taxonomic treatments of the scleractinian order were carried out by Cairns (1982, 1995) and continue to this day (Kitahara and Cairns 2021).

It was also in the 1970s and 1980s that global economic interests shifted towards the deep ocean and human impacts were recognized shortly thereafter (Rogers 1999). Fisheries were expanding into deeper waters as shallow-water fish stocks became depleted (Morato et al. 2006; Mackay et al. 2014). Oil and gas development transitioned from being entirely on land, to significant production from offshore platforms in the Gulf of Mexico and North Sea (Gass and Roberts 2006; Cordes et al. 2016b). By the 1990s, corals became common in fisheries bycatch, particularly in the Southwest Pacific and North Atlantic (Anderson and Clark 2003) and were frequently observed in the seafloor surveys of the energy industry (Cordes et al. 2016b).

The proliferation of deep-sea sampling and observation capacity, along with the advent of multibeam echosounders and deployment of long-term moored observatories in the early 2000s, led to a dramatic increase in the fundamental knowledge of the deep ocean in general, and CWCs in particular. The recent discoveries of CWC reefs off the coast of Southwest Africa (e.g., (Hanz et al. 2019; Tamborrino et al. 2019), on seamounts in the North and Southwest Pacific (e.g., Baco et al. 2017; Rowden et al. 2017), in areas of the Mediterranean (Taviani et al. 2005; Orejas et al. 2009; Fabri et al. 2014; Angeletti et al. 2020), and elsewhere have enhanced our understanding of the broad distribution and significant ecological role of cold-water coral reefs in the global ocean, and many of these are covered in the subsequent chapters of this volume.

1.1.2 Cold-Water Coral Biology

The coral species that form deep-water reefs and contribute to the complexity of the structure belong to a variety of taxonomic groups, but a small group of scleractinians are the only reef-building corals in cold aphotic waters. The most well-known of these species is *Lophelia pertusa*, also referred to as *Desmophyllum pertusum*. This proposed name change is based on a combination of genetic (microsatellite) and morphological (primarily the arrangement and number of septa in the calyx of the polyp) data, which revealed close taxonomic similarity with the solitary coral species *Desmophyllum dianthus* (Addamo et al. 2016). However, *L. pertusa* remains an iconic species name that requires substantial evidence to revise, and there continues to be significant debate over this name change (see comments in WoRMS by Dr. Steven Cairns in 2019). There are substantial efforts currently underway to determine the validity of the revised taxonomic position of this species, including the publication of the genome assembly of *L. pertusa* (Herrera and Cordes 2023). Thus, we will be using the original name in this introduction and have suggested (but not insisted) that the authors of the other chapters retain the use of *L. pertusa* throughout.

Lophelia pertusa is a common species of scleractinian coral that forms most of the cold-water coral reefs in the Atlantic Ocean (Roberts et al. 2009). The long history of CWC research, dating back to the 1700s, is primarily based on this species, and therefore much of the existing knowledge of CWCs and most of the assumptions about their life history, ecological role, distribution, and physiological responses come from this species. It is relatively fast-growing compared to other CWC species, with linear extension rates typically in the 0.5–3.4 cm $year^{-1}$ range (Hall-Spencer et al. 2002; Roberts et al. 2009; Orejas et al. 2011a, b). However, growth rate data from other locations and other species of scleractinians show wide variability (e.g., *M. oculata*, see Sabatier et al. (2012) and Orejas et al. (2021), likely due to species-specific variability in physiology and variable (often suboptimal) habitat conditions. In addition, individual coral colonies can be very long lived; in the Southwest Pacific, a 1-m diameter colony of the commonly occurring scleractinian *Solenosmilia variabilis* was estimated to be over 2000 years old (Fallon et al. 2014).

The main source of nutrition for CWCs on cold-water reefs derives from the photic zone. Depending on polyp size, among both individual polyps and among species, corals are capable of feeding on a broad variety of food sources from marine snow to zooplankton that are a few centimeters long (Duineveld et al. 2007; Becker et al. 2009; Mueller et al. 2014; Maier et al. 2019) and can take advantage of material that is resuspended by deep-water currents (Duineveld et al. 2007). In addition, most corals can supplement these food sources with the uptake of nutrients and free amino acids from their environment (Gori et al. 2014). Because the coral reefs discussed here live below the photic zone, there are no photosynthetic symbionts (azooxanthellate), but there is a diverse community of microbial associates that appear to form a "core microbiome" in many, if not all, species (Weinbauer et al. 2012; Appah et al. 2022). Some of these microbes are capable of chemosynthesis (Middelburg et al. 2015), although it is only in a few rare instances where this appears to provide a significant supplement to their nutrition in situ (Osman et al. 2023).

Reproductive strategies in CWC, like their warm water counterparts, include species that are gonochoristic (separate sexes), hermaphrodites, broadcast spawners, and brooders (Waller and Tyler 2005). These strategies are species- or taxon-specific, but other aspects such as fecundity or the timing of reproduction can be driven by factors such as food supply and quantity (Pires et al. 2014). When conditions are appropriate and they have sufficient energy reserves, corals will undergo sexual reproduction. In *L. pertusa*, it takes approximately 1 year for eggs to develop and reproduction appears to be seasonal, with spawning in February-March in Norway (Brooke and Jarnegren 2013) and September-October in the western Atlantic (Brooke and Sogluizzo 2017). Recent work in the Mediterranean Sea supports the seasonal spawning of *L. pertusa* but showed *M. oculata* to be a continuous spawner (Chemel et al. 2023).

In *Goniocorella* from the Southwest Pacific, larvae appear to be brooded for extended periods of time (Waller et al. 2023). In other species, including *L. pertusa*, planula larvae may have extended periods of longevity in their planktotrophic larval stage, varying somewhere between 3 and 6 weeks, resulting in long dispersal capability and broad realized connectivity among populations (Larsson et al. 2014; Pires et al. 2014; Fox et al. 2016; Stromberg and Larsson 2017). Vertical swimming ability may augment the long-distance dispersal capacity of CWCs by allowing them to spend time in shallower, faster-moving portions of the water column (Fagerstrom et al. 2022). The settlement cues for CWC larvae remain unknown, but for some species, such as *L. pertusa*, they appear to be relatively broad since they have been observed growing on all types of substrata (Wheeler et al. 2007), including the steel legs of oil platforms (Bell and Smith 1999; Gass and Roberts 2006).

1.1.3 Controls on Cold-Water Coral Reef Formation

There are sufficient data on the distribution of reef features coupled with both short- and long-term oceanographic observations that the present-day distribution of CWC and reef habitats can be predicted with some confidence. Some of the most extensive CWC reefs (as many as 1500 of them) occur along the Norwegian coast (Fossa et al. 2002). The shallowest communities are found within the fjords where offshore Atlantic water is forced into the shallows, bringing abundant coral larvae and food (Fossa et al. 2002). In the eastern North Atlantic off Europe (between Ireland and the Iberian Peninsula), where the oceanographic conditions are highly suitable, CWC reefs and mounds can be found at depths near the intersection of the Eastern North Atlantic Water and the Mediterranean Outflow Water (White and Dorschel 2010; Raddatz et al. 2014; Wienberg et al. 2020). These reefs consist of scleractinians (primarily *L. pertusa*) along with diverse octocorals and sponges (De Mol et al. 2002; Roberts et al. 2008) on banks just above the shelf break, on the continental slope, and within canyons off the edge of the shelf (Stewart et al. 2014). Over geological timescales, cold-water corals have formed giant carbonate mounds, in some cases reaching >150 m above the seabed, through successive phases of reef development (Thierens et al. 2013). For example, the Challenger Mound in the Porcupine Seabight, offshore W Ireland, was dated to be 2.6 Ma old (Kano et al. 2007; Thierens et al. 2013).

Within the Mediterranean Sea, there are many CWC habitats, but reef-formers may be at the end of their temperature tolerance in this warm and semi-enclosed sea with relatively little deep-water exchange (Naumann et al. 2013). Further to the south, off the coast of West Africa, the potential limiting factor to CWC reef development appears to be dissolved oxygen concentrations—in particular in the extended oxygen minimum zones off Mauritania, Angola, and Namibia—rather than temperature or productivity (Wienberg et al. 2018; Hanz et al. 2019; Tamborrino et al. 2019). However, even in these areas of low oxygen availability, the reefs and mounds extend over hundreds of kilometers (Wienberg et al. 2018; Tamborrino et al. 2019; Hebbeln et al. 2020; Tamborrino et al. 2022) and have persisted for at least 200,000 years (Wienberg et al. 2018) and potentially for 600,000 years (Wienberg, unpub data).

Across the South Atlantic, off the coast of Brazil, there is relatively high dissolved oxygen and availability of the carbonate ions that have supported the development of reefs throughout most of the Holocene, but warmer waters may be the limiting factor (Bernardino et al., this volume). Many sampled CWCs on the Brazilian margin, including *Lophelia*, *Solenosmilia*, *Enallopsammia*, and *Madrepora*, occur over hard substrata created by carbonate structures on the seafloor and on the edges of submarine canyons (Almada and Bernardino 2017; Gaurisas and Bernardino 2023). Off of the Southeast USA and beneath the Gulf Stream is one of the largest of the CWC mound provinces, referred to as the Million Mounds, with nearly continuous CWC mounds stretched across at least 300 km and potentially for 1000 km from far offshore of the states of North and South Carolina to the southern tip of Florida. Within the Gulf of Mexico, there are CWC mounds and reefs along the Campeche Bank (Hebbeln et al. 2014) and Florida Platform (Ross et al. 2017) and on authigenic carbonates in the northern Gulf of Mexico (Cordes et al. 2008). Here, they have persisted for at least 300,000 years (Roberts and Kohl 2018) but they are presently near the limits of their tolerance for high temperatures, low oxygen concentrations, and low pH (Georgian et al. 2014; Lunden et al. 2014).

In the Pacific Ocean, reef development is limited by low oxygen and pH along the coasts of the Americas, but patchy reefs occur in southern California (Gómez et al. 2018) and in some of the fjords of Chile (Häussermann and Försterra 2007; Jantzen et al. 2013; Forsterra et al. 2014). There are numerous reefs on the seamounts of the northern and central Pacific supported by the high productivity concentrated in some areas, and these discoveries in low-pH waters have challenged our thinking of where reefs can develop (Baco et al. 2017). In the Southwest Pacific, CWC reefs occur on the seamounts and ridges of the region where the temperature, aragonite saturation state, and flux of particulate organic matter are able to support reef development (Anderson et al. 2015, 2016). Part of the distribution of *S. variabilis* occurs below the aragonite saturation horizon (ASH) in the Southwest and North Pacific indicating that this species appears to be more resilient to undersaturation of aragonite than other colonial reef-coral species (Bostock et al. 2015; Baco et al. 2017).

There have been numerous models of the distribution and niches of CWC species, including scleractinians (Davies et al. 2008; Tittensor et al. 2009; Davies and Guinotte 2011; Anderson et al. 2016; Rowden et al. 2017; Georgian et al. 2019; Morato et al. 2020; Williams et al. 2020; Gasbarro et al. 2022), octocorals (Yesson et al. 2017; Georgian et al. 2020), and black corals (Yesson et al. 2017; Etnoyer et al. 2018). However, the distribution of the CWC reef habitat has only been modeled directly at fine (~100 m) scales in the Northeast Atlantic for *L. pertusa* (Howell et al. 2011), and so it is not entirely clear what environmental conditions favor the development of reefs within the broad oceanographic conditions that support their component species globally.

In general, field observations suggest that scleractinian corals will grow anywhere there is hard substrate and sufficient food supply within the individual species' range of environmental conditions (e.g., carbonate chemistry, temperature, oxygen, pH). Often these are sites where topography interacts with hydrography leading to hydrodynamic processes (e.g., internal tides, downwelling) that increase turbulence, and hence, food delivery (Duineveld et al. 2007; Davies et al. 2009; Hebbeln et al. 2016; Lim et al. 2020), with many individual colonies surviving even in marginal habitats (Wienberg et al. 2009; Georgian et al. 2014; Lim et al. 2020). In modeling exercises, locally elevated bathymetric highs (measured by the bathymetric position index), temperature, pH (or aragonite saturation state), and dissolved oxygen concentrations are the most common limiting factors for any individual species (Howell et al. 2011; Anderson et al. 2016; Georgian et al. 2019). However, the deep-water reef structures formed by scleractinians along with many other coral and sponge species have been observed only within a specific subset of the locations predicted to be suitable for scleractinian growth (Howell et al. 2011; Rowden et al. 2017; Williams et al. 2020).

As part of this introductory chapter, we collected all of the existing information on CWC distribution to predict where scleractinian taxa and reefs would be found in the World Ocean. In this process, we also set out to test for differences in habitat preference among and within reef-forming CWC taxa in different regions. All data used in this study come from publicly available and regularly updated datasets in order to create a reproducible workflow aligned with the principles of open data (Reichman et al. 2011) that will allow for iterative improvements of these models as new biogeographic and ocean model information become available. Taken together, these analyses have revealed the subset of ideal conditions for reef growth beyond simple coral survival.

1.2 Methods

1.2.1 Presence and Pseudoabsence Data

Two sets of data were compiled for habitat suitability modeling efforts. Occurrence data for individual reef-building scleractinian genera, and occurrence data for areas specifically considered to be "reefs." Presence locations for five reef-building scleractinia genera (*Lophelia*, *Madrepora*, *Enallopsammia*, *Solenosmilia*, and *Goniocorella*) were extracted from Freiwald et al. (2021), an expert-curated and iteratively refined dataset of global CWC occurrences maintained by the United Nations Environment Programme

World Conservation Monitoring Centre (see Freiwald et al. 2004). The high taxonomic and positional quality of this dataset reduces common data errors that arise from raw database occurrence data. Records that were duplicated within a grid cell (1/12° resolution across a global grid) or shallower than 200 m were filtered for each taxon to avoid including shallow-water records and reduce the influence of local sampling intensity. The final occurrence dataset contained 1428, 682, 205, 279, and 138 distinct grid cells, respectively, for *Lophelia, Madrepora, Enallopsammia, Solenosmilia, and Goniocorella* (Fig. S1.1). While *Oculina* can also form CWC reefs, this taxon was not included because their distribution extends well into the photic zone and currently known oculinid reefs are geographically limited to the southwest Atlantic (Reed 2002).

For CWC reef occurrences we began by sourcing reef occurrence records from the primary and gray (e.g., research cruise and government reports) literature with 223 observations returned (Fig. S1.2), 203 of which occurred below 200 m. This excluded, for example, emergent fjordic reefs whose environmental conditions are not well constrained by large-scale oceanographic models. This list of "known reefs" also included occurrences provided by each of the lead authors of the chapters in this book to ensure that the list is up-to-date. We note that this is still a non-exhaustive occurrence list, as some published studies, especially non-biological ones, were agnostic to whether a CWC habitat is a "reef" or did not include adequate positional information, and thus were not included.

To help to account for the incomplete knowledge of reefs, the occurrence list described above was augmented with additional reef presence grid cells if they met some of the following criteria:

(i) The grid cell is at or above the 95th percentile of a "Visitation Ratio" (VR). VR was defined as the per-grid-cell ratio of framework-forming CWC occurrences in the UNEP-WCMC dataset to occurrences from a pooled faunal dataset that contained all presences of annelids, echinoderms, and poriferans in the Ocean Biogeographic System in January 2021 ($n = {\sim}1.75 \times 10^6$ occurrences). A high VR (a high number of CWC observations per total faunal observations in a given grid cell) is considered here as indicative of reef-building CWC presence, because CWC reefs typically support a high diversity of other fauna (Mortensen and Buhl-Mortensen 2005; Henry and Roberts 2007; Cordes et al. 2008).
(ii) There were > 10 framework-forming CWC occurrences within that grid cell.
(iii) The grid cell is within a 2° buffer of a known CWC reef sourced from the literature.

The presence points in each of these categories are shown in Fig. S1.2. A total of 63 and 213 grid cells met the first two criteria, respectively. To be included as an additional reef occurrence in the model input, a grid cell had to meet either of the first two criteria along with the third criterion, which limited additional presences to within 2° of known reefs. In total, this process added an additional 32 reef occurrences for a total of 235 grid cells for CWC reef occurrences (Fig. S1.3).

While accurate presence data are accessible for scleractinian CWCs, reliable absence data are not available at the resolution and scale of the current study. Therefore, 10,000 pseudoabsences (or "background" points) were generated for each taxon. While the use of pseudoabsence data is common in the HSM literature, randomly distributed pseudoabsences can cause spatial biases that affect model performance (Phillips et al. 2009). Therefore, we generated pseudoabsences within a geodesic distance of 10–500 km from known presence points. The lower limit of this distance-based selection limits spatial correlation with presence points while the upper limit recognizing the capability for long-distance dispersal that scleractinian CWCs likely possess due to their long larval durations and ability to disperse within relatively fast-moving upper layers of the water column (Larsson et al. 2014; Stromberg and Larsson 2017; Fagerstrom et al. 2022).

1.2.2 Oceanographic and Bathymetry-Derived Terrain Data

In the deep sea, oceanographic data are typically sparse and consist of point samples with limited temporal and spatial coverage (Clark et al. 2016). Consequently, we used modeled data from the EU Copernicus Mercator model for oceanographic input variables and the General Bathymetric Chart of the Oceans (GEBCO 2021) for the bathymetric data (Fig. S1.3) from which the terrain variables were derived; both provide the requisite continuous environmental coverage of the seafloor and near-bottom oceanography (but see discussion in Anderson et al. 2016). Our framework can be updated with additional data as more comprehensive multibeam bathymetry becomes available for the global ocean (Mayer et al. 2018). The Copernicus Marine Environment Monitoring Service (CMEMS) data include their Global Physics Reanalysis (GLOBAL_MULTIYEAR_PHY_001_030 (Lellouche et al. 2021)) and Global Biogeochemical Hindcast (GLOBAL_MULTIYEAR_BGC_001_029) products that are available at 1/12° and ¼° native resolutions with 50 and 75 vertical levels, respectively. The former is an eddy-resolving ocean model (GLORYS12V1; Lellouche et al. 2021) that incorporates both model data and assimilated observations from satellites and in situ samples. The Global Biogeochemical Hindcast product simulates nutrient, biochemical, and lower trophic levels using the Pelagic Interactions Scheme for Carbon and Ecosystem Studies (PISCES) biogeochemical model (Aumont et al. 2015) initialized with World Ocean Atlas conditions. Full details on each of these products are available at https://data.marine.copernicus.eu/.

Network Common Data Form (NetCDF) files containing monthly means from 1993 to 2020 for both surface and near-bottom (i.e., the deepest vertical level containing data) variables were downloaded. Annual means were calculated for each variable and averaged into long-term means, and intra-annual variance (standard deviation of 1993–2020 means for each month) variables were created for the ocean physics variables (Table 1.1). The use of intra-annual variables represents a departure from many other broad-scale CWC modeling studies (Davies et al. 2008; Morato et al. 2020; Gasbarro et al. 2022; Tong et al. 2022) where only long-term means were considered. All NetCDF file processing was completed with the "Nctoolkit" (Wilson 2020) package in Python and the command-line "Climate Data Operators" tool (Schulzweida 2019). Where necessary, oceanographic variables were bilinearly interpolated to a common 1/12° (i.e., ~8 km at the equator) grid.

Terrain variables (Table 1.1) derived from 15 arc-second (~450 m at the equator) horizontal resolution GEBCO (2021) bathymetry included slope, aspect (converted into "northness" and "eastness" components), and Bathymetric Position Index (BPI), which is a measure of elevation relative to a given grid-cell neighborhood, at both "fine" (8-cell radius) and "broad" (49 cell radius) scales. The R package "raster" (Hijmans 2023) was used to create these variables according to the methodology of Wilson et al. (2007) and then resampled to match the 1/12° resolution of the oceanographic variables.

Pairs of variables with Pearson's correlation coefficient > 0.7 (Fig. S1.4) were selected and one of the two variables was manually removed according to their ecological relevance to CWCs. For example, oceanographic variables reliably shown to affect CWC distributions such as those related to temperature and carbonate chemistry were prioritized over nutrient variables such as iron or phosphate, and variables representing near-bottom conditions were prioritized over surface variables. Then, the vifstep() function in the "usdm" package (Naimi 2015) was used to ensure that no two variables had a variance inflation factor (VIF) > 5. Summaries of all variables are presented in Table 1.1.

1.2.3 Niche Differentiation by Region and Taxon

We used a Principal Component Analysis (PCA) ordination to visualize the environmental conditions at known occurrences (i.e., their realized niche) for each taxon and between regions. Environmental variables (see Table 1.1) were scaled to zero mean and unit variance prior to PCA, which was implemented with the "ade4" R package (Thioulouse et al. 2018). There were ten different biogeographic regions delineated a priori to examine differences in the reef habitats modeled. These regions generally correspond to the nine regions covered in the chapters of this book (see Table of Contents), except that an additional category was created for the reefs discovered around Greenland and Iceland because of the distinct habitat characteristics that have been observed in these locations.

To test whether the centroids and/or variance of the environmental variables in multivariate space varied between regions/taxa, we employed permutational multivariate analysis of variance (PERMANOVA (Anderson 2001)) and analysis of multivariate homogeneity of group dispersions using the *pairwise.adonis() and betadisper()* functions in the R packages "pairwiseAdonis" (Arbizu 2017) and "vegan" (Oksanen et al. 2022). Thus, we tested the complementary null hypotheses that each pairwise region/taxa combination experiences equivalent environmental conditions and that those conditions are equally variable between regions/taxa. We note that observed differences between centroids can be driven by differences in multivariate dispersions, and so the former should be interpreted with caution when there are significant pairwise differences in centroids between taxa/regions. Both analyses were run to test for differences between each of the six modeled taxa and for *L. pertusa* (because only this species had sufficient data) between regions. Note that the Southwest Pacific region was not included in these analyses because the one *L. pertusa* presence record that has been documented in this region was shown to be erroneous (Tracey et al. 2011). Prior to analyses, all variables were scaled to a 0–1 range without altering their internal structure using their respective minima and maxima. PERMANOVAs were run with 999 permutations and pairwise differences in environment between occurrence points were represented by Bray-Curtis dissimilarities. The significance of taxonomic/regional differences in group variances calculated with the *betadisper()* function was tested with ANOVAs and post-hoc pairwise comparisons were conducted with Tukey's Honestly Significant Differences (Tukey's HSD) tests.

Table 1.1 Environmental variable summaries (mean ± SE) at CWC occurrence locations by taxon. Variables used in habitat suitability models are in bold

Type	Short name	Variable [units]	Lophelia	Madrepora	Enallopsammia	Solenosmilia	Goniocorella	Reefs
Physics	**sst**	Surface temperature [°C]	14.6 ± 0.17	20.8 ± 0.26	20.6 ± 0.43	16.3 ± 0.41	15 ± 0.35	16.4 ± 0.48
	sss	Surface salinity [PSU]	35.4 ± 0.03	35.6 ± 0.04	35.4 ± 0.05	35.3 ± 0.05	34.9 ± 0.05	35.4 ± 0.07
	nv_surf	Northward current velocity [m s^{-1}]	0.094 ± $7e^{-3}$	0.1 ± 0.01	0.022 ± $7.8e^{-3}$	0.016 ± 0.01	$-5.92e^{-3}$ ± $9.1e^{-3}$	0.14 ± 0.02
	ev_surf	Eastward current velocity [m s^{-1}]	$4.56e^{-2}$ ± $3.5e^{-3}$	0.026 ± $6.4e^{-3}$	$2.11e^{-3}$ ± $8.1e^{-3}$	0.028 ± 0.01	0.053 ± $9.4e^{-3}$	0.053 ± 0.01
	mld	Mixed-layer depth defined by sigma-theta [m]	73.2 ± 1.4	49.4 ± 1.6	47.3 ± 1.8	69 ± 2.5	57 ± 2.1	62.7 ± 2.7
	t_btm	Bottom temperature [°C]	8.47 ± 0.07	8.78 ± 0.15	6.56 ± 0.29	4.99 ± 0.2	8.19 ± 0.24	8.13 ± 0.21
	s_btm	Bottom salinity [PSU]	35.5 ± 0.02	35.4 ± 0.04	34.8 ± 0.03	34.8 ± 0.03	34.6 ± 0.02	35.3 ± 0.06
	nv_btm	Bottom northward current velocity [m s^{-1}]	0.041 ± $2.5e^{-3}$	0.029 ± $3.4e^{-3}$	0.01 ± $3.4e^{-3}$	0.016 ± $4.5e^{-3}$	$5.65e^{-3}$ ± $3.5e^{-3}$	0.052 ± $7.4e^{-3}$
	ev_btm	Bottom eastward current velocity [m s^{-1}]	$5.56e^{-3}$ ± $2.1e^{-3}$	$-2.31e^{-6}$ ± $3e^{-3}$	$-5.61e^{-3}$ ± $3.5e^{-3}$	$4.1e^{-4}$ ± $4.3e^{-3}$	0.01 ± $2.4e^{-3}$	$4.89e^{-3}$ ± $4.9e^{-3}$
	sst_sd	Standard deviation of monthly mean sst	2.22 ± 0.02	2.11 ± 0.04	1.85 ± 0.06	1.82 ± 0.05	1.96 ± 0.04	2.23 ± 0.05
	sss_sd	Standard deviation of monthly mean sss	0.16 ± $6.8e^{-3}$	0.16 ± $9.6e^{-3}$	0.097 ± $8.7e^{-3}$	0.11 ± 0.01	0.074 ± 0.01	0.18 ± 0.02
	nv_surf_sd	Standard deviation of monthly mean nv_surf	0.027 ± $7e^{-4}$	0.036 ± $1.2e^{-3}$	0.032 ± $2.4e^{-3}$	0.028 ± $2.7e^{-3}$	0.016 ± $1.1e^{-3}$	0.031 ± $2.1e^{-3}$
	ev_surf_sd	Standard deviation of monthly mean ev_surf	0.025 ± $5.7e^{-4}$	0.037 ± $1.3e^{-3}$	0.03 ± $1.9e^{-3}$	0.031 ± $2.2e^{-3}$	0.018 ± $1.8e^{-3}$	0.027 ± $1.4e^{-3}$
	mld_sd	Standard deviation of monthly mean mld	65.1 ± 1.48	38.8 ± 1.72	31.3 ± 1.74	55.2 ± 2.64	43.6 ± 1.85	52.8 ± 3.08
	t_btm_sd	Standard deviation of monthly mean t_btm	0.19 ± $3.7e^{-3}$	0.12 ± $5.1e^{-3}$	0.1 ± 0.01	0.073 ± $4.7e^{-3}$	0.11 ± $6.5e^{-3}$	0.16 ± $9.9e^{-3}$
	s_btm_sd	Standard deviation of monthly mean s_btm	0.02 ± $5.3e^{-4}$	0.016 ± $7.5e^{-4}$	0.012 ± $9e^{-4}$	0.012 ± $1.1e^{-3}$	0.018 ± $1.1e^{-3}$	0.022 ± $1.8e^{-3}$
	nv_btm_sd	Standard deviation of monthly mean nv_btm	$9.68e^{-3}$ ± $2.2e^{-4}$	$8.12e^{-3}$ ± $2.7e^{-4}$	$8.78e^{-3}$ ± $7.2e^{-4}$	$7.58e^{-3}$ ± $3.6e^{-4}$	$5.2e^{-3}$ ± $3.6e^{-4}$	0.01 ± $5.7e^{-4}$
	ev_btm_sd	Standard deviation of monthly mean ev_btm	$8.99e^{-3}$ ± $2.2e^{-4}$	$7.8e^{-3}$ ± $3.1e^{-4}$	$8.67e^{-3}$ ± $6.8e^{-4}$	$7.74e^{-3}$ ± $3.5e^{-4}$	$4.73e^{-3}$ ± $2.8e^{-4}$	$9.15e^{-3}$ ± $4.7e^{-4}$
Biogeochem	**si_surf**	Surface silicate [mole conc.]	2.87 ± 0.03	3.34 ± 0.14	3.35 ± 0.27	4.86 ± 0.36	3.63 ± 0.28	2.84 ± 0.09
	spco2_surf	Surface CO_2 [partial pressure]	36.1 ± 0.05	37.2 ± 0.09	36.4 ± 0.11	36.7 ± 0.11	35.6 ± 0.12	36.2 ± 0.15
	ph_surf	Surface pH	8.08 ± $5e^{-4}$	8.06 ± $8.3e^{-4}$	8.07 ± $1.2e^{-3}$	8.07 ± $1.1e^{-3}$	8.08 ± $1.4e^{-3}$	8.08 ± $1.4e^{-3}$
	phyc_surf	Surface phytoplankton carbon [mole conc.]	2.05 ± 0.02	1.77 ± 0.05	1.47 ± 0.03	1.68 ± 0.04	2.16 ± 0.08	1.87 ± 0.03
	po4_surf	Surface phosphate [mole conc.]	0.26 ± $5.4e^{-3}$	0.18 ± 0.01	0.27 ± 0.03	0.45 ± 0.03	0.53 ± 0.03	0.23 ± 0.01
	fe_surf	Surface dissolved iron [mole conc.]	$8.32e^{-4}$ ± $1.8e^{-5}$	$1.1e^{-3}$ ± $3.6e^{-5}$	$6.16e^{-4}$ ± $4.5e^{-5}$	$5e^{-4}$ ± $3e^{-5}$	$4.62e^{-4}$ ± $4.5e^{-5}$	$8.33e^{-4}$ ± $5.5e^{-5}$
	nppv_surf	Surface net primary production [C unit $vol.^{-1}$]	9.36 ± 0.27	8.88 ± 0.59	5.53 ± 0.41	7.18 ± 0.4	11.3 ± 0.98	8.63 ± 0.43
	o2_surf	Surface dissolved oxygen [mole conc.]	259 ± 0.79	231 ± 1.11	232 ± 1.98	251 ± 1.97	258 ± 1.82	251 ± 2.17

	no3_surf	Surface nitrate [mole conc.]	3.37 ± 0.07	1.97 ± 0.15	2.59 ± 0.37	5.39 ± 0.41	5.62 ± 0.47	2.83 ± 0.17
	chl	Surface chlorophyll-a [mass conc.]	$0.41 \pm 6.1e^{-3}$	0.29 ± 0.01	0.23 ± 0.01	0.32 ± 0.01	0.46 ± 0.02	0.36 ± 0.01
	si_btm	Bottom silicate [mole conc.]	12 ± 0.39	31.6 ± 1.35	65.6 ± 3.43	47.9 ± 2.19	33.8 ± 2.85	25.6 ± 2.49
	ph_btm	Bottom pH	$7.99 \pm 2.1e^{-3}$	$7.91 \pm 2e^{-3}$	$7.85 \pm 4.4e^{-3}$	$7.9 \pm 4e^{-3}$	$7.93 \pm 5.7e^{-3}$	$7.95 \pm 6.9e^{-3}$
	po4_btm	Bottom phosphate [mole conc.]	1 ± 0.01	1.46 ± 0.03	2.01 ± 0.04	1.76 ± 0.03	1.68 ± 0.04	1.24 ± 0.04
	fe_btm	Bottom iron [mole conc.]	$1.24e^{-3} \pm 1.5e^{-5}$	$1.1e^{-3} \pm 1.5e^{-5}$	$7.81e^{-4} \pm 1.8e^{-5}$	$7.58e^{-4} \pm 1.2e^{-5}$	$1.31e^{-3} \pm 5e^{-5}$	$1.22e^{-3} \pm 4.4e^{-5}$
	o2_btm	Bottom dissolved oxygen [mole conc.]	227 ± 1.29	178 ± 2.21	177 ± 3.55	209 ± 2.39	208 ± 3.2	209 ± 3.49
	no3_btm	Bottom nitrate [mole conc.]	14.9 ± 0.16	21.4 ± 0.36	29.2 ± 0.57	25.8 ± 0.45	23.4 ± 0.6	18.4 ± 0.64
Bathymetry-derived terrain	depth	Depth [m]	656 ± 15.4	916 ± 28	1240 ± 61	1450 ± 49.3	634 ± 37.5	735 ± 36.3
	slope	8-cell slope in [°]	2.62 ± 0.08	4.08 ± 0.15	5.87 ± 0.32	5.96 ± 0.25	2.77 ± 0.25	3.7 ± 0.28
	bpi_fine	8-cell radius BPI	0.087 ± 0.01	0.16 ± 0.03	0.49 ± 0.09	0.35 ± 0.06	0.17 ± 0.06	0.29 ± 0.07
	bpi_broad	49-cell radius BPI	18.8 ± 2.13	27 ± 5.38	96.9 ± 15.2	59.9 ± 9.16	34.9 ± 8.83	53.2 ± 11.8
	roughness	Roughness	48.7 ± 1.54	81.2 ± 3.05	117 ± 6.46	116 ± 5.06	54 ± 4.87	72.5 ± 5.73
	northness	Cosine of aspect [°]	$6.79e^{-3} \pm 1.28e^{-3}$	$0.015 \pm 2e^{-3}$	$0.015 \pm 2.6e^{-3}$	$5.53e^{-3} \pm 2.3e^{-3}$	$7.59e^{-3} \pm 4.7e^{-3}$	$1.84e^{-3} \pm 2.5e^{-3}$
	eastness	Sine of aspect [°]	$-9.99e^{-3} \pm 1.3e^{-2}$	$-6.76e^{-3} \pm 1.6e^{-3}$	$-4.7e^{-3} \pm 2.6e^{-3}$	$-8.98e^{-3} \pm 2e^{-3}$	$8.7e^{-3} \pm 5.6e^{-3}$	$-0.01 \pm 3.2e^{-3}$

1.2.4 Habitat Suitability Modeling

Where independent survey data are not available to evaluate model performance, a portion of data can be set aside for this purpose while the rest of the data are used to train the model. However, randomly splitting presence/pseudoabsence data into testing and training datasets can exaggerate spatial biases (Hao et al. 2019) and thus, often is not the optimal cross-validation strategy (Valavi et al. 2019; Hao et al. 2020). Therefore, we split our data into five spatial block cross-validation folds (Fig. S1.5) using the *spatialBlock()* function in the "blockCV" package (Valavi et al. 2019). Approximately evenly dispersed (i.e., containing roughly the same amount of presences and pseudoabsences within each fold) 1000 km blocks were created that encompassed all presence and pseudoabsence data for each taxon. The block size was selected using the *spatialAutoRange()* function within "blockCV" based on the median range of spatial autocorrelation within the bottom temperature data, which we hypothesized to be a key predictor based on previous studies (Georgian et al. 2019; Morato et al. 2020; Gasbarro et al. 2022). Blocks were sequentially assigned to folds 1–5. Habitat suitability models (HSMs) were iteratively trained with four of the spatial folds while the remaining fold was withheld as testing data.

HSMs were fitted using a common procedure for each taxon. For each of the five testing-training folds, habitat suitability index (HSI) scores were computed with each of Random Forest (RF), Gradient Boosting Machine (GBM), and generalized additive model (GAM) algorithms, creating 15 models which were later combined into ensemble models. The use of ensemble models from multiple algorithms reduces biases from single methods (Buisson et al. 2010; Hao et al. 2019), and thus we fit and ensembled models from both tree-based (RF, GBM) and regression-based (GLM) approaches. Individual model performance was assessed with the True Skill Statistic (TSS; Lawson et al. 2014) and area under the receiver operating curve (AUC) metrics. Relative variable importance scores (Guisan et al. 2017) and variable response curves (Elith et al. 2005) were also generated for each algorithm.

The parameters of each algorithm were tuned using the *BIOMOD_tuning()* function in "biomod2" (Thuiller et al. 2021). Model tuning is often overlooked in ensemble distribution models, but it is an important step as the use of default parameters can lead to potentially poorer performance than individual, tuned models (Hao et al. 2020). The tuning function was programmed to automatically set parameters to maximize model sensitivity and specificity, as measured by TSS. Only individual models with TSS scores > 0.7 were retained for ensemble models for each taxon. Ensemble means were weighted by model TSS, unweighted mean, median, and cross-validation models and were computed and projected into geographic space. All model fitting, projection, and evaluation were accomplished with the "biomod2" package (v.3.5.1; Thuiller et al. 2021) in R (v4.1.2; R Core Team 2021) using Temple University's High-Performance Computing cluster.

1.3 Results

The overall PCA showed distinctions in environmental conditions among the six CWC taxa and the ten reef regions. The first two axes of the PCA explained 15.2% and 13.7% of the variation in environmental conditions, respectively (Fig. 1.3). Variable loadings on the first axis were primarily driven by differences in sea-surface temperature and bottom oxygen, while loadings on the second were driven by a mixture of bottom temperature, the bathymetry-derived terrain variables (e.g., slope, BPI), and monthly variance in bottom temperature and salinity (Fig. 1.4). Loadings for monthly variance in surface currents, bottom pH, mixed-layer depth, and bottom iron indicate roughly equal contributions to both of the first two PCA axes (Fig. 1.4). Axes three and four of the PCA had lesser, but still notable contributions (9.7% and 7.6%, respectively); axis three was driven primarily by bottom salinity and temperature and variance in sea-surface temperatures, while axis four was driven primarily by the BPI variables and surface current variables (Fig. S1.6). Clear regional distinctions were apparent in both plots of PCA axes 1–2 (Fig. 1.3) and 3–4 (Fig. S1.6). Well-oxygenated regions with lower surface temperatures (e.g., Norway) were clearly distinguishable from regions with opposite conditions (e.g., Gulf of Mexico and West Africa; Fig. 1.3). The Mediterranean was notably distinct from all other regions on the third and fourth PCA axes due to relatively high bottom salinities (Fig. S1.6). The Southeast USA was also distinct in the high velocities of surface currents, due to the influence of the Gulf Stream (Fig. S1.6).

Statistical tests for differences in environmental niche centroids between taxa revealed significant differences that varied in their strength (Table S1). All six taxa had significantly different niche centroids (PERMANOVAs, $p < 0.05$), but with low explanatory power ($R^2 < 0.1$) for all pairwise comparisons except between *Enallopsammia* and *Goniocorella* ($R^2 = 0.12$; Table S1). Interestingly, the difference between *Lophelia* and "Reefs" was the smallest between all pairwise comparisons ($F = 10.29$, $R^2 = 0.006$), suggesting only a slight degree of niche centroid separation between the two at the scale of this study. These relatively slight differences in niche centroids between taxa may have been driven by differences in multivariate dispersions of environmental niches, which were significant between many of the taxa (Table S2). The largest pairwise differences

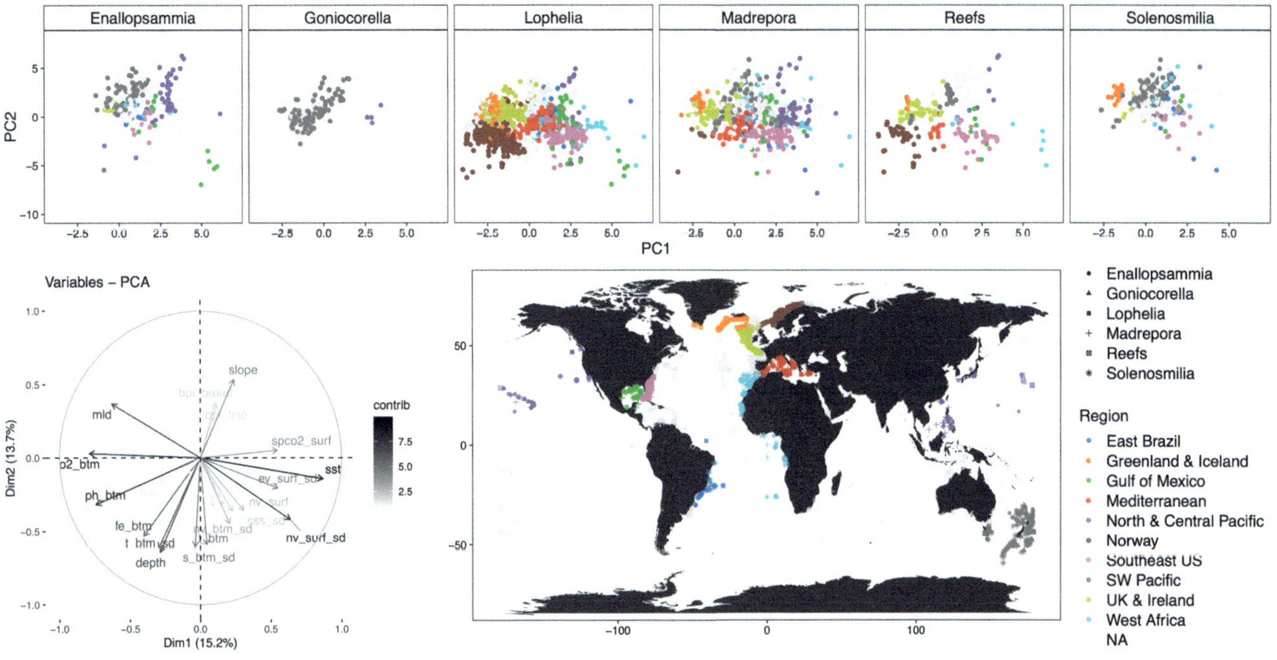

Fig. 1.3 PCA results and map showing locations of occurrence points in each region/taxon combination. The influence of individual variables and their relative contributions to these axes can be seen in the eigenvector plot (top row)

were between *Goniocorella-Madrepora*, with the former significantly less varied. No differences in dispersions were found between *Solenosmilia-Lophelia*, *Solenosmilia-Enallopsammia*, Reefs-*Enallopsammia*, and Reefs-*Solenosmilia* ($p > 0.05$; Table S2).

Intra-regional niche centroid differences for *L. pertusa* were more pronounced than those between taxa (Table S3), ranging from relatively small differences between East Brazil and West Africa ($F = 11.85$, $R^2 = 0.12$) to strong distinctions between Greenland & Iceland and the Mediterranean ($F = 721.66$, $R^2 = 0.75$). Strong differences were also found between *L. pertusa* environments in the Southeast USA versus Norway ($F = 616.56$, $R^2 = 0.56$), versus Greenland & Iceland ($F = 607.31$, $R^2 = 0.68$), versus the Mediterranean ($F = 501.86$, $R^2 = 0.68$), and versus the UK & Ireland ($F = 508.05$, $R^2 = 0.55$). The Mediterranean and Norway also had a relatively high degree of difference ($F = 623.39$, $R^2 = 0.59$).

Multivariate dispersions of *L. pertusa* environmental variables varied between regions, with a number of regions showing no differences in dispersions (Table S4). *L. pertusa* sites in the North & Central Pacific region had the greatest dispersions in environmental conditions, with significantly higher dispersion than all other regions they were tested against, while sites in the Mediterranean were the least dispersed (Table S4). See Table S4 for all pairwise comparisons of multivariate dispersions in environmental conditions for *L. pertusa*.

1.3.1 Habitat Suitability Model Evaluation

Habitat suitability model performance, as assessed through five-fold spatial cross-validation, varied among taxa and algorithms but generally revealed, for each taxon, good to excellent spatial discrimination capacity in a number of the models (Fig. S1.7). In general, the tree-based algorithms (i.e., RF and GBM) outperformed the regression-based GAM algorithms, which varied more in their performance and in the shape of important variable response curves (Figs. 1.4 and S1.7). There was high correspondence in variable responses between RF and GBM algorithms (Fig. 1.4). In general, most models had TSS > 0.7 and thus were retained for ensemble modeling and projection. However, models for the taxa with relatively few presences (i.e., *Goniocorella* and Reefs) performed worse on average, and a number of individual models for these taxa were not retained. No GAM models were retained for *Goniocorella*, for example. Scores for *Enallopsammia* were generally high despite the relatively few number of presences for this taxon. AUC scores were generally high (>0.9), indicating excellent spatial discrimination capacity for all taxa except for *Goniocorella*, where only a few of the taxon's 15 models scored in this range.

1.3.2 Taxon Distributions

The combination of the different oceanographic variables in the models revealed patterns in the environmental controls of

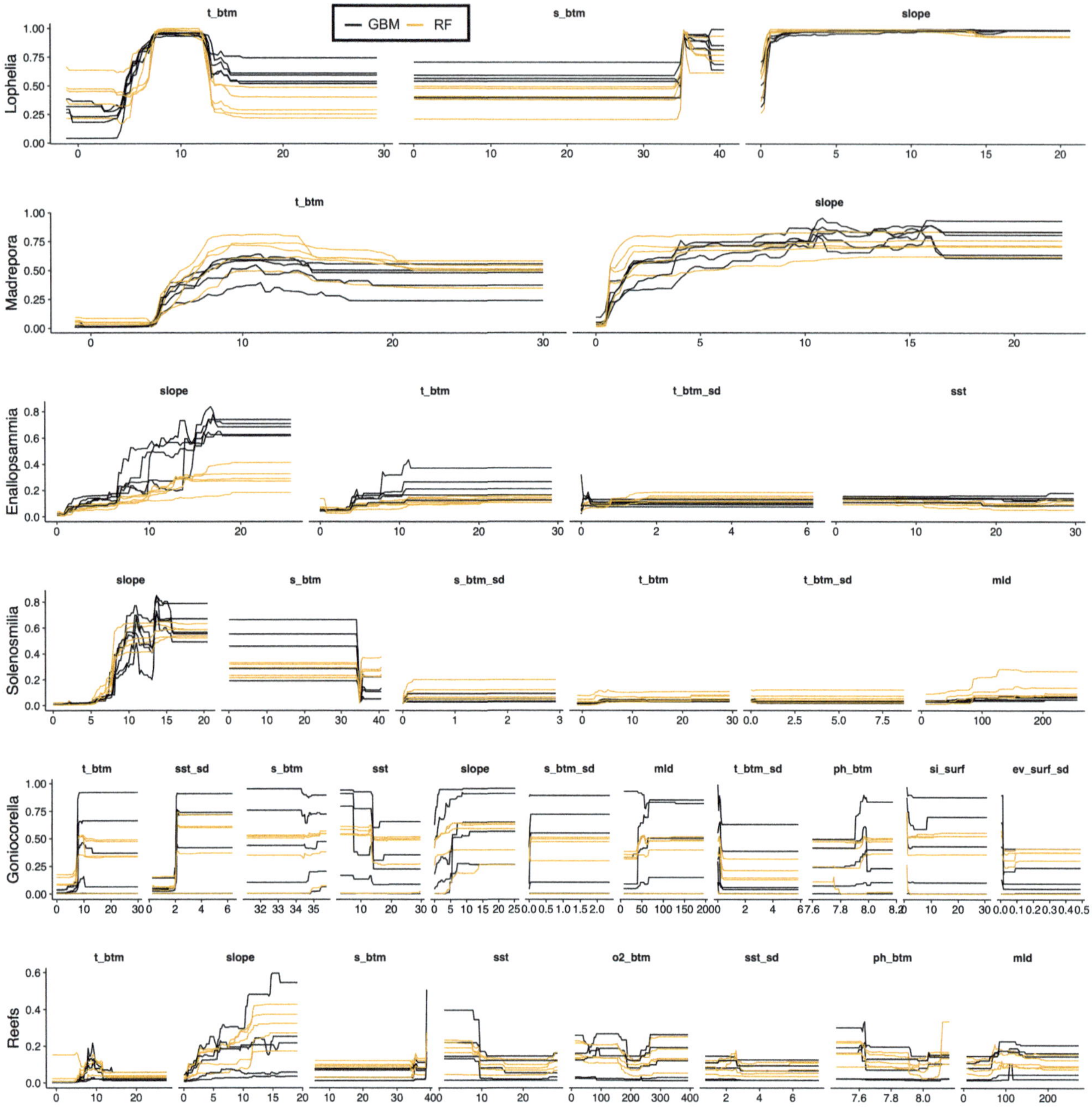

Fig. 1.4 Response curves showing the probability of occurrence (y-axis) for each modeled taxon (rows) across the range of each predictor variable for GBM & RF algorithms (line colors). Variables with relative importance scores greater than 0.1 are shown and arranged left to right in order of decreasing importance (see this figure)

the different reef-forming taxa in the deep ocean. The global distributions of most of the taxa (i.e., all except for *Enallopsammia* and *Solenosmilia*) were primarily controlled by the average bottom temperature (Fig. 1.5). For *L. pertusa*, the range of bottom temperatures experienced was similar to the other taxa, although the sea-surface temperature was the lowest for this species, and the variance in SST was the highest (Table 1.1). For *M. oculata*, bottom temperature accounted for the highest relative importance of any variable for any species (Fig. 1.5), and *M. oculata* also showed the highest average bottom and surface temperature of any species where it was distributed (Table 1.1). *Enallopsammia* sp. was present in areas with a relatively low average seafloor temperature, but in areas where the surface temperature was among the highest (Table 1.1). *Solenosmilia* appears to have the most tolerance for low bottom temperatures, and where the temperature has the lowest variance, which coincides with its distribution at the deepest depths (avg = 1450 m)

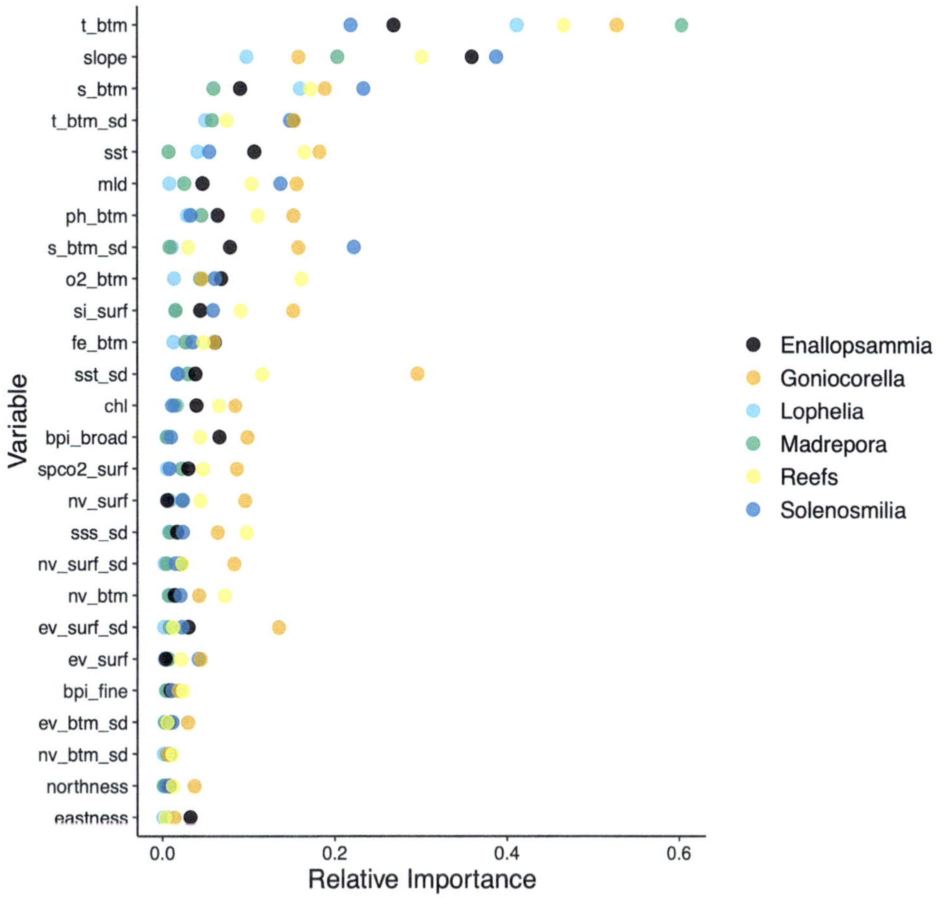

Fig. 1.5 Relative importance of variables included in habitat suitability models for each taxon

examined here. In contrast, *Goniocorella* had the shallowest depth distribution (avg = 634 m) and the variables with the highest explanatory power were bottom temperature, variance in sea-surface temperature, bottom salinity, and sea-surface temperature (Fig. 1.5).

The second most important variable overall was the slope of the seafloor (calculated at the native 15 arc-second horizontal resolution of the bathymetry data, equivalent to ~450 m at the equator, before interpolation to the 1/12° model resolution). Slope was the single most important variable for *Enallopsammia* and *Solenosmilia*, surpassing bottom temperature (Fig. 1.5). It is also notable that the slope of the seafloor was less important for *L. pertusa* than it was for any of the other taxa modeled (Fig. 1.5). *Enallopsammia* also occurs at the highest BPI (both broad and fine scale) of any of the corals, with the most extreme values from the seamounts of the North and Central Pacific where it is one of the dominant reef-forming species (Table 1.1).

If the CWC reef habitats are considered together, at the exclusion of the points where corals are present but apparently not forming reefs, some additional trends are revealed. Similar to many of the individual taxon models, the most important variables in the predictive model for Reefs are bottom temperature, slope, bottom salinity, sea-surface temperature, and dissolved oxygen concentration (Fig. 1.5). When the relative importance of the variables is compared between the reef model and the individual coral taxa (Fig. 1.5), reefs occur in areas where there is elevated surface and bottom current velocity, and these variables are more important to the predicted distribution of reefs than surface productivity. The dissolved oxygen concentration is also more important for reef formation than it is for individual coral taxa distribution. Reefs are also supported in areas with higher BPI than exhibited by the individual taxa in the Atlantic, but not as compared to *Enallopsammia* or *Goniocorella*. Interestingly, reefs are also distributed in areas with lower concentrations of surface silica than where individual taxa are found.

1.3.3 Regional Trends in Coral Habitats

This global analysis of CWC and reef distribution has also provided a broad perspective on the prevailing environmental conditions at reef habitats and distinctions between different regions where CWC reefs occur (Table 1.1). The CWC reef habitats lying at the center of the PCA of environmental characteristics are in the Mediterranean and off of the Southeast U.S. coast (Fig. 1.3). On the Norwegian reefs, the highest pH reported values are recorded along with high iron concentrations. In other areas of the North Atlantic, the reef habitats are typified by high dissolved oxygen concentrations, a deep mixed-layer depth, relatively high pH, and low sea-surface temperatures. This is in contrast to the other sites in the South Atlantic and the Gulf of Mexico that show high sea-surface temperatures and lower average dissolved oxygen concentrations at depth, along with high variability in surface velocity.

1.3.4 Predicted Reef Distribution

One of the most powerful and useful outcomes of the modeling effort is in the precision of reef distribution projections (Fig. 1.6). For example, the majority of the known coral habitats in Norwegian waters are inshore and in fjords, but the model points to large areas on the outer continental shelf that appear to be suitable for reef development (Fig. 1.6c). These suitable areas continue south along the shelf break to the Hebrides Islands and to sites along the Irish margin around the Porcupine Seabight and the Wyville-Thomson Ridge (Fig. 1.6b). High suitability is also predicted for areas further south in the canyon systems off the coast of France, Spain, and Portugal (Fig. 1.6). In the Mediterranean, large portions of the Aegean Sea appear to be suitable for CWC reef and mound development (Fig. 1.6f).

Off of the West African coast, the model suggests that the areas along the mound province of Mauritania are suitable for coral presence at the current time, but not reef and mound development (Fig. 1.6h). However, conditions appear more favorable to the north on the edge of the continental shelf near the Canary Islands, where there are known reefs (Zibrowius 1980), but also at the edge of the exclusive economic zones of Western Sahara and Morocco where only relic CWC mounds have been sampled to date (Hebbeln et al. 2019). Further offshore, CWC reefs may be found topping some of the seamounts of the Guinea Rise and on the Romanche Transform Fault of the Mid-Atlantic Ridge. The known sites off of Angola (Le Guilloux et al. 2009) also appear highly suitable for reef development in the model predictions, as do some of the sites off of Namibia (Hanz et al. 2019) despite their relatively low dissolved oxygen concentrations. Across the Atlantic, many of the known CWC areas off of Brazil (Viana et al. 1998) are suitable for coral growth, with a few hotspots of potential reef development (Fig. 1.6g). This includes an area offshore Brazil north of the Amazon river outflow and extending into French Guiana, Suriname, and Guyana where CWC reefs are predicted to be found.

The areas off of the Southeast USA (Fig. 1.6d) all appear to be highly suitable for reef development, and this area is well known from previous work (Reed et al. 2006), although individual reef structures continue to be found (Gasbarro et al. 2022). The model predicts that additional CWC reefs will be present throughout the Bahamas, which has had less deep-water work completed over the years. Within the Gulf of Mexico (Fig. 1.6a), there are patchy locations suitable for reef development, including the entire Yucatan peninsula, not just the known sites of the Campeche coral mounds (Matos et al. 2017).

In the Pacific Ocean, the model predicts scattered areas of potential reef development on many of the seamounts of the region. (Fig. 1.6). The waters surrounding Taiwan in the East and South China Seas are highly probable areas for reef development (Fig. 1.7). In the south Pacific, the Macquarie Ridge off of New Zealand is predicted to be good coral reef habitat (Fig. 1.6i). These suitable areas continue to the Antarctic Peninsula and the Weddell Sea. Large areas of the Indian Ocean also appear to be suitable for reef development including the area North of Madagascar, around the Seychelles, off the Horn of Africa, the north shore of the Arabian Sea and the Bay of Bengal, and the Andaman Sea (Fig. 1.7).

1.4 Discussion

1.4.1 Predicted Distribution of Coral Reefs in the Deep Ocean

The results of the predictive habitat modeling indicate a high likelihood that coral reefs are present across large areas of the disphotic and aphotic, cold, deep oceans between approximately 200 and 1200 m and at temperate and tropical latitudes, even above the Arctic Circle. However, habitat suitability models tend to over-predict, and there are a number of limitations to the models presented in this study. Despite the relatively high resolution of the source data that served as inputs to our models relative to past global- and basin-scale CWC modeling efforts (Davies et al. 2008; Morato et al. 2020; Tong et al. 2022) there is still a wide body of evidence that more localized (i.e., <100 m) factors play an important role in CWC reef formation. Specifically, geomorphological structures (e.g., ridges and escarpments) at this scale have been shown to be highly predictive of reef structures (Wheeler et al. 2007; Howell et al. 2011; Rowden et al. 2017; Williams et al. 2020), while vertical walls, which can host extensive "hanging reefs," are not often mapped correctly and are therefore often overlooked (Robert et al. 2020). This highlights the need for continued exploration and high-resolution mapping of the seafloor.

The models presented in this study are also hampered by the data limitations that are consistent across all large-scale CWC modeling studies—namely, that we have relatively few records on the distribution of CWCs and CWC reefs in particular, and discoveries continue to be made that augment our knowledge of the breadth of their environmental niches.

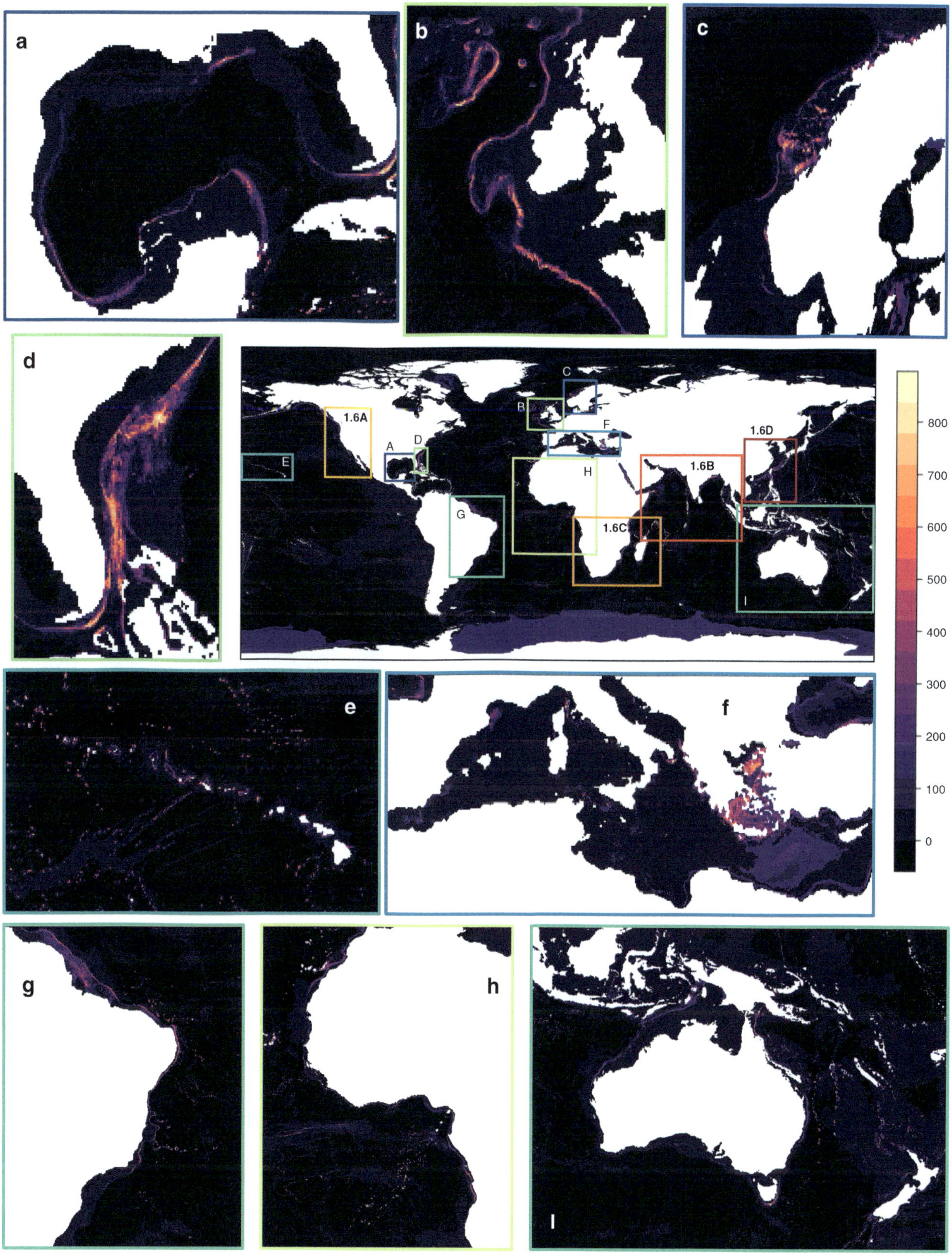

Fig. 1.6 Map of ensemble weighted mean habitat suitability index (HSI) scores for CWC reefs and reef-forming taxa known to occur within the regions of the world covered by the chapters of this volume (**a**) Gulf of Mexico, (**b**) UK & Ireland, (**c**) Norway, (**d**) Southeast USA, (**e**) Hawaii, (**f**) Mediterranean Sea, (**g**) East Brazil, (**h**) West Africa, and (**i**) Southwest Pacific CWC reef regions. Additional areas not covered in the book are shown in figures **a**–**d**. Major land masses are shown in white

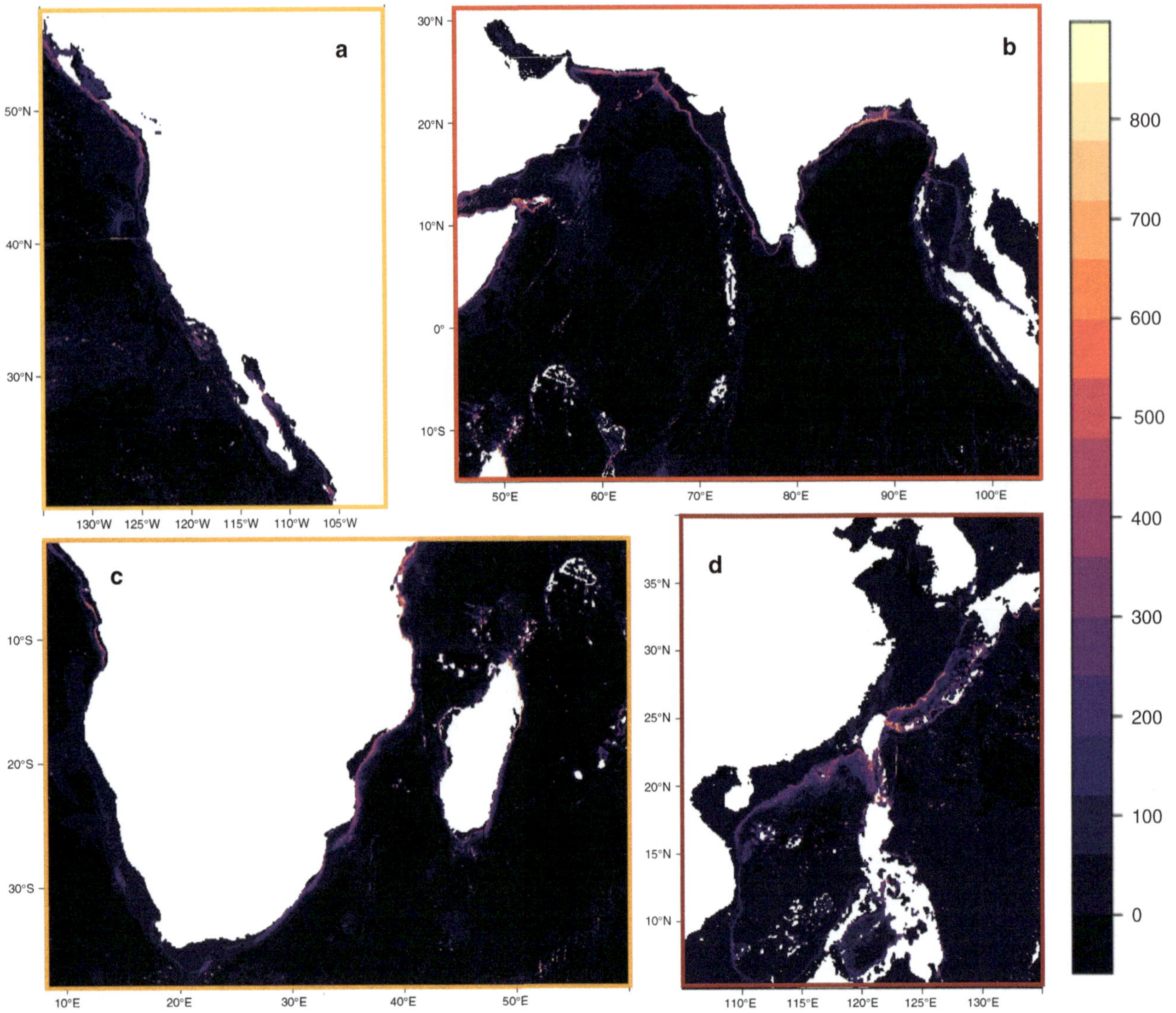

Fig. 1.7 Map of ensemble weighted mean habitat suitability index (HSI) scores for CWC reefs and reef-forming taxa known to occur within largely unexplored areas of the (**a**) Pacific coast of North America, (**b**) Indian Ocean, (**c**) Southern Africa, and (**d**) Eastern Pacific, South China Sea, and East China Sea. Major land masses are shown in white

The decision was made to limit the input presence data to locations within 2° of known reefs. This excludes large areas of the World Ocean where our other criteria were met, namely frequent observations of framework-forming CWC taxa and co-occurrence of these observations with collections of a diverse reef-associated fauna. However, the strength of this approach was revealed in the outcome of the model, which predicted high habitat suitability for CWC reef formation in many of the areas that were not included in the model input, including the parts of the Indian Ocean, central Pacific, and South China Sea (Figs. 1.6 and 1.7).

Another limitation is that these models are also only capable of capturing the present-day conditions that the reef and mound structures experience, while the onset of reef development could be millennia earlier. While the Norwegian reefs are all post-glacial in age, the initiation of development of the Challenger Mound in the eastern Porcupine Seabight was dated at 2.6 million years ago (Li et al. 2011), while the detailed stratigraphy suggested that the North Atlantic has been conducive to CWC growth since the Late Pliocene (Thierens et al. 2013). This agrees with the dating of a CWC mound in the northern Gulf of Mexico at over 300,000 years (Roberts and Kohl). Overall, the NE Atlantic CWC reefs and mounds started their most recent reef growth period some 11,000 years ago on substrata as small as drop stones from the last period of glaciation in the region (Lopez Correa et al. 2012; Raddatz et al. 2014; Wienberg et al. 2020; O'Reilly et al. 2022).

Despite these limitations, predictive models such as the ones presented here are necessary to guide future conservation and exploration efforts for CWC reefs in concert with more localized information on the size and extent of CWC reefs. Taken together, the model predicts that the distribution of CWC reefs is more limited than the distribution of colonial, deep-sea scleractinian taxa, and that reefs will generally occur within the habitable niche of the species that form them where there are relatively elevated currents, elevated topography, and high dissolved oxygen concentrations (Fig. 1.5), although there are always exceptions.

It is also interesting that the modeling effort revealed that reefs tend to be found in areas with lower surface silica concentration than individual coral species. This is potentially related to the distribution of diatoms in surface waters, as their abundance tends to track silica concentrations (Villareal et al. 2012). There is an abundance of data that the corals themselves are feeding on small zooplankton (Duineveld et al. 2004; Kiriakoulakis et al. 2005; Becker et al. 2009) and the diatom-derived fatty acids that have been sampled in *L. pertusa* are likely coming from herbivorous zooplankton (Dodds et al. 2009). It is possible that if diatoms are at the base of the food web, then there is less energy in the exported productivity to depth because of the generally low nutritional value of diatoms for zooplankton (Jones and Flynn 2005) and therefore, presumably, for corals. By extension, high concentrations of silica would indicate lower food quality at depth, which would provide insufficient energy for the corals to elevate their growth rates to form large, complex reef structures. This is highly speculative, but is an interesting hypothesis for future research.

Even in areas where there has been an abundance of research effort over the years, there may still be reefs to be discovered. Some of the most recent discoveries of CWC reef structures have been in the North and Central Pacific seamounts (Baco et al. 2017), many of which are threatened by commercial fishing (Baco et al. 2020) and the potential for deep-sea mining in the prime cobalt crust zone (Miller et al. 2018). Our models predicted additional discoveries of CWC reefs may be made in this region (Fig. 1.6), although it did not do a good job of recreating the boundary of reef development hypothesized to be a result of differential food availability and current direction north of Academician seamount in the NW Hawaiian Islands (Baco et al. 2017). In the East Pacific, there have been recent discoveries offshore British Columbia and California (Gómez et al. 2018) that suggest CWC reef development in other locations that are only marginally suitable in the model (Fig. 1.7a). Together, these discrepancies point to a significant need for exploration and environmental characterization of CWC habitats in this region.

In the Atlantic (Fig. 1.6), the model is largely in agreement with previous modeling efforts for *Lophelia pertusa* and other species of reef-forming scleractinians (Davies et al. 2008; Tittensor et al. 2009; Davies and Guinotte 2011). However, additional oceanographic variables were included here that were not commonly included in previous models, including sea-surface and bottom-water oceanography and current speeds along with their variability. For *L. pertusa*, sea-surface temperature over the reef locations was the lowest of any of the species, while the variance in SST was the highest. This finding may be related to the fact that the mixed layer was the deepest where *L. pertusa* was the primary reef-forming species present. At the scale of the modeling effort, this is primarily due to the frequency of *L. pertusa* reefs in the extreme northern Atlantic and in the area under the influence of the rapid and dynamic Gulf Stream in the western Atlantic. At the more local scale, it is also possible that the coral mound structures themselves contribute to the increased depth of the overlying mixed layer through their hydrodynamic interactions (Wheeler et al. 2007) and the generation of a "topographically-enhanced carbon pump" that leads to a positive feedback between mound growth and food supply (Soetaert et al. 2016).

In the South Atlantic, there is intensifying pressure from offshore energy activities that makes exploration in these regions an urgent need. The model output indicates that large areas offshore northern Brazil, north of the Amazon river outflow, and extending into French Guiana, Suriname, and Guyana (Fig. 1.6g), where there are only scattered accounts of CWCs (Barbosa et al. 2020), are highly suitable for CWC reef development. Along the Angola margin, north of the known coral sites (Le Guilloux et al. 2009), in the Gulf of Guinea near the island of Sao Tome, and further offshore on the St. Helena seamount chain there are numerous hotspots of potential reef development in this area of heavy offshore drilling activity (Fig. 1.6h). Exploration of these predicted CWC reefs is essential for the proper management of this industrial activity (Cordes and Levin 2018).

One of the largest gaps in our knowledge of global CWC and CWC reef distribution is in the Indian Ocean (Figs. S1.2 and S1.3). The model results highlight areas off of South Africa, north of Madagascar (Fig. 1.7c), off the Horn of Africa, the entire Indian continental shelf break, and large areas of the Andaman Sea off of Burma and Malaysia as key locations for potential reef development (Fig. 1.7b). There is a desperate need for further exploration in the Indian Ocean that will reveal the true distribution of CWC reefs and their relative importance to the function and ecosystem services of this Ocean.

Finally, there is a large area to the north and south of the island of Taiwan that is highly suitable for CWC reef development (Fig. 1.7d). In the East China Sea, many of the areas around the Okinawa Trough between mainland China and Taiwan, Okinawa, and Kyushu appear to be highly suitable for reef development. Although there has been active research on the cold seeps of this area, some of which mention octocoral colonies (Cao et al. 2019) and *Enallopsammia rostrata* (Dong et al. 2021), we are not aware of records of CWC reefs from the area. There are also isolated reports of CWC from the South China Sea, including the framework-forming *Enallopsammia* sp. (Xu et al. 2019), that indicate that reef formation could be occurring under the right conditions. In the area of this observation, there are numerous seep-related authigenic carbonates that could provide ample substrate and rugose bathymetry that could support locally elevated currents and food delivery (Xu et al. 2019), all of the necessary precursors for reef development.

1.4.2 Role of Cold-Water Coral Reefs

The tremendous size of the area potentially occupied by CWC reefs has implications for the role they play as essential components of a high-functioning global ocean. CWC reefs provide significant ecosystem services including direct provisioning through fisheries exploitation and unique marine genetic resources, indirect supporting services such as habitat creation and nutrient regeneration, regulating services including carbon sequestration on geological time scales, and cultural services derived from the beauty and awe inspired by the seascapes they create (Thurber et al. 2014; Cordes et al. 2021). We are just beginning to be able to quantify these services, and this is an active area of current and future research (Armstrong et al. 2012; Cordes et al. 2021).

Cold-water coral reefs are most closely tied to fisheries production in the areas of the world where fisheries have expanded into deeper waters. The utilization of CWC gardens and reefs for habitat, shelter, nursery, and feeding grounds for numerous fish and shrimp species has been well established to the point where CWCs are considered "Essential Fish Habitat," which is managed directly in the USA and elsewhere (Peterson et al. 2000; Andrews et al. 2002). CWC reefs across the globe harbor a distinct community from the surrounding deep sea, and this has been explicitly demonstrated in the Southwest Pacific (O'Hara et al. 2008), Gulf of Mexico (Cordes et al. 2008; Lessard-Pilon et al. 2010; Caesar et al. 2021), Southeast USA (Ross and Quattrini 2009), North Atlantic (Roberts et al. 2008; Linley et al. 2017; Price et al. 2019), Mediterranean (D'Onghia et al. 2019), and off West Africa (Colman et al. 2005). The highest diversity of fauna on the reefs appears to be on dead, standing coral structure, which provides complex habitat without the direct predation by the suspension feeding corals, although elevated biodiversity is still found on live coral and coral rubble (Mortensen and Fosså 2006; Cordes et al. 2008; Price et al. 2019), and in a "halo" around the reefs for over 100 m distance (Henry and Roberts 2007; Demopoulos et al. 2014).

The role of CWC reefs in supporting fisheries goes well beyond habitat provision. The deep sea has long been understood to be the source of nutrients that drives high productivity in upwelling regions. It is in recent years that we have better defined the prominent role that CWC reefs play in nitrogen and phosphorus cycling. In a series of studies from the waters of the UK and Ireland, the metabolic rates of cold-water corals and the communities they support revealed them to be hotspots of secondary productivity and biogeochemical cycling in the deep ocean (Van Oevelen et al. 2009; Cathalot et al. 2015; Rix et al. 2018; de Froe et al. 2019; Maier et al. 2021).

The role of cold-water corals in carbon sequestration remains a key area for future research. CWC reefs might have the potential to be important blue carbon sinks. It is well understood that carbon is stored in carbonate mounds on geological time scales (Dorschel et al. 2007; Titschack et al. 2009, 2015, 2016). The dynamics of mound formation are best understood on the UK and Irish margins where a number of mounds have been cored (Mienis et al. 2009; Frank et al. 2011; Thierens et al. 2013; Wienberg et al. 2020; O'Reilly et al. 2022). In these waters, there are episodic periods of mound growth, but the majority of these records only go back as far as the last glaciation event. However, an International Ocean Drilling Program core of a 155-m deep section of the Challenger Mound in the Porcupine Seabight revealed that coral mounds extend back 2.6 Mya, but with significant hiatuses in mound accumulation (Kano et al. 2007; Thierens et al. 2010). In the central Mediterranean Sea, the earliest period of mound development appears to have begun nearly 400 kya, with glacial-interglacial periodicity of mound development corresponding to the shift in the depth of the intersection between Atlantic (AW) and Levantine Intermediate Waters (LIW) (Corbera et al. 2022b), while mound development in the western Mediterranean probabaly began as early as 1 Ma ago and showed a remarkably coherent pattern with precession-driven changes in African hydroclimate (Wienberg et al. 2022). Off the West African coast, the mounds have persisted on similar time scales but the bathymetric distribution of suitable habitat for coral growth has shifted over time such that there are roughly parallel series of mounds separated by 150–200 m depth off of Mauritania (Wienberg et al. 2018) and Morocco (Hebbeln et al. 2019). Over these time scales, and over the spatial domain that the models presented here now reveal to be suitable for reef formation, these carbonate mounds can accumulate massive volumes of carbon.

There is a wide body of evidence that tide–topography interactions enhance food delivery at CWC reef locations via internal waves, turbidity, and focused downwelling (Mienis et al. 2007; Dullo et al. 2008; Davies et al. 2009, 2010; Mohn et al. 2014; van Haren et al. 2014; De Clippele et al. 2018; de Froe et al. 2019; Hanz et al. 2019; Wang et al. 2019) and may promote positive feedbacks between coral growth and these oceanographic features that enhance CWC growth, persistence, and reef/mound development in the long-term (Van der Kaaden et al. 2021). Evidence from cores across the North Atlantic suggest that consistent, high food delivery relative to off-reef areas may be a primary driver of reef persistence through time (Portilho-Ramos et al. 2022). While this study presents evidence that monthly variability in oceanographic conditions plays a role in CWC distributions, pulses of nutrient-rich waters on timescales of days-to-weeks have been observed by benthic landers at CWC reefs (Mienis et al. 2012; Lim et al. 2020) but these are not resolved by global ocean models. Getting a better understanding of these dynamics will have a significant effect on the estimates for carbon cycling on the reefs.

It remains to be seen how carbon is processed on the reefs over shorter time scales. It is clear that there is strong benthic-pelagic coupling in many of these continental margin systems. Dense aggregations of mid-water, diel vertical migrators (DVM) have been seen accumulating over topographic highs, including CWC mounds, along continental margins (Davies et al. 2010; Mienis et al. 2012; Van Engeland et al. 2019). It is clear that the zooplankton in these communities are among the food sources of the corals as well as the galatheids (Schnabel 2020, 2022) and ophiuroids living on the reefs. If these processes are significant, a large carbon subsidy is provided to the reefs. It is estimated that the Norwegian reefs account for up to 35% of total benthic respiration of the entire Norwegian shelf, cycling up to 5% of the total productivity of the region, with the majority of particulate organic matter retained on the reef as mucus (White et al. 2012). Furthermore, with the recent finding that we may have been over-estimating rates of microbial carbon decomposition in the deep sea (Amano et al. 2023), much of this carbon may remain within the reef structure rather than being cycled back to dissolved inorganic carbon. Further in situ measurements of carbon degradation are required to test this hypothesis.

1.4.3 Threats to Cold-Water Coral Reefs

CWC reefs have already been subjected to human impacts and there are multiple looming threats to their health, the effects of which are compounded by our incomplete knowledge of their distribution and dynamic role in global ocean function. The first documented interactions with CWC corals came from fishermen in the North Atlantic, and the early species descriptions were from these samples (Roberts et al. 2009). In the present-day North Atlantic, many of the impacts from the fisheries are now unable to be quantified because of the long history of bottom trawling, but it was the impact on CWC reefs (Wheeler et al. 2005) that helped drive the push to ban bottom trawling in deep waters throughout the European Union (EU) (De Santo and Jones 2007). Trawling is banned in areas of known reef structures in Norway and 18 marine protected areas have been established to protect them. Because CWCs can regrow from fragments, protecting impacted sites can still have benefits (Huvenne et al. 2016). There is now a deep-sea cabled observatory, the LoVe Observatory, that is monitoring the recovery of these populations in Norwegian marine protected areas (Godø et al. 2014).

In the South Pacific, there is a long history of deep-water fisheries, and it was on the seamounts off of Tasmania that the impacts from fisheries on cold-water coral reefs were first reported in the primary scientific literature (Koslow et al. 2001). There was a documented decline in coral cover from 20 to 50% before fishing to 1–5% after less than 10 trawls (Clark and Tittensor 2010). In the 20 years since those observations, there has been minimal recovery and the majority of the impacts still persist (Clark et al. 2022). The North Pacific seamounts similarly have a long history of trawl fisheries as well as CWC fisheries for precious corals (Grigg 2002; Clark and Koslow 2007; Victorero et al. 2018). Trawling continues on high seas seamounts in this region in areas with CWC reefs, with documented damage to the reefs and other CWC from fisheries (Baco et al. 2020). However, no areas of North Pacific CWC reefs have yet been explicitly protected, although some of the large marine protected areas of the Pacific (U.S. Pacific Remote Islands National Marine Sanctuary, Papahānaumokuākea Marine National Monument, Phoenix Islands Protected Area) include areas of CWC reefs within them.

In the Atlantic off the Southeast USA there is intense fishing activity, but less use of bottom-contact gear and 60,000 km^2 of deep coral habitat (primarily *Lophelia pertusa* and *Enallopsammia profunda* reefs) have been designated as "Habitat Areas of Particular Concern." The first CWC reefs to be protected (1984) in the USA were the *Oculina* Banks off the east coast of Florida, constructed by the stony coral *Oculina varicosa*. Due to lack of enforcement, these reefs were subsequently decimated and have not recovered (Reed 2002). In Brazil, the deep-water shrimp fishery has targeted upper-slope areas for decades, and many areas were shown to harbor CWC reefs (Perez et al. 2009), but impacts are poorly documented. Similarly, in West Africa, there is more bottom trawling in deeper waters in recent years, but little documentation of any impacts to the living or relict reefs in the region (Wienberg et al. this issue).

Significant impacts from the oil and gas industry have also been documented on CWC reefs. The most pronounced of these is the *Deepwater Horizon* incident in the deep Gulf of Mexico in 2010. Impacts from this explosion and subsequent spill were documented to deep-sea coral gardens up to 22 km away (Fisher et al. 2014), and it is highly probable that there were additional impacts that went undocumented (Georgian et al. 2020). Smaller-scale impacts from normal operations are also apparent in the highly industrialized deep Gulf of Mexico. The pace of new offshore drilling has slowed in the Gulf of Mexico and the North Sea, but continues to expand in the Caribbean, South America, and off West Africa where it threatens known and as-yet undiscovered CWC reefs.

On the immediate horizon is another threat from a developing industry, deep-sea mining (Washburn et al. 2019; Simon-Lledó et al. 2020). This industry will target manganese nodules on the seafloor, seafloor massive sulfides on active and extinct hydrothermal vents, cobalt and nickel crusts on seamounts, and phosphate deposits in areas of the continental shelf near CWC reefs. The most significant overlap with CWC reef distribution will be on the seamounts of the Central Pacific, which lie in the prime cobalt crust zone. This is an area of active deep-sea exploration, but CWC reefs have appeared in areas that we would not have predicted a few short years ago, and new discoveries will continue to come.

Beyond these direct industrial impacts, the pervasive effects of global climate and ocean change will continue to put pressure on CWC reefs in the future. Predictions from the North Atlantic suggest changes in the distribution of suitable habitat for CWC reefs to favor deeper waters and higher latitudes (Morato et al. 2020; Gasbarro et al. 2022), while models for CWC off New Zealand predicted that for most taxa there will be substantial shifts in the location of the most suitable habitat and decreases in the area of such habitat by the end of the twenty-first century, driven primarily by decreases in seafloor oxygen concentrations, shoaling of aragonite and calcite saturation horizons, and increases in nitrogen concentrations (Anderson et al. 2022). Whether CWC species can shift their distribution through larval dispersal into these new areas remains to be seen (Morato et al. 2020; Anderson et al. 2022; Gasbarro et al. 2022). Corals from the reefs of the southeastern USA have been shown to be resilient to temperature fluctuations, although this comes at a cost with increased respiration and excretion supported by a shift to protein-dominated catabolism (Gómez et al. 2022). Additional stressors, such as ocean acidification and deoxygenation, will only exacerbate these responses and potentially push coral reefs, which includes both the live coral and the dead coral framework, to a tipping point (Barnhill et al. 2023). This has been observed in the paleontological record where multiple stressors have led to mound failure on a large scale off West Africa (Hebbeln et al. 2019). Once the coral framework is no longer protected by living coral tissue, ocean acidification can lead to the dissolution and decomposition of the CWC reef structure (Hennige et al. 2020). The models presented here suggest that CWC reefs are at the edge of their environmental niche in the Gulf of Mexico, the Eastern Pacific, the Mediterranean Sea, and other areas of the world. Some may already be past that niche, much of the reef structure in the North Pacific is already below the aragonite saturation horizon (ASH) (Baco et al. 2017), potentially due to shoaling of the ASH that has occurred in that region (Feely et al. 2012; Carter et al. 2017). Without continued efforts to protect CWC reefs from multiple stressors, restore CWC habitats where we can, and address the issue of global change directly, we may lose large areas of CWC reefs before we know that they are there.

1.4.4 Future Work

There is so much additional work to be done on CWC reefs that it is difficult to know where to begin. Before we can study the details of how these habitats function, we need to continue to explore the world's oceans for CWC reef habitats. The model presented here should help to guide future exploratory efforts, although new discoveries await us wherever we go. As we continue to make these discoveries, we improve our understanding of the fundamental CWC and CWC reef niche, which will lead to improvements in our predictive capacity for CWC distribution. We hope that this study and the regional information contained in the chapters of this volume stimulate additional discoveries that will, in turn, lead to more complete future iterations of global CWC modeling efforts.

One of the biggest needs in CWC research is the documentation of ecosystem services (Armstrong et al. 2014; Cordes et al. 2021). It is apparent that CWC reefs cover large areas of the world ocean, therefore the processes carried out by these ecosystems must play a significant role in global biogeochemical cycling. Better quantitative estimates of biomass, benthic respiration, calcification, nitrogen cycling, and phosphorus cycling are critical to the understanding of CWC reef systems. These assessments need to be conducted at a regional scale because of the wide range of environmental conditions encountered in different coral habitats as revealed by this study and many others, and the genetic and niche differentiation that has followed form this differential selective pressure. To ensure that these ecosystems are maintained, better continuous monitoring of reef habitats is required, particularly where there are threats from industrial activity or where the CWCs are near the edge of their fundamental niche. More information on the reproduction, larval dispersal, and connectivity of CWC habitats will improve our ability to manage these systems where they are threatened.

Advances in technology continue to drive research innovation especially in the seabed mapping where 3D habitat characterization and the use of autonomous technologies is rapidly advancing, primarily driven by increased computational power (de Oliveira et al. 2021, 2022; Thornton et al. 2021). Nevertheless, the global capacity for deep-sea research needs to be improved. Many of the areas where exploration and documentation of CWC reef distribution are most desperately needed are where our capacity is at the lowest, particularly in the South Pacific, South Atlantic, and the Indian Ocean. There are efforts underway as part of the UN Ocean Decade, in particular the Challenger-150 Program (Howell et al. 2020), to bring greater capacity for oceanographic and deep-sea research to the global South and developing nations, but these efforts require large amounts of resources. Improved collaborations among industry, government, and academia could facilitate this transfer of capacity and bring essential resources to bear on these critical questions. The next few years will reveal whether we have been able to obtain the knowledge required to manage the industrial development of the deep ocean and avoid some of these potential impacts on the Cold-Water Coral Reefs of the World.

Acknowledgments The authors wish to thank all of the ocean explorers who have come before us and revealed the abundance of the cold-water coral reefs of the world. This work would not have been possible without the officers and crew of the ships who have explored the global ocean and the technicians and pilots operating the sampling gear and vehicles that made these observations possible. These are too numerous to possibly list here, but their stories live on through the discoveries they have made.

References

Addamo AM, Vertino A, Stolarski J, Garcia-Jimenez R, Taviani M, Machordom A (2016) Merging scleractinian genera: the overwhelming genetic similarity between solitary Desmophyllum and colonial Lophelia. BMC Evol Biol 16. https://doi.org/10.1186/s12862-016-0654-8

Almada G, Bernardino AF (2017) Conservation of deep-sea ecosystems within offshore oil fields on the Brazilian margin, SW Atlantic. Biol Conserv 206:92–101. https://doi.org/10.1016/j.biocon.2016.12.026

Amano C, Reinthaler T, Sintes E, Varela MM, Stefanschitz J, Kaneko S, Nakano Y, Borchert W, Herndl GJ, Utsumi M (2023) A device for assessing microbial activity under ambient hydrostatic pressure: The in situ microbial incubator (ISMI). Limnol Oceanogr Methods 21(2): 69–81. https://doi.org/10.1002/lom3.10528

Anderson MJ (2001) A new method for non-parametric multivariate analysis of variance. Austral Ecol 26(1):32–46. https://doi.org/10.1111/j.1442-9993.2001.01070.pp.x

Anderson OF, Clark MR (2003) Analysis of bycatch in the fishery for orange roughy, Hoplostethus atlanticus, on the South Tasman Rise. Mar Freshw Res 54(5):643–652. https://doi.org/10.1071/mf02163

Anderson KD, Heron SF, Pratchett MS (2015) Species-specific declines in the linear extension of branching corals at a subtropical reef, Lord Howe Island. Coral Reefs 34(2):479–490. https://doi.org/10.1007/s00338-014-1251-1

Anderson OF, Guinotte JM, Rowden AA, Tracey DM, Mackay KA, Clark MR (2016) Habitat suitability models for predicting the occurrence of vulnerable marine ecosystems in the seas around New Zealand. Deep Sea Res Part I Oceanogr Res Pap 115:265–292. https://doi.org/10.1016/j.dsr.2016.07.006

Anderson OF, Stephenson F, Behrens E, Rowden AA (2022) Predicting the effects of climate change on deep-water coral distribution around New Zealand-Will there be suitable refuges for protection at the end of the 21st century? Glob Chang Biol 28(22):6556–6576. https://doi.org/10.1111/gcb.16389

Andrews AH, Cordes EE, Mahoney MM, Munk K, Coale KH, Cailliet GM, Heifetz J (2002) Age, growth and radiometric age validation of a deep-sea, habitat-forming gorgonian (Primnoa resedaeformis) from the Gulf of Alaska. Hydrobiologia 471(1):101–110. https://doi.org/10.1023/A:1016501320206

Angeletti L, Castellan G, Montagna P, Remia A, Taviani M (2020) The "Corsica channel cold-water coral province" (Mediterranean Sea). Front Mar Sci 7. https://doi.org/10.3389/fmars.2020.00661

Appah JKM, Lynch SA, Lim A, Riordan RO, O'Reilly L, de Oliveira L, Wheeler AJ (2022) A health survey of the reef forming scleractinian cold-water corals Lophelia pertusa and Madrepora oculata in a remote submarine canyon on the European continental margin, NE Atlantic. J Invertebr Pathol 192. https://doi.org/10.1016/j.jip.2022.107782

Arbizu PM (2017) pairwiseAdonis: pairwise multilevel comparison using Adonis. R package version 041

Armstrong CW, Foley NS, Tinch R, van den Hove S (2012) Services from the deep: Steps towards valuation of deep sea goods and services. Ecosyst Serv 2:2–13. https://doi.org/10.1016/j.ecoser.2012.07.001

Armstrong CW, Foley NS, Kahui V, Grehan A (2014) Cold water coral reef management from an ecosystem service perspective. Mar Policy 50:126–134. https://doi.org/10.1016/j.marpol.2014.05.016

Aumont O, Ethe C, Tagliabue A, Bopp L, Gehlen M (2015) PISCES-v2: an ocean biogeochemical model for carbon and ecosystem studies. Geosci Model Dev 8(8):2465–2513. https://doi.org/10.5194/gmd-8-2465-2015

Baco AR, Morgan N, Roark EB, Silva M, Shamberger KEF, Miller K (2017) Defying dissolution: discovery of deep-sea scleractinian coral reefs in the North Pacific. Sci Rep 7. https://doi.org/10.1038/s41598-017-05492-w

Baco AR, Morgan NB, Roark EB (2020) Observations of vulnerable marine ecosystems and significant adverse impacts on high seas seamounts of the northwestern Hawaiian Ridge and Emperor Seamount Chain. Mar Policy 115. https://doi.org/10.1016/j.marpol.2020.103834

Barbosa RV, Davies AJ, Sumida PYG (2020) Habitat suitability and environmental niche comparison of cold-water coral species along the Brazilian continental margin. Deep Sea Res Part I Oceanogr Res Pap 155. https://doi.org/10.1016/j.dsr.2019.103147

Barnhill KA, Roberts JM, Myers-Smith I, Williams M, Dexter KG, Ryan C, Wolfram U, Hennige SJ (2023) Incorporating dead material in ecosystem assessments and projections. Nat Clim Change 13(2): 113–115. https://doi.org/10.1038/s41558-022-01565-5

Becker EL, Cordes EE, Macko SA, Fisher CR (2009) Importance of seep primary production to Lophelia pertusa and associated fauna in the Gulf of Mexico. Deep Sea Res I Oceanogr Res Pap 56(5): 786–800

Bell N, Smith J (1999) Coral growing on North Sea oil rigs. Nature (London) 402:601

Bostock HC, Tracey DM, Currie KI, Dunbar GB, Handler MR, Fletcher SEM, Smith AM, Williams MJM (2015) The carbonate mineralogy and distribution of habitat-forming deep-sea corals in the southwest pacific region. Deep Sea Res Part I Oceanogr Res Pap 100:88–104. https://doi.org/10.1016/j.dsr.2015.02.008

Brooke S, Jarnegren J (2013) Reproductive periodicity of the scleractinian coral Lophelia pertusa from the Trondheim Fjord,

Norway. Mar Biol 160(1):139–153. https://doi.org/10.1007/s00227-012-2071-x

Brooke S, Sogluizzo A (2017) Chapter 11: Invertebrate reproductive biology. In: CSA Ocean Sciences (ed) Exploration and research of mid-Atlantic deepwater hard bottom habitats and shipwrecks with emphasis on canyons and coral communities, vol OCS Study BOEM 2017-060. U.S. Dept. of the Interior, Bureau of Ocean Energy Management, Atlantic OCS Region, pp 537–577

Buhl-Mortensen L, Vanreusel A, Gooday AJ, Levin LA, Priede IG, Buhl-Mortensen P, Gheerardyn H, King NJ, Raes M (2010) Biological structures as a source of habitat heterogeneity and biodiversity on the deep ocean margins. Mar Ecol 31(1):21–50. https://doi.org/10.1111/j.1439-0485.2010.00359.x

Buisson L, Thuiller W, Casajus N, Lek S, Grenouillet G (2010) Uncertainty in ensemble forecasting of species distribution. Glob Chang Biol 16:1145–1157. https://doi.org/10.1111/j.1365-2486.2009.02000.x

Caesar L, McCarthy GD, Thornalley DJR, Cahill N, Rahmstorf S (2021) Current Atlantic meridional overturning circulation weakest in last millennium. Nat Geosci 14(3):118. https://doi.org/10.1038/s41561-021-00699-z

Cairns SD (1982) A new subfamily of operculate stylasterine (Coelenterata, Hydrozoa) from the Sub-Antarctic. J Nat Hist 16(1): 71–81. https://doi.org/10.1080/00222938200770051

Cairns SD (1995) New records of azooxanthellate stony corals (Cnidaria, Scleractinia and Stylasteridae) from the neogene of Panama and Costa-Rica. Proc Biol Soc Wash 108(3):533–550

Cairns SD (2007) Deep-water corals: an overview with special reference to diversity and distribution of deep-water scleractinian corals. Bull Mar Sci 81:311–322

Cao H, Wu N, Liu W, Liu C, Jiang Z, Geng W, Zhang X, Wang L, Zhai B, Jiang X, Liu L, Li X (2019) Mineralogical and geochemical records of seafloor cold seepage history in the Northern Okinawa Trough, East China Sea. Deep Sea Res I Oceanogr Res Pap 155: 103165. https://doi.org/10.1016/j.dsr.2019.103165

Carter BR, Feely RA, Mecking S, Cross JN, Macdonald AM, Siedlecki SA, Talley LD, Sabine CL, Millero FJ, Swift JH, Dickson AG, Rodgers KB (2017) Two decades of Pacific anthropogenic carbon storage and ocean acidification along Global Ocean Ship-lebased Hydrographic Investigations Program sections P16 and P02. Global Biogeochem Cycles 31(2):306–327. https://doi.org/10.1002/2016gb005485

Cathalot C, Van Oevelen D, Cox TJS, Kutti T, Lavaleye M, Duineveld G, Meysman FJR (2015) Cold-water coral reefs and adjacent sponge grounds: hotspots of benthic respiration and organic carbon cycling in the deep sea. Front Mar Sci 2. https://doi.org/10.3389/fmars.2015.00037

Chemel M, Meistertzheim AL, Mouchi V, Lartaud F (2023) Reproductive biology of the two main reef-building cold-water coral species (Desmophyllum pertusum and Madrepora oculata) in the Mediterranean Sea. Deep Sea Res I Oceanogr Res Pap 195:103984. https://doi.org/10.1016/j.dsr.2023.103984

Clark MR, Koslow JA (2007) Impacts of fisheries on seamounts. In: Pitcher TJ, Morato T, Hart PJB, Clark MR, Haggan N, Santos RS (eds) Seamounts: ecology, fisheries & conservation. Blackwell, Oxford, pp 413–441. https://doi.org/10.1002/9780470691953.ch19

Clark MR, Tittensor DP (2010) An index to assess the risk to stony corals from bottom trawling on seamounts. Mar Ecol Evol Perspect 31:200–211. https://doi.org/10.1111/j.1439-0485.2010.00392.x

Clark MR, Consalvey M, Rowden AA (2016) Biological sampling in the deep sea. Wiley, New York. https://doi.org/10.1002/9781118332535

Clark MR, Bowden DA, Stewart R, Rowden AA, Goode SL (2022) Seamount recovery: analysis of 20 years of time-series data from the Graveyard Knolls, Chatham Rise, New Zealand. New Zealand Aquatic Environment and Biodiversity Report No. 292

Colman JG, Gordon DM, Lane AP, Forde MJ, Fitzpatrick JJ (2005) Carbonate mounds off Mauritania, Northwest Africa: status of deep-water corals and implications for management of fishing and oil exploration activities. In: Freiwald AaR JM (ed) Cold-water corals and ecosystems. Springer, Berlin, pp 417–441

Corbera G, Lo Iacono C, Simarro G, Grinyo J, Ambroso S, Huvenne VAI, Mienis F, Carreiro-Silva M, Martins I, Mano B, Orejas C, Larsson A, Hennige S, Gori A (2022a) Local-scale feedbacks influencing cold-water coral growth and subsequent reef formation. Sci Rep 12(1). https://doi.org/10.1038/s41598-022-24711-7

Corbera G, Lo Iacono C, Standish CD, Gracia E, Ranero C, Huvenne VAI, Anagnostou E, Foster GL (2022b) Glacial-aged development of the Tunisian Coral Mound Province controlled by glacio-eustatic oscillations and changes in surface productivity. Mar Geol 446. https://doi.org/10.1016/j.margeo.2022.106772

Cordes EE, Levin LA (2018) Exploration before exploitation. Science 359(6377):719. https://doi.org/10.1126/science.aat2637

Cordes EE, McGinley MP, Podowski EL, Becker EL, Lessard-Pilon S, Viada ST, Fisher CR (2008) Coral communities of the deep Gulf of Mexico. Deep Sea Res Part I Oceanogr Res Pap 55(6):777–787. https://doi.org/10.1016/j.dsr.2008.03.005

Cordes EE, Arnaud-Haond S, Bergstad O-A, Da Costa Falcao AP, Freiwald A, Roberts JM (2016a) Chapter 42: Cold-water corals. In: Inniss L, Simcock A (eds) The first global integrated marine assessment: world ocean assessment I. United Nations, New York

Cordes EE, Jones DOB, Schlacher TA, Amon DJ, Bernardino AF, Brooke S, Carney R, DeLeo DM, Dunlop KM, Escobar-Briones EG, Gates AR, Genio L, Gobin J, Henry LA, Herrera S, Hoyt S, Joye M, Kark S, Mestre NC, Metaxas A, Pfeifer S, Sink K, Sweetman AK, Witte U (2016b) Environmental impacts of the deep-water oil and gas industry: a review to guide management strategies. Front Environ Sci 4. https://doi.org/10.3389/fenvs.2016.00058

Cordes EE, Clark MR, Evans K, Hennige S, Kazanidis G (2021) Chapter 7E: Cold water corals. In: Ruwa R, Simcock A (eds) The second world ocean assessment. United Nations, New York

Davies AJ, Guinotte JM (2011) Global habitat suitability for framework-forming cold-water corals. PLoS One 6:e18483. https://doi.org/10.1371/journal.pone.0018483

Davies AJ, Wisshak M, Orr JC, Roberts MJ (2008) Predicting suitable habitat for the cold-water coral Lophelia pertusa (Scleractinia). Deep Sea Res I Oceanogr Res Pap 55(8):1048–1062. https://doi.org/10.1016/j.dsr.2008.04.010

Davies AJ, Duineveld GCA, Lavaleye MSS, Bergman MJN, van Haren H, Roberts JM (2009) Downwelling and deep-water bottom currents as food supply mechanisms to the cold-water coral Lophelia pertusa (Scleractinia) at the Mingulay Reef complex. Limnol Oceanogr 54(2):620–629

Davies AJ, Duineveld GCA, Van Weering TCE, Mienis F, Quattrini AM, Seim HE, Bane JM, Ross SW (2010) Short-term environmental variability in cold-water coral habitat at Viosca Knoll, Gulf of Mexico. Deep-Sea Res I Oceanogr Res Pap 57(2):199–212

De Clippele LH, Huvenne VAI, Orejas C, Lundalv T, Fox A, Hennige SJ, Roberts JM (2018) The effect of local hydrodynamics on the spatial extent and morphology of cold-water coral habitats at Tisler Reef, Norway. Coral Reefs 37(1):253–266. https://doi.org/10.1007/s00338-017-1653-y

de Froe E, Rovelli L, Glud RN, Maier SR, Duineveld G, Mienis F, Lavaleye M, van Oevelen D (2019) Benthic oxygen and nitrogen exchange on a cold-water coral reef in the North-East Atlantic Ocean. Front Mar Sci 6(665). https://doi.org/10.3389/fmars.2019.00665

De Mol B, Van Rensbergen P, Pillen S, Van Herreweghe K, Van Rooij D, McDonnell A, Huvenne V, Ivanov M, Swennen R, Henriet JP (2002) Large deep-water coral banks in the Porcupine Basin, southwest of Ireland. Mar Geol 188(1–2):193–231

Demopoulos AWJ, Bourque JR, Frometa J (2014) Biodiversity and community composition of sediment macrofauna associated with deep-sea Lophelia pertusa habitats in the Gulf of Mexico. Deep Sea Res Part I Oceanogr Res Pap 93:91–103. https://doi.org/10.1016/j.dsr.2014.07.014

de Oliveira LMC, Lim A, Conti LA, Wheeler AJ (2021) 3D classification of cold-water coral reefs: a comparison of classification techniques for 3D reconstructions of cold-water coral reefs and seabed. Front Mar Sci 8. https://doi.org/10.3389/fmars.2021.640713

de Oliveira LMC, De Oliveira PA, Lim A, Wheeler AJ, Conti LA (2022) Developing mobile applications with augmented reality and 3D photogrammetry for visualisation of cold-water coral reefs and deep-water habitats. Geosciences 12(10). https://doi.org/10.3390/geosciences12100356

De Santo EM, Jones PJS (2007) Offshore marine conservation policies in the North East Atlantic: Emerging tensions and opportunities. Mar Policy 31(3):336–347. https://doi.org/10.1016/j.marpol.2006.10.001

Dias V, Oliveira F, Boavida J, Serrao EA, Goncalves JMS, Coelho MAG (2020) High coral bycatch in bottom-set gillnet coastal fisheries reveals rich coral habitats in Southern Portugal. Front Mar Sci 7

Dong D, Li X, Yang M, Gong L, Li Y, Sui J, Gan Z, Kou Q, Xiao N, Zhang J (2021) Report of epibenthic macrofauna found from Haima cold seeps and adjacent deep-sea habitats, South China Sea. Mar Life Sci Technol 3(1):1–12. https://doi.org/10.1007/s42995-020-00073-9

D'Onghia G, Sion L, Capezzuto F (2019) Cold-water coral habitats benefit adjacent fisheries along the Apulian margin (central Mediterranean). Fish Res 213:172–179. https://doi.org/10.1016/j.fishres.2019.01.021

Dons C (1944) Norges korallrev. Det kongelige norske videnskabers selskabs forhandlinger 16:37–82

Dodds LA, Black KD, Orr H, Roberts JM (2009) Lipid biomarkers reveal geographical differences in food supply to the coldwater coral Lophelia pertusa (Scleractinia). Mar Ecol Prog Ser 397:113–124

Dorschel B, Hebbeln D, Rüggeberg A, Dullo C (2007) Carbonate budget of a cold-water coral carbonate mound: propeller mound, porcupine seabight. Int J Earth Sci 96:73–83

Duineveld GCA, Lavaleye MSS, Berghuis EM (2004) Particle flux and food supply to a seamount cold water coral community (Galicia Bank, NW Spain). Mar Ecol Prog Ser 277:13–23

Duineveld GCA, Lavaleye MSS, Bergman MIN, De Stigter H, Mienis F (2007) Trophic structure of a cold-water coral mound community (Rockall Bank, NE Atlantic) in relation to the near-bottom particle supply and current regime. Bull Mar Sci 81(3):449–467

Dullo C, Floegel S, Rüggeberg A (2008) Cold-water coral growth in relation to the hydrography of the Celtic and Nordic European Continental Margin. Mar Ecol Prog Ser 371:165–176

Duncan PM (1870) Deep sea corals. Nature 1(10):267

Elith J, Ferrier S, Huettmann F, Leathwick JR (2005) The evaluation strip: A new and robust method for plotting predicted responses from species distribution models. Ecol Model 186:280–289. https://doi.org/10.1016/j.ecolmodel.2004.12.007

Etnoyer PJ, Wagner D, Fowle HA, Poti M, Kinlan B, Georgian SE, Cordes EE (2018) Models of habitat suitability, size, and age-class structure for the deep-sea black coral Leiopathes glaberrima in the Gulf of Mexico. Deep Sea Res Part II Topical Stud Oceanogr 150: 218–228. https://doi.org/10.1016/j.dsr2.2017.10.008

Fabri MC, Pedel L, Beuck L, Galgani F, Hebbeln D, Freiwald A (2014) Megafauna of vulnerable marine ecosystems in French mediterranean submarine canyons: Spatial distribution and anthropogenic impacts. Deep Sea Res Part II Topical Stud Oceanogr 104:184–207. https://doi.org/10.1016/j.dsr2.2013.06.016

Fagerstrom V, Brostrom G, Larsson AI (2022) Turbulence affects larval vertical swimming in the cold-water coral Lophelia pertusa. Front Mar Sci 9. https://doi.org/10.3389/fmars.2022.1062884

Fallon SJ, Thresher RE, Adkins J (2014) Age and growth of the cold-water scleractinian Solenosmilia variabilis and its reef on SW Pacific seamounts. Coral Reefs 33(1):31–38. https://doi.org/10.1007/s00338-013-1097-y

Feely RA, Sabine CL, Byrne RH, Millero FJ, Dickson AG, Wanninkhof R, Murata A, Miller LA, Greeley D (2012) Decadal changes in the aragonite and calcite saturation state of the Pacific Ocean. Global Biogeochem Cycles 26. https://doi.org/10.1029/2011gb004157

Fisher CR, Hsing PY, Kaiser CL, Yoerger DR, Roberts HH, Shedd WW, Cordes EE, Shank TM, Berlet SP, Saunders MG, Larcom EA, Brooks JM (2014) Footprint of Deepwater Horizon blowout impact to deep-water coral communities. Proc Natl Acad Sci U S A 111(32): 11744–11749. https://doi.org/10.1073/pnas.1403492111

Forsterra G, Haeussermann V, Laudien J, Jantzen C, Sellanes J, Munoz P (2014) Mass die-off of the cold-water coral Desmophyllum dianthus in the Chilean Patagonian fjord region. Bull Mar Sci 90(3): 895–899. https://doi.org/10.5343/bms.2013.1064

Fossa JH, Mortensen PB, Furevik DM (2002) The deep-water coral Lophelia pertusa in Norwegian waters: distribution and fishery impacts. Hydrobiologia 471:1–12. https://doi.org/10.1023/a:1016504430684

Fox AD, Henry LA, Corne DW, Roberts JM (2016) Sensitivity of marine protected area network connectivity to atmospheric variability. Roy Soc Open Sci 3(11). https://doi.org/10.1098/rsos.160494

Frank N, Freiwald A, López Correa M, Wienberg C, Eisele M, Hebbeln D, Van Rooij D, Henriet JP, Colin C, van Weering T, de Haas H, Buhl-Mortensen P, Roberts JM, De Mol B, Douville E, Blamart D, Hatte C (2011) Northeastern Atlantic cold-water coral reefs and climate. Geology 39:743–746

Frederiksen R, Jensen A, Westerberg H (1992) The distribution of the scleractinian coral *Lophelia pertusa* around the Faeroe islands and the relation to internal tidal mixing. Sarsia 77:157–171

Freiwald A, Henrich R, Paetzold J (1997) Anatomy of a deep-water coral reef mound from Stjernsund, West Finnmark, Northern Norway. In: James NP, Clarke JAD (eds) Cool-water carbonates, vol 56. Special publication SEPM (Society for Sedimentary Geology), Tulsa, pp 141–162

Freiwald A, Fossâ JH, Grehan A, Koslow T, Roberts JM (2004) Cold-water coral reefs: out of sight-no longer out of mind, UNEP-WCMC biodiversity series, vol 22. UNEP-WCMC, Cambridge

Freiwald A, Rogers A, Hall-Spencer J, Guinotte JM, Davies AJ, Yesson C, Martin CS, Weatherdon LV (2021) Global distribution of cold-water corals (version 5.1). Fifth update to the dataset in Freiwald et al. 2004 by UNEP-WCMC, in collaboration with Andre Freiwald and John Guinotte. https://doi.org/10.34892/72x9-rt61

Gasbarro R, Sowers D, Margolin A, Cordes EE (2022) Distribution and predicted climatic refugia for a reef-building cold-water coral on the southeast US margin. Glob Chang Biol 28(23):7108–7125. https://doi.org/10.1111/gcb.16415

Gass SE, Roberts JM (2006) The occurrence of the cold-water coral Lophelia pertusa (Scleractinia) on oil and gas platforms in the North Sea: Colony growth, recruitment and environmental controls on distribution. Mar Pollut Bull 52(5):549–559

Gaurisas DY, Bernardino AFP (2023) Benthic biogeographic patterns on the deep Brazilian margin. PeerJ 11:e14585. https://doi.org/10.7717/peerj.14585

GEBCO (2021) Grid, *GEBCO Compilation Group*, 2021. https://doi.org/10.5285/c6612cbe-50b3-0cff-e053-6c86abc09f8f

Georgian SE, Shedd W, Cordes EE (2014) High-resolution ecological niche modelling of the cold-water coral Lophelia pertusa in the Gulf of Mexico. Mar Ecol Prog Ser 506:145–154. https://doi.org/10.3354/meps10816

Georgian SE, Anderson OF, Rowden AA (2019) Ensemble habitat suitability modeling of vulnerable marine ecosystem indicator taxa to inform deep-sea fisheries management in the South Pacific Ocean. Fish Res 211:256–274. https://doi.org/10.1016/j.fishres.2018.11.020

Georgian SE, Kramer K, Saunders M, Shedd W, Roberts H, Lewis C, Fisher C, Cordes E (2020) Habitat suitability modelling to predict the spatial distribution of cold-water coral communities affected by the Deepwater Horizon oil spill. J Biogeogr 47(7):1455–1466. https://doi.org/10.1111/jbi.13844

Godø OR, Johnsen S, Torkelsen T (2014) The LoVe ocean observatory is in operation. Mar Technol Soc J 48(2):24–30

Gómez CE, Wickes L, Deegan D, Etnoyer PJ, Cordes EE (2018) Growth and feeding of deep-sea coral Lophelia pertusa from the California margin under simulated ocean acidification conditions. PeerJ 6: e5671. https://doi.org/10.7717/peerj.5671

Gómez CE, Gori A, Weinnig AM, Hallaj A, Chung HJ, Cordes EE (2022) Natural variability in seawater temperature compromises the metabolic performance of a reef-forming cold-water coral with implications for vulnerability to ongoing global change. Coral Reefs 41(4):1225–1237. https://doi.org/10.1007/s00338-022-02267-2

Gori A, Grover R, Orejas C, Sikorski S, Ferrier-Pagès C (2014) Uptake of dissolved free amino acids by four cold-water coral species from the Mediterranean Sea. Deep Sea Res II Top Stud Oceanogr 99:42–50. https://doi.org/10.1016/j.dsr2.2013.06.007

Grigg RW (2002) Precious corals in Hawaii: discovery of a new bed and revised management measures for existing beds. Mar Fish Rev 64: 13–20

Guisan A, Thuiller W, Zimmermann NE (2017) Habitat suitability and distribution models: with applications in R. Ecology, biodiversity and conservation. Cambridge University Press, Cambridge. https://doi.org/10.1017/9781139028271

Gunnerus JE (1768) Om nogle Norske Coraller. Konglige Norske Videnskabers Selskabs Skrifter 4:38–73

Hall-Spencer J, Allain V, Fossa JH (2002) Trawling damage to Northeast Atlantic ancient coral reefs. Proc R Soc B Biol Sci 269(1490): 507–511. https://doi.org/10.1098/rspb.2001.1910

Hanz U, Wienberg C, Hebbeln D, Duineveld G, Lavaleye M, Juva K, Dullo WC, Freiwald A, Tamborrino L, Reichart GJ, Flögel S, Mienis F (2019) Environmental factors influencing benthic communities in the oxygen minimum zones on the Angolan and Namibian margins. Biogeosciences 16(22):4337–4356. https://doi.org/10.5194/bg-16-4337-2019

Hao T, Elith J, Guillera-Arroita G, Lahoz-Monfort JJ (2019) A review of evidence about use and performance of species distribution modelling ensembles like BIOMOD. Divers Distrib 25(5): 839–852. https://doi.org/10.1111/ddi.12892

Hao T, Elith J, Lahoz-Monfort JJ, Guillera-Arroita G (2020) Testing whether ensemble modelling is advantageous for maximising predictive performance of species distribution models. Ecography 43(4):549–558. https://doi.org/10.1111/ecog.04890

Häussermann V, Försterra G (2007) Large assemblages of cold-water corals in Chile: a summary of recent findings and potential impacts. Bull Mar Sci 81(3):195–207

Hebbeln D, Wienberg C, Wintersteller P, Freiwald A, Becker M, Beuck L, Dullo C, Eberli GP, Glogowski S, Matos L, Forster N, Reyes-Bonilla H, Taviani M, Party MSMSS (2014) Environmental forcing of the Campeche cold-water coral province, southern Gulf of Mexico. Biogeosciences 11(7):1799–1815. https://doi.org/10.5194/bg-11-1799-2014

Hebbeln D, Van Rooij D, Wienberg C (2016) Good neighbours shaped by vigorous currents: Cold-water coral mounds and contourites in the North Atlantic. Mar Geol 378:171–185. https://doi.org/10.1016/j.margeo.2016.01.014

Hebbeln D, Portilho-Ramos R, Wienberg C, Titschack J (2019) The fate of cold-water corals in a changing world: a geological perspective. Front Mar Sci 6. https://doi.org/10.3389/fmars.2019.00119

Hebbeln D, Wienberg C, Dullo WC, Freiwald A, Mienis F, Orejas C, Titschack J (2020) Cold-water coral reefs thriving under hypoxia. Coral Reefs 39(4):853–859. https://doi.org/10.1007/s00338-020-01934-6

Heezen BC, Tharp M (1977) World ocean floor panorama. Full color, painted by H Ber Mercator Projection, Scale 1 (23,230,300)

Hennige SJ, Wolfram U, Wickes L, Murray F, Roberts JM, Kamenos NA, Schofield S, Groetsch A, Spiesz EM, Aubin-Tam ME, Etnoyer PJ (2020) Crumbling reefs and cold-water coral habitat loss in a future ocean: evidence of "coralporosis" as an indicator of habitat integrity. Front Mar Sci 7. https://doi.org/10.3389/fmars.2020.00668

Henry LA, Roberts JM (2007) Biodiversity and ecological composition of macrobenthos on cold-water coral mounds and adjacent off-mound habitat in the bathyal Porcupine Seabight, NE Atlantic. Deep Sea Res Part I Oceanogr Res Pap 54(4):654–672. https://doi.org/10.1016/j.dsr.2007.01.005

Herrera S, Cordes EE (2023) Genome assembly of the deep-sea coral Lophelia pertusa. GigaByte. https://doi.org/10.46471/gigabyte.78

Hijmans RJ (2023) Geographic data analysis and modeling [R package raster version 3.6-13]

Hovland M, Croker PF, Martin M (1994) Fault-associated seabed mounds (carbonate knolls?) off western Ireland and north-west Australia. Mar Petroleum Geol 11(2):232–246

Howell KL, Holt R, Endrino IP, Stewart H (2011) When the species is also a habitat: Comparing the predictively modelled distributions of Lophelia pertusa and the reef habitat it forms. Biol Conserv 144(11): 2656–2665. https://doi.org/10.1016/j.biocon.2011.07.025

Howell KL, Hilario A, Allcock AL, Bailey DM, Baker M, Clark MR, Colaco A, Copley J, Cordes EE, Danovaro R, Dissanayake A, Escobar E, Esquete P, Gallagher AJ, Gates AR, Gaudron SM, German CR, Gjerde KM, Higgs ND, Le Bris N, Levin LA, Manea E, McClain C, Menot L, Mestre NC, Metaxas A, Milligan RJ, Muthumbi AWN, Narayanaswamy BE, Ramalho SP, Ramirez-Llodra E, Robson LM, Rogers AD, Sellanes J, Sigwart JD, Sink K, Snelgrove PVR, Stefanoudis PV, Sumida PY, Taylor ML, Thurber AR, Vieira RP, Watanabe HK, Woodall LC, Xavier JR (2020) A blueprint for an inclusive, global deep-sea ocean decade field program. Front Mar Sci 7. https://doi.org/10.3389/fmars.2020.584861

Huvenne VAI, Bett BJ, Masson DG, Le Bas TP, Wheeler AJ (2016) Effectiveness of a deep-sea cold-water coral Marine Protected Area, following eight years of fisheries closure. Biol Conserv 200:60–69. https://doi.org/10.1016/j.biocon.2016.05.030

Jantzen C, Haussermann V, Forsterra G, Laudien J, Ardelan M, Maier S, Richter C (2013) Occurrence of a cold-water coral along natural pH gradients (Patagonia, Chile). Mar Biol 160(10):2597–2607. https://doi.org/10.1007/s00227-013-2254-0

Jeffreys J (1869) The deep-sea dredging expedition in H.M.S. Porcupine. Nature 1:135–137

Jones RH, Flynn KJ (2005) Nutritional status and diet composition affect the value of diatoms as copepod prey. Science (New York, NY) 307(5714):1457–1459. https://doi.org/10.1126/science.1107767

Kano A, Ferdelman TG, Williams T, Henriet JP, Ishikawa T, Kawagoe N, Takashima C, Kakizaki Y, Abe K, Sakai S, Browning EL, Li X, Scientists IE (2007) Age constraints of the origin and growth history of a deep-water coral mound in the northeast Atlantic drilled during Integrated Ocean Drilling Program Expedition 307. Geology 35(11):1051–1054

Kazanidis G, Henry LA, Roberts JM (2021) Hidden structural heterogeneity enhances marine hotspots' biodiversity. Coral Reefs 40(5): 1615–1630. https://doi.org/10.1007/s00338-021-02114-w

Kiriakoulakis K, Fischer E, Wolff GA, Freiwald A, Grehan A, Roberts JM (2005) Lipids and nitrogen isotopes of two deep-water corals from the North-East Atlantic: initial results and implications for their nutrition. In: Freiwald A, Roberts JM (eds) Cold-water corals and ecosystems. Springer, Berlin, pp 715–729

Kitahara M, Cairns S (2021) Azooxanthellate scleractinia (Cnidaria, Anthozoa) from New Caledonia

Koslow JA, Gowlett-Holmes K, Lowry JK, O'Hara T, Poore GCB, Williams A (2001) Seamount benthic macrofauna off southern Tasmania: community structure and impacts of trawling. Mar Ecol Prog Ser 213:111–125. https://doi.org/10.3354/meps213111

Larsson AI, Jarnegren J, Stromberg SM, Dahl MP, Lundalv T, Brooke S (2014) Embryogenesis and larval biology of the cold-water coral Lophelia pertusa. PLoS One 9(7). https://doi.org/10.1371/journal.pone.0102222

Lawson CR, Hodgson JA, Wilson RJ, Richards SA (2014) Prevalence, thresholds and the performance of presence–absence models. Methods Ecol Evol 5(1):54–64. https://doi.org/10.1111/2041-210X.12123

Le Danois E (1948) Les profondeurs de la mer. Payot, Paris, pp 1–303

Le Guilloux E, Olu K, Bourillet JF, Savoye B, Iglésias SP, Sibuet M (2009) First observations of deep-sea coral reefs along the Angola margin. Deep Sea Res II Top Stud Oceanogr 56(23):2394–2403. https://doi.org/10.1016/j.dsr2.2009.04.014

Lellouche J-M, Eric G, Romain B-B, Gilles G, Angélique M, Marie D, Clément B, Mathieu H, Olivier LG, Charly R, Tony C, Charles-Emmanuel T, Florent G, Giovanni R, Mounir B, Yann D, Pierre-Yves LT (2021) The Copernicus global 1/12° oceanic and sea ice GLORYS12 reanalysis. Front Earth Sci 9. https://doi.org/10.3389/feart.2021.698876

Lessard-Pilon SA, Podowski EL, Cordes EE, Fisher CR (2010) Megafauna community composition associated with Lophelia pertusa colonies in the Gulf of Mexico. Deep Sea Res Part II Topical Stud Oceanogr 57(21–23):1882–1890. https://doi.org/10.1016/j.dsr2.2010.05.013

Li XH, Takashima C, Kano A, Sakai S, Chen YH, Xu BL, Scientists IE (2011) Pleistocene geochemical stratigraphy of the borehole 1317E (IODP Expedition 307) in Porcupine Seabight, SW of Ireland: applications to palaeoceanography and palaeoclimate of the coral mound development. J Quat Sci 26(2):178–189. https://doi.org/10.1002/jqs.1441

Lim A, Wheeler AJ, Price DM, O'Reilly L, Harris K, Conti L (2020) Influence of benthic currents on cold-water coral habitats: a combined benthic monitoring and 3D photogrammetric investigation. Sci Rep 10(1):19433. https://doi.org/10.1038/s41598-020-76446-y

Linley TD, Lavaleye M, Maiorano P, Bergman M, Capezzuto F, Cousins NJ, D'Onghia G, Duineveld G, Shields MA, Sion L, Tursi A, Priede IG (2017) Effects of cold-water corals on fish diversity and density (European continental margin: Arctic, NE Atlantic and Mediterranean Sea): Data from three baited lander systems. Deep Sea Res Part II Topical Stud Oceanogr 145:8–21. https://doi.org/10.1016/j.dsr2.2015.12.003

Lopez Correa M, Montagna P, Joseph N, Ruggeberg A, Fietzke J, Flogel S, Dorschel B, Goldstein SL, Wheeler A, Freiwald A (2012) Preboreal onset of cold-water coral growth beyond the Arctic Circle revealed by coupled radiocarbon and U-series dating and neodymium isotopes. Quat Sci Rev 34:24–43. https://doi.org/10.1016/j.quascirev.2011.12.005

Lowenstam H (1950) Niagaran reefs of the Great Lakes area. J Geol 58(4):430–487. https://doi.org/10.1086/625755

Lunden JJ, McNicholl CG, Sears CR, Morrison CL, Cordes EE (2014) Acute survivorship of the deep-sea coral Lophelia pertusa from the Gulf of Mexico under acidification, warming, and deoxygenation. Front Mar Sci 1. https://doi.org/10.3389/fmars.2014.00078

Mackay K, Rowden A, Bostock H, Tracey D (2014) Revisiting squires' coral coppice, Campbell Plateau, New Zealand. N Z J Mar Freshw Res 48(4):507–523

Maier SR, Kutti T, Bannister RJ, van Breugel P, van Rijswijk P, van Oevelen D (2019) Survival under conditions of variable food availability: Resource utilization and storage in the cold-water coral Lophelia pertusa. Limnol Oceanogr 64(4):1651–1671. https://doi.org/10.1002/lno.11142

Maier SR, Mienis F, de Froe E, Soetaert K, Lavaleye M, Duineveld G, Beauchard O, Kaaden ASV, Koch BP, van Oevelen D (2021) Reef communities associated with 'dead' cold-water coral framework drive resource retention and recycling in the deep sea. Deep Sea Res Part I Oceanogr Res Pap 175. https://doi.org/10.1016/j.dsr.2021.103574

Matos L, Wienberg C, Titschack J, Schmiedl G, Frank N, Abrantes F, Cunha MR, Hebbeln D (2017) Coral mound development at the Campeche cold-water coral province, southern Gulf of Mexico: Implications of Antarctic Intermediate Water increased influence during interglacials. Mar Geol 392:53–65. https://doi.org/10.1016/j.margeo.2017.08.012

Mayer L, Jakobsson M, Allen G, Dorschel B, Falconer R, Ferrini V, Lamarche G, Snaith H, Weatherall P (2018) The Nippon Foundation-GEBCO Seabed 2030 Project: The quest to see the world's oceans completely mapped by 2030. Geosciences 8(2). https://doi.org/10.3390/geosciences8020063

McKnight D, Estcourt I (1978) Benthic ecology of Caswell and Nancy sounds. N Z Oceanogr Inst Memoir 79:85–94

Middelburg JJ, Mueller CE, Veuger B, Larsson AI, Form A, van Oevelen D (2015) Discovery of symbiotic nitrogen fixation and chemoautotrophy in cold-water corals. Sci Rep 5. https://doi.org/10.1038/srep17962

Mienis F, De Stigter HC, White M, Duineveld G, De Haas H, Van Weering TCE (2007) Hydrodynamic controls on cold-water coral growth and carbonate-mound development at the SW and SE Rockall Trough Margin, NE Atlantic Ocean. Deep Sea Res I Oceanogr Res Pap 54(9):1655–1674

Mienis F, Van der Land C, De Stigter HC, Van de Vorstenbosch M, De Haas H, Richter T, Van Weering TCE (2009) Sediment accumulation on a cold-water carbonate mound at the Southwest Rockall Trough margin. Mar Geol 265(1–2):40–50

Mienis F, Duineveld GCA, Davies AJ, Ross SW, Seim H, Bane J, van Weering TCE (2012) The influence of near-bed hydrodynamic conditions on cold-water corals in the Viosca Knoll area, Gulf of Mexico. Deep Sea Res Part I Oceanogr Res Pap 60:32–45. https://doi.org/10.1016/j.dsr.2011.10.007

Miller KA, Thompson KF, Johnston P, Santillo D (2018) An overview of seabed mining including the current state of development, environmental impacts, and knowledge gaps. Front Mar Sci 4:418

Mohn C, Rengstorf A, White M, Duineveld G, Mienis F, Soetaert K, Grehan A (2014) Linking benthic hydrodynamics and cold-water coral occurrences: A high-resolution model study at three cold-water coral provinces in the NE Atlantic. Prog Oceanogr 122:92–104. https://doi.org/10.1016/j.pocean.2013.12.003

Moore DR, Bullis HR Jr (1960) A deep-water coral reef in the Gulf of Mexico. Bull Mar Sci 10(1):125–128

Morato T, Watson R, Pitcher TJ, Pauly D (2006) Fishing down the deep. Fish Fish 7(1):24–34. https://doi.org/10.1111/j.1467-2979.2006.00205.x

Morato T, Gonzalez-Irusta JM, Dominguez-Carrio C, Wei CL, Davies A, Sweetman AK, Taranto GH, Beazley L, Garcia-Alegre A, Grehan A, Laffargue P, Murillo FJ, Sacau M, Vaz S, Kenchington E, Arnaud-Haond S, Callery O, Chimienti G, Cordes E, Egilsdottir H, Freiwald A, Gasbarro R, Gutierrez-Zarate C, Gianni M, Gilkinson K, Hayes VWE, Hebbeln D, Hedges K, Henry LA, Johnson D, Koen-Alonso M, Lirette C, Mastrototaro F, Menot L, Molodtsova T, Munoz PD, Orejas C, Pennino MG, Puerta P, Ragnarsson SA,

Ramiro-Sanchez B, Rice J, Rivera J, Roberts JM, Ross SW, Rueda JL, Sampaio I, Snelgrove P, Stirling D, Treble MA, Urra J, Vad J, van Oevelen D, Watling L, Walkusz W, Wienberg C, Woillez M, Levin LA, Carreiro-Silva M (2020) Climate-induced changes in the suitable habitat of cold-water corals and commercially important deep-sea fishes in the North Atlantic. Glob Chang Biol 26(4): 2181–2202. https://doi.org/10.1111/gcb.14996

Mortensen P, Buhl-Mortensen L (2005) Morphology and growth of the deep-water gorgonians Primnoa resedaeformis and Paragorgia arborea. Mar Biol 147(3):775–788. https://doi.org/10.1007/s00227-005-1604-y

Mortensen PB, Fosså JH (2006) Species diversity and spatial distribution of invertebrates on deep-water Lophelia reefs in Norway. In: Proceedings of 10th international coral reef symposium. ICRS Okinawa, p 1868

Mortensen PB, Hovlund M, Brattegard T, Farestveit R (1995) Deep water bioherms of the scleractinian coral *Lophelia pertusa* (L.) at 64° N on the Norwegian shelf: structure and associated megafauna. Sarsia 80:145–158

Mortensen PB, Buhl-Mortensen L, Gebruk AV, Krylova EM (2008) Occurrence of deep-water corals on the Mid-Atlantic Ridge based on MAR-ECO data. Deep Sea Res Part II Topical Stud Oceanogr 55(1–2):142–152. https://doi.org/10.1016/j.dsr2.2007.09.018

Mueller CE, Larsson AI, Veuger B, Middelburg JJ, van Oevelen D (2014) Opportunistic feeding on various organic food sources by the cold-water coral Lophelia pertusa. Biogeosciences 11(1): 123–133. https://doi.org/10.5194/bg-11-123-2014

Mullins HT, Vanburen HM, Davis RA (1981) Anatomy of modern open-ocean windward carbonate slope—Northern Little Bahama Bank. AAPG Bull Am Assoc Pet Geol 65(5):962–963

Naimi B (2015) Usdm: Uncertainty analysis for species distribution models 1:1–12

Naumann MS, Orejas C, Ferrier-Pages C (2013) High thermal tolerance of two Mediterranean cold-water coral species maintained in aquaria. Coral Reefs 32(3):749–754. https://doi.org/10.1007/s00338-013-1011-7

Oksanen J, Simpson G, Blanchet FG, Kindt R, Legendre P, Minchin P, Hara R, Solymos P, Stevens H, Szöcs E, Wagner H, Barbour M, Bedward M, Bolker B, Borcard D, Carvalho G, Chirico M, De Cáceres M, Durand S, Weedon J (2022) Vegan community ecology package version 2.6-2 April 2022

O'Hara TD, Rowden AA, Williams A (2008) Coldwater coral habitats on seamounts: do they have a specialist fauna? Divers Distrib 14:925–934

O'Reilly L, Fentimen R, Butschek F, Titschack J, Lim A, Moore N, O'Connor OJ, Appah J, Harris K, Vennemann T, Wheeler AJ (2022) Environmental forcing by submarine canyons: Evidence between two closely situated cold-water coral mounds (Porcupine Bank Canyon and Western Porcupine Bank, NE Atlantic). Mar Geol 454. https://doi.org/10.1016/j.margeo.2022.106930

Orejas C, Jimenez C (2019) Mediterranean cold-water corals: past, present and future. Springer, Cham. doi:https://doi.org/10.1007/978-3-319-91608-8

Orejas C, Gori A, Lo Iacono C, Puig P, Gili JM, Dale MRT (2009) Cold-water corals in the Cap de Creus canyon, northwestern Mediterranean: spatial distribution, density and anthropogenic impact. Mar Ecol Prog Ser 397:37–51. https://doi.org/10.3354/meps08314

Orejas C, Ferrier-Pages C, Reynaud S, Gori A, Beraud E, Tsounis G, Allemand D, Gili JM (2011a) Long-term growth rates of four Mediterranean cold-water coral species maintained in aquaria. Mar Ecol Prog Ser 429:57–65. https://doi.org/10.3354/meps09104

Orejas C, Ferrier-Pages C, Reynaud S, Tsounis G, Allemand D, Gili JM (2011b) Experimental comparison of skeletal growth rates in the cold-water coral Madrepora oculata Linnaeus, 1758 and three tropical scleractinian corals. J Exp Mar Biol Ecol 405(1–2):1–5. https://doi.org/10.1016/j.jembe.2011.05.008

Orejas C, Wienberg C, Titschack J, Tamborrino L, Freiwald A, Hebbeln D (2021) *Madrepora oculata* forms large frameworks in hypoxic waters off Angola (SE Atlantic). Sci Rep 11(1). https://doi.org/10.1038/s41598-021-94579-6

Osman EO, Vohsen SA, Girard F, Cruz R, Glickman O, Bullock LM, Anderson KE, Weinnig AM, Cordes EE, Fisher CR, Baums IB (2023) Capacity of deep-sea corals to obtain nutrition from cold seeps aligned with microbiome reorganization. Glob Chang Biol 29(1):189–205. https://doi.org/10.1111/gcb.16447

Pérès J-M, Picard J (1964) Nouveau manuel de bionomie benthique de la mer Méditerranée. Station Marine d'Endoume 31(47):5–137

Perez JA, Pezzuto PR, Wahrlich R, de Souza Soares AL (2009) Deep-water fisheries in Brazil: history, status and perspectives. Lat Am J Aquat Res 37:513–541

Peterson CH, Summerson HC, Thomson E, Lenihan HS, Grabowski J, Manning L, Micheli F, Johnson G (2000) Synthesis of linkages between benthic and fish communities as a key to protecting essential fish habitat. Bull Mar Sci 66:759–774

Phillips SJ, Dudík M, Elith J, Graham CH, Lehmann A, Leathwick J, Ferrier S (2009) Sample selection bias and presence-only distribution models: implications for background and pseudo-absence data. Ecol Appl 19(1):181–197. https://doi.org/10.1890/07-2153.1

Pires DO, Silva JC, Bastos ND (2014) Reproduction of deep-sea reef-building corals from the southwestern Atlantic. Deep Sea Res Part II Topical Stud Oceanogr 99:51–63. https://doi.org/10.1016/j.dsr2.2013.07.008

Pontoppidan E (1753) Det første Førsog paa Norges Naturlige Historie, førestillende dette kongeriges Luft, Grund, Fielde, Bande, Baexter, Metaller, Mineralier, Steen-Arter, Dyr, Fugle, Fiske og omsider Indbyggernes Naturel, samt Gaedvaner og Levemaade. Berlingske Arvingers Bogtrykkerie, Kobenhaven

Portilho-Ramos R, Titschack J, Wienberg C, Rojas MGS, Yokoyama Y, Hebbeln D (2022) Major environmental drivers determining life and death of cold-water corals through time. PLoS Biol 20(5). https://doi.org/10.1371/journal.pbio.3001628

Price DM, Robert K, Callaway A, Lo Lacono C, Hall RA, Huvenne VAI (2019) Using 3D photogrammetry from ROV video to quantify cold-water coral reef structural complexity and investigate its influence on biodiversity and community assemblage. Coral Reefs 38(5): 1007–1021. https://doi.org/10.1007/s00338-019-01827-3

Quelch JJ (1886) Report on the reef-corals collected by HMS Challenger during the years 1873-76, vol 16, no 3. HM Stationery Office

R Core Team (2021) R: a language and environment for statistical computing. R Foundation for Statistical Computing, Vienna, Austria. https://www.R-project.org/

Raddatz J, Ruggeberg A, Liebetrau V, Foubert A, Hathorne EC, Fietzke J, Eisenhauer A, Dullo WC (2014) Environmental boundary conditions of cold-water coral mound growth over the last 3 million years in the Porcupine Seabight, Northeast Atlantic. Deep Sea Res Part II Topical Stud Oceanogr 99:227–236. https://doi.org/10.1016/j.dsr2.2013.06.009

Ralph PM, Squires DF (1962) The extant scleractinian corals of New Zealand. Department of Zoology, Victoria University of Wellington

Reed JK (2002) Deep-water Oculina coral reefs of Florida: biology, impacts, and management. Hydrobiologia 471:43–55. https://doi.org/10.1023/a:1016588901551

Reed JK, Weaver DC, Pomponi SA (2006) Habitat and fauna of deep-water Lophelia pertusa coral reefs off the southeastern US: Blake Plateau, Straits of Florida, and Gulf of Mexico. Bull Mar Sci 78:343–375

Reichman OJ, Jones MB, Schildhauer MP (2011) Challenges and opportunities of open data in ecology. Science 331(6018):703–705. https://doi.org/10.1126/science.1197962

Rix L, de Goeij JM, van Oevelen D, Struck U, Al-Horani FA, Wild C, Naumann MS (2018) Reef sponges facilitate the transfer of coral-

derived organic matter to their associated fauna via the sponge loop. Mar Ecol Prog Ser 589:85–96. https://doi.org/10.3354/meps12443
Robert K, Jones DOB, Georgiopoulou A, Huvenne VAI (2020) Cold-water coral assemblages on vertical walls from the Northeast Atlantic. Divers Distrib 26(3):284–298. https://doi.org/10.1111/ddi.13011
Roberts HH, Kohl B (2018) Temperature control of cold-water coral (Lophelia) mound growth by climate-cycle forcing, Northeast Gulf of Mexico. Deep Sea Res Part I Oceanogr Res Pap 140:142–158. https://doi.org/10.1016/j.dsr.2018.08.002
Roberts JM, Henry LA, Long D, Hartley JP (2008) Cold-water coral reef frameworks, megafaunal communities and evidence for coral carbonate mounds on the Hatton Bank, north east Atlantic. Facies 54(3):297–316. https://doi.org/10.1007/s10347-008-0140-x
Roberts JM, Wheeler A, Freiwald A, Cairns S (2009) Cold-water corals: the biology and geology of deep-sea coral habitats. Cambridge University Press, Cambridge. https://doi.org/10.1017/CBO9780511581588
Rogers AD (1999) The biology of Lophelia pertusa (Linnaeus 1758) and other deep-water reef-forming corals and impacts from human activities. Int Rev Hydrobiol 84:315–406
Ross SW, Quattrini AM (2009) Deep-sea reef fish assemblage patterns on the Blake Plateau (Western North Atlantic Ocean). Mar Ecol 30: 74–92
Ross SW, Rhode M, Brooke S (2017) Deep-sea coral and hardbottom habitats on the west Florida slope, eastern Gulf of Mexico. Deep Sea Res Part I Oceanogr Res Pap 120:14–28. https://doi.org/10.1016/j.dsr.2016.12.005
Rossi S, Bramanti L, Gori A, Orejas C (2017) Marine animal forests: the ecology of benthic biodiversity hotspots, 1st edn. Springer, Cham
Rowden AA, Anderson OF, Georgian SE, Bowden DA, Clark MR, Pallentin A, Miller A (2017) High-resolution habitat suitability models for the conservation and management of vulnerable marine ecosystems on the Louisville Seamount Chain, South Pacific Ocean. Front Mar Sci 4. https://doi.org/10.3389/fmars.2017.00335
Sabatier P, Reyss JL, Hall-Spencer JM, Colin C, Frank N, Tisnerat-Laborde N, Bordier L, Douville E (2012) Pb-210-Ra-226 chronology reveals rapid growth rate of Madrepora oculata and Lophelia pertusa on world's largest cold-water coral reef. Biogeosciences 9(3):1253–1265. https://doi.org/10.5194/bg-9-1253-2012
Sars M (1865) Om de i Norge forekommende fossile dyrelevninger fra Quartærperioden. Universitetets program for første halvaar 1864. Christiania
Schnabel KE (2020) The marine fauna of New Zealand: Squat lobsters (Crustacea, Decapoda, Chirostyloidea). National Institute of Water and Atmospheric Research (NIWA)
Schnabel K (2022) The marine fauna of New Zealand. Squat lobsters (Crustacea, Decapoda, Chirostyloidea)
Schulzweida U (2019) CDO User Guide (1.9.8). Zenodo. https://doi.org/10.5281/zenodo.3539275
Simon-Lledó E, Pomee C, Ahokava A, Drazen J, Leitner A, Flynn A, Parianos J, Jones D (2020) Multi-scale variations in invertebrate and fish megafauna in the mid-eastern Clarion Clipperton Zone. Prog Oceanogr 187:102405. https://doi.org/10.1016/j.pocean.2020.102405
Soetaert K, Mohn C, Rengstorf A, Grehan A, van Oevelen D (2016) Ecosystem engineering creates a direct nutritional link between 600-m deep cold-water coral mounds and surface productivity. Sci Rep 6. https://doi.org/10.1038/srep35057
Squires DF (1959) Deep sea corals collected by the Lamont Geological Observatory. 1, Atlantic corals. American Museum novitates; no. 1965
Squires DF (1964) Fossil coral thickets in Wairarapa, New Zealand. J Paleontol 904–915
Stetson TR, Squires DF, Pratt RM (1962) Coral banks occurring in deep water on the Blake Plateau. American Museum novitates; no. 2114
Stewart HA, Davies JS, Guinan J, Howell KL (2014) The Dangeard and Explorer canyons, South Western Approaches UK: Geology, sedimentology and newly discovered cold-water coral mini-mounds. Deep Sea Res Part II Topical Stud Oceanogr 104:230–244. https://doi.org/10.1016/j.dsr2.2013.08.018
Stromberg SM, Larsson AI (2017) Larval behavior and longevity in the cold-water coral Lophelia pertusa indicate potential for long distance dispersal. Front Mar Sci 4. https://doi.org/10.3389/fmars.2017.00411
Tamborrino L, Wienberg C, Titschack J, Wintersteller P, Mienis F, Schröder-Ritzrau A, Freiwald A, Orejas C, Dullo WC, Haberkern J, Hebbeln D (2019) Mid-Holocene extinction of cold-water corals on the Namibian shelf steered by the Benguela oxygen minimum zone. Geology 47(12):1185–1188. https://doi.org/10.1130/g46672.1
Tamborrino L, Titschack J, Wienberg C, Purkis S, Eberli GP, Hebbeln D (2022) Spatial distribution and morphometry of the Namibian coral mounds controlled by the hydrodynamic regime and outer-shelf topography. Front Mar Sci 9:877616
Taviani M, Remia A, Corselli C, Freiwald A, Malinverno E, Mastrototaro F, Savini A, Tursi A (2005) First geo-marine survey of living cold-water Lophelia reefs in the Ionian Sea (Mediterranean Basin). Facies 50:409–417
Teichert C (1958) Cold- and deep-water coral banks. AAPG Bull 42(5): 1064–1082. https://doi.org/10.1306/0bda5af5-16bd-11d7-8645000102c1865d
Thierens M, Titschack J, Dorschel B, Huvenne VAI, Wheeler AJ, Stuut JB, O'Donnell R (2010) The 2.6 Ma depositional sequence from the Challenger cold-water coral carbonate mound (IODP Exp. 307): Sediment contributors and hydrodynamic palaeo-environments. Mar Geol 271(3–4):260–277
Thierens M, Browning E, Pirlet H, Loutre MF, Dorschel B, Huvenne VAI, Titschack J, Colin C, Foubert A, Wheeler AJ (2013) Cold-water coral carbonate mounds as unique palaeo-archives: the Plio-Pleistocene Challenger Mound record (NE Atlantic). Quat Sci Rev 73:14–30. https://doi.org/10.1016/j.quascirev.2013.05.006
Thioulouse J, Dray S, Dufour A-B, Aurélie S, Jombart T, Pavoine S (2018) Multivariate analysis of ecological data with ade4. Springer, New York. https://doi.org/10.1007/978-1-4939-8850-1
Thornton B, Bodenmann A, Yamada T, Stanley D, Massot-Campos M, Huvenne V, Durden J, Bett B, Ruhl H, Newborough D (2021) Visualizing multi hectare seafloor habitats with BioCam. Oceanography 34(4):92–93. https://doi.org/10.5670/oceanog.2021.supplement.02-34
Thuiller W, Georges D, Gueguen M, Engler R, Breiner F (2021) biomod2: ensemble platform for species distribution modeling. R package version 351
Thurber AR, Sweetman AK, Narayanaswamy BE, Jones DOB, Ingels J, Hansman RL (2014) Ecosystem function and services provided by the deep sea. Biogeosciences 11(14):3941–3963. https://doi.org/10.5194/bg-11-3941-2014
Titschack J, Thierens M, Dorschel B, Schulbert C, Freiwald A, Kano A, Takashima C, Kawagoe N, Li X (2009) Carbonate budget of a cold-water coral mound (Challenger Mound, IODP Exp. 307). Mar Geol 259(1–4):36–46
Titschack J, Baum D, De Pol-Holz R, Lopez Correa M, Forster N, Flögel S, Hebbeln D, Freiwald A (2015) Aggradation and carbonate accumulation of Holocene Norwegian cold-water coral reefs. Sedimentology 62(7):1873–1898. https://doi.org/10.1111/sed.12206
Titschack J, Fink HG, Baum D, Wienberg C, Hebbeln D, Freiwald A (2016) Mediterranean cold-water corals—an important regional carbonate factory? Depositional Rec 2(1):74–96. https://doi.org/10.1002/dep2.14
Tittensor DP, Baco AR, Brewin PE, Clark MR, Consalvey M, Hall-Spencer J, Rowden AA, Schlacher T, Stocks KI, Rogers AD (2009) Predicting global habitat suitability for stony corals on seamounts. J

Biogeogr 36(6):1111–1128. https://doi.org/10.1111/j.1365-2699.2008.02062.x
Tong R, Davies AJ, Purser A, Liu X, Liu F (2022) Global distribution of the cold-water coral Lophelia pertusa. IOP Conf Seri Earth Environm Sci 1004(1):012010. https://doi.org/10.1088/1755-1315/1004/1/012010
Tracey DM, Rowden AA, Mackay KA, Compton T (2011) Habitat-forming cold-water corals show affinity for seamounts in the New Zealand region. Mar Ecol Prog Ser 430:1–U59. https://doi.org/10.3354/meps09164
Valavi R, Elith J, Lahoz-Monfort JJ, Guillera-Arroita G (2019) blockCV: An r package for generating spatially or environmentally separated folds for k-fold cross-validation of species distribution models. Methods Ecol Evol 10(2):225–232. https://doi.org/10.1111/2041-210X.13107
Van der Kaaden AS, Mohn C, Gerkema T, Maier SR, de Froe E, van de Koppel J, Rietkerk M, Soetaert K, van Oevelen D (2021) Feedbacks between hydrodynamics and cold-water coral mound development. Deep Sea Res Part I Oceanogr Res Pap 178. https://doi.org/10.1016/j.dsr.2021.103641
Van Engeland T, Godø OR, Johnsen E, Duineveld GCA, van Oevelen D (2019) Cabled ocean observatory data reveal food supply mechanisms to a cold-water coral reef. Prog Oceanogr 172:51–64. https://doi.org/10.1016/j.pocean.2019.01.007
van Haren H, Mienis F, Duineveld GCA, Lavaleye MSS (2014) High-resolution temperature observations of a trapped nonlinear diurnal tide influencing cold-water corals on the Logachev mounds. Prog Oceanogr 125:16–25. https://doi.org/10.1016/j.pocean.2014.04.021
Van Oevelen D, Duineveld G, Lavaleye M, Mienis F, Soetaert K, Heip CHR (2009) The cold-water coral community as a hot spot for carbon cycling on continental margins: A food-web analysis from Rockall Bank (northeast Atlantic). Limnol Oceanogr 54(6):1829–1844
Viana AR, Faugeres JC, Kowsmann RO, Lima JAM, Caddah LFG, Rizzo JG (1998) Hydrology, morphology and sedimentology of the Campos continental margin, offshore Brazil. Sediment Geol 115(1):133–157. https://doi.org/10.1016/S0037-0738(97)00090-0
Victorero L, Watling L, Deng Palomares ML, Nouvian C (2018) Out of sight, but within reach: a global history of bottom-trawled deep-sea fisheries from >400 m depth. Front Mar Sci 5. https://doi.org/10.3389/fmars.2018.00098
Villareal TA, Brown CG, Brzezinski MA, Krause JW, Wilson C (2012) Summer diatom blooms in the North Pacific subtropical gyre: 2008–2009. PloS One 7(4):e33109. https://doi.org/10.1371/journal.pone.0033109
Waller RG, Tyler PA (2005) The reproductive biology of two deep-water, reef-building scleractinians from the NE Atlantic Ocean. Coral Reefs 24:514–522
Waller RG, Goode S, Tracey D, Johnstone J, Mercier A (2023) A review of current knowledge on reproductive and larval processes of deep-sea corals. Mar Biol 170(5):58. https://doi.org/10.1007/s00227-023-04182-8
Wang HZ, Lo Iacono C, Wienberg C, Titschack J, Hebbeln D (2019) Cold-water coral mounds in the southern Alboran Sea (western Mediterranean Sea): Internal waves as an important driver for mound formation since the last deglaciation. Mar Geol 412:1–18. https://doi.org/10.1016/j.margeo.2019.02.007
Washburn TW, Turner PJ, Durden JM, Jones DO, Weaver P, Van Dover CL (2019) Ecological risk assessment for deep-sea mining. Ocean Coast Manag 176:24–39
Weinbauer MG, Ogier J, Maier C (2012) Microbial abundance in the coelenteron and mucus of the cold-water coral Lophelia pertusa and in bottom water of the reef environment. Aquat Biol 16(3):209–216. https://doi.org/10.3354/ab00443
Wheeler AJ, Kozachenko M, Beyer A, Foubert A, Huvenne V, Klages M, Masson DG, Olu-Le Roy K, Thiede J (2005) Sedimentary processes and carbonate mounds in the Belgica Mound province, Porcupine Seabight, NE Atlantic. In: Freiwald A, Roberts MJ (eds) Cold-water corals and ecosystems. Springer, Berlin, pp 571–603
Wheeler A, Beyer A, Freiwald A, De Haas H, Huvenne V, Kozachenko M, Olu-Le Roy K, Opderbecke J (2007) Morphology and environment of cold-water coral carbonate mounds on the NW European margin. Int J Earth Sci 96:37–56
White M, Dorschel B (2010) The importance of the permanent thermocline to the cold water coral carbonate mound distribution in the NE Atlantic. Earth Planet Sci Lett 296(3–4):395–402. https://doi.org/10.1016/j.epsl.2010.05.025
White M, Wolff GA, Lundalv T, Guihen D, Kiriakoulakis K, Lavaleye M, Duineveld G (2012) Cold-water coral ecosystem (Tisler Reef, Norwegian Shelf) may be a hotspot for carbon cycling. Mar Ecol Prog Ser 465:11–23. https://doi.org/10.3354/meps09888
Wienberg C, Hebbeln D, Fink HG, Mienis F, Dorschel B, Vertino A, Correa ML, Freiwald A (2009) Scleractinian cold-water corals in the Gulf of Cadiz-First clues about their spatial and temporal distribution. Deep Sea Res Part I Oceanogr Res Pap 56(10):1873–1893. https://doi.org/10.1016/j.dsr.2009.05.016
Wienberg C, Titschack J, Freiwald A, Frank N, Lundalv T, Taviani M, Beuck L, Schröder-Ritzrau A, Krengel T, Hebbeln D (2018) The giant Mauritanian cold-water coral mound province: Oxygen control on coral mound formation. Quat Sci Rev 185:135–152. https://doi.org/10.1016/j.quascirev.2018.02.012
Wienberg C, Titschack J, Frank N, De Pol-Holz R, Fietzke J, Eisele M, Kremer A, Hebbeln D (2020) Deglacial upslope shift of NE Atlantic intermediate waters controlled slope erosion and cold-water coral mound formation (Porcupine Seabight, Irish margin). Quat Sci Rev 237. https://doi.org/10.1016/j.quascirev.2020.106310
Wienberg C, Krengel T, Frank N, Wang H, Hebbeln D (2022) Cold-water coral mounds in the western Mediterranean Sea: new insights into their initiation and development since the Mid-Pleistocene in response to changes in African hydroclimate. Quat Sci Rev 293:107723
Williams A, Althaus F, Green M, Maguire K, Untiedt C, Mortimer N, Jackett CJ, Clark M, Bax N, Pitchers R, Schlacher T (2020) True size matters for conservation: a robust method to determine the size of deep-sea coral reefs shows they are typically small on seamounts in the Southwest Pacific Ocean. Front Mar Sci 7. https://doi.org/10.3389/fmars.2020.00187
Wilson JB (1979) The distribution of the coral Lophelia pertusa (L.) [L. prolifera (pallas)] in the north-east Atlantic. Mar Biol Assoc UK 59:149–164
Wilson R (2020) r4ecology/nctoolkit: v0.2.1 (v0.2.1). Zenodo. https://doi.org/10.5281/zenodo.4034527
Wilson MF, O'Connell B, Brown C, Guinan JC, Grehan AJ (2007) Multiscale terrain analysis of multibeam bathymetry data for habitat mapping on the continental slope. Mar Geod 30:3–35
Wood R (1998) The ecological evolution of reefs. Annu Rev Ecol Syst 29:179–206. https://doi.org/10.1146/annurev.ecolsys.29.1.179
Xu A, Chen Z, Qu Y, Tian Y, Shu C, Zheng X, Li G, Yan W, Zhao M (2019) Cold-water corals in a cold seep area on the northern continental slopes of the South China sea and their isotopic characteristics. Deep-Sea Res I Oceanogr Res Pap 152:103043
Yesson C, Bedford F, Rogers AD, Taylor ML (2017) The global distribution of deep-water Antipatharia habitat. Deep Sea Res Part II Topical Stud Oceanogr 145:79–86. https://doi.org/10.1016/j.dsr2.2015.12.004
Zibrowius H (1980) Les Scléractiniaires de la Méditerranée et de l'Atlantique nord-oriental. Mémoires de l'Institut océanographique, Monaco

Biology, Ecology, and Threats to Cold-Water Corals on Brazil's Deep-Sea Margin

2

Angelo F. Bernardino, Daniela Y. Gaurisas, and Paulo Y. G. Sumida

Abstract

In the region of Brazil, cold-water corals are distributed from the Equatorial to the Southern Brazilian margins and on nearby seamounts, mainly between 250 and 1200 m water depth, forming reefs, carbonate banks, and octocoral gardens. Larger reef structures formed by *Lophelia pertusa* are only known from the Southeast-Southern (SE-S) slope, with depth ranges limited by high temperatures (>12 °C) and the lower aragonite saturation state of lower bathyal water masses. Besides *L. pertusa*, three other reef-forming scleractinians occur in larger numbers, *Solenosmilia variabilis*, *Enallopsammia rostrata*, and *Madrepora oculata*. On the SE Brazilian continental margin, these species are associated with mounds formed by authigenic carbonates related to seabed pockmarks with active seepage. Particulate organic carbon flux may be important to the development of Brazilian scleractinian deep reefs, but Octocorallia and Antipatharia are apparently not restricted to similar environmental filtering as the calcifying species and are common at depths below 1200 m. The limited information on the ecology and economic importance of cold-water coral ecosystems in Brazil prevent any measure of economic losses associated with impacts by the deep-water fisheries and the offshore oil and gas industry. In the next decades, climate change effects will likely decrease habitat suitability for cold-water corals on the upper slope by changing patterns of primary productivity, with higher temperatures and decreased pH of the oceans. Brazil potentially harbors rich deep-water coral reefs similar to other provinces in the North Atlantic, but limited government investment and management will keep these treasures in the dark.

Keywords

Continental margin · South Atlantic · Scleractinia · Octocorallia · Ecology · Climate change · Impacts

2.1 Introduction

The Brazilian continental margin extends for over 6000 km, from the equatorial Atlantic near the Amazon river fan (6°N) to the Southern Atlantic on the Argentinean Basin (35°S). The deep-sea margin, considered here at depths below 200 m, includes the main continental slope and the expansions of the Brazilian Economic Exclusive Zone (EEZ) through territorial sovereignty in the North and South Atlantic of the oceanic archipelagos of St. Peter and St. Paul (0°55′N; 29°20′W), Fernando de Noronha (3°52′S; 32°26′W), and Trindade and Martim Vaz (20°30′S; 29°19′W; Fig. 2.1). The large extent of the Brazilian margin includes a number of important geomorphological seafloor features, including submarine canyons, seamounts, seeps, pockmarks, carbonate mounds, and large sedimentary basins along the slope (Fig. 2.1; Sumida et al. 2020a). The distribution of these features will be discussed here in light of potential habitats (hard substrata) for deep-sea corals, and also considering the deep margin as a major area of economic and biological interest in the South Atlantic.

The initial recognition of the taxonomic diversity and distribution of cold-water corals along Brazil's continental margin started with the global oceanographic expedition of the H.M.S. Challenger (1872–1876) and several others that followed until the decades of the 1940s through 1960s (Pires 2007). In the late 1990s and early 2000s, Brazilian research programs and increasing offshore oil and gas exploration in deep waters increased the discovery of new deep-water coral species along Brazil's deep-sea margin. However, the

A. F. Bernardino (✉) · D. Y. Gaurisas
Grupo de Ecologia Bêntica, Departamento de Oceanografia, CCHN, Universidade Federal do Espirito Santo, Goiabeiras, Vitória, ES, Brazil
e-mail: angelo.bernardino@ufes.br

P. Y. G. Sumida
Departamento de Oceanografia Biológica, Instituto Oceanográfico, Universidade de São Paulo, Cidade Universitária, São Paulo, SP, Brazil

E. Cordes, F. Mienis (eds.), *Cold-Water Coral Reefs of the World*, Coral Reefs of the World 19,
https://doi.org/10.1007/978-3-031-40897-7_2

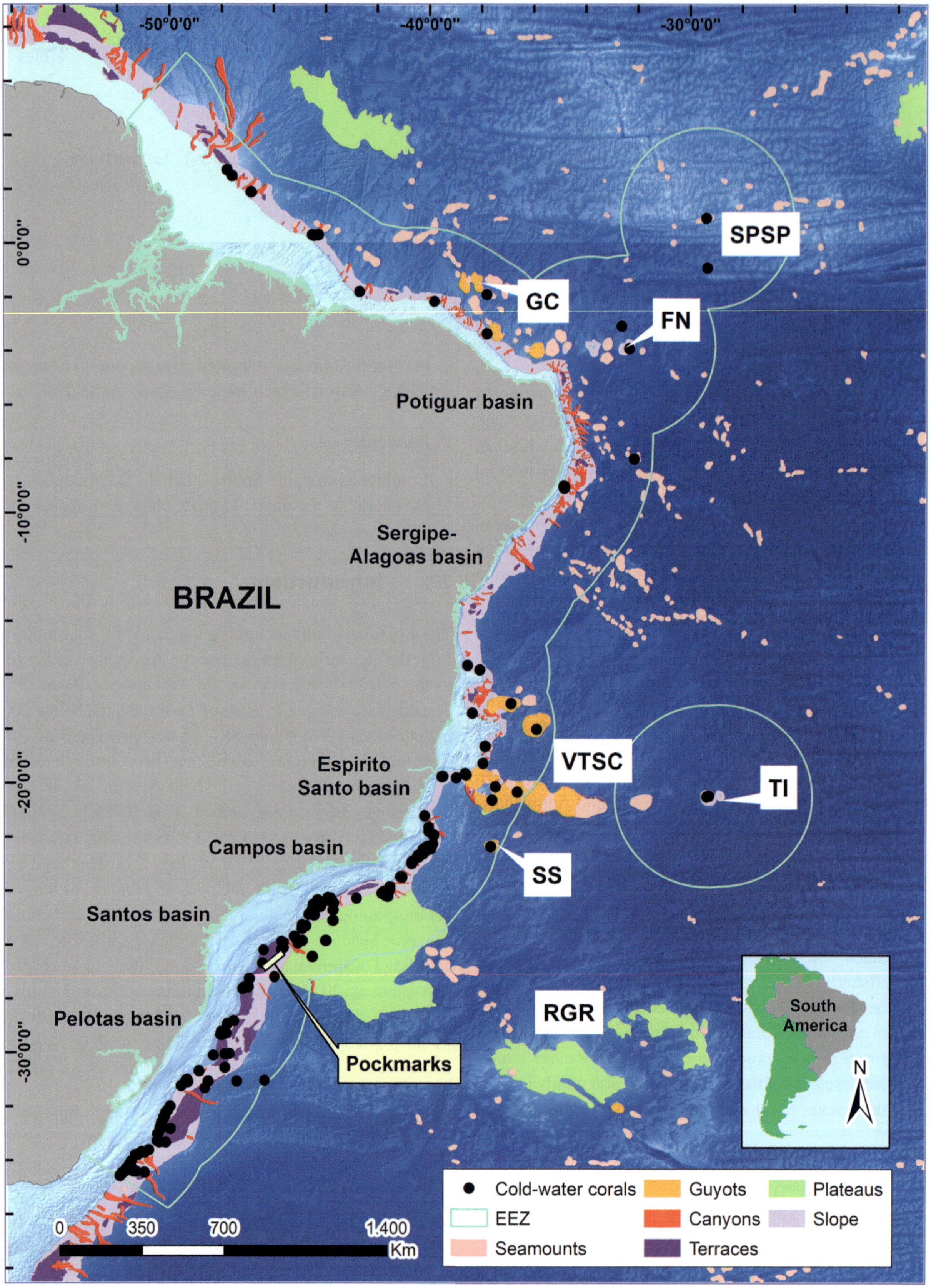

Fig. 2.1 Distribution of current cold-water coral records (black dots) over the seabed geomorphic features on the slope (bathyal) and continental rise (abyssal) of the Brazilian margin. Main sedimentary basins along the margin are indicated. Brazil's EEZ is delimited by the green line. Seamounts and oceanic islands: SPSP, St. Peter St. Paul; CG, Ceará guyot; FN, Fernando de Noronha Archipelago; VTSC, Vitória-Trindade seamount chain; TI, Trindade Island; SS, Saldanha seamount; RGR, Rio Grande Rise. GIS database from Blue Habitats (UNEP), Marine Regions, and General Bathymetric Chart of the Oceans (GEBCO 2020)

reported diversity and distribution of deep-sea corals on Brazil's margin is still limited in sampling effort, and spatially skewed to intense sampling from the oil and gas industry in basins of commercial interest in Southeastern (SE) Brazil.

In this chapter we review current knowledge about the environmental setting and the distribution, ecology, and threats to cold-water coral ecosystems of the deep Brazil margin. With the logistical and research funding constraints that have historically limited the exploration of deep-sea ecosystems of biological interest on Brazil's margin, including cold-water corals, we further explore potential areas of occurrence and vulnerability to deep-sea economic exploration and climate change effects. We then discuss the sparse recognition of the distribution, habitat sizes, ecological processes, and diversity of cold-water corals along the deep Brazil margin, its oceanic islands and seamounts.

2.2 Oceanography of the Deep Brazil Margin

The Atlantic Ocean is bathed by water masses with higher oxygen content and deeper aragonite compensation depth when compared to the Pacific and Indian Oceans (Levin and Gooday 2003; Jiang et al. 2015). These parameters are important for the biology and distribution of calcifying organisms such as cold-water corals, as these organisms need adequate oxygen levels to grow and high concentrations of carbonate ions to make up their skeletons (i.e., scleractinian corals). There are four deep-water masses over the slope and abyss off Brazil (Table 2.1), which mostly originate from well-oxygenated polar regions in the North Atlantic and Southern Ocean. Those water masses can be distinguished by slightly different temperatures, salinities, and nutrient contents that result from their distinct geographical areas of formation and mixing (De Madron and Weatherly 1994; Stramma and England 1999; Silveira et al. 2020).

The Antarctic Bottom Water (AABW) originates off the coast of Antarctica and is the deepest water mass flowing in the South Atlantic Ocean. The AABW is colder and thus has a higher density when compared to deep-water masses formed in the North Atlantic (Morozov et al. 2010). The Antarctic Bottom Water flows toward the Equator through lower bathyal and abyssal depths along Brazil's margin below depths of 3500 m. It fills abyssal basins in the Eastern and Western South Atlantic, including the Brazil Basin, with high oxygen content (>5 ml l^{-1}), low temperatures (<0.5 °C), and salinity near to 34.7. Above the AABW and flowing in the opposite direction, the North Atlantic Deep Water (NADW, 1000–3500 m depth) originates from a number of regions in the North Atlantic and flows southward into the South Atlantic meandering along the western boundary of South America. When it reaches the Equator, the NADW propagates into different layers and branches, partly flowing to the East (African continent) and partly flowing south through the western boundary currents along the Brazilian margin. The NADW marks a deep oxygen maximum layer (>6.0 ml l^{-1}) along Brazil's margin and it is separated into three different vertical layers (upper, central, and lower) due to their different densities and area of origin in the North Atlantic (De Madron and Weatherly 1994). Located above the main bathyal (NADW) and abyssal (AABW) water masses, the top 1000 m pelagic layer of the Southwestern Atlantic along Brazil's margin includes the surface tropical water (0–150 m) and two sub-surface water masses. The latter are the South Atlantic Central Water (SACW, 150–500 m) and the Antarctic Intermediate Water (AAIW, 500–1000 m). The SACW has higher temperatures and salinities when compared to the AAIW underneath (Table 2.1) and partially originates from central waters of the Indian Ocean brought into the Atlantic by the Agulhas Current (Stramma and England 1999). On the other hand, the Antarctic Intermediate Water originates from spills of the Antarctic Circumpolar water near the Drake Passage and the Falklands. The AAIW has very low concentrations of carbonate ions and has low aragonite saturation (Table 2.1). Both the SACW and AAIW form central water masses at depths of 150–1000 m, but have opposite flow directions between the NE-Equatorial margin and SE-S Brazil margin. This division occurs in the Subtropical-Tropical boundary of

Table 2.1 Depth range, temperature, salinity, dissolved oxygen, carbonate and aragonite saturation state of deep-water masses along Brazil's margin. Values represent means or ranges within each water mass

Depth (m)	Water mass	Temperature (°C)/salinity	Oxygen (ml l^{-1})	Carbonate (CO_3^{-2}; μmol kg^{-1})[1] and Ω_{arag} range	Remarks
150–500	South Atlantic Central Water (SACW)	18–6 °C/35.8	5.1	>150; Ω_{arag} = 1.2–2.1	Oxygen minimum layer. Flows Equatorward from 21°S
500–1000	Antarctic Intermediate Water (AAIW)	6–2 °C/34.2	5.3	<100; Ω_{arag} = 1.2–0.9	Salinity minimum, upper aragonite lysocline (600–1300 m)
1000–3500	North Atlantic Deep Water (NADW)	4–2 °C/34.9	6.2	>100; Ω_{arag} < 1.0	Deep oxygen maximum, lower limit to reefs (~1200 m)
3500–6000	Antarctic Bottom Water (AABW)	<2 °C/34.7	5.5	<100; Ω_{arag} < 1.0	2 °C is the "top" of the AABW and marks transition to the NADW

Data from De Madron and Weatherly (1994); Gerhardt and Henrich (2001); Jiang et al. (2015)

the South Atlantic located at the Vitoria-Trindade Seamount Chain (21° S). As a result, these central water masses flow equatorward north of the Vitoria-Trindade seamounts transported by the North Brazil Current toward the Caribbean Sea. To the south of Vitoria-Trindade seamounts, there is poleward flow of central water masses along the subtropical Brazil Current until reaching the Argentine Basin (Silveira et al. 2020).

The temperature and oxygen signatures of central, bathyal, and abyssal water masses along Brazil's margin are markedly distinct from the ones in the Pacific and Indian Oceans (Levin and Gooday 2003). The upper bathyal depths on Brazil's margin (500–1000 m) have an oxygen minimum layer with dissolved O_2 higher than 5 ml l^{-1}, which increases toward lower bathyal and abyssal depths. In contrast, both the Indian and Pacific Oceans have extensive sub-surface oxygen minimum zones (OMZ) that may cover vast areas of the upper slope down to 1000 m (Demopoulos et al. 2003; Smith and Demopoulos 2003). Bottom dissolved oxygen concentrations within OMZs are frequently below 0.5 ml l^{-1}, which may have a significant effect on deep-sea habitats and fauna distribution. OMZs do not currently occur on the deep-sea Brazilian margin indicating that oxygen is not currently a limiting factor to cold-water corals, although future climate change will likely lead to decreased sub-surface oxygenation in the Atlantic Ocean and a consequent impact in cold-water corals (Morato et al. 2020).

The carbonate ion concentrations (CO_3^{-2}) and the aragonite saturation state of deep-water masses are also of additional importance for cold-water corals as a decrease in CO_3^{-2} concentrations leads to chemical dissolution of coral skeletons (Orr et al. 2005). The aragonite saturation depth thus indicates a chemical threshold where formation of skeleton structure by cold-water corals would be unfavorable, and global patterns indicate that the South Atlantic has a much shallower saturation depth (~1200 m) than the North Atlantic (~2500 m, Feely et al. 2004; Jiang et al. 2015). This shallowing of the aragonite saturation depth is explained by the intermediate (depths of 1000–1500 m) inflow of Antarctic waters in the South Atlantic. The carbonate concentration of deep-water masses in the South Atlantic indicates that the Antarctic Intermediate Water (AAIW) and the Antarctic Bottom Waters (AABW) are highly corrosive due to low carbonate concentrations (Gerhardt and Henrich 2001, Table 2.1). The upper carbonate lysocline occurs in a depth range within the AAIW that has minimum CO_3^{-2} concentrations and may be related to regional processes of pelagic organic matter decomposition (Gerhardt and Henrich 2001). Therefore, the upper lysocline associated with the AAIW at depths of 600–1200 m would be stronger in areas of the margin close to inputs of continental nutrients (e.g., Amazon basin), and in areas of the margin with higher primary productivity and a weaker thermocline toward southern Brazil. Below the AAIW, the mixing with the upper layers (~1200 m) of the North Atlantic Deep Water slightly increases the dissolved CO_3^{-2} concentrations. The bathymetric pattern of carbonate concentrations in deep-water masses offshore Brazil's margin suggests that upper bathyal depths offer better conditions for the occurrence of calcifying corals (Scleractinia), which is supported by global and South Atlantic habitat suitability models based on oceanographic data (Tittensor et al. 2009; Barbosa et al. 2020).

2.3 Geological Setting and Reef Structures

2.3.1 Geological Setting

Brazil has a passive continental margin that is predominantly composed of soft sedimented slopes, plateaus, and continental rises (Fig. 2.1). However, as in other continental margins, there is a range of habitats created by topographic, geological, physical, and biological processes that provide an opportunity for colonization by a range of organisms, including cold-water corals. The diversity of seafloor habitats is highly variable at regional and local spatial scales along continental margins (Levin and Sibuet 2012), and Brazil's margin is no exception. The geomorphology of seafloor hard substrate habitats along the Brazilian continental margin suggests a wide distribution of cold-water corals, as these organisms often rely on hard substrate for larval settlement and development (Roberts et al. 2006). As a result, potential areas with adequate substrate for development of cold-water corals are common along the slope, submarine canyons, seamounts, and oceanic islands that occur throughout the margin (Fig. 2.1).

Brazil's margin has large sedimentary basins throughout its slope that host major deep-sea hydrocarbon reservoirs from the Paleozoic and Mesozoic (Fig. 2.1, Mohriak 2003). These basins are relatively well mapped due to industrial interest in offshore exploration, although very few have been thoroughly sampled at depths below 200 m. The seafloor at slope depths is typically covered by soft sediments of marine and terrestrial origin as a result of downslope and lateral transport of continental inorganic and organic matter with varying contributions of marine organic sources. The deep equatorial margin is closer to the Mid-Atlantic Ridge and exhibits seafloor features associated with transform faults and volcanism, including many seamounts and oceanic archipelagos, which also occur on the SE Brazilian margin (Vitória-Trindade seamount chain). Throughout the margin, submarine canyons, seabed pockmarks, seeps, and other biogenic or thermogenic hard-bottom habitats formed by ancient coral reefs or authigenic carbonate formations have been discovered during the last two decades (Fig. 2.1).

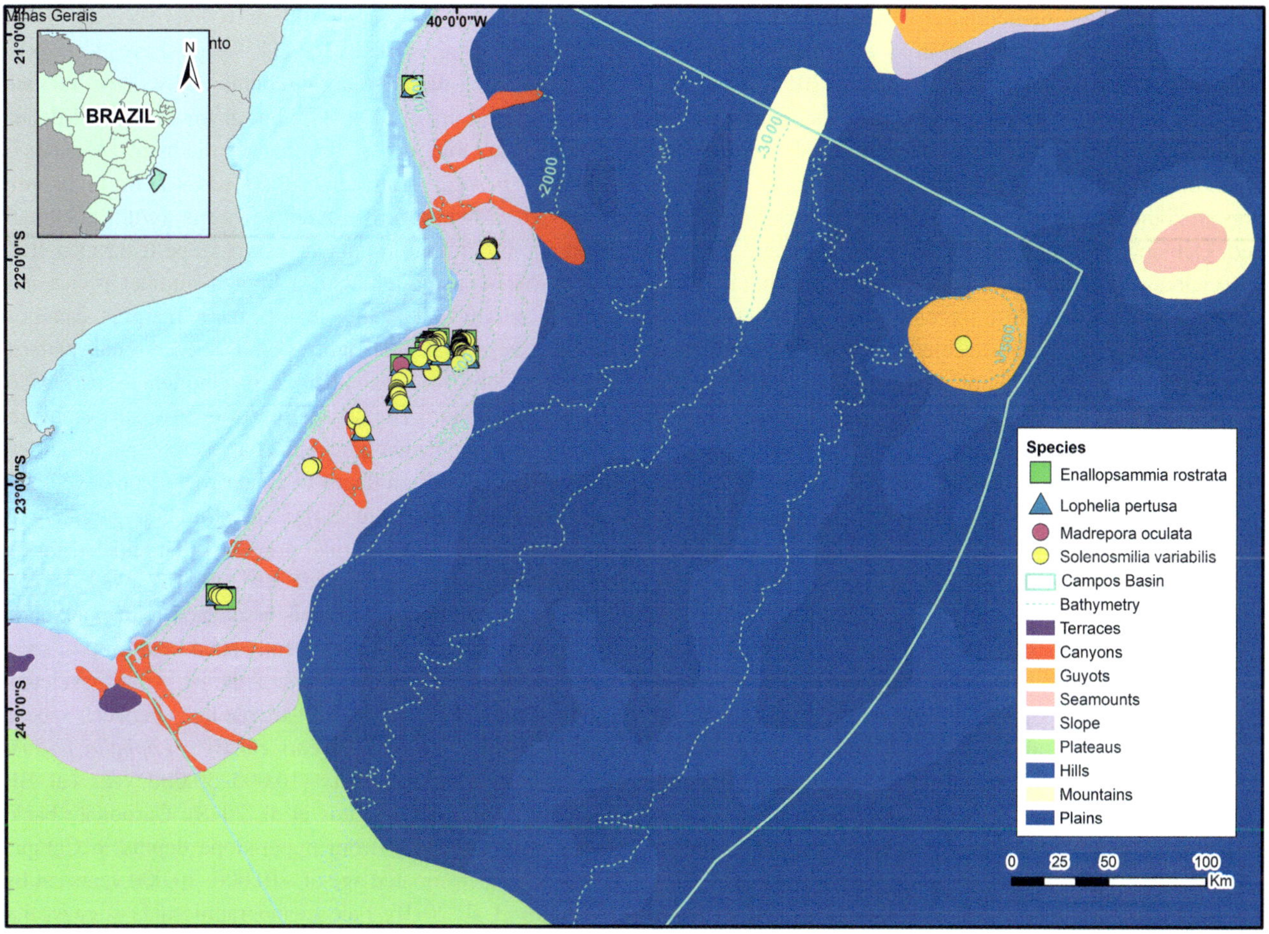

Fig. 2.2 Overlap of occurrence of top four-ranked cold-water coral-reef species over the seabed geomorphic features on the slope of Campos Basin (delimited within the blue line). GIS database from Blue Habitats (UNEP) and General Bathymetric Chart of the Oceans (GEBCO 2020)

Brazil's deep margin has a number of distinct geomorphological features and a complex topography that work in concert with oceanographic variability to support distinct habitats for cold-water corals. The continental slope typically has a low slope angle (0–5°, Almeida and Kowsmann 2014) and is frequently interrupted by submarine canyons that promote mass transport and create additional hard-bottom topography across the slope. Submarine canyons are a particularly abundant feature of Brazil's margin, with over 161 mapped canyons occupying an area of over 50,000 km^2, which is 1.4% of the EEZ at bathyal (200–3000 m) depths (De Leo et al. 2020; Sumida et al. 2020b). Canyon sizes and types vary widely across the margin, with some extending for over 42 km across the slope and others presenting a U-shaped morphology of over 8 km between canyon walls (Almeida and Kowsmann 2014). These canyons also vary geomorphologically, with both shelf-incised and blind (slope) canyons commonly occurring along the margin. There is a very limited understanding of biological communities associated with submarine canyons across the deep Brazilian margin, although cold-water corals have been observed growing on canyon walls (Bernardino et al. 2016, 2019; Cavalcanti et al. 2017). A number of canyons in Campos Basin support coral-reef structures and mapped reefs may be over 700 m long (Cavalcanti et al. 2017). However, these reports are mostly qualitative and there is limited overlap between the reported occurrence of cold-water coral species and other canyons. From the main reef-building corals, only *Solenosmilia variabilis* Duncan, 1873 and *Madrepora oculata* Linnaeus, 1758 have occurrence data that suggest an association with submarine canyons in Campos Basin (Fig. 2.2). Most records of cold-water reefs in Brazil result from ROV surveys over carbonated structures on slope areas or from nearby offshore infrastructure. So there is limited recognition of cold-water corals associated with submarine canyons due to limited benthic surveys and sampling along Brazil's margin. The global occurrence of corals thriving on upper-slope canyon walls and the high number of submarine canyons suggest they offer potential habitats for cold-water corals across the Brazilian margin (Almada and Bernardino 2017).

Seeps and pockmarks are other types of slope ecosystems with potential for harboring cold-water corals, but are also understudied along Brazil's margin. Sumida et al. (2004) presented the first evidence of cold-water corals (Scleractinia and Octocorallia) and shells of chemosynthetic bivalves associated with pockmark structures at 800 m depth in Santos Basin (SE Brazil). More recently, detailed geophysical mapping in Santos Basin revealed that over 980 pockmarks of variable sizes (few tens of meters to over a kilometer in diameter) occur in a 100 km strip between depths of 300 and 700 m within a region named the "Santos Basin pockmark province" (Mahiques et al. 2017; Fig. 2.1). Mahiques et al. (2017) suggested that this province may extend further north along the Campos and Espirito Santo Basins. Pockmarks have also been identified on the southern Brazil margin at Pelotas Basin and on the Uruguayan margin (Carranza et al. 2012; Giongo et al. 2016), where associations of corals with seep chemosynthetic fauna potentially occur. The extremely high densities of pockmarks and the outstanding presence of carbonate structures on Santos Basin support the close association of hydrocarbon seepage and coral development over the upper slope (Fig. 2.3). In addition, past periods of active and passive methane seepage associated with coral growth may have caused the build-up of an extensive (187 km^2, 150 m high) carbonate ridge recently discovered at depths between 450 and 1250 m in Santos Basin (Maly et al. 2019). The limited extension of mapped seafloor by multibeam and seismic data currently limits our understanding of the seep and cold-water coral occurrences along the slope, but indirect evidence extends the pockmark-seep province to over a third of Brazil's continental margin.

The Pelotas, Santos, Campos, and Espirito Santo deep-water hydrocarbon provinces expand for over 700,000 km^2 within the Brazilian EEZ, where the lower depth ranges of these pockmarks (700 m) suggest an exceptional area for the occurrence of cold-water corals. The wide occurrence of pockmarks in Santos Basin may be related to ancient or recent gas seepage from hydrocarbon resources (Mahiques et al. 2017). Salt diapir-related faults facilitate the upward gas release on the upper-slope seafloor, and geochemical data from pockmarks in Santos Basin suggest a relatively recent (<200 years BP) seepage activity (Santos et al. 2018). The seepage of methane gas on the seafloor surface leads to microbially-mediated oxidation of CH_4 and the formation of authigenic carbonate structures that serve as hard substrate for coral settlement (Levin 2005; Cordes et al. 2008).

Seafloor carbonate structures are often observed along Pelotas, Santos, and Campos Basin (S-SE Brazil), and carbonate mounds composed of dead coral rubble were found supporting vast densities of living colonies of cold-water corals between 500 and 1200 m (Giongo et al., 2016; Cavalcanti et al. 2017). Asphalt seeps have also been observed along the SE margin, colonized by octocorals and sponges (Fujikura et al. 2017), but at lower bathyal depths that are unsuitable for reef-forming scleractinians. Mounds of *Lophelia pertusa* (Linnaeus, 1758) 35–67 m high were observed growing next to active seeps (with pockmark depressions) on the Uruguayan upper slope (Carranza et al. 2012). So even though with a limited number of oceanographic surveys, it seems that Brazil's margins can host similar seep-coral associations and successional patterns than were observed on other deep continental margins of the North Atlantic. The succession of deep-sea assemblages from a seep stage when pockmarks are active and sustaining chemosynthetic assemblages to a later reef stage developing over seafloor authigenic carbonate structures is likely to occur over a larger scale than previously anticipated on the SE-S Brazilian margin (Sumida et al. 2004).

Radiocarbon dating of corals associated with pockmarks on the Brazilian margin has recently shed light on past seep-coral associations in the South Atlantic. Coral skeletons sampled over pockmarks had variable ages between species of *Caryophyllia* sp. (1120 years cal BP), *Lophelia pertusa* and *Solenosmilia variabilis* (10,000–30,000 year cal BP, Mangini et al. 2010; Santos et al. 2018). Carbonate banks and *L. pertusa* corals from upper-slope depths in Campos Basin have an estimated age of ~10,000–20,000 years cal BP (Mangini et al. 2010). Paleoceanographic data suggest that the onset of growth of cold-water corals on the deep Brazil margin first occurred before 25,000 years BP and was associated with significant cooling of sea surface temperatures that promoted the sinking and downslope transport of surface waters on the slope. The Atlantic Meridional Overturning Circulation then later changed (<20,000 years BP) to the predominance of sub-surface deep-water masses that mirror current conditions of the SACW, AAIW, and NADW (Mangini et al. 2010). These conditions then led to the predominant cold-water coral-reef development during the Holocene. Recent work on cold-coral reef mounds from Campos Basin (800 m) has reported corals with estimated ages of over 150,000 years BP, with colonies of *S. variabilis* showing variable aggradation rates (2–80 cm ka^{-1}) with peaks during glacial periods and high food supply (Raddatz et al. 2020). This dataset supports an ancient origin of the cold-water coral-reef structures and suggests a marked link between coral growth and food supply that has yet to be studied in modern reefs off Brazil.

Oceanic islands, seamounts, plateaus, and continental rises are also major tectonic features of the Brazilian margin. There are over 100 seamounts with summits shallower than 1500 m in the SW Atlantic suggesting wide potential for the occurrence of cold-water corals (Clark et al. 2010; Yesson et al. 2011). However, only a handful of seamounts have ever

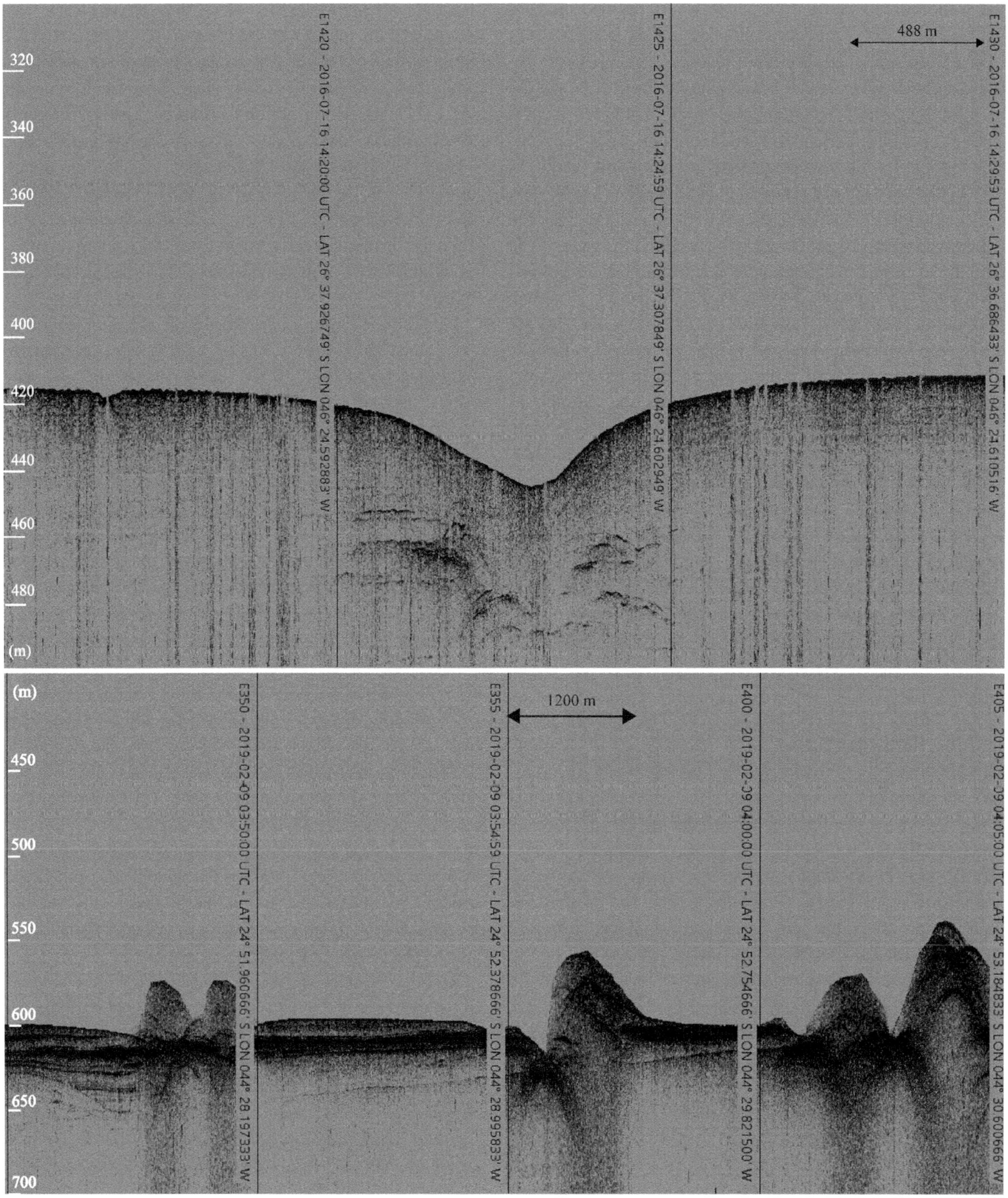

Fig. 2.3 Single-channel 3.5 kHz CHIRP seismic reflection profiles from the Santos Basin slope (400–600 m depth). Upper panel: Seafloor pockmarks with sub-surface reflections suggesting carbonate concretions associated with corals or authigenic origin. Lower panel: Cold-water coral carbonate mounds up to 50 m high near seafloor faults with potential inactive seepage activity. Images gently provided by R. Ramos; M. Maly and M. Mahiques, IOUSP

been thoroughly sampled by an ROV and detailed imagery and virtually all records of cold-water corals in these areas come from dredging. There are records of corals over seamount chains and oceanic islands on the Equatorial and SE margin and most records come from shallower depths (<500 m) in the Vitoria-Trindade seamount chain (~20°50′ S; Fig. 2.4). The Vitória-Trindade seamount chain is associated with a transform fault and seafloor uplift with volcanic origin. The summits of seamounts are typically shallower than 100 m, covered by benthic calcifying algae and the steep slopes of seamounts down to 400 m are covered by carbonate deposits 300 m thick (O'Hara et al. 2010; Skolotnev et al. 2010). Lower-slope flanks of seamounts are covered by volcanic rocks and 1–2 cm thick Fe-Mn deposits (Skolotnev et al. 2010). Additional potential hard substrate habitats for corals on Brazil's margin occur on seamounts and guyots with summit depths shallower than 1000 m that are distributed along the Brazilian Equatorial margin. The flat tops of those seamounts (e.g., Ceará seamount) are commonly covered by sediments, but deeper seamount flanks over the Brazilian margin offer a wide range of habitats for corals (Skolotnev et al. 2010; Jovane et al. 2016). On the southern margin, the Rio Grande Rise (~30°S) is part of the southernmost seamount chain at ~1000 km distance from Southern Brazil, with potential habitat for corals (Montserrat et al. 2019; Barbosa et al. 2020).

2.3.2 Reef Structures

The structure of cold-water coral reefs on Brazil's margin is only known at the Campos Basin due to the decades of intense offshore oil and gas exploration at depths of over 2000 m. *Solenosmilia variabilis*, *Lophelia pertusa*, *Enallopsammia rostrata* (Pourtalès, 1878), and *Madrepora oculata* are the four main coral-reef builders currently identified in Brazil's deep-sea margin, with the family Caryophylliidae the most abundant (Castro et al. 2006; Pires 2007; Barbosa et al. 2020). Most records of reef-building species and reef structures were sampled from the SE-Southern Brazil's deep-sea margin (Kitahara 2007; Cavalcanti et al. 2017; Lavrado et al. 2017), with very limited sampling of the NE-Northern equatorial margin (see also Cordeiro et al. 2020).

In Campos Basin (SE Brazil), the first reports of deep-water coral reefs indicated reef structures of 10–15 m high extending for over 40 km at depths of 500–800 m (Viana et al. 1998). Recent observations reported a wider occurrence of cold-water corals, with several reefs 1 km long and 10–30 m high occurring in Campos Basin (Cavalcanti et al. 2017), which are similar to reef structures reported by Sumida et al. (2004) in Santos Basin. There are also numerous reports (>3500 reflective targets, Cavalcanti et al. 2017) of patches of hard substrate along the soft sediment slope with potentially smaller reefs (10s to 100m wide and 1–7 m high), some of them potentially related to authigenic carbonates from inactive seep sites. Most reef structures from Campos Basin exhibit similar growth directions (based on 2D sonar images) facing predominant bottom currents across the slope (Cavalcanti et al. 2017). Bottom currents are in the range of 0.5–1.0 m s^{-1}, with occasional current velocities up to 3 m s^{-1} at slope depths and are generated by internal waves, cold fronts and eddies occurring throughout the SE-S margin (Viana et al. 1998; Silveira et al. 2020).

The Campos Basin scleractinian reefs are abundant on the upper slope from 400 to 800 m, with smaller reef patches distributed down to 1200 m. These patches of coral reefs growing over hard carbonate substrates seem similar to the ones present in the Gulf of Mexico and other continental margins worldwide (Cordes et al. 2008, 2016a; Becker et al. 2009; Davies et al. 2010). The patchy distribution of cold-water coral reefs on the southern and southeastern slope of Brazil suggests a similar association to clusters of carbonate mounds, with potential links to inactive seep ecosystems. Box-corer sampling over the Santos Basin carbonate ridge revealed dead and living reef-building corals (*L. pertusa*, *E. rostrata*, and *M. oculata*) along with other associated benthic assemblages (Maly et al. 2019). As substrate availability would be key for initial settlement and development of reef structures (Roberts et al. 2006), the wide occurrence of these corals on the southern-southeastern slope may be facilitated by the high density of patches of carbonate mounds along the upper-slope depths. Therefore, the SE-S Brazilian upper slope is likely a key province for cold-water corals in a similar way to the rich coral provinces from the NE Atlantic (Roberts et al. 2006). The potentially remarkable association between the formation of seeps and the later development of suitable habitat for cold-water corals needs to be explored within Brazil's deep-sea margin.

Records of scleractinians and octocorals on seamounts and oceanic islands on Brazil's margin are mostly limited to scattered occurrence records from opportunistic sampling by dredging over the years (Fig. 2.4), and there is no reporting of reef structures from those areas. The main reef builders *Madrepora oculata*, *Lophelia pertusa*, and *Solenosmilia variabilis* have all been sampled on the Vitoria-Trindade seamount chain and over the Northern Equatorial margin on St. Peter and St. Paul islands at upper bathyal depths (Fig. 2.4, Castro et al. 2006; Kitahara 2007; Pires 2007). In the Rio Grande Rise seamounts, submersible dives have recorded isolated scleractinian colonies (*Caryophyllia* sp.) over the seamount plateau between 1000 and 1200 m depth (Perez et al. 2018). Barbosa et al. (2020) predicted the Rio Grande Rise as a suitable habitat for scleractinians and, in fact, recent sampling in the area detected colonies of

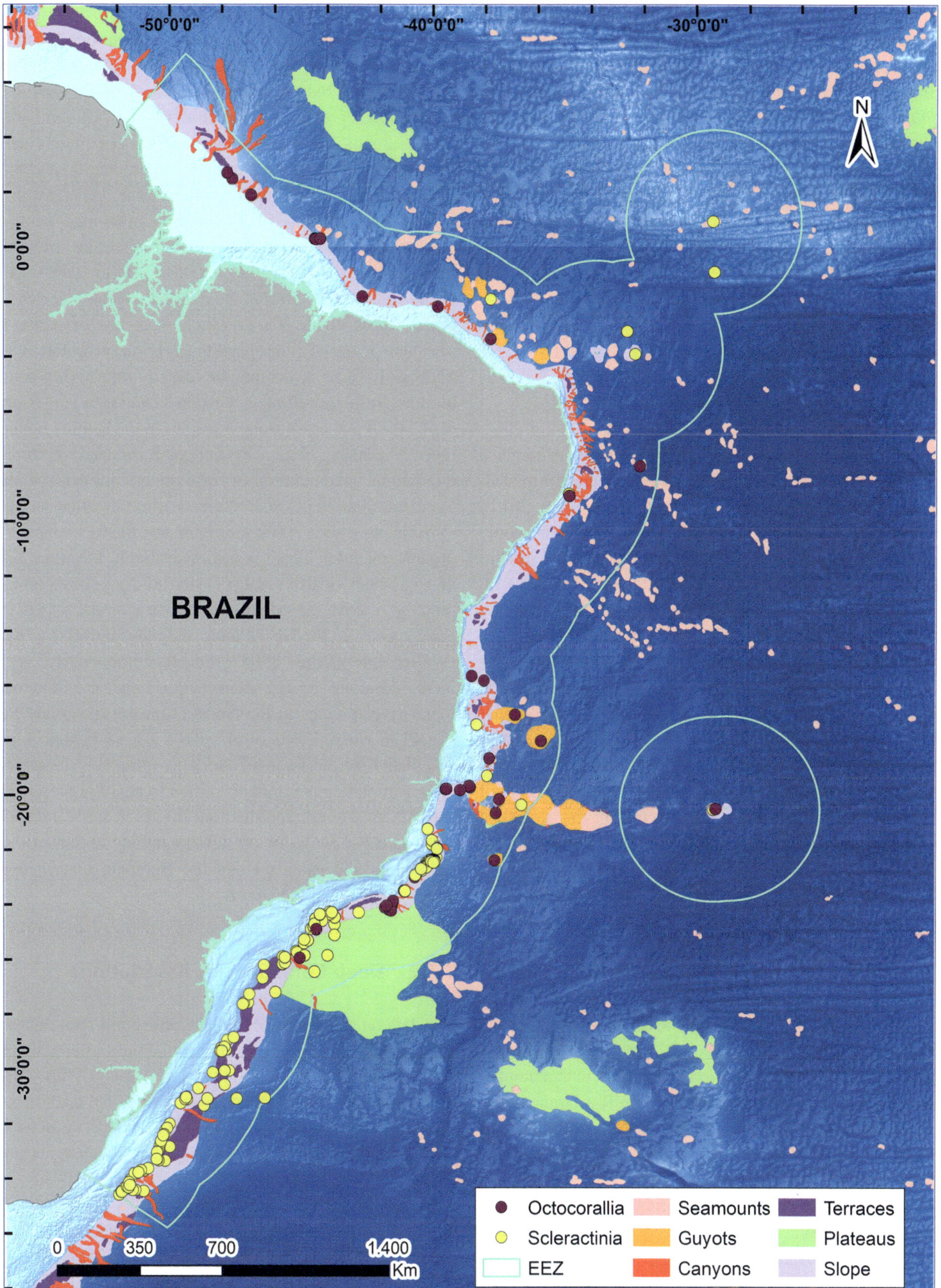

Fig. 2.4 Distribution of cold-water scleractinians and octocorals over the seabed geomorphic features along Brazil's margin (delimited within the blue line). GIS database from Blue Habitats (UNEP), Marine Regions, and General Bathymetric Chart of the Oceans (GEBCO 2020)

S. variabilis (unpublished data). The seamount flanks may offer a better suitable habitat for corals and occasionally create extensive reef structures similar to slope or canyon environments. Reef-forming species including *Solenosmilia variabilis and Enallopsammia* sp. have been observed at depths of 1300–1500 m in the Equatorial Atlantic seamounts to the West (9°16′N; Victorero et al. 2018), which are bathed by similar water masses to those described for Brazil's margin. However, Brazilian research programs have never sampled seamounts and oceanic archipelagos within the EEZ at depths below 500 m using ROVs.

2.4 Reef Biology and Ecology

2.4.1 Taxonomy and Reproduction

There are 63 species of azooxanthellate scleractinian corals in Brazil's EEZ, which include a mixture of species from shallow water (<200 m, Pires 2007; Cordeiro et al. 2014). Octocorallia has a similar number of species (~70) currently identified from deep coral reefs (Pérez et al. 2011; Cordeiro et al. 2015). Twenty-six species of black corals (Antipatharia) from the Brazilian margin and seamounts are currently described, which is nearly 1/3 of the group's diversity in the entire Atlantic Ocean (Lima et al. 2019). Deep-water coral species are widely distributed across the Brazilian deep margin, seamounts, and oceanic islands, with higher diversity in areas with higher sampling effort. Scleractinian cold-water coral-reef structures have been mainly mapped on the SE-S upper-slope depths, but a high diversity of octocorals and antipatharians also occurs on the slope, on isolated seamounts (e.g., Vitoria-Trindade, Almirante Saldanha) and on the Rio Grande Rise down to 1600 m depth (Castro et al. 2006; Perez et al. 2018; Lima et al. 2019). In general, the Brazilian deep cold-water coral fauna (Scleractinia, Octocorallia, and Antipatharia) includes a range of species with similar composition as the adjacent Caribbean, Equatorial Eastern Atlantic, and Southern Oceans, but also cosmopolitan species (Cairns 2007; Kitahara et al. 2020). Black corals may have the largest biogeographic ranges on the Brazilian margin, with a number of sampled species on seamounts and the Rio Grande Rise from 300 to 2300 m that are widely distributed across the North Atlantic, East and Equatorial Atlantic, and in the Pacific Ocean (Lima et al. 2019).

There is very little information on reproductive patterns of deep-sea corals from Brazil's margin. Data from only one octocoral species (*Anthoptilum murrayi* Kölliker, 1880) and the four dominant scleractinian reef-building corals from Campos Basin are available (Pires et al. 2009, 2014). Gametogenesis in *A. murrayi* (sampled between 1300 and 1800 m depth) and coral colonies of reef species *Madrepora oculata*, *Lophelia pertusa*, *Solenosmilia variabilis, and Enallopsammia rostrata*, sampled at 600 m depth, suggests that all species are gonochoristic with rare (<2%) occurrences of hermaphroditic colonies of *L. pertusa* and *M. oculata* reported by Pires et al. (2014). All coral species from Campos Basin had predominantly continuous gamete development suggesting continuous reproduction and broadcast spawning with lecithotrophic larvae. As a result, the reproductive patterns of Brazilian scleractinian corals are conservative and comparable to those species from other deep-sea basins (Waller 2005). Only two species (*L. pertusa* and *S. variabilis*) exhibited seasonal reproduction peaks (Pires et al. 2014), which were out of phase with seasonal peaks of phytoplanktonic primary productivity in SE Brazil. These peaks may be related to the spring-summer blooms from upwelling events in Campos Basin (Sumida et al. 2005) and cross-slope transport in the benthic boundary layer. The spring blooms in Campos Basin lead to prolonged interannual accumulation of fresh organic material of planktonic, zooplanktonic, and bacterial origin in slope sediments (Yoshinaga et al. 2008; Cordeiro et al. 2018). This fresh organic material likely becomes available through sinking of aggregates or from resuspension to deep corals and may potentially drive reproductive cycles in a similar way to cold-coral species in the NE Atlantic and other upwelling regions (Waller and Tyler 2005). Pelagic productivity is also increased along the Brazilian margin under the influence of major river plumes to the North (Amazon river) and South (La Plata river, Ciotti et al. 2010). These pelagic blooms stimulated by the input of nutrients ultimately increase the export of particulate organic carbon to adjacent slope depths and influence coral reproduction. However, it is yet unclear if cold-water coral reefs of non-upwelling regions or over seamounts and islands along the Brazilian margin exhibit distinct reproductive patterns.

2.4.2 Deep Coral Species Associations

The data on species associations within coral-reef clusters on the Brazilian margin are mainly limited to individual counts from dredged samples on the slope from Campos Basin (Castro et al. 2006; Arantes et al. 2009) and to some observations using ROV (Frensel et al. 2010). Arantes et al. (2009) analyzed cold-water coral species associations among clusters of stations from the upper- (50–800 m), mid- (1100–1400 m), and lower-slope (1300–1800 m) depths and found distinct species groups. They found that a number of octocorals had wide bathymetric distributions from the shelf to the upper slope, whereas the octocoral *A. murrayi* was mainly sampled at depths below 1000 m. Dominant scleractinian corals (*L. pertusa*, *E. rostrata,* and *S. variabilis*) were mostly sampled at mid-slope depths

(Arantes et al. 2009), although they also occur on the upper slope since 500 m depth (Cavalcanti et al. 2017). In general, Arantes et al. (2009) suggested that species associations between octocoral and scleractinian species change with depth across the slope. In a similar work, Castro et al. (2006) also found a distinct set of octocoral species on the upper slope (<500 m) of Campos Basin and argued that higher octocoral diversity was potentially associated with scleractinian reefs, particularly in the vicinity of the Vitoria-Trindade and Almirante Saldanha seamounts.

Observations during a deep-sea coral banks survey at the Campos Basin from 700 to 1100 m using an ROV revealed associations of brittle star species living on deep-sea gorgonian octocorals (Frensel et al. 2010). Ophiuroids of the genera *Asterochema* and *Ophiacantha* have been associated with deep-sea octocorals, being commonly found perched on corals using the currents to feed (Dearborn 1977; Emson and Woodley 1987). It is believed that ophiuroids occur as epizoic suspension feeders on gorgonians, antipatharians, pennatulacea, and alcyonaceans (Mcknight 2000), but the nature of the association between brittle stars and deep-sea corals is yet unclear. Nevertheless, Frensel et al. (2010) collected *Asteroschema arenosum* Lyman, 1878 on gorgonians from Plexauridae and Primnoidae families at 1100 m at Campos Basin and reported a high abundance of ophiacanthids in the southern area of the basin, possibly related to deep-sea coral banks.

Most cold-water coral programs in Brazil to date had a very limited sampling effort and the nature of dredged material precludes detailed information about species associations. This is because bottom trawls sample over multiple bottom types and do not capture benthic invertebrates over hard bottoms or soft sediments with similar efficiency. Sampling efficiency between hard and soft corals is also unlikely to be similar, so overall, we have a very limited picture of assemblage structure and species associations from cold-water coral reefs in Brazil. Cold-water corals have been shown to attract a rich and abundant fauna (Demopoulos et al. 2014; Bourque and Demopoulos 2018; Rueda et al. 2019), but there are only qualitative reports of benthic organisms in association with reef-forming and octocoral species on Brazil's margin. The association of other non-coral deep-sea species, mainly polynoid polychaetes, with cold-water coral reefs from Campos Basin has been reported for octocorals and antipatharians (Miranda and Brasil 2014; De Assis et al. 2019). In general, sampling of deep-water corals from slope depths has resulted in a number of new associated-species descriptions, with similar observations of commensal associations between annelids and cold-water corals detected globally. Although yet with limited sampling, it appears that corals host a number of species that are not typically sampled in background slope sediments and thus may offer critical habitat for the deep-sea biodiversity on Brazil's margin.

2.4.3 Ecology and Distribution

New species occurrences for deep-water corals along Brazil's margin will continue to occur with sampling at depths below 200 m, especially at the poorly sampled slopes of the Northeast and Northern margins and on seamounts and oceanic islands. Currently, inferences on the biology and ecological processes driving the distribution of deep-water coral assemblages on Brazil's margin largely rely on overlapping sampling sites with the bathymetric distribution of deep-water masses. To this end, distribution patterns and habitat suitability (or ecological niche) models based on species occurrence datasets have attempted to circumvent the limited deep-sea sampling and provide potential biogeographic distribution of cold-water corals (Barbosa et al. 2020; Gaurisas and Bernardino 2023).

On the Brazilian margin, the habitat niche modeling of cold-water coral records with deep-water mass variables indicates that corals have a wide potential distribution (the whole margin) and wide bathymetric range for all taxonomic groups (Davies et al. 2008; Barbosa et al. 2020). In general, reef-forming species have peak densities at depths within the aragonite saturation horizons (ASH) above 1.0 (Barbosa et al. 2020), but models suggest they potentially span far below the ASH in to the NADW water masses (to 2500 m depth). Carbonate ion concentrations (CO_3^{-2}) and the ASH of NADW are unlikely to provide adequate conditions for reef growth, which would create stress to calcification and reef development below 1200 m depth in most regions of the Brazilian margin. This depth is also the lower limit for larger cold-water reefs on Caribbean seamounts that are under similar deep-water masses, although corals have been observed scattered down to over 2000 m (Auscavitch et al. 2020). The mapping of reefs in Campos Basin suggests that the ASH may indeed be a key limit to the deeper (downslope) distribution of scleractinian reefs, reflected by the number of large reefs mapped at upper to mid-slope depths between 500 and 1200 m (Fig. 2.5).

Dissolved oxygen, temperature, and particulate organic matter flux were found to be additionally important to the habitat preference for deep-water corals on Brazil's margin (Barbosa et al. 2020). However, dissolved oxygen and temperature gradients across intermediate and lower-slope water masses (AAIW, NADW) are very limited when compared to harsher bottom water conditions ($<O_2$ and $>>T$) where scleractinian and octocoral associations have been sampled in other basins. For example, *L. pertusa* reefs have been found thriving under dissolved oxygen concentrations

Fig. 2.5 Deep-water corals from 600 to 800 m depth in Campos Basin, SE Brazilian Margin. (**a**) *Lophelia pertusa* (white) and *Madrepora oculata* reef. Note two small colonies of *Solenosmilia variabilis* on the left of the picture; (**b**) dead scleractinian colonies serving as substrate for several species of octocorals; (**c**) live *Lophelia pertusa* reef; (**d**) small colony of *Madrepora oculata* with a majid crab walking over it; (**e**) a goniasterid sea star associated with a deep-water reef; (**f**) a primnoid coral serving as substrate for a large sea anemone. Images courtesy of Petrobrás Petróleo Brasileiro S.A.

(DO) lower than 3.0 ml l^{-1} in the Gulf of Mexico (Davies et al. 2010) and under hypoxia (DO < 1.5 ml l^{-1}) on the Namibian margin (Hebbeln et al. 2020). As a result, it is unlikely that dissolved oxygen concentrations limit current growth of coral-reef builders over Brazil's margin. It remains to be determined if Brazilian scleractinian corals would be less tolerant to future low O_2 conditions given their acclimation to highly oxygenated water masses. In any case, the presence of tolerant reef species may have important implications for the survival of deep-water reefs worldwide.

On the Namibian margin, the *Lophelia* reefs are also under exceptional high temperatures (14.2 °C) and productivity regimes (>3.6 g C m^2 day^{-1}). Thermal stress (and higher metabolic demand) would be compensated by an increased food input (Hebbeln et al. 2020). On the Brazilian margin, temperatures below 14 °C are typically encountered below 400 m northward of 35°S, but shallower (200 m) on the Equatorial margin (5°S and 5°N; Mémery et al. 2000). As a result, it is likely that higher temperatures (>12–14 °C) limit the upper depth range of cold-water coral reefs along the Brazilian margin, with marked regional differences between the Equatorial and SE-S margin. Although coral species have been sampled at shallower depths, large reef structures (100s m) are unlikely to thrive under continuous warmer temperatures, which would be similar to ecological patterns observed in the North Atlantic (Freiwald 2002). These upper distribution limits along the 12 °C thermocline are within the SACW and vary through mixing with other deep-water masses along Brazil's margin.

Another hypothesis for changes in the bathymetric range of cold-water reefs is that acclimation of cold-water corals to shallower (200–300 m) and higher temperature waters would occur toward the southern Brazilian margin as this region is highly productive (NPP > 4 g C m^2 $year^{-1}$; Muller-Karger et al. 2005). In fact, large *Lophelia* reefs have already been found at around 250 m depth on the SE margin (P. Sumida, unpublished data). Therefore, the higher food input to cold-water corals on the southern Brazilian margin would compensate stress and metabolic demands at higher temperatures. The Brazilian Equatorial margin, which is less productive (NPP < 2 g C m^2 $year^{-1}$), would have cold-water coral reefs at deeper depth ranges due to limited energy availability for temperature acclimation. However, it must be noted that cold-water coral reefs are often present in areas with apparent low habitat suitability (high temperature, low salinity, low POC flux; Mienis et al. 2014), and habitat models are largely limited by restricted sampling and long-term observations. POC flux had a low power to indicate adequate environmental niches for cold-water corals over large spatial scales on the Brazilian margin (Barbosa et al. 2020), but as suggested before, regional differences may be more significant. Both current and geological records suggest that food input is key to coral acclimation to extreme temperatures and to lower aragonite saturation state, and periods of higher growth were marked by periods of high ocean productivity, which are regionally different along Brazil's margin. These regional oceanographic differences may be highly important and give clues to the future state of cold-water coral reefs along the deep Brazilian margin under climate change.

2.5 Threats and Outlook

2.5.1 Fisheries

Bottom trawling is one of the major threats to deep-water coral reefs globally. In the South Atlantic, fisheries catches have historically increased toward the upper slope from the 1970s to early 2000s, increasing the mean longevity of captured fish by fishing more longer-lived and vulnerable species, and thus, removing deep fish stocks (Morato et al. 2006). By the late 1990s, the fish landings from the upper (200–500 m) and mid-slope depths between 800 and 1000 m increased by over two times compared to the previous decades (Morato et al. 2006). These operations were later intensified by international chartered ships that heavily trawled mid-slope depths over the Equatorial, the SE-S margin (Campos and Santos Basin), and on seamounts (Perez et al. 2009, 2020). Perez et al. (2009) reported that nearly 21,000 trawl hauls have taken place over upper-slope depths (200–800 m) between 2000 and 2007. Based on an average commercial fishing by individual trawl shots duration and gear (10–15 min, Clark and Rowden 2009), we estimate that deep-sea trawling swept an upper-slope area of 2940 km^2 on the Brazilian margin, which largely overlaps with suitable habitats for cold-water coral reefs. The seamount area swept by trawling during the same period is close to 135 km^2 and was concentrated on seamounts from the Equatorial margin and the Vitoria-Trindade Chain.

The Brazilian deep-water fisheries targeted high-value commercial shrimp species *Aristaeopsis edwardsiana* (J.Y. Johnson, 1868) (Aristeidae; Scarlet shrimp) and *Aristaeomorpha foliacea* (Risso, 1827) (Aristeidae; Giant red shrimp) from upper-slope depths (500–800 m), with bycatch reports of corals and vulnerable deep-sea fish species including Macrouridae, Berycidae, and Trachichthyidae families (Perez et al. 2009). Based on these reports and on the overlap with records of cold-water coral species (Kitahara 2009), it is likely that the trawling caused significant impacts to deep-water reefs. The targeted shrimp fisheries likely exploited stocks that relied on deep-water coral reefs. *Lophelia pertusa* reefs have been found to support both high densities and gravid female shrimps that could potentially be associated with reef structures (Henry et al. 2013; Purser et al. 2013). In addition, fish bycatch included deep-sea Macrouridae, Berycidae (Alfonsino), and Trachichthyidae (Orange roughy), which are typically associated with seamounts and cold-water coral reefs (Clark 1999). Although trawling over the slope has likely caused widespread damage to cold-water corals on the Equatorial and on the SE-S provinces and nearby seamounts, there are no scientific assessments of damage or of the health status of reef ecosystems from the Brazilian deep-sea margin.

2.5.2 Offshore Industrial Impacts

The offshore oil and gas exploration on the Brazilian deep-sea margin is one of the largest in the world. The industry supports a significant portion of the country's gross domestic product (GDP, Bernardino et al. 2020), and over 80% of the hydrocarbon production comes from deep-sea offshore platforms (Almada and Bernardino 2017; Bernardino and Sumida 2017). There are multiple threats from regular offshore operations to cold-water corals, which include direct physical impacts and chemical contamination during exploration and production operations (Cordes et al. 2016b). The processes, scales, and effects of contaminants from the oil and gas industry to the cold-water coral reefs have been reviewed by Cordes et al. (2016b) and largely apply to the Brazilian margin. The acute and long-lasting effects of catastrophic spills over deep-sea ecosystems must be also considered, as the blowout of the Macondo well in the Gulf of Mexico (GoM) in April 2010 has shown. The Macondo spill caused significant and long-term (>10 years) biological and ecological impacts to benthic assemblages, including corals, over large regions in the GoM (Fisher et al. 2014; Girard and Fisher 2018). These observations indicate that accidental spills may be especially severe to cold-water coral fauna and will have lasting effects on the biology and ecology of coral reefs. There are no reports of impacts on deep-sea corals from regular offshore operations or accidental spills on Brazil's margin, which is notable after over 60 years of exploration. Although the industry is bound to environmental compliance rules, it is also clear that a framework of independent scientific impact assessments would be critical to evaluate these impacts.

Regulatory frameworks to minimize impacts of accidental spills on cold-water coral reefs are limited to installation setback practices, which suggest the industry to operate at least 100s m away from identified deep-water coral reefs. In practice, this is not enforced and it would be difficult to monitor without independent scientific assessments. During industrial prospecting and research activities, hundreds to thousands of exploratory corers and drilling operations impact seafloor assemblages within leased blocks, with further impacts on the deep-sea slope during installation and operation procedures. There are no standards for environmental impact assessments during those operations, and most areas leased to the offshore industry lack consistent mapping showing occurrence of cold-water coral reefs. Additional potential impacts to cold-water coral reefs include potential long-term effects of exploitation on seepage of fluids on the seabed, which create substrate (carbonate mounds) for settlement and growth of coral reefs (Baco et al. 2010; Cordes et al. 2016b). These effects may be potentially significant over time frames relevant to coral growth and over areas with abundant carbonate mounds such as Campos and Santos Basins and need to be properly assessed.

The lack of independent scientific monitoring programs leads to a great uncertainty on the health status of cold-water coral reefs and potential impacts from offshore operations on Brazil's deep margin. There is a significant overlap between offshore leasing areas and deep-sea ecosystems in the main cold-water coral provinces on Brazil's margin (Fig. 2.6). Most coral records fall within offshore production or exploration areas, which is a result of the nearly exclusive survey programs carried out by oil and gas companies within their areas of interest. The wide expanse of production and leased offshore oil and gas areas on Brazil's margin raises concerns on the status of reefs within and nearby those areas. Initiatives to foster good practices of spatial management have shown viable ways to minimize threats to these ecosystems (Almada and Bernardino 2017). However, spatial conservation planning efforts would need to be followed by in-situ independent impact assessments and basin-wide management practices that include the high number of potentially harmful oil and gas platforms. These initiatives are still absent from Brazil's regulatory framework and are unlikely to change in the near future.

The seamounts of Vitória-Trindade and Rio Grande Rise are the main areas with potential mining interest on the Brazilian margin given the occurrence of Fe-Mn crusts (Skolotnev et al. 2010; Montserrat et al. 2019). The potential adverse effects of deep-sea mining on vulnerable ecosystems including cold-water corals have raised intense global concerns of its viability (Levin et al. 2016). As indicated previously, the Brazilian seamount fauna has been undersampled and only a few scattered records of cold-water corals exist (Fig. 2.4). There are a number of biological and ecological patterns in the resident fauna over those seamounts that would need to be properly addressed if effective mining would be proposed (Smith et al. 2008b). Current data suggest that seamounts offer critical habitat to a number of cold-water coral species on the Brazilian margin (Castro et al. 2006), which would need to be the top scientific evidence against any proposal of deep-sea mining exploration in those ecosystems.

2.5.3 Climate Change

Climate change during the next century will lead to major changes in biogeochemical conditions of the deep oceans. In the South Atlantic along the Brazilian margin, scenarios indicate significant warming and an increased acidity, as the upper ocean layers absorb the excess heat and CO_2 from the atmosphere (Sweetman et al. 2017). As a result of ocean warming and acidification, models suggest a marked decrease in dissolved oxygen concentrations, lower pH, and lower food supply to benthic ecosystems by the year 2100 (Sweetman et al. 2017). The upper ocean (1000 m) of the tropical and subtropical South Atlantic warmed on average 0.1–0.2 °C per decade during the past 50 years, decreasing the solubility of oxygen and potentially increasing respiration rates and biological demand for oxygen (Schmidtko et al. 2017). Models indicate that the velocity of changes in these upper mesopelagic layers will be faster when compared to other layers in the global deep ocean (Brito-Morales et al. 2020). On the Brazilian margin, upper-slope (200–1000 m) ecosystems have experienced an average 50% deoxygenation during the last 50 years due to decreased solubility and higher biological demand, which were stronger in subtropical regions (~40 °S, Schmidtko et al. 2017). Ocean warming has also significant effects on pelagic productivity and on

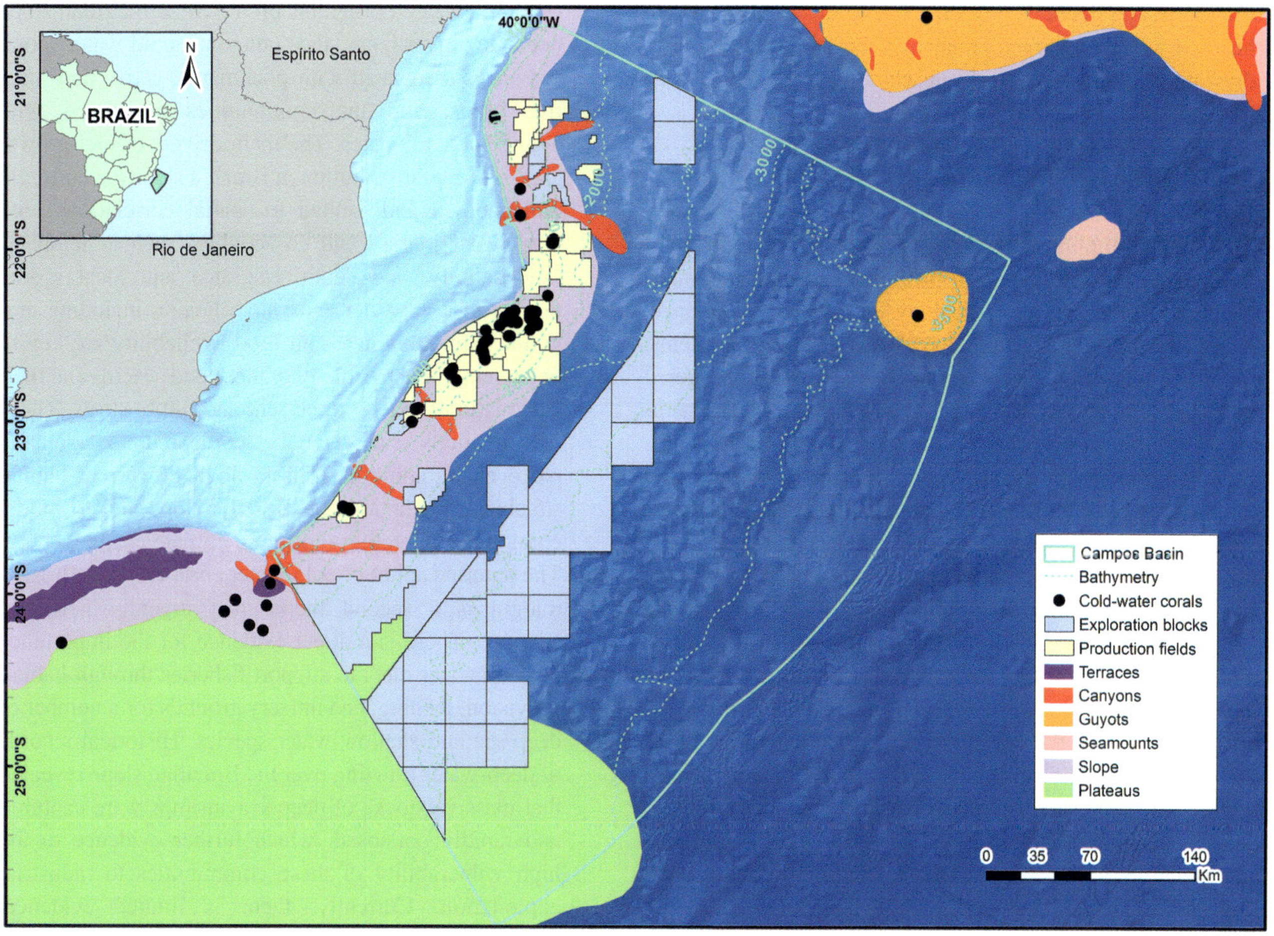

Fig. 2.6 Distribution of cold-water corals over the seabed geomorphic features and within oil and gas exploration (blue blocks) and production (yellow blocks) fields in Campos Basin (delimited within the blue line). GIS database from Blue Habitats (UNEP), General Bathymetric Chart of the Oceans (GEBCO 2020), and Agência Nacional de Petróleo, Gás Natural e Biocombustíveis (ANP)

the transfer of organic material to the deeper areas, with negative feedback to benthic assemblages and ecological processes on the seafloor (Smith et al. 2008a).

Based on the past and projected changes for the next century, cold-water scleractinian corals at bathyal depths in the South Atlantic will likely experience significant stress due to a combination of lower oxygen, lower food input, and lower aragonite saturation in intermediate water masses. Deep-sea coral reefs exposed to higher acidity may still be able to cope with changes at the cost of weaker calcification, for example (Hennige et al. 2015). The processes of physiological acclimation to lower aragonite saturation or hypoxia have been suggested in experimental and in field observations showing that corals can thrive under conditions that seem unfit. A number of studies suggest that cold-water corals may regulate their metabolism to acclimate to stressful conditions (Mienis et al. 2014). In some cases, corals may rely on an abundant food input to meet their metabolic demands during increased temperatures for example, but there is still very limited knowledge on how these processes will evolve with time and under multiple stressors.

In addition to warming and acidification, the heterotrophic deep-water corals will have to cope with lower pelagic organic input, which is predicted to decrease by up to 36% in the South Atlantic by the year of 2100 at bathyal depths (Sweetman et al. 2017). The Brazilian Equatorial margin will likely see higher deficits in organic matter flux to the seafloor as a result of stronger upper ocean stratification (Smith et al. 2008a). Upwelling regions such as Campos Basin may also experience lower productivity scenarios as summer temperatures become warmer and shutdown the upwelling of cold sub-surface and nutrient-rich SACW, and there is strong evidence for the past and future decreases in pelagic productivity throughout Brazil's margin (Behrenfeld et al. 2006). From the available climate change models, sub-surface water masses of the Brazilian margin (e.g., Subtropical and Equatorial margin) have experienced and will continue to show faster deoxygenation, lower aragonite saturation, and declines in food input to the benthos, where the cold-water coral-reef fauna may be subjected to early physiological and ecological impacts.

2.6 Future Directions

The limited sampling of biological communities over most of the Brazilian deep-sea margin precludes an essential understanding of the diversity, ecology, economic importance, health, and future resilience of cold-water coral ecosystems. Currently, data from cold-water corals in Brazil are scattered and mostly limited to nearly 2000 presence or absence taxonomic records throughout the deep margin and seamounts. There is also a general absence of distribution and reef structure records based on geophysical surveys and ROV imagery, which is only available from restricted access databases and from technical reports produced by the oil and gas industry. The institutional and legal arrangement requires impact assessments during a number of stages throughout the offshore oil and gas operations. However, these surveys are frequently carried out by either contracted private firms or by the industry itself, which in general fail to use standard scientific protocols and hold the "sensitive environmental" data restricted under confidentiality. As a result of decades of limited scientific funding and logistical platforms and technology to support deep-sea research, the scientific impact of research partnerships between the oil and gas industry is very low. Recent scientific advances have led to descriptions of age and growth of reef structures at one site of the Brazilian margin. Recent academic (public Universities) geophysical studies highlighted spectacular pockmark provinces and carbonate ridges over the upper slope of Santos Basin. However, there are yet limited biological and ecological datasets, and several key questions that are fundamental to advance our understanding of Brazilian cold-water coral ecosystems remain.

Within the objectives of this chapter, we proposed a number of research questions and hypotheses that should advance and match global efforts in understanding cold-water coral ecosystems:

(i) We need to expand the knowledge on habitat distribution and species association of cold-water coral reefs, with increased effort on poorly sampled and heterogeneous seafloor over submarine canyons, seamount chains, and the Equatorial margin slope. A consolidated understanding of geomorphological and biological species association will allow an initial recognition of areas of higher biological and ecological importance for cold-water coral reefs on the deep-sea margin that need to be protected and managed. As suggested earlier in this chapter, research programs on the SE-S Brazilian slope need to investigate the potential cold-water coral provinces associated with pockmarks, active and inactive seeps, and the communities associated with authigenic carbonates. The main drivers of reef growth and coral-reef distribution in Brazil's slope are currently understudied and limited to spatial correlations with deep-water masses and co-variables such as temperature, salinity, and depth. Regional and local spatial drivers of cold-water ecosystem health, including aragonite saturation state and food availability are key to understanding the ecology of those ecosystems and their acclimation capacity to current and future stress. These investigations can then be compared to available paleoceanographic datasets to advance current habitat suitability predictions on Brazil's slope and to model coral-reef vulnerability to future climate change.

(ii) The natural capital of cold-water coral reefs on Brazil's margin is a second theme that urgently needs to advance. There is indirect evidence for the importance of deep-water reefs to support fisheries through habitat provision, feeding, and nursery grounds for a number of deep-sea and shallow-water species. Historical records of deep-water trawling over the Brazilian slope revealed that massive stocks of deep-sea shrimps were captured and rapidly collapsed, which further evidence to the high vulnerability of these communities to industrial exploitation. Currently, there is limited evidence, supported by bycatch records, that cold-water coral reefs along the Brazilian margin and seamounts function in a similar way to other slope reefs and seamount ecosystems worldwide by offering key supporting services to commercial fishes. As the future climate warms and the oceans become increasingly less productive, these deep-sea reefs may become greatly important to support fisheries as the shallow-water stocks are already overexploited. The links to other sources of pressures to reefs, including climate and the offshore oil and gas industry will also be critical to explore and will need to be included toward effective management.

(iii) Implementing an effective funding and research agreement between industry, government, and scientists is the third major need to advance cold-water coral-reef research. It is unlikely that government resources alone will break the barriers of the expensive research needed to advance the exploration of cold-water coral reefs along Brazil's margin. Although some initiatives exist where the industry funds research in Brazilian universities, the limited scope of individual projects (e.g., taxonomy and species occurrence) will be insufficient to address the research questions raised above. Such initiatives would require a better rationalization of funding to increase independent scientific assessments without increasing costs to the offshore oil and gas industry in Brazil. As government institutions regulate the industry to meet environmental performance, these institutions also need to recognize and support a change in environmental management toward science-based projects. These arrangements exist in other major hydrocarbon provinces around the globe and will only strengthen the quality of routine surveys and impact assessment studies by using the same capital investments. In addition, management of monitoring programs in partnership with scientific committees will allow basin-wide strategic management of hydrocarbon provinces that are explored simultaneously by dozens of offshore platforms owned by different companies or consortium agreements.

Acknowledgments A.F.B. and P.Y.G.S. are supported by CNPq, CAPES, FAPES, and FAPESP. D.Y.G. acknowledges a FAPES graduate scholarship (Projeto iAtlântico). This project has received funding from the European Union's Horizon 2020 research and innovation programme under grant agreement No 818123 (iAtlantic). This output reflects only the author's view and the European Union cannot be held responsible for any use that may be made of the information contained therein.

References

Almada G, Bernardino AF (2017) Conservation of deep-sea ecosystems within offshore oil fields on the Brazilian margin, SW Atlantic. Biol Conserv 206:92–101

Almeida AG, Kowsmann RO (2014) Geomorfologia do talude continental e do Platô de São Paulo. In: Kowsmann RO (ed) Geologia e Geomorfologia, vol 1. Elsevier. Habitats, Rio de Janeiro, pp 33–66

Arantes RCM, Castro CB, Pires DO, Seoane JCS (2009) Depth and water mass zonation and species associations of cold-water octocoral and stony coral communities in the southwestern Atlantic. Mar Ecol Prog Ser 397:71–79. https://doi.org/10.3354/meps08230

Auscavitch SR, Lunden JJ, Barkman A, Quattrini AM, Demopoulos AWJ, Cordes EE (2020) Distribution of deep-water scleractinian and stylasterid corals across abiotic environmental gradients on three seamounts in the Anegada Passage. PeerJ 8:e9523. https://doi.org/10.7717/peerj.9523

Baco AR, Rowden AA, Levin LA, Smith CR, Bowden DA (2010) Initial characterization of cold seep faunal communities on the New Zealand Hikurangi margin. Mar Geol 272(1):251–259

Barbosa RV, Davies AJ, Sumida PYG (2020) Habitat suitability and environmental niche comparison of cold-water coral species along the Brazilian continental margin. Deep Sea Res I 155:103147

Becker EL, Cordes EE, Macko SA, Fisher CR (2009) Importance of seep primary production to Lophelia pertusa and associated fauna in the Gulf of Mexico. Deep Sea Res I 56(5):786–800. https://doi.org/10.1016/j.dsr.2008.12.006

Behrenfeld MJ, O'Malley RT, Siegel DA, McClain CR, Sarmiento JL, Feldman GC, Milligan AJ, Falkowski PG, Letelier RM, Boss ES (2006) Climate-driven trends in contemporary ocean productivity. Nature 444:752–755. https://doi.org/10.1038/nature05317

Bernardino AF, Sumida PYG (2017) Deep risks from offshore development. Science 358(6361):312. https://doi.org/10.1126/science.aaq0779

Bernardino AF, Berenguer V, Ribeiro-Ferreira VP (2016) Bathymetric and regional changes in benthic macrofaunal assemblages on the deep Eastern Brazilian margin, SW Atlantic. Deep Sea Res I 111:110–120. https://doi.org/10.1016/j.dsr.2016.02.016

Bernardino AF, Gama RN, Mazzuco ACA, Omena EP, Lavrado HP (2019) Submarine canyons support distinct macrofaunal assemblages on the deep SE Brazil margin. Deep Sea Res I 149:103052. https://doi.org/10.1016/j.dsr.2019.05.012

Bernardino AF, Cordes EE, Schlacher TA (2020) The natural capital of offshore oil, gas, and methane hydrates in the World Ocean. In: Baker M, Llodra ER, Tyler P (eds) Natural capital and exploitation of the deep ocean. Oxford University Press, Oxford, pp 111–125. https://doi.org/10.1093/:oso/9780198841654.003.0006

Bourque JR, Demopoulos AWJ (2018) The influence of different deep-sea coral habitats on sediment macrofaunal community structure and function. PeerJ 6:e5276. https://doi.org/10.7717/peerj.5276

Brito-Morales I, Schoeman DS, Molinos JG, Burrows MT, Klein CJ, Arafeh-Dalmau N, Kaschner K, Garilao C, Kesner-Reyes K, Richardson AJ (2020) Climate velocity reveals increasing exposure of deep-ocean biodiversity to future warming. Nat Clim Change 10:576–581. https://doi.org/10.1038/s41558-020-0773-5

Cairns SD (2007) Deep-water corals: an overview with special reference to diversity and distribution of deep-water Scleractinia. Bull Mar Sci 81:311–322

Carranza A, Recio AM, Kitahara M, Scarabino F, Ortega L, López G, Franco-Fraguas P, De Mello C, Acosta J, Fontan A (2012) Deep-water coral reefs from the Uruguayan outer shelf and slope. Mar Biodivers 42:411–414. https://doi.org/10.1007/s12526-012-0115-6

Castro CB, Pires DO, Medeiros MS, Loiola LL, Arantes RCM, Thiago CM, Berman E (2006) Cnidaria. In: Lavrado HP, Ignacio BL (eds) Biodiversidade bentica da costa central da Zona Econômica Exclusiva Brasileira, vol 14. Museu Nacional, Rio de Janeiro, Livros, pp 147–192

Cavalcanti GH, Arantes RCM, Falcão APC, Curbelo-Fernandez MP, Silveira MAS, Politano AT, Viana AR, Hercos CM, Brasil ACS (2017) Ecossistemas de corais de águas profundas da Bacia de Campos. In: Curbelo-Fernandez MP, Braga AC (eds) Comunidades Demersais e Bioconstrutores: caracterização ambiental regional da Bacia de Campos, Atlântico Sudoeste, vol 4. Elsevier. Habitats, Rio de Janeiro, pp 43–85

Ciotti AM, Garcia CAE, Jorge DSF (2010) Temporal and meridional variability of Satellite-estimates of surface chlorophyll concentration over the Brazilian continental shelf. Pan Am J Aquat Sci 5(2):236–253

Clark MR (1999) Fisheries for orange roughy (Hoplostethus atlanticus) on seamounts in New Zealand. Oceanologica Acta 22:593–602. https://doi.org/10.1016/S0399-1784(00)88950-1

Clark MR, Rowden AA (2009) Effect of deepwater trawling on the macro-invertebrate assemblages of seamounts on the Chatham Rise, New Zealand. Deep Sea Res I 56:1540–1554. https://doi.org/10.1016/j.dsr.2009.04.015

Clark MR, Rowden AA, Schlacher T, Williams A, Consalvey M, Stocks KI, Rogers AD, O'Hara TD, White M, Shank TM, Hall-Spencer JM (2010) The ecology of seamounts: structure, function, and human impacts. Annu Rev Mar Sci 2:253–278

Cordeiro RTS, Kitahara MV, Amaral FD (2014) New records and range extensions of azooxanthellate scleractinians (Cnidaria: Anthozoa) from Brazil. Mar Biodivers Rec 5:e35. https://doi.org/10.1017/S175526721200019X

Cordeiro RTS, Castro C, Pérez CD (2015) Deep-water octocorals (Cnidaria: Octocorallia) from Brazil: Family Chrysogorgiidae Verrill, 1883. Zootaxa 4058(1):81–100

Cordeiro LGMS, Wagener ALR, Carreira RS (2018) Organic matter in sediments of a tropical and upwelling influenced region of the Brazilian Continental margin (Campos Basin, Rio de Janeiro). Org Geochem 120:86–98. https://doi.org/10.1016/j.orggeochem.2018.01.005

Cordeiro RTS, Neves BM, Kitahara MV, Arantes RCM, Perez CD (2020) First assessment on Southwestern Atlantic euqatorial deep-sea coral communities. Deep-Sea Res I 163:103344

Cordes EE, McGinley MP, Podowski EL, Becker EL, Lessard-Pilon S, Viada ST, Fisher CR (2008) Coral communities of the deep Gulf of Mexico. Deep Sea Res I 55:777–787. https://doi.org/10.1016/j.dsr.2008.03.005

Cordes EE, Arnaud-Haond S, Bergstad O, Falcao APC, Freiwald A, Roberts JM, Bernal P (2016a) Cold-water corals. In: Group of Experts of the Regular Process (ed) World Ocean Assessment I. United Nations General Assembly, Chapter 42, pp 1–28

Cordes EE, Jones DOB, Schlacher TA, Amon DJ, Bernardino AF, Brooke S, Carney R, DeLeo DM, Dunlop KM, Escobar-Briones E, Gates AR, Genio L, Gobin J, Henry L-A, Herrera S, Hoyt S, Joye S, Kark S, Mestre NC, Metaxas A, Pfeifer S, Sink K, Sweetman AK, Witte U (2016b) Environmental impacts of the deep-water oil and gas industry: a review to guide management strategies. Front Environ Sci 4:58. https://doi.org/10.3389/fenvs.2016.00058

Davies AJ, Wisshak M, Orr JC, Roberts JM (2008) Predicting suitable habitat for the cold-water coral Lophelia pertusa (Scleractinia). Deep Sea Res I 55(8):1048–1062. https://doi.org/10.1016/j.dsr.2008.04.010

Davies AJ, Duineveld GCA, van Weering TCE, Mienis F, Quattrini AM, Seim HE, Bane JM, Ross SW (2010) Short-term environmental variability in cold-water coral habitat at Viosca Knoll, Gulf of Mexico. Deep Sea Res I 57(2):199–212. https://doi.org/10.1016/j.dsr.2009.10.012

Dearborn JH (1977) Food and feeding characteristics of Antarctic asteroids and ophiuroids. In: Llano GA (ed) Adaptations within Antarctic ecosystems. Gulf Publishing, Houston, pp 293–326

De Assis JE, Souza JRB, Lima MM, Lima GV, Cordeiro RTS, Pérez CD (2019) Association between deep-water scale-worms (Annelida: Polynoidae) and black corals (Cnidaria: Antipatharia) in the Southwestern Atlantic. Zoologia 36:1–13. https://doi.org/10.3897/zoologia.36.e28714

De Leo FC, Bernardino AF, Sumida PYG (2020) Chapter 3: Continental slope and submarine canyons: benthic biodiversity and human impacts. In: Sumida PYG, Bernardino AF, De Leo FC (eds) Brazilian deep-sea biodiversity, Brazilian marine biodiversity. Springer, Cham. https://doi.org/10.1007/978-3-030-53222-2_3

De Madron XD, Weatherly G (1994) Circulation, transport and bottom boundary layers of the deep currents in the Brazil Basin. J Mar Res 52:583–638

Demopoulos AWJ, Smith CR, Tyler PA (2003) The deep Indian ocean floor. In: Tyler PA (ed) Ecosystems of the deep oceans. Ecosystems of the world, vol 28. Elsevier, Amsterdam, pp 219–237

Demopoulos AWJ, Bourque JR, Frometa J (2014) Biodiversity and community composition of sediment macrofauna associated with deep-sea Lophelia pertusa habitats in the Gulf of Mexico. Deep Sea Res I 93:91–103. https://doi.org/10.1016/j.dsr.2014.07.014

Emson RH, Woodley JD (1987) Submersible and laboratory observations on Asteroschema tenue, a long-armed euryaline brittle star epizoic on gorgonians. Mar Biol 96:31–45

Feely RA, Sabine CL, Lee K, Berelson W, Kleypas J, Fabry VJ, Miller FJ (2004) Impact of anthropogenic CO2 the CaCO3 system in the oceans. Science 305:362–366. https://doi.org/10.1126/science.1097329

Fisher CR, Hsing P, Kaiser CL, Yoerger DR, Roberts HH, Shedd WW, Cordes EE, Shank TM, Berlet SP, Saunders MG, Larcom EA, Brooks JM (2014) Footprint of Deepwater Horizon blowout impact to deep-water coral communities. Proc Natl Acad Sci USA 111(32): 11744–11749. https://doi.org/10.1073/pnas.1403492111

Freiwald A (2002) Reef forming cold water corals. In: Wefer G, Billett D, Hebbeln D, Jorgensen BB, Schluter M, Van Weering T (eds) Ocean margin systems. Springer, Berlin, pp 365–385

Frensel R, Barboza CAM, Moura RB, Campos LS (2010) Southwest Atlantic deep-sea brittle stars (Echinodermata: Ophiuroidea) from Campos Basin, Brazil. In: Durham–Harris et al (eds) Echinoderms. Taylor & Francis, London, pp 173–180. ISBN 978-0-415-40819-6.

Fujikura K, Yamanaka T, Sumida PYG, Bernardino AF, Pereira OS, Kanehara T, Nagano Y, Nakayama C, Nobrega M, Pellizari V, Shigeno S, Yoshida T, Zhang J, Kitazato H (2017) Discovery of asphalt seeps in the deep Southwest Atlantic of Brazil. Deep Sea Res II 146:35–44. https://doi.org/10.1016/j.dsr2.2017.04.002

Gaurisas DY, Bernardino AF (2023) Benthic biogeographic patterns on the deep Brazilian margin. PeerJ 11:e14585. https://doi.org/10.7717/peerj.14585

GEBCO Compilation Group (2020) GEBCO 2020 Grid. https://doi.org/10.5285/a29c5465-b138-234d-e053-6c86abc040b9

Gerhardt S, Henrich R (2001) Shell preservation of Limacina inflata (Pteropoda) in surface sediments from the Central and South Atlantic Ocean: a new proxy to determine the aragonite saturation state of water masses. Deep Sea Res I 48:2051–2071

Giongo A, Haag T, Simão TLL, Medina-Silva R, Utz LRP, Bogo MR et al (2016) Discovery of a chemosynthesis-based community in the western South Atlantic Ocean. Deep Sea Res I 112:45–56

Girard F, Fisher CR (2018) Long-term impact of the Deepwater Horizon oil spill on deep-sea corals detected after seven years of monitoring. Biol Conserv 225:117–127. https://doi.org/10.1016/j.biocon.2018.06.028

Hebbeln D, Wienberg C, Dullo W, Freiwald A, Mienis F, Orejas C, Titschack J (2020) Cold-water coral reefs thriving under hypoxia. Coral Reefs 39:853–859. https://doi.org/10.1007/s00338-020-01934-6

Hennige SJ, Wicks LC, Kamenos NA, Perna G, Findlay HS, Roberts JM (2015) Hidden impacts of ocean acidification to live and dead coral framework. Proc R Soc B 282:20150990. https://doi.org/10.1098/rspb.2015.0990

Henry L-A, Moreno-Navas J, Roberts JM (2013) Multi-scale interactions between local hydrography, seabed topography, and community assembly on cold-water coral reefs. Biogeosciences 10: 2737–2746. https://doi.org/10.5194/bg-10-2737-2013

Jiang L-Q, Feely RA, Carter BR, Greeley DJ, Gledhill DK, Arzayus KM (2015) Climatological distribution of aragonite saturation state in the global oceans. Global Biogeochem Cycles 29:1656–1673. https://doi.org/10.1002/2015GB005198

Jovane L, Figueiredo JJP, Alves DPV, Iacopini D, Giorgioni M, Vannucchi P, Moura DS, Bezerra FHR, Vital H, Rios ILA, Molina EC (2016) Seismostratigraphy of the Ceará Plateau: clues to Decipher the Cenozoic evolution of Brazilian equatorial margin. Front Earth Sci 4:90. https://doi.org/10.3389/feart.2016.00090

Kitahara MV (2007) Species richness and distribution of azooxanthellate Scleractinia in Brazilian waters. Bull Mar Sci 81(3):497–518

Kitahara MV (2009) The deep-sea demersal fisheries and the azooxanthellate corals from southern Brazil. Biota Neotrop 9(2): 36–43. http://www.biotaneotropica.org.br/v9n2/en/abstract?article+bn00409022009

Kitahara MV, Cordeiro RTS, Barbosa RV, Pires DO, Sumida PYG (2020) Chapter 4: Brazilian deep-sea corals. In: Sumida PYG, Bernardino AF, De Leo FC (eds) Brazilian deep-sea biodiversity, Brazilian marine biodiversity. Springer, Cham. https://doi.org/10.1007/978-3-030-53222-2_4

Lavrado HP, Bernardino AF, Omena EP (2017) Distribuição da comunidade megabêntica ao longo da plataforma e talude continental da bacia de campos. In: Curbelo-Fernandez MP, Braga AC (eds) Comunidades Demersais e Bioconstrutores: caracterização ambiental regional da Bacia de Campos, Atlântico Sudoeste, vol 4. Elsevier. Habitats, Rio de Janeiro, pp 139–166

Levin LA (2005) Ecology of cold seep sediments: interactions of fauna with flow, chemistry and microbes. Oceanogr Mar Biol Annu Rev 43:1–46

Levin LA, Gooday AJ (2003) The deep Atlantic Ocean. In: Tyler PA (ed) Ecosystems of the deep oceans, Ecosystems of the world, vol 28. Elsevier, Amsterdam, pp 111, 582p–178

Levin LA, Sibuet M (2012) Understanding continental margin biodiversity: a new imperative. Annu Rev Mar Sci 4:79–112. https://doi.org/10.1146/annurev-marine-120709-142714

Levin LA, Mengerink K, Gjerde KM, Rowden AA, Van Dover CL, Clark MR, Ramirez-Llodra E, Curries B, Smith CR, Sato KN, Gallo N, Sweetman AK, Lily H, Armstrong CW, Brider J (2016) Defining "serious harm" to the marine environment in the context of deep-seabed mining. Mar Policy 74:245–259. https://doi.org/10.1016/j.marpol.2016.09.032

Lima MM, Cordeiro RTS, Pérez CD (2019) Black Corals (Anthozoa: Antipatharia) from the Southwestern Atlantic. Zootaxa. https://doi.org/10.11646/zootaxa.4692.1.1

Mahiques MM, Schattner U, Lazar M, Sumida PYG, Souza LAP (2017) An extensive pockmark field on the upper Atlantic margin of Southeast Brazil: spatial analysis and its relationship with salt diapirism. Heliyon 3:e00257. https://doi.org/10.1016/j.heliyon.2017.e00257

Maly M, Schattner U, Lobo FJ, Dias RJS, Ramos RB, Couto DM, Sumida PYG, Mahiques MM (2019) The Alpha Crucis Carbonate Ridge (ACCR): Discovery of a giant ring-shaped carbonate complex on the SW Atlantic margin. Sci Rep 9:18697. https://doi.org/10.1038/s41598-019-55226-3

Mangini A, Godoy JM, Godoy ML, Kowsmann R, Santos GM, Ruckelshausen M, Schroeder-Ritzrau A, Wacker L (2010) Deep sea corals off Brazil verify a poorly ventilated Southern Pacific Ocean during H2, H1 and the Younger Dryas. Earth Planet Sc Lett 293:269–276. https://doi.org/10.1016/j.epsl.2010.02.041

McKnight DC (2000) The marine fauna of New Zealand: Basket-stars and Snake-stars (Echinodermata: Ophiuroidea: Euryalinida). NIWA Biodiversity memoir. ISSN 0083-7903:115

Mémery L, Arhan M, Alvarez-Salgado XA, Messias MJ, Mercier H, Castro CG, Rios AF (2000) The water masses along the western boundary of the south and equatorial Atlantic. Prog Oceanogr 47: 69–98

Mienis F, Duineveld GCA, Davies AJ, Lavaleye MMS, Ross SW, Seim H, Bane J, van Haren H, Bergman MJN, de Haas H, Brooke S, van Weering TCE (2014) Cold-water coral growth under extreme environmental conditions, the Cape Lookout area, NW Atlantic. Biogeosciences 11:2543–2560. https://doi.org/10.5194/bg-11-2543-2014

Miranda VR, Brasil ACS (2014) Two new species and a new record of Scale-worms (Polychaeta) from Southwest Atlantic deep-sea coral mounds. Zootaxa 3856(2):211–226

Mohriak WU (2003) Sedimentary basins of the Brazilian continental margin. In: Bizzi LA, Schobbenhaus C, Vidotti RM, Gonçalves JH (eds) Geologia, Tectônica e Recursos Minerais do Brasil. CPRM, Brasília, pp 87–94

Montserrat F, Guilhon M, Corrêa PVF, Bergo NM, Signori CN, Tura PM, Maly ML, Moura D, Jovane L, Pellizari V, Sumida PYG, Brandini FP, Turra A (2019) Deep-sea mining on the Rio Grande Rise (Southwestern Atlantic): A review on environmental baseline, ecosystem services and potential impacts. Deep Sea Res I 145:31–58. https://doi.org/10.1016/j.dsr.2018.12.007

Morato T, Watson R, Pitcher TJ, Pauly D (2006) Fishing down the deep. Fish Fish 7:24–34

Morato T, Gonzalez-Irusta J-M, Dominguez-Carrió C, Wei C-L, Davies A et al (2020) Climate-induced changes in the suitable habitat of cold-water corals and commercially important deep-sea fishes in the North Atlantic. Glob Chang Biol 26:2181–2202

Morozov EG, Demidov AN, Tarakanov RY, Zenk W (2010) Deep water masses of the South and North Atlantic. In: Abyssal channels in the Atlantic Ocean. Springer, Dordrecht, pp 25–50. https://doi.org/10.1007/978-90-481-9358-5_2

Muller-Karger FE, Varela R, Thunell R, Luerssen R, Hu C, Walsh JJ (2005) The importance of continental margins in the global carbon cycle. Geophys Res Lett 32:L01602. https://doi.org/10.1029/2004GL021346

O'Hara TD, Consalvey M, Lavrado HP, Stocks KI (2010) Environmental predictors and turnover of biota along a seamount chain. Mar Ecol 31(S1):1–11. https://doi.org/10.1111/j.1439-0485.2010.00379.x

Orr J, Fabry V, Aumont O, Bopp L, Doney SC, Feely RA, Gnanadesikan A, Gruber N, Ishida A, Joos F, Key RM, Lindsay K, Maier-Reimer E, Matear R, Monfray P, Mouchet A, Najjar RG, Plattner G-K, Rodgers KB, Sabine CL, Sarmiento JL, Schlitzer R, Slater RD, Totterdell IJ, Weirig M-F, Yamanaka Y, Yool A (2005) Anthropogenic ocean acidification over the twenty-first century and its impact on calcifying organisms. Nature 437:681–686. https://doi.org/10.1038/nature04095

Perez JAA, Pezzuto PR, Wahrlich R, Soares ALS (2009) Deep-water fisheries in Brazil: history, status and perspectives. Lat Am J Aquat Res 37(3):513–541. https://doi.org/10.3856/vol37-issue3-fulltext-18

Pérez CD, Neves BM, Oliveira DH (2011) New records of octocorals (Cnidaria: Anthozoa) from the Brazilian coast. Aquat Biol 13:203–214. https://doi.org/10.3354/ab00363

Perez JAA, Kitazato H, Sumida PYG, Sant'Ana R, Mastella AM (2018) Benthopelagic megafauna assemblages of the Rio Grande Rise (SW Atlantic). Deep Sea Res I 134:1–11. https://doi.org/10.1016/j.dsr.2018.03.001

Perez JAA, Abreu JGN, Lima AOS, Silva MAC, Souza LP, Bernardino AF (2020) Chapter 8: Living and non-living resources in Brazilian deep waters. In: Sumida PYG, Bernardino AF, De Leo FC (eds) Brazilian deep-sea biodiversity, Brazilian marine biodiversity. Springer, Cham. https://doi.org/10.1007/978-3-030-53222-2_8

Pires DO (2007) The azooxanthellate coral fauna of Brazil. In: George RY, Cairns SD (eds) Conservations and adaptative management of seamount and deep-sea coral ecosystems. Rosenstiel School of Marine and Atmospheric Science, University of Miami, Miami, pp 265–272

Pires DO, Castro CB, Silva JC (2009) Reproductive biology of the deep-sea pennatulacean Anthoptilum murrayi (Cnidaria, Octocorallia). Mar Ecol Prog Ser 397:103–112

Pires DO, Silva JC, Bastos ND (2014) Reproduction of deep-sea reef-building corals from the southwestern Atlantic. Deep Sea Res II 99: 51–63

Purser A, Orejas C, Gori A, Tong R, Unnithan V, Thomsen L (2013) Local variation in the distribution of benthic megafauna species associated with cold-water coral reefs on the Norwegian margin. Cont Shelf Res 54:37–51. https://doi.org/10.1016/j.csr.2012.12.013

Raddatz J, Titschack J, Frank N, Freiwald A, Conforti A, Osborne A, Skornitzke S, Stiller W, Ruggeberg A, Voigt S, Albuquerque ALS, Vertino A, Schroder-Ritzrau A, Bahr A (2020) Solenosmilia variabilis-bearing cold-water coral mounds off Brazil. Coral Reefs 39(1):69–83. https://doi.org/10.1007/s00338-019-01882-w

Roberts JM, Wheeler AJ, Freiwald A (2006) Reefs of the deep: the biology and geology of cold-water coral ecosystems. Science 312: 543–547. https://doi.org/10.1126/science.1119861

Rueda JL et al (2019) 29 Cold-water coral associated fauna in the Mediterranean Sea and adjacent areas. In: Orejas C, Jiménez C (eds) Mediterranean cold-water corals: past, present and future. Coral reefs of the world, vol 9. Springer, Cham. https://doi.org/10.1007/978-3-319-91608-8_29

Santos RFD, Nagaoka D, Ramos RB, Salaroli AB, Taniguchi S, RCL F, Bícego MC, Lobo FJ, Schattner U, Mahiques MM (2018) Metal/Ca ratios in pockmarks and adjacent sediments on the SW Atlantic slope: Implications for redox potential and modern seepage. J Geochem Explor 192:163–173

Schmidtko S, Stramma L, Visbeck M (2017) Decline in global oceanic oxygen content during the past five decades. Nature 542:335–339. https://doi.org/10.1038/nature21399

Silveira IC, Napolitano DC, Farias IU (2020) Water masses and oceanic circulation of the Brazilian Continental Margin and adjacent abyssal plain. In: Sumida PYG, Bernardino AF, De Leo FC (eds) Brazilian deep sea biodiversity. Springer, Cham, pp 7–36

Skolotnev SG, Peyve AA, Turko NN (2010) New data on the structure of the Vitoria–Trindade seamount chain (Western Brazil Basin, South Atlantic). Dokl Akad Nauk 431(2):435–440. https://doi.org/10.1134/S1028334X10040057

Smith CR, Demopoulos AWJ (2003) The deep Pacific Ocean floor. In: Tyler PA (ed) Ecosystems of the deep oceans. Ecosystems of the world, vol 28. Elsevier, Amsterdam, pp 179–218

Smith CR, De Leo FC, Bernardino AF, Sweetman AK, Arbizu PM (2008a) Abyssal food limitation, ecosystem structure and climate change. Trends Ecol Evol 23(9):518–528

Smith CR, Gaines SD, Friedlander A, Morgan C, Thurnherr AM, Mincks S, Watling L, Rogers AD, Clark MR, Baco-Taylor AR, Bernardino AF, De Leo FC, Dutrieux P, Rieser A, Kittinger J, Padilla-Gamino J, Prescott R, Srsen P (2008b) Preservation reference areas for nodule mining in the Clarion-Clipperton Zone: rationale and recommendations to the International Seabed Authority. Expert Participants in a Workshop to Design Marine Protected Areas for Seamounts and the Abyssal Nodule Province in the Pacific High Seas. University of Hawaii at Manoa

Stramma L, England M (1999) On the water masses and mean circulation of the South Atlantic Ocean. J Geophys Res Oceans 104(C9): 20863–20883

Sumida PYG, Yoshinaga MY, Madureira LAS, Hovland M (2004) Seabed pockmarks associated with deepwater corals off SE Brazilian continental slope, Santos Basin. Mar Geol 207:159–167

Sumida PYG, Yoshinaga MY, Ciotti AM, Gaeta SA (2005) Benthic response to upwelling events off the SE Brazilian coast. Mar Ecol Prog Ser 291:35–42

Sumida PYG, Bernardino AF, De Leo FC (2020a) Brazilian deep-sea biodiversity. Brazilian marine biodiversity. Springer, Cham. 273p. https://doi.org/10.1007/978-3-030-53222-2

Sumida PYG, Bernardino AF, De Leo FC (2020b) Chapter 1: An introduction to the Brazilian deep-sea biodiversity. In: Sumida PYG, Bernardino AF, De Leo FC (eds) Brazilian deep-sea biodiversity, Brazilian marine biodiversity. Springer, Cham. https://doi.org/10.1007/978-3-030-53222-2_1

Sweetman AK, Thurber AR, Smith CR, Levin LA, Mora C, Wei C-L, Gooday AJ, Jones DOB, Rex M, Yasuhara M, Ingels J, Ruhl HA, Frieder CA, Danovaro R, Wurzberg L, Baco A, Grupe BM, Pasulka A, Meyer KS, Dunlop KM, Henry L, Roberts JM (2017) Major impacts of climate change on deep-sea benthic ecosystems. Elem Sci Anth 5:4. https://doi.org/10.1525/elementa.203

Tittensor DP, Baco AR, Brewin PE, Clark MR, Consalvey M, Hall-Spencer J, Rowden AA, Schlacher T, Stocks KI, Rogers AD (2009) Predicting global habitat suitability for stony corals on seamounts. J Biogeogr 36:1111–1128

Viana AR, Faugeres JC, Kowsmann RO, Lima JAM, Caddah LFG, Rizzo JG (1998) Hydrology, morphology and sedimentology of the Campos continental margin, offshore Brazil. Sediment Geol 115:133–157

Victorero L, Robert K, Robinson LF, Taylor ML, Huvenne VAI (2018) Species replacement dominates megabenthos beta diversity in a remote seamount setting. Sci Rep 8:4152. https://doi.org/10.1038/s41598-018-22296-8

Waller RG (2005) Deep-water Scleractinia (Cnidaria: Anthozoa): current knowledge of reproductive processes. In: Freiwald A, Roberts JM (eds) Cold-water corals and ecosystems. Springer, Berlin, pp 691–700

Waller RG, Tyler PA (2005) The reproductive biology of two deep-water, reef-building scleractinians from the NE Atlantic Ocean. Coral Reefs 24:514–522

Yesson C, Clark MR, Taylor ML, Rogers AD (2011) The global distribution of seamounts based on 30 arc seconds bathymetry data. Deep Sea Res I 58:442–453. https://doi.org/10.1016/j.dsr.2011.02.004

Yoshinaga MY, Sumida PYG, Wakeham SG (2008) Lipid biomarkers in surface sediments from an unusual coastal upwelling area from the SW Atlantic Ocean. Org Geochem 39:1385–1399

Cold-Water Corals of the World: Gulf of Mexico

3

Sandra Brooke, Amanda Demopoulos, Harry Roberts, Jay Lunden, Tracey Sutton, and Andrew Davies

Abstract

The Gulf of Mexico is a semi-enclosed sea that borders the USA and Mexico and covers approximately 1.5 million square kilometers. The northern Gulf is topographically complex and is a rich source of oil and gas deposits, which has led to a great deal of research on benthic ecosystems from the coastal zone to the deep sea. While not fully explored, the distribution of cold seeps and deep corals in the northern Gulf is reasonably well described. The eastern Gulf has a moratorium on energy industry development and consequently less exploration and research has been conducted in this region; however, recent explorations have revealed deep scleractinian reefs on the west Florida slope and extensive octocoral gardens on the deep escarpment. The Gulf is a productive sea with lucrative fisheries in addition to oil and gas. Exploitation of natural resources and potential climate change impacts threaten vulnerable ecosystems in the Gulf, including those in the deep sea. This chapter describes the oceanography and geology of the Gulf of Mexico, presents the current state of the knowledge of cold-water coral distribution, physiology and ecology, and provides an assessment of the threats to these vulnerable ecosystems.

Keywords

Coral reef · Deep sea · Geology · Salt tectonics · Loop current · Scleractinian · Coral biology · Microbiome · Deep-sea ecology

3.1 Introduction

The Gulf of Mexico (GoM) is a semi-enclosed marginal sea of the Atlantic Ocean, located at the southern extent of the North American continent. To the east, north, and north-west the GoM is bounded by the southern United States of Florida, Alabama, Mississippi, Louisiana, and Texas, and the south west to southeast is bounded by the Mexican states of Tamaulipas, Veracruz, Tabasco, Campeche, Yucatan, and Quintana Roo. The GoM covers an area of approximately 1.5 million square kilometers and contains many ecologically and economically valuable resources. The northern GoM has a highly complex geological formation, driven by salt tectonics, gas seepage and production of irregular carbonate formations as a result of bacterial methane metabolism (CSA 2007; Brooks et al. 2015). The eastern GoM (Florida and the Yucatan Peninsula) comprises massive carbonate platforms, with wide gentle slopes leading to steep escarpments with complex erosional gullies and canyons. In addition to natural habitats, the GoM has many shipwrecks and oil platforms that provide artificial habitats for marine communities.

The geological complexity and variety of the GoM provides a wide diversity of cold-water coral (CWC, also known as deep-sea coral) habitats; however, CWC ecosystem types in the GoM can be broadly divided into coral mounds that are created by stony corals and coral gardens that are hard substrates colonized by different coral community types. Coral communities have been defined as *assemblages of*

S. Brooke (✉)
Coastal and Marine Lab, Florida State University, St. Teresa, FL, USA
e-mail: sbrooke@fsu.edu

A. Demopoulos
US Geological Survey, Wetland and Aquatic Research Center, Gainesville, FL, USA

H. Roberts
Louisiana State University, Baton Rouge, LA, USA

J. Lunden
Department of Biology, Temple University, Philadelphia, PA, USA

T. Sutton
Oceanic Ecology Lab, Nova Southeastern University, Dania Beach, FL, USA

A. Davies
Department of Biological Sciences, University of Rhode Island, Kingston, RI, USA

Graduate School of Oceanography, University of Rhode Island, Narragansett, Rhode Island, USA

E. Cordes, F. Mienis (eds.), *Cold-Water Coral Reefs of the World*, Coral Reefs of the World 19,
https://doi.org/10.1007/978-3-031-40897-7_3

structure-forming deep corals and other associated species, such as sedentary and motile invertebrates and demersal fishes (Lumsden et al. 2007). Structure-forming coral taxa include not only the reef-building stony corals, but also many different "tree" corals (some species of octocorals and black corals). Larger hydrocorals (Stylasteridae) and some soft corals (e.g., Nephtheidae) can be locally abundant and shelter smaller invertebrates, but do not appear to contribute significantly to habitat structure (Brooke and Schroeder 2007).

The northern GoM has historically been the focus of exploration and research, as this is the most active region for energy industry activities. The West Florida Platform is under a moratorium for offshore drilling, so has been less studied than the northern region. The areas that border Mexico are the least studied of all the regions, despite active drilling for oil in the southwest region. Over the past decade, research on CWC in the GoM has increased substantially, primarily through federal funding by the Bureau of Ocean Energy Management (BOEM), National Oceanographic and Atmospheric Administration (NOAA) and United States Geological Survey (USGS). Because some CWCs colonize seep-derived authigenic carbonate, earlier discoveries of CWCs were incidental to studies of chemosynthetic ecosystems. The first targeted study of CWC was a partnership of the Minerals Management Service (subsequently re-named BOEM) and USGS. The study focused on *L. pertusa* ecosystems offshore of Mississippi and Louisiana and greatly expanded our understanding of the distribution, biology, and ecology of these systems in the GoM (CSA 2007; Cordes et al. 2008; Sulak et al. 2008). The sequel to these studies commenced in 2007 and was a collaboration between BOEM, USGS, and NOAA (Brooks et al. 2015; Demopoulos et al. 2017). This study included coral communities associated with natural and artificial habitats, and the geographic scope extended eastward to the West Florida Slope. Subsequent studies by NOAA Deep Sea Coral Research and Technology Program (DSCRTP) and NOAA Office of Ocean Exploration and Research (OER) further expanded our knowledge of coral communities on the west Florida platform (Boland et al. 2017). The Deepwater Horizon (DWH) oil well explosion in 2010 precipitated a significant increase in research effort as a result of the federally mandated Natural Resource Damage Assessment (NRDA), which funded 5 CWC focused cruises in the aftermath of the spill (White et al. 2012; Fisher et al. 2014a, b). Subsequently, several additional cruises and research programs were initiated through the Gulf of Mexico Research Initiative (GOMRI) to understand the effects of the oil on deepwater fauna, including corals (Cordes et al. 2014). Between 2012 and 2014, nine additional cruises were conducted by NOAA and partners, as well as the Schmidt Ocean Institute. The result of these intensive, multidisciplinary, and geographically expansive efforts has been a rapid increase in our understanding of CWC systems over the past decade and a recognition of the vulnerability of these ancient systems to anthropogenic impact such as destructive fishing, energy industry activities, and climate change, among others (Joye et al. 2016; Schwing et al. 2020; Halanych et al. 2021).

Place-based management measures have been implemented by BOEM to protect corals and other sensitive habitats from fossil fuel extraction and by the GoM Fisheries Management Council (GMFMC) to prevent damage from fishing gear. These kinds of measures will not protect the GoM marine ecosystems from wide-ranging impacts such as oil spills and climate change, which require a different type of management approach (Cordes et al. 2016a, b). As we continue to progress in our understanding of these systems, new approaches to their protection will hopefully develop, so they can survive future ocean conditions.

3.2 Oceanography

3.2.1 Major Current Systems

The GoM is predominately a semi-enclosed basin that is narrowly connected to the Caribbean Sea through the Yucatan Strait and the Atlantic Ocean through the Florida Straits. Circulation in the GoM is facilitated by multiple physical processes, including the Loop Current (Hofmann and Worley 1986), anticyclonic rings propagated from the Loop Current (Welsh and Inoue 2000), and cyclonic eddies that shed from the Loop Current and its rings (Hamilton et al. 1999). At the surface, circulation is primarily governed by the Loop Current, which enters from the Yucatan Strait and exits through the Florida Straits. Loop Current rings strengthen surface circulation and typically move westward at intervals of 3–17 months (Sturges and Leben 2000; Sturges and Lugo-Fernandez 2005). Compared to surface seawater circulation in the GoM, deepwater circulation is less understood, though deep anticyclone-cyclone pairs are hypothesized to migrate westward along the seafloor in conjunction with surface rings driven by the Loop Current (Welsh and Inoue 2000).

Physical properties within the GoM, including dissolved oxygen, salinity, and temperature, tightly reflect the properties of the water masses within it. Water in the Loop Current is primarily composed of North Atlantic Central Water (NACW), which ranges in temperature from 8 to 19 °C, salinity from 35.1 to 36.7, and dissolved oxygen from 2.5 to 3.5 mL L^{-1} (Schmitz and McCartney 1993; Mienis et al. 2012). Subtropical Underwater (SUW) is also present within the Loop Current and drives the formation of a salinity maximum of 36.5 below the mixed layer in the GoM (Jochens and DiMarco 2008). In deeper waters, there is evidence of two additional water masses, including Antarctic

Intermediate Water (AAIW) and Tropical Atlantic Central Water (TACW). The AAIW is characterized by high nutrient concentrations and a salinity minimum of 34.9 (Jochens and DiMarco 2008; Sturges 2005), and the TACW is believed to influence the GoM's oxygen minimum layer of 600 m (Morrison et al. 1983). In addition to the water masses discussed above, riverine inputs originating from the Mississippi–Atchafalaya River System discharge large amounts of freshwater into the Northern GoM, along with sediment, organic material, and nutrients such as phosphates and nitrates (Milliman and Meade 1983; Balsam and Beeson 2003; Lunden 2013).

3.2.2 Environmental Setting

A range of abiotic and physical variables are critical to the growth and survival of CWC (Davies and Guinotte 2011; Yesson et al. 2012). Across the global ocean, reef-forming CWC occur within a generally narrow range of such variables including temperature, dissolved oxygen, density, and carbonate saturation state; particularly aragonite, which stony corals use to create their skeletons. Physical factors such as current speed also exert control over cold-water coral distribution in relation to larval delivery and food supply, sediment transport, and particle capture rates (Davies et al. 2009; Mienis et al. 2019; Orejas et al. 2016).

The ranges of major abiotic factors at cold-water coral reefs in the GoM are generally similar to those observed in other regions of the world. Temperature typically ranges from approximately 8 to 10 °C at reefs dominated by *Lophelia pertusa* throughout the global ocean (Davies and Guinotte 2011; Davies et al. 2008). Long- and short term Eulerian observations at the Viosca Knoll 826 site (Fig. 3.1) revealed temperatures between 6.5 and 11.6 °C (Davies et al. 2010; Mienis et al. 2012) and CTD observations within the coral ecosystems of the Campeche cold-water coral province suggest a range between 7.5 and 9.5 °C (Hebbeln et al. 2014). Habitat suitability modeling efforts support similar ranges, with a temperature range centered on a mean of 10 °C (6–16 °C) for the black coral *Leiopathes glaberrima* (Etnoyer et al. 2018) and 11.5 °C for reef-forming scleractinians (Hu et al. 2020). Salinity at GoM coral reefs also appears to follow the typical range for many species, centered around means of approximately 35 on the practical salinity scale with reported ranges of 34.95–35.4 (Mienis et al. 2012) and 34.9–35.1 (Hebbeln et al. 2014). Current velocity observations suggest a fairly moderate habitat, again bearing strong resemblance to other sites around the world, average 8 cm s^{-1} and peak current speeds of up to 38 cm s^{-1} (Mienis et al. 2012). Oxygen, in contrast, is generally lower than observations in the North Atlantic, with values of 2.7–2.8 mL L^{-1} reported in both the Campeche coral province and Viosca Knoll (Davies et al. 2010; Mienis et al. 2012; Hebbeln et al. 2014). Regarding the aragonite saturation state, Ω_{arag}, values at the GoM cold-water coral reef are slightly lower than other major reef areas and range from approximately 1.2 to 1.6 (Lunden et al. 2013). At the Viosca Knoll reefs, total alkalinity is slightly elevated over the reef structures relative to the surrounding water column, which may slow the rate of dissolution of reef framework caused by ocean acidification (Georgian et al. 2016a).

3.3 Geological Setting

The present GoM occupies an area of over 1.5×10^6 km^2 and has a maximum water depth of ~3750 m (Salvador 1991). Continental margins of the Gulf vary from carbonates of the West Florida Escarpment and the Yucatan Platform to terrigenous clastic sediments delivered by rivers of the northwestern and northern Gulf. These regional lithological differences reflect the Gulf's geologic history starting with the opening of the Gulf Basin by rifting during late Triassic (~237–201 myr BP; Walker et al. 2018) and deposition of thick salt (Louann Salt) during mid-to-late Jurassic (175–145 myr BP; Buffler 1991). After salt deposition, crustal stretching continued and separated the Louann Salt into the Louisiana-Texas salt province in the north and the Mexican or Campeche salt province in the south (Fig. 3.2a). These events were accompanied by a general deepening of the Gulf as the central basin cooled and slowly subsided (Marton and Buffler 2010). From mid-Jurassic to early Cretaceous (~175–130 myr BP) the Gulf was a shallow "ramp" setting, with widespread deposition of carbonates, which are acknowledged as source rocks for much of the petroleum found in the present Gulf and its onshore extensions. As the basin subsided, rudist reefs and other carbonates built the basin margins.

By the end of the Cretaceous (~ 95 myr BP) significant non-carbonate deposits started showing up along the northern and northwestern margins of the Gulf (Galloway et al. 2011). A recent detrital-Zircon study by Blum et al. (2017) indicates that by late Cretaceous (~90 myr BP) an ancestral Tennessee-Alabama River system, which drained the Appalachian Mountains, contributed most of the sediment to the Gulf. Also, at this time the Chicxulub impact event marked the end of the Cretaceous and caused large-scale collapse along both the carbonate margins of Yucatan and Florida. However, by the beginning of Cenozoic times (~65 myr BP) drainage of southern North America had reorganized so that significant terrigenous sediments were being contributed from the western mountain ranges as well as the Appalachians. These sediments were delivered primarily to the northwestern Gulf (Fig. 3.2b). By Oligocene times (~30 myr BP) the Rocky Mountains provided sediments from the

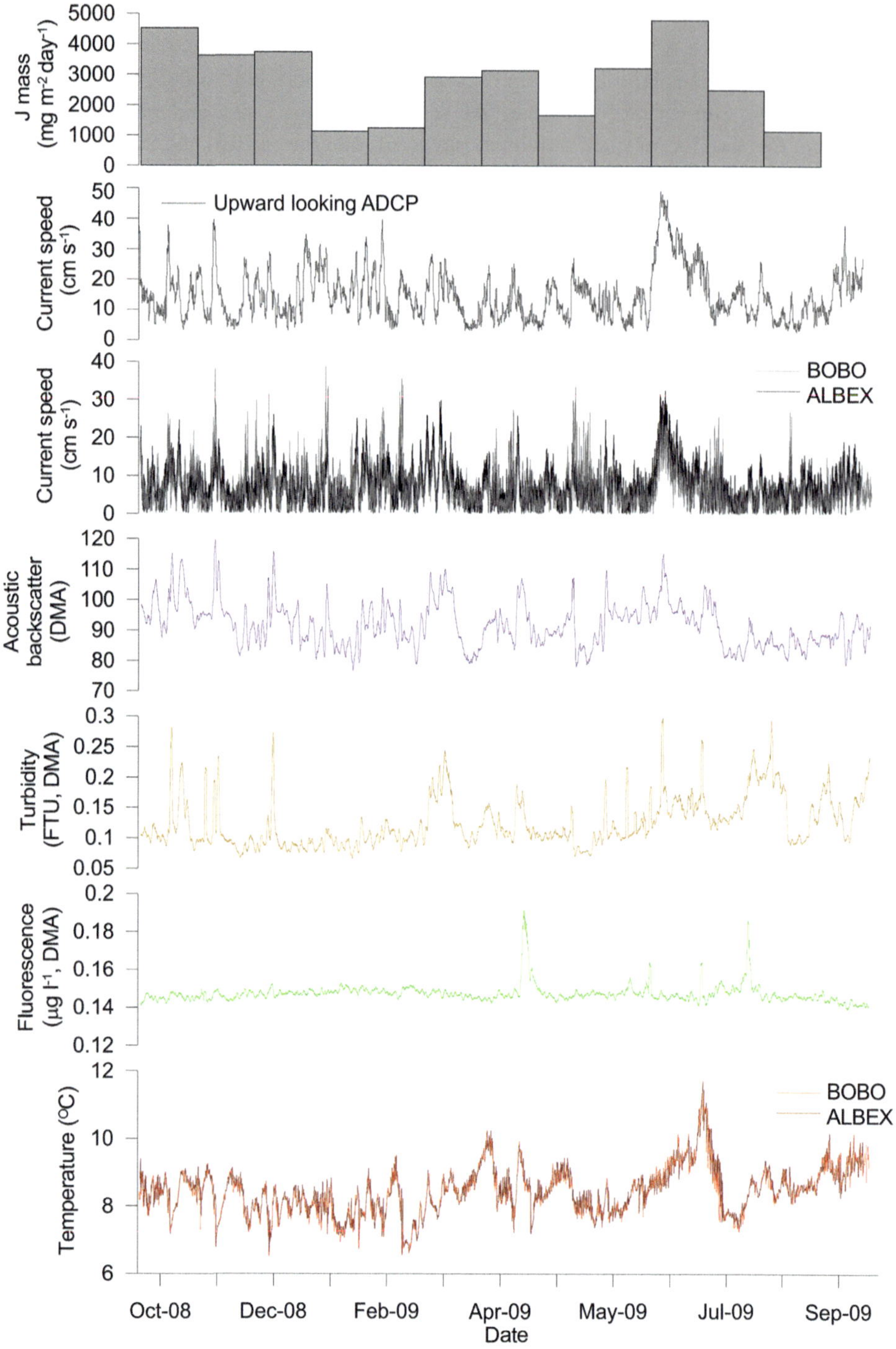

Fig. 3.1 Data on oceanographic variability at Viosca Knoll 826, taken from almost one year of benthic observation using two lander systems (BOBO and ALBEX). Variables recorded include temperature at 15-min intervals, daily moving averages of fluorescence, turbidity, and acoustic backscatter. Near bottom current speeds, depth averaged current speed from the upward looking ADCP (0–130 mab), and mass flux from sediment trap data (collecting interval 27 days). Reproduced from Mienis et al. (2012)

west to the Mississippi Embayment by way of the paleo-Red, Arkansas, and Mississippi Rivers. From mid-Tertiary times (Oligocene) to the present, massive volumes of sediment have been delivered to the northern and northwestern margins of the Gulf (Fig. 3.2b). The interaction between sediment transported to the Gulf by one of the world's great continental drainage systems, now represented by the Mississippi River, has created a present continental slope with over 90 sediment-filled intraslope basins surrounded by a variety of salt-cored features that have up to 150 m relief.

The events outlined above have set the stage for seafloor characteristics we observe today throughout the deepwater (>200 m) Gulf from the West Florida Platform to the Yucatan Platform. Each major sector of the deepwater Gulf has its own substrate type and oceanographic setting.

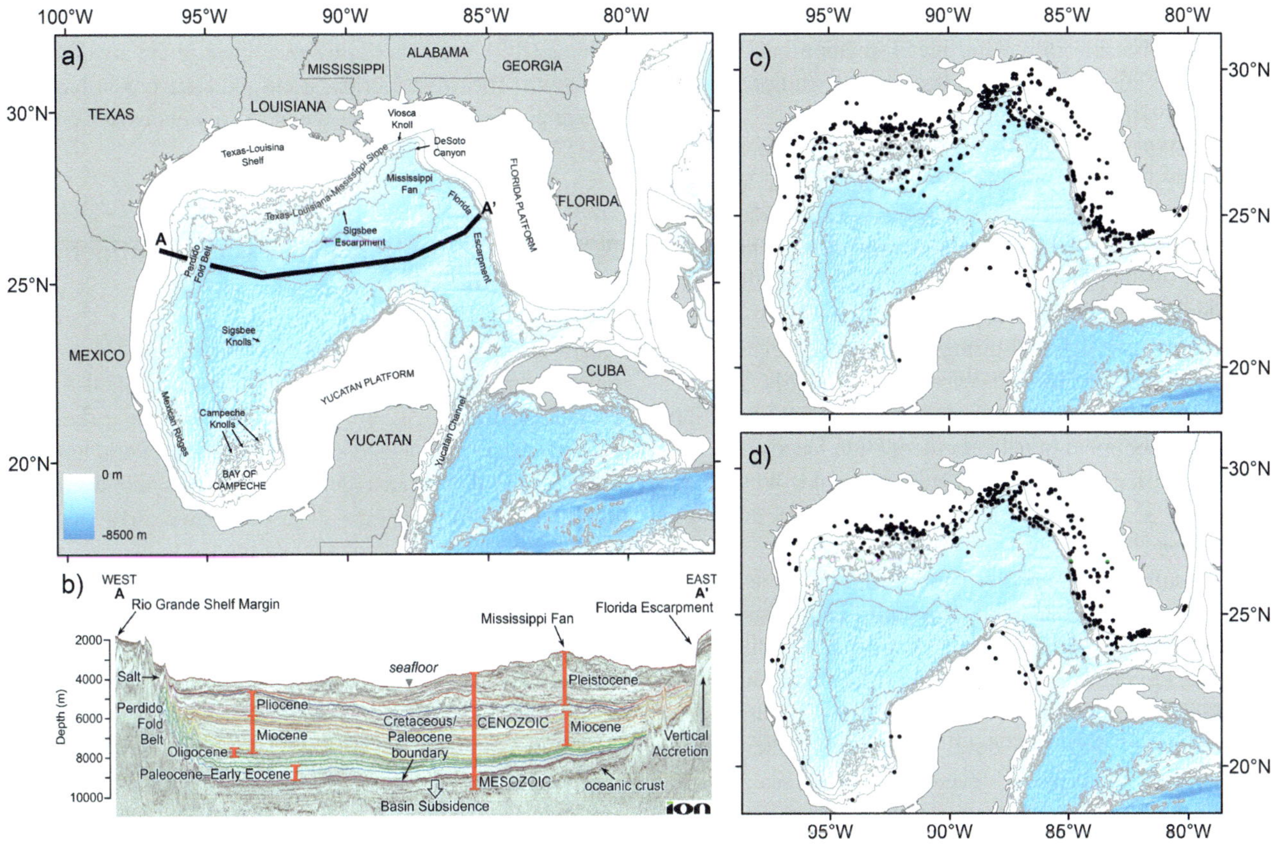

Fig. 3.2 (**a**) Map of the Gulf of Mexico showing submarine features and political boundaries; (**b**) cross section A–A[1]: regional seismic line across the middle of the Gulf showing thickness and age of sediments (modified from Blum et al. 2017); (**c**) locations of cold-water octocorals; (**d**) locations of cold-water stony corals (coral data from NOAA National Database for Deep-Sea Corals and Sponges, 2020). Underlying bathymetry is from the GEBCO 2020 grid, with 500 m contours (GEBCO Compilation Group 2020 Gebco 2020 Grid)

3.3.1 West Florida Slope-Escarpment

The West Florida Slope and Escarpment is separated from the northern Gulf's Texas-Louisiana-Mississippi continental slope by the DeSoto Canyon (Fig. 3.2a). This feature is not a true steep-walled submarine canyon which has been eroded into the continental margin (Shepard 1963; Shepard and Dill 1966). Rather, many geoscientists think it is the incompletely filled Gulfward extension of the South Georgia Rift Basin formed during Jurassic times (Smith and Lord 1997; Hine 2019). Coleman et al. (1991) characterized the DeSoto Canyon as simply the intersection of the river-borne sediment province of the northern Gulf rim and the West Florida carbonate province. Recent research by Dunn (2016) reveals that sediments spilling into DeSoto Canyon from the west are part of a large contourite-sediment drift system which started in mid-Miocene times (~15 myr BP) and was driven eastward by the paleo-Loop Current. In fact, the origin of DeSoto Canyon appears related to along-slope depositional processes rather than downslope or up-slope erosional processes associated with typical canyon formation.

Along the western edge of DeSoto Canyon, a series of channel-like features with a maximum relief of ~100 m deliver sediment to the canyon. As long as the Loop Current is operative, these sediment pathways will continue to fill the canyon from the west while the eastern side of the canyon will remain characterized by processes of erosion. The erosional east side of the canyon exposes truncated Florida Platform carbonates typical of the remaining West Florida Platform margin to the south.

The east side of the DeSoto Canyon transitions southward from a low gradient continental slope into a continental margin characterized by limestone outcrops, steep-to-overhanging limestone cliffs, slope failures of various scales, and large re-entrants to true submarine canyons. These carbonate rocks are inherited from platform accretion under shallow-water conditions during the Jurassic and early Cretaceous times followed by rapid platform drowning during the mid-Cretaceous. Seismic profiles from the deep Gulf onto the Florida Platform coupled with drilling data indicate near-vertical accretion of the West Florida Platform by shallow-water carbonate-evaporite facies while the central Gulf subsided (Corso et al. 1988). This period produced the steep platform margin of several kilometers relief thought to be mainly attributable to rudistid reef formation backed to the east by various other shallow-water carbonates and evaporites (Brooks and Holmes 2011).

Global tectonics in the mid-Cretaceous reduced ocean basin-volume causing rapid and sustained sea level rise. Combined with steady subsidence, the West Florida Platform was flooded along with many similar features worldwide. As the platform fell beneath the photic zone, reef-building and shallow-water carbonate deposition was replaced by deep-water carbonate deposits primarily composed of foraminifera, pteropods, and other calcareous microfossils. This transition was accompanied by the beginning of significant input of riverine sediment into the northern Gulf. Deposition of the deep-water sediments resulted in a ramp-like West Florida Terrace or slope located stratigraphically above the Florida Escarpment.

Exposure of low energy platform interior limestones on the Escarpment face and a missing rudist reef facies suggests that as much as 1–5 km of retreat has occurred in the central and southern parts of the Escarpment (Freeman-Lynde 1983). Erosion of the Escarpment resulted in failure of large blocks of limestone into the deep Gulf where they largely disintegrated and were covered by deep-sea fan and hemipelagic deposits. Collapse and retreat of the Florida Escarpment may be at least partially related to the Chicxulub impact event on the Yucatan Peninsula which occurred at the end of the Cretaceous. In addition, Mullins et al. (1986) identified a mid-Miocene slide scar 120 km across and up to 30 km wide that removed a massive amount of material from the platform margin. These large-scale collapse events have been followed by smaller-scale gravity-driven slumping. The results of these failures are large re-entrants that now dominate the morphology of the central part of the West Florida Escarpment and smaller-scale steep scalloped features. The re-entrants transition to the south into sizeable submarine canyons, up to 25 km long, eroded into the margin.

Using DSV Alvin, Paull et al. (1984) and Paull and Neumann (1987) observed methane and sulfide-rich hypersaline water seeping from the limestones at the Escarpment base. When combined with seawater this acidic solution dissolves limestone. It is thought that the sulfide-rich water originates from dissolution of evaporites in the shallow platform facies. Undercutting, collapse, and general retreat of the carbonate margin may be in part related to limestone dissolution processes (Paull et al. 1990).

The West Florida Slope and Escarpment from DeSoto Canyon to the Florida Strait represents a nearly continuous series of carbonate outcrops, steep cliffs, and displaced limestone blocks. This huge hard bottom area, which is impacted by the Loop Current, offers CWC an extremely favorable environment for attachment and food for growth. Despite many human occupied vehicle (HOV) and remotely operated vehicle (ROV) studies of this area, most of the area remains unknown. The West Florida Slope and Escarpment have the characteristics to be perhaps the most continuous and productive CWC province in the U.S. GoM.

3.3.2 Texas-Louisiana-Mississippi Continental Slope

Deepwater exploration and production of hydrocarbons was pioneered in the Gulf, starting in 1978 with the Shell-Amoco Cognac platform (~312 m water depth). Since the early 1970s no other deepwater oil and gas province in the world has been the subject of as much data collection as the deepwater northern GoM. These datasets have convincingly documented a geologic complexity unique to the continental slope, with its variety of knolls, intraslope basins, and various salt features that are unmatched in today's oceans.

Complexity of the continental slope is related to an interaction between sedimentation and salt deformation. Progradation of the shelf edge and slope has occurred primarily during periods of lowered sea level when fluvial systems migrated across the continental shelf to form deltas at the shelf edge and to deposit sediment directly onto the continental slope. Over-steepening of the shelf margin by rapid and localized fluvial deposition has led to slope failures that transfer sediment to the deep slope and create canyons that deliver sediment to the deep basin floor. Trainor et al. (1988) estimate that there have been as many as 176 high frequency sea level changes since the beginning of the Cenozoic (~65 myr BP) that have forced shelf margin progradation. Massive sediment deposition on the continental slope has caused the formation of locally thickened intraslope basins. Sediment loading of underlying mobile salt has forced the formation of a variety of salt structures from vertical columns attached to the parent salt layer at depth, detached tear-shaped masses, and sheet-like structures (McBride et al. 1998). Coincident with these dynamic processes of salt deformation is faulting which is constantly being reactivated by salt movement. Faults provide fluid and gas migration pathways from the deep subsurface to the modern seafloor (McBride et al. 1998).

Because of the northern Gulf's long history of river-borne sediment input, the natural and slow processes of lithification leading to non-carbonate rock formation are far outpaced by rate of deposition. Therefore, hard substrates available for CWC attachment are limited to three primary types: (1) highly compacted clay-rich sediments displaced upward to the seafloor by faulting and/or salt movement, (2) sands and gravels deposited during lowered sea level periods and exposed by salt movement, or (3) authigenic Ca-Mg carbonates formed as a by-product of microbial oxidation of hydrocarbons at seep sites (Roberts and Aharon 1994; Roberts et al. 2010). Of these three, current data indicate that seep-related carbonates are by far the most common deepwater substrates of the Texas-Louisiana-Mississippi continental slope. Seep-related carbonates are generally found around the faulted and elevated margins of intraslope basins, above and on the flanks of shallow salt masses, where deep-cutting faults intersect the modern seafloor, and above slowly decomposing gas hydrate deposits. Their formation is an ongoing process. These hardgrounds take the form of (1) large and complex outcrops of which chemosynthetic organisms are an integral part, (2) boulders, (3) slabs-pavements, and (4) nodular masses and shells of chemosynthetic organisms (e.g., clams) within host sediments (Roberts et al. 2010). All forms may function as good substrates for CWC and the complex elevated margins of intraslope basins project into currents that flow across the slope, supplying food to sessile organisms.

3.3.3 East Mexico Slope (Mexican Ridge and Bay of Campeche)

The Mexican Ridges of East Mexico Slope occur between the Texas-Louisiana-Mississippi salt diapir province to the north and the Bay of Campeche to the south (Fig. 3.2a). First described by Bryant et al. (1968) and later by Buffler et al. (1979) the Mexican Ridges are a series of anticlinal folds which transition from the similar, but smaller Perdido Fold Belt to the north. The ridges roughly parallel the present coastline, have a length of up to ~500 km, and are spaced 10–12 km apart (Bryant et al. 1991). Jurassic salt, interpreted from seismic lines, appears relatively undeformed beneath the ridges, indicating that the subparallel folds are not caused by salt tectonics. They occur above a detachment zone in upper Cretaceous-Lower Cenozoic shale (Buffler et al. 1979). Numerous thrust faults associated with the folds suggest compressional stresses acting in an east-west direction. Building on previous studies, Pew (1982) interpreted the southern Mexican Ridges as the result of massive gravity sliding, which started in the mid-Miocene and folded Plio-Pleistocene and Holocene strata indicating that the gravity-driven downslope displacement is an ongoing process. Starting in the Miocene, sedimentation in the troughs of the folds appears to have forced the flow of plastic shale into the cores of the anticlinal structures much like the salt-cored anticlines of the northern Gulf.

Until recently details of the modern seafloor of the Mexican Ridge were completely unknown. Interest in oil and gas potential of this part of the Gulf has initiated a comprehensive acquisition of new generation seismic data and multibeam bathymetry coupled with gravity coring and geochemistry. The results of these studies are still mostly unpublished, but recent research confirms the faulted nature of the ridges, the occurrence of numerous local slope failures, and hard bottom areas closely associated with hydrocarbon seeps along the faulted ridge crests (Teas et al. 2017). The slope failures are accompanied by large blocks of displaced sediment that now reside in the downslope depositional sites and may offer good habitats for CWC. Coring has revealed that seeps have chemosynthetic communities and authigenic carbonates much like the many seep sites of the northern Gulf (Perez-Drago et al. 2019). The Mexican Ridges are virtually unexplored, but new datasets suggest there are numerous habitats for CWC in this Gulf province. The ridges transition to the south into the Bay of Campeche.

The Bay of Campeche is an area of hummocky topography, known as the Campeche Knolls, (Fig. 3.2a), a reflection of salt tectonics. The salt responsible for the erratic seafloor character of this region is from the same parent salt unit that underlies the northern Gulf's continental slope. The Campeche Knolls and Sigsbee Knolls to the northeast, first reported by Ewing et al. (1958) and later discussed in more detail by Garrison and Martin (1973), are similar salt-cored and highly faulted dome-like features. Some of these features attain as much as 1500 m relief (Bryant et al. 1991). The first seafloor photographs of the Campeche Knolls were collected by Bohrmann and Schenck (2004). They discovered lush chemosynthetic communities, authigenic carbonate pavements, and unusual asphalt deposits that had the appearance of lava flows. Follow-up studies by Brüning et al. (2010), Sahling et al. (2016), and Hsu et al. (2019) extended observations in the Campeche Basin as well as among the Sigsbee Knolls. High gravity oil seepage-to-flows and gas emissions were found to be common to the knolls of both areas. Considering that carbonate hardgrounds were also found to be abundant at each site, these areas should be ideal settings for CWC even though comparatively few sites have been visited so far.

3.3.4 Yucatan Platform and Escarpment

Much like the geologic evolution of the Florida Platform, the Yucatan Platform is a carbonate province that started significant vertical accretion in the Lower Cretaceous. The present platform is surrounded by steep slopes of the Campeche Escarpment (Fig. 3.2a). This Escarpment is inferred to be originally the product of rudist reef-building as with the Florida Escarpment. The rudist reefs were backed by a variety of shallow-water carbonate facies and evaporites leading to slow overall vertical accretion and platform formation. Platform accretion kept up with or outpaced subsidence of the Gulf Basin until the Mid-Cretaceous drowning event. As in the case of the Florida Platform, this drowning event ended the rudist reef-building phase and introduced Tertiary deep-water, foraminifera-rich deposits that draped across the platform. However, the broad terrace (~200 km wide and slope of

~5°) at the northeastern part of the platform is internally different than other parts of the platform. Seismic data indicate that this area has a stratigraphic architecture that indicates progradation, probably related to sediments being driven across the platform margin by strong currents from the Yucatan Channel (Bryant et al. 1991).

An early study by Locker and Buffler (1983) noted that seismic data suggested that the Campeche Escarpment was considerably more complex than the Florida Escarpment. This observation was made before the geologic features caused by the Chicxulub impact event at the Cretaceous-Paleogene boundary were well known. More recently, Paull et al. (2014) revealed the intricate erosional nature of the Campeche Escarpment using high quality multibeam bathymetry. They identified the Cretaceous-Paleogene boundary as a dramatic slope change on the Escarpment face, average slopes of >25° below the boundary and ~5° above. This steep lower part is sculpted by ~80 box submarine canyons with near-vertical cliff-like sides ~500 m high (Paull et al. 2014). The canyon heads occur in a water depth of 2000–2500 m. These submarine canyons plus less steep gullied features to the northeast suggest significant collapse of the Yucatan Platform and retreat of the Escarpment face during the late Cretaceous, the time of the Chicxulub impact which occurred only ~230 km away on the Yucatan Peninsula. At the base of the Escarpment are huge km-scale blocks of material derived from the Escarpment face and debris flow deposits that probably traveled well out into the GoM basin. Paull et al. (2014) suggest that the Escarpment was massively altered during the Chicxulub impact and may have retreated as much as 11 km.

The present Campeche Escarpment is a vast, mostly unexplored province for CWC. Steep-sided gullies, vertical to near-vertical walls of submarine canyons, truncated carbonate rocks of the platform interior, and huge blocks of displaced material in the debris fans at the base of the Escarpment provide many hard substrate opportunities for coral attachment and growth. The vertical nature of the submarine canyon walls and gullies minimize accumulation of pelagic and hemipelagic sediment while leaving hard substrates exposed. Molinari and Morrison (1988) and Merino (1997) also identify upwelling in the northeast sector of the Campeche Bank, which suggests increased productivity and an abundant food source for corals. This is the area where the largest coherent CWC mound province in the Gulf has been discovered (Hubscher et al. 2010; Hebbeln et al. 2014). *Enallopsammia profunda* and *Lophelia pertusa* dominate this coral community. Strong bottom currents from the Yucatan Channel provide supportive environmental conditions for these thriving corals. However, most of the Yucatan Platform margin remains unstudied, but habitat conditions for CWC appear to be excellent.

3.4 Structure-Forming Corals in the Gulf of Mexico

3.4.1 Scleractinians (Anthozoa, Hexacorallia, Scleractinia)

Stony corals are common in the deep GoM (Fig. 3.2d), with 72 known species at depths >200 m (Hourigan et al. 2017); however, few of these species form significant structure as most are solitary (cup) corals (Cairns et al. 2009). The primary reef-building species in the GoM is *Lophelia pertusa*, which forms well-developed thickets and bioherms at ~400–700 m (Fig. 3.3a). Another colonial species, *Madrepora oculata* (Fig. 3.3b) does not form monospecific stands but occurs on coral mounds with *L. pertusa*, and as individual colonies on exposed hard substrates. Other structure-forming scleractinians such as *M. carolina*, *Enallopsammia profunda*, *E. rostrata* and *Solenosmilia variabilis* have been observed in the GoM (Cairns et al. 2009), but the extent of any of these species is unclear.

3.4.2 Antipatharians (Anthozoa, Hexacorallia, Antipatharia)

Black corals are ubiquitous across the deep GoM, but not highly diverse with only 16 documented species found >200 m depth (Etnoyer and Cairns 2017). Some species form branching colonies that provide structure and refuge for other organisms, while others have one or few branches. Arguably the most important of these species is *Leiopathes glaberrima* (Fig. 3.3c) which can form very large (>2 m), ancient, complex colonies and is moderately abundant throughout the GoM (Boland et al. 2017; Prouty et al. 2011; Brooke and Schroeder 2007). Most other black coral species are smaller than *L. glaberrima*, but mobile invertebrates such as galatheid crabs are often observed within their branches. The single branched *Stichopathes* sp. can occur as dense populations, collectively creating shelter for other species.

3.4.3 Octocorals (Anthozoa, Octocorallia, Alcyonacea)

Octocorals (which include gorgonians and the true soft corals) are a large, diverse, and widely distributed taxa in the GoM (Fig. 3.2c), with ~118 species documented from >200 m (Quattrini et al. 2014; Hourigan et al. 2017). This is probably an underestimate given the dominance of octocorals in under-explored deeper (>1000 m) areas of the GoM, and the presence of taxonomically similar or cryptic species

Fig. 3.3 Images of coral habitat from different regions of the Gulf of Mexico. (**a**) *Lophelia pertusa* with golden crabs at 500 m on the West Florida Slope. Image credit: Lophelia II 2010, USGS; (**b**) *Madrepora oculata* colonies on a carbonate boulder at 1050 m in Atwater Valley. Image credit: Lophelia II, BOEM/NOAA; (**c**) White *Leiopathes glaberrima* colony surrounded by anemones at Viosca Knoll. Image credit: Lophelia II, 2008 BOEM/NOAA; (**d**) The octocoral *Callogorgia* sp with Euryalid ophiuroids (*Asteroschema* sp.) entwined in the branches at Mississippi Canyon. Image credit: ECOGIG (2017); (**e**) collection of corals (*Paramuricea* sp. and *Lophelia pertusa*) with anemones and starfish (Brissingidae) at Viosca Knoll. Image credit: ECOGIG (2013); (**f**) large unidentified bamboo coral at 2000 m on the Florida Escarpment. Image credit: Brooke et al. (2019), NOAA Ocean Exploration and Research

(Hourigan et al. 2017). Octocoral species distributions may be driven by their adaptation to environmental conditions such as temperature that allows them to occupy specific depth zones, and by more random processes such as recruitment. Quattrini et al. (2017) identified a zone of octocoral species' turnover between 800 and 1200 m in the GoM, with different species occupying the shallower (250–800 m) and deeper (1200–2500 m) depth zones. Octocorals can be large, with complex branching, and provide habitat for numerous other species (Edinger et al. 2007; Husebo et al. 2002; Miller et al. 2012). Significant structure-forming species are found within several octocoral families, including Paragorgiidae, Coralliidae, Chrysogorgiidae, Primnoidae, Plexauridae, and Isididae. Some members of the family Paragorgiidae, commonly known as bubblegum corals, can span several meters (Smith 2001), but the species found in the GoM (*Sibogagorgia cauliflora, Paragorgia johnsoni, P. regalis*) do not reach such massive colony sizes. The family Coralliidae contains the precious pink and red corals. In the GoM, members of this family (*Hemicorallium niobe* and *Corallium medea*) are rarely observed as they are only found at depths >2000 m (Brooks et al. 2015; Etnoyer and Cairns 2017). Recent explorations of the deep West Florida Escarpment revealed abundant large colonies of Corallidae and several unidentified species of *Iridogorgia* sp. and *Chrysogorgia* sp. (family Chrysogorgiidae) at >1800 m depth (Brooke et al. 2020). Two species of *Iridogorgia* (*I. splendens* and *I. magnispiralis*) and six species of *Chrysogorgia* were also found in the deep northern GoM (Brooks et al. 2015; Etnoyer and Cairns 2017). The Primnoidae and Plexauridae are both speciose families, with a range of sizes and structural complexity. There are four species of the common Primnoid genera *Callogorgia* in the GoM: *C. gracilis*, *C. americana*, *C. linguimaris*, and the endemic *C. delta* (Quattrini et al. 2013; Etnoyer and Cairns 2017). These species occupy different depth zones with *C. gracilis* occurring at the shallowest sites (82–514 m), and *C. delta* the deepest (366–913). These species are large and often occur as dense populations, providing significant habitat for other fauna (Fig. 3.3d). The family Plexauridae contains many of the branching, structure-forming octocorals in the GoM, including the common genera *Swiftia*, *Thesea*, and *Paramuricea* (Fig. 3.3e). The deepest known member of the Plexauridae is *P. biscaya*, which was discovered near the DWH wellhead at ~1000–1500 m depth. The Isididae, or bamboo corals are unique amongst the octocorals for their skeletal structure of alternating gorgonin nodes and calcium carbonate internodes. Some species, such as *Keratoisis flexibilis*, can become very large and highly branched (Fig. 3.3f), and others (e.g., *Acanella* sp.) can form dense monotypic stands. Bamboo corals are ubiquitous members of the GoM CWC community and can be very abundant in some locations. Soft corals in the genus *Anthomastus* are common in the GoM and will colonize rocky substrates as well as dead *L. pertusa* skeleton, and the Nephtheid *Pseudodrifa nigra* is locally abundant on the West Florida Slope coral mounds. None of the soft coral species contribute significantly to habitat structure since they neither form large colonies nor have solid skeletons.

3.4.4 Hydrocorals (Hydrozoa, Hydroidolina, Stylasteridae)

Of the nine species of hydrocorals found in the deep GoM (Etnoyer and Cairns 2017), only two are found in the northern Gulf; most are restricted to the less sediment-laden habitats of the West Florida Slope. These colonies are usually small and isolated, but occasionally they may reach moderate size and be locally abundant. Under these circumstances, their highly branched complex colony morphology can provide structure for other species.

3.5 Cold-Water Coral Biology

3.5.1 Feeding and Nutrition

Cold-water corals are heterotrophic suspension feeders, relying on surface-derived organic carbon and zooplankton for food. Corals are considered "living sediment traps," because their tissues integrate the biogeochemical fingerprint of recently exported particulate organic matter (POM) which is then incorporated into skeletal structure (Prouty et al. 2017). Higher growth of CWC occurs in areas that have enhanced food delivery and access, typically in areas of high current speeds, associated with topographic highs and/or in areas that promote mixing (e.g., via internal waves, Davies et al. 2010). Water masses have been shown to play a role in influencing the supply of nutrients to deepwaters. For example, the Antarctic Intermediate Water from the Loop Current provides high nutrient waters to the depth zones that *L. pertusa* occupies (500–1000 m, Davies et al. 2010).

Flux and availability of organic carbon to the seafloor are major factors that influence deep-sea coral distribution. Export productivity in particular is a strong predictor in habitat suitability modeling for CWC (Georgian et al. 2014). Organic matter flux to CWC (e.g., *Lophelia pertusa*) in the GoM is best known at VK826 (Davies et al. 2010; Mienis et al. 2012), where deep-sea landers with sediment traps have amassed long-term data on the oceanographic conditions coupled with mass flux data. Sediment traps at VK826 collected fine-grained aggregates with high organic content. The source of this material to VK826 may originate from the Mississippi Outflow, which delivers high concentrations of nutrients and terrestrial organic matter, which in turn promotes high primary production in the surface ocean. This material descends to the seafloor and reaches the corals. A large proportion of the organic material was derived from the water column (transported rather than resuspended), with stable carbon isotope data reflecting its phytodetrital origin (–21.5 to –20.4‰), and a typical range for marine organic matter.

Reinforcing the tight coupling between surface production and deep-sea coral habitats, peaks in fluorescence measured by the landers corresponded to surface blooms visible in satellite images. Increased primary production in the northern GoM occurs during July, largely driven by upwelling associated with the reversal of the prevailing wind direction to the south-southwest, forcing water to the east (Mienis et al. 2012). During this period, mass fluxes recorded at VK826 (1120–4479 mg m^{-2} day^{-1}) were larger than from NE Atlantic and were highest during periods of westward flow of warm water (21 days). While high fluxes recorded in sediment traps were comparable to other deep-sea coral locations around the globe (Mienis et al. 2012), particle load and source of organic matter at VK826 differs from other deep-sea coral environments (Mienis et al. 2012). Organic carbon fluxes at VK826 were 2–4× higher than observations from the Rockall and Galicia Banks in the North Atlantic Ocean. Sediment trap material was composed of high organic carbon content, high C:N ratios, and low $CaCO_3$ content, consistent with fluvial origin. Near bottom currents are sufficiently high (sometimes >15 cm s^{-1}, east to west) to resuspend and transport sediment and aggregates. From October to January, high flux corresponded to resuspended material with low organic content in the traps, associated with peaks in backscatter and turbidity. In contrast, low mass fluxes occurred in February and March. During April and June, fluorescence peaks observed near the seabed were consistent with the arrival of phytodetritus. Corals thriving at VK826, therefore, experience a very dynamic system, with stark contrasts in sediment delivery/mass flux, turbidity, and temperature with shifting current direction. Currents shift from west (more turbid, warmer water conditions) to east (clearer, cooler waters), which may ameliorate any impact from sedimentation during the westward flow conditions, since high sediment load can limit coral growth. In addition to sinking phytodetritus, acoustic Doppler current profiler (ADCP) data revealed diel vertical migration of zooplankton over short and long time periods (Davies et al. 2010; Mienis et al. 2012), suggesting another possible consistent source of food, with the presence of zooplankton recorded in sediment traps throughout the year (Mienis et al. 2012).

In order to reveal dominant food resources for CWC, stable carbon and nitrogen isotopes, ($\delta^{13}C$, $\delta^{15}N$), have been used to reconstruct carbon sources and characterize food webs, because the animal tissues integrate these isotopes from their diets over time (Peterson and Fry 1987). For *Lophelia pertusa*, $\delta^{13}C$ values indicate that they derive their nutrition from surface-derived phytodetritus (–22.3 to –18.0‰), rather than chemosynthesis-derived organic carbon from seeps (< –25‰, Becker et al. 2009). In terms of $\delta^{15}N$ (8.7–10.8‰), *L. pertusa* does not appear to be a primary consumer, rather it may rely on a mixture of phytodetritus and higher trophic level organisms, such as zooplankton (Becker et al. 2009; Brooks et al. 2015). Likewise, the black corals, *Leiopathes* spp., also feed on fresh organic carbon that is recently exported from surface waters (–21.0 to –19.0‰; Prouty et al. 2011) and may also consume a mixture of zooplankton and phytodetritus, based on tissue and skeletal $\delta^{15}N$ values (7.4–10.8‰). Tracking $\delta^{15}N$ over the life of these black corals indicates increased contributions of terrestrial-derived effluent to the deep sea (Williams et al. 2007) within the last few centuries.

Most deep-sea octocorals (e.g., *Paramuricea*) have isotope results indicative of utilizing surface-derived organic matter with one exception (Brooks et al. 2015). *Callogorgia delta,* known to live in proximity to seep environments, had the broadest range and lighter than typical $\delta^{13}C$ and $\delta^{15}N$ values (range: –33.1 to –20.2‰, 2.1–11.5‰, for $\delta^{13}C$ and $\delta^{15}N$ values, values respectively), consistent with some assimilation of seep-derived primary production (Brooks et al. 2015). Possible alternative sources of organic carbon to *C. americana delta* include free-living bacteria suspended by bottom currents or the presence of microbial symbionts within the coral tissues. Other octocoral species, including congeneric species of *Callogorgia*, plus *Chrysogorgia* sp. *Swiftia exserta*, *Paramuricea biscaya*, *Paramuricea* B3, and *Paramuricea* sp. E, had stable isotope values that fell within the range of expected values for surface-derived photosynthetic production. *Paramuricea* colonies were generally located on topographic highs, including the tops of carbonate outcroppings and ridges (Doughty et al. 2014), enabling them access to variable flow and enhancing access to food. *Paramuricea* is most likely to consume surface-derived POM and zooplankton for food (Prouty et al. 2016). Their congeners are known to feed on microzooplankton (Coma et al. 1994; Ribes et al. 1999), inhabiting areas with enhanced food supply. Food supply also was an important variable influencing habitat suitability modeling of *Paramuricea* (Georgian et al. 2019), and occurrence of these corals was predicted to be the highest where there were high flux rates. Additionally, analyses of environmental filtering controls on octocoral structure identified three traits associated with food availability and selectivity, including polyp density, size, and interpolyp distance, all of which changed as a function of depth (Quattrini et al. 2017). Increased polyp size with depth could facilitate particle capture where food is limited, whereas variable polyp size and distance may enable selection for different sized particles, allowing for the coexistence of different species of corals (Quattrini et al. 2017).

3.5.2 Age and Growth

One of the most notable characteristics of the CWC is their longevity; they are among the longest-lived marine organisms (Roberts et al. 2009). The growth of CWC is extremely slow compared to that of their shallow-water relatives and highly variable among species. Using a variety of methods, including direct linear measurements and isotopic techniques in the laboratory and from in situ samples, growth rates of *L. pertusa* in the North Sea have been constrained to a range of 9.4 to 20–25 mm $year^{-1}$ (Mortensen 2001; Mikkelsen et al. 1982; Mortensen and Rapp 1998). In the GoM, in situ growth rates of *L. pertusa* fall toward the lower end of this growth spectrum, with average growth rates ranging from 2.43 to 3.76 mm $year^{-1}$ (Brooke and Young 2009). On human-made structures including deepwater energy installations and shipwrecks in the GoM, growth rates of *L. pertusa* ranged from 3.2 to 32.3 mm $year^{-1}$ (Larcom et al. 2014). The elevated growth rates of corals on artificial substrates may be attributed to favorable access to currents and associated food particles (Mortensen 2001; Larcom et al. 2014).

Various approaches are used to estimate the ages of CWC and the structures they form, including visual, stratigraphic, and isotopic dating techniques (Roberts et al. 2006, 2009). Select groups within the Octocorallia, Antipatharia, and Zoanthidea clades lay down annual growth bands as the axis develops, and these bands may be enumerated to infer age and growth rate if the chronology of banding is validated by isotopic methods (Andrews et al. 2002; Sherwood et al. 2005). Isotopic techniques such as uranium/thorium, ^{14}C, and ^{210}Pb dating of skeletal material can provide age estimates of various specimens, although these methods are most often applied to obtain ages of reefs produced by scleractinians (e.g., Roberts et al. 2009; López Correa et al. 2012). Over the past decade, sophisticated aging techniques have revealed a latitudinal shift in optimal growth conditions for CWC mounds (Frank et al. 2011). During the Holocene (up to 11,650 ybp) coral provinces along the NW European margin have undergone rapid growth (100–1500 cm ka^{-1}) (López Correa et al. 2012; Wienberg et al. 2018). Further south, CWC mounds have occurred on the Mauritian margin for the past 120,000 years, but ceased growing with the onset of the Holocene (Wienberg et al. 2018). The authors concluded this shift in CWC growth was related to changes in water mass structure, which currently exposes CWC on the Mauritian margin to low dissolved oxygen conditions.

In the northern GoM a CWC mound ("Roberts Reef") showed successive periods of coral growth and non-growth over a period of ~300 ka. Five growth periods corresponded with interglacials, while growth ceased during glacial periods when low sea level stands increased bottom temperatures beyond the thermal limits for the corals (Roberts and Kohl 2018). The larger Campeche Bank coral mound province in the southern GoM showed similar patterns of growth and non-growth over the past 140 ka, also corresponding to interglacial and glacial periods, respectively (Matos et al. 2017); however, the authors attributed the cessation of growth to diminished food supply rather than temperature.

The age of individual coral colonies, however, is less understood for the GoM. Following the DWH Spill in 2010, Prouty et al. (2014) reported ages for individual colonies of the octocorals *Paramuricea biscaya* and *Chrysogorgia* sp. collected from nearby the impact site. Using radiocarbon dating, *P. biscaya* individuals were aged to a maximum of 600 years and *Chrysogorgia* sp. to a maximum of 525 years. The black coral, *L. glaberrima*, has been aged over 2K years based on radiocarbon dating (Prouty et al. 2011). However, despite the power of radiocarbon dating, visual surveys can also produce robust age estimates when coupled with size-class models and radiocarbon dating data. This approach was employed to estimate the age of individual colonies of *L. glaberrima*, where individual colonies reached a maximum age of 909 years (Etnoyer et al. 2018). The results of these studies contribute broadly to the understanding of CWC as extremely long-lived organisms.

The longevity of CWC, coupled with their widespread distribution, allows them to serve as paleo-environmental archives (Roberts et al. 2009). Biological processes such as calcification produce "vital effects," alterations to the isotopic and trace element composition of the skeleton (Weber and Woodhead 1972), and have been observed in CWC (Emiliani et al. 1978). Isotopes such as ^{18}O, ^{13}C, and ^{11}B, as well as Mg/Ca and Sr/Ca ratios, are commonly employed in the examination of vital effects in cold-water coral skeletons. Trends including changes in seawater temperature (Smith et al. 2000), circulation patterns (Adkins et al. 1998), and pH (Trotter et al. 2011) can be identified using these techniques and may help reconstruct Earth's climate history. Examples of such studies in the GoM are limited; however, an examination of the skeletons of *Leiopathes glaberrima* detected signals of nutrient enrichment over the past ~200 years that may be linked to anthropogenically-driven terrestrial runoff (Prouty et al. 2014). The anthropogenic release of CO_2 from fossil fuel combustion was also detected in skeletons of *L. glaberrima* via a decline in ^{13}C from specimens from the GoM (Williams et al. 2007). These studies support the continued use of CWC as archives of environmental conditions, particularly during the current period of rapid environmental changes in ocean temperature, oxygen concentration, and pH.

3.5.3 Ecophysiology

3.5.3.1 Environmental Tolerances

The range of variables that cold-water coral reefs are exposed to is generally informed by studies that characterize local conditions near the seafloor. A series of studies in the GoM has examined the ranges of environmental variables that are coincident with CWC reefs, including temperature (e.g., Davies et al. 2010), dissolved oxygen (e.g., Schroeder 2002), sedimentation (e.g., Brooke et al. 2009), and aragonite saturation state (e.g., Lunden et al. 2013; Georgian et al. 2016a) (described above in Sect. 3.2.2). Further exploration of the niche of *Lophelia pertusa* revealed a significant contribution of depth, local topography, and presence of hard substrate to the distribution of this framework-forming scleractinian (Georgian et al. 2014). While observational studies are useful in defining the natural ranges of abiotic factors that cold-water coral reefs experience, they lack the ability to fully capture the maximum limits of physiological tolerance.

Manipulative experiments provide useful context for understanding the physiological tolerance of CWC. In the GoM, a number of studies have employed both field-based and laboratory-based approaches to understand the tolerance limits of reef-forming taxa, particularly *L. pertusa*. A major objective of these studies included assessing the tolerance of *L. pertusa* within the context of ocean warming and acidification as these ongoing phenomena may threaten to reduce the niche of CWC (Roberts and Cairns 2014). Two different laboratory-based studies identified an upper-level thermal maximum of ~14–15 °C (Brooke et al. 2013; Lunden et al. 2014) after 7 days of exposure; however, individual corals may be able to sustain shorter exposures to warmer conditions. Regarding reductions in pH due to ocean acidification, *L. pertusa* from the GoM appear to be able to maintain net positive calcification at aragonite saturation states near 1.0 (Lunden et al. 2014), but this appears to be linked to individual genotypes (Kurman et al. 2017). Additionally, *L. pertusa* from the GoM exhibits different responses to ocean acidification than *L. pertusa* from Tisler Reef in the North Atlantic, where Tisler Reef individuals appeared to be more tolerant of reduced pH which may be due to increased respiration and prey capture rates (Georgian et al. 2016b). *Lophelia pertusa* from the GoM exhibited a mortality threshold in oxygen concentration near 1.5 mL L^{-1} (Lunden et al. 2014), suggesting potentially negative consequences of hypoxia due to ocean warming and nutrient loading in the GoM. When *L. pertusa* colonies were exposed to a combination of multiple stressors specifically, increased temperature, decreased seawater pH, and deoxygenation, net calcification decreased with reduced pH conditions, and survivorship declined with increased temperature and deoxygenation (Lunden et al. 2014). However, some *L. pertusa* genotypes could calcify at saturation states as low as 1.0, which suggests that their sensitivity may be influenced at the genetic level (Lunden et al. 2014). Impacts of decreased pH also resulted in reduction of respiration and prey capture rates. These responses represent common stress responses in marine invertebrates (Georgian et al. 2016b). Additional exposure experiments indicated responses differed with exposure time (Kurman et al. 2017). Over the short term, no significant changes in net calcification occurred across different pH treatments. After 6 months, there was overall net dissolution of the skeletons in all except one genotype, which maintained positive net calcification at lowest pH, suggesting that some genotypes may be more resilient than others (Kurman et al. 2017).

Cold-water coral communities in the GoM were impacted by the DWH oil spill in April 2010 (White et al. 2012, Hsing et al. 2013). While the immediate impacts of the spill were primarily constrained to ~20 km of the Macondo wellhead site (Fisher et al. 2014a, b), the incident itself prompted a series of laboratory investigations to test the physiological tolerance of different CWC species to oil and dispersant exposure (DeLeo et al. 2016, 2018; Frometa et al. 2017). Experiments conducted on three species of CWC: *Paramuricea* type B3, *Callogorgia delta*, and *Leiopathes glaberrima* indicated that dispersant use led to the degradation of coral tissues, and exposure produced a toxic effect on CWC (DeLeo et al. 2016). The most significant health declines occurred in the dispersant-only and mixed dispersant-oil treatments. Follow-up research tracking the sub-lethal effects of the DWH event on corals examined gene expression in *Paramuricea biscaya* (DeLeo et al. 2018). Specific genes were differentially expressed in the impacted corals, including those associated with oxidative stress, immunity, wound repair, tissue regeneration, and metabolism of xenobiotics. The corals displayed stress responses that manifested across the genome, which has implications for using next-generation sequencing as a monitoring tool for future impacts. Pollution effects to *L. pertusa* were assessed with the inclusion of temperature as a stress (Weinnig et al. 2020). The dispersant-only treatment yielded the most significant decline in coral health after 24 h exposure under ambient temperature, with polyp health returning to the normal, pre-exposure health state. However, with increased temperature, coral recovery declined (72 h) following the dispersant treatment. Thus, increased temperature resulting from climate change coupled with pollution, may influence the resilience of CWC to environmental stress. The octocorals, *Swiftia exserta* exhibited a similar stress response to the oil-dispersant treatment or dispersant alone, with mortality occurring within 48 h (Frometa et al. 2017). The lethal concentration (LC50) for Corexit 9500 was calculated to be 51.17 mg L^{-1}, thus identifying toxicity thresholds, not previously examined in GoM CWC.

Our ability to measure stress responses at the genetic or metabolic level has greatly increased our understanding of how animals may be compromised, before they manifest extreme responses such as respiratory distress, tissue death, or mortality. Metabolomic diversity of four cold-water coral species from the northern GoM were assessed for diversity of metabolomics (Vohsen et al. 2019); these included a stony coral (*L. pertusa*), two black corals (*L. glaberrima*, *Stichopathes* sp.), and an octocoral (*C. a. delta*). The study showed that each coral species had a distinct metabolomic richness, including some ions that were unique to each species or taxonomic group, and others that were present in all samples. *Lophelia pertusa* had the highest ion diversity, and *C. a. delta* the lowest. They also discovered differences within populations of *C. a. delta* from different locations. While the function of many of these metabolites is not clear, some are well-known. For example, diterpenes and diterponoids have antimicrobial and anti-inflammatory properties and in corals may function as predator deterrents and anti-fouling agents. Other ions may come from the coral diet; for example, an algal-derived steroid was found in the cold-water octocoral *Acanthogorgia*, which (like other CWC) consumes marine snow (Bonini et al. 1983). Samples of *C. a. delta* from different locations had distinct profiles, which may be driven by differences in environmental conditions (Costa-Lotufo et al. 2018), or by underlying genetic differences (Gong et al. 2013). The metabolite diversity and species-specific ions may contribute to their abilities to withstand natural and anthropogenic stress (Vohsen et al. 2019).

3.5.3.2 Metabolic Rates and Energy Budgets

Metabolism in CWC is required to produce ATP to be allocated to biological functions such as calcification, reproduction, and cellular maintenance. Measurements of metabolic rate in CWC in situ are generally lacking, but recent evidence suggests the metabolism of *Lophelia pertusa* changes seasonally over the year as a result of variability in food input, which includes surface-derived POM and zooplankton (Maier et al. 2020). Accordingly, the metabolic performance of *L. pertusa* is likely adapted to pulses of surface productivity linked to temperature, nutrient, and light regimes (Baussant et al. 2017). Relative to CWC from the North Atlantic Ocean, limited data exist on the metabolic rates of CWC from the GoM and are primarily derived from experiments in the laboratory. Under a set of manipulated pH exposures, *L. pertusa* exhibited a linear decline in metabolic rate from ambient pH (pH 7.9, respiration rate 6.9 μmol O_2 gTW^{-1} h^{-1}) to reduced pH (pH 7.6, respiration rate 2.6 μmol O_2 gTW^{-1} h^{-1}), suggesting that *L. pertusa* from the GoM is metabolically-sensitive to ocean acidification (Georgian et al. 2016b). Interestingly, this same study identified a robust metabolic response in *L. pertusa* from Tisler Reef in Norwegian waters, suggesting that metabolic rates in CWC may be controlled by local conditions.

3.5.4 Reproduction and Early Life History

3.5.4.1 Reproductive Strategies

Corals exhibit several different reproductive strategies, which can be broadly categorized as asexual and sexual. The former includes fragmentation, and less commonly, parthenogenesis, which is the production of asexual larvae (Kahng et al. 2011), and polyp bailout, where the polyp or tentacles detach and settle apart from the main colony (Sammarco 1982; Rakka et al. 2019). Coral sexual reproductive strategies include gonochorism (colonies are separate sexes), simultaneous or sequential hermaphroditism (both sexes occur in a single colony), broadcast spawning of gametes, and internal fertilization followed by brooding of embryos and larvae. Coral larvae (planulae) may stay in the water column for several weeks or settle and metamorphose after a few days. Some reproductive strategies are constrained within a taxon; for example, hydrocorals (Anthoathecatae) have internal fertilization and release well-developed planulae (Brooke and Stone 2007), whereas black corals (Antipatharia) are (with few exceptions) gonochoric broadcast spawners (Wagner et al. 2012). However, octocorals (Alcyonacea; Kahng et al. 2011) and stony corals (Scleractinia; Harrison 2011) have diverse reproductive strategies. Reproductive traits have a significant effect on the ecology of a species, influencing the quantity, timing, and dispersal of propagules, and their potential to recolonize impacted habitats and recruit into new areas. Understanding reproductive ecology requires sampling of multiple specimens of target species across different seasons, and this is challenging for deep-sea species. Consequently, there is no information available for most deep-sea species and geographically limited or superficial data for most others.

The complete reproductive cycle of *L. pertusa* was first described, using samples collected from a fjord in Norway (Brooke and Jarnegren 2013). Unlike offshore areas, where inclement weather precludes sampling during the winter, the more projected fjord allowed for sampling throughout the year. Like the other reef-forming stony corals studied to date (Burgess and Babcock, 2005; Waller and Tyler 2005), *L. pertusa* is a gonochoristic broadcast spawning species, with spawning occurring annually, but over a protracted period (Brooke and Jarnegren 2013). In the northeastern Atlantic this species spawns from January to March with a peak in February. In the GoM, information on gametogenic cycles is incomplete, but summer and fall samples captured the latter part of the cycle, which included vitellogenesis and spawning (CSA 2007; Demopoulos et al. 2017), with spawning occurring in mid-October (Brooke pers. obs.).

Populations of *L. pertusa* from the eastern and western Atlantic showed a difference of ~5 months in the timing of gametogenesis and spawning (Brooke unpubl. data). This difference correlates approximately to the regional timing of organic material flux from the surface, which may drive the onset of energetically expensive gametogenesis. Timing of food delivery to the seafloor was hypothesized decades ago (Tyler 1988), to drive reproductive seasonality in deep-sea species, and there are indications that this may be the case for CWCs also (e.g., Sun et al. 2010b; Mercier et al. 2011; Brooke and Jarnegren 2013).

The black coral *L. glaberrima* is an important structure-forming species in the southeastern U.S. A phylogenetic study of this species in the northern GoM (Ruiz-Ramos et al. 2015) concluded that different reproductive strategies were dominant in different locations. Fragmentation (asexual reproduction) dominated in isolated sites in the Viosca Knoll region, whereas sexual reproduction was predominant across all other sampling sites in the northern GoM. This mixed reproductive strategy may enable species to maintain populations in isolated areas but increase genetic diversity across broader areas.

The Plexaurid octocoral genus *Paramuricea* is abundant in the GoM and can form large colonies that provide habitat for other fauna (Doughty et al. 2014). Oil from the Deepwater Horizon spill damaged several large colonies of *P. biscaya* in the vicinity of the wellhead. Evaluation of the long-term impacts of the damage was limited by the virtual absence of information on this species. Potential for future impacts prompted an investigation of the population dynamics of this species and other congeners in the GoM (Doughty et al. 2014). Most *Paramuricea* colonies occurred on outcrops or steep ridges, with a generally patchy and clumped distribution. This distribution pattern has also been observed in the mid-Atlantic canyons (Brooke et al. 2017). The reproductive biology of the GoM *Paramuricea* is not known, but information from the western Atlantic indicates this genus is gonochoristic with large eggs and an absence of internal brooding (Brooke and Sogluizzo 2017). The Mediterranean congener *P. clavata* broods embryos externally, resulting in settlement close to the parent colony rather than long range dispersal. This mode of reproduction explains the observed distribution of this genus in the GoM and elsewhere and may be interpreted as a characteristic of the genus, rather than from exogenous influence.

While there is little information on reproduction of other coral taxa in the GoM, some inferences can be made from studies of the same or related taxa in other regions. Kahng et al. (2011), Wagner et al. (2012), and Waller and Tyler (2005) provide reviews of octocoral, antipatharian, and cold-water scleractinian reproductive biology, respectively.

3.5.4.2 Early Development

Deep coral early development has been described in very few species as obtaining and rearing cold-water coral larvae is time-sensitive and challenging. So far, information on post-spawning development is limited to three scleractinians: *Oculina varicosa* (Brooke and Young 2003), *L. pertusa* (Larsson et al. 2014), and *Flabellum angulare* (Mercier et al. 2011); four species of soft corals *Drifa glomerata, Drifa* sp. (Sun et al. 2010a, b), *Duva florida*, and *Gersemia fruticosa* (Sun et al. 2011); and a deep-sea anemone, *Allantactis parasitica* (Mercier and Hamel 2009). None of these studies were conducted in the GoM, although congeners or closely related species occur in the Gulf.

Spawning in *L. pertusa* was asynchronous amongst colonies and did not correlate with any obvious environmental condition. Small eggs (~160 μm) developed into swimming planulae ~3 days post-fertilization at 8 °C. Larvae remained in the water column for 3–4 weeks then began probing the bottom of the aquaria, indicating they were ready to settle and metamorphose. Larval settlement was not observed, and some larvae remained alive for up to 57 days (Larsson et al. 2014). Little is known of CWC settlement cues; an experiment by Lavaleye and Duineveld (2001) deployed various substrates adjacent to live coral thickets and showed settlement on coral skeleton, with none on ceramic tile or oyster shells. However, *L. pertusa* (and other corals) also settles on substrates that have no adult corals nearby, such as oil rigs and shipwrecks (Gass and Roberts 2006), which indicates that other factors may be used as settlement cues. In 2010, a series of calcium carbonate settlement plates were impregnated with four different strains of bacteria (*Pseudoalteromonas*) and deployed near deep-coral habitat in the northern GoM (Kellogg et al. 2017a, b). The plates were recovered 12 months later and examined microscopically for coral settlement. Hydroids and foraminifera were the most common colonizers of the plates; three cnidaria polyps were observed on the control (no bacterial treatment) plates, but morphological and genetic analyses confirmed they were not *L. pertusa*. This avenue of research may have potential to reveal bacterial settlement cues, but field work of this nature requires considerable time and resources. The early development of *O. varicosa* was similar to *L. pertusa*, but development was completed faster (2–3 weeks until bottom probing began), which may be explained by phylogenetic differences and/or warmer culture temperatures. A few of the larvae settled and metamorphosed but died shortly thereafter and settlement cues were not identified (Brooke and Young 2005). The scleractinian cup coral *F. angulare* was also a gonochoristic broadcast spawner that released gametes over a protracted period (Mercier et al. 2011). Fertilization was external, with large oocytes

(900–1200 μm) sticking to the tentacles until they either dropped to the bottom or moved into the water column. Embryos developed into large (2–3 mm) larvae that did not settle, again possibly due to the lack of appropriate cues.

Octocorals in the genus *Drifa* brooded planulae in the polyps and released them over several months (Sun et al. 2010b). Most planulae were large (3–5 mm long) and were primarily demersal until settlement (1–30 days after release). Settlement trials showed higher settlement rates on rough natural surfaces that had been pre-conditioned with a biofilm. Two other octocorals from the NW Atlantic, *G. fruticosa* and *D. florida*, are also brooding species (Sun et al. 2011). Several of these species have common traits such as extended spawning period and long-lived larvae, which confers some resilience; however, these may be counteracted by highly selective settlement substrates/cues and slow growth rates, which increases vulnerability (Sun et al. 2010b)

3.5.5 Population Connectivity

3.5.5.1 Population Genetic Structure

The degree of connectivity among spatially separated populations influences the vulnerability of a system. A high degree of population connection confers some resilience, whereas genetic isolation increases probability of extinction from disturbance (Cowen et al. 2007). Genetic connectivity is maintained by dispersal of sexually produced larvae, which is influenced by biological and physical factors such as pelagic larval duration (PLD) and hydrodynamic patterns. Larval dispersal and recruitment among populations cannot be measured directly; however, it can be inferred from studies of population genetic structure. Highly variable molecular markers, for example, microsatellites (Morrison et al. 2008) or Single Nucleotide Polymorphisms (SNPs; Bracco et al. 2019) are used to examine allele frequencies over a range of spatial scales, which can then be interpreted as degree of differentiation among populations. Large-scale population structure of *L. pertusa* showed that GoM populations were distinct from western and eastern North Atlantic populations (Morrison et al. 2011), but the study did not resolve the location of the geographic break. A subsequent study (Demopoulos et al. 2017) examined population structure of *L. pertusa* across ten locations within the GoM, spanning ~900 km from Texas to the West Florida Slope. Genetic analysis of the samples showed some structuring but mixing among populations was sufficient to maintain a coherent regional population. The Loop Current would appear to provide an ideal conduit for larvae from the GoM into the Atlantic but genetic analysis indicates restricted gene flow between these regions (Morrison et al. 2011; Demopoulos et al. 2017). Hydrodynamic studies at Viosca Knoll *L. pertusa* mounds in the northern GoM show occasional reversals to the prevailing eastward Loop Current flow, which would enhance mixing among GoM populations (Davies et al. 2010; Mienis et al. 2012) and explain the observed weak genetic structure across GoM *L. pertusa* populations. Clonal species such as *L. pertusa* complicate genetic analysis as the usual assumptions about random mating and non-overlapping generations are confounded by asexual reproduction (Hey and Machado 2003). Additional approaches, such as network analyses (Rozenfeld et al. 2007), would quantify genetic structure of each population and clarify the role that each plays in the overall GoM system. The more recent application of hydrodynamic modeling of larval dispersal coupled with genetic connectivity estimates is also a powerful approach (e.g., Bracco et al. 2019).

The black coral *L. glaberrima* has several color morphotypes (red, orange, white), but genetic analysis has shown these do not represent species differentiation (Brooks et al. 2015; Ruiz-Ramos et al. 2015). The variation in color as well as branching pattern and polyp size suggests the possibility of undescribed species of *Leiopathes* in the GoM (Brugler 2011). To understand the reproductive patterns and population genetic structure of this important species, samples from three regions in the GoM were analyzed using microsatellites: the West Florida Slope, northeastern GoM, and northwestern GoM. Two distinct lineages were identified from this study; one was found across all three regions. A second was found exclusively at the Viosca Knoll (VK) lease block 906, and the nearby VK826 site was composed of a mixture of both lineages, some of which appeared to be hybrids. The lineages reflected differing reproductive patterns, with the second lineage showing higher levels of inbreeding than the first, which the authors attributed to the relative isolation of the VK906 site. High clonal levels within a site can be caused by predominantly asexual reproduction. Locally well-adapted genotypes of shallow stony corals can successfully dominate communities (Boulay et al. 2014); this may explain how an apparently isolated lineage of *L. glaberrima* can persist at VK906 and in a limited section of VK826.

Cairns and Bayer (2009) listed 162 octocoral species in the GoM, with 77 species found deeper than 200 m. More recent research (Brooks et al. 2015) increased this number by 12 species not previously known from the GoM, and 3 new endemic sub-species (Quattrini et al. 2014). In the GoM, one of the most widespread and abundant large octocorals belongs to the genus *Callogorgia* (Family Primnoidae). Molecular studies (Brooks et al. 2015) revealed three species of *Callogorgia*: *C. gracilis*, *C. americana americana* (<425 m), and *C. americana delta* (425–1000 m) in the GoM. A population genetic analysis of *C. a. delta* in the northern GoM revealed within-species population structuring with depth, but not with geographic distance (Quattrini et al.

2015; Bracco et al. 2019). Species in the genus *Paramuricea* (Family Plexuaridae) are also large, abundant octocorals that can occur up to 2400 m depth. A comprehensive molecular study of this genus across the northern GoM (Doughty et al. 2014) identified seven haplotypes and five putative species of *Paramuricea* that showed strong depth-structured distributions with little overlap: *P.* haplotype H (<260 m), *P.* haplotype E (280–400 m), *P.* haplotype A (443–550 m), *P.* haplotype B3, (837–1040 m), and *P. biscaya* B1, B1a, B2 (860–2400 m). There was no apparent geographic structuring within this genus, and the GoM haplotypes were found throughout the north Atlantic and Caribbean (Thoma et al. 2009; Doughty et al. 2014). In summary, the species and putative species within these genera were broadly distributed across the GoM, with each occupying a specific depth range (Doughty et al. 2014; Brooks et al. 2015). This pattern has been observed for other octocorals in the deep GoM (Quattrini et al. 2015; Bracco et al. 2019) and elsewhere (Eytan et al. 2009; Baco and Cairns 2012). This pattern could be the result of speciation along an environmental gradient correlated with depth, as occurs in other corals (Prada and Hellberg 2013), or niche adaptations to specific depth-related environmental conditions. Genetic analysis provides information on contemporary population structure, but larval dispersal modeling, which integrates information on reproduction, early life histories, and ocean currents can provide insight into the mechanisms of population dynamics (Bracco et al. 2019).

3.5.5.2 Larval Dispersal Modeling

The limitation of these models is usually the lack of data on reproduction and early life history characteristics (larval lifespan, environmental tolerances, behavior) of deep-sea species. Most models treat larvae as passive particles and estimate larval distributions by integrating ocean currents with a specified planktonic larval duration (PLD). While imperfect, these models can be used to estimate potential dispersal pathways and map these trajectories on the biogeographic distribution of species to gain insight into their recruitment dynamics (Young et al. 2012). A thorough study of dispersal of deep-sea species was conducted by Young et al. (2012) using seven species (three echinoderms, two mollusks, one annelid, and one sipunculid) from sites in the GoM, Bahamas, and Barbados. The spawning time and PLD values for the target species were derived from spawning and rearing of larvae in the lab. Larval trajectories were estimated by "releasing" particles from the appropriate location during the known spawning period for each species and incorporating their respective PLD into the selected ocean circulation model. Model outputs for GoM species released off the Louisiana slope show that larvae of the seep tubeworm (*Lamellibrachia luymesi*), with a PLD of 21 days, were retained within the northern GoM. Those with longer PLD (up to 660 days) could travel throughout the Gulf and into the north Atlantic, although the majority were retained within the GoM with an underlying directional movement from west to east, then south following the Loop Current. Estimates of larval dispersal for the seep mussel, *Bathymodiolus childressi*, were supported by the discovery of shared mitochondrial haplotypes in adults of this species from the GoM and mid-Atlantic canyons (Coykendall et al. 2019). The models also show that if larvae move into a shallower depth zone (e.g., for planktotrophic feeding), they are dispersed further than remaining at their release depth. A larval dispersal model for *L. pertusa* (CSA 2007) estimated a PLD of 21 days, based on data from deep populations of *Oculina varicosa* (Brooke and Young 2003). Subsequent studies of *L. pertusa* larvae indicated a similar time to competence (Larsson et al. 2014). Model predictions for larvae released from two known coral sites (VK826 and Green Canyon 234) show similar trajectories with some particles moving west, but most moving eastward, in a similar pattern to *L. luymesi* (Young et al. 2012). In both cases, the authors concluded that oceanic conditions in the GoM can facilitate some retention near the release sites, but dispersal was primarily eastern, then southern following the along isobath current flows. Dispersal is strongly influenced by PLD, with longer lived larvae dispersing far into the north Atlantic. Increased information on reproduction and ecology of early life histories can refine larval dispersal models and improve our understanding of mechanisms driving biogeographic patterns.

3.5.6 Microbial Ecology

The consortia of corals and their associated microbial assemblages are termed a "holobiont" (Rohwer et al. 2002), and the microorganisms in the holobiont can include bacteria, archaea, and eukarya (e.g., fungi, protists) (Rosenberg et al. 2007a), as well as viruses (Thurber and Correa 2011). Microorganisms associated with shallow corals may have a probiotic function that prevents biofouling of tissue and skeleton (Ritchie 2006; Reshef et al. 2006) or help cycle nutrients and carbon (Wegley et al. 2007; Yang et al. 2013; Zhang et al. 2015; Silveira et al. 2017). Some of these microbes comprise a "core" community that is necessary for coral health (Rosenberg et al. 2007b; Shade and Handelsman 2012), and others can change in response to shifts in environmental conditions; for example, pathogenicity may increase with higher temperatures (Bruno et al. 2007; Ben-Haim et al. 2003). There is less information on cold-water coral microbial assemblages and their function within the holobiont, but several studies have focused on understanding these relationships in the stony coral *L. pertusa* and several deep-sea octocorals (Kellogg 2019). The first study to culture bacteria from *L. pertusa* was conducted in 2004–2007 (Galkiewicz et al. 2011) and revealed high abundance of *Vibrio* spp. and *Pseudoalteromonas* spp. These are commonly found in cultures of microbionts of corals (Chimetto et al. 2008; Raina et al. 2009) and sponges (Menezes et al. 2009) and may be symbiotic but their ecological and metabolic roles may vary. Both *Pseudoalteromonas* spp and *Vibrio* spp. produce compounds that are bioactive and antibiotic (Dobretsov and Qian 2004; Ritchie 2006), which may confer some protection to the coral from biofouling and bacterial infection. In addition, *Vibrio* spp. may have a role in enhancing coral nutrition through cycling of nitrogen and refractory carbon (Chimetto et al. 2008). Phylogenetic studies of *L. pertusa* microbionts in Norway (Neulinger et al. 2008) and the Gulf of Mexico (Kellogg et al. 2009) supported the existence of cold-water coral-specific bacterial groups and also concluded that a subset of these bacteria appeared to play a role in the nutrition of their hosts. Western Atlantic cold-water coral ecosystems (including the GoM) are dominated by a white phenotype of *L. pertusa*, with very rare occurrences of an orange phenotype that is moderately common in the eastern Atlantic. The mixotrophic *Rhodobacteraceae* can exploit low food levels and was one of the dominant taxa in the white coral microbiont, but less abundant in the orange phenotype (Neulinger et al. 2008; Kellogg et al. 2017a, b). The authors posited that these bacteria gave the corals a nutritional advantage and explained the prevalence of the white phenotype in less food-rich environments. More recent work has highlighted the bacterial community variation at the regional and individual *L. pertusa* colony levels (Kellogg et al. 2017a, b). Other microbial associates such as fungi have also been documented in the cold-water coral holobiont. The first fungi cultured from *L. pertusa* came from corals in the northern and eastern GoM and the Atlantic coast of Florida (Galkiewicz et al. 2012). Corals from these different sites showed phylogenetically similar fungal isolates, indicating consistency of association across broad geographic distances, but differences in amount and composition of the two most common taxa: Ascomycetes and Basidiomycetes. Culturing techniques can provide insight into functional roles of microbial species, but genetic analysis provides a more comprehensive understanding of the microbial community (Staley and Konopka 1985; Raghukumar 2006). Applying a combination of these techniques is a powerful approach to understanding the roles and relationships within the coral holobiont.

Community composition of the deep-sea stony coral *Madrepora oculata* from the eastern Atlantic (Hansson et al. 2009) and the Mediterranean (Meistertzheim et al. 2016) was dominated by the bacterial group *Endozoicomonas* (previously *Spongiobacter*) which is associated with many marine species and may have adaptive mutualistic functions in the holobiont (Neave et al. 2017). Comparisons between two dominant stony corals, *M. oculata* and *L. pertusa*, that occupy the same habitats and geographic range, show distinct differences between their microbial communities (Hansson et al. 2009; Meistertzheim et al. 2016; Weinbauer et al. 2019). While *M. oculata* maintained a consistent (host-structured) core microbiome, communities associated with *L. pertusa* were much more variable temporally and spatially, suggesting more influence of local environmental conditions or some other variable factor such as nutrition (Meistertzheim et al. 2016; Galand et al. 2020). These data suggest contrasting strategies relative to the microbiome between two species that are found co-existing across an extensive geographic area.

Microbial associations of a number of different deep-sea octocoral species from different ocean basins supported the concept of a core microbial community associated with each host species or genus, and an additional microbial community that varied amongst individuals (Kellogg et al. 2016; Lawler et al. 2016; Goldsmith et al. 2018). Congeners of several of the species studied from other locations also occur in the GoM so the general observations may be applied with caution. Recent work has examined the ecology of eukaryotic microbial associates known as corallicolids in Gulf of Mexico octocorals and black corals (Vohsen et al. 2020).

In shallow-water corals, the coral microbiome includes three components: a small "core" community, a species or site-specific community, and a large and variable community that reflects fluctuations in local conditions (Hernandez-Agreda et al. 2016, 2017; van de Water et al. 2017). These components also appear to occur in CWC (Meistertzheim et al. 2016; Kellogg et al. 2017a, b; Goldsmith et al. 2018; Kellogg 2019; Weinbauer et al. 2019). The relationship between host and microbial associates has revealed potential benefits to the coral host, but these associates also include pathogenic bacteria. Understanding these relationships may provide insight into coral health status and resilience to disease and other stressors.

3.6 Cold-Water Coral Ecosystems

3.6.1 Community Structure

3.6.1.1 Structure-Forming Coral Communities

Structure-forming corals are those that create significant three-dimensional structure, either as large individual colonies or as assemblages that provide habitat for other species. These include not only the reef-building stony corals such as *Lophelia pertusa* but also large black corals such as

Leiopathes glaberrima or octocorals such as *Callogorgia* sp. and smaller species that occur in large numbers and collectively create habitat.

3.6.1.1.1 West Florida Slope and Escarpment

The massive carbonate platform that dominates the eastern GoM supports a variety of cold-water coral habitats. A distinct but irregular escarpment runs almost parallel to the west Florida coastline for ~88 km (Ross et al. 2017a); the average depth of the top of the feature is ~440 m, with a variable vertical drop (7–85 m) to the wide, gentle slope. The escarpment comprises a rocky ledge with scattered boulders at the base. Submersible surveys (Reed and Wright 2004; Reed et al. 2006) revealed small colonies of *L. pertusa* on the ledges, but the dominant fauna were various species of black corals (including large *L. glaberrima*), bamboo corals (Isididae), and other octocorals, plus several species of sponge. Mounds (lithoherms) of phosphorite-encrusted boulders were capped with *L. pertusa* thickets and other sessile fauna such as octocorals, stylasterid hydrocorals, black corals (*Antipathes* sp. and *Cirrhipathes* sp.), and sponges (Reed et al. 2006). More recent explorations (2009–2012) to the west of the escarpment have revealed intermittent groups of coral bioherms at 450–500 m that appeared to have formed over a base of large rocks, possibly broken off the escarpment. These span approximately 20 km (between 26° 20′N and 26° 30N), with individual mounds ranging from 5 to 15 m tall (Ross et al. 2017a). Dead *L. pertusa* skeleton provided the underlying structure, with live coral cover increasing to >50% near the tops of the mounds (Ross et al. 2017a). Colonies of *Madrepora oculata* were also observed on the bioherms. The cup coral *Thecopsammia socialis* formed large patches on dead coral rubble or rocky habitats, and the soft nephtheid octocorals were also locally abundant on the rubble zones. Further south (~25.6°N; 84.5°W), a site called Okeanos Ridge (450–600 m) is comprised of a curving escarpment (620–825 m), and a long ridge (~10 km) to the east. Surveys of this site revealed relatively high coral diversity, with large colonies of *L. pertusa*, together with *M. oculata* and abundant black corals (e.g., *Bathypathes* sp and *L. glaberrima*), gorgonians (including a large Primnoid, *Paracalytrophora carinata*), and stylasterine hydrocorals.

The Florida Escarpment is between 200 and 400 km offshore of the west Florida coast and stretches almost the length of the Florida peninsula as a steep, rugged slope incised with gulleys and canyons. The Escarpment is one of the least accessible places in the GoM; it is far from shore, very deep (1800–3200 m) and is subject to strong currents generated by the Loop Current. Less than 20 surveys have been conducted along the 600 km length of the Escarpment (Okeanos Explorer cruises Ex1202, Ex1403, Ex1711, Brooks et al. 2015; Brooke et al. 2020), but these have revealed dense coral communities that are not found at shallower depths, including octocorals in the families Chrysogorgiidae, Corallidae, Paragorgiidae, Plexauridae, and Isididae. The stony corals common at slope depths, such as *L. pertusa* and *M. oculata*, were not observed on the Escarpment, but large individual colonies of *E. rostrata* were moderately abundant. Another notable difference were the number of mobile invertebrates and fishes, which were far fewer on the deep Escarpment than on the shallower slope habitats, possibly due to food limitations at extreme depths

3.6.1.1.2 Northern Gulf of Mexico

Deep-coral sites in the northern GoM are designated following the BOEM lease block numbers in which they occur. Distribution of cold-water coral communities in this region is driven in part by the availability of exposed hard substrate. Unlike the eastern carbonate platform, hard substrate in the northern deep GoM is primarily authigenic carbonate, which can be extensive but is sporadic rather than continuous.

Some of the best known cold-water coral communities in the GoM occur on a mound in the southwest corner of the Viosca Knoll (VK) lease block 826 on the upper DeSoto slope. Abundant authigenic carbonate supports *L. pertusa* communities (Schroeder 2002; Brooke and Schroeder 2007), which range from small individual colonies to large aggregated thickets, interspersed with sediment and coral rubble/shell hash. To the northeast of the main crest, the colonies form thickets, with an understory of dead coral matrix capped by live corals. This is the initial phase of bioherm formation and is rarely observed in the northern GoM. The eastern flank of the knoll is also colonized by dense colonies of *L. pertusa*, *L. glaberrima,* and octocorals (e.g., *Callogorgia* sp.). To the east of the main knoll is a ridge of almost continuous *L. pertusa* thickets. Active methane seepage occurs on this feature, with communities of chemosynthetic tubeworms and vesicomyid clams. Corals live near these seeps and do not appear to be affected by them.

To the west of VK826 is another well-known cold-water coral site (VK862/906), which comprises a submarine canyon and numerous coral bioherms. The shallow (320–350 m) northern section of this site near the canyon head (VK862) has abundant rugged exposed carbonate, often colonized with extremely high densities of white anemones, as well as sponges, large *L. glaberrima* colonies, *L. pertusa* and octocorals such as *Muriceides* sp. Sandy areas support dense populations of *Acanella* sp., an unusual bamboo coral that lives in soft sediment (Brooke and Schroeder 2007; Brooks et al. 2015). At the southern end of the canyon (VK906) are four mounds that were discovered in 2009 and are the only known coral bioherms in the northern GoM. The largest mound (~50 m tall) was named after Dr. Harry Roberts (LSU), who was instrumental in the discovery. The base of Roberts Reef was comprised of coral rubble with small *L. pertusa* colonies, glass sponges, comatulid crinoids, and echinoderms. The size and abundance of *L. pertusa* colonies increased on the slopes and were interspersed with colonies of *L. glaberrima*. At the crest of the mound were dense thickets of live *L. pertusa*. Octocorals associated with these features include bamboo corals, *Muriceides* cf *hirta*, *Acanthogorgia aspera*, *Paramuricea* sp., and *C. a. delta*.

Several lease blocks in this area have abundant authigenic carbonate formations, which provide hard substrate for colonization by sessile benthic fauna. Deep-coral communities have been documented within several of the Mississippi Canyon (MC) lease blocks, but others probably exist in unexplored parts of this region. Within lease block MC 751 is an oblong area of approximately 20 m relief that is composed of large blocks, slabs, and outcroppings of authigenic carbonate (Brooks et al. 2015). The carbonate is colonized by isolated colonies of *L. pertusa* and a diverse suite of large gorgonians such as *C. a. delta* (commonly found near seep areas), *Paramuricea* sp, *M.* cf *hirta*, *Swiftia exserta*, *Chelidonisis aurantiaca mexicana*, and a species of bubblegum coral *Paragorgia johnsoni*, which is uncommon in the GoM. Black corals such as *Bathypathes* sp. and *Stichopathes* sp. were also present. Lease block MC 885 is deeper (620–650 m) and has a series of small (30 m tall) mounds, which are colonized by both *L. pertusa* and *M. oculata*, as well as *L. glaberrima*. The gorgonian *C. a. delta* was also common at this site, but overall coral diversity was low (CSA International Inc. 2007). Lease block MC118 is the deepest (880 m) of the well-known sites in this region. In addition to cold seeps and methane hydrates, dense coral assemblages colonized the authigenic carbonates (Brooks et al. 2015), including the deep octocoral *Chrysogorgia* sp., many large colonies of *Paramuricea biscaya*, and unusually high densities of *M. oculata* that formed thickets similar to *L. pertusa*.

During post oil-spill exploration, one of the largest bathyal coral communities in the northern GoM was discovered in Atwater Valley (AT) lease block 357, near the mouth of the Mississippi Canyon (Brooks et al. 2015). The main part of the site consists of carbonate slabs covered with dense thickets of *M. oculata* (Fig. 3.3b) large colonies of *Paramuricea* sp. and other deep octocoral genera such as *Anthothela* and *Chrysogorgia*.

The Green Canyon (GC) area also has abundant cold-water coral communities, some of which were discovered during exploration and research of methane seeps. Within lease block GC140, a large salt diapir rises steeply from ~400 to 230 m and has a very rugged top comprised of carbonate slabs colonized by large colonies of the black coral, *L. glaberrima* and the octocorals *Callogorgia gracilis* and *Paramuricea* sp., plus several species of bamboo corals, but *L. pertusa* was absent from this site. GC 234 is a linear carbonate ridge at 450–500 m depth and has abundant live and dead colonies of *L. pertusa*, but octocorals dominated this site, including *C. a. delta* with their characteristic ophiuroid associates (*Asteroschema* sp.), *Paragorgia johnsoni*, *Paramuricea* sp., *Chelidonisis* sp., *M.* cf. *hirta*, *Acanthogorgia aspera*, *Thesea* sp., and *Scleracis* sp. At GC 354, live and dead *L. pertusa* and large sponges cap a large salt diapir (580–525 m). Carbonate boulders on the sides of the mound supported diverse communities of octocorals including *A. aspera*, *M.* cf *hirta*, *Nicella* sp., *Paramuricea* sp., *S. exserta*, *Chelidonisis a. mexicana*, and *Paracalyptrophora carinata* (CSA International Inc. 2007; Brooks et al. 2015). Within lease block GC852 is a long, wide ridge that rises from the base at 1600 m to a rugged area at 1400 m, colonized by a dense coral assemblage that was different from those at the shallower sites (Brooks et al. 2015). Stony corals included *Solenosmilia variabilis*, *E. rostrata*, and *M. oculata*. Octocoral communities were dominated by bamboo corals (Keratoisidinae), with several other octocorals, including *Iridogorgia splendens*. *Paramuricea biscaya*, *Swiftia pallida*, *Narella pauciflora*, and the precious coral *Corallium* sp.

The Garden Banks (GB) region lies offshore of the Texas-Louisiana border and is the western-most area in the northern GoM region. Within lease block GB535 (500–550 m) is a long ridge with 50 m relief and exposed carbonate along the crest, colonized by a diverse coral and sponge community. Large mounds of *L. pertusa* with significant dead framework suggested long-term colonization and growth. Dominant fauna included large white demosponges, antipatharian whip corals (*Stichopathes* sp.), and three species of octocorals (*Narella* sp., *Nicella* sp. and *Scleracis* sp.). The diversity of octocorals was high for this region and included a new species of *Anthothela* sp., as well as *A. aspera*, *M.* cf. *hirta*, *C. a. mexicana*, and *Paracalyptrophora carinata* (Brooks et al. 2015). Lease block GB299 (340–500 m) lies on a NE-SW trending feature with a rugged top and a steep southern slope (Brooks et al. 2015). The substrate was colonized by black corals and octocorals with *L. glaberrima* and *C. a. delta* being unusually abundant, particularly along the north-facing slope at ~365 m depth. Other octocorals included *Nidalia dissidens*, *M.* cf *hirta*, *Paramuricea* sp., and bamboo corals.

3.6.1.1.3 Campeche Bank

The Campeche Bank in the southern GoM is flanked by the Campeche Basin in the west and the Yucatan Straits in the east. In the northeastern region, the shelf slopes gently to ~550 m, where it terminates in a sharp shelf edge and steep scarp. East of the shelf edge, the slope is covered with hundreds of steep V-shaped ridges, 20–40 m high and 500–100 m long (Hebbeln 2012). An area of ~50 km^2 has been mapped (~23° N, –87° W), and several of the features have been identified as CWC reefs, dominated by *L. pertusa* and *E. profunda* (Hubscher et al. 2010; Hebbeln 2012). If all of the structures within the mapped area are coral reefs, the Campeche Bank would represent the largest continuous CWC province in the GoM; however, most of this vast rugged area remains unexplored. To the west of the coral ridges lies a steep, narrow canyon which incises the bank at approximately 450 m depth and extends to the base of the Campeche Escarpment at ~3000 m. This canyon has never been explored. Most shelf-incised canyons are conduits of biogenic material, and characteristically strong currents remove sediment to expose underlying hard substrate that allows colonization of sessile benthic fauna such as corals and sponges. This canyon therefore may support as yet undiscovered CWC communities.

3.6.1.1.4 Artificial Habitats

There are 53 energy platforms and shipwrecks deeper than 300 m in the northern GoM (Boland et al. 2017), and some support dense communities of CWC and sponges. In 2003, *L. pertusa* was documented for the first time on the Pompano platform in the GoM at 400 m depth (Boland et al. 2017), and cold-water stony corals and octocorals have been observed on several other shipwrecks and platforms (Larcom et al. 2014). Particularly large, dense colonies of *L. pertusa* were documented on two shipwrecks: the Gulf Penn (Church et al. 2007) and The Gulfoil (Brooks et al. 2015). Artificial substrates such as shipwrecks and energy platforms provide habitat for some CWC and may represent stepping-stones among populations of some species (Brooks et al. 2015). However, the communities associated with artificial habitats may not completely represent those on natural substrate (Boland et al. 2017) so it is unclear how much they contribute to the overall ecology of CWC ecosystems in the GoM.

3.6.1.2 Coral-Associated Invertebrate Communities

Within the GoM, only a few CWC environments have been examined in detail, including quantitative descriptions, collections, and identifications of the associated coral communities. These sites include the West Florida Slope, Campeche coral province, and the northern GoM: Viosca Knoll (VK826, 862, 906), Mississippi Canyon (MC751), and Green Canyon (GC852).

The West Florida Slope CWC habitats include extensive and complex reef environments, primarily composed of living and dead *L. pertusa* framework (Newton et al. 1987; Reed et al. 2006; Hubscher et al. 2010; Ross et al. 2017a). In addition to a diverse coral assemblage, dominant sessile fauna includes anemones (Actiniaria), hydroids (Hydrozoa), and sponges (*Heterotella* spp., *Phakellia* spp., Corallistidae, Pachastrellidae, Petrosiidae, Astrophorida). Other fauna includes various bivalves and gastropods (Mollusca), sea cucumbers (Holothuroidea), crinoids (Crinoidea), and decapod crustaceans including squat lobsters (*Gastroptychus* spp., *Munidopsis* spp., *Munidopsis turkayi, Eumunida picta*), crabs (*Chaceon fenneri, Bathynectes* sp., *Rochinia* sp.), and shrimp (*Periclimenes* sp. *Eugonatonotus* sp.) (Reed et al. 2006; Macpherson et al. 2016; Nizinski and Ames 2017). Several new records have been observed from this region, including the squat lobsters *Leiogalathea agassizii* (previously recorded off Cuba and in the Caribbean) and *Munida sanctipauli* (eastern GoM) (Nizinski and Ames 2017), as well as scale worms (e.g., *Leanira robusta*) found associated with CWC habitats (Barnich et al. 2013). The squat lobster, *Munidopsis glabra* and the anemone *Adamsia obvolva* (which live on the shells of hermit crabs) were collected from cold-water coral habitats on the West Florida Slope. Both of these species are endemic to the GoM but neither are restricted to coral ecosystems.

The relationship between corals and their associates has been examined for the northern and eastern GoM, based on 720 coral-associated invertebrates (brittle stars, shrimp, amphipods, anemones, barnacles, and crabs; Brooks et al. 2015). Coral-associated composition varied with depth and biogeographic location. The highest diversity of associates was found with live *L. pertusa* and *Paramuricea* sp., with the highest number of morphospecies occurring on *L. pertusa*, possibly due to the mixture of live and dead coral framework. Overall, there was a low number of obligate species, which included the ophiuroid *Asteroschema clavigerum* on *Paramuricea biscaya*, and the shrimp (*Pathypalaemonella serratipalma*) on *Chrysogorgia* spp. and *Iridogorgia* spp. Arthropods represented the most common and diverse group (38 morphospecies) and were dominated by galatheid and chyrostylid crabs, followed by echinoderms (30 different morphospecies), which were dominated by ophiuroids.

In the northern GoM, collections made with otter trawls, ROVs, HOVs, and traps, coupled with quantitative image analysis have revealed abundant and diverse megafaunal communities within and adjacent to CWC habitats. Trawl collections around VK826 indicate high species diversity, with the crustaceans representing the most abundant and speciose group, comprising 69% of the total individuals collected, within 74 species (Nizinski and Ames 2017). The most abundant species included the brachyuran crab *Bathyplax typhla*, the squat lobster *Munida* spp., and the blind shrimp *Polycheles* sp. Golden crabs (*Chaceon fenneri*), which are harvested in the southeastern U.S., are also often found in *L. pertusa* habitats.

High-resolution photo mosaics, collected using the HOV *Johnson Sea-link* (Harbor Branch Oceanographic Inst.), and ROV *Jason* (Woods Hole Oceanographic Inst.), were used to quantify megafaunal communities associated with multiple deep-sea coral habitats across the northern GoM (Lessard-Pilon et al. 2010; Brooks et al. 2015). High beta diversity occurred within deep-sea coral sites and across the region

(Lessard-Pilon et al. 2010) and overall communities were more similar within regions rather than among regions. Diversity and community composition of coral-associated megafauna is a function of the proportion of live and dead coral framework (Lessard-Pilon et al. 2010; Brooks et al. 2015), with higher diversity of fish and invertebrates associated with dead thickets of *L. pertusa*, rather than live colonies. Overall, thirty-three taxa were distinguishable in the image mosaics, with common taxa including squat lobsters (*Eumunida picta*, *Munidopsis* spp.), shrimp (*Odontozona edwardsi*), cup corals (*Caryophyllia*), and tube-dwelling anemones (*Cerianthus* sp.). Species' dominance varied among sites and many of the fauna were non-randomly distributed across substrata. Abundance of *Munidopsis* and *E. picta* was correlated with dead coral and rubble, but *E. picta* is also commonly observed within live *L. pertusa* colonies on natural and artificial habitats, where the coral may be providing refuge from predation, feeding habitat, or some combination therein (Kilgour and Shirley 2008). Other taxa associated with live coral included the gastropod, *Coralliophila* sp., crinoids (*Comatonia cristata*), and urchins (*Echinus* spp.).

Communities associated with the interstitial framework of *L. pertusa* and *M. oculata* habitats have been quantified using a bushmaster and coral pot collection devices (Cordes et al. 2008; Brooks et al. 2015). Specifically, 17 collections from *L. pertusa* and 2 collections within *M. oculata* framework were examined at 12 different sites across the NGoM, in depths from 388 to 2445 m. At least 90 species (>2 mm) were found associated with *L. pertusa* and eight species were associated with *M. oculata* (Cordes et al. 2008; Brooks et al. 2015). Sponges (Porifera) and hydroids (Hydrozoa) remain unresolved to species. Community collections yielded several potentially transient associates as well as some potential coral-endemic species. The most common and abundant species were the epifaunal sabellid polychaetes (*Euratella* sp.), as well as encrusting sponges and hydroids. Other polychaetes (*Glycera tesselata* and *Eunice* sp.) and shrimp (*Periclemes* sp.) were also abundant. Other common coral associates included *Coralliophila* sp. (corallivorous gastropod), squat lobsters (*Eumunida* sp., *Munidopsis* sp.), and the ophiuroid *Ophiotreta valenciennesi rufescens*, plus another unidentified ophiuroid. The polychaete *Eunice* sp. is similar to *E. norvegica*, which is known to accrete pieces of coral to its tubes, which are then calcified by *L. pertusa*. Community similarities among collections were largely due to depth and proportion of live coral. Additional variance was also explained by the surface area:volume ratio and distance between collections, where greater similarity occurred within than among sites.

Diversity from the community collections was inversely proportional to amount of live coral present (Cordes et al. 2008; Brooks et al. 2015). Diversity (in terms of total number of species and Shannon-Weiner, H') was also a function of surface area of the coral in the collections. The authors hypothesized that the more diverse communities may preferentially use the dead coral framework over live polyps, which have nematocysts and mesenterial filaments that prohibit larval settlement. The animal associates may also be contributing to coral tissue mortality through abrasion and/or direct consumption, enhancing the overall availability of dead coral. Most of the taxa found within the coral framework are part of the regional species pool within the GoM continental slope. The three-dimensional, heterogeneous substrate created by CWCs provides shelter and food for associated taxa. There was also a high level of beta diversity with overall low similarities within and across sites, which was consistent with the results from the image mosaics. For example, the two *M. oculata* collections had no species in common. For *L. pertusa*, VK826 had high beta diversity, whereas sites in the western GoM (GB535, GC354, GC234) clustered together, exhibiting higher similarity overall and lower beta diversity across those sites.

Dominant species associated with octocoral colonies included commensal ophiuroids (White et al. 2012; Girard et al. 2016, 2018). Specifically, the euryalid ophiuroids, *A. clavigerum*, were observed closely associated with their host corals, *Paramuricea* sp. Image analysis following the DWH oil spill revealed that recovery and hydroid colonization of damaged *Paramuricea* colonies was negatively correlated with distance from commensal ophiuroids. Branches near ophiuroids were more likely to recover, suggesting that the corals benefit from the association, possibly through the removal of deposited material and prevention of hydroid settlement by ophiuroids. Other taxa observed with *Paramuricea* include anemones, forams, hermit crabs, aplacophorans, hormatiid anemones, and fishes (Girard et al. 2016, 2018). Small ophiuroid individuals were found on small coral colonies; the authors speculate that they may settle directly on the corals early in their development (Girard et al. 2016).

The Campeche coral province is located in the southern GoM, off of the Yucatan peninsula. *Lophelia pertusa* is the dominant scleractinian reef-building coral (Hebbeln et al. 2014), but *Enallopsammia profunda* also contributes to the reef structure. A variety of megafaunal invertebrates use the reef as habitat, including the sponges *Aphrocallistes* sp. and *Hyalonema*, actiniarian anemones, and squat lobsters. There appear to be discrete communities associated with the availability of live coral. Within the live coral zone several decapod species (*Bathynectes longispina*, *Eumunida picta*, *Chaceon fenneri*, *Munidopsis penescabra*, *M. turkayi*, and *Rochinia crassa*), echinoids (*Cidaris* sp. and *Gracilechinus* sp.), seastars (*Hippasteria*), and the gastropod, *Coralliophila richardi*, were observed. Crinoids were primarily found on the high points of live coral branches. On bare branches of the

L. pertusa, zoanthids and anemones were present. In contrast to live *L. pertusa* in the NGoM, *Eunice* polychaetes were rare within the Campeche coral province. The exposed coral framework within the dead coral zone had a diverse community of taxa, including flytrap anemones (~*Actinoscyphia* sp.), bamboo (*Keratoisis* sp.), solitary corals (*Desmophyllum dianthus, Javania cailetti, Stenocyathus vermiformis, Trochopsammia infundibulum*), and glass sponges attached to framework (*Aphrocallistes* sp.).

3.6.1.3 Coral-Associated Fish Communities

Information on fish assemblages associated with CWC is extremely limited compared to that of shallow-reef associations owing to the limited suite of tools available for fish surveys. SCUBA-diving, the primary census technique in shallow-waters, is not feasible at depths occupied by CWC. Likewise, trawling, the primary census technique for deep-pelagic and deep-benthic habitats, is not practical (or allowed) over deep-coral habitat. Thus, the primary tools used for fish surveying over CWC are ROVs and HOVs, both of which are highly limited in spatiotemporal coverage, and both prone to avoidance by certain fish taxa. That said, the information that does exist suggests that CWC provide important habitat for diverse assemblages of fishes. Here we review the available information regarding fish assemblages associated with CWC in the GoM and adjacent waters (Straits of Florida, southeast Florida coast), with notes on the ecology of selected species when such information exists.

The faunal structure of fish assemblages over cold-water coral reefs in the Gulf differs substantially from those of shallow reefs. Shallow-water coral reef assemblages are dominated numerically by damselfishes, wrasses, angelfishes, grunts and surgeonfishes (Pomacentridae, Labridae, Pomacanthidae, Haemulidae and Acanthuridae, respectively), whereas deepwater assemblages tend to be dominated by small-bodied groupers (e.g., Serranidae [Anthiinae]), scorpaenoid fishes (Scorpaenidae, Sebastidae), hakes (Moridae, Phycidae), roughies (Gibberichthyidae, Trachichthyidae), rattails (Macrouridae), and eels (Synaphobranchidae, Congridae). Specific fishes that dominate various GoM deep-reef systems include thorny tinselfish (*Grammicolepus brachiusculus*) (Fig. 3.4a), blackbelly rosefish (*Helicolenus dactylopterus*) (Fig. 3.4b) roughtongue bass (*Pronotogrammus martinicensis*) (Fig. 3.4d), red barbier (*Hemanthias vivanus*), creole-fish (*Paranthias furcifer*), yellowfin bass (*Anthias nicholsi*), big roughy (*Gephyroberyx darwini*) (Fig. 3.4e), swallowtail bass (*Anthias woodsi*) (Fig. 3.4f), threadnose bass (*Choranthias tenuis*), tattler (*Serranus phoebe*), various hakes (*Laemonema barbatulum, L. goodebeanorum, Urophycis cirrata,* and *U. floridana*), yellowtail reeffish (*Chromis enchrysura*), brown chromis (*C. multilineata*), yellow goatfish (*Mulloidichthys martinicus*), deepbody boarfish (*Antigonia capros*), Atlantic roughy (*Hoplostethus occidentalis*), rattails (*Nezumia* sp.), and cutthroat eels (e.g., *Dysommina rugosa*) (Weaver et al. 2006a, b; Streich et al. 2017; Ross et al. 2017b).

Large, predatory fishes observed at cold-water reef sites in the Gulf include Warsaw grouper (*Epinephelus nigritus*), scamp (*Mycteroperca phenax*), snowy grouper (*E. niveatus*), speckled hind (*E. drummondhayi*), yellowedge grouper (*E. flavolimbatus*), Atlantic wreckfish (*Polyprion americanus*), barrelfish (*Hyperoglyphe perciformis*), queen snapper (*Etelis oculatus*) (Fig. 3.4c), silk snapper (*Lutjanus vivanus*), blueline tilefish (*Caulolatilus microps*), smalltooth sandtiger shark (*Odontaspis ferox*), and swordfish (*Xiphias gladius*).

At deep *L. pertusa* reefs, 53 taxa of demersal fishes are associated with prime and transition reef habitats in the northern Gulf (Viosca Knoll), with the thorny tinselfish (*Grammicolepis brachiusculus*) (Fig. 3.4a) tightly associated with *L. pertusa* (Sulak et al. 2007). Sulak et al. (2007) and several other studies highlight depth as the primary determinant of faunal composition over deep reefs, though depth itself is a multivariate parameter encompassing variation in temperature, pressure, solar light, and physical oceanography. In the environments adjacent to *L. pertusa* reefs, several additional species not observed during submersible dives (either HOV or ROV) contribute to the overall number of taxa sampled in proximity to the reefs (Brooks et al. 2015; Ross et al. 2017b), which is comparable to the species richness documented in the southeastern U.S. reef environments (Ross and Quattrini 2007). Based on submersible surveys reviewed here, there is an apparent association between complex reef habitats and certain fish species, including lophiiform fishes. Specific obligate associates of reef environments include the fishes *Conger oceanicus, Hoplostethus occidentalis, Dysommina rugosa, Nezumia sclerorhynchus, Physiculus kerrerae, Benthocometes robustus, Beryx* spp., *Idiastion kyphos*, and *Anthias woodsi*. While difficult to collect, cryptic reef associates have been observed within the coral matrix, including the new brotula species, *Bellotia robusta* (Nielsen et al. 2009). Notably, trawl and trap samples did not collect some fishes commonly observed near the reefs: *Nezumia sclerorhynchus, Lophoides* spp., *Beryx* spp., and *Benthocometes robustus*, which may indicate that these species do not venture far from the reef. Along the west coast of Florida, the extensive lithotherms serve as habitats for a variety of fish species, including anthiine groupers, *Chlorophthalmus agassizi* (shortnose greeneye), black belly rosefish (*H. dactylopterus*), codlings (*Laemonema* spp.), beardfish (*Polymixia* spp.), and hake (*Urophycis* spp.) Reed et al. 2006), which is similar in species composition to those found within the northern GoM reef environments.

Fig. 3.4 Fishes commonly associated with cold-water corals in the Gulf of Mexico: (**a**) thorny tinselfish, *Grammicolepus brachiusculus*; (**b**) blackbelly rosefish, *Helicolenus dactylopterus*; (**c**) queen snapper, *Etelis oculatus*; (**d**) roughtongue bass, *Pronotogrammus martinicensis*; (**e**) big roughy, *Gephyroberyx darwinii*; (**f**) swallowtail bass, *Anthias woodsi*. Image credits: (**a**) NOAA Southeast Deep Coral Initiative and Pelagic Research Services; (**b**) NOAA Office of Exploration and Research (OER), Deep Connections 2019; (**c, d**) NOAA OER, Exploring Deep-sea Habitats off Puerto Rico and the U.S. Virgin Islands; (**e**) NOAA OER, Gulf of Mexico 2017; (**f**) NOAA OER, Windows to the Deep 2018

Generally, cruising predators are abundant over reefs located between depths of 300–450 m but are nearly absent at reefs below 500 m. For example, Benfield et al. (2019) listed barrelfish, amberjack, big roughy, snowy grouper, misty grouper (*Hyporthodus mystacinus*), and American conger (*C. oceanicus*) as dominant cruising predators near the base of a *Lophelia*-covered oil rig at 305 m depth in the northern Gulf. Reed et al. (2006) listed 58 species of fishes from the Gulf, Straits of Florida, and off southeastern Florida. In addition to species already listed, the red bream (*Beryx decadactylus*) and hagfish (*Myxine glutinosa*) were commonly observed. Reed et al. (2014) listed 62 taxa of fishes along the Pourtalès Terrace reef complex (Florida Keys), with several taxa showing tight association with coral. Weaver et al. (2006a, b) listed 50 species of fishes at the Miller's Ledge (Florida Keys) deep-reef habitat. Ross et al. (2015) listed 18 species of fishes associated with *Lophelia pertusa* colonies off northeastern Florida. Fish assemblages between 150 and 350 m were distinct from those occurring below 450 m. Harter et al. (2009) listed 62 species of fishes associated with the Oculina Bank off eastern Florida. No fishes were exclusive to coral, but grouper densities were higher at sites with more and higher stands of *Oculina*. Ross and Quattrini (2007) listed 99 taxa of fishes associated with deep reefs off the southeastern U.S. coast (Florida to North Carolina). Close reef associations were reported for swallowtail bass, red bream, Atlantic conger and a cutthroat eel (*D. rugosa*). Taxa reported were similar to those reported from the Gulf, with the additional observations of mesopelagic fish (e.g., *Diaphus dumerilii, Maurolicus weitzmani, Polyipnus clarus*) aggregations near coral habitat, highlighting benthopelagic coupling at the deep reef/oceanic interface. In these studies, records were derived using ROV and HOV observation. It is likely that the relative abundances of fishes listed may be skewed toward more conspicuous fishes such as hovering zooplanktivores and large cruising predators. Cryptic, epibenthic fishes such as gobies, blennies, small scorpaenoid fishes, and eels are likely underrepresented in all surveys (discussed in Sulak et al., 2007). Finally, the invasion (circa 2010) of lionfish (*Pterois volitans/miles*), known in the Gulf to depths over 300 m, will likely cause faunal shifts in deep-coral communities as lionfish populations expand (Reed et al. 2019).

Within the Campeche bank coral province, there appeared to be differences in fish assemblages based on the availability of live coral, much like the invertebrate communities (Hebbeln et al. 2014). Fishes associated with the exposed coral framework, including *H. dactylopterus* and *Nettenchelys exoria*, differed from those observed on the rubble near the mounds, *Chaunax suttkusi*, *Chlorophthalmus agassizi*, *Laemonema* spp., *Nezumia* spp., Phycidae and Rajidae.

In addition to providing habitat for a unique fish assemblage structure, there is evidence that cold-water coral habitat in the Gulf provides nursery habitat for some species. Etnoyer and Warrenchuk (2007) reported high densities of catshark (Scyliorhinidae) egg cases attached to primnoid octocorals (*Callogorgia a. delta*) between 300 and 350 m depth in the Viosca Knoll region. The authors hypothesized that the elevated location above the seafloor increases survivorship of shark embryos via enhanced water circulation and reduced predation. Weaver et al. (2006a, b) reported bicolor-phase Warsaw grouper at Miller's Ledge, indicating spawning readiness. Baumberger et al. (2010) reported a large spawning aggregation of beardfish (*Polymixia* sp.) in a sinkhole on the Pourtalès Terrace. In the Mediterranean, deep-coral habitat (*Lophelia-Madrepora* mounds, 350–1100 m depth) is a reported spawning area for blackbelly rosefish (D'Ongia et al. 2010), a numerically dominant species in the Gulf, as well as nursery habitat for that and four other species. Given the under-sampling/under-observation of deep-reef fishes in the Gulf, it is reasonable to assume that new species, new records, and essential linkages between cold-water coral habitat and fish life histories will be discovered as observation is expanded (e.g., Quattrini et al. 2004).

3.6.2 Trophic Relationships

While CWC have been shown to primarily utilize surface-derived primary production as a food source (see Sect. 3.5.1), food resources for animals associated with deep-sea coral habitats are less known. In addition to surface-derived primary production, animals residing in corals near seeps could also utilize seep-derived organic carbon. Through the application of stable isotope analysis, it was determined that fauna associated with *L. pertusa* had $\delta^{13}C$ values ranging from –25 to –13‰, $\delta^{15}N$ from 6 to 14‰, and $\delta^{34}S$ from 10 to 25‰, consistent with reliance on surface POM as a primary energy source (Becker et al. 2009; Brooks et al. 2015). There were some notable exceptions, including the gastropod *Provanna sculpta* ($\delta^{13}C$: –30.5‰, $\delta^{15}N$: 3.9‰, $\delta^{34}S$: –5.4‰), the cup coral *Caryophyllia berteriana* ($\delta^{13}C$: –27‰, $\delta^{15}N$: 0.76‰, $\delta^{34}S$: 27‰), one *Odontozona* specimen (–27.6‰ for $\delta^{34}S$), polychaete worms (Hesionidae; $\delta^{13}C$: –28.5‰), and hexactinellid "lollipop sponges" ($\delta^{13}C$: –27.6 to –26.1‰). *Provanna sculpta* may be utilizing seep-derived primary production, either by feeding on free-living microbes or they might contain symbionts (Becker et al. 2009). Because only a few specimens ($n < 10$) had isotope results that suggest seep-derived energy, more collections would be necessary to quantify how widespread this food source is within deep-sea coral habitats in proximity to seeps.

Lophelia pertusa-associated communities encompassed multiple trophic levels. Several taxa represented suspension feeders, including hydroids and *Euratella* polychaetes, with isotope values reflecting POM and possibly zooplankton as possible food sources. Animals with the highest $\delta^{15}N$ values represented several top predators (e.g., *Coralliophila* sp., *Munidopsis sp.*, *O. edwardsi*, *Eunice* sp., *Periclimenes* sp., *Chloeia* sp., *Bathynectes longispina*, and unidentified galatheid) consistent with occupying high trophic levels (Becker et al. 2009; Brooks et al. 2015). For example, the gastropod *Coralliophila*, a putative coral predator (Becker et al. 2009), had isotope values consistent with consuming *L. pertusa*. Becker et al. (2009) suggest that *Munidopsis* sp. and *Eunice* may select hydroids and *Euratella* polychaetes for food. However, none of the other mobile predators exhibited a clear feeding relationship with coral associates, which indicates that either the study did not sufficiently capture the food resources for these taxa or they move off reef to feed and their reliance on *L. pertusa* is primarily as a refuge and habitat.

Commensal ophiuroids were examined to evaluate their trophic reliance on their host octocorals. For ophiuroids associated with *Paramuricea*, stomach contents were devoid of coral polyps, suggesting no direct reliance on corals for food. Stable isotope analysis revealed a significant linear relationship between corals and commensal ophiuroid $\delta^{13}C$

values. Likewise, similar $\delta^{15}N$ values between the host corals and associated ophiuroids suggest that the corals are providing access to food sources to the ophiuroids that are similar to what is consumed and assimilated by the corals (Brooks et al. 2015).

Diet analysis (based on examination of gut contents) of deep-sea fishes found associated with corals and surrounding environments has been conducted on several species (Ross et al. 2017b). Thirty species of demersal fishes, encompassing 926 specimens, were collected by trawl, baited traps, or submersibles. Overall, various crustaceans dominated the gut content assemblages for several groups, including the Macrouridae, Moridae, Percophidae, Chaunacidae, Peristeidiidae, Phycidae, and Epigonidae. Other fishes, including those tightly associated with the reefs, had a variety of food sources. For black bellied rosefish (*H. dactylopterus*), tunicates were a dominant food item in stomach contents. Crustaceans also dominated the diets of *H. occidentalis* and *L. goodebeanorum*. However, diets for several fishes closely associated with deep-reef environments in the GoM were not examined, including *Beryx* spp., *C. oceanicus*, *H. perciformis*, and *G. brachiusculus* due to insufficient sample sizes, empty stomachs, or inability to capture these species.

3.6.3 Interactions with Surrounding Ecosystems

Cold-water corals are ecosystem engineers, creating three-dimensional structure for a variety of organisms, but also modifying the biotic and abiotic conditions in the surrounding seafloor environments. Infaunal communities adjacent to CWC differ from those found in soft sediments (Demopoulos et al. 2014, 2016; Fisher et al. 2014a, b; Bourque and Demopoulos 2018). Within *L. pertusa* sediments, 86 discrete taxa were identified from near coral and background cores (away from the reef), with 16 taxa occurring near coral habitats, 14 taxa unique to background sediments, and several taxa overlapping the two habitats. Diversity was higher near corals at one site, but taxon richness was similar across habitats. Building on this work, Bourque and Demopoulos (2018) examined coral-associated infaunal communities and their functional traits near multiple coral species (scleractinians [*L. pertusa, Madrepora oculata*] and octocorals [*Chrysogorgia*, *Paramuricea*, *Callogorgia*]), at depths ranging from 263 to 1095 m. While infaunal densities were highest at *L. pertusa* habitats, diversity and evenness were highest at the octocoral sites. Infaunal community composition also differed among coral habitats, with *L. pertusa* infauna most distinct from *M. oculata* and octocoral communities in terms of composition and functional traits. Specific environmental drivers that helped explain the observed patterns included mud content, depth, and organic carbon, whereas geographic location was the primary driver of functional traits. Octocoral communities had distinct functional traits from *L. pertusa* and *M. oculata* habitats, due to lower abundances of motile burrowers and subsurface-deposit feeders within octocoral communities. *Lophelia pertusa* had higher densities of discretely motile tube dwellers, and surface and suspension feeders compared to *M. oculata* habitats.

For background soft sediments in the northern GoM, density declines with depth, but diversity exhibits a mid-depth (1100–1300 m) maximum (Rowe and Kennicutt 2009). In contrast, diversity and density in sediments adjacent to CWCs were decoupled from these patterns observed in soft sediments; thus, corals are structuring adjacent sediment communities in unique ways. Patch size may influence coral sediment communities (Demopoulos et al. 2014; Bourque and Demopoulos 2018), likely a function of the local hydrodynamic regime. Deep-sea coral habitats examined represented a range in sizes and associated complexities, with *L. pertusa* representing the largest habitats, followed by *M. oculata* the intermediate sizes, and octocorals having the smallest habitats. Significant taxa turnover with distance from the coral environment was observed, indicating that the reef extent, and associated local environmental conditions, may be important in structuring adjacent sediment communities. Thus, the sphere of influence of CWC extends beyond the corals themselves.

3.7 Threats to Cold-Water Corals in the Gulf of Mexico

While the deep sea may appear remote and isolated from anthropogenic stress, CWC are not immune to disturbances. Specific threats include fisheries, oil and gas exploration and extraction, seafloor mining, and climate change (e.g., ocean acidification and warming). Given slow growth rates, patchy distribution, and low recruitment rates, CWC are highly sensitive to disturbance (Prouty et al. 2014; Doughty et al. 2014). While fishery regulations and protected areas can prevent fishing damage, they are unable to protect corals from ocean acidification, warming temperatures, or oil spills. Understanding cold-water coral tolerance and resilience to these large-scale stressors will improve our understanding of their sensitivity to environmental change and further anthropogenic threats.

3.7.1 Fishing

Bottom-contact fishing activities threaten CWC globally, through direct impacts from bottom-tending trawls, longlines and traps, and indirect damage from suspension of sediments that can smother the feeding apparatus of corals and other suspension feeders (Clark and Koslow 2007; Brooks et al. 2015; Rooper et al. 2017; Victorero et al. 2018). In the GoM, bottom longlines and bottom trawls have impacted some shelf-depth coral habitats (Boland et al. 2017; Etnoyer et al. 2016), but there are currently few fishing threats to CWCs below 200 m. Deep bottom trawling is limited to a small royal red shrimp (*Pleoticus robustus*) fishery (400–500 m depth) that occurs off Texas, Florida, and Alabama. Damage by this fishery to coral habitats is unknown, given the shrimp live in soft sediment environments, and the heavy shrimp trawls are typically not deployed in rugged reef environments given the high risk of losing gear. However, if the fishing is close to coral habitats (for example, royal red shrimp fishing occurs near VK862/906), there is potential for accidental damage. This is especially likely in high current environments, where sediments disturbed by trawls can be transported great distances before settling out, possibly smothering far-off corals.

There are no bottom longline fisheries that target slope depths (>200 m), but fishing line has been observed ensnared on octocoral colonies at ~1000 m depth at multiple sites in the northern GoM (Fisher et al. 2014a, b). This was likely discarded pelagic longline, given the depth and location of the observation. Potential target fish taxa that are found near CWC in the GoM include blackbelly rosefish (*H. dactylopterus*), alfonsinos (*Beryx* spp.), roughies (*Gephyroberyx* and *Hoplostethus*), wreckfish (*Polyprion americanus*), barrelfish (*Hyperoglyphe perciformis*), and deeper snapper and grouper (e.g., *Epinephelus niveatus, E. nigritus*). Blackbelly rosefish are currently caught commercially in the eastern Gulf of Mexico, in low quantities (1547 kg in 2019; National Marine Fisheries Service), but recreational fishing for this species has fluctuated from 43 kg in 2000 to a maximum of 92,600 kg in 2015 (all Florida landings). While the preferred deepwater targets of recreational fishers are typically snapper and grouper, blackbelly rosefish are targeted when these other species are unavailable. Roughies and alfonsinos occur near deep-coral habitats in the GoM and are intensively harvested elsewhere (Branch 2001; Victorero et al. 2018) using midwater and bottom trawls. However, given their relatively low numbers in the GoM, they are unlikely to support a commercial fishery.

Golden crabs (*Chaceon fenneri*) are commonly observed in and near to cold-water coral reefs, which they appear to use for feeding and shelter. They are fished commercially using rectangular wire traps that are tied together in lines (~1 km long), often deployed in the vicinity of coral habitat, where they may remain for several days. Traps are recovered using a grappling hook that catches the line and the traps are hauled to the surface. Golden crabs were historically fished in the eastern GoM in the 1980s (Crosson et al. 2013), but because of several quality control and market problems (which have now been solved), the fishery did not develop. Currently, golden crabs are not a managed fishery in the GoM, but in November 2019 the GMFMC approved a 2-year permit to conduct an experimental fishery to determine viability of a golden crab fishery off west Florida. In the southeastern U.S., the golden crab annual catch limit (2 million pounds) is significantly higher than the average annual catch (< 800,000 pounds; NMFS), and there is interest by fishers to expand their allowable fishing areas within the Coral Habitat Areas of Particular Concern (HAPC) under the South Atlantic Fishery Management Council Golden Crab Amendment 10 (in progress). This fishery operates near coral habitats, in areas of rugged topography and strong currents, using strings of weighted traps, so proposed expansions are cause for concern regarding potential damage to fragile coral ecosystems.

3.7.2 Energy Industry

The northern GoM contains the largest offshore oil and gas operation in the USA, with several thousand active platforms and approximately 50% of the U.S. petroleum refining and natural gas processing occurring along the Gulf coast. Oil and gas production from deep wells (>300 m) has surpassed shallow production (Murawski et al. 2020), and drilling is moving into increasingly deeper water in search of new resources.

Fossil fuel operations may impact CWC and adjacent environments during all phases from exploration to platform decommissioning. These include infrastructure installation (anchor chains, pipelines), release and deposition of drill fluids and cuttings, and sediment resuspension (CSA 2006). Deployment of pipelines and cables have the potential to damage benthic communities as they drag along the seafloor (Cordes et al. 2016a). A large trough surrounded by damaged *L. pertusa* colonies was observed at the VK826 coral habitat (Schroeder 2002) and was possibly caused by a wire anchor cable which was deployed during energy operations (Schroeder 2002, 2007).

Drill cuttings may also have a toxic effect on sensitive species, but these are challenging to separate in situ from the effect of sediment alone. Experiments on the effects of sediment and drill cuttings, and observations of cold-water coral reefs in close proximity to drilling operations, indicate that *L. pertusa* can tolerate a moderate level of exposure to these conditions (Lepland and Mortensen 2008; Brooke et al. 2009; Purser and Thomsen 2012; Larsson et al. 2013); however, effects on other CWC species are less understood. Larvae of *L. pertusa* were exposed to a range of concentrations of drill cuttings and drill fluid components (barite and bentonite) for different time periods (Jarnegren et al. 2017, 2020). Mortality occurred after larval cilia became clogged with material, and responses were influenced by larval age and exposure material. Bentonite had a greater effect than the other materials, as the fine particles clogged the larval cilia at lower concentrations than the other materials.

In their review of stressors to CWC in the GoM, Brooke and Schroeder (2007) identified oil and gas development as a medium level threat, the highest relative level for human activities in the GoM. This assessment was based not only on impacts from routine operations, but also the risk of widespread damage in the event of a large oil spill. An updated report in 2017 (Hourigan et al. 2017) maintained the "medium" threat level for the energy industry. With the increase in energy demand, there is a corresponding increased risk for oil spills. There are an estimated 1–3 spills per week within the U.S. Exclusive Economic Zone (EEZ, NOAA Office of Response and Restoration), although most are small and nearshore. However, the GoM was the site of the two largest marine oil spills in history, the DWH in 2010 and Ixtoc I spill in 1979–1980. DWH risk frequency was estimated 8–91 years (once every 17 years, Eckle et al. 2012). See Case Study for additional information on the DWH oil spill.

3.7.3 Climate Change

The deep sea is not isolated from the effects of climate change (e.g., increased temperature, CO_2, decreased oxygen concentration and pH), and warming has already been recorded in the deep ocean. While mobile benthos can move to more hospitable environments including deeper, cooler depths, sessile species such as corals are limited by available substrate and specific environmental thresholds. CWC are sensitive to warming conditions, given their upper thermal limits. Clues into how corals will respond to future change can be gleaned from retrospective studies using piston cores collected within framework-forming scleractinians. Analysis from these cores provide clarity on the conditions under which the corals formed and the range of environmental conditions under which they thrive (Hebbeln et al. 2019; Roberts and Kohl 2018). For example, past warming conditions in bottom water environments may have limited *L. pertusa* development (growth and mound accretion), while current conditions in the GoM are hospitable for coral growth.

Ocean acidification is a consequence of increased atmospheric CO_2 being absorbed by the ocean, reducing seawater pH (Le Quéré et al. 2010). Seawater carbonate saturation state is a function of temperature and is lower in cold, deep, waters. Thus, CWC reside closer to the saturation horizon (the depth below which carbonate minerals dissolve) than shallow-water corals (Cordes et al. 2016b). The saturation horizon will continue to shoal (become shallower) as ocean acidification progresses, resulting in more CWC being exposed to under-saturated carbonate conditions (Guinotte et al. 2006). Potential impacts to CWC include weakening of the skeletal structure, increased energy demand for constructing the skeleton, and dissolution of dead coral framework. This may lead to decrease in available habitat for coral-associated fauna, but much remains to be learned regarding how coral ecosystems will respond to future conditions.

Studies on the effects of ocean acidification (OA) on CWC in the GoM have centered on Viosca Knoll (VK), where *Lophelia pertusa* is currently close to its environmental thresholds for temperature (8.6–13.6 °C), dissolved oxygen, and carbonate parameters (Georgian et al. 2016a). Some *Lophelia* reefs in the GoM occur close to the aragonite saturation horizon, with minimum observed saturation state of 1.25 (Lunden et al. 2013). Differential response at the colony level indicated that certain genotypes may be more resilient to OA. These coral mounds also have elevated total alkalinity (TA) and dissolved inorganic carbon (DIC), possibly due to dissolution and remineralization of dead coral understory. This may ameliorate the low aragonite saturation states, but dissolution of dead coral skeleton undermines the foundation of cold-water coral reefs (Hennige et al. 2015, 2020). As OA increases, it is unclear whether dissolution of dead coral skeleton will be faster than the rate of accretion from coral growth (Georgian et al. 2016a, b). Acidified waters are reducing Ω_{arag} in other parts of the world's oceans (e.g., between 0.0021 and 00.48 units per year in the North Atlantic; Georgian et al. 2016a); similar declines in the GoM would lead to under-saturation of *L. pertusa* reefs over just a few decades. Increased spatial and temporal resolution of oceanographic data would clarify carbonate conditions within the GoM and help constrain deep-sea coral responses to changing conditions (Georgian et al. 2016a, b).

3.8 Conservation and Management

3.8.1 Management of Fishing Impacts

Federal fishery management authority resides with the Secretary of Commerce, NOAA Fisheries, and eight regional fishery management councils. The Magnuson-Stevens Fishery Conservation and Management Act (MSA) is the primary law governing marine fisheries management in federal waters of the USA. The MSA contain several mechanisms that can be employed to protect deep-coral ecosystems: designation as essential fish habitat (EFH; MSA 303[a][7]), amendments to minimize bycatch (MSA 301[a][9]), amendments to fishery management plans (FMP) and through a discretionary provision of the MSA (303[b][2][B]) enacted during the 2006

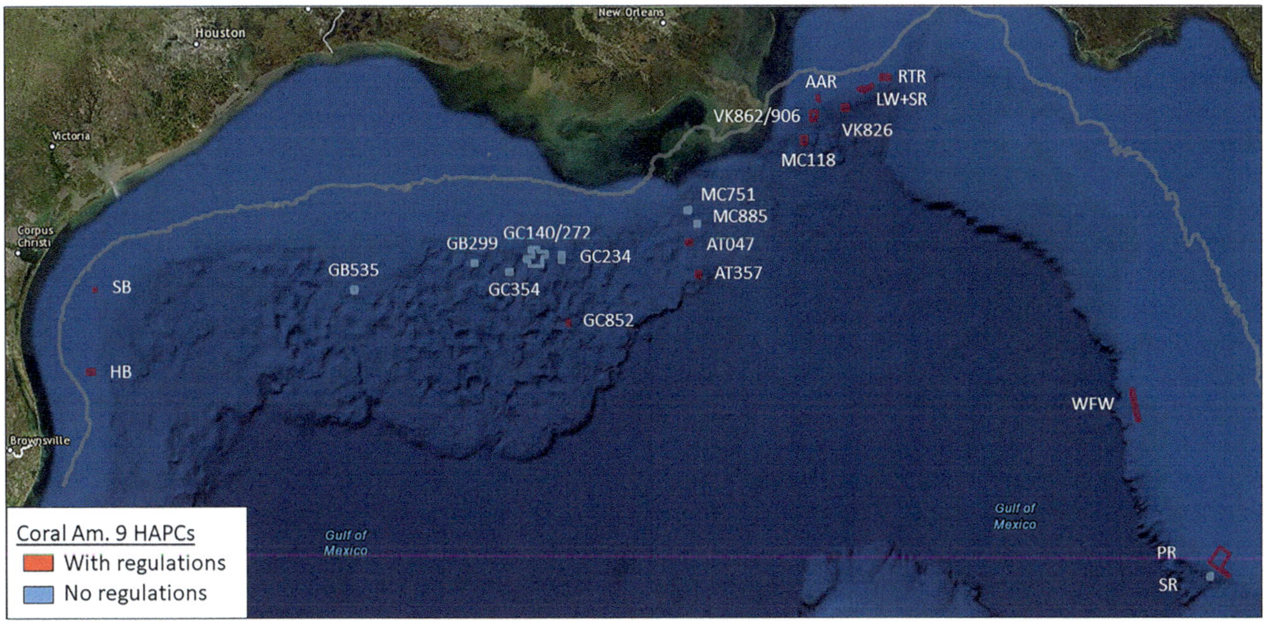

Fig. 3.5 Map showing locations of the HAPCs established under Coral Amendment 9 to the Coral and Coral Reefs FMP. HAPCs with regulations: Pulley Ridge (PR), West Florida Wall (WFW), Roughtongue Reef (RTR), L+W pinnacles and Scamp Reef (LW+SR), Viosca Knoll 826 (VK826), Alabama Alps Reef (AAR), Viosca Knoll (VK826, VK862/906), Mississippi Canyon (MC118), Atwater Valley (AT, 047, AT357), Green Canyon (GC852), Southern Bank (SB), Harte Bank (HB). HAPCs without regulations: South Reed (SR), MC751, MC885, GC234, GC140/272, GC354, Garden Banks (GB299, GB535)

MSA reauthorization, which also established the deep-sea coral research and technology program (DSCRTP).

The Gulf of Mexico Fishery Management Council (GMFMC) is responsible for fisheries management in Federal waters of the GoM. Existing protections for cold-water corals in the GoM include the prohibition on harvest of black corals and stony corals, unless authorized for scientific research, exempted fishing permit activity or educational activity. Octocorals are not part of the fishery management unit (FMU) because they are harvested in state waters off the coast of Florida and are managed by the Florida Fish and Wildlife Conservation Commission (GMFMC 2011). While these regulations were initially put in place for shallow reefs, they also apply to deeper coral communities. Spatial protections for a number of shallow and shelf edge habitats was implemented through the Flower Garden Banks National Marine Sanctuary (1992), the Madison Swanson and Steamboat Lumps Marine Reserves (2000), and thirteen Habitat Areas of Particular Concern, including the Florida Middle Grounds (1982) and Pulley Ridge (2006). In 2009, the South Atlantic Fishery Management Council (SAFMC) implemented five Coral HAPC areas, encompassing ~65,000 km^2 of known cold-water coral habitat in the southeast region, and other FMCs followed with extensive spatial protections established for deep-coral habitats (Hourigan et al. 2017). In 2014, the GMFMC convened a Special Coral Working Group that identified 47 cold-water coral areas of varying size but all in >50 m water depths that they felt qualified as potential Habitat Areas of Particular Concern. After an extensive review and input process, a subset of 21 of these recommended areas were formally proposed by the GMFMC to the Department of Commerce, for HAPC designation. The Final Rule of *Amendment 9 to the FMP for Coral and Coral Reef Resources in the GOM* (Federal register 85-201, 50 CFR 622, 635) designated 21 HAPC sites (Fig. 3.5), totaling 1254 km^2, effective November 16, 2020. Thirteen of these HAPCs have fishing gear restrictions that prohibit bottom trawls, bottom longlines, traps, and anchors. The expanded Pulley Ridge area is exempted for bottom longliners that have historically fished the area, and royal red shrimp trawlers are allowed to enter the boundaries of the VK862/906, but only with nets kept off bottom. These exemptions were granted because of undue hardships to the fishers. Hook and line fishing is still permissible because of perceived limited impact to the benthos. Eight of the new sites do not have any fishing restrictions, and the HAPC status does not protect any of the sites from non-fishing activities. However, GMFMC may recommend additional restrictions (avoidance, mitigation) or offset negative effects of activities that are under the jurisdiction of federal or state agencies at these sites, including oil and gas exploration and extraction. Prevailing marine protected area (MPA) theory argues that MPAs, such as HAPCs, help increase ecosystem resilience to impacts that are not managed by boundaries (e.g., oil spills, pollution, climate change), and expanding and implementing protection prior to fishery investments is an effective way to conserve these environments, while minimizing economic impact.

3.8.2 Management of Energy Industry Activities

Seabed energy and mineral resources on the Outer Continental Shelf (OCS) are managed by BOEM. Through its Environmental Studies Program, BOEM develops, funds, and manages needed scientific research to guide OCS resource development decisions. There are dozens of regulations, executive orders, and policies that guide BOEM's stewardship decisions. These include the Outer Continental Shelf Lands Act (OCSLA-43 U.S.C. 1301-1356), which requires BOEM to examine and consider marine (as well as coastal and human) impacts when determining effective steps to promoting energy independence, environmental protection, and economic development. Other policies include the National Environmental Policy Act (NEPA), and as stated above, highly relevant to CWC are the Magnuson-Stevens Act and National Historic Preservation Act (16 U.S.C. 1531), since CWC have been shown to colonize shipwrecks (Brooks et al. 2015). BOEM makes use of the best available scientific information about natural resources and impact producing factors in order to perform environmental assessments and publish NEPA-required environmental impact statements to evaluate potential impacts of BOEM's proposed actions. Assessments typically include compilation of regional baseline data, documentation of environmental sensitivities, and recommendations to policy makers about where future energy exploration could safely occur or should be avoided. BOEM consults with state (Coastal Zone Management Act Consistency Determinations) and federal (e.g., Essential Fish Habitat consultations with NOAA) agencies that are also tasked with protecting resources under their jurisdiction.

Within the Gulf of Mexico OCS Region's Office of the Environment, the Biological Sciences Unit is responsible for evaluating specific sensitive biological resources, including deepwater benthic communities such as seeps, CWC, and soft bottom environments. BOEM is responsible for identifying sensitive habitats and avoiding or minimizing harm from energy industry activities. BOEM's notices to lessees (NTL) define information requirements needed to delineate biologically sensitive habitats and restrict certain activities in their vicinity. For deep-sea environments greater than 300 m water depth, separation distances are specified between bottom disturbing activities and known communities. Drill muds and cuttings discharge points must be sited at least 2000 ft (609.6 m) from the sensitive communities and at least 250 ft (76.2 m) for all other proposed seafloor disturbances (including those caused by anchors, anchor chains, wire ropes, seafloor template installation, and pipeline construction). New recommendations proposed by Cordes et al. (2016a) suggest increasing the current setback distance to 2000 m for muds and cuttings discharge points and 200 m for all other seafloor disturbing activities.

In their NTLs, BOEM provides specific guidance to operators about site clearance and potential mitigation measures in deepwaters. For deepwater (>300 m) benthic communities. NTL-2009-G40 stipulates that oil and gas industry operators submit maps showing bathymetry, geological features, and potential sensitive biological areas that could be impacted by the proposed activities. Areas with potential habitats for deep-sea coral and chemosynthetic organisms habitat (e.g., methane seeps) have been identified (https://www.boem.gov/oil-gas-energy/mapping-and-data/map-gallery/seismic-water-bottom-anomalies-map-gallery) based on reviews of seismic survey data. If a well is drilled near an identified hard substrate habitat considered to be potential sensitive habitat, then visual surveys would be required prior to start of any seafloor disturbing activity, such as infrastructure emplacement or drilling. If the visual surveys show the presence of chemosynthetic or coral communities, the operator is required to report their occurrence and submit copies of the imagery and maps for BOEM subject matter expert (SME) review. Special conditions of approval requiring benthic community impact avoidance or mitigation measures can be issued along with the activity permits during the assessment process.

3.8.2.1 Lessons Learned from the Deepwater Horizon Oil Spill

Despite the wealth of information from previous studies in the GoM (Rowe and Menzel, 1971; Cordes et al. 2006, 2008; Rowe and Kennicutt 2009; Demopoulos et al. 2014; Quattrini et al. 2014), essential baseline data were lacking in proximity to the spill location and major components of the ecosystem (microbial communities and processes). The impact to injured coral colonies could have changed the demography of affected species (see Cordona et al. 2016). For example, tissue loss may have led to declined reproductive capacity and larval supply, limiting the recovery of injured corals. Given the limited baseline information on distribution of CWC, it is possible that the overall footprint of impact was underestimated (Georgian et al. 2019) and is a function of the broader connectivity among populations, which is poorly known.

Following DWH, major changes to oversight and regulation of the oil and gas industry were implemented. The Minerals Management Service was divided into BOEM and

BSEE to separate the regulatory function from the leasing arm. New policies were enacted that have implications for deep-sea communities, including corals. The updated NTL-2015-N01 requires operators to demonstrate preparation for blowout and worst-case scenario discharge. Potential future actions include expanding the no activity zones in proximity to densely spaced habitat features adjacent to existing banks, which include potentially sensitive biological features.

Given the spatial extent of active leasing in the GoM, it is impractical to establish buffer zones around all mitigation areas that potentially contain high density communities, because that would necessitate excluding 98% of the active lease blocks. One alternative approach would be to protect representative areas of sufficient size, while allowing for industrial activity to continue in other areas (Cordes et al. 2016a). Additional changes could include establishing the setback distance for infrastructure with no planned discharges to at least 200 m from sensitive communities and increasing that distance to 2 km when discharge is planned (Cordes et al. 2016a).

3.9 Future Directions

Energy interests in the northern GoM have facilitated extensive research into deep-sea communities, including CWC. Recent efforts have also increased our understanding of other regions, particularly the West Florida Slope, but there remain some large unmapped and/or virtually unexplored areas that are known or potential cold-water coral habitats. These include the Florida Escarpment, the Mexican Ridge and Bay of Campeche, and the Yucatan platform and Escarpment. Seafloor bathymetry is available for most of the U.S. GoM (NOAA National Centers for Environmental Information, and BOEM); while this is a significant resource, coverage is incomplete (especially in Mexican waters), and some data are low resolution. The importance of seafloor mapping has been highlighted by the Seabed 2030 project (https://seabed2030.org) which was initiated in 2016, with the ambitious goal of having 100% of the ocean mapped by 2030. Within the USA, EEZ, expanding high-resolution bathymetric data may lead to the discovery of new seafloor features and their associated communities, identify marine hazards (such shelf slumping), and improve our understanding of ocean circulation.

The southern regions of the GoM may support a different (more tropical) fauna than those further north, but it is unclear if, or where, biogeographic breaks exist and how diversity changes with latitude, region, or depth in the deep GoM. Exploration of cold-water coral habitats in Mexican waters is limited to a few ROV dives on the Campeche Bank mounds; these revealed the most extensive coral mound province in the GoM. Further explorations of these and other features will undoubtedly significantly increase coral distribution information. Limited explorations into deeper regions (>2000 m) of the GoM have revealed very different species assemblages from those found at slope depths (Brooks et al. 2015; Boland et al. 2017). There are vast areas of potential cold-water coral habitat along deep escarpments of the Florida and Yucatan Peninsulas, and we know virtually nothing about the biology or ecology of their associated communities. Additional information on cold-water coral distribution would enhance our understanding of the larger system and expand the application of management decision support tools, such as predictive habitat models, beyond U.S. waters.

It is also important to understand the effects of stressors on cold-water coral ecosystems, to prevent (if possible) or mitigate for anthropogenic threats. A significant number of studies have focused on understanding the effects of natural and anthropogenic stressors on *L. pertusa*, which is one of the most widespread reef-building CWC species, and is tolerant of a wide range of abiotic conditions (see Sect. 3.5.3). This species is also easy to manipulate and maintain in aquaria, so lends itself well to experimental research. Very little work of this kind has been performed on other CWC species. Best practices in resource management accommodate the less resilient species, not just the most tolerant, so this lack of data on other important species is a significant information gap.

Regions of the GoM are linked by the powerful Loop Current, which passes through unexplored regions off Mexico, before moving into the northern and eastern GoM. Currents facilitate population connectivity through larval dispersal, where downstream populations may be dependent on the health of those residing upstream. Currents also carry pollutants such as oil and may be conduits of other unfavorable conditions such as thermal stress. Nature does not recognize political boundaries, and the U.S. and Mexican regions do not operate in isolation from each other. Therefore, it is important to understand how regions of the GoM Large Marine Ecosystem interact with each other in order to effectively manage CWC environments within the broader region.

3.10 Case Study: Impacts of the Deepwater Horizon Oil Spill on Cold-Water Corals

In April 2010, the *Deepwater Horizon* (DWH) oil drilling rig wellhead exploded, leading to the release of 4.1 million barrels of oil at 1500 m depth over a period of 87 days (McNutt et al. 2012). Approximately half of the oil formed a surface slick (Fig. 3.6), while the rest of the gas and oil remained in deepwater (~1100 m) and extended out 50 km

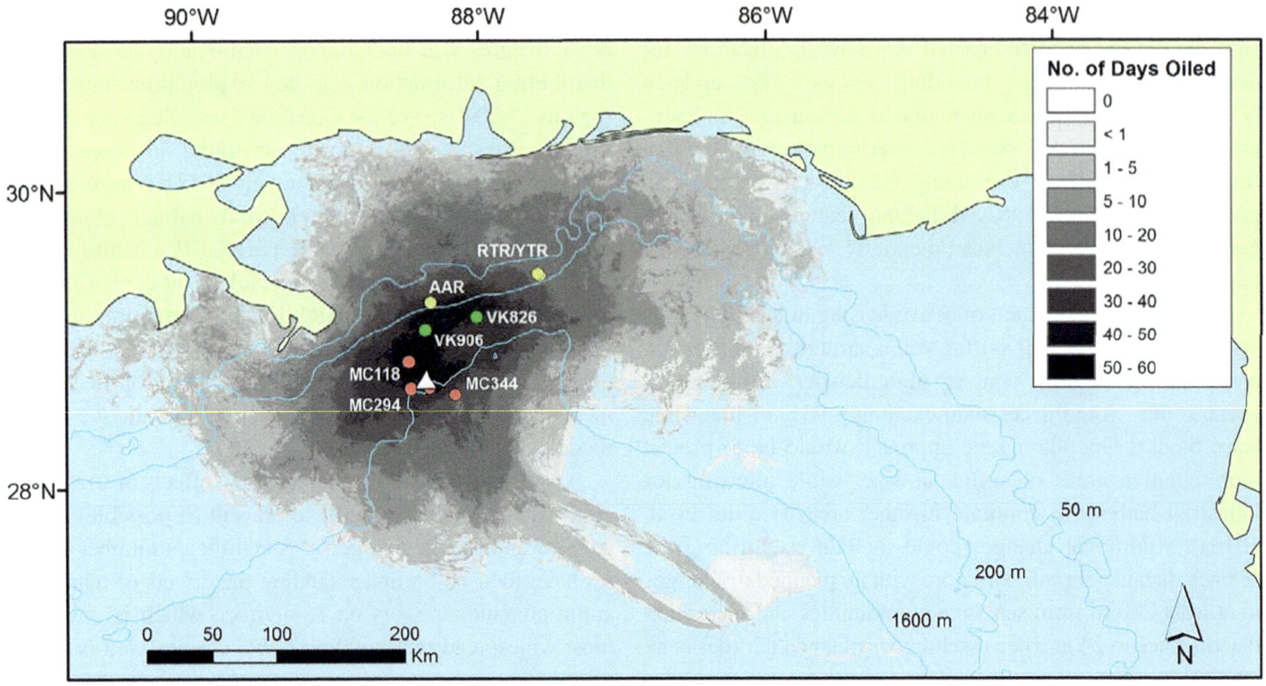

Fig. 3.6 Map showing known cold-water coral habitats relative to the extent and duration of the surface oil slick produced by the Deepwater Horizon blowout at the Macondo wellhead (white triangle). Yellow circles: mesophotic coral habitats at Alabama Alps Reef (AAR); Yellowtail Reef (YTR), Roughtongue Reef (RTR). Green circles: *Lophelia* habitats at VK826, 906. Red circles: octocoral gardens at MC118, 294, 344. Image credit: Hourigan et al. (2017)

from the wellhead (Camilli et al. 2010). To facilitate oil degradation, 0.77 million gallons of chemical dispersant (Corexit 9527 and 9500A) were injected directly into the leaking oil, and 1.4 million gallons were applied to the surface slick (Hemmer et al. 2011; Kujawinski et al. 2011). Oiled marine snow sank and deposited a layer on the seafloor over a large area (~3200 km^2, Chanton et al. 2014; Valentine et al. 2014). The surface oil slick from the DWH spill remained over the shelf edge off Mississippi and Alabama for up to 39 days. Several rocky mound features occur within this area at 70–100 m depth (Fig. 3.6, yellow dots). These sites are colonized by dense and diverse communities of octocorals, which were exposed to the DWH oil (Silva et al. 2016). These features were surveyed in 2011, and four octocoral species (*Swiftia exserta*, *Hypnogorgia pendula*, *Thesea* spp., and *Placogorgia* spp.) showed tissue discoloration and loss, and overgrowth by hydroids (Etnoyer et al. 2016). Elevated polycyclic aromatic hydrocarbons (PAH) were also found in coral tissues and sediments (Silva et al. 2016). As with the deeper species, the injured colonies had further declined by 2014 (Etnoyer et al. 2016)

Visible damage to CWC was documented 11 km from the wellhead in November 2010 (White et al. 2012), where branches of several large octocorals, including *Paramuricea biscaya, Swiftia pallida, and Paragorgia regalis*, were covered in brown flocculent material (Fig. 3.7), which contained DWH oil, components of Corexit (White et al. 2012, 2014), and oil-degrading bacteria (Simister et al. 2016). The damaged corals exhibited tissue loss, excess mucus production, retracted polyps, bare skeleton, enlarged sclerites, and discolored or bleached ophiuroid associates (*Asteroschema clavigerum*). Impacts extended out to 22 km from the well at depths up to 1950 m, which was greater than the depth of the wellhead (Hsing et al. 2013; Fisher et al. 2014a, b). Either the oil-dispersant plume extended deeper than models predicted or oil/dispersant marine snow was deposited from the water column. Coral damage was greater near the wellhead, with less and more patchy damage at sites further away (Fisher et al. 2014a, b), and healthy and injured colonies co-occurring. While some recovery occurred over a period of 17 months, the more severely impacted colonies continued to decline, and hydroids colonized the exposed dead branches (Hsing et al. 2013; Fisher et al. 2014a, b). In laboratory studies, *P. biscaya* exhibited physiological decline and mortality, and a genome-wide stress response when exposed to dispersant and oil treatments (DeLeo et al. 2016, 2018). Sub-lethal effects could influence energy allocation to other physiological functions, including reproduction, which could hinder future recruitment and recovery. Recovery from DWH was estimated to occur on timescales of centuries to millennia given the slow growth and recolonization rates of deepwater corals (Fisher et al. 2014a, b). When considering the gene expression results, sub-lethal effects could persist for decades or longer (DeLeo et al. 2018).

Fig. 3.7 Large colony of *Paramuricea biscaya* with commensal ophiuroids. Healthy coral tissue is yellow, and brown discoloration is from the oil released by the Deepwater Horizon accident. Image credit: Chuck Fisher, Pennsylvania State University

The Natural Resource Damage Assessment and Gulf of Mexico Research Initiative, impact assessments were limited to a few sites that contained deep-sea coral habitats. Recent habitat suitability modeling for the genus *Paramuricea* indicated that the spatial extent of injury may be higher than previously shown (Georgian et al. 2019). Within the 2291 km^2 area affected by the spill, several thousand *Paramuricea* colonies may have been affected. However, these sites require further ground truthing in order to not only confirm the presence of *Paramuricea*, but also assess the magnitude of impact from the DWH spill.

References

Adkins JF, Cheng H, Boyle EA et al (1998) Deep-sea coral evidence for rapid change in ventilation history of the deep north Atlantic 15,400 years ago. Science 280:725–728

Andrews AH, Cordes EE, Mahoney MM et al (2002) Age, growth, and radiometric age validation of a deep-sea, habitat-forming gorgonian (*Primnoa resedaeformis*) from the Gulf of Alaska. Hydrobiologia 471:101–110

Baco AR, Cairns SD (2012) Comparing molecular variation to morphological species designations in the deep-sea coral *Narella* reveals new insights into seamount coral ranges. PLos One 7:e45555

Balsam WL, Beeson JP (2003) Sea-floor sediment distribution in the Gulf of Mexico. Deep Sea Res Pt I 50:1421–1444

Barnich R, Beuck L, Freiwald A (2013) Scale worms (Polychaeta: Aphroditiformia) associated with cold-water corals in the eastern Gulf of Mexico. J Mar Biol Assoc U K 93:2129–2143

Baumberger RE, Brown-Peterson NJ, Reed JK et al (2010) Spawning aggregation of beardfish, *Polymixia lowei*, in a deep-water sinkhole off the Florida Keys. Copeia 1:41–46

Baussant T, Nilsen M, Ravagnan E et al (2017) Physiological responses and lipid storage of the coral *Lophelia pertusa* at varying food density. J Toxicol Env Health A 80:266–284

Becker EL, Cordes EE, Macko SA, Fisher CR (2009) Importance of seep primary production to *Lophelia pertusa* and associated fauna in the Gulf of Mexico. Deep Sea Res Pt I 56:786–800

Benfield MC, Kupchik MJ, Palandro DA et al (2019) Documenting deep-water habitat utilization by fishes and invertebrates associated with *Lophelia pertusa* on a petroleum platform on the outer continental shelf of the Gulf of Mexico using a remotely operated vehicle. Deep Sea Res Pt I 149:103045

Ben-Haim Y, Thompson F, Thompson MC et al (2003) Vibriocoralliilyticus sp. nov, a temperature-dependent pathogen of the coral *Pocilloporadamicornis*. Int J Syst Evol Microbiol 53:309–315

Blum MD, Milliken KT, Pecha MA et al (2017) Detrital-Zircon records of Cenomanian, Peleocene, and Oligocene Gulf of Mexico drainage integration and sediment routing: Implications for scales of basin-floor fans. Geosphere 13:1–37

BOEM. Northern Gulf of Mexico Deep-water Bathymetry Grid from 3D Seismic. https://www.boem.gov/oil-gas-energy/mapping-and-data/map-gallery/boem-northern-gulf-mexico-deepwater-bathymetry-grid-3d

Bohrmann G, Schenck S (2004) RV SONNE cruise report SO174, OTEGA II, Balboa-Corpus Christi - Miami (1 October–12 November 2003), GEOMAR Report 117

Boland GS, Etnoyer PJ, Fisher CR, Hickerson E (2017) State of deep-sea coral and sponge ecosystems in the Gulf of Mexico Region: Texas to the Florida Straits. In: Hourigan TF, Etnoyer PJ, Cairns SD (eds) The state of deep-sea coral and sponge ecosystems of the

United States. NOAA Technical Memorandum NMFS-OHC-4, Silver Spring, MD, pp 321–378
Bonini C, Kinnel RB, Li M et al (1983) Minor and trace sterols in marine invertebrates 38: Isolation, structure elucidation and partial synthesis of papakusterol, a new biosynthetically unusual marine sterol with a cyclopropyl-containing side chain. Tetrahedron Lett 24(3):227–280
Boulay JN, Cortés J, Hellberg M, Baums IB (2014) Unrecognized coral species diversity masks differences in functional ecology. Proc Roy Acad Sci Ser B 281:20131580
Bourque JR, Demopoulos AWJ (2018) The influence of different deep-sea coral habitats on sediment macrofaunal community structure and function. PeerJ Aquat Biol 6:e5276
Bracco A, Liu G, Galaska MP et al (2019) Integrating physical circulation models and genetic approaches to investigate population connectivity in deep-sea corals. J Mar Syst 198:103189
Branch TA (2001) A review of orange roughy Hoplostethus atlanticus fisheries, estimation methods, biology and stock structure. S Afr J Mar Sci 23(1):181–203
Brooke S et al. (2019) HydroSMAC Daily Log. https://marinelab.fsu.edu/archive/hydrosmac/daily-blog/
Brooke S, Jarnegren J (2013) Reproductive periodicity of the deep-water scleractinian coral, *Lophelia pertusa* from the Trondheim Fjord, Norway. Mar Biol 160:139–153
Brooke S, Schroeder WW (2007) Chapter 7: State of deep coral ecosystems in the Gulf of Mexico region: Texas to the Florida Straits. In: Lumsden SE, Hourigan TF, Bruckner AW (eds) The state of deep coral ecosystems of the United States. NOAA Technical Memorandum NOS-CRCP-3, Silver Spring, MD, pp 271–306
Brooke S, Sogluizzo A (2017) Chapter 11: Invertebrate reproductive biology. In: CSA Ocean Sciences (ed) Exploration and research of mid-Atlantic deepwater hard bottom habitats and shipwrecks with emphasis on canyons and coral communities. U.S. Dept. of the Interior, Bureau of Ocean Energy Management, Atlantic OCS Region. OCS Study BOEM 2017-060, pp 537–577
Brooke S, Stone R (2007) Reproduction of deep-water hydrocorals (Family Stylasteridae) from the Aleutian Islands, Alaska. Bull Mar Sci 81(3):519–532
Brooke S, Young CM (2003) Reproductive ecology of a deep-water scleractinian coral, *Oculina varicosa,* from the southeast Florida shelf. Cont Shelf Res 23:847–858
Brooke S, Young CM (2005) Embryogenesis and larval biology of the ahermatypic scleractinian *Oculina varicosa*. Mar Biol 146(4): 665–675
Brooke SD, Young CM (2009) In situ measurement of survival and growth of *Lophelia pertusa* in the northern Gulf of Mexico. Mar Ecol Prog Ser 397:153–161
Brooke SD, Holmes M, Young CM (2009) Sediment tolerance of two different morphotypes of the deep-sea coral *Lophelia pertusa* from the Gulf of Mexico. Mar Ecol Prog Ser 390:137–144
Brooke S, Ross SW, Bane JM et al (2013) Temperature tolerance of the deep-sea coral *Lophelia pertusa* from the southeastern United States. Deep Sea Res Pt II 92:240–248
Brooke SD, Watts MW, Heil AD et al (2017) Distributions and habitat associations of deep water corals in Norfolk and Baltimore canyons of the mid-Atlantic Bight, USA. Deep Sea Res pt II 137:131–147
Brooke S, Ingels J, Baco-Taylor A et al (2020) Cruise Report for NA180AR0110285: Combining habitat suitability and physical oceanography for targeted discovery of new benthic communities on the west Florida slope. Submitted to NOAA Office of Exploration and Research, p 18
Brooks GR, Holmes CW (2011) West Florida continental slope. In: Buster NA and Holmes CW (eds) Gulf of Mexico, origin, waters, and biota. Texas A&M Press, College Station, TX (Geology), pp 129–139
Brooks JM, Fisher C, Roberts H et al (2015) Exploration and research of northern Gulf of Mexico deepwater natural and artificial hard-bottom habitats with emphasis on coral communities: Reefs, rigs, and wrecks—"Lophelia II" Final report. U.S. Dept. Int., Bureau of Ocean Energy Management, Gulf of Mexico OCS Region, New Orleans, LA. OCS Study BOEM 2016-021. 628 p
Brugler MR (2011) Molecular evolution in black corals (Cnidaria: Anthozoa:Hexacorallia). Implications for Antipatharian systematics. University of Lousiana at Lafayette, Lafayette, LA
Brüning M, Sahling H, MacDonald IR et al (2010) Origin, distribution, and alteration of asphalts at Chapopote Knoll, Southern Gulf of Mexico. Mar Petrol Geol 27:1093–1106
Bruno JF, Selig ER, Casey KS et al (2007) Thermal stress and coral cover as drivers of coral disease outbreaks. PLoS Biol 5:e124
Bryant WR, Antoine JW, Ewing M, Jones B (1968) Structure of the Mexican continental shelf and slope, Gulf of Mexico. Am Assoc Petrol Geol Bull 52:1204–1228
Bryant WR, Cordova C, Cordova JL, Salvador A (1991) Physiography and bathymetry. In: Salvador A (ed) The Gulf of Mexico Basin. Geol Soc of America J, Boulder, CO, pp 13–30
Buffler RT (1991) Early evolution of the Gulf of Mexico. In: Goldthwaite D (ed) An introduction to central Gulf Coast geology. New Orleans Geological Society, New Orleans, LA, pp 1–15
Buffler RT, Shaub J, Watkins JS, Worzel JL (1979) Anatomy of the Mexican Ridges, southwestern Gulf of Mexico. In: Watkins JS (ed) Geological and geophysical investigations of continental margins, vol 29. American Assoc Petrol Geol Mem, Tulsa, OK, pp 319–327
Burgess S, Babcock RC (2005) Reproductive ecology of three reef-forming, deep-sea corals in the New Zealand region. In: Freiwald A, Roberts JM (eds) Cold-water corals and ecosystems. Springer, New York, pp 701–713
Cairns SD, Bayer FM (2009) Chapter 13: Octocorallia (Cnidaria) of the Gulf of Mexico. In: Felder DL, Camp DK (eds) Gulf of Mexico, origins, waters and biota: Volume 1, Biodiversity. Texas A and M University Press, College Station, TX, pp 321–331
Cairns SD, Jaap WC, Lang JC (2009) Ch14: Scleractinia (Cnidaria) of the Gulf of Mexico. In: Felder DL, Camp DK (eds) Gulf of Mexico origin, waters, and biota. Volume 1: Biodiversity. Harte Research Institute for Gulf of Mexico Studies Series. Texas A&M University Press, College Station, TX, pp 333–347
Camilli R et al (2010) Tracking hydrocarbon plume transport and biodegradation at Deepwater Horizon. Science 330:201–204
Chanton J, Zhao T, Rosenheim BE et al (2014) Using natural abundance radiocarbon to trace the flux of petrocarbon to the seafloor following the Deepwater Horizon oil spill. Environ Sci Technol 49:847–854. https://doi.org/10.1021/es5046524
Chimetto LA, Brocchi M, Thompson CC et al (2008) Vibrios dominate as culturable nitrogen-fixing bacteria of the Brazilian coral *Mussismilia hispida*. Syst Appl Microbiol 31:312–319
Church R, Cullimore WD, Johnston R et al (2007) Archaeological and biological analysis of World War II shipwrecks in the Gulf of Mexico: artificial reef effect in deep water. U.S. Department of the Interior, Minerals Management Service, Gulf of Mexico OCS Region, New Orleans, LA. OCS Study MMS 2007-015
Clark MR, Koslow JA (2007) Impacts of fisheries on seamounts. In: Pitcher TJ, Morato T, Hart PJB, Clark MR, Haggan N, Santos R (eds) Seamounts: ecology, fisheries & conservation. Blackwell, Oxford, pp 413–441
Coleman JM, Roberts HH, Bryant WR (1991) Late quaternary sedimentation. In: Salvador A (ed) The Gulf of Mexico Basin. The Geology of North America. Geol Soc America J, pp 325–352
Coma R, Gili JM, Zabala M, Riera T (1994) Feeding and prey capture cycles in the aposymbiotic gorgonian *Paramuricea clavata*. Mar Ecol Prog Ser 155:257–270
Continental Shelf Associates, Inc. (CSA) (2006) Effects of oil and gas exploration and development at selected continental slope sites in the Gulf of Mexico, Vol. II. Technical report, U.S. Department of the

Interior, Minerals Management Service, Gulf of Mexico OCS Region, New Orleans, LA

Cordes EE, Bergquist DC, Predmore BL et al (2006) Alternate unstable states: convergent paths of succession in hydrocarbon-seep tubeworm-associated communities. J Exp Mar Biol Ecol 339:159–176

Cordes EE, McGinley M, Podowski EL et al (2008) Coral communities of the deep Gulf of Mexico. Deep Sea Res Pt I 55(6):777–787

Cordes EE, Berlet SP, Cardman Z et al (2014) Exploring deep-sea coral communities and the effects of oil and gas inputs to the Gulf of Mexico. Oceanography 27:28–29. https://doi.org/10.5670/oceanog.2014.supplement.01

Cordes EE, Jones DOB, Schlacher TA et al (2016a) Environmental impacts of the deep-water oil and gas industry: a review to guide management strategies. Front Environ Sci 4:58. https://doi.org/10.3389/fenvs.2016.00058

Cordes EE, Arnaud-Haond S, Bergstad O-A et al (2016b) Chapter 42: Cold-water corals. In: Inniss L, Simcock A (eds) The first global integrated marine assessment: world ocean assessment I, under the auspices of the United Nations General Assembly

Cordona Y, Ruiz-Ramos DV, Baums IB, Bracco A (2016) Potential connectivity of cold-water black coral communities in the Northern Gulf of Mexico. PLoS One 11(5):e0156257. https://doi.org/10.1371/journal.pone.0156257

Corso W, Austin JA Jr, Buffler RT (1988) The early Cretaceous platform off northwest Florida: Controls on morphological development of carbonate margins. Mar Geol 86:1–14

Costa-Lotufo LV, Carnevale-Neto F, Trindade-Silva AE et al (2018) Chemical profiling of two congeneric sea mat corals along the Brazilian coast: Adaptive and functional patterns. Chem Commun 54(16):1952–1955

Cowen RK, Gawarkiewicz G, Pineda J et al (2007) Population connectivity in marine systems: An overview. Oceanography 20(3):14–21

Coykendall DK, Cornman RS, Prouty NG et al (2019) Molecular characterization of *Bathymodiolus* mussels and gill symbionts associated with chemosynthetic habitats from the U.S. Atlantic margin. PLoS One 14(3):e0211616

Crosson S, Yandle T, Stoffle B (2013) Renegotiating property rights in the Florida golden crab fishery. Int J Commons 7(2):521–548

CSA International Inc. (2007) Characterization of northern Gulf of Mexico deepwater hard bottom communities with emphasis on *Lophelia* coral. U.S. Department of the Interior, Minerals Management Service, Gulf of Mexico OCS Region, New Orleans, LA. OCS Study MMS 2007-044

Davies A, Guinotte J (2011) Global habitat suitability for framework-forming cold-water corals. PLoS One 6(4):e18483

Davies AJ, Wisshak M, Orr JC, Roberts JM (2008) Predicting suitable habitat for the cold-water reef framework-forming coral *Lophelia pertusa* (Scleractinia). Deep Sea Res Pt I 55:1048–1062

Davies AJ, Duineveld GCA, Lavaleye MSS et al (2009) Downwelling and deep-water bottom currents as food supply mechanisms to the cold-water coral *Lophelia pertusa* (Scleractinia) at the Mingulay Reef Complex. Limnol Oceanogr 54:620–629

Davies AJ, Duineveld GCA, van Weering TCE et al (2010) Short-term environmental variability in cold-water coral habitat at Viosca Knoll, Gulf of Mexico. Deep Sea Res Pt I 57:199–212

DeLeo DM, Ruiz-Ramos DV, Baums IB et al (2016) Response of deep-water corals to oil and chemical dispersant exposure. Deep Sea Res Pt II 129:137–147

DeLeo DM, Herrera S, Lengyel SD et al (2018) Gene expression profiling reveals deep-sea coral response to the Deepwater Horizon oil spill. Mol Ecol 27:4066–4077

Demopoulos AWJ, Bourque JR, Frometa J (2014) Biodiversity and community composition of sediment macrofauna associated with deep-sea *Lophelia pertusa* habitats in the Gulf of Mexico. Deep Sea Res Pt I 93:91–103. https://doi.org/10.1016/j.dsr.2014.07.014

Demopoulos AWJ, Bourque JR, Cordes E, Stamler KM (2016) Impacts of the Deepwater Horizon oil spill on deep-sea coral-associated sediment communities. Mar Ecol Prog Ser 561:51–68. https://doi.org/10.3354/meps11905

Demopoulos AWJ, Ross SW, Kellogg CA et al (2017) Deepwater program: Lophelia II, continuing ecological research on deep-sea corals and deep-reef habitats in the Gulf of Mexico. U.S. Geological Survey Open-File Report 2017-1139, p 269

Dobretsov S, Qian PY (2004) The role of epibotic bacteria from the surface of the soft coral *Dendronephthya sp.* in the inhibition of larval settlement. J Exp Mar Biol Ecol 299(1):35–50

D'Onghia G, Maiorano P, Sion L et al (2010) Effects of deep-water coral banks on the abundance and size structure of the megafauna in the Mediterranean Sea. Deep Sea Res Pt II 57(5–6):397–411

Doughty CD, Quattrini AM, Cordes EE (2014) Insights into the population dynamics of the deep-sea coral genus *Paramuricea* in the Gulf of Mexico. Deep Sea Res II 99:71–82

Dunn SC (2016) Miocene contourite deposition (along-slope) near DeSoto Canyon, Gulf of Mexico: A product of an enhanced paleo-Loop Current: PhD Dissertation, College of Marine Science, University of South Florida, 139 p

Eckle P, Burgherr P, Michaux E (2012) Risk of large oil spills: a statistical analysis in the aftermath of Deepwater Horizon. Environ Sci Techonol 46:13002–13008. https://doi.org/10.1016/j.marpol.2013.12.002

ECOGIG (2013) https://ecogig.org/expeditions?id=36

ECOGIG (2017) https://ecogig.org/expeditions?id=36

Edinger EN, Wareham VE, Haedrich RL (2007) Patterns of groundfish diversity and abundance in relation to deep-sea coral distributions in Newfoundland and Labrador waters. Bull Mar Sci 81:101–122

Emiliani C, Hudson JH, Shinn EA et al (1978) Oxygen and carbon isotopic growth records in a reef coral from the Florida Keys and a deep-sea coral from the Blake Plateau. Science 202:627–629

Etnoyer PJ, Cairns SD (2017) Deep-sea coral taxa in the U.S. Gulf of Mexico: depth and geographical distribution. In: Hourigan TF, Etnoyer PJ, Cairns SD (eds) The state of deep-sea coral and sponge ecosystems of the United States. NOAA Tech Memo NMFS-OHC-4

Etnoyer P, Warrenchuk J (2007) A catshark nursery in a deep gorgonian field in the Mississippi Canyon, Gulf of Mexico. Bull Mar Sci 81(3):553–559

Etnoyer PJ, Wickes LN, Silva M et al (2016) Decline in condition of gorgonian octocorals on mesophotic reefs in the northern Gulf of Mexico: before and after the Deepwater Horizon oil spill. Coral Reefs 35:77–90

Etnoyer PJ, Wagner D, Fowle HA et al (2018) Models of habitat suitability, size, and age-class structure for the deep-sea black coral *Leiopathes glaberrima* in the Gulf of Mexico. Deep Sea Res Pt II 150:218–228

Ewing M, Erickson DB, Heezen BC (1958) Sediments and topography of the Gulf of Mexico. In: Weeks LG (ed) Habitat of oil. American Assoc Petrol Geol, Tulsa, OK, pp 995–1053

Eytan RI, Hayes M, Arbour-Reily P et al (2009) Nuclear sequences reveal mid-range isolation of an imperiled deep-water coral population. Mol Ecol 18:2375–2389

Fisher CR, Hsing P-Y, Kaiser CL et al (2014a) Footprint of Deepwater Horizon blowout impact to deep-water coral communities. Proc Natl Acad Sci USA 111:11744–11749. https://doi.org/10.1073/pnas.1403492111

Fisher CR, Demopoulos AWJ, Cordes EE et al (2014b) Deep-sea coral communities as indicators of ecosystem level impacts resulting from the Deepwater Horizon oil spill. BioScience 64(9):796–807. https://doi.org/10.1093/biosci/biu129

Frank N, Freiwald A, Lopez-Correa M et al (2011) Northeastern Atlantic cold-water coral reefs and climate. Geology 39:743–746

Freeman-Lynde RR (1983) Cretaceous and Tertiary samples dredged from the Florida Escarpment, eastern Gulf of Mexico. Gulf Coast Assoc Geol Soc Trans 33:91–99

Frometa J, DeLorenzo ME, Pisarski EC, Etnoyer PJ (2017) Toxicity of oil and dispersant on the deep water gorgonian octocoral *Swiftia exserta*, with implications for the effects of the Deepwater Horizon oil spill. Mar Poll Bull 122(1–2):91–99

Galand PE, Remize M, Meistertzheim AL et al (2020) Diet shapes cold-water corals bacterial communities. Environ Microbiol 22(1): 354–368

Galkiewicz JP, Pratte ZA, Gray MA, Kellogg CA (2011) Characterization of culturable bacteria isolated from the cold-water coral *Lophelia pertusa*. FEMS Microb Ecol 77(2):333–346

Galkiewicz JP, Stellick SH, Gray MA, Kellogg CA (2012) Cultured fungal associates from the deep-sea coral *Lophelia pertusa*. Deep Sea Res I 67:12–20

Galloway WE, Whitaker TL, Ganey-Curry PR (2011) History of Cenozoic North American drainage basin evolution, sediment yield, and accumulation in the Gulf of Mexico basin. Geosphere 7:938–973. https://doi.org/10.1130/GES00647.1

Garrison LE, Martin RG Jr (1973) Geologic structure in the Gulf of Mexico. U.S. Geological Survey Professional Paper 773, 85 pp

Gass SE, Roberts JM (2006) The occurrence of the cold-water coral *Lophelia pertusa* (Scleractinia) on oil and gas platforms in the North Sea: colony growth, recruitment and environmental controls on distribution. Mar Poll Bull 52:549–559

GEBCO Compilation Group (2020) GEBCO 2020 Grid. https://doi.org/10.5285/a29c5465-b138-234d-e053-6c86abc040b9

Georgian SE, Shedd W, Cordes EE (2014) High-resolution ecological niche modelling of the cold-water coral *Lophelia pertusa* in the Gulf of Mexico. Mar Ecol Prog Ser 506:145–161

Georgian S, DeLeo D, Durkin A et al (2016a) Oceanographic patterns and carbonate chemistry in the vicinity of cold-water coral reefs in the Gulf of Mexico: implications for resilience in a changing ocean. Limnol Oceangr 61:648–665

Georgian SE, Dupont S, Kurman M et al (2016b) Biogeographic variability in the physiological response of the cold-water coral *Lophelia pertusa* to ocean acidification. Mar Ecol 37:1345–1359

Georgian SE, Kramer K, Saunders M et al (2019) Habitat suitability modelling to predict the spatial distribution of cold-water coral communities affected by the Deepwater Horizon oil spill. J Biogeogr 47:1455–1466

Girard F, Fu B, Fisher CR (2016) Mutualistic symbiosis with ophiuroids limited the impact of the Deepwater Horizon oil spill on deep-sea octocorals. Mar Ecol Prog Ser 549:89–98

Girard F, Katriona S, Fisher C (2018) Projecting the recovery of a long-lived deep-sea octocoral species after the Deepwater Horizon oil spill using structured population models. J Appl Ecol 55:1812–1822

GMFMC (2011) Generic annual catch limits/accountability measures amendment for the Gulf of Mexico Fishery Management Council's red drum, reef fish, shrimp, coral and coral reefs fishery management plans. Gulf of Mexico Fishery Management Council, Tampa, FL. 378 pp. http://gulfcouncil.org/wp-content/uploads/Final-Generic-ACL-AM-Amendment-September-9-2011-v.pdf

Goldsmith DB, Kellogg CA, Morrison CL et al (2018) Comparison of microbiomes of cold-water corals *Primnoa pacifica* and *Primnoa resedaformis*, with possible link between microbiome composition and host genotype. Sci Rep 8:12383

Gong L, Chen W, Gao Y et al (2013) Genetic analysis of the metabolome exemplified using a rice population. Proc Natl Acad Sci USA 110(50):20320–20325

Guinotte JM, Orr J, Cairns S et al (2006) Will human-induced changes in seawater chemistry alter the distribution of deep-sea scleractinian corals? Front Ecol Environ 4:141–146

Halanych KM, Ainsworth CH, Cordes EE et al (2021) Effects of petroleum by-products and dispersants on ecosystems. Oceanography 34:152–163. https://doi.org/10.5670/oceanog.2021.123

Hamilton P, Fargion GS, Biggs DC (1999) Loop current eddy paths in the western Gulf of Mexico. J Phys Oceanogr 29:1180–1207

Hansson L, Agis M, Maier C et al (2009) Community composition of bacteria associated with cold-water coral *Madrepora oculata*: within and between colony variability. Mar Ecol Prog Ser 397:89–102

Harrison PL (2011) Sexual reproduction of scleractinian corals. In: Dubinsky Z, Stambler N (eds) Coral reefs: An ecosystem in transition. Springer, Dordrecht, pp 59–85

Harter SL, Ribera MM, Shepard AN, Reed JK (2009) Assessment of fish populations and habitat on Oculina Bank, a deep-sea coral marine protected area off eastern Florida. Fish Bull 107(2):195–206

Hebbeln D (2012) West Atlantic cold-water coral ecosystems: the west side story. RV Maria S. Merian, Cruise Report MSM20/L4, 2012, 57 pp., DFG Senatskommission für Ozeanographie

Hebbeln D, Wienberg C, Wintersteller P et al (2014) Environmental forcing of the Campeche cold-water coral province, southern Gulf of Mexico. Biogeosciences (BG) 11:1799–1815

Hebbeln D, Portilho-Ramos RC, Wienberg C, Titschack J (2019) The fate of cold-water corals in a changing world: a geological perspective. Front Mar Sci 6:119. https://doi.org/10.3389/fmars.2019.00119

Hemmer MJ, Barron MG, Greene RM (2011) Comparative toxicity of eight oil dispersants, Louisiana sweet crude oil (LSC), and chemically dispersed LSC to two aquatic test species. Environ Toxicol Chem 30(10):2244–2252

Hennige SJ, Wicks LC, Kamenos NA et al (2015) Hidden impacts of ocean acidification to live and dead coral framework. Proc R Soc B 282:20150990. https://doi.org/10.1098/rspb.2015

Hennige SJ, Wolfram U, Wickes L et al (2020) Crumbling reefs and cold-water coral habitat loss in a future ocean: evidence of "coralporosis" as an indicator of habitat integrity. Front Mar Sci 7: 668. https://doi.org/10.3389/fmars.2020.00668

Hernandez-Agreda A, Leggat W, Bongaerts P et al (2016) The microbial signature provides insight into the mechanistic basis of coral success across reef habitats. Am Soc Microbiol 7:1–10

Hernandez-Agreda A, Gates RD, Ainsworth TC (2017) Defining the core microbiome in corals' soup. Trends Microbiol 25:125–140

Hey J, Machado CA (2003) The study of structured populations: New hope for a difficult and divided science. Nat Rev Genet 4(7):535–543

Hine AC (2019) Geologic history of Florida. University Press of Florida, Gainesville, FL. 230 p. ISBN: 978-0-8130-6412-3

Hofmann EE, Worley SJ (1986) An investigation of the circulation of the Gulf of Mexico. J Geophys Res 91:14221–14236

Hourigan TF, Etnoyer PJ, Cairns SD (2017) The state of deep-sea coral and sponge ecosystems of the United States. NOAA Tech Memo NMFS-OHC-4, 467p

Hsing P-Y, Fu B, Larcom EA et al (2013) Evidence of lasting impact of the Deepwater Horizon oil spill on a deep Gulf of Mexico coral community. Elementa Sci Anthropocene 1:52

Hsu CW, MacDonald IR, Römer M et al (2019) Characteristics and hydrocarbon seepage at the Challenger Knoll in the Sigsbee Basin, Gulf of Mexico. Geo Mar Lett 39:391–399

Hu Z, Hu J, Hu H, Zhou Y (2020) Predictive habitat suitability modeling of deep-sea framework-forming scleractinian corals in the Gulf of Mexico. Sci Tot Environ 742:140562

Hubscher C, Dullo C, Flogel S et al (2010) Contourite drift evolution and related coral growth in the eastern Gulf of Mexico and it's gateways. Int J Earth Sci 99(1):S191–S206

Husebo A, Nottestad L, Fossa JH et al (2002) Distribution and abundance of fish in deep-sea coral habitats. Hydrobiologia 471:91–99

Jarnegren J, Brooke S, Jensen H (2017) Effects of drill cuttings on larvae of the cold-water coral *Lophelia pertusa*. Deep Sea Res Pt II 137: 454–462

Jarnegren J, Brooke S, Jensen H (2020) Effects and recovery of the cold-water coral *Lophelia pertusa* (*Desmophyllum pertusum*) exposed to suspended bentonite, barite and drill cuttings. Mar Environ Res 158: 104996

Jochens AE, DiMarco SF (2008) Physical oceanographic conditions in the deep-water Gulf of Mexico in summer 2000-2002. Deep Sea Res Pt II 55:2541–2554

Joye SB, Bracco A, Ozgokmen T et al (2016) The Gulf of Mexico ecosystem, six years after the macondo oil well blowout. Deep Sea Res II 129:4–19. https://doi.org/10.1016/j.dsr2.2016.04.018

Kahng SE, Benayahu Y, Lasker HR (2011) Sexual reproduction in octocorals. Mar Ecol Prog Ser 443:265–283

Kellogg CA (2019) Microbiomes of stony and soft deep-sea corals share rare core bacteria. Microbiome 7(1):90

Kellogg CA, Lisle JT, Galkiewicz JP (2009) Culture-independent characterization of bacterial communities associated with the cold-water coral *Lophelia pertusa* in the northeastern Gulf of Mexico. Appl Env Microbiol 75(8):2294–2303

Kellogg CA, Ross SW, Brooke SD (2016) Bacterial community diversity of the deep sea octocoral *Paramuricea placomus*. PeerJ 4:e2529

Kellogg CA, Galkiewicz JP, Gray MA (2017a) Microbial ecology and functional metagenomics of *Lophelia pertusa* in the Gulf of Mexico. In: Demopoulos AWJ et al (eds) Deepwater Program: *Lophelia* II, continuing ecological research on deep-sea corals and deep-reef habitats in the Gulf of Mexico. SGS Open File Report 2017-1139

Kellogg CA, Goldsmith DB, Gray MA (2017b) Biogeographic comparison of *Lophelia* associated bacterial communities in the western Atlantic reveals conserved core microbiome. Front Microbiol 8:796

Kilgour MJ, Shirley TC (2008) *Eumunida picta* S.I. Smith, 1883, and *Lophelia pertusa* (Linnaeus, 1758): a relationship or just good friends? Crustaceana 81:587–593

Kujawinski EB, Kido Soule MC, Valentine DL et al (2011) Fate of dispersants associated with the Deepwater Horizon oil spill. Environ Sci Technol 45(4):1298–1306

Kurman MD, Gomez CE, Georgian SE et al (2017) Intra-specific variation reveals potential for adaptation in a cold-water coral from the Gulf of Mexico. Front Mar Sci 4:111

Larcom EA, McKean DL, Brooks JM et al (2014) Growth rates, densities, and distribution of *Lophelia pertusa* on artificial structures in the Gulf of Mexico. Deep Sea Res Pt I 85:101–109

Larsson AI, van Oevelen D, Purser A, Thomsen L (2013) Tolerance to long-term exposure of suspended benthic sediments and drill cuttings in the cold-water coral *Lophelia pertusa*. Mar Poll Bull 70(1–2):176–188

Larsson AI, Järnegren J, Strömberg SM et al (2014) Embryogenesis and larval biology of the cold water coral *Lophelia pertusa*. PLoS One 9(7):e102222

Lavaleye MSS, Duineveld GCA (2001) Coral recruitment experiments. ACES Deliverable, p 205

Lawler SN, Kellogg CA, France SC et al (2016) Coral associated bacterial diversity is conserved across two deep-sea *Anthothela* species. Front Microbiol 7:458

Lepland A, Mortensen PB (2008) Barite and barium in sediments and coral skeletons around the hydrocarbon exploration drilling site in the Traena Deep, Norwegian Sea. Environ Geol 56:119–129. https://doi.org/10.1007/s00254-007-1145-4

Le Quéré C, Takahashi T, Buitenhuis ET et al (2010) Impact of climate change and variability on the global oceanic sink of CO2. Glob Biogeochem Cycles 24(4):GB4007

Lessard-Pilon SA, Podowski EL, Cordes EE, Fischer CR (2010) Megafauna community composition associated with *Lophelia pertusa* colonies in the Gulf of Mexico. Deep Sea Res Pt II 57(21–23): 1882–1890

Locker S, Buffler RT (1983) Comparison of lower Cretaceous carbonate shelf margins, northern Campeche Escarpment and northern Florida Escarpment, Gulf of Mexico. Am Assoc Petrol Geol Stud Geol 15: 123–128

López Correa M, Montagna P et al (2012) Preboreal onset of cold-water coral growth beyond the Arctic Circle revealed by coupled radiocarbon and U-series dating and neodymium isotopes. Quat Sci Rev 34: 24–43

Lumsden SE, Hourigan TF, Bruckner AW (eds) (2007) The state of deep coral ecosystems of the United States. NOAA Technical Memorandum NOS-CRCP-3, Silver Spring, MD

Lunden JJ (2013) Ocean acidification and the cold-water coral *Lophelia pertusa* in the Gulf of Mexico. PhD Thesis, Temple University, 134 p

Lunden JJ, Georgian SE, Cordes EE (2013) Aragonite saturation states at cold-water coral reefs structured by *Lophelia pertusa* in the northern Gulf of Mexico. Limnol Oceanogr 58:354–362

Lunden JJ, McNicholl CM, Sears C et al (2014) Acute survivorship of the deep-sea coral *Lophelia pertusa* from the Gulf of Mexico under acidification, warming, and deoxygenation. Front Mar Sci 1:78

Macpherson E, Beuck L, Freiwald A (2016) Some species of *Munidopsis* from the Gulf of Mexico, Florida Straits and Caribbean Sea (Decapoda: Munidopsidae), with the description of two new species. Zootaxa 4137:405–416

Maier SR, Kutti T, Bannister RJ et al (2020) Recycling pathways in cold-water coral reefs: Use of dissolved organic matter and bacteria by key suspension feeding taxa. Sci Rep 10:9942

Marton G, Buffler RT (2010) Jurassic reconstruction of the Gulf of Mexico Basin. Int Geol Rev 36(6):545–586. https://doi.org/10.1080/00206819409465475

Matos L, Wienberg C, Titschack J et al (2017) Coral mound development at the Campeche cold-water coral province, southern Gulf of Mexico: Implications of Antarctic Intermediate Water increased influence during interglacials. Mar Geol 392:53–65

McBride BC, Weimer P, Rosen MG (1998) The effect of allochthonous salt on the petroleum systems of northern Green Canyon and Ewing Bank (offshore Louisiana), Northern Gulf of Mexico. Am Assoc Petrol Geol Bull 82:1083–1112

McNutt MK, Camilli R, Crone TJ et al (2012) Review of flow rate estimates of the Deepwater Horizon oil spill. Proc Natl Acad Sci USA 109(50):20260–20267

Meistertzheim A-L, Lartaud F, Arnaud-Haond S et al (2016) Patterns of bacteria host associations suggest different ecological strategies between two reef building cold-water coral species. Deep Sea Res Pt I 114:12–22

Menezes CBA, Bonugli-Santos RC, Miqueletto PB et al (2009) Microbial diversity associated with algae, ascidians and sponges from the north coast of Sao Paulo state, Brazil. Microbiol Res 165:466–482

Mercier A, Hamel J-F (2009) Reproductive periodicity and hostspecific settlement and growth of a deep-water symbiotic sea anemone. Can J Zool 87:967–980

Mercier A, Sun Z, Hamel J-F (2011) Reproductive periodicity, spawning and development of the deep-sea scleractinian coral *Flabellum angulare*. Mar Biol 158:371–380

Merino M (1997) Upwelling on the Yucatan Shelf: hydrographic evidence. J Mar Syst 13:101–121

Mienis F, Duineveld GCA, Davies AJ et al (2012) The influence of near-bed hydrodynamic conditions on cold-water corals in the Viosca Knoll area, Gulf of Mexico. Deep Sea Res Pt I 60:32–45

Mienis F, Bouma TJ, Witbaard R et al (2019) Experimental assessment of the effects of coldwater coral patches on water flow. Mar Ecol Prog Ser 609:101–117

Mikkelsen N, Erlenkeuser H, Killingley JS et al (1982) Norwegian corals; radiocarbon and stable isotopes in *Lophelia pertusa*. Boreas 11:163–171

Miller RJ, Hocevar J, Stone RP, Fedorov DV (2012) Structure-forming corals and sponges and their use as fish habitat in Bering Sea submarine canyons. PLoS One 7:e33885

Milliman JD, Meade RH (1983) World-wide delivery of river sediment to the oceans. J Geol 91(1):1–21. https://doi.org/10.1086/628741
Molinari RL, Morrison J (1988) The separation of the Yucatan Current from the Campeche Bank and intrusion of the Loop Current into the Gulf of Mexico. J Geophys Res 93:10645–10654
Morrison JM, Merrell WJ Jr, Key RM et al (1983) Property distributions and deep chemical measurements within the Western Gulf of Mexico. J Geophys Res 88:2601–2608
Morrison CL, Eackles MS, Johnson RL, King TL (2008) Characterization of 13 microsatellite loci for the deep-sea coral, *Lophelia pertusa* (Linnaeus 1758), from the western North Atlantic Ocean and Gulf of Mexico. Mol Ecol Res 8:1037–1039
Morrison CL, Ross SW, Nizinski MS et al (2011) Genetic discontinuity among regional populations of *Lophelia pertusa* in the North Atlantic Ocean. Conserv Genet 12:713–729
Mortensen PB (2001) Aquarium observations on the deep-water coral *Lophelia pertusa* (L., 1758) (Scleractinia) and selected associated invertebrates. Ophelia 54:83–104
Mortensen PB, Rapp HT (1998) Oxygen and carbon isotope ratios related to growth linepatterns in skeletons of *Lophelia pertusa* (L) (Anthozoa, Scleractinia): implications for determinations of linear extension rates. Sarsia 83:433–446
Mullins HT, Gardulski AF, Hine AC (1986) Catastrophic collapse of the West Florida Platform margin. Geology 14:167–170
Murawski SA, Hollander DJ, Gilbert S, Gracia A (2020) Deepwater oil and gas production in the Gulf of Mexico and related global trends. In: Murawski S et al. Scenarios and responses to future deep oil spills. Springer, Cham. https://doi.org/10.1007/978-3-030-12963-7_2
Neave M, Mitchell C, Apprill A et al (2017) *Endozoicomonas* genomes reveal functional adaptation and plasticity in bacterial strains symbiotically associated with diverse marine hosts. Sci Rep 7:40579
Neulinger SC, Järnegren J, Ludvigsen M et al (2008) Phenotype-specific bacterial communities in the cold-water coral *Lophelia pertusa* (Scleractinia) and their implications for the coral's nutrition, health, and distribution. Appl Environ Microbiol 74(23):7272–7285
Newton CR, Mullins HT, Gardulski AF, Hine AC, Dix GR (1987) Coral mounds on the West Florida slope: unanswered questions regarding the development of deep-water banks. Palaios 2:359–367
Nielsen JG, Ross SW, Cohen DM (2009) Atlantic occurrence of the genus *Bellottia* (Teleostei, Bythitidae) with two new species from the Western North Atlantic. Zootaxa 2018(1):45–57
Nizinski M, Ames C (2017) Distribution, abundance, and species composition of the megafaunal invertebrate assemblages associated with deep-sea coral habitats in the Gulf of Mexico. In: Demopoulos AWJ, Ross SW, Kellogg CA et al (eds) Deepwater Program: Lophelia II, continuing ecological research on deep-sea corals and deep-reef habitats in the Gulf of Mexico. U.S. Geological Survey Open-File Report 2017-1139, p 269
Orejas C, Gori A, Rad-Menéndez C et al (2016) The effect of flow speed and food size on the capture efficiency and feeding behaviour of the cold-water coral *Lophelia pertusa*. J Exp Mar Biol Ecol 481:34–40
Paull CK, Neumann AC (1987) Continental margin brine seeps: Their geological consequences. Geology 15:545–548. https://doi.org/10.1130/0091-7613(1987)15<545:CMBSTG>2.0.co;2
Paull CK, Hecker B, Commeau R et al (1984) Biological communities at the Florida Escarpment resemble hydrothermal vent taxa. Science 226:965–967. https://doi.org/10.1126/science.226.4677.965
Paull CK, Spiess FN, Curray JR, Twichell D (1990) Origin of Florida Canyon and the role of spring sapping on the formation of submarine box canyons. Geol Soc Am Bull 102:502–515
Paull CK, Caress DW, Gwiazda R et al (2014) Cretaceous-Paleogene boundary exposed. Campeche Escarpment, Gulf of Mexico. Mar Geol 357:392–400
Perez-Drago G, Perez-Cruz G, Chenet P (2019) Origin of natural seabed seepage along Mexico: Petroleum system implications and potential. In: Proceedings 2nd EAGE workshop on deepwater exploration in Mexico, April 2019, pp 1–5. https://doi.org/10.3997/2214-4609.201900356
Peterson BJ, Fry B (1987) Stable isotopes in ecosystem studies. Ann Rev Ecol Syst 18:293–320
Pew E (1982) Seismic structural analysis of deformation in the southern Mexico Ridges. MA thesis, University of Texas Libraries, The University of Texas at Austin, 102 p
Prada C, Hellberg ME (2013) Long pre-reproductive selection and divergence by depth in a Caribbean candelabrum coral. Proc Natl Acad Sci USA 110(10):3961–3966
Prouty NG, Roark EB, Buster NA, Ross SW (2011) Growth-rate and age distribution of deep-sea black corals in the Gulf of Mexico. Mar Ecol Prog Ser 423:101–115
Prouty NG, Roark EB, Koenig AE et al (2014) Deep-sea coral record of human impact on watershed quality in the Mississippi River Basin. Global Biogeochem Cycles 28:29–43. https://doi.org/10.1002/2013GB004754
Prouty N, Fisher CR, Demopoulos AWJ, Druffel ERM (2016) Growth rates and ages of deep-sea corals impacted by the Deepwater Horizon oil spill. Deep Sea Res Pt II 129:196–212
Prouty NG, Roark EB, Andrews AH et al (2017) Age, growth rates, and paleoclimate studies in deep-sea corals of the U.S. In: Hourigan TF, Etnoyer PJ, Cairns SD (eds) State of deep-sea coral and sponge ecosystems of the United States. NOAA Technical Memorandum NMFS-OHC-4, Silver Spring, MD
Purser A, Thomsen L (2012) Monitoring strategies for drill cutting discharge in the vicinity of cold-water coral ecosystems. Mar Pollut Bull 64:2309–2316. https://doi.org/10.1016/j.marpolbul.2012.08.003
Quattrini AM, Ross SW, Sulak KJ et al (2004) Marine fishes new to continental United States waters, North Carolina, and the Gulf of Mexico. Southeastern Nat 3(1):155–172
Quattrini AM, Georgian SE, Byrnes L et al (2013) Niche divergence by deep-sea octocorals in the genus Callogorgia across the continental slope of the Gulf of Mexico. Mol Ecol 22:4123–4140. https://doi.org/10.1111/mec.12370
Quattrini AM, Etnoyer PJ, Doughty C et al (2014) A phylogenetic approach to octocoral community structure in the deep Gulf of Mexico. Deep Sea Res Pt II 99:92–102
Quattrini AM, Baums IB, Shank TM, Morrison CL, Cordes EE (2015) Testing the depth differentiation hypothesis in a deepwater octocoral. Proc R Soc B 282:20150008
Quattrini AM, Gomez CE, Cordes EE (2017) Environmental filtering and neutral processes shape octocoral community assembly in the deep sea. Oecologia 183:221–236
Raghukumar S (2006) Marine microbial eukaryotic diversity, with particular reference to fungi: lessons from prokaryotes. Indian J Mar Sci 35(4):388–398
Raina JB, Tapiolas D, Willis BL, Bourne DG (2009) Coral associated bacteria and their role in the biogeochemical cycling of sulfur. Appl Environ Microbiol 75:3492–3501
Rakka M, Bilan M, Godinho A et al (2019) First description of polyp-bailout in cold-water octocorals under aquaria maintenance. Coral Reefs 38:15–20
Reed JK, Wright A (2004) Submersible and scuba collections on deep-water reefs off the east coast of Florida, including the Northern and Southern Straits of Florida and Florida Marine Keys National Sanctuary for biomedical and biodiversity research of the benthic communities with emphasis on the Porifera and Gorgonacea, May 20-June 2, 2004. Final cruise report. Conducted by the Center of Excellence, HBOI and FAU, p 54
Reed JK, Weaver DC, Pomponi SA (2006) Habitat and fauna of deep-water *Lophelia pertusa* coral reefs off the southeastern US: Blake Plateau, Straits of Florida, and Gulf of Mexico. Bull Mar Sci 78(2):343–375

Reed JK, Harter S, Farrington S, David A (2014) Characterization and interrelationships of deepwater coral/sponge habitats and fish communities off Florida. In: Interrelationships between corals and fisheries. CRC, Boca Raton, FL, pp 51–82

Reed JK, Farrington S, David A et al (2019) Pulley Ridge, Gulf of Mexico, USA. In: Loya Y, Puglise KA, Bridge TCL (eds) Mesophotic coral ecosystems. Springer, Cham, pp 57–69

Reshef L, Koren O, Loya Y et al (2006) The coral probiotic hypothesis. Environ Microbiol 8:2068–2073

Ribes M, Coma R, Gili J (1999) Heterogenous feeding in benthic suspension feeders: the natural diet and grazing rate of the temperate gorgonian *Paramuricea clavata* (Cnidaria: Octocorallia) over a year cycle. Mar Ecol Prog Ser 183:125–137

Ritchie KB (2006) Regulation of microbial populations by coral surface mucus and mucus-associated bacteria. Mar Ecol Prog Ser 322:1–14

Roberts HH, Aharon P (1994) Hydrocarbon-derived carbonate buildups of the northern Gulf of Mexico continental slope—a review of submersible investigations. Geo Mar Lett 14:135–148

Roberts JM, Cairns SD (2014) Cold-water corals in a changing ocean. Curr Opin Environ Sustain 7:118–126

Roberts HH, Kohl B (2018) Temperature control of cold-water coral (*Lophelia*) mound growth by climate-cycle forcing, Northeast Gulf of Mexico. Deep Sea Res Pt I 140:142–158

Roberts JM, Wheeler AJ, Freiwald A (2006) Reefs of the deep: The biology and geology of cold-water coral ecosystems. Science 312: 543–547

Roberts JM, Wheeler AJ, Freiwald A et al (2009) Cold-water corals: the biology and geology of deep-sea coral habitats. Cambridge University Press, New York

Roberts HH, Feng D, Joye SB (2010) Cold-seep carbonates of the middle and lower continental slope. Deep Sea Res Pt II 57:2040–2054

Rohwer F, Seguritan V, Azam F, Knowlton N (2002) Diversity and distribution of coral-associated bacteria. Mar Ecol Prog Ser 243:1–10

Rooper CN, Etnoyer PJ, Stierhoff KL, Olson JV (2017) Effects of fishing gear on deep-sea corals and sponges in U.S. waters. In: Hourigan TF, Etnoyer PJ, Cairns SD (eds) The State of deep-sea coral and sponge ecosystems of the United States. NOAA Technical Memorandum NMFS-OHC-4, Silver Spring, MD. 36 p

Rosenberg E, Kellogg CA, Rohwer F (2007a) Coral microbiology. Oceanography 20(2):114–122

Rosenberg E, Koren O, Reshef L et al (2007b) The role of microorganisms in coral health, disease and evolution. Nat Rev Microbiol 5:355–362

Ross SW, Quattrini AM (2007) The fish fauna associated with deep coral banks off the southeastern United States. Deep Sea Res Pt I 54(6):975–1007

Ross SW, Brooke S, Quattrini AM et al (2015) A deep-sea community, including *Lophelia pertusa*, at unusually shallow depths in the western North Atlantic Ocean off northeastern Florida. Mar Biol 162(3):635–648

Ross SW, Rhode M, Brooke SD (2017a) Deep sea coral and hardbottom habitats on the west Florida slope, eastern Gulf of Mexico. Deep Sea Res Pt II 120:14–28

Ross SW, Quattrini AM, Rhode M (2017b) Fishes associated with deep-sea coral habitats in the Gulf of Mexico In: Demopoulos AWJ, Ross SW, Kellogg CA et al (eds) Deepwater Program: Lophelia II, continuing ecological research on deep-sea corals and deep-reef habitats in the Gulf of Mexico. U.S. Geological Survey Open-File Report 2017-1139, p 269

Rowe GT, Kennicutt II MC, (eds) (2009) Northern Gulf of Mexico continental slope habitats and benthic ecology study: Final report, U.S. Dept. of the Interior, Minerals Management Service, Gulf of Mexico OCS Region, New Orleans, LA

Rowe GT, Menzel DW (1971) Quantitative benthic samples from the deep Gulf of Mexico with some comments on the measurement of deep-sea biomass. Bull Mar Sci 21:556–566

Rozenfeld AF, Arnaud-Haond S, Hernández-García E (2007) Spectrum of genetic diversity and networks of clonal organisms. J R Soc Interface 4(17):1093–1102

Ruiz-Ramos DV, Saunders M, Fisher CR, Baums IB (2015) Home bodies and wanderers: sympatric lineages of the deep-sea black coral *Leiopathes glaberrima*. PLoS One 10(10):e0138989

Sahling H, Borowski C, Escobar-Briones E et al (2016) Massive asphalt deposits, oil seepages, and gas venting support abundant chemosynthetic communities at the Campeche Knolls Southern Gulf of Mexico. Biogeosciences 13:4491–4512. https://doi.org/10.5194/bg-13-4491-2016

Salvador A (1991) Introduction. In: Salvador A (ed) The Gulf of Mexico Basin. The Geology of North America. Geol Soc America J, pp 1–12

Sammarco PW (1982) Polyp bail-out: an escape response to environmental stress and a new means of reproduction in corals. Mar Ecol Prog Ser 10:57–65

Schmitz WJ, McCartney MS (1993) On the North Atlantic circulation. Rev Geophys 31:29–49

Schroeder WW (2002) Observations of *Lophelia pertusa* and the surficial geology at a deep-water site in the northeastern Gulf of Mexico. Hydrobiologia 471:29–33

Schroeder WW (2007) Seafloor characteristics and distribution patterns of *Lophelia pertusa* and other sessile megafauna at two upper-slope sites in the northwestern Gulf of Mexico. U.S. Dept. of the Interior, Minerals Management Service, Gulf of Mexico OCS Region, New Orleans, LA. OCS Study MMS 2007-035

Schwing P, Montagna PA, Joye SB et al (2020) Deep benthic ecosystem impacts of the *Deepwater Horizon* oil spill: Assembling the record of species and community change. Front Mar Sci 7:560012. https://doi.org/10.3389/fmars.2020.560012

Shade A, Handelsman J (2012) Beyond the Venn diagram: the hunt for a core microbiome. Environ Microbiol 14:4–12

Shepard FP (1963) Submarine geology. Harper and Row, New York. 557 p

Shepard FP, Dill RF (1966) Submarine canyons and other sea valleys. Rand McNally, Chicago, IL. 381 p

Sherwood OA, Heikoop JM, Scott DB et al (2005) Stable isotope composition of deep sea gorgonian corals *Primnoa* spp.: a new archive of surface processes. Mar Ecol Prog Ser 301:135–148

Silva M, Etnoyer PJ, MacDonald I (2016) Coral injuries observed at mesophotic reefs after the Deepwater Horizon oil discharge. Deep Sea Res Pt II 129:96–107

Silveira CB, Cavalcanti GS, Walter JM et al (2017) Microbial processes driving coral reef organic carbon flow. FEMS Microbiol Rev 41: 575–595

Simister RL, Antzis EW, White HK (2016) Examining the diversity of microbes in a deep-sea coral community impacted by the Deepwater Horizon oil spill. Deep Sea Res Pt II 129:157–166

Smith PJ (2001) Managing biodiversity: Invertebrate bycatch in seamount fisheries in the New Zealand exclusive economic zone (a case study). UNEP workshop on managing global fisheries for biodiversity, Victoria

Smith DL, Lord KM (1997) Tectonic evolution and geophysics of the Florida Basement. In: Randazzo AF, Jones DS (eds) The Geology of Florida. University of Florida Press, Gainesville, FL, pp 13–37

Smith JE, Schwarcz HP, Risk MJ et al (2000) Paleotemperatures from deep-sea corals: overcoming 'vital effects'. Palaios 15:25–32

Staley JT, Konopka A (1985) Measurement of in situ activities of nonphotosynthetic microorganisms in aquatic and terrestrial habitats. Ann Rev Microbiol 39(1):321–346

Streich MK, Ajemian MJ, Wetz JJ, Stunz GW (2017) A comparison of fish community structure at mesophotic artificial reefs and natural banks in the western Gulf of Mexico. Mar Coast Fish Dyn Manag Ecosyst Sci 9:170–189

Sturges W (2005) Deep-water exchange between the Atlantic, Caribbean, and Gulf of Mexico. In: Sturges W, Lugo-Fernandez A (eds) Circulation in the Gulf of Mexico. American Geophysical Union, Washington, DC, pp 263–278

Sturges W, Leben R (2000) Frequency of ring separations from the Loop Current in the Gulf of Mexico: a revised estimate. J Phys Oceanogr 30:1814–1819

Sturges W, Lugo-Fernandez A (2005) Circulation in the Gulf of Mexico: Observations and models. American Geophysical Union Geophys Monog Ser 161, Washington, DC. https://doi.org/10.1029/GM161

Sulak KJ, Brooks RA, Luke KE et al (2007) Demersal fishes associated with *Lophelia pertusa* coral and hard-substrate biotopes on the continental slope, northern Gulf of Mexico. Bull Mar Sci 81(3): 65–92

Sulak KJ, Randall MT, Luke KE et al (eds) (2008) Characterization of Northern Gulf of Mexico deepwater hard bottom communities with emphasis on *Lophelia* coral. *Lophelia* reef megafaunal community structure, biotopes, genetics, microbial ecology, and geology (2004–2006): U.S. Geological Survey Open-File Report 2008–1148; OCS Study MMS 2008–015. https://doi.org/10.3133/ofr20081148

Sun Z, Hamel J-F, Edinger E, Mercier A (2010a) Reproductive biology of the deep-sea octocoral *Drifa glomerata* in the Northwest Atlantic. Mar Biol 157:863–873

Sun Z, Hamel J-F, Edinger E, Mercier A (2010b) Planulation periodicity, settlement preferences and growth of two deep-sea octocorals from the northwest Atlantic. Mar Ecol Prog Ser 410:71–87

Sun Z, Hamel J-F, Edinger E, Mercier A (2011) Planulation, larval biology and early growth of the deep-sea soft corals *Gersemia fruticosa* and *Duva florida* (Octocorallia: Alcyonacea). Invert Biol 130(2):91–99

Teas P, Orange D, Decker J, Brumley K (2017) High-resolution multibeam reveals surficial character of the Perdido, Mexican Ridges, and Campeche Fold Belt, deep water Gulf of Mexico. In: AAPG annual convention and exhibition, April 2–5, 2017, Houston, TX

Thoma JN, Pante E, Brugler MR, France SC (2009) Deep-sea octocorals and antipatharians show no evidence of seamount-scale endemism in the NW Atlantic. Mar Ecol Prog Ser 397:25–35

Thurber RLV, Correa AM (2011) Viruses of reef-building scleractinian corals. J Exp Mar Biol Ecol 408(1–2):102–113

Trainor DM, Williams DF, Lerche I (1988) Refinement and spectral analysis of Plio-Pleistocene oxygen and carbon isotope records. Gulf Coast Assoc Geol Soc Trans 38:422–435

Trotter JA, Montagna P, McCulloch MT et al (2011) Quantifying the pH 'vital effect' in the temperate zooxanthellate coral *Cladocora caespitosa*: Validation of the boron seawater pH proxy. Earth Planet Sci Lett 303:163–173

Tyler PA (1988) Seasonality in the deep sea. Oceanogr Mar Biol Annu Rev 26:227–258

Valentine DL, Fisher GB, Bagby SC et al (2014) Fallout plume of submerged oil from Deepwater Horizon. Proc Nat Acad Sci USA 111:15906–15911. https://doi.org/10.1073/pnas.1414873111

van de Water JA, Melkonian R, Voolstra CR et al (2017) Comparative assessment of Mediterranean gorgonian-associated microbial communities reveals conserved core and locally variant bacteria. Microb Ecol 73(2):466–478

Victorero L, Watling L, Deng Palomares ML, Nouvian C (2018) Out of sight, but within reach: a global history of bottom-trawled deep-sea fisheries from >400m depth. Front Mar Sci 5(98):1–17

Vohsen SA, Fisher CR, Baums IB (2019) Metabolomic richness and fingerprints of deep-sea coral species and populations. Metabolomics 15(3):34

Vohsen SA, Anderson KE, Gade AM et al (2020) Deep-sea corals provide new insight into the ecology, evolution, and the role of plastids in widespread apicomplexan symbionts of anthozoans. Microbiome 8(1):1–15

Wagner D, Luck DG, Toonen RJ (2012) The biology and ecology of black corals (Cnidaria: Anthozoa: Hexacorallia: Antipatharia). Adv Mar Biol 63:67–132. Academic

Walker JD, Geissman JW, Bowring SA et al (2018) Geologic time scale v. 5.0. Geological Society of America. https://doi.org/10.1130/2018.CTS005R3C

Waller RG, Tyler PA (2005) The reproductive biology of two deep-water, reef-building scleractinians from the NE Atlantic Ocean. Coral Reefs 24(3):514

Weaver DC, Hickerson EL, Schmahl GP (2006a) Deep reef fish surveys by submersible on Alderdice, McGrail, and Sonnier Banks in the Northwestern Gulf of Mexico. In: Emerging technologies for reef fisheries research and management. NOAA, Seattle, WA, pp 69–87 (NOAA Professional Paper NMFS,5)

Weaver DC, Naar DF, Donahue BT (2006b) Deepwater reef fishes and multibeam bathymetry of the Tortugas South Ecological Reserve. Florida Keys National Marine Sanctuary, Florida

Weber JN, Woodhead PM (1972) Temperature dependence of oxygen-18 concentration in reef coral carbonates. J Geophys Res 77:463–473

Wegley L, Edwards R, Rodriguez-Brito B et al (2007) Metagenomic analysis of microbial community associated with the coral *Porites astreoides*. Environ Microbiol 9(11):2707–2719

Weinbauer MG, Oregioni D, Groskurth A et al (2019) 33 Diversity of Bacteria Associated with the Cold Water Corals *Lophelia pertusa* and *Madrepora oculata*. In: Orejas C, Jiménez C (eds) Mediterranean cold-water corals: past, present and future. Coral reefs of the world, vol 9. Springer, Cham

Weinnig AM, Gomez CE, Hallaj A et al (2020) Cold-water coral (*Lophelia pertusa*) response to multiple stressors: High temperature affects recovery from short-term pollution exposure. Sci Rep 10: 1768

Welsh SE, Inoue M (2000) Loop Current rings and the deep circulation in the Gulf of Mexico. J Geophys Res 105(C7):16951–16959

White HK, Hsing P-Y, Cho W et al (2012) Impact of the Deepwater Horizon oil spill on a deep-water coral community in the Gulf of Mexico. Proc Natl Acad Sci USA 109:20303–20308. https://doi.org/10.1073/pnas.1118029109

White HK, Lyons SL, Harrison SJ et al (2014) Long-term persistence of dispersants following the Deepwater Horizon oil spill. Environ Sci Technol Lett 1:295–299. https://doi.org/10.1021/ez500168r

Wienberg C, Titschack J, Freiwald A et al (2018) The giant Mauritanian cold-water coral mound province: Oxygen control on coral mound formation. Quat Sci Rev 185:135–152

Williams B, Risk MJ, Ross SW et al (2007) Stable isotope data from deep-water antipatharians: 400-year records from the southeastern coast of the United States of America. Bull Mar Sci 81:437–447

Yang S, Sun W, Zhang F, Li Z (2013) Phylogenetically diverse denitrifying and ammonia-oxidizing bacteria in corals *Alcyonium gracillimum* and *Tubastraea coccinea*. Mar Biotechnol 15:540–551

Yesson C, Taylor ML, Tittensor DP et al (2012) Global habitat suitability of cold-water octocorals. J Biogeogr 39:1278–1292

Young CM, He R, Emlet RB et al (2012) Dispersal of deep-sea larvae from the Intra-American Seas: Simulations of trajectories using ocean models. Int Comp Biol 52(4):483–496

Zhang Y, Ling J, Yang Q et al (2015) The functional gene composition and metabolic potential of coral-associated microbial communities. Sci Rep 5:16191

Cold-Water Coral Reefs of the Southeastern United States

4

Andrea M. Quattrini, Martha S. Nizinski, Jay J. Lunden, Furu Mienis, Cheryl L. Morrison, Leslie Sautter, Harvey Seim, Robert E. Todd, and John Reed

Abstract

The southeastern United States (SEUS), from North Carolina through southern Florida, has the most extensive cold-water coral reefs, including coral mounds and coral gardens, in the US Exclusive Economic Zone. In fact, a region extending from off central Florida to Georgia has been named the "Million Mounds" area and is estimated to contain tens-of-thousands of coral mounds. Oceanographic patterns in this region are dominated by the Gulf Stream and associated oceanographic events (i.e., eddies, meanders, intrusions). Therefore, corals in the SEUS often experience strong currents and rapid changes in environmental conditions, such as temperature and salinity. The Gulf Stream also serves as a conduit for dispersal of larvae, and thus coral populations throughout the area are highly connected, although there are some signatures of differentiation across a bathymetric gradient. The faunal community in the region is diverse and includes a characteristic deep-reef fauna that differs from areas off reef. While resource extraction activities in deep water are present in the region, the most significant anthropogenic threat to cold-water corals in the SEUS is climate change. Future research efforts should focus on our understanding of the interplay between changing environmental conditions and coral development, growth, physiology, and persistence. In addition, we need to better understand both abiotic and biotic processes that govern cold-water coral ecosystems in the region.

Keywords

Deep sea · Habitat area of particular concern · Essential fish habitat · Gulf Stream · Coral mound province · Geomorphology · Oceanographic time series · Coral reproduction · Coral connectivity

4.1 Introduction

The southeastern U.S. (SEUS) region has the most extensive cold-water coral (CWC) areas in the U.S. Exclusive Economic Zone (EEZ; Ross and Nizinski 2007). CWCs are defined as azooxanthellate (non-photosymbiotic) heterotrophic corals that primarily live in cold and/or deep (<12 °C, >50 m) environments (Roberts et al. 2009). A diverse and abundant CWC fauna populates the deep seafloor off the SEUS from Cape Hatteras, NC to the Florida Straits

Andrea M. Quattrini and Martha S. Nizinski contributed equally with all other contributors.

A. M. Quattrini (✉)
Department of Invertebrate Zoology, Smithsonian Institution, National Museum of Natural History, Washington, DC, USA
e-mail: quattrinia@si.edu

M. S. Nizinski
NOAA, National Marine Fisheries Service, National Systematics Lab, National Museum of Natural History, Washington, DC, USA

J. J. Lunden
Department of Biology, Temple University, Philadelphia, PA, USA

F. Mienis
Department of Ocean Systems, NIOZ Royal Netherlands Institute for Sea Research and Utrecht University, Den Burg, The Netherlands

C. L. Morrison
US Geological Survey, Leetown Science Center, Kearneysville, WV, USA

L. Sautter
Department of Geology and Environmental Geosciences, College of Charleston, Charleston, SC, USA

H. Seim
Department of Earth, Marine and Environmental Sciences, University of North Carolina at Chapel Hill, Chapel Hill, NC, USA

R. E. Todd
Department of Physical Oceanography, Woods Hole Oceanographic Institution, Woods Hole, MA, USA

J. Reed
Cooperative Institute of Ocean Research, Exploration and Technology, Harbor Branch Oceanographic Institute, Florida Atlantic University, Fort Pierce, FL, USA

E. Cordes, F. Mienis (eds.), *Cold-Water Coral Reefs of the World*, Coral Reefs of the World 19,
https://doi.org/10.1007/978-3-031-40897-7_4

(Fig. 4.1). At least 202 species of corals, representing antipatharians (black corals), octocorals (sea fans, soft corals), scleractinians (stony corals), and hydrocorals (lace corals) inhabit depths ranging from mesophotic (~40 m) to greater than 2000 m (Ross and Nizinski 2007; Hourigan et al. 2017b). Corals colonize hard-bottom habitat common on the Blake Plateau (BP) and Florida-Hatteras Slope (Fig. 4.1), often forming "forests" or "gardens" on the seafloor. Perhaps more remarkable, however, is the abundance of coral-built mounds and reef structures that are formed throughout the region. Although a few species of framework-forming or reef-building corals occur in the SEUS, *Lophelia pertusa*, which commonly occurs at depths of ~200–1000 m, is the most common framework-forming species in the region. *Lophelia pertusa* forms mounds, banks, ridges, and reef structures that extend 10s–100s of km in length and can rise off the seafloor to heights up to 100 m.

Historically, investigators have surveyed deep waters off the SEUS since the late 1800s, with missions dedicated to exploring CWC habitats beginning in the 1950s (see review in Ross and Nizinski 2007). With the exception of the *Oculina* reefs off northeastern Florida (Reed 2002a, b), the majority of research efforts have focused on CWC habitats occurring at depths of ~400–800 m (Fig. 4.1). This is in part due to the discovery and documentation of corals and coral mound features in earlier studies at these depths on the BP.

Stetson et al. (1962), for example, used single-beam echo soundings to identify and describe numerous mound features in an area now known as Stetson Banks. In the decade to follow, several studies using echosounders noted many more mounds on the Florida-Hatteras slope and the BP (e.g., Uchupi and Tagg 1967; Uchupi 1967; Zarudzki and Uchupi 1968). Using seismic profiling, Uchupi (1967) documented the occurrence of the northernmost coral mounds in the region, off Cape Lookout, NC. Although dredge and trawl samples confirmed coral presence on some mounds in the 1960s, the fact that many of these mounds were composed of coral was not confirmed until decades later (see Ross and Nizinski 2007).

Increased access to the tools needed to survey these areas of rugged topography and extreme currents has given researchers the opportunity to explore more areas where CWCs likely occur. For example, numerous (>100) dives using human occupied vehicles (HOVs; e.g., *Johnson-Sea-Link, Alvin*) and remotely operated vehicles (ROVs; e.g., *Jason*) have been conducted in the region since the 1980s. While a few towed and drop cameras had previously provided glimpses of CWC habitats in the region (see Ross and Nizinski 2007), in situ observations obtained by ROVs and HOVs provided the first holistic views of CWC habitats and their characteristic reef fauna. Since the early 2000s, numerous multidisciplinary research efforts targeting CWC habitat have been conducted. Results from these studies have increased our knowledge of geographic and bathymetric distributions, species composition, and community structure of corals and associated fishes and invertebrates.

Molecular techniques have added considerably to the amount and type of information gathered on CWCs. DNA barcoding of single genes in CWCs has aided identification of species that can be difficult to identify from morphology alone. Often, DNA barcoding studies have revealed the presence of cryptic species (e.g., species morphologically similar but genetically divergent) as well as previously undescribed species (e.g., Herrera et al. 2010; Quattrini et al. 2014). Microsatellites (e.g., repetitive elements of DNA) have illuminated population structure of CWCs in the North Atlantic (Le Goff-Vitry et al. 2004; Dahl et al. 2012), including the SEUS (Morrison et al. 2011).

Advanced technologies have also facilitated data acquisition of the geological and physical environments of the SEUS region. The advent of high-resolution multibeam sonar has enabled much of the region to be mapped in detail. Multibeam mapping has revealed 10s of 1000s of mounds and reef structures that were previously uncharacterized (e.g., Sowers 2020). In addition, benthic lander (autonomous observational platforms) deployments, moorings, and glider surveys in the region have facilitated long-term (8–16 months) acquisition of environmental data (e.g., current speed and direction, temperature) in the water column and near the seafloor at CWC habitats (e.g., Mienis et al. 2014; Prouty et al. 2017). Recently, new programs have created deep-water observatory networks (e.g., ADEON, https://adeon.unh.edu) capable of recording oceanographic and deep-water soundscape data over a time scale of months to years. In combination, technologies such as submergence vehicles, multibeam sonar, benthic landers, and observatories that can host a suite of instrumentation will continue to advance our understanding of CWC habitat in the SEUS region and elsewhere.

This chapter focuses primarily on framework-forming CWCs (herein referred to as CWCs), such as *L. pertusa*, that construct reef, ridge, and mound structures and colonize hard bottoms on the upper to middle continental slope (200–1000 m) of the SEUS. Because many of these framework-forming corals also co-occur with other corals (black corals, octocorals, solitary stony corals) that increase habitat complexity, we refer to these multi-species assemblages as "CWC habitats" throughout this chapter. Here, we document: (1) geology of the BP and Florida-Hatteras Slope that serve as the setting for CWC habitat development, (2) oceanographic patterns and processes that contribute to larval dispersal, population connectivity, and nutrient supply, (3) development and geomorphology of coral structures (e.g., reefs, mounds, and bioherms) in notable areas throughout the SEUS, (4) biology of corals that enable them to persist in the region, (5) the structure and function of

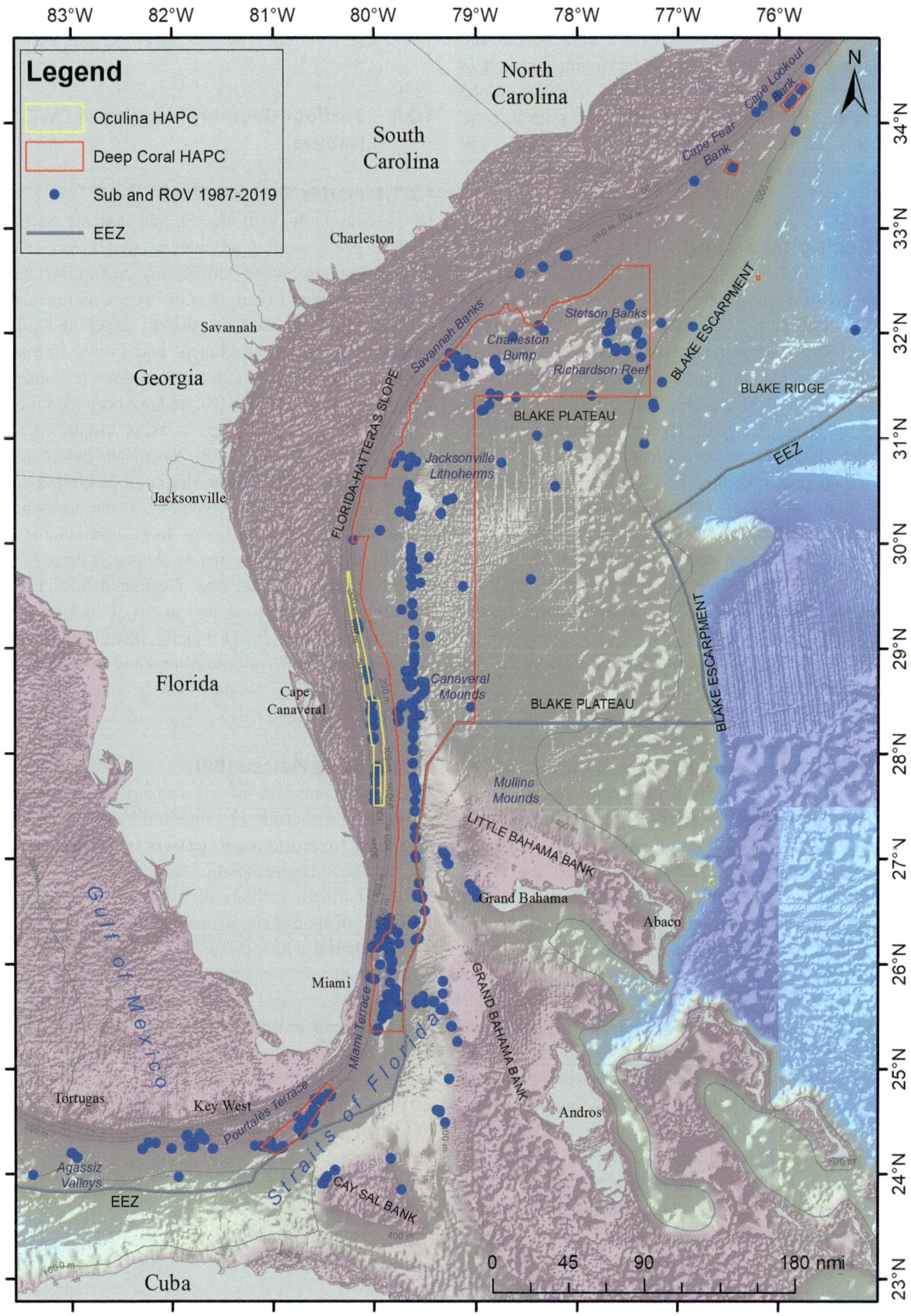

Fig. 4.1 Southeastern U.S. study region. Available locations of dives conducted in cold-water coral habitats from 1987 to 2019 with various ROVs (e.g., *Deep Discoverer, Jason*) and submersibles (e.g., *Johnson-Sea-Link, Clelia, Alvin*). HAPC, habitat area of particular concern; ROV, remotely operated vehicle; EEZ, exclusive economic zone

ecosystems and the communities that they support, (6) current and potential anthropogenic disturbances to coral habitat, and (7) new initiatives and suggestions for future research to increase knowledge of CWCs and CWC habitats in the SEUS region. We also highlight three exceptional areas (*North Carolina Mounds, Richardson Reef Complex*, and *Blake Plateau West or "Million Mounds"*) as case studies.

4.2 Geological Setting

The continental margin off the SEUS coast is unlike most passive (i.e., tectonically inactive) margins. Although the typical shelf-slope-rise sequence is found north of Cape Hatteras, NC, the margin geomorphology from the tip of Cape Hatteras, NC to Miami, FL differs dramatically from the typical sequence (Appendix 1). Beyond the shelf, the BP (Figs. 4.1 and 4.2) is the dominant feature in the region. The BP is a broad, relatively flat, 127,700 km^2 physiographic province (Pratt and Heezen 1964) that extends as far south as the Bahama Banks. The BP is wider at its southern reaches, where the width is as broad as 300 km. The feature narrows north of 31°N, intersecting the continental shelf off Cape Fear, NC. Depths on the BP range from approximately 400 to 1250 m.

The continental shelf inshore of the BP is broader than its counterpart off northern North Carolina. The seaward edge of the shelf begins to steepen at depths between 80 and 100 m creating a narrow, steeper physiographic province between the continental shelf edge and BP known as the Florida-Hatteras Slope (Paull and Dillon 1979; Figs. 4.1 and 4.2). This province extends southward to the Florida Keys.

The eastern edge of the BP steepens rapidly along a prominent and unusually steep feature called the Blake Escarpment. These steep areas include numerous cliff-like scarps, submarine canyons (Gardner et al. 1996), and exposed rock-layer edges (Pratt and Heezen 1964). The escarpment includes the large eastward-protruding Blake Spur, which is located well over 300 km from shore, outside the U.S. EEZ. Beyond the Blake Escarpment and north of the Blake Spur lies the triangular-shaped Blake Ridge with giant bedforms at depths ranging from 2500 to >5000 m (Appendix 1). Specifically CWC habitats on SEUS continental margin are found on the Florida-Hatteras Slope, across the full depth range of the BP, and on the uppermost Blake Escarpment. The portions of the seafloor along the SEUS margin that have been mapped with high-resolution multibeam echosounder bathymetric data show significant variations in geomorphology. The regional geologic structure, variety of complex seafloor features, and substrate character (sediments vs. rocky seabed) offer a wide range of benthic habitats. Habitat conditions are influenced by their proximity to the Gulf Stream (GS), which provides variable temperatures, current velocities, and food availability.

4.2.1 Seafloor Geomorphologies of CWC Habitat

4.2.1.1 Florida-Hatteras Slope

The geomorphology of the Florida-Hatteras Slope is the result of gently sloping sedimentary rock layers covered by modern unconsolidated terrigenous and carbonate sands. Along its seaward extent, the GS sweeps sedimentary-strata outcrops clean of sediments, thereby exposing hard-bottom conducive to CWC attachment and growth. From North Carolina to southern Florida, at least three prominent CWC areas in deep water (~350–550 m) have been identified on the Florida-Hatteras Slope: Cape Lookout Mounds, Cape Fear Mound (both off NC), and Savannah Banks (off GA) (Fig. 4.3, areas 1–3). Off the east coast of Florida, at depths of 70–100 m, 1–2 m colonies of the azooxanthellate scleractinian *Oculina varicosa* form coral mounds up to 30 m in height (Fig. 4.3, area 4). Although the mean annual bottom temperature on the *Oculina* banks is 15 °C, temperatures can drop as low as 10 °C during upwelling events (Reed 1980, 1981). Finally, along the western edge of the Florida Straits are the Miami and Pourtalès Terraces, which support CWC habitats in deep waters (~200–500 m) (Fig. 4.3, areas 5 and 6).

4.2.1.2 Blake Plateau (BP)

The BP is composed of a foundation of a thick layered sequence of carbonate (i.e., limestone) rock, overlain with a veneer of unconsolidated pelagic sediments (Pratt and Heezen 1964). High-resolution mapping surveys of the western and northern sections of the BP have resulted in the discovery of thousands of mound-like features likely to be CWC bioherms. Other dominant seafloor features of the BP include dominant, irregularly eroded, elongate depressions cut into the carbonate layers, with steep cliff-like walls, or scarps. Several multibeam mapping cruises conducted by NOAA ship *Okeanos Explorer* in 2018 and 2019 provided details of several major scarp features across large sections of the BP. Several of these scarps have vertical relief exceeding 100 m. Due to the large geographic extent of the BP and for the purposes of subsequent discussion, we refer to areas within the Plateau as BP North (BPN), BP West (BPW), and BP Central (BPC) (Fig. 4.3).

BP North (BPN) The large, triangular section of BPN, above the 31°30′N parallel, has a complex variety of seafloor features including elongate ridges with large cliff-like scarps, circular depressions, and mound features, as well as broad, flat-lying expanses. Three areas within the BPN are known to

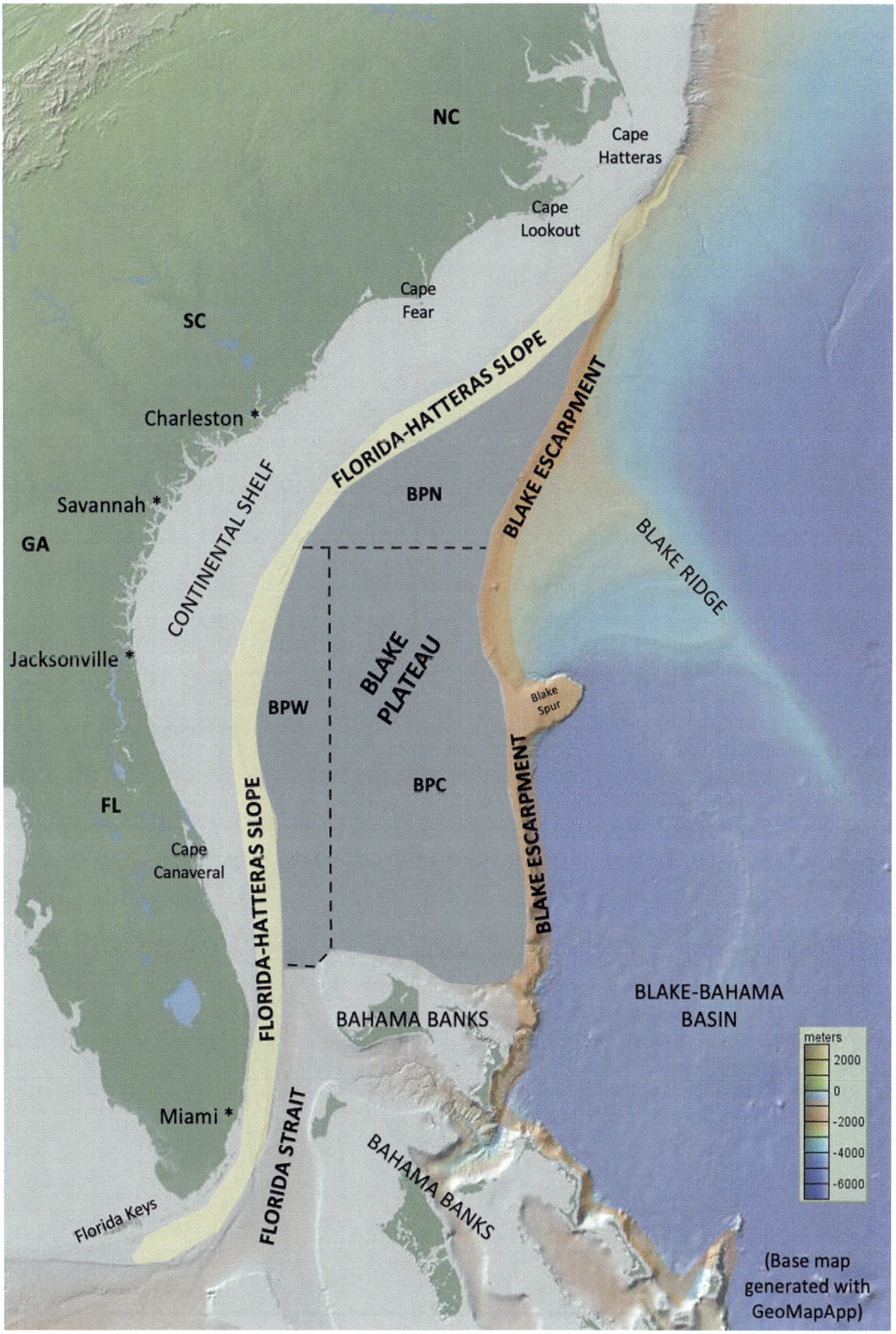

Fig. 4.2 The southeast U.S. continental margin, showing the unique physiographic provinces of the Florida-Hatteras Slope, BP (North, West, and Central regions; BPN, BPW, and BPC, respectively), and Blake Escarpment, all of which include CWC habitat discussed in this chapter. The depth scale color range (0–1500 m) has been modified to highlight variable topography on the relatively flat BP. All dark purple areas show depths >1500 m

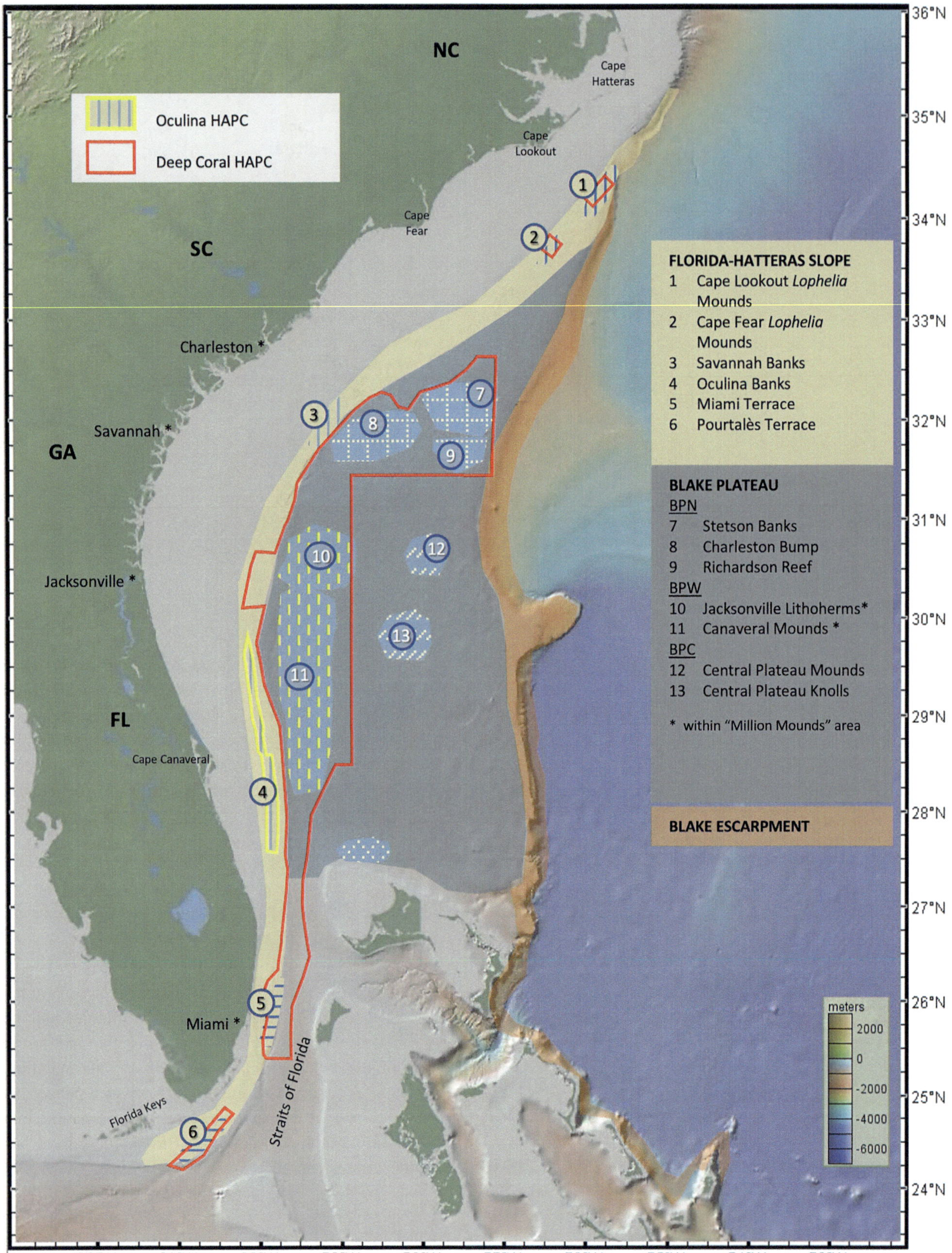

Fig. 4.3 CWCs and CWC habitats discussed in this chapter, shown on their respective physiographic provinces

host CWC habitat (Fig. 4.3, areas 7, 8 and 9). A portion of BPN is included within the region's deep-water Coral Habitat Area of Particular Concern (CHAPC), a protected area designated by the South Atlantic Fisheries Management Council (SAFMC) (Figs. 4.1 and 4.2). On the western side of BPN, at the base of the Florida-Hatteras Slope lies the Charleston Bump, a large geologic feature with exposed rocky surfaces at 400–700 m depth (Sedberry 2001). Here, the GS begins to deflect northeastward (Popenoe and Manheim 2001). East of the Charleston Bump, the seafloor is dominated by both elongate and undulating cliffs that create wide, flat depressions, the likely result of erosion into the flat-lying carbonate rocks caused by bottom currents (Paull and Dillon 1979). Recent exploration of several bioherms along a scarp edge confirmed the existence of a 150-km reef tract referred to as the Richardson Reef Complex Gasbarro et al. 2022. Stetson Banks is the northernmost locality hosting large aggregations of CWC habitats within BPN; however, much of the seafloor north of Stetson Banks remains to be mapped in high resolution.

BP West (BPW) Located directly beneath the main axis of the GS, BPW area (also referred to as the Stetson Mesa; Sautter et al. 2019; Cantwell et al. 2020) contains thousands of mound features, some verified to be CWC bioherms (Sowers 2020). Most bioherms in the BPW are found on relatively flat seabed at 650–850 m depth and can have vertical reliefs ranging from 10 to 80 m off the seafloor (Cantwell et al. 2020). Preliminary evidence (Reed 2002a, b) that the BPW area contained several bioherms with abundant CWC communities at their crests prompted the SAFMC to include this area in the CHAPC. However, the sheer number and extent of mounds in this region were not fully realized until recently (Cantwell et al. 2020; Sowers 2020). The overall extent, high abundance, and high density of CWC bioherms (>25,000 mounds) have earned this area the nickname "*Million Mounds*." Within the "*Million Mounds*" area, two discrete locations, Jacksonville Lithoherms and Cape Canaveral Mounds, have been named because of prior investigations at these locations (Fig. 4.3, areas 10 and 11). In the northern portion of BPW, several major north-south oriented depressions with steep, cliffed walls include apparent but unexplored mound-like structures.

BP Central (BPC) BPC lies east of the main axis of the GS. At present, less than half of this area has been mapped in high resolution. The southeastern section has the sparest coverage with only exploratory and transit lines of high-resolution sonar data available (Appendix 2). Where high-resolution maps are available, high concentrations of individual CWC mounds were discovered on the broad flat seabed of BPC (Cantwell et al. 2020; Fig. 4.3, areas 12 and 13). Although not in detail herein, mounds were notably also discovered in Bahamian waters along the northern slope of the Little Bahama Banks (Mullins et al. 1981), marking the southernmost mounds discovered on the BP (Fig. 4.3). In contrast, the seabed is gently sloping and surprisingly featureless just west of the Blake Spur.

4.2.1.3 Blake Escarpment

Where the BP begins to steepen at the Blake Escarpment, outcropping edges of the rock layers form stair-step features or intra-slope terraces (Sautter et al. 2019). These exposed, hard-bottom terraces provide numerous small scarps for CWC growth. Portions of the Blake Escarpment have been mapped using high-resolution multibeam sonar, however, due to depths ranging approximately 1200–2000 m, relatively little of the escarpment has been explored. Recent ROV dives on intra-slope terraces north of Blake Spur (Fig. 4.3, areas 14, 15) documented vertical scarps within the terraces hosting octocorals, black corals, and scattered solitary cup corals.

4.3 Oceanography

4.3.1 Circulation and Water Masses

The regional oceanic circulation plays a key role in regulating the physical and biogeochemical properties of near-bottom water throughout the SEUS, thereby influencing growth and survival of CWC habitats and larval dispersal. The circulation in the Florida Straits is dominated by the subtropical western boundary current of the North Atlantic, which flows through the Caribbean, into the Gulf of Mexico as the Loop Current, and then enters the Florida Straits south of the Florida Keys as the Florida Current. The flow follows the deep channel, turning counterclockwise to flow northward along the east coast of Florida, where it exits the Straits and becomes the GS (Appendix 3). There is some inflow to the Florida Current through the passages in the Bahamas platform, principally through the Old Bahamas Channel (Rousset and Beal 2014). This inflow causes the average transport to increase from 24–30 Sv where the current enters the Gulf of Mexico through the Yucatan Channel (Athie et al. 2015; Sheinbaum et al. 2002) to 32 Sv at 27°N just before it leaves the Florida Straits (Meinen et al. 2010). The transport continues to increase as the GS flows toward Cape Hatteras, NC, due to inflow from the east, first from the Antilles Current, which enters just north of the Bahamas (Meinen et al. 2019), then from more diffuse inflow, reaching a transport of 57 Sv in the upper 1000 m as the GS moves into deep water northeast of Cape Hatteras (Heiderich and Todd 2020). For convenience, we will refer to this main flow feature as the

GS, though as pointed out above, its name varies along its course.

The spatial structure of the flow in the GS is that of a half-jet, with peak currents at the surface, decreasing with depth and horizontally away from the jet axis (e.g., Appendix 4). The GS is 50–100 km wide and extends to the bottom at depths of 500–700 m in the Florida Straits and 1000 m over the western half of the BP (e.g., Appendix 4; Heiderich and Todd 2020). The path of the GS is constrained by the bathymetry of the Florida Straits, though it does vary in response to the Loop Current path in the southern Florida Straits (Androulidakis et al. 2020; Fratantoni et al. 1998); north of the Straits the flow generally follows the upper slope as it crosses the BP (e.g., Appendix 4). An exception occurs at the Charleston Bump, a topographic feature on the BP at approximately 31.5°N (see the 500-m isobath in Appendix 4, a), where the GS deflects offshore, sometimes weakly, sometimes strongly (Bane and Dewar 1988; Gula et al. 2015; Legeckis 1979; Zeng and He 2016). Strong near-bottom flows (Appendix 4) produce large bottom boundary layers, up to 100 m thick, and excite energetic lee waves in the overlying stratified flow (Todd 2017).

The shoreward side of the GS is prone to instability (e.g., Todd et al. 2016) and forms meanders, with along-stream scales of 100–200 km (Lee and Atkinson 1983), and amplitudes that vary along its path. Within the Florida Straits, meander amplitudes are relatively small, but meander amplitudes increase as the GS flows over the BP (Lee et al. 1991). Inshore of the meander troughs (offshore deflections of the GS), frontal eddies often form cold-core cyclonic circulation features that can drive upwelling onto the shelf at their leading edge and downwelling from the shelf at their trailing edge (Lee et al. 1991). The frontal eddies can draw fluid from upstream meander crests (onshore deflections of the GS) around their shoreward edge, forming warm filaments, surface-trapped linear features, that are typically found over the shelf break. These features propagate downstream with speeds of 0.5–0.7 m s^{-1} on average (Lee and Atkinson 1983).

Within the southern portion of the Florida Straits, recirculations develop near Tortugas or Cuba, depending on the path of the Loop Current. If the Loop Current extends far into the Gulf of Mexico, the Tortugas cyclonic gyre can form to the north of the GS (Fratantoni et al. 1998). If the Loop Current turns anticyclonically out of the Yucatan Channel into the Florida Straits, then an anticyclonic circulation is established north of Cuba (Androulidakis et al. 2020). These recirculation features can be long-lived, surviving for months (shown schematically in Appendix 3). Meander amplitudes over the BP vary systematically, growing in amplitude downstream of Cape Canaveral, before decaying as the GS approaches Palm Beach, FL, between 30°N and 32°N. Meander amplitudes grow again downstream of the Charleston Bump, then decay from 33°N to Cape Hatteras. When the GS is strongly deflected at the Charleston Bump, the Charleston gyre, another recirculation feature, can develop on its shoreward side (Bane et al. 2001). Strong equatorward flows along the upper slope help close the recirculation (Seim and Edwards 2019).

The GS transports waters with a variety of physical and biogeochemical properties through the region (see Heiderich and Todd 2020; Szuts and Meinen 2017). Of particular relevance to CWC habitats are the near-bottom water masses in and near the GS (Fig. 4.4), which exhibit two distinctive tails in potential temperature-salinity space at potential densities greater than 27 kg m^{-3} (Fig. 4.5). Antarctic Intermediate Water (AAIW), the fresher of the two near-bottom waters, is typically high in nutrients and silica (Tsuchiya 1989) and low in oxygen (typically around 3 mg m^{-3}, Fig. 4.5; Atkinson 1983). The fresh signature of AAIW spreads northward from the Florida Straits and over the BP (Fig. 4.2; Atkinson 1983), but the volume transport of AAIW decreases toward Cape Hatteras (Heiderich and Todd 2020) due to mixing with surrounding water masses that results from the strong near-bottom flow (Fig. 4.4c) encountering topography (Todd 2017). AAIW is found at shallower depths on the western side of the Florida Straits and BP due to the tilting of isopycnals associated with the geostrophic flow of the GS (Szuts and Meinen 2017). The other major near-bottom water mass over the BP is upper Labrador Sea Water (uLSW; Pickart and Smethie 1993), which is entrained into the GS over the BP and northeast of Cape Hatteras (Heiderich and Todd 2020). Found at densities greater than 27.4 kg m^{-3}, uLSW is saltier than AAIW of similar density (Fig. 4.4) and, with transit times of a few years from formation regions in the Labrador Sea (Le Bras et al. 2017), is rich in oxygen (Fig. 4.5; Pickart 1992).

4.3.2 Near-Bed Environmental Conditions

Benthic ecosystems like CWC reefs and mounds are governed by interactions between the organisms and the physical and chemical environment. In general, *Lophelia* CWC mounds and reefs occur in areas with strong bottom currents (Mienis et al. 2007; Roberts et al. 2006), a temperature range between 4 and 12 °C and well-oxygenated waters (Freiwald 2002). It also has been shown that CWCs can withstand daily fluctuations of temperature (up to 3 °C) and current speed, which are often linked to internal tides or downwelling (Davies et al. 2009; van Haren et al. 2014).

When compared to other CWC reef areas, reefs and mounds in the SEUS region (Fig. 4.1) are living under extreme environmental conditions characterized by the largest natural temperature range (e.g., Cape Lookout, 5.6–15.2 °C, Richardson Reef Complex 3.8–10.6 °C) recorded to date

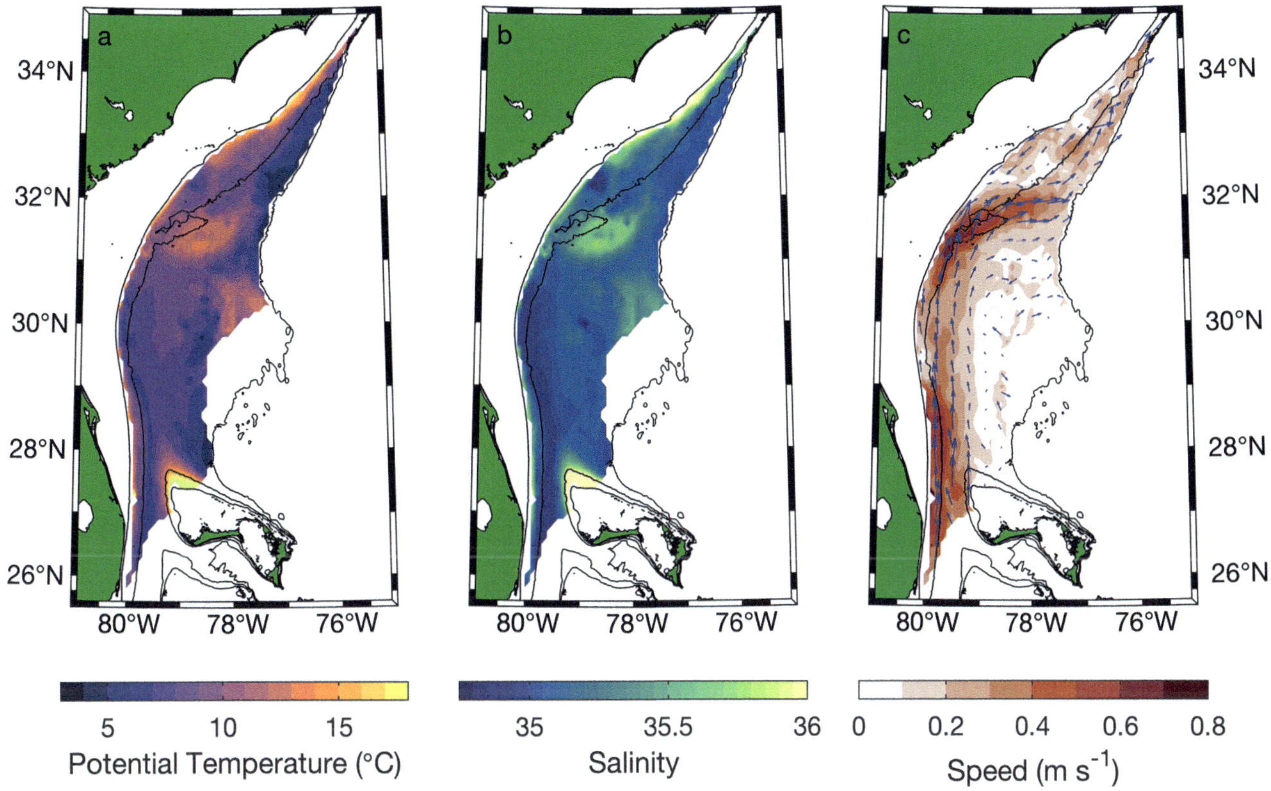

Fig. 4.4 Maps of mean near-bottom properties over the BP estimated from glider observations (Todd and Owens 2016; Todd 2020) averaged as in Todd (2021). (**a**) Potential temperature, (**b**) salinity, and (**c**) velocity in the deepest averaging bins at each 0.1° × 0.1° grid point. The 100-, 500-, and 1000-m isobaths are drawn in black. Note that color bars are saturated at the high ends in order to emphasize properties in deeper water

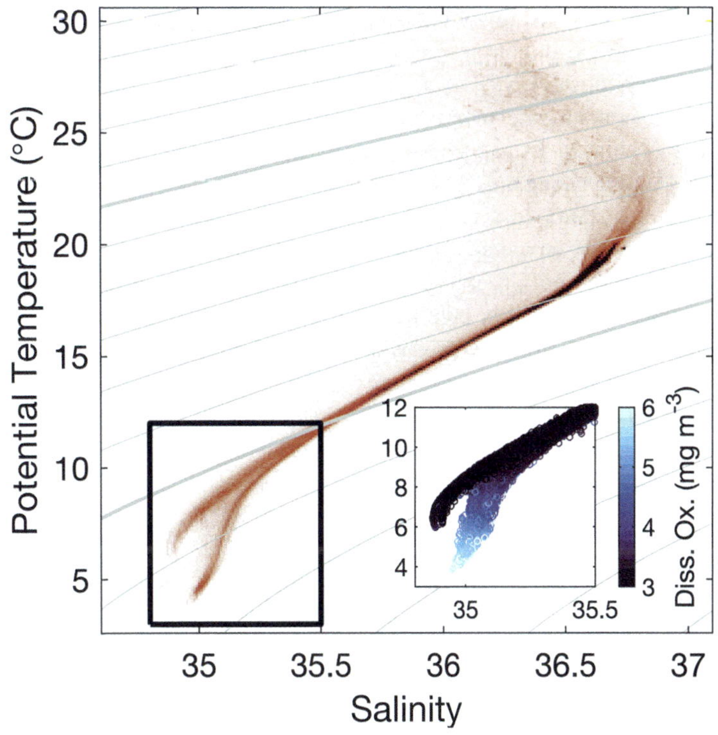

Fig. 4.5 Potential temperature-salinity distribution for waters over the BP, shown as a probability density function derived from all Spray glider observations in the region. Potential density is contoured in gray with a contour interval of 0.5 kg m^{-3} and the 24.0 and 27.0 kg m^{-3} isopycnals bold. The two distinct modes of the distribution at densities greater than 27 kg m^{-3} are due to fresher AAIW and saltier uLSW. The inset shows dissolved oxygen measured by gliders in color as a function of potential temperature and salinity for the near-bottom waters over the BP

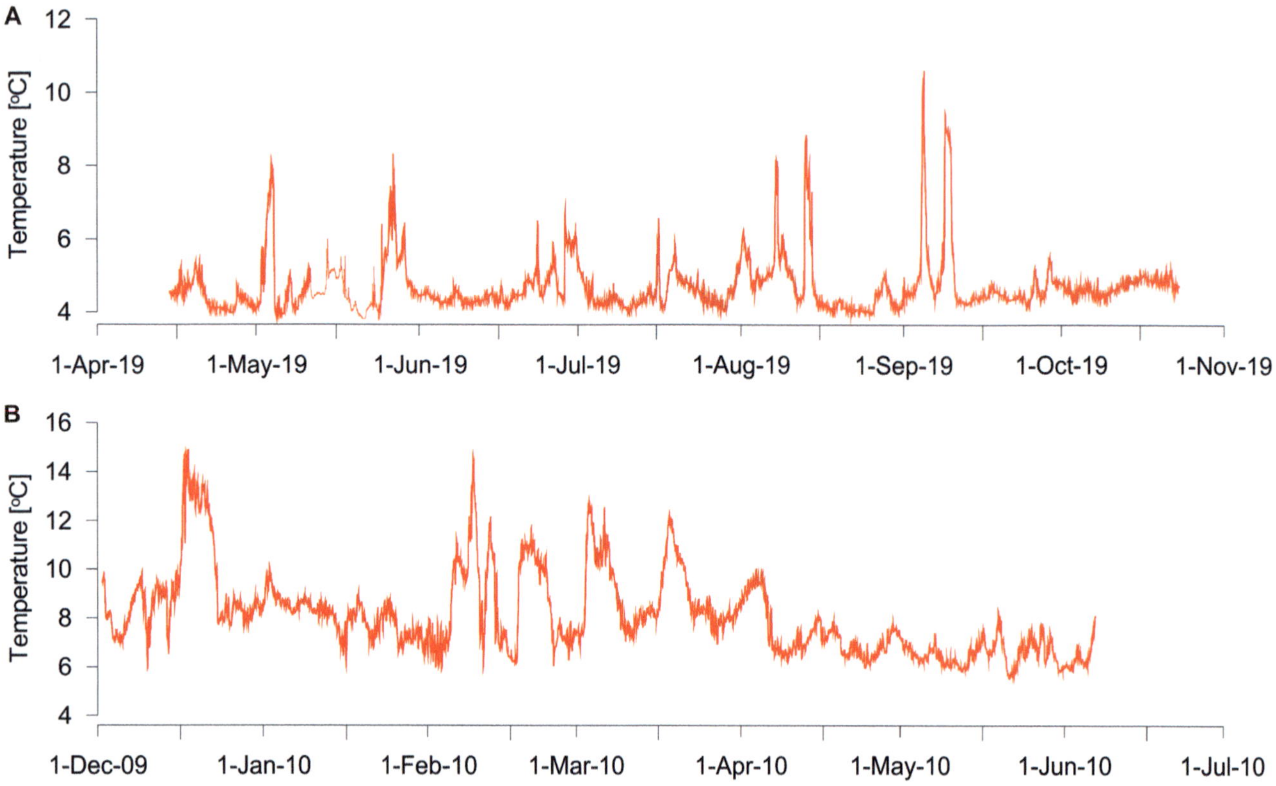

Fig. 4.6 Long-term deployment of benthic landers shows the presence of the Gulf Stream in the CWC reef areas as shown by increases in bottom-water temperature. (**a**) Richardson reef complex, (**b**) Cape Lookout

(Mienis et al. 2014). Long-term data collected with benthic landers in the CWC areas near Richardson Reef Complex and Cape Lookout have shown that CWCs thrive in extreme conditions, related to the occasional presence of warm, oxygen poor, and oligotrophic GS water (Fig. 4.6). Coastward meanders of the GS in the Cape Lookout area lead to rapid rises in bottom-water temperature of up to 9 °C and salinity of 1.18 within 24 h. In addition, upper temperature limits of 15.2 °C are among the highest temperatures ever observed for *L. pertusa* (Brooke et al. 2013; Mienis et al. 2014), while bottom-water temperatures as low as 3.8 °C outside the influence of the GS are among the lowest in the range. Overall, more than six pronounced high temperature events with a duration from several days up to a week were observed over a period of 6 months near Cape Lookout in 2010 (Mienis et al. 2014). These extremely large fluctuations in temperature could have consequences on coral metabolic and calcification processes (see Dodds et al. 2007; Dorey et al. 2020).

Near-bed current patterns at Richardson Reef Complex and Cape Lookout (Fig. 4.6) are also driven by the meandering of the GS. When the pathway of the GS is more offshore, current speeds are relatively low <0.1 m s^{-1}. In contrast, the onset of warm events related to inshore movement of the GS increases near-bottom current speeds (Mienis et al. 2014). Near-bottom current measurements showed, on average, higher mean current speeds (>0.2 m s^{-1}) (Grasmueck et al. 2006; Mienis et al. 2014; Paull et al. 2000) compared to CWC reefs in the NE Atlantic and Gulf of Mexico (Davies et al. 2009; Mienis et al. 2007, 2012). For example, at the Great Bahama Banks, corals living between 600 and 800 m depth experienced sustained Florida Current speed strengths of 0.2 m s^{-1} with peak speeds up to 0.5 m s^{-1}. Current direction in this area, related to semi-diurnal tidal flow, changed between a north and southward direction (Grasmueck et al. 2006). In the Cape Lookout mound area, GS presence was characterized by a northward flow with peak current speeds up to 0.6 m s^{-1}. During these high current speed events acoustic backscatter also increased, related to the resuspension of bottom sediments (Mienis et al. 2014). Similar observations were made at the Richardson Reef Complex where peak current speeds reached up to 0.8 m s^{-1}. It has been suggested that GS meanders influence biological primary production, related to upwelling intrusion events, influencing phytoplankton and bacterioplankton production (Lee et al. 1991; Matrai et al. 1996). However, only one fluorescence peak was recorded near the seabed in March, which was related to a spring bloom (Leterme and Pingree 2008).

One of the major sets of environmental parameters suggested to limit CWC reef formation at the near-bed includes factors related to the saturation of calcium carbonate in seawater, including pH and the aragonite saturation state (Guinotte et al. 2006; Davies and Guinotte 2011). Indeed, recent work in the eastern North Atlantic revealed new information on the interplay between water mass circulation and limitations of calcium carbonate precipitation of reef-forming CWCs (Fontela et al. 2020). Generally, reef-forming CWCs inhabit areas where the aragonite saturation state approaches 1.0 (Form and Riebesell 2012; Lunden et al. 2013), but *L. pertusa* has been observed at saturation states as low as 0.81 in the Southern California Bight (Gómez et al. 2018) although no reefs or mounds are present in this region. In the SEUS, measured values for aragonite saturation are lacking overall. However, surveys of the Richardson Reef Complex in 2018 identified a range of 1.49–1.59 in aragonite saturation state and pH values of 7.68–7.81 (Gasbarro et al. 2022). While no large mounds or reefs are present in the northern Mid-Atlantic Canyons, Brooke and Ross (2014) reported aragonite saturation values of 1.41–1.44 and pH of 7.94–7.97 at Baltimore Canyon, providing evidence that other factors drive reef and mound formation other than aragonite saturation state. As increasing emissions of CO_2 continue to alter the carbonate chemistry of the oceans, these values may change in the future as the aragonite saturation horizon shoals and exposes the reefs to undersaturated conditions. However, some evidence exists that the U.-S. South Atlantic Bight may serve as a refugia for cold-water coral reefs below 700 m depth (Morato et al. 2020; Gasbarro et al. 2022).

4.4 The Reef Structure

CWC habitats extend across much of the northern and western portions of the BP ranging from off North Carolina southward through the Florida Straits. These CWC habitats include coral mounds, rocky ridges, reefs, and escarpments; various hard-bottom habitats including the Miami Terrace and Pourtalès Terrace in the southern Florida Straits; and deep-water canyons (Agassiz and Tortugas Valleys) off of the Tortugas (Reed et al. 2013b) and off North Carolina (Partyka et al. 2007; Ross and Nizinski 2007). In addition, extensive CWC habitat occurs in the eastern Florida Straits on the deep island slopes off western Bahamas and possibly off northern Cuba. Notably, the SEUS harbors some of the most abundant CWC reefs and mounds as compared to nearby regions (e.g., Gulf of Mexico, Caribbean, northeastern U.S.). The dominant structure-forming scleractinian corals are *L. pertusa, Enallopsammia profunda*, and *Madrepora oculata*. Other foundation species include stylasterid corals, octocorals, black corals, and sponges (see Sect. 4.6).

4.4.1 Mound Formation and Geomorphology

Framework-forming CWCs have been defined as coral reefs, coral banks, coral mounds, bioherms, or lithoherms (Teichert 1958; Stetson et al. 1962; Neumann et al. 1977; Wilson 1979; Paull et al. 2000; Reed et al. 2005a, b, 2006, 2013b). Rogers (1999) suggested that such topographic features fall within the definition of a "coral reef" based on their physical and biological characteristics. A bioherm is a deep-water coral bank that, over centuries, has formed a mound of unconsolidated sediment and coral debris and is often capped with thickets of live coral (Reed 2002a, b, 2006). Lithoherms are high-relief, lithified carbonate mounds, rather than unconsolidated sediment mounds, which also may be covered with thickets of coral (Neumann et al. 1977). Both lithoherms and bioherms may occur as isolated individual features or as connected chains of clustered mounds. Seismic profiles of CWC mounds typically do not show their internal structure (Stetson et al. 1962; Mullins et al. 1981). However, cores recently extracted from coral mounds support the hypothesis that bioherms are accumulations of coral debris and sediment that may or may not be built upon a hard substrate (William et al. 2006; Messing et al. 2008). Ayers and Pilkey (1981) examined a number of coral banks on the Florida-Hatteras Slope and BPW and found that dead-coral samples ranged from 5000 to 44,000 years old.

The formation of a CWC reef or mound may progress through the following hypothetical sequence as proposed in part by Squires (1964), Wilson (1979), and Mullins et al. (1981): (1) coral larvae initially settle and develop into isolated colonies on rock pavement or outcrops; (2) a coral thicket forms as other colonies grow nearby either by sexual reproduction or by branch fragmentation and regrowth; (3) a coppice stage or mound develops from trapped sediment and coral debris; and finally (4) the coppice develops into a coral bank, a large structure composed of unconsolidated coral debris and sediment and capped with live coral. A final mature phase may result in which the mantle of living coral is relatively negligible to the large volume of dead coral (Newton et al. 1987). This succession may explain the numerous extinct, relic CWC mounds that are common in the Atlantic and Gulf of Mexico. However, Newton et al. (1987) suggested that paleoclimate fluctuations may also contribute to the deterioration of many of these senescent reefs. Glacial-interglacial conditions may cause CWC ecosystems to die or recolonize in certain areas (e.g., Wienberg et al. 2009; Hebbeln et al. 2019).

Several studies have indicated that the northward flowing Florida Current and GS control the abundance and shape of coral mounds in the Florida Straits (e.g., Correa et al. 2012a, b). In general, mounds influenced by the GS and Florida Current are often teardrop shaped, with the steepest peak and densest live coral on the south slope, facing the strongest current (e.g., Neumann et al. 1977; Messing et al. 1990; Quattrini et al. 2012). However, coral mounds at the base of the Miami Terrace escarpment have a different geomorphology. Unlike most coral mounds on the BP and in the Florida Straits, these coral ridges are aligned perpendicular to a locally persistent southward-flowing bottom current, resulting in the dense framework of *L. pertusa* and *E. profunda* to occur on the north slopes of these mounds (Neumann and Ball 1970; Correa et al. 2012a, b). The SEUS region also includes other hard-bottom habitats such as rocky escarpments, ridges, boulders, and pavement, as well as karstic topographic features including deep-water sinkholes. All of these habitats provide hard substrate for a variety of corals and sponges.

4.4.2 CWC Reefs of the SEUS

4.4.2.1 Florida-Hatteras Slope North: North Carolina Mounds and Savannah Banks

Historically, Uchupi (1967) reported the first record of coral mounds in the late 1960s. Two years later, Rowe and Menzies (1969) referred to discontinuous banks of *L. pertusa* along the 450-m isobath off North Carolina based on echosounder recordings (Messing et al. 2008). No CWC mounds are known north of Cape Lookout in U.-S. waters, although *L. pertusa* colonies occur on steep mud walls in several submarine canyons north of Cape Hatteras (Quattrini et al. 2015; Brooke et al. 2017).

Case Study—Box 4.1: North Carolina Mounds

The northernmost coral-built mounds in the region are located off North Carolina at depths of 320–550 m (Mienis et al. 2014). These CWC mounds, known as the Cape Lookout and Cape Fear Mounds, are composed primarily of *Lophelia pertusa* with few, scattered colonies of *Madrepora oculata*. At least ten large (>30 m high) and several smaller mounds occur off Cape Lookout (over a ~20 km^2 area) whereas only one isolated mound (~100 m tall, 0.7 km^2) has been documented off Cape Fear (Ross and Quattrini 2009; Mienis et al. 2014). The Cape Lookout and Cape Fear mounds are formed by successive coral growth followed by collapse and sediment entrapment by the coral framework. The bases and slopes of the mounds are covered with dead-coral rubble, and the tops are often capped with extensive live-coral cover particularly on the south-southwest sides facing the currents. In contrast, the Cape Fear mound is composed of mostly dead coral (Ross and Quattrini 2009; Mienis et al. 2014, Fig. 4.7a). Off Cape Lookout, mound development appears to have started in the middle Holocene (~6K years ago, Matos et al. 2015). Coral mounds in this region can experience extreme shifts in current directions and speeds, salinity and temperature in a single day due to Gulf Stream meanders (Mienis et al. 2014).

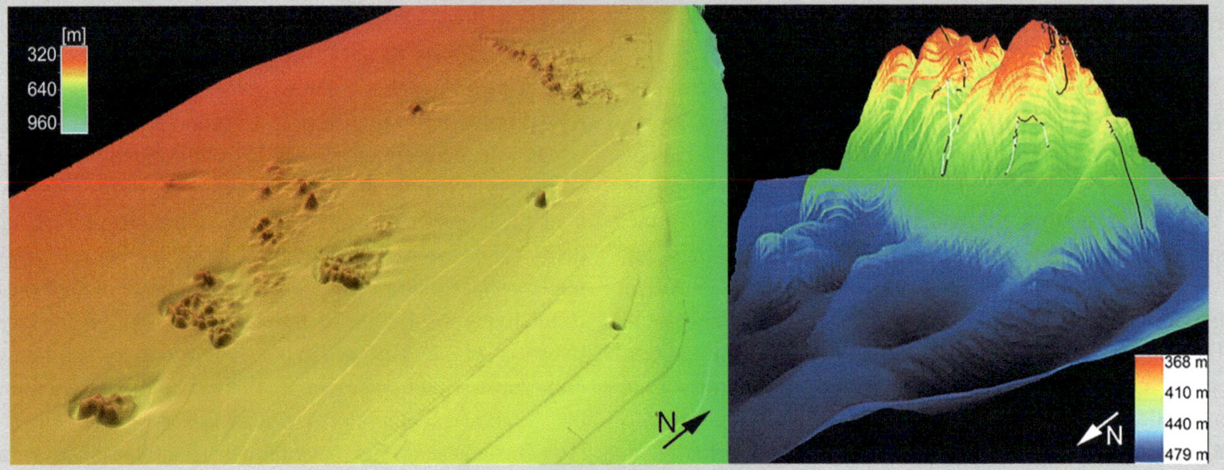

Cape Lookout Mounds (left, photo adapted from Mienis et al. 2014) and Cape Fear Mound (right, photo adapted from Quattrini et al. 2012)

Fig. 4.7 Cold-water coral habitats of the southeastern U.S. (**a**) Cape Fear Mound, conger eel *Conger oceanicus*, sea anemones Hormathiidae, 386 m (*Johnson-Sea-Link* sub JSL I-4896, photo from Partyka et al. 2007); (**b**) Richardson Ridge, *Lophelia pertusa* mound, BP, 778 m (ROV *Deep Discoverer* EX1806-07, photo courtesy of NOAA Office of Ocean Exploration and Research); (**c**) Blake Plateau Knolls, *Lophelia pertusa* mound, 827 m (ROV *Deep Discoverer* EX1903-04, photo courtesy of NOAA Office of Ocean Exploration and Research); (**d**) Cape Canaveral *Lophelia pertusa* mounds, Reed Peak #339, 491 m (JSLII-09-Atl-3716, photo courtesy of S. Ross); (**e**) Miami Terrace, Reed Peak #BU-4, Stylasteridae, *Leiopathes* sp., 284 m (JSL I-4666, photo courtesy of J. Reed, Hourigan et al. 2017b); (**f**) Pourtalès Terrace lithoherms, Reed Peak #311, Stylasteridae, *Madracis asperula*, 167 m (JSL II-3597, photo courtesy of J. Reed)

The Cape Lookout and Cape Fear mounds, residing within the boundaries of the Deepwater Coral HAPC (Ross and Quattrini 2007), appear to be the northernmost bioherms in the SEUS region (Box 4.1) (Partyka et al. 2007; Ross and Nizinski 2007; Roberts et al. 2009; Quattrini et al. 2012). These features consist of 80–100 m tall mounds and ridges at depths of 366–463 m (see Box 4.1). The mounds consist of a sediment and coral rubble matrix, capped with mostly monotypic stands (up to 3 m tall) of live and dead *L. pertusa* with lesser amounts of other colonial scleractinians, including *M. oculata* and *E. profunda*, also present (Fig. 4.7a). In the North Carolina mounds area, current patterns seem to have shaped the mounds, which have teardrop-shaped trenches at the base. A deeper CWC bank, the Agassiz Coral Hills, dominated by solitary corals was reported off North Carolina at depths of 650–750 m (George 2002), although its exact location requires verification (Ross and Nizinski 2007).

The Savannah Banks, located on the western side of the Charleston Bump, consists of numerous lithoherms and bioherms at depths of 490–555 m (Reed et al. 2006; Partyka et al. 2007; Messing et al. 2008). Wenner and Barans (1990, 2001) described a variety of 15–23 m tall mounds. These mounds tend to be covered mostly with dead *L. pertusa* rubble and standing thickets of live *L. pertusa* colonies (15–30 cm diameter) intermixed with small colonies of *M. oculata* and stylasterid hydrocorals and sponges (Reed 2002a, b; Reed et al. 2006; Ross and Quattrini 2007).

4.4.2.2 Florida-Hatteras Slope South: Miami Terrace and Pourtalès Terrace

Off southeastern FL, the continental slope is interrupted by two terraces at intermediate depths (200–600 m): the Miami and Pourtalès terraces. These Miocene age terraces have complex karst-like topography of Tertiary limestones, which form high-relief rock ridges, rock mounds, scarps, slabs, and rock pavement that host significant CWC habitat. The aerial cover of CWC habitat at these two regions is ~2329 and 5823 km^2, respectively (Reed et al. 2013b). Although CWCs occur here, they generally do not accumulate as topographically recognizable bioherms.

The Miami Terrace is a 65-km long carbonate platform in the northern Florida Straits that lies between Boca Raton and South Miami, Florida at depths of 200–400 m (Reed and Wright 2004). Along the eastern edge is the Miami Terrace Escarpment, which extends as a steep slope from the top edge of the terrace at 275 m to the eastern base at 550–600 m. The slope consists of a series of terrace-like steps of rock slabs and ledges with 30–45° slopes and sharp vertical escarpments up to 90 m in height, with sinkholes in some areas (Neumann and Ball 1970; Ballard and Uchupi 1971; Reed et al. 2006). This hard-bottom habitat is capped with scattered colonies and thickets of *L. pertusa*, stylasterid hydrocorals (Stylasteridae), bamboo corals (Isididae), and various other octocorals and sponges (Fig. 4.7e; Reed and Wright 2004; Reed et al. 2006).

The Pourtalès Terrace occurs in the southern Florida Straits and parallels the Florida Keys. Along its 213 km extent, Pourtalès Terrace has a maximum width of 32 km and ranges in depths of 200–450 m (Reed et al. 2005a). High-relief, hard-bottom, topographic features consist of a chain of sinkholes extending ~100 km along the southwest margin and, numerous high-relief knolls and ridges on the mid-terrace (Jordan et al. 1964; Malloy and Hurley 1970; Gomberg 1976; Land and Paull 2000; Reed et al. 2005a, 2014). The Marathon sinkhole is massive, measuring 900 m in diameter, 527 m at the base, and 445 m at the rim. High densities of sponges (1–80 colonies per m^2), stylasterid hydrocorals (9–96 colonies per m^2), and octocorals (16–48 colonies per m^2) were observed (Reed et al. 2005a), especially on the rims and slopes of the plateaus and terraces, and rims of the sinkholes (Fig. 4.7f). CWC reefs disappear in the southern Florida Straits off the Florida Keys. Except for one *Lophelia* mound found off Marathon Key, scleractinians are sparse and consist of small colonies of *E. profunda*, *L. pertusa*, *Madrepora* sp., and *Dendrophyllia* sp.

4.4.2.3 Blake Plateau North: Stetson Banks, Richardson Reef Complex, Charleston Bump

Found off the coasts of South Carolina and Georgia, Stetson Banks, Richardson Reef Complex, and Charleston Bump represent major areas of high-relief, hard-bottom topography with significant coral habitat, including *L. pertusa* mounds (Partyka et al. 2007; Ross and Quattrini 2009). Stetson Banks, a large region near the northeastern edge of the BP, is an area of extremely rugged topography with a variety of substrates and geomorphologies (Stetson et al. 1962; Reed 2002a, b, 2006; Messing et al. 2008; Sedberry et al. 2010). Additionally, Stetson Banks is one of the deeper and more complex CWC habitats on the BP and supports a variety and often high density of corals and sponges (Stetson et al. 1962; Reed 2002a, b; Reed et al. 2006; Ross and Quattrini 2007). *Enallopsammia profunda* dominates throughout with *L. pertusa* concentrated on mound crests. This area includes numerous mounds up to 146 m high spread over a 6174-km^2 area (Ross and Nizinski 2007). One feature, known as Stetson's Pinnacle, rising nearly 150 m from the base to a peak at 627 m (Reed et al. 2006), has complex and rugged topography consisting of 60–90° rock walls and 3–9 m tall rock outcrops, with live *L. pertusa* colonies up to 60 cm tall becoming more common and completely covering some rock ledges (Reed et al. 2006). Although *L. pertusa* dominates, *E. profunda* and *S. variabilis* also occur.

The Charleston Bump is located on the northwestern region of the BP at depths of 400–700 m (Sedberry 2001). Although coral mounds have not been documented here, the Charleston Bump provides a variety of other hard substrates conducive to formation of CWC habitat, including precipitous rocky slopes and escarpments with up to 100 m relief and flat phosphoritic-manganese pavement (Sedberry et al. 1994; Popenoe and Manheim 2001).

Recent expeditions (2018–2019, e.g., R/V *Atlantis*, R/V *Ron Brown*, NOAA *Okeanos Explorer*) have mapped and surveyed (using ROV and HOV) the northern region of the BP and Blake Escarpment. These expeditions discovered mounds dominated by *L. pertusa*, at depths of ~800–900 m, that appear to be true bioherms. Additionally, on the northeastern BP at the outer edge of the GS, the Richardson Reef Complex was discovered (Gasbarro et al. 2022). The Richardson Reef Complex comprises numerous coral mounds that appear solely covered with coral rubble and colonies of live *L. pertusa* and *M. oculata* (Fig. 4.7b; Box 4.2). Documenting that these elongate, steep ridge areas are actually coral mounds was a significant finding, as these features occur in many areas in the BPN region. Although few surveys have been conducted at deeper depths, no framework-forming CWCs have been found on the Blake Escarpment (1736 m) and Blake Ridge (3424 m). However, solitary stony corals, black corals, octocorals, and sponges were commonly observed attached to the manganese encrusted rock.

Case Study—Box 4.2: Richardson Reef Complex

Although CWCs were known to exist in this area (Stetson et al. 1962; Popenoe 1994; Popenoe and Manheim 2001), recent multibeam sonar surveys conducted by NOAA Ships *Okeanos Explorer* and *Ronald H. Brown* as well as R/V *Atlantis* revealed over 3000 CWC features that often form contiguous reef tracts. Richardson Reef Complex, comprising 150 km of reef tracts, lies at the eastern edge of the BP, approximately 180 km off South Carolina, at 700–900 m depth (Gasbarro et al. 2022). This deep-reef complex ranks among the largest in the world (Gasbarro et al. 2022). Video imagery from two HOV *Alvin* and five ROV (*Jason-II* and *Deep Discoverer*) dives verified that the reef consists primarily of the scleractinian coral *Lophelia pertusa*, with fewer colonies of *Madrepora oculata* and *Enallopsammia profunda*. Typical of CWC reefs in the region, the slopes of these features are draped in coral rubble, transitioning to higher coverage of live-coral colonies toward reef crests. Oceanographic conditions at the Richardson Reef Complex are also noteworthy, where some of the highest current speeds and lowest temperatures and pH levels for deep reefs were recorded. Continued investigations aim to reveal the influence of the Gulf Stream on this extensive cold-water coral reef habitat.

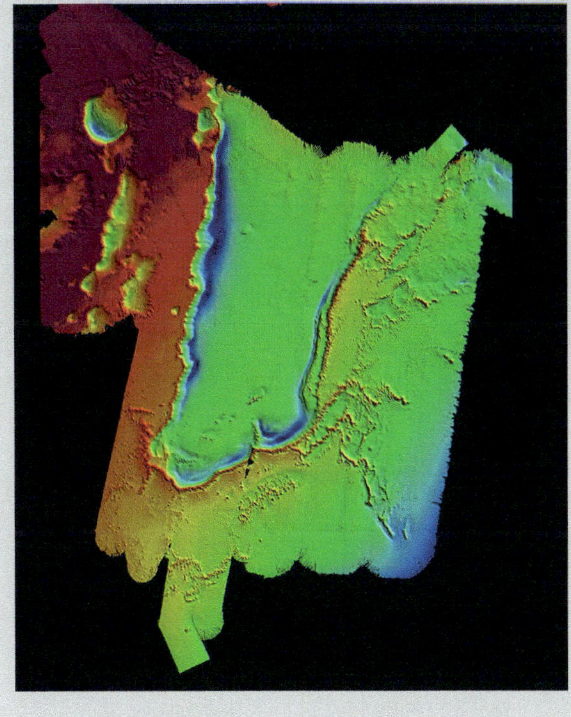

Case Study—Box 4.2 (continued)

Multibeam bathymetry map of the Richardson Reef Complex. Coral habitat features including linear reefs, mounds, and ridges are observable

4.4.2.4 Blake Plateau West: Jacksonville Lithoherms, Canaveral Mounds, "Million Mounds"

High-relief coral mounds and ridges, located off Florida from Jacksonville to Cape Canaveral at 200–900 m depth, are nearly continuous in the BPW region (e.g., Paull et al. 2000; Reed 2002a, b; Reed et al. 2006, 2013a; Hourigan et al. 2017b). Paull et al. (2000) estimated the abundance of coral lithoherms off northern Florida to be over 40,000. In 2013, the extent of CWC habitat off eastern Florida was estimated at over 13,400 km^2 (Reed et al. 2013b). Based on recent multibeam surveys, Sowers (2020) estimated that >25,000 individual peaks, likely coral mounds, occur in this region that has become known as "*Million Mounds*" (Box 4.3). The exact nature and extent of the peaks in this region will require further validation, including high-resolution multibeam mapping and groundtruthing of live and/or dead CWCs. Nevertheless, this region likely contains the largest CWC province in U.S. waters. Additionally, the shallowest recorded *L. pertusa* bioherm in the western North Atlantic occurs on the BPW (Ross et al. 2015). Discovered off Jacksonville at a depth of 210 m (Ross et al. 2015), this mound is significantly shallower than previous records of live *L. pertusa* in the region (Ross and Nizinski 2007).

Many of the mounds off the Jacksonville region appear to be primarily lithoherms, carbonate rock structures capped with coral thickets (Reed et al. 2006). The lithoherms are generally 100–1000 m long with a ridge crest generally oriented perpendicular to the northerly flowing GS. One feature off Jacksonville, described as a "massive lithoherm," was reported to extend 5.7-km (oriented in a N–S direction) and consists of at least seven individual peaks with heights of 30–60 m, rising from a base at 701 m, and total vertical relief of 157 m (Reed et al. 2006). These mounds are composed predominantly of *L. pertusa* but some are intermixed with *E. profunda*. *Madrepora oculata* is not a major component of these mounds and occurs only as isolated, small colonies.

The features, south of St. Augustine to central Florida, appear to be primarily unconsolidated sediment and coral rubble mounds capped with dense 1-m tall thickets of *L. pertusa* on the peaks. On the southern slopes, *L. pertusa* is intermixed with *E. profunda*, scattered *M. oculata*, and varying amounts of coral debris and dead standing coral. The mounds off Cape Canaveral support some of the densest living *L. pertusa* thickets observed on the BPW. One *L. pertusa* mound, Reed Peak 402, is 1600 m wide at the base with its peak at 410 m and an overall vertical relief of 44 m. One to three meter-tall thickets of *L. pertusa* completely cover the upper south slope and peak (Fig. 4.7d, Box 4.3).

Case Study—Box 4.3: BP West Mounds ("Million Mounds")

Recent multibeam sonar mapping surveys revealed an abundance of mound-like structures on the southwestern BP (Sowers 2020). These mounds, located primarily off the coast of Florida from Jacksonville to Cape Canaveral, occur at 400–900 m depth. An additional 300 mounds, ranging from 8 to 168 m above the seafloor, were discovered along a 222-km stretch from Jacksonville to Jupiter, Florida (Reed et al. 2006). Due to the high density of mounds, this region has been referred to as "Million Mounds." Habitat suitability models predicted a high likelihood of CWC presence in this region (Kinlan et al. 2013). Reed et al. (2013b) estimated that CWC habitat covers over 13,400 km^2 off eastern Florida. These features have been surveyed extensively with the *Johnson-Sea-Link* submersible, and the ROV *Deep Discoverer* (Paull et al. 2000; Reed 2002a, b; Reed et al. 2006, 2013a; Partyka et al. 2007; Ross and Quattrini 2007; Messing et al. 2008). Many of the mounds appear to be composed primarily of unconsolidated sediment and coral rubble, capped with dense thickets of *Lophelia pertusa* on the peaks, and *L. pertusa* intermixed with *Enallopsammia profunda*, *E. rostrata*, and scattered *Madrepora oculata* on the south slopes.

Case Study—Box 4.3 (continued)

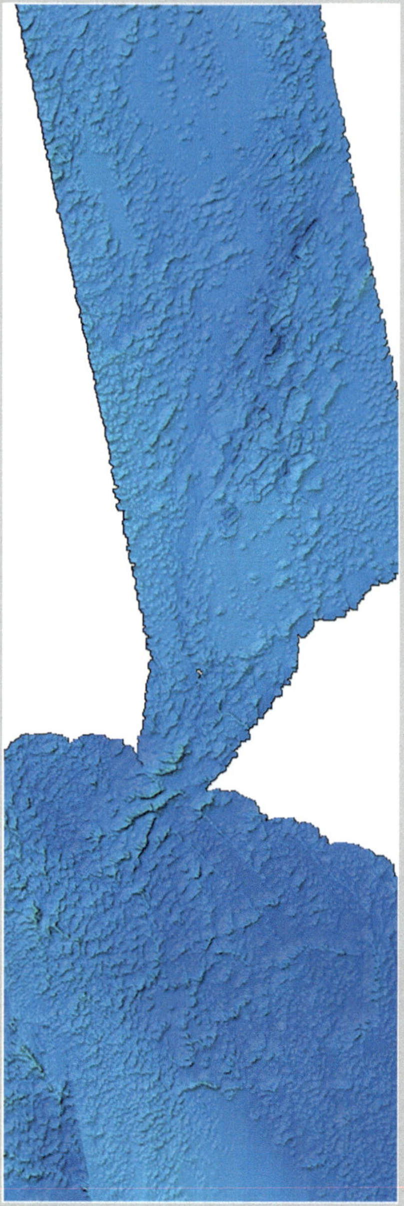

Multibeam bathymetry of the Million Mounds area off the eastern coast of Florida. It is likely that each feature is a CWC mound, although relatively few have been groundtruthed. Data from NOAA Ocean Exploration Program.

4.5 Coral Biology

4.5.1 Coral Reproduction

Understanding basic biological attributes such as timing and seasonality of reproduction, larval behavior (e.g., swimming, vertical migrations, feeding), and larval lifespans is important because they affect larval survivorship, dispersal distances, and hence connectivity. Yet, limited information exists regarding reproduction and early life histories for most CWCs (Waller 2005). Corals have a wide range of reproductive strategies including separate sexes (gonochorism) or intersexual (hermaphroditism); broadcast spawning or brooding; and continuous, seasonal, or periodic gamete release. While most shallow-water scleractinian corals studied to date are hermaphroditic (Harrison 2011), many CWC species are gonochoric broadcast spawners (Waller 2005), which might allow for greater dispersal potential (Szmant 1986). Additionally, larvae can be either lecithotrophic (non-feeding, generally shorter longevity) or planktotrophic (feeding, longer-lived, and possibly higher dispersal potential).

Of all CWCs, reproduction and larval biology are best known for *L. pertusa*, but most studies have been conducted on populations outside of the SEUS region. *L. pertusa* has an annual reproductive cycle, with broadcast spawning followed by embryogenesis and larval development in the water column (Larsson et al. 2014). An offset in timing of spawning has been noted, with spawning occurring over 6–8 weeks beginning in late January in the eastern North Atlantic (Brooke and Jarnegren 2013) and in either October (Gulf of Mexico) or December (mid-Atlantic Canyons, Brooke unpubl) in the western North Atlantic. Larvae of *L. pertusa* appear to be planktotrophic, actively swim upward, and may spend 3–5 weeks in the water column; all characteristics that suggest high dispersal potential (Larsson et al. 2014). Another CWC framework-forming species, *M. oculata* appears to be a gonochoristic species with periodic spawning and lecithotrophic larvae (Waller and Tyler 2005).

4.5.2 Thermal Tolerance

Even though CWCs can survive variable environmental conditions, CWCs along the SEUS margin naturally tolerate higher temperature ranges than CWCs in other regions due to the meandering of the GS. Periods of high temperatures at the upper limits of thermal tolerance (~12–14 °C, see Brooke et al. 2013) for *L. pertusa* in the SEUS often last for multiple days, which can affect respiration, calcification, and growth rates, as well as compromise juvenile and larval growth (Brooke et al. 2013). Experiments have shown that *L. pertusa* can survive short temperature spikes, but prolonged exposure can lead to increased mortality (Brooke et al. 2013; Lunden et al. 2014). A rise of almost 10 °C will lead to a three- to fivefold increase in respiration (Dodds et al. 2007; Dorey et al. 2020). Therefore, prolonged periods of high temperatures will require an increase in food availability to compensate for the stressor (Büscher et al. 2017). Temperature variability also will impact reef-building coral species differently. Naumann et al. (2014) demonstrated that calcification and respiration rates of *L. pertusa* are less affected by temperature variability compared to *M. oculata*; acclimation times differed also. Differences in temperature tolerance will likely influence coral distribution patterns and might help explain why *L. pertusa* is the dominant species on contemporary coral mounds on the BP (Mienis et al. 2014; Paull et al. 2000; Quattrini et al. 2012).

4.5.3 Genetics and Connectivity

Recently, genetic and genomic techniques have been increasingly applied to a variety of CWC taxa in support of sound taxonomy (e.g., delimiting species, Erickson et al. 2021), refining understanding of evolutionary relationships (Herrera and Shank 2016; Quattrini et al. 2020) and elucidating patterns of genetic structuring and gene flow across seascapes (e.g., Bracco et al. 2019). In general, few studies have harnessed sufficient genetic data from CWC taxa common in the SEUS. To date, only *L. pertusa* has been used to address questions regarding genetic differentiation in the SEUS. Utilizing nine microsatellite loci, studies of connectivity patterns of *L. pertusa* in the North Atlantic detected regional structuring, with the SEUS distinct from the Gulf of Mexico, New England Seamounts, the northeastern Atlantic (Fig. 4.8, Morrison et al. 2011; Lunden et al. 2014). Similarly, significant differences in genetic composition between ocean regions have been documented in other CWCs such as the bamboo corals (Smith et al. 2004), bubblegum coral *P. arborea*, and the solitary scleractinian coral *D. dianthus* (Addamo et al. 2021).

Within the SEUS, weak genetic structuring was detected among *L. pertusa* populations from the Florida-Hatteras Slope (Miami Terrace, Savannah Banks, North Carolina mounds; 287–511 m) and the BP (Canaveral Mounds, Jacksonville Lithoherms, and Stetson Banks; 430–740 m). When samples from Norfolk and Baltimore canyons in the mid-Atlantic region were added to the analysis, the canyons were distinctive from SEUS populations (Fig. 4.8; Morrison et al. 2017).

Phylogeographic (or biogeographic) breaks, or connectivity barriers, often occur at the boundaries of biogeographical provinces. Across such ocean transition zones, water masses that differ sharply in physical characteristics (often

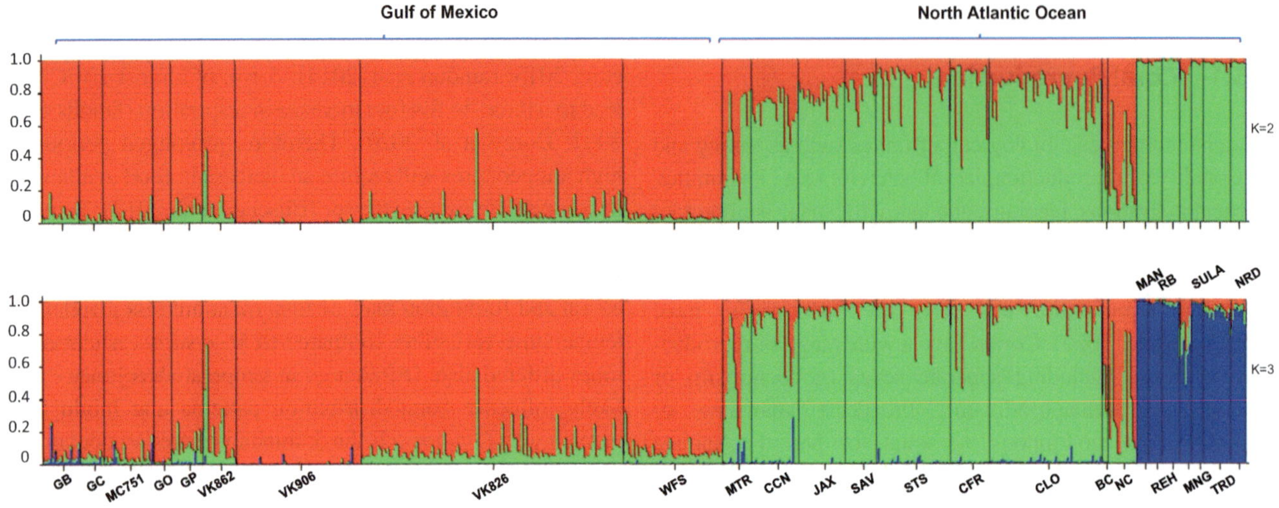

Fig. 4.8 Structure plot of 617 *Lophelia pertusa* individuals from the Gulf of Mexico and the North Atlantic Ocean based upon eight microsatellite loci. Individuals are represented by a vertical bar partitioned into sections with lengths proportional to estimated probability of membership into *K* clusters. Significant genetic structuring was detected in the dataset; the most likely number of clusters was $K = 3$. Sample site abbreviations were Gulf of Mexico: GB, Garden Banks; GC, Green Canyon; MC751, Mississippi Canyon 751; GO, Gulf Oil; GP, Gulfpenn; VK862, VK906, VK826, Viosca Knoll 862, 906 and 826; WFS, West Florida Slope; SEUS: MTR, Miami Terrace; CCN, Cape Canaveral; JAX, Jacksonville; SAV, Savannah; STS, Stetson Banks; CFR, Cape Fear; CLO, Cape Lookout; mid-Atlantic Canyons: BC, Baltimore Canyon; NC, Norfolk Canyon; New England Seamounts: MAN, Manning Seamount; REH, Rehoboth Seamount; northeastern Atlantic: RB, Rockall Banks; MNG, Mingulay; SULA, Sula Ridge; NRD, Nordleska; TRD, Trondheim Fjord (from Morrison et al. 2017)

temperature) may create ecological and/or physical gradients that limit larval exchange. Two phylogeographic breaks exist within the region and connectivity data for *L. pertusa* suggest that these breaks may extend to waters deep enough to affect CWC larvae. A transition between the Caribbean and Carolinian zoogeographic provinces occurs in the southern portion of the SEUS in waters off central Florida near Cape Canaveral (Briggs 1974). Phylogeographic breaks in shallow-water marine organisms have been recognized in this area for decades (reviewed by Avise 1992, 2000). Miami Terrace and the Canaveral Mounds exhibited transitional population structure in *L. pertusa*, with admixture between Gulf of Mexico and SEUS (Fig. 4.8). At the northern edge of the SEUS region, colder waters from the Virginian Province meet warmer waters from the Carolinian Province (Briggs 1974) at Cape Hatteras, NC. Genetic data has detected breaks in connectivity at Cape Hatteras for some species [the toadfish, *Opsanus tau* (Avise et al. 1987) and black sea bass, *Centropristis striata* (McCartney et al. 2013)]. Similarly, *L. pertusa* populations from above (mid-Atlantic Canyons) and below the transitional boundary (Cape Lookout) are differentiated despite being separated by only 307 km. Although phylogeographic breaks exist at the northern and southern extent in the SEUS, *L. pertusa* populations between these transition zones follow the paradigm observed for most deep-sea species in that populations at similar depths appear well connected over 100s of kilometers (Taylor and Roterman 2017).

Population structure across depth has been observed even at smaller spatial scales (100s–1000 m; Taylor and Roterman 2017; Etter and Bower 2015). Like phylogeographic breaks, layering of water masses with differing physical characteristics may create environmental gradients that limit gene flow and structure populations (Radice et al. 2016; Bracco et al. 2019). For example, an isolation-by-depth pattern of differentiation was documented in *Callogorgia delta* populations in the Gulf of Mexico (Quattrini et al. 2015; Bracco et al. 2019). Recent analyses of microsatellite data including samples from the shallowest *L. pertusa* population (off Jacksonville, FL, 214 m, Ross et al. 2015) and the deeper Richardson Reef Complex (~780 m) suggest limited connectivity among these shallow and deep populations. In addition, populations inhabiting depths less than 420 m show limited connectivity, with increased connectivity among deeper populations (Morrison, unpublished data); a pattern also observed in deep-sea bivalves (Etter et al. 2005). This pattern may be indicative of a more consistent influence of the GS at deeper sites, whereas the shoreward side is prone to instability (Todd et al. 2016, see above). Additional analyses including higher resolution genomic data are underway and, combined with physical oceanographic data, will help confirm or refute these findings.

4.6 Reef Ecosystems

4.6.1 Coral Species Composition

Corals associated with CWC habitat in the SEUS consist of a diversity of azooxanthellate (non-photosymbiotic) species. In the region, there are approximately 202 CWC species currently described (Hourigan et al. 2017a). Orders Alcyonacea (soft corals, sea fans) and Scleractinia (both colonial and solitary stony corals) are the most diverse with 83 and 77 species, respectively. Orders Antipatharia (black corals), Pennatulacea (sea pens), and Anthoathecata (lace corals) are equally diverse with 16, 11, and 15 species, respectively. However, 38 of these species (six species of antipatharians, 10 scleractinians, 17 alcyonaceans, four pennatulaceans, and one stylasterid) commonly occur at depths shallower than 200 m. New species continue to be discovered, with 12 species described since 2000 (Hourigan et al. 2017a) and others awaiting description. Many CWC species, particularly structure-forming scleractinians, alcyonaceans, and antipatharians, are foundation species, thereby providing shelter, food, and nursery habitat for a variety of fishes and invertebrates (Ross and Quattrini 2007; Etnoyer and Warrenchuk 2007; Baillon et al. 2012).

Only a few CWCs (Table 4.1, *L. pertusa*, *S. variabilis*, *Oculina* spp., *Madrepora* spp., *Enallopsammia* spp., *Madracis* spp.) can be considered ecosystem engineers in the deep sea. Ecosystem engineers not only create habitat, but they also modify the physical environment, thereby directly and indirectly regulating the availability of resources in an area (Jones et al. 1994). *Lophelia pertusa* is by far the most common ecosystem engineer in the SEUS (Reed 2002a, b, 2006; Hourigan et al. 2017b). But in the more southern parts of the SEUS region, *E. profunda* contributes more to the coral framework, particularly at deeper depths. In fact, some deeper reefs are composed entirely of *E. profunda* (J. Reed, pers. observ.). *Solenosmilia variabilis*, usually observed in the region at depths deeper than 700 m, occurs only as isolated colonies and not as part of larger reef-like structures. Although considered a structure-forming scleractinian, *M. oculata* occurs mostly as small colonies and does not contribute significantly to the reef framework (Hourigan et al. 2017b). The SEUS region is unique in that azooxanthellate colonies of the scleractinian *O. varicosa* form an extensive reef system, similar in structure to *Lophelia* banks, but at mesophotic depths (70–100 m) along the shelf edge off central eastern Florida (Reed 2002a, b).

4.6.2 Community Structure

Live and dead CWCs provide a variety of microhabitats (Buhl-Mortensen and Mortensen 2005; Mortensen and Fosså 2006; Buhl-Mortensen et al. 2010) and therefore support a diverse faunal assemblage of recreationally, commercially, and ecologically-important species. Interestingly, presence of live coral does not appear to be the main factor driving these associations (Buhl-Mortensen et al. 2010; Hourigan et al. 2017b). Rather it is the hard substrate and the variety of microhabitats provided by dead-coral skeletons that facilitate the high biodiversity associated with reef-forming CWCs (see Mortensen and Fosså 2006; Cordes et al. 2008; Lessard-Pilon et al. 2010). Several invertebrate taxa are often observed living around, on, or in the matrix formed by CWCs. These include crustaceans (crabs, shrimps, amphipods, barnacles), echinoderms (sea stars, brittle stars, sea urchins, crinoids), cnidarians (octocorals, anemones, hydroids, zoanthids), sponges, mollusks (squid, octopus, gastropods), pycnogonids (sea spiders), and sipunculan and polychaete worms (Fig. 4.9, Table 4.2). Most studies are broad surveys focused primarily on the corals but also mention or provide an annotated species list of associated fauna,

Table 4.1 Colonial deep-water Scleractinia off southeastern US

Species	Location	Depth (m)
Cladocora debilis (Milne Edwards and Haime, 1849)	FS	11–400
Lophelia pertusa (Linnaeus, 1758)	NC-FS(PT)	146–895
Solenosmilia variabilis (Duncan, 1873)	GA-FS	220–1383
Dendrophyllia alternata (Pourtalès, 1880)	SEFL(BP)	276–900
Enallopsammia profunda (Pourtalès, 1867)	NC(BP)FS(PT, AT)	305–1748
Enallopsammia rostrata (Pourtalès, 1878)	SC	300–1646
Madrepora carolina (Pourtalès, 1871)	NC-FL	53–1003
Madrepora oculata (Linnaeus, 1758)	NC-FL	80–1500
Oculina tenella (Pourtalès, 1871)	SEFL, FS	25–159
Oculina varicosa (Lesueur, 1821)	NC-FL	3–150
Madracis asperula (Milne Edwards and Haime, 1849)	FS	24–311
Madracis myriaster (Milne Edwards and Haime, 1849)	NC-FS(PT)	20–1220
Madracis pharensis f. pharensis (Heller, 1868)	NC	11–333

Modified from Hourigan et al. (2017a)

Fig. 4.9 Images of reef-associates. (**a**) Sea urchins *Echinus tylodes* (top) and *Cidaris* sp. (bottom) on live *Lophelia pertusa*, Richardson Ridge, 873 m depth (ROV *Deep Discoverer* EX1806-07), (**b**) large venus flytrap sea anemones (*Actinoscyphia aurelia*), barnacles, crinoids, and squat lobsters (*Gastroptychus* sp.) on a black coral colony, Stetson Mesa South Mounds, 760 m depth (ROV *Deep Discoverer* EX1903-02), (**c**) several large squat lobsters (*Eumunida picta*), pink anemones, large octocorals (*Paramuricea* sp.), including one with ophiuroid associate, sponges, and a swimming crab (*Bathynectes* sp.) on dead *L. pertusa*, Cape Fear Mound, 454 m depth, (ROV *Deep Discoverer* EX1806-10), (**d**) Western roughy *Hoplostethus occidentalis*, Million Mounds Area, 775 m depth (ROV *Deep Discoverer* EX1907-02), (**e**) goosefish *Lophiodes beroe*, Million Mounds Area, 775 m depth (ROV *Deep Discoverer* EX1907-02), (**f**) Swallowtail bass *Anthias woodsi*, off North Carolina, 350 m (ROV *Deep Discoverer* EX1806-13). All photos courtesy of NOAA Office of Ocean Exploration and Research

particularly larger megafauna (Messing et al. 1990, 2008; Reed et al. 2005a, b; 2006; Ross and Nizinski 2007; Quattrini et al. 2012; Ross et al. 2015). The few detailed studies of invertebrate associates reflect the interests of the investigator (e.g., crustaceans: Reed et al. 1982, molluscs: Reed and Mikkelson 1987, sea stars: Mah 2020; Mah et al. 2010, hydroids: Henry et al. 2008).

Assemblages of organisms can vary depending on the primary underlying structure. For example, live tissue of deep-water scleractinian corals seems to prevent attachment of sessile species (Buhl-Mortensen et al. 2010). Thus, mobile megafauna such as squat lobsters (e.g., *Eumunida picta*, *Munidopsis* spp.) and sea urchins (e.g., *Echinus tylodes*) are more frequently observed perched on live branches of *L. pertusa*. Crustaceans (shrimps, amphipods) and ophiuroids are commonly observed on large, structure-forming gorgonians. In contrast, a more diverse assemblage of sessile and mobile invertebrates utilize dead-coral skeletons. For

Table 4.2 List of mobile megafauna commonly associated with cold-water coral reefs (>200 m depth) in the region

Megafauna	Regional common name
Fishes	
Anthias woodsi	Swallowtail bass
Beryx decadactylus	Red bream
Conger oceanicus	American conger eel
Dysommina rugosa	Cutthroat eel
Rubicundus lopheliae	Coral hagfish
Helicolenus dactylopterus	Blackbelly rosefish
Hoplostethus occidentalis	Western roughy
Laemonema barbatulum	Short beard codling
Laemonema melanurum	Reef codling or Coral hake
Nettenchelys exoria	Exterior pike conger
Nezumia sclerorhynchus	Roughtip grenadier
Polyprion americanus	Wreckfish
Scyliorhinus retifer	Chain catshark
Synaphobranchus spp.	Cutthroat eels
Trachyscorpia cristulata	Atlantic thornyhead
Crustaceans	
Bathynectes longispina	Bathyal swimming crab
Chaceon fenneri	Golden crab
Eumunida picta	Squat lobster
Rochinia crassa	Inflated spiny crab
Gastroptychus salvadori	Squat lobster
Uroptychus spp.	Squat lobster
Munida spp.	Squat lobster
Munidopsis spp.	Squat lobster
Echinoderms	
Ceramaster grenadensis	Seastar
Chondraster grandis	Seastar
Cidaris rugosa	Pencil urchin
Cladaster rudis	Seastar
Echinus tylodes	Sea urchin
Evoplosoma spp.	Seastar
Gilbertaster caribaea	Seastar
Gracilechinus gracilis	Sea urchin
Novodinia antillensis	Brisingid
Ophiacanthia bidentata	Brittle star
Ophidiaster guildingi	Seastar
Plinthaster dentatus	Seastar
Solaster spp.	Seastar
Sthenaster emmae	Seastar
Stylocidaris spp.	Seastar

References: Reed et al. (2006); Ross and Quattrini (2007, 2009); Ross and Nizinski (2007); Mah (2020)

example, coral rubble habitats, found at the base of coral reefs and mounds and escarpments, support large aggregations of brittle stars (*Ophiacantha bidentata* off North Carolina). Associations between corals and the invertebrate fauna can range from facultative, occasional use to obligate species-specific relationships. Overall, many of the associations between SEUS CWCs and their associated invertebrates appear to be facultative, suggesting that the structure of the coral and not the coral itself is the predominant factor driving the association (Quattrini et al. 2012; Hourigan et al. 2017b). Our understanding of these symbiotic relationships will continue to increase as data from recent expeditions to the region (e.g., NOAA ship *Okeanos Explorer*, Deep Search) provide new insights into the diversity and abundance of invertebrates associated with CWCs and CWC habitats.

CWC serve as fish habitat by providing food, shelter, and/or nursery grounds for a variety of fish species, although the degree to which the structure built by corals or the corals themselves is important to fishes remains elusive (Auster 2005; Costello et al. 2005; Ross and Quattrini 2007).

Nevertheless, new species, including a deep-water snake eel (McCosker and Ross 2007) and a hot-pink hagfish (Fernholm and Quattrini 2008) described from CWC habitats in the SEUS, highlight the close association of fishes to these habitats. In addition, observations of large pelagic fishes (e.g., swordfish, Ross and Quattrini 2007; Gasbarro et al. 2022; wreckfish, Ross and Quattrini 2007) indicate that top predators and highly-migratory species utilize CWC habitat; however, it remains unknown how long these species reside in CWC habitats and to what extent they utilize (e.g., feeding, reproduction) the habitats. In the SEUS, a study by Ross and Quattrini (2007) found that fishes observed in CWC habitats were different than those found in soft-sediment habitats (Table 4.2, Fig. 4.9); few species were found to be generalists to both habitats. This study also documented the close association of fishes to CWCs, as several fish species were observed hiding within, moving through, or resting on the coral framework (both live and dead). In a follow-up study, depth and latitude were shown to drive changes in fish assemblages across the SEUS region (Ross and Quattrini 2009). Observations of fishes at a deep-water shipwreck in the 2009 study showed that any high profile, complex structure in the deep sea likely meets habitat requirements for at least some fish species (e.g., *Beryx decadactylus*). In combination, the studies by Ross and Quattrini (2007, 2009) provide evidence for a characteristic "reef fish" fauna at CWC habitats in the SEUS. Similarly, the azooxanthellate *Oculina* Banks harbor a characteristic reef fish fauna (Harter et al. 2009) that is more typical of mesophotic reefs at similar depths (e.g., Stefanoudis et al. 2019) and is essential fish habitat for fisheries including the snapper-grouper complex (Koenig et al. 2000).

4.6.3 Trophic Dynamics

Whether fishes and invertebrates are facultative, obligate, or transient members of CWC habitats, they all contribute to trophic dynamics and function of CWC ecosystems. Organic particles derived from surface productivity fuel the food web, consisting of primary consumers to top-level predators, in CWC ecosystems. Thus, CWCs often occur in areas of high productivity and strong currents; areas conducive to sessile filter feeders. Marine snow and detritus brought by currents supports not only dense coral growth and biomass, but a diverse and abundant infauna and detritivore assemblage. For example, anemones and brisingid sea stars are frequently found perched, elevated off the seafloor, on coral substratum, presumably to gain better access to food items transported by these strong currents, such as the GS. Associated and transient mobile predators add complexity to the trophic web. These predators may drive community assembly and/or serve as conduits for energy exchange among ecosystems. Based on recent in situ observations, knowledge of predator–prey interactions and feeding behavior in the SEUS has increased substantially. For example, several species of sea stars (e.g., *Cladaster rudis*, *Sthenaster emmae*, *Gilbertaster caribaea*) are documented corallivores having been observed feeding on octocorals or scleractinians (Mah et al. 2010; Watling et al. 2011; Mah 2020). Additionally, other species (e.g., *Chondraster* sp.) are suspected to feed on the organic surface layer of *L. pertusa* and not necessarily on the coral itself (Mah 2020). In situ observations have also illustrated that some crustaceans, particularly squat lobsters, are not strictly detritivores or scavengers as once thought, but are also predators on a variety of pelagic organisms such as fishes (M. Nizinski, pers. observ.).

CWC ecosystems on the seafloor are connected with those on the surface of the ocean not only through the surface primary productivity that falls to the seafloor, but also through the mesopelagic species that make diel vertical migrations between both environments. Dense and persistent aggregations of mesopelagic fishes have been noted from CWC mounds in the region (Gartner et al. 2008). Gartner et al. (2008) suggested that these species are not only exploiting food resources at the seafloor, but they are likely serving as a regular food source to bottom-dwelling species. By examining stomach contents in fishes collected from hard-bottom habitats on the Charleston Bump, Weaver and Sedberry (2001) found that vertically migrating species support dense aggregations of large predators (e.g., wreckfish). Squat lobsters have also been observed feeding on mesopelagic species (e.g., fishes, pyrosomes, pers. obs.). Thus, diel vertical migrators transfer energy from the surface to the seafloor, and perhaps from the seafloor back to the surface. Flexibility in feeding modes by a variety of invertebrates provides an additional conduit for transfer of energy between pelagic and benthic environments. *Ophiacantha bidentata*, a brittle star that occurs in dense aggregations at the North Carolina CWC mounds, can switch from deposit feeding to suspension feeding, depending upon resource availability (Litvinova 1980; Gallagher et al. 1998). Feeding plasticity of squat lobsters allows them to feed on resources ranging from primary producers to dead organic matter (Lovrich and Thiel 2011). Thus, these organisms play an important role in energy flow and secondary production through the utilization, processing, and redistribution of organic matter and subsequent transfer of organic material to higher trophic levels (Pearson and Gage 1984; Pape-Lindstrom et al. 1997; Summers and Nybakken 2000; Lovrich and Thiel 2011).

4.7 Threats and Outlook

Resource extraction (e.g., fisheries, mining, oil, and gas), marine debris, pollution, invasive species, and climate change threaten all marine ecosystems, including those in the deep sea. Unlike other regions, resource extraction is minimal in deep waters of the SEUS, and limited mostly to fishing. However, resource extraction activities in the region will likely increase in the future and disrupt CWC ecosystems. Discussions of opening areas in the region to energy extraction activities are ongoing, fisheries continually expand into deep water, and there is a global push to mine the deep seafloor in regions across the world.

Commercial fisheries for golden crab (*Chaceon fenneri*) and royal red shrimp (*Pleoticus robustus*) are the primary managed fisheries in the SEUS that can impact CWC reefs (>200 m). While neither of these fisheries target coral areas directly, there is overlap between CWC reef habitat and fishery operations (Hourigan et al. 2017b). Golden crabs are most common at depths of ~400–500 m (Reed and Farrington 2010; Reed et al. 2017) and have been observed at several CWC mounds nestled in both live and dead-coral framework (Reed et al. 2017, pers. observ). The relatively small golden crab fishery (11 active permits and 5–6 active vessels; NOAA Fisheries 2020; Hourigan et al. 2017b) operates off eastern Florida within designated access areas of the deep-water CHAPC (Fig. 4.1, Reed et al. 2017). Damage to coral habitat from traps is minimal overall, as traps are apparently deployed in soft-sediment habitats adjacent to the coral mounds. However, there are instances when CWC reefs might be impacted, including when strong currents carry traps into coral habitats or when traps are towed behind the fishing vessel upon retrieval (Reed et al. 2017).

Royal red shrimp have also been observed in CWC habitat (Reed and Farrington 2010) but in contrast to golden crabs, the trawl fishery for royal red shrimp has greater potential to impact CWC reefs. Royal reds currently represent only a small portion of the SEUS shrimp industry with fewer than ten active fishing vessels (Stiles et al. 2007); however, the fishery can operate in the designated access area within the CHAPC. The largest aggregations of royal red shrimp occur primarily at 250–550 m depth off northeast Florida with most trawling activity at 250–400 m depth off eastern Florida (Anderson and Lindner 1971; Reed et al. 2017). Damage to CWCs has been minimal since most of the trawling apparently occurs in soft-sediment habitat.

There is potential for development of new fisheries in the region that could impact CWC reefs. For example, several finfish species (e.g., *Helicolenus dactylopterus*) are closely associated with CWC habitat in the region. Although currently not fished in the SEUS, these species are heavily fished elsewhere in the world (Table 4.2). Due to the presence of the CHAPCs in the region and a proactive fishery management council, it is likely that the development of new fisheries will occur in conjunction with regional fishery management plans. The historical actions by the SAFMC to protect CHAPCs, including the mesophotic *Oculina* CHAPC (Fig. 4.1), have set a precedent for fishery management efforts in the region. Although illegal fishing for snappers, groupers, and perhaps rock shrimp have recently been illuminated in *Oculina* CHAPC (Reed et al. 2007), regulations for vessel monitoring systems have reduced impacts of fishing gear on *Oculina* in the region (see Hourigan et al. 2017b).

Currently, energy extraction activities are minimal in the SEUS region, although there is potential for their development. Early studies indicated that certain areas may contain significant oil and gas reserves (BOEM 2014). In the 1970s and 80s, a few exploratory wells were drilled off of Florida for natural gas, however, they were abandoned as flow rates were variable and declined over time (BOEM 2020). At present, there are no active oil and gas leases in the SEUS. Renewable energy projects (e.g., wind, currents) are underway in the region and could potentially be developed in offshore areas, including the Florida Straits, that harbor CWC habitats. However, the few commercial wind leases that exist are closer to shore and in shallower depths.

Rapid and growing interest in mineral extraction in the deep sea presents significant risks to midwater and benthic habitats (Drazen et al. 2020), including CWCs. Mining activities target, for example, sulfide deposits on hydrothermal vents, polymetallic nodules on the deep seafloor, and cobalt-rich crusts on seamounts (Lusty and Murton 2018). Physical damage and noise produced by collector vehicles and subsequent sediment resuspension, which has the potential to disperse 10s–100s of km, can directly impact benthic communities (e.g., burial, smothering, etc.; Wedding et al. 2013; Drazen et al. 2020). Furthermore, sediment plumes that are released in the midwater and/or travel upwards in the water column can impact midwater fauna. Because of bentho-pelagic coupling at CWC mounds (Gartner et al. 2008), the impacts of mining to either benthic or midwater communities may have drastic effects to CWC ecosystem functioning (Drazen et al. 2020). Currently, there are no mining activities in the SEUS or the U.S. EEZ in general, but there is potential for development in the region. Large volumes of lower-grade ferromanganese crusts, which can be colonized by CWCs, and manganese nodules occur in the U.S. EEZ on the BP off Florida and South Carolina (Commeau et al. 1984; Manheim 1986). Therefore, if mining were to open in this region, CWC reefs would be severely impacted.

Several additional anthropogenic disturbances can impact CWC ecosystems. Marine debris has become a pervasive problem. Approximately 50% of plastic debris sinks to the seafloor (Galgani et al. 2015), making the deep sea a likely sink for substantial quantities of microplastic debris (Woodall et al. 2014), which can potentially be incorporated into the diets of CWCs (Taylor et al. 2016). Derelict fishing gear is a significant problem because it can destroy reef structure or become entangled in the coral matrix (Quattrini et al. 2015). Submarine cable deployment can also impact CWC habitat, and areas off southeastern Florida are a major hub for communication cables (Hourigan et al. 2017b). In addition to marine debris, runoff from land might be an additional source of pollutants to the deep sea (Prouty et al. 2014). Williams et al. (2007) found that nitrogen enrichment in black coral skeletons increased over the past ~100 years, suggesting a terrestrial source (e.g., sewage or manure effluent) for the nutrients impacting the deep sea.

Finally, invasive species could impact ecosystem functioning of CWC reefs by eliminating native species and disrupting trophic webs. To the best of our knowledge, the only invasive species in deep waters of the region is the Indo-Pacific lionfish (*Pterois volitans*). This species has been reported from mesophotic depths off the SEUS coast (Meister et al. 2005), including the *Oculina* CHAPC (Harter et al. 2020). Lionfish, reported as deep as 300 m in the western North Atlantic (Muñoz et al. 2011; Gress et al. 2017), overlap the depth distributions of several coral species, including coral mounds built by *L. pertusa* and *O. varicosa.*

Extending beyond the threats discussed above, one emergent and global challenge to CWC reefs is the increased accumulation of CO_2 in the atmosphere due to society's dependence on fossil fuels. As CO_2 continues to accumulate in the atmosphere, the oceans will be impacted in three major ways: (1) water temperatures will increase as the oceans absorb excess atmospheric heat; (2) oxygen saturation will decrease due to higher water temperatures; and (3) the acidity of the oceans will increase, thus compromising the stability and persistence of CWC reefs (Hebbeln et al. 2019; Hennige et al. 2020). Further weakening of the GS in the region (Caesar et al. 2021) due to global climate change (Yang et al. 2016) may also impact CWC growth and survival. For the North Atlantic Ocean in general (including habitat within the SEUS region), modelling studies predict a 50% decrease in habitat suitability for scleractinian corals by 2100 resulting from climate change impacts (Morato et al. 2020).

4.8 Future Directions

Research and exploration efforts in CWC habitats in the SEUS have been extensive compared to other U.S. regions. However, like many areas in the deep sea, additional work is needed to gain a more holistic understanding of CWC habitats in the region. Fortunately, advances in new technologies and analytical methodologies are enabling a rapid pace of discovery in the deep sea. Here, we outline a few research directions and exploration targets that will advance our knowledge and understanding of CWC habitats in the SEUS.

Numerous areas in the SEUS region remain poorly characterized, particularly in deep (>1000 m) waters as well as mesophotic and rariphotic depths (~40–300 m, with the exception of the *Oculina* CHAPC). The following studies would provide much needed information about poorly characterized SEUS areas: (1) further exploration and characterization of mesophotic and rariphotic depths throughout the region where rocky hard-bottoms might harbor new coral species and unique CWC habitats (>200 m off FL, see Ross et al. 2015), (2) multibeam mapping and additional surveys of the Blake Escarpment, including the Blake Spur, and the central BP, and (3) additional surveys at deeper depths along the Florida-Hatteras slope and western BP (e.g., "Million Mounds") where numerous coral mounds have been discovered yet await to be surveyed (Fig. 4.1, Appendix 4). Further characterization in these areas would elucidate the diversity and extent of CWC communities in this region. In addition, the recent discovery of Richardson Reef Complex highlights the critical need for high-resolution multibeam mapping throughout the entire region to realize the full extent of CWC habitats. Prioritizing efforts to map more areas in the SEUS, particularly on the BP, would provide vital information to elucidate CWS habitats.

While a few studies have increased our knowledge of community ecology at CWC reefs in the SEUS, more focused efforts, including experiments and well-designed field studies, are needed to advance our understanding of the processes that shape community structure. Abiotic factors (e.g., temperature, oxygen) are known to influence CWC distribution and associated communities, but biotic factors (e.g., competition, predation, and symbioses) have rarely been studied at CWC habitats. Recent observations of asteroids feeding on cold-water octocorals and scleractinians (Mah 2020) raise the question of whether predation pressure by sea stars is strong enough to impact the structure of coral assemblages. In addition, few studies (e.g., Girard et al. 2016) have investigated how invertebrate-CWC mutualistic relationships influence community structure. In fact, the degree to which invertebrates specifically associate with different CWC species or microhabitats (e.g., rubble, live or dead coral) rarely has been quantified.

Investigating how CWCs influence demographics (e.g., population size, age structure, fecundity) of commercially- and ecologically-important fish species would help to better understand their role as essential fish habitat. Finally, documenting whether CWC habitats have higher functional diversity due to higher productivity and/or complexity

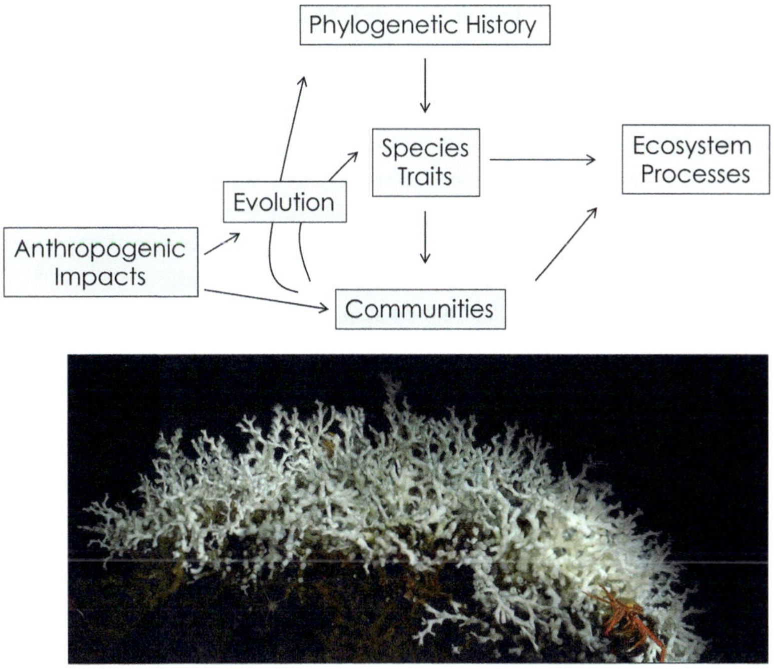

Fig. 4.10 Evolution and ecology are intrinsically linked. At the center of this association is the connection between the species that occupy ecosystems and the traits they possess. Organismal traits (e.g., reproductive output, dispersal potential, longevity, physiological traits) arise as innovations on phylogenies and can influence the distribution and ecological success of organisms. Anthropogenic impacts have the capacity to disrupt both ecology and evolution of CWC habitats. Illustration adapted from Cavender-Bares et al. (2009). Image credit to *Lophelia* II program (BOEM/USGS)

compared with other deep-sea habitats is essential to understanding ecosystem processes in the deep sea. Combining phylogenetic studies with trait-based ecology can help to elucidate the multiple dimensions of diversity, community dynamics, and ecosystem processes at CWC habitats, and the subsequent changes occurring in response to anthropogenic change (Fig. 4.10).

Not only are data lacking on processes at the community level, little is known about broad-scale ecosystem processes at CWC habitats in the SEUS. For example, Weaver and Sedberry (2001) demonstrated that trophic pathways in benthic habitats are linked to pelagic prey, but it is unknown whether (and to what extent) mesopelagic species are feeding on the benthos. Future studies should focus on the trophic connections among resident and transient as well as benthic and pelagic members of CWC habitats. In addition, more data are needed to determine the sources of primary productivity supporting CWC habitats. For example, future studies should examine the amount of seasonal and annual primary productivity that reaches the benthos from the surface; the amount of material that becomes bioavailable from sediment resuspension; the input of chemosynthetic-derived material to CWC habitats (particularly on the Blake Escarpment and BP); and the transport mechanisms of this material among CWC habitats. Finally, elucidating the role of microbes in nutrient acquisition or recycling in CWCs is critically needed. Middelburg et al. (2015) suggested CWC microbionts are enabling chemoautotrophy, nitrogen fixation, and nitrogen recycling, but indicated that follow-up studies are necessary to understand whether these interactions are critical for ecosystem functioning at CWC habitats.

Advances in molecular techniques and high-throughput sequencing will facilitate conservation and restoration of CWC habitats in the SEUS region and beyond. First, techniques such as targeted sequence capture and RADseq will improve species delimitation and identification of cryptic species, enabling a more complete documentation of species' distributions and improved predictive habitat suitability models (e.g., Herrera and Shank 2016; Quattrini et al. 2019; Arrigoni et al. 2020; Erickson et al. 2021). Second, genomic techniques may help detect genes that are under selection, and thus enable a greater understanding of cellular and physiological adaptations to environmental conditions. Third, population genomics may reveal dispersal pathways and barriers among disjunct CWC habitats within the SEUS (e.g., Leydet et al. 2018; Bracco et al. 2019) and among the SEUS and other regions. This information would improve our understanding of historical and contemporary sources and sinks of larvae, helping to identify isolated deep-sea ecosystems as management units.

A particularly promising approach couples population genomics with biophysical transport models that may refine understanding of the processes shaping connectivity patterns (Bracco et al. 2019). To date, the only population connectivity study on CWCs in the region was on *L. pertusa* (Morrison et al. 2011, 2017). Thus, efforts should be made to examine connectivity patterns of other structure-forming scleractinians, octocorals, and black corals in the SEUS, along with associated fishes and invertebrates. Shared patterns of genetic connectivity across co-distributed species may indicate that oceanographic features affect connectivity in similar ways, while contrasting patterns may indicate that differences in life histories are driving resultant dispersal patterns (Cunningham and Collins 1998). To that end, biological data on longevity, fecundity, reproductive timing and mode, and pelagic larval durations, are critically needed to refine analyses of connectivity.

As detailed above, several anthropogenic disturbances threaten CWC ecosystems. Thus, improving our knowledge of the impacts of climate change and other stressors on CWCs is crucial. A clear understanding of baseline environmental conditions is an essential first step. Only then can we increase our understanding of how carbonate chemistry (pH, alkalinity), temperature, dissolved oxygen, particulate organic carbon (POC) or matter (POM) at depth in this region will change under climate change scenarios. In addition, improved knowledge of contemporary geographic and bathymetric distributions of CWCs with concurrent data of their stressors and physiological tolerances will allow us to predict impacts of environmental changes on growth, survival, and connectivity. This information is critical for CWC habitats in the SEUS, particularly because these corals are already living under extreme environmental conditions close to their tolerance limits. Additional climatic shifts may expose CWCs to more stressful conditions beyond their adaptive capability, and hence pose a serious threat to their health, survival, and the ecosystem services these CWCs provide. Effective management strategies to mitigate the potentially detrimental effects of these changes on CWCs in the SEUS are urgently needed.

Acknowledgments S. Farrington, W. Sautter, and A. Reft are gratefully acknowledged for preparing Fig. 4.1, Appendix 3, and Fig. 4.10, respectively. J. Chaytor and R. Gasbarro produced the maps of Richardson Reef complex. Harbor Branch Oceanographic Institution, Florida Atlantic University, (HBOI-FAU), S. Ross, and NOAA Office of Ocean Exploration and Research provided images for Figs. 4.9 and 4.10. Bureau of Ocean Energy Management, Environmental Studies Program (DeepSearch, Contract Number M17PC00009), NOAA Cooperative Institute for Ocean Exploration, Research, and Technology at HBOI-FAU, the Smithsonian Institution and the US Geological Survey provided partial funding and support for data collection and salaries. Extensive surveys conducted in the region by NOAA Ship *Okeanos Explorer* provided data that led to recent discoveries. The South Atlantic Fishery Management Council is gratefully acknowledged for assuming the task of protecting these deep-water coral reefs. Any use of trade, firm, or product names is for descriptive purposes only and does not imply endorsement by the U.S. Government.

Appendices

Appendix 1

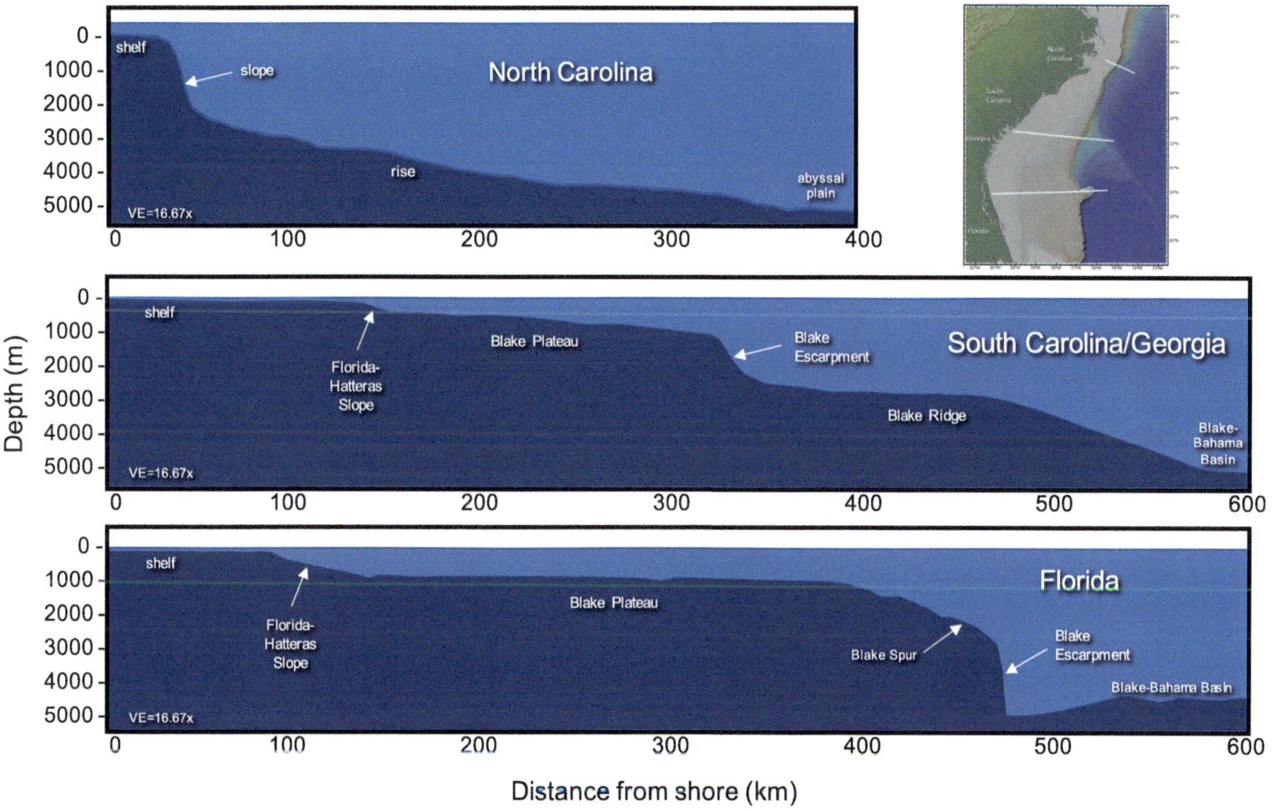

Depth profile comparison of North Carolina, South Carolina/Georgia, and Florida continental margins and their distinctive features, shown at the same scale with vertical exaggeration of 16.67×. Inset map shows profile locations (map and profiles generated using GeoMapApp)

Appendix 2

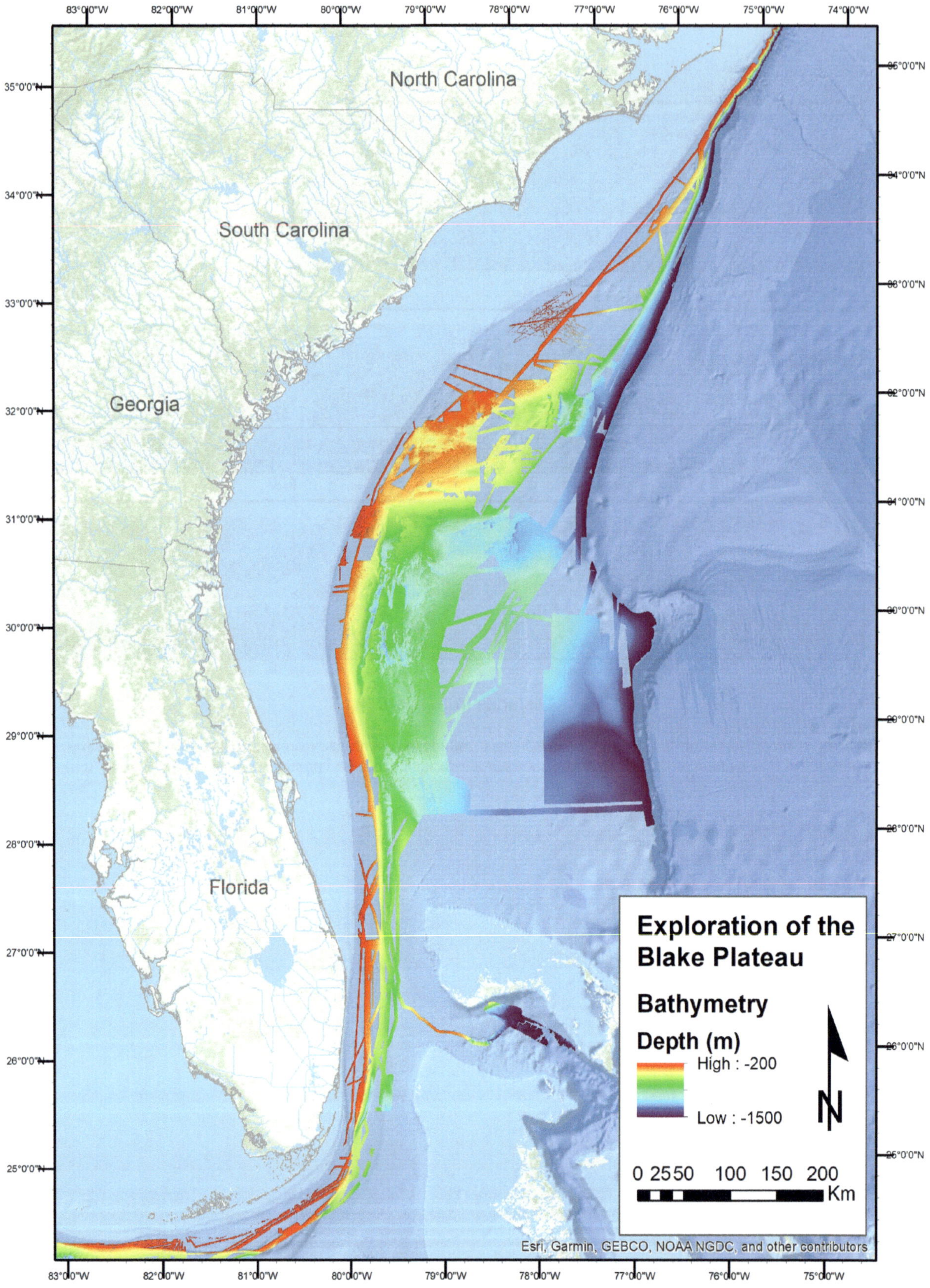

High-resolution (30 m) multibeam sonar bathymetry of the SEUS continental margin as of December 2019. Only depths between 200 and 1500 m are illustrated (overlain on the GEBCO bathymetry). Large areas of the southeastern BP remain unmapped at high resolution

Appendix 3

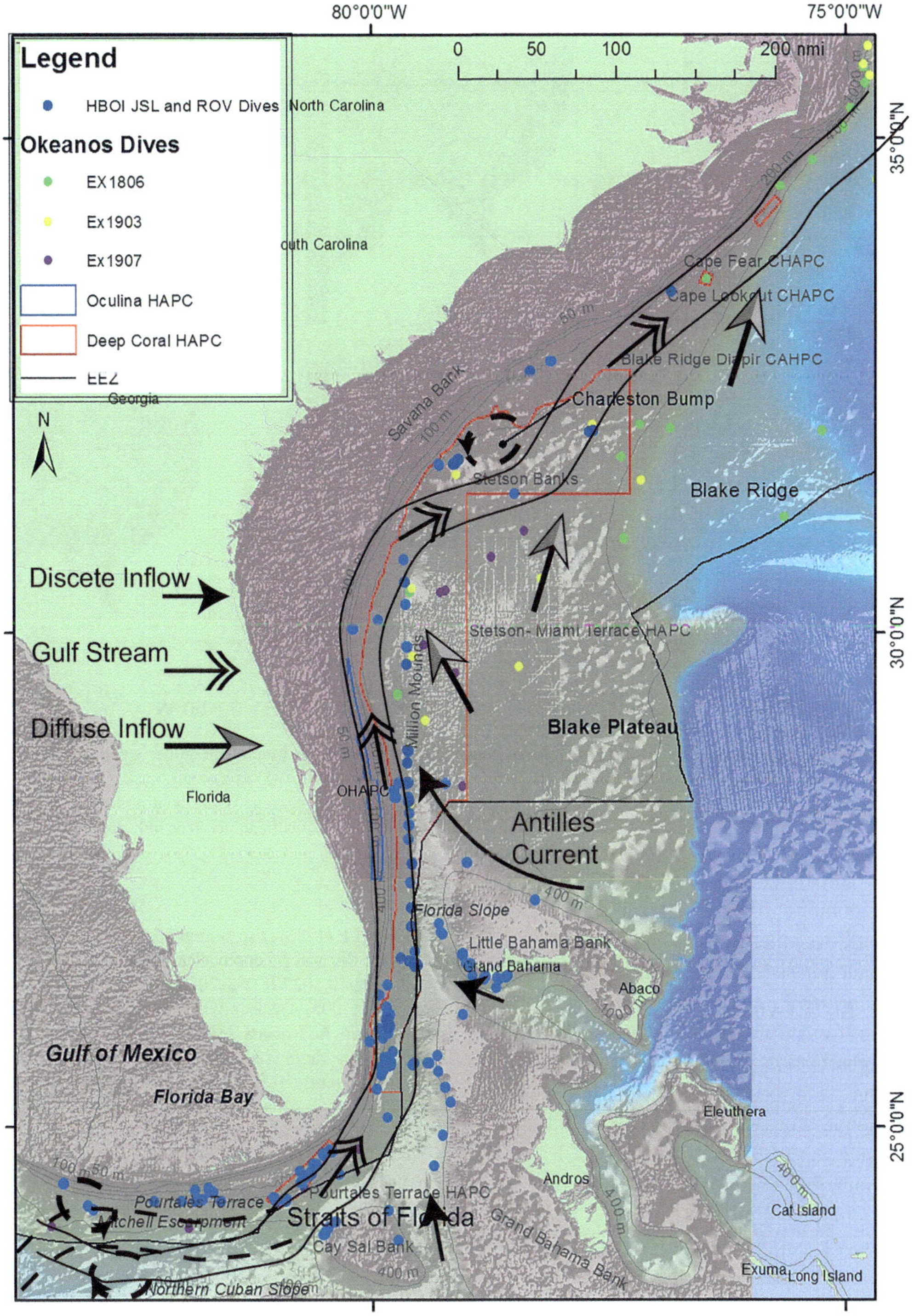

A schematic of circulation. Black solid lines mark the typical location of the GS; dashed lines in the southwest corner denote an alternate path when the Loop Current is retracted. Quasi-stationary recirculations are marked with dashed ellipses, and arrows denote different inflows

Appendix 4

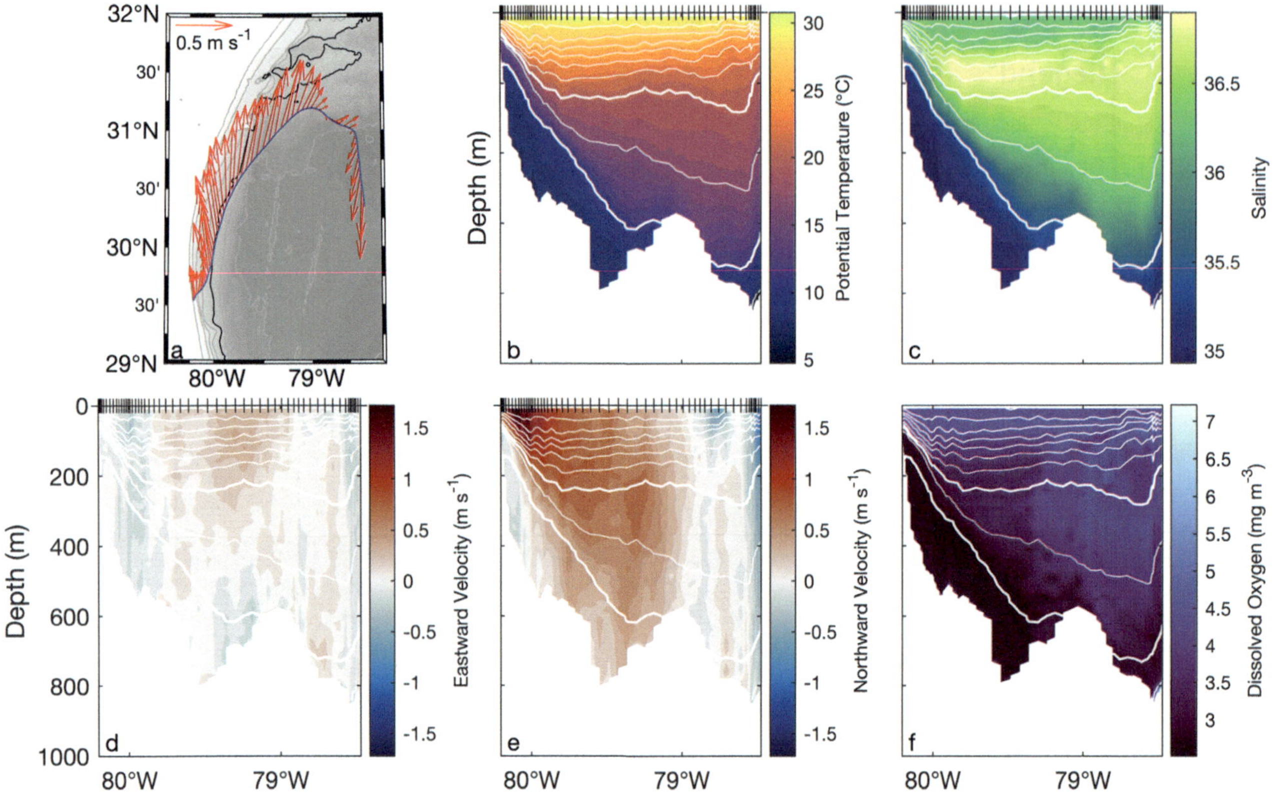

Transect across the Gulf Stream over the BP collected by a Spray glider in August 2019. (**a**) Map of the glider's path (blue) with measured depth average currents (red) and bathymetry contoured; contours are drawn every 100 m with the 500-m isobath black. (**b–f**) Sections of (**b**) potential temperature, (**c**) salinity, (**d**) eastward velocity, (**e**) northward velocity, and (**f**) dissolved oxygen. White contours in (**b–f**) are isopycnals with a contour interval of 0.5 kg m^{-3} and the 26.0 and 27.0 kg m^{-3} isopycnals bold, and tick marks on the upper axes indicate locations of individual glider profiles

References

Addamo AM, Miller KJ, Häussermann V, Taviani M, Machordom A (2021) Global-scale genetic structure of a cosmopolitan cold-water coral species. Aquat Conserv 31(1):1–14. https://doi.org/10.1002/aqc.3421

Anderson W, Lindner M (1971) Contributions to the biology of the royal red shrimp, *Hymenopenaeus robustus* Smith. Fish Bull 69: 313–336

Androulidakis Y, Kourafalou V, Le Henaff M, Kang HS, Ntaganou N, Hu CM (2020) Gulf Stream evolution through the Straits of Florida: the role of eddies and upwelling near Cuba. Ocean Dyn 70(9): 1281–1281. https://doi.org/10.1007/s10236-020-01393-1

Arrigoni R, Berumen ML, Mariappan KG, Beck PS, Hulver AM, Montano S, Pichon M, Strona G, Terraneo TI, Benzoni F (2020) Towards a rigorous species delimitation framework for scleractinian corals based on RAD sequencing: The case study of *Leptastrea* from the Indo-Pacific. Coral Reefs 39:1001–1025

Athie G, Sheinbaum J, Leben R, Ochoa J, Shannon MR, Candela J (2015) Interannual variability in the Yucatan Channel flow. Geophys Res Lett 42(5):1496–1503. https://doi.org/10.1002/2014gl062674

Atkinson LP (1983) Distribution of Antarctic Immediate Water over the Blake Plateau. J Geophys Res Oceans 88(NC8):4699–4704. https://doi.org/10.1029/JC088iC08p04699

Auster PJ (2005) Are deep-water corals important habitats for fishes? In: Freiwald A, Roberts JM (eds) Cold-water corals and ecosystems. Erlangen earth conference series. Springer, Berlin, pp 747–760. https://doi.org/10.1007/3-540-27673-4_39

Avise JC (1992) Molecular population structure and the biogeographic history of a regional fauna: a case history with lessons for conservation biology. Oikos 63:62–76

Avise JC (2000) Phylogeography: the history and formation of species. Harvard University Press, Cambridge, MA

Avise JC, Reeb CA, Saunders NC (1987) Geographic population structure and species differences in mitochondrial DNA of mouthbrooding marine catfishes (Ariidae) and demersal spawning toadfishes (Batrachoididae). Evol 41:991–1002

Ayers MW, Pilkey OH (1981) Piston core and surficial sediment investigations of the Florida-Hatteras Slope and inner Blake Plateau. In: Popenoe P (ed) Environmental geologic studies on the southeastern Atlantic outer continental shelf. USGS Open File Rept. 81-582-A. pp 5.1–5.89

Baillon S, Hamel JF, Wareham VE, Mercier A (2012) Deep cold-water corals as nurseries for fish larvae. Front Ecol Environ 10(7):351–356

Ballard RD, Uchupi E (1971) Geological Observations of Miami Terrace from Submersible Ben-Franklin. Mar Technol Soc J 5(2):43

Bane JM, Dewar WK (1988) Gulf Stream bimodality and variability downstream of the Charleston Bump. J Geophys Res Oceans 93 (C6):6695–6710. https://doi.org/10.1029/JC093iC06p06695

Bane JM, Atkinson LP, Brooks DA (2001) Gulf Stream physical oceanography at the Charleston Bump: deflection, bimodality, meanders, and upwelling. In: Sedberry GR (ed) Island in the stream: oceanography and fisheries of the Charleston Bump, Amer Fish Soc Symp 25, Bethesda, MD, pp 25–36

BOEM (2014). https://www.boem.gov/sites/default/files/oil-and-gas-energy-program/Resource-Evaluation/Resource-Assessment/Assessment-of-Oil-and-Gas-Resources-2014-Update.pdf. Accessed 1 Sept 2020

BOEM (2020). https://www.boem.gov/oil-gas-energy/atlantic-oil-and-gas-information. Accessed 1 Sept 2020

Bracco A, Liu G, Galaska MP, Quattrini AM, Herrera S (2019) Integrating physical circulation models and genetic approaches to investigate population connectivity in deep-sea corals. J Mar Syst 198:103189

Briggs JC (1974) Marine zoogeography. McGraw-Hill, New York

Brooke S, Jarnegren J (2013) Reproductive periodicity of the deep-water scleractinian coral, *Lophelia pertusa* from the Trondheim Fjord, Norway. Mar Biol 160:139–153

Brooke S, Ross SW (2014) First observations of the cold-water coral *Lophelia pertusa* in mid-Atlantic canyons of the USA. Deep Sea Res Pt II 104:245–251

Brooke S, Ross SW, Bane JM, Seim HE, Young CM (2013) Temperature tolerance of the deep-sea coral *Lophelia pertusa from* the southeastern United States. Deep Sea Res Pt II 92:240–248

Brooke SD, Watts MW, Heil AD, Rhode M, Mienis F, Duineveld GCA, Davies AJ, Ross SW (2017) Distributions and habitat associations of deep-water corals in Norfolk and Baltimore Canyons, Mid-Atlantic Bight, USA. Deep Sea Res Pt II 137:131–147. https://doi.org/10.1016/j.dsr2.2016.05.008

Buhl-Mortensen L, Mortensen PB (2005) Distribution and diversity of species associated with deep-sea gorgonian corals off Atlantic Canada. In: Freiwald A, Roberts JM (eds) Cold-water corals and ecosystems. Erlangen earth conference series. Springer, Berlin, pp 849–879

Buhl-Mortensen L, Vanreusel A, Gooday AJ, Levin LA, Priede IG, Buhl-Mortensen P, Gheerardyn H, King NJ, Raes M (2010) Biological structures as a source of habitat heterogeneity and biodiversity on the deep ocean margins. Mar Ecol 31:21–50. https://doi.org/10.1111/j.1439-0485.2010.00359.x

Büscher JV, Form AU, Riebesell U (2017) Interactive effects of ocean acidification and warming on growth, fitness and survival of the cold-water coral *Lophelia pertusa* under different food availabilities. Front Mar Sci 4:101

Caesar L, McCarthy GD, Thornalley DJ, Cahill N, Rahmstorf S (2021) Current Atlantic meridional overturning circulation weakest in last millennium. Nat Geosci 14(3):118–120

Cantwell K, Sautter L, Morrison C, Sowers DC, Bowman A (2020) EX-18-06: Windows to the deep 2018 cruise report. OER Expedition Rep. 18-06. https://doi.org/10.25923/50hx-3p68

Cavender-Bares J, Kozak KH, Fine PVA, Kembel SW (2009) The merging of community ecology and phylogenetic biology. Ecol Lett 12(7):693–715

Commeau R, Clark A, Manheim F, Aruscavage P, Johnson C, Lane CM (1984) Ferromanganese crust resources in the Pacific and Atlantic Oceans. In: Oceans 1984. IEEE, pp 421–430. https://doi.org/10.1109/OCEANS.1984.1152203

Cordes EE, McGinley M, Podowski EL, Becker EL, Lessard-Pilon S, Viada S, Fisher CR (2008) Coral communities of the deep Gulf of Mexico. Deep Sea Res I 55:777–787

Correa TBS, Eberli GP, Grasmueck M, Reed JK, Correa AMS (2012a) Genesis and morphology of cold-water coral ridges in a unidirectional current regime. Mar Geol 326:14–27. https://doi.org/10.1016/j.margeo.2012.06.008

Correa TBS, Grasmueck M, Eberli GP, Reed JK, Verwer K, Purkis S (2012b) Variability of cold-water coral mounds in a high sediment input and tidal current regime, Straits of Florida. Sedimentology 59(4):1278–1304. https://doi.org/10.1111/j.1365-3091.2011.01306.x

Costello MJ, McCrea M, Freiwald A, Lundälv T, Jonsson L, Bett BJ, van Weering TC, de Haas H, Roberts JM, Allen D (2005) Role of cold-water *Lophelia pertusa* coral reefs as fish habitat in the NE Atlantic. In: Freiwald A, Roberts JM (eds) Cold-water corals and ecosystems. Erlangen earth conference series. Springer, Berlin, pp 771–805

Cunningham CW, Collins TM (1998) Beyond area relationships: extinction and recolonization in molecular marine biogeography. In: DeSalle R, Schierwater B (eds) Molecular approaches to ecology and evolution. Birkhäuser, Basel, pp 297–321

Dahl MP, Pereyra RT, Lundälv T, André C (2012) Fine-scale spatial genetic structure and clonal distribution of the cold-water coral *Lophelia pertusa*. Coral Reefs 31(4):1135–1148

Davies AJ, Guinotte JM (2011) Global habitat suitability for framework-forming cold-water corals. PLoS One 6(4):e18483

Davies AJ, Duineveld GC, Lavaleye MS, Bergman MJ, van Haren H, Roberts JM (2009) Downwelling and deep-water bottom currents as food supply mechanisms to the cold-water coral *Lophelia pertusa* (Scleractinia) at the Mingulay Reef Complex. Limnol Oceanogr 54(2):620–629

Dodds LA, Roberts JM, Taylor AC, Marubini F (2007) Metabolic tolerance of the cold-water coral *Lophelia pertusa* (Scleractinia) to temperature and dissolved oxygen change. J Exp Mar Biol Ecol 349(2):205–214

Dorey N, Gjelsvik Ø, Kutti T, Büscher JV (2020) Broad thermal tolerance in the cold-water coral *Lophelia pertusa* from Arctic and Boreal Reefs. Front Phys 10:1636

Drazen JC, Smith CR, Gjerde KM et al (2020) Opinion: Midwater ecosystems must be considered when evaluating environmental risks of deep-sea mining. PNAS 117(30):17455–17460. https://doi.org/10.1073/pnas.2011914117

Erickson KL, Pentico A, Quattrini AM, McFadden CS (2021) New approaches to species delimitation and population structure of anthozoans: two case studies of octocorals using ultraconserved elements and exons. Mol Ecol Res 21(1):78–92

Etnoyer P, Warrenchuk J (2007) A catshark nursery in a deep gorgonian field in the Mississippi Canyon, Gulf of Mexico. Bull Mar Sci 81(3): 553–559

Etter RJ, Bower AS (2015) Dispersal and population connectivity in the deep North Atlantic estimated from physical transport processes. Deep Sea Res Pt I 104:159–172

Etter RJ, Rex MA, Chase MC, Quattro JM (2005) Population differentiation decreases with depth in deep-sea bivalves. Evol 59(7): 1479–1491

Fernholm B, Quattrini AM (2008) A new species of hagfish (Myxinidae: *Eptatretus*) associated with deep-sea coral habitat in the western North Atlantic. Copeia 2008(1):126–132

Fontela M, Pérez FF, Carracedo LI, Padín XA, Velo A, García-Ibañez MI, Lherminier P (2020) The Northeast Atlantic is running out of excess carbonate in the horizon of cold-water corals communities. Sci Rep 10(1):1–10

Form AU, Riebesell U (2012) Acclimation to ocean acidification during long-term CO2 exposure in the cold-water coral *Lophelia pertusa*. Global Change Biol 18(3):843–853

Fratantoni PS, Lee TN, Podesta GP, Muller-Karger F (1998) The influence of loop current perturbations on the formation and evolution of Tortugas eddies in the southern Straits of Florida. J Geophys Res Oceans 103(C11):24759–24779. https://doi.org/10.1029/98jc02147
Freiwald A (2002) Reef-forming cold-water corals. In: Wefer G, Billeit D, Hebbeln D, Jørgensen BB, Schlüter M, Van Weering T (eds) Ocean margin systems. Springer, Berlin, pp 365–385
Galgani F, Hanke G, Maes T (2015) Global distribution, composition and abundance of marine litter. In: Bergmann M, Gutow L, Klages M (eds) Marine anthropogenic litter. Springer, Cham
Gallagher ML, Ambrose WG Jr, Renaud PE (1998) Comparative studies in biochemical composition of benthic invertebrates (bivalves, ophiuroids) from the Northeast Water (NEW) Polynya. Polar Biol 19:167–171
Gardner JV, Field ME, Twichell DC (eds) (1996) Geology of the United States' seafloor: the view from GLORIA. Cambridge University Press
Gartner JV, Sulak KJ, Ross SW, Necaise AM (2008) Persistent near-bottom aggregations of mesopelagic animals along the North Carolina and Virginia continental slopes. Mar Biol 153(5):825–841
George RY (2002) Ben Franklin temperate reef and deep sea 'Agassiz Coral Hills' in the Blake Plateau off North Carolina. Hydrobiologia 471(1–3):71–81. https://doi.org/10.1023/A:1016545102459
Girard F, Fu B, Fisher CR (2016) Mutualistic symbiosis with ophiuroids limited the impact of the Deepwater Horizon oil spill on deep-sea octocorals. Mar Ecol Prog Ser 49:89–98
Gomberg DN (1976) Geology of the Pourtalès Terrace, Straits of Florida. Dissertation, University of Miami, Florida
Gómez CE, Wickes L, Deegan D, Etnoyer PJ, Cordes EE (2018) Growth and feeding of deep-sea coral *Lophelia pertusa* from the California margin under simulated ocean acidification conditions. PeerJ 6: e5671
Gasbarro R, Sowers D, Margolin A, Cordes EE (2022) Distribution and predicted climatic refugia for a reef-building cold-water coral on the southeast US margin. Glob Chang Biol 28(23):7108–7125
Grasmueck M, Eberli GP, Viggiano DA, Correa T, Rathwell G, Luo JG (2006) Autonomous underwater vehicle (AUV) mapping reveals coral mound distribution, morphology, and oceanography in deep water of the Straits of Florida. Geophys Res Lett 33(23):6. https://doi.org/10.1029/2006gl027734
Gress E, Andradi-Brown DA, Woodall L, Schofield PJ, Stanley K, Rogers AD (2017) Lionfish (*Pterois* spp.) invade the upper-bathyal zone in the western Atlantic. PeerJ 5:e3683
Guinotte JM, Orr J, Cairns S, Freiwald A, Morgan L, George R (2006) Will human-induced changes in seawater chemistry alter the distribution of deep-sea scleractinian corals? Front Ecol Environ 4(3): 141–146
Gula J, Molemaker MJ, McWilliams JC (2015) Gulf Stream dynamics along the southeastern US seaboard. J Phys Oceanogr 45(3): 690–715. https://doi.org/10.1175/jpo-d-14-0154.1
Harrison P (2011) Sexual reproduction of scleractinian corals. In: Dubinsky Z, Stambler N (eds) Coral reefs: an ecosystem in transition. Springer, New York, pp 59–85
Harter S, Ribera M, Shepard A, Reed J (2009) Assessment of fish populations and habitat on *Oculina* Bank, a deep-sea coral marine protected area off eastern Florida. Fish Bull 107(2):195–206
Harter S, Reed J, Farrington S, David A (2020) South Atlantic MPAs and *Oculina* HAPC: Characterization of benthic habitat and biota. NOAA Ship Pisces Cruise 19-02. NOAA CIOERT Cruise Report, pp 388
Hebbeln D, Portilho-Ramos RD, Wienberg C, Titschack J (2019) The fate of cold-water corals in a changing world: a geological perspective. Front Mar Sci 6:119
Heiderich J, Todd RE (2020) Along-stream evolution of Gulf Stream volume transport. J Phys Oceanogr 50(8):2251–2270. https://doi.org/10.1175/jpo-d-19-0303.1
Hennige SJ, Wolfram U, Wickes L, Murray F, Roberts JM, Kamenos NA, Schofield S, Groetsch A, Spiesz EM, Aubin-Tam ME, Etnoyer PJ (2020) Crumbling reefs and cold-water coral habitat loss in a future ocean: evidence of "Coralporosis" as an indicator of habitat integrity. Front Mar Sci 7:668
Henry LA, Nizinski MS, Ross SW (2008) Occurrence and biogeography of hydroids (Cnidaria: Hydrozoa) from deep-water coral habitats off the southeastern United States. Deep Sea Res Pt I 55(6):788–800
Herrera S, Shank TM (2016) RAD sequencing enables unprecedented phylogenetic resolution and objective species delimitation in recalcitrant divergent taxa. Mol Phyl Evol 100:70–79
Herrera S, Baco A, Sánchez JA (2010) Molecular systematics of the bubblegum coral genera (Paragorgiidae, Octocorallia) and description of a new deep-sea species. Mol Phylogenet Evol 55(1):123–135
Hourigan TF, Cairns S, Reed J, Ross S (2017a) Deep-sea coral taxa in the U.S. Southeast Region: depth and geographical distribution. Online resource: https://deepseacoraldata.noaa.gov/library/2017-state-of-deep-sea-corals-report
Hourigan TF, Reed J, Pomponi S, Ross S, David A, Harter S (2017b) State of deep-sea coral and sponge ecosystems of the Southeast United States. In: Hourigan T, Etnoyer P, Cairns S (eds) The state of deep-sea coral and sponge ecosystems of the United States. NOAA Tech Memo NMFS-OHC-4, Silver Spring, MD, p 60
Jones CG, Lawton JH, Shachak M (1994) Organisms as ecosystem engineers. In: Samson FB, Knopf FL (eds) Ecosystem management. Springer, New York, pp 130–147
Jordan GF, Malloy RJ, Kofoed JW (1964) Bathymetry and geology of Pourtalès Terrace, Florida. Mar Geol 1(3):259–287. https://doi.org/10.1016/0025-3227(64)90063-5
Kinlan BP, Poti M, Hourigan T, Dorfman D, Caldow C (2013) Digital data: Predictive models of deep-sea coral habitat suitability in the U.S. Southeast region. Released August 2013. Available at: https://nccospublicstor.blob.core.windows.net/projects-attachments/35/DSC_PredictiveModeling_US_Southeast.zip
Koenig CC, Coleman FC, Grimes CB, Fitzhugh GR, Scanlon KM, Gledhill CT, Grace M (2000) Protection of fish spawning habitat for the conservation of warm-temperate reef-fish fisheries of shelf-edge reefs of Florida. Bull Mar Sci 66(3):593–616
Land LA, Paull CK (2000) Submarine karst belt rimming the continental slope in the Straits of Florida. Geo Mar Lett 20(2):123–132. https://doi.org/10.1007/s003670000041
Larsson AI, Järnegren J, Strömberg SM, Dahl MP, Lundälv T, Brooke S (2014) Embryogenesis and larval biology of the cold-water coral *Lophelia pertusa*. PLoS One 9(7):e102222
Le Bras IA, Yashayaev I, Toole JM (2017) Tracking Labrador sea water property signals along the deep western boundary. Curr J Geophys Res Oceans 122:5348–5366. https://doi.org/10.1002/2017JC012921
Lee TN, Atkinson LP (1983) Low frequency current and temperature variability from Gulf Stream frontal eddies and atmospheric forcing along the Southeast United States outer continental shelf. J Geophys Res Oceans 88:4541–4567. https://doi.org/10.1029/JC088iC08p04541
Lee TN, Yoder JA, Atkinson LP (1991) Gulf Stream frontal eddy influence on productivity of the Southeast US continental shelf. J Geophys Res 96:22191–22205
Legeckis RV (1979) Satellite observations of the influence of bottom topography on the seaward deflection of the Gulf Stream off Charleston, South Carolina. J Phys Oceanogr 9:483–497
Le Goff-Vitry MC, Pybus OG, Rogers AD (2004) Genetic structure of the deep-sea coral *Lophelia pertusa* in the northeast Atlantic revealed by microsatellites and internal transcribed spacer sequences. Mol Ecol 13(3):537–549
Lessard-Pilon S, Podowski EL, Cordes EE, Fisher CR (2010) *Lophelia pertusa* as a habitat resource in the deep Gulf of Mexico. Deep Sea Res II 57:1882–1890

Leterme SC, Pingree RD (2008) The Gulf Stream, rings and North Atlantic eddy structures from remote sensing (Altimeter and SeaWiFS). J Mar Syst 69:177–190

Leydet KP, Grupstra CGB, Coma R, Ribes M, Hellberg ME (2018) Host-targeted RAD-Seq reveals genetic changes in the coral *Oculina patagonica* associated with range expansion along the Spanish Mediterranean coast. Mol Ecol 27:2529–2543

Litvinova NM (1980) Ways of feeding in some species of brittle stars. Zool Zh 59:239–247

Lovrich GA, Thiel M (2011) Ecology, physiology, feeding and trophic role of squat lobsters. In: Poore GCB, Ahyong ST, Taylor J (eds) The biology of squat lobsters. CSIRO, Melbourne, pp 183–222

Lunden JJ, Georgian SE, Cordes EE (2013) Aragonite saturation states at cold-water coral reefs structured by *Lophelia pertusa* in the northern Gulf of Mexico. Limnol Oceanogr 58:354–362

Lunden JJ, McNicholl CG, Sears CR, Morrison CL, Cordes EE (2014) Acute survivorship of the deep-sea coral *Lophelia pertusa* from the Gulf of Mexico under acidification, warming, and deoxygenation. Front Mar Sci 1(78):1–12

Lusty PA, Murton BJ (2018) Deep-ocean mineral deposits: metal resources and windows into earth processes. Elements 14(5): 301–306

Mah CL (2020) New species, occurrence records and observations of predation by deep-sea Asteroidea (Echinodermata) from the North Atlantic by NOAA ship Okeanos Explorer. Zootaxa 4766(2): 201–260. https://doi.org/10.11646/zootaxa.4766.2.1

Mah CL, Nizinski M, Lundsten L (2010) Phylogenetic revision of the Hippasterinae (Goniasteridae; Asteroidea): systematics of deep sea corallivores, including one new genus and three new species. Zool J Linn Soc 160:266–301

Malloy RJ, Hurley RJ (1970) Geomorphology and geologic structure: Straits of Florida. Geol Soc Am Bull 81(7):1947–1972

Manheim FT (1986) Marine cobalt resources. Science 232(4750): 600–608

Matos L, Mienis F, Wienberg C, Frank N, Kwiatkowski C, Groeneveld J, Thil F, Abrantes F, Cunha MR, Hebbeln D (2015) Interglacial occurrence of cold-water corals off Cape Lookout (NW Atlantic): first evidence of the Gulf Stream influence. In: Deep sea research part I: oceanographic research papers, vol 105. Vancouver, pp 158–170

Matrai PA, Cooper DJ, Saltzman ES (1996) Frontal enhancement of dimethylsulfide concentrations across a Gulf Stream meander. J Mar Syst 7(1):1–8

McCartney MA, Burton ML, Lima TG (2013) Mitochondrial DNA differentiation between populations of black sea bass (*Centropristis striata)* across Cape Hatteras, North Carolina (USA). J Biogeogr 40: 1386–1398

McCosker JE, Ross SW (2007) A new deepwater species of the snake eel genus *Ophichthus* (Anguilliformes: Ophichthidae) from North Carolina. Copeia 2007(4):783–787

Meinen CS, Baringer MO, Garcia RF (2010) Florida Current transport variability: an analysis of annual and longer-period signals. Deep Sea Res Pt I 57(7):835–846. https://doi.org/10.1016/j.dsr.2010.04.001

Meinen CS, Johns WE, Moat B, Smith RH, Johns EM, Rayner D, Frajka-Williams E, Garcia RF, Garzoli SL (2019) Structure and variability of the Antilles Current at 26.5°N. J Geophys Res Oceans 124(6):3700–3723. https://doi.org/10.1029/2018jc014836

Meister HS, Wyanski DM, Loefer JK, Ross SW, Quattrini AM, Sulak KJ (2005) Further evidence for the invasion and establishment of *Pterois volitans* (Teleostei: Scorpaenidae) along the Atlantic coast of the United States. Southeast Nat 4(2):193–206

Messing CG, Neumann AC, Lang JC (1990) Biozonation of deep-water lithoherms and associated hardgrounds in the northeastern Straits of Florida. Palaios 5(1):15–33

Messing CG, Reed JK, Brooke SD, Ross SW (2008) Deepwater coral reefs of the United States. In: Riegl B, Dodge RE (eds) Coral reefs of the USA. Coral reefs of the world, vol 1. Springer, Dordrecht, pp 767–791. https://doi.org/10.1007/978-1-4020-6847-8_21

Middelburg JJ, Mueller CE, Veuger B, Larsson AI, Form A, Oevelen DV (2015) Discovery of symbiotic nitrogen fixation and chemoautotrophy in cold-water corals. Sci Rep 5(1):17962

Mienis F, De Stigter HC, White M, Duineveld G, De Haas H, Van Weering TCE (2007) Hydrodynamic controls on cold-water coral growth and carbonate-mound development at the SW and SE Rockall Trough Margin, NE Atlantic Ocean. Deep Sea Res Pt I 54(9):1655–1674

Mienis F, Duineveld GCA, Davies AJ, Ross SW, Seim H, Bane J, Van Weering TCE (2012) The influence of near-bed hydrodynamic conditions on cold-water corals in the Viosca Knoll area, Gulf of Mexico. Deep Sea Res Pt I 60:32–45

Mienis F, Duineveld GCA, Davies AJ, Lavaleye MMS, Ross SW, Seim H, Bane J, Van Haren H, Bergman MJN, De Haas H, Brooke S (2014) Cold-water coral growth under extreme environmental conditions, the Cape Lookout area, NW Atlantic. Biogeosciences 11(9):2543–2560

Morato T, González-Irusta JM, Dominguez-Carrió C, Wei CL, Davies A, Sweetman AK, Taranto GH, Beazley L, García-Alegre A, Grehan A, Laffargue P (2020) Climate-induced changes in the suitable habitat of cold-water corals and commercially important deep-sea fishes in the North Atlantic. Glob Chang Biol 26(4): 2181–2202. https://doi.org/10.1111/gcb.14996

Morrison CL, Ross SW, Nizinski MS, Brooke S, Waller RG, Johnson RL, King TL (2011) Genetic discontinuity among regional populations of *Lophelia pertusa* in the North Atlantic Ocean. Conserv Genet 12(3):713–729

Morrison CL, Coykendall DK, Springmann MJ, Shroades KM, Sanders LR, Waller RG, Ross SW, Brooke SD (2017) Scleractinian coral biodiversity and patterns of inter-canyon connectivity among four coral species. In: CSA Ocean Sciences Inc, Ross S, Brooke S et al (eds) Exploration and research of Mid-Atlantic Deepwater hard bottom habitats and shipwrecks with emphasis on canyons and coral communities: Atlantic deepwater canyons study. OCS Study BOEM 2017-060. U.S. Department of the Interior, Bureau of Ocean Energy Management, Sterling, VA

Mortensen PB, Fosså JH (2006) Species diversity and spatial distribution of invertebrates on deep-water *Lophelia* reefs in Norway. In: Proceedings of 10th international coral reef symposium 2006, pp 1849–1868

Mullins HT, Newton CR, Heath K, Vanburen HM (1981) Modern deep-water coral mounds north of Little Bahama Bank—criteria for recognition of deep-water coral bioherms in the rock record. J Sediment Petrol 51(3):999–1013

Muñoz RC, Currin CA, Whitfield PE (2011) Diet of invasive lionfish on hard bottom reefs of the Southeast USA: insights from stomach contents and stable isotopes. Mar Ecol Prog Ser 432:181–193

Naumann MS, Orejas C, Ferrier-Pagès C (2014) Species-specific physiological response by the cold-water corals *Lophelia pertusa* and *Madrepora oculata* to variations within their natural temperature range. Deep Sea Res Pt II 99:36–41

Neumann AC, Ball MM (1970) Submersible observations in the Straits of Florida: geology and bottom currents. Geol Soc Am Bull 81(10): 2861–2874

Neumann AC, Kofoed J, Keller GH (1977) Lithoherms in the Straits of Florida. Geology 5(1):4–10

Newton CR, Mullins HT, Gardulski AF, Hine AC, Dix GR (1987) Coral mounds on the West Florida slope: unanswered questions regarding the development of deep-water banks. Palaios 2(4):359–367

NOAA Fisheries (2020). https://www.fisheries.noaa.gov/management-plan/south-atlantic-golden-crab-fishery-management-plan. Accessed 1 Sept 2020

Pape-Lindstrom PA, Feller RJ, Stancyk SE, Woodin SA (1997) Sublethal predation: field measurements of arm tissue loss from the ophiuroid *Microphiopholis gracillima* and immunochemical identification of its predators in North Inlet, South Carolina, USA. Mar Ecol Prog Ser 156:131–140

Partyka ML, Ross SW, Quattrini AM, Sedberry GR, Birdsong TW, Potter J, Gottfried S (2007) Southeastern United States deep-sea corals (SEADESC) initiative: a collaboration to characterize areas of habitat forming deep-sea corals. Silver Spring, MD

Paull CK, Dillon WP (1979) The subsurface geology of the Florida-Hatteras slope and inner Blake Plateau. US Geol Surv Open-File Rept No 79-448. https://doi.org/10.3133/ofr79448

Paull C, Neumann A, Am Ende BA, Ussler W III, Rodriguez N (2000) Lithoherms on the Florida–Hatteras slope. Mar Geol 166(1):83–101

Pearson M, Gage JD (1984) Diets of some deep-sea brittle stars in the Rockall Trough. Mar Biol 82(3):247–258

Pickart RS (1992) Water mass components of the North Atlantic deep western boundary current. Deep Sea Res Pt A 39(9):1553–1572. https://doi.org/10.1016/0198-0149(92)90047-W

Pickart RS, Smethie WM (1993) How does the deep western boundary current cross the Gulf Stream. J Phys Oceanogr 23(12):2602–2616

Popenoe P (1994) Bottom character map of the northern Blake Plateau. US Geol Surv Open-File Rept No. 93-724. https://doi.org/10.3133/ofr93724

Popenoe P, Manheim FT (2001) Origin and history of the Charleston Bump-geological formations, currents, bottom conditions, and their relationship to wreckfish habitats on the Blake Plateau. In: Sedberry GR (ed) Island in the stream: oceanography and fisheries of the Charleston Bump, Amer Fish Soc Symp 25, Bethesda, MD, pp 43–94

Pratt RM, Heezen BC (1964) Topography of the Blake Plateau. Deep Sea Res Pt II 11(5):721–728

Prouty NG, Roark EB, Koenig AE, Demopoulos AWJ, Batista FC, Kocar BD, Selby D, McCarthy MD, Mienis F, Ross SW (2014) Deep-sea coral record of human impact on watershed quality in the Mississippi River Basin. Global Biogeochem Cycles 28:29–43. https://doi.org/10.1002/2013GB004754

Prouty NG, Mienis F, Campbell-Swarzenski P, Roark EB, Davies AJ, Robertson CM, Duineveld G, Ross SW, Rhode M, Demopoulos AWJ (2017) Seasonal variability in the source and composition of particulate matter in the depositional zone of Baltimore Canyon, U.S. Mid-Atlantic Bight. Deep Sea Res Pt I 127:77–89. https://doi.org/10.1016/j.dsr.2017.08.004

Quattrini AM, Ross SW, Carlson MCT, Nizinski MS (2012) Megafaunal-habitat associations at a deep-sea coral mound off North Carolina, USA. Mar Biol 159(5):1079–1094

Quattrini AM, Etnoyer PJ, Doughty C, English L, Falco R, Remona N, Rittinghouse M, Cordes EE (2014) A phylogenetic approach to octocoral community structure in the deep Gulf of Mexico. Deep Sea Res Pt II 99:92–102. https://doi.org/10.1016/j.dsr2.2013.05.027

Quattrini AM, Baums IB, Shank TM, Morrison CL, Cordes EE (2015) Testing the depth-differentiation hypothesis in a deepwater octocoral. Proc R Soc B 282:20150008

Quattrini AM, Wu T, Soong K, Jeng M-S, Benayahu Y, McFadden CS (2019) A next generation approach to species delimitation reveals the role of hybridization in a cryptic species complex of corals. BMC Evol Biol 19:116. https://doi.org/10.1186/s12862-019-1427-y

Quattrini AM, Rodríguez E, Faircloth BC, Cowman PF, Brugler MR, Farfan GA, Hellberg ME, Kitahara MV, Morrison CL, Paz-García DA, Reimer JD, McFadden CS (2020) Palaeoclimate ocean conditions shaped the evolution of corals and their skeletons through deep time. Nat Ecol Evol. https://doi.org/10.1038/s41559-020-01291-1

Radice VZ, Quattrini AM, Wareham VE, Edinger EN, Cordes EE (2016) Vertical water mass structure in the North Atlantic influences the bathymetric distribution of species in the deep-sea coral genus *Paramuricea*. Deep Sea Res Pt I 116:253–263

Reed JK (1980) Distribution and structure of deep-water *Oculina varicosa* coral reefs off central eastern Florida. Bull Mar Sci 30(3): 667–677

Reed JK (1981) In situ growth rates of the scleractinian coral *Oculina varicosa* occurring with zooxanthellae on 6-m reefs and without on 80-m banks. Proc Fourth Int Coral Reef Symp 2:201–206

Reed JK (2002a) Comparison of deep-water coral reefs and lithoherms off southeastern USA. Hydrobiologia 471(1–3):57–69. https://doi.org/10.1023/A:1016593018389

Reed JK (2002b) Deep-water *Oculina* coral reefs of Florida: biology, impacts, and management. Hydrobiologia 471(1–3):43–55. https://doi.org/10.1023/A:1016588901551

Reed JK (2006) Deep-water *Oculina* reefs of Florida: summary of the state of knowledge of the habitat, fauna, geology, and physical processes of the ecosystem. Port Canaveral, Florida South Atlantic Fishery Management Council Oculina Evaluation Team Workshop, p 27

Reed JK, Farrington S (2010) Distribution of deep-water commercial fisheries species-golden crab, tilefish, royal red shrimp-in deep-water habitats off eastern Florida from submersible and ROV dives. South Atlantic Fishery Management Council and NOAA National Marine Fisheries Service Report

Reed JK, Mikkelson PM (1987) The molluscan community associated with the scleractinian coral *Oculina varicosa*. Bull Mar Sci 40(1): 99–131

Reed JK, Wright A (2004) Final cruise report. Submersible and scuba collections on deep-water reefs off the east coast of Florida, including the Northern and Southern Straits of Florida and Florida Keys National Marine Sanctuary for biomedical and biodiversity research of the benthic communities with emphasis on the Porifera and Gorgonacea, May 20–June 2, 2004. Center of Excellence, HBOI and FAUs, p 54

Reed JK, Gore RH, Scotto LE, Wilson KA (1982) Community composition, structure, aereal and trophic relationships of decapods associated with shallow- and deep-water *Oculina varicosa* coral reefs. Bull Mar Sci 32:761–786

Reed JK, Pomponi SA, Weaver D, Paull CK, Wright AE (2005a) Deep-water sinkholes and bioherms of south Florida and the Pourtalès Terrace—habitat and fauna. Bull Mar Sci 77(2):267–296

Reed JK, Shepard AN, Koenig CC, Scanlon KM, Gilmore RG Jr (2005b) Mapping, habitat characterization, and fish surveys of the deep-water *Oculina* coral reef Marine Protected Area: a review of historical and current research. In: Freiwald A, Roberts JM (eds) Cold-water corals and ecosystems. Erlangen earth conference series. Springer, Berlin, pp 443–465

Reed JK, Weaver DC, Pomponi SA (2006) Habitat and fauna of deep-water *Lophelia pertusa* coral reefs off the southeastern US: Blake Plateau, Straits of Florida, and Gulf of Mexico. Bull Mar Sci 78(2): 343–375

Reed JK, Koenig CC, Shepard AN (2007) Impacts of bottom trawling on a deep-water *Oculina* coral ecosystem off Florida. Bull Mar Sci 81:481–496

Reed J, Harter S, Farrington S, David A (2013a) South Atlantic MPAs and deepwater coral HAPCs: Characterization of Benthic Habitat and Biota. HBOI Tech Rept No. 144

Reed JK, Messing C, Walker BK, Brooke S, Correa TBS, Brouwer M, Udouj T, Farrington S (2013b) Habitat Characterization, Distribution, and Areal Extent of Deep-sea Coral Ecosystems off Florida, Southeastern U.S.A. Caribb J Sci 47(1):13–30. https://doi.org/10.18475/cjos.v47i1.a3

Reed JK, Harter S, Farrington S, David A (2014) Characterization and interrelationships of deepwater coral/sponge habitats and fish communities off Florida, USA. In: Bortone S (ed) Coral habitat and fish interrelationships. CRC, New York, pp 49–80

Reed JK, Farrington S, Messing C, David A (2017) Distribution and habitat use of the golden crab *Chaceon fenneri* off eastern Florida based on in situ submersible and ROV observations and potential for impacts to deep water coral/sponge habitat. Gulf Caribb Res 28:1–14. https://doi.org/10.18785/gcr.2801.03

Roberts JM, Wheeler AJ, Freiwald A (2006) Reefs of the deep: The biology and geology of cold-water coral ecosystems. Science 312(5773):543–547. https://doi.org/10.1126/science.11(1986)1

Roberts JM, Wheeler A, Freiwald A, Cairns S (2009) Cold-water corals: the biology and geology of deep-sea coral habitats. Cambridge University Press, Cambridge

Rogers AD (1999) The biology of *Lophelia pertusa* (Linnaeus 1758) and other deep-water reef-forming corals and impacts from human activities. Int Rev Hydrobiol 84(4):315–406

Ross S, Nizinski M (2007) State of deep coral ecosystems in the southeastern United States Region: Cape Hatteras to the Florida In: Lumsden SE, Hourigan TF, Bruckner AW, Dorr G (eds) The state of deep coral ecosystems of the United States. NOAA Tech Memo CRCP-3, Silver Spring, MD, pp 233–270

Ross SW, Quattrini AM (2007) The fish fauna associated with deep coral banks off the southeastern United States. Deep Sea Res Pt I 54(6):975–1007. https://doi.org/10.1016/j.dsr.2007.03.010

Ross SW, Quattrini AM (2009) Deep-sea reef fish assemblage patterns on the Blake Plateau (Western North Atlantic Ocean). Mar Ecol 30(1):74–92

Ross SW, Brooke S, Quattrini AM, Rhode M, Watterson JC (2015) A deep-sea community, including *Lophelia pertusa*, at unusually shallow depths in the western North Atlantic Ocean off northeastern Florida. Mar Biol 162(3):635–648. https://doi.org/10.1007/s00227-015-2611-2

Rousset C, Beal LM (2014) Closing the transport budget of the Florida Straits. Geophys Res Lett 41(7):2460–2466. https://doi.org/10.1002/2014gl059498

Rowe GT, Menzies RJ (1969) Zonation of large benthic invertebrates in the deep-sea off the Carolinas. Deep-Sea Res Oceanogr Abstr 16(5): 531–537

Sautter LR, Morrison C, Cantwell K, Sowers D, Lobecker E (2019) Windows to the deep 2018: exploration of the Southeast US Continental Margin. Oceanography 32(1, Suppl):82–87

Sedberry GR (ed) (2001) Island in the stream: oceanography and fisheries of the Charleston Bump. Am Fish Soc Symp 25, Bethesda, MD

Sedberry GR, Ulrich GF, Applegate A (1994) Development and status of the fishery for wreckfish *(Polyprion americanus)* in the southeastern United States. Proc Gulf Caribb Fish Inst 43:168–192

Sedberry G, Andrus C, Barans C, Crowe S, Fiore C, Griffin S, Jutte PC, Sautter L, Schobernd C (2010) Mounds, depressions, scarps and caves: exploration of habitat and species diversity on the Charleston Bump and Blake Plateau. Final Project Report. Ocean Exploration, p 62

Seim H, Edwards C (2019) Upper-slope jets and Gulf Stream filaments inshore of the Charleston Bump during winter 2012. J Phys Oceanogr 49:1423–1438. https://doi.org/10.1175/JPO-D-18-0205.1

Sheinbaum J, Candela J, Badan A, Ochoa J (2002) Flow structure and transport in the Yucatan Channel. Geophys Res Lett 29:1040. https://doi.org/10.1029/2001GL013990

Smith PJ, McVeagh SM, Mingoia T, France SC (2004) Mitochondrial DNA sequence variation in deep-sea bamboo coral (Keratoisidinae) species in the southwest and northwest Pacific Ocean. Mar Biol 144: 253–261

Sowers DC (2020) Utilizing extended continental shelf (ECS) and ocean exploration mapping data for standardized marine ecological classification of the US Atlantic Margin (2020.28260193) Dissertation, University of New Hampshire. ProQuest Dissertations

Squires DF (1964) Fossil coral thickets in Wairarapa, NZ. J Paleontol 38(5):904–915

Stefanoudis PV, Gress E, Pitt JM, Smith SR, Kincaid T, Rivers M, Andradi-Brown DA, Rowlands G, Woodall LC, Rogers AD (2019) Depth-dependent structuring of reef fish assemblages from the shallows to the rariphotic zone. Front Mar Sci 6:307. https://doi.org/10.3389/fmars.2019.00307

Stetson TR, Squires DF, Pratt RM (1962) Coral banks occurring in deep water on the Blake Plateau. Am Mus Novit 2114:1–39

Stiles ML, Harrould-Kolieb E, Faure P, Ylitalo-Ward H, Hirshfield MF (2007) Deep-sea trawl fisheries of the southeast US and Gulf of Mexico: rock shrimp, royal red shrimp, calico scallops. Oceana Rept, May 2007. https://www.coris.noaa.gov/activities/resourceCD/resources/trawl_report_bm.pdf. Accessed 1 Sept 2020

Summers AC, Nybakken J (2000) Brittle star distribution patterns and population densities on the continental slope off central California (Echinodermata: Ophiuroidea). Deep Sea Res Pt II 47(5–6): 1107–1137. https://doi.org/10.1016/S0967-0645(99)00138-1

Szmant AM (1986) Reproductive ecology of Caribbean reef corals. Coral Reefs 5:43–54

Szuts ZB, Meinen CS (2017) Florida Current salinity and salinity transport: mean and decadal changes. Geophys Res Lett 44(20): 10495–10503. https://doi.org/10.1002/2017gl074538

Taylor ML, Roterman CN (2017) Invertebrate population genetics across Earth's largest habitat: the deep-sea floor. Mole Ecol 26: 4872–4896

Taylor ML, Gwinnett C, Robinson LF, Woodall LC (2016) Plastic microfibre ingestion by deep-sea organisms. Sci Rep 6:33997. https://doi.org/10.1038/srep33997

Teichert C (1958) Cold-and deep-water coral banks. AAPG Bull 42(5): 1064–1082

Todd RE (2017) High-frequency internal waves and thick bottom mixed layers observed by gliders in the Gulf Stream. Geophys Res Lett 44(12):6316–6325. https://doi.org/10.1002/2017gl072580

Todd RE (2020) Spray glider observations in support of PEACH [Data set]. Scripps Institution of Oceanography, Instrument Development Group. https://doi.org/10.21238/S8SPRAY0880

Todd RE (2021) Gulf Stream mean and eddy kinetic energy: three-dimensional estimates from underwater glider observations. Geophys Res Lett 48(6):e2020GL090281. https://doi.org/10.1029/2020GL090281

Todd RE, Owens B (2016) Gliders in the Gulf Stream [Data set]. Scripps Institution of Oceanography, Instrument Development Group. https://doi.org/10.21238/S8SPRAY2675

Todd RE, Owens WB, Rudnick DL (2016) Potential vorticity structure in the North Atlantic Western Boundary Current from underwater glider observations. J Phys Oceanogr 46(1):327–348. https://doi.org/10.1175/jpo-d-15-0112.1

Tsuchiya M (1989) Circulation of the Antarctic intermediate water in the North-Atlantic Ocean. J Mar Res 47(4):747–755. https://doi.org/10.1357/002224089785076136

Uchupi E (1967) The continental margin south of Cape Hatteras, North Carolina: shallow structure. Southeast Geol 8(7):155–177

Uchupi E, Tagg R (1967) Microrelief of the continental margin south of Cape Lookout, North Carolina. Geol Soc Amer Bull 77:427–430

Van Haren H, Mienis F, Duineveld GCA, Lavaleye MSS (2014) High-resolution temperature observations of a trapped nonlinear diurnal tide influencing cold-water corals on the Logachev mounds. Prog Oceanogr 125:16–25. https://doi.org/10.1016/j.pocean.2014.04.021

Waller RG (2005) Deep-water Scleractinia (Cnidaria: Anthozoa) current knowledge of reproductive processes. In: Friewald A, Roberts JM (eds) Cold water corals and ecosystems. Springer, Berlin, pp 691–700

Waller RG, Tyler PA (2005) The reproductive biology of two deep-water, reef-building scleractinians from the NE Atlantic Ocean. Coral Reefs 24(3):514–522

Watling L, France SC, Pante E, Simpson A (2011) Biology of deep-water octocorals. Adv Mar Biol 60:41–122. https://doi.org/10.1016/B978-0-12-385529-9.00002-0

Weaver DC, Sedberry GR (2001) Trophic subsidies at the Charleston Bump: food web structure of reef fishes on the continental slope of the southeastern United States. In: Sedberry GR (ed) Island in the stream: oceanography and fisheries of the Charleston Bump, Amer Fish Soc Symp 25, Bethesda, MD, pp 137–152

Wedding LM, Friedlander AM, Kittinger JN, Watling L, Gaines SD, Bennett M, Hardy SM, Smith CR (2013) From principles to practice: a spatial approach to systematic conservation planning in the deep sea. Proc R Soc B 280:20131684. https://doi.org/10.1098/rspb.2013.1684

Wenner E, Barans C (1990) In situ estimates of density of golden crab, *Chaceon fenneri*, from habitats on the continental slope, southeastern US. Bull Mar Sci 46(3):723–734

Wenner E, Barans C (2001) Benthic habitats and associated fauna of the upper- and middle-continental slope near the Charleston Bump. In: Sedberry GR (ed) Island in the stream: oceanography and fisheries of the Charleston Bump, Amer Fish Soc Symp 25, Bethesda, MD, pp 161–176

Wienberg C, Hebbeln D, Fink HG, Mienis F, Dorschel B, Vertino A, Correa ML, Freiwald A (2009) Scleractinian cold-water corals in the Gulf of Cadiz-First clues about their spatial and temporal distribution. Deep Sea Res Pt I 56:1873–1893

William T, Kano A, Ferdelman T, Henriet JP, Abe K, Andres MS, Bjerager M, Browning EL, Cragg BA, De Mol B (2006) Cold-water coral mounds revealed. Eos Trans Am Geophys Union 87(47):525–526

Williams B, Risk MJ, Ross SW, Sulak KJ (2007) Stable isotope data from deep-water antipatharians: 400-year records from the southeastern coast of the United States of America. Bull Mar Sci 81(3):437–447

Wilson J (1979) Patch development of the deep-water coral *Lophelia pertusa* (L.) on Rockall Bank. J Mar Biol Assoc UK 59(1):165–177

Woodall LC, Sanchez-Vidal A, Canals M, Paterson GLJ, Coppock R, Sleight V, Calafat A, Rogers AD, Narayanaswamy BE, Thompson RC (2014) The deep sea is a major sink for microplastic debris. R Soc Open Sci 1:140317. https://doi.org/10.1098/rsos.140317

Yang H, Lohmann G, Wei W, Dima M, Ionita M, Liu J (2016) Intensification and poleward shift of subtropical western boundary currents in a warming climate. J Geophys Res Oceans 121:4928–4945

Zarudzki EF, Uchupi E (1968) Organic reef alignment on the continental margin south of Cape Hatteras. Geol Soc Am Bull 79:1867–1870

Zeng X, He R (2016) Gulf Stream variability and a triggering mechanism of its large meander in the South Atlantic Bight. J Geophys Res Oceans 121(11):8021–8038. https://doi.org/10.1002/2016jc012077

Norwegian Coral Reefs

5

Pål Buhl-Mortensen and André Freiwald

Abstract

Cold-water corals (CWC) in Norwegian waters have been known for more than two centuries, but direct studies of CWC reefs were first enabled with the introduction of new ROV (Remotely Operated Vehicle) technology in the 1980s. This provided new knowledge about reef ecology and how fisheries have impacted these fragile habitats. In this chapter we present the status of research on CWC reefs formed by the scleractinian *Desmophyllum pertusum* in Norway. More than 1500 reef locations have been reported from Norway. These reefs can be up to 40 m high and 1 km long. Their bathymetric distribution displays three peaks (90–120 m, 200–280 m, and 310–370 m depth), reflecting three different settings: (1) fjords and coastal zone, (2) continental shelf, and (3) shelf break and upper slope. Below 370 m depth, reefs are much less common and the lower distribution limit at around 500 m fits with the depth influenced by the Arctic Intermediate Water with temperatures below 2 °C. Linear extension rates for *D. pertusum* vary within colonies, as well as within individual polyps over time. Most young polyps generate new buds each second year, following a period with high growth (linear extension of 2–3 cm per year). Small coral fragments may survive severe impact if the live polyps are not covered with sediments. CWC ecosystems have a great bioengineering role and are important in a wider ecosystem perspective. The reefs represent biodiversity hotspots, and complex habitats that may function as shelter, feeding ground and nursery for a few commercially interesting fish species. Reefs represent large and complex structures that significantly increase habitat heterogeneity. CWC reef locations have long been appreciated as good fishing grounds by local long-line fishers. Bottom fishing is destructive for CWC reefs and especially bottom trawling has led to severe damage. Norwegian authorities have therefore implemented regulations to protect CWC reefs from destructive practices. Norwegian fisheries act prohibits trawling on all known coral reefs, and special marine protected areas have been established for 18 areas.

Keywords

Deep sea · Seabed mapping · Norwegian Current · Coral reefs · Coral growth · Coral feeding · Carbon cycling · Biodiversity · Symbiosis · Fishing pressure · Offshore oil and gas

5.1 Introduction

Cold-water corals in Norwegian waters have been known for more than two centuries (Linné 1758), but reefs built by the scleractinian *Desmophyllum pertusum* (Linnaeus, 1758) (syn. *Lophelia pertusa*) were first studied in detail by Dons (1944). With the introduction of new ROV (Remotely Operated Vehicle) technology in the 1980s (Hovland et al. 1994; Mortensen et al. 1995; Freiwald et al. 1997; Fosså et al. 2002), new *in-situ* insights were provided to the reef ecology and how fisheries have impacted these fragile habitats.

Globally, *D. pertusum* is most common in the North Atlantic, and especially the Northeast Atlantic. On the western North American margin, *D. pertusum* reefs are found from North Carolina, southward along Florida, and into the Gulf of Mexico (e.g., Messing et al. 2008). A single reef occurrence in the mouth of the Laurentian Channel in Atlantic Canada is an exception to this. This reef occurs in the southward flowing warm water that has passed the southern coast of Greenland after branching off the Gulf stream south of Iceland. The recently discovered reef off SW Greenland (Kenchington et al. 2017) occurs in the same water mass.

P. Buhl-Mortensen (✉)
Institute of Marine Research, Bergen, Norway
e-mail: paal.mortensen@hi.no

A. Freiwald
Marine Research Department, Senckenberg am Meer, Wilhelmshaven, Germany

E. Cordes, F. Mienis (eds.), *Cold-Water Coral Reefs of the World*, Coral Reefs of the World 19,
https://doi.org/10.1007/978-3-031-40897-7_5

D. pertusum is also found along the mid-Atlantic ridge, but living reefs have not been confirmed and live coral is only represented by small, scattered colonies. Large patches of coral rubble, however, indicate that proliferate reefs occurred in the past.

The Northeast Atlantic can be divided into three main reef provinces based on geography and environmental similarities: (1) The Nordic occurrences, including Sweden, Norway, Faroe Island, and Iceland, (2) Irish-British margins, and (3) Franco-Iberian margin (Wheeler et al. 2007). The cold-water coral reefs in Norwegian waters consist mainly of *Desmophyllum pertusum*, but other species such as *Madrepora oculata, Paragorgia arborea, Paramuricea placomus,* and *Primnoa resedaeformis* are commonly found in association with the reefs, along with a rich fauna including various species of fish and invertebrates (Mortensen and Fosså 2006). Rare scleractinians are *Desmophyllum dianthus* (Fosshagen and Høisæter 1992) and *Stenocyathus vermiformis* (Freiwald and Mortensen 2000). In Norway, individual reefs can be up to 40 m high, and a total of 1500 reef locations have been reported (Sundahl et al. 2020). The distribution of the reefs is concurrent with the influx of Atlantic Water (Norwegian Current) on the Norwegian shelf and in fjords. They are found in various geologic settings, including areas of sedimentary as well as crystalline bedrock (Fig. 5.1).

5.2 Oceanography and Geological Settings

5.2.1 Geological Settings

Desmophyllum pertusum can use all kinds of hard substrate as a foundation, even human-made structures such as legs of oil platforms in the North Sea (Gass and Roberts 2006). The substrate can be a shell or a pebble, and as soon as one colony is present it provides new hard substrate for subsequent colonization. In Norwegian waters *D. pertusum* reefs are found on finer sediments mixed with gravel as well as directly on bedrock. Occurrences on bedrock are found in fjords and coastal areas. Further from the coast, on the continental shelf and slope, bedrock is rare and morainic material with gravel and boulder on banks and edges of troughs is the most common reef foundation substrate (Mortensen et al. 2001). Near the shelf break, and on the upper part of the slope morainic material and ledges formed by prehistoric submarine slides are common landscape elements suitable for *D. pertusum*. One example of this is the area called Aktivneset, part of Storegga, shown in Fig. 5.2.

During the last 0.6 Ma, the Norwegian shelf has been overrun numerous times by ice-sheets, eroding and transporting sediments to the shelf edge (Henrich and Baumann 1994). The direction of ice flow has varied, as observed by the orientation of flow-parallel lineations in different sub-seafloor horizons. One example is found in the Trænadjupet Trough where the direction of ice-flow during the last ice-advance was parallel to the trough (Ottesen et al. 2005). The last deglaciation of the shelf occurred between 16.5 ky and 10 ky BP (Dahlgren and Vorren 2003), and seafloor habitats found on the shelf are thus younger than this. Cold-water coral growth commenced after the Pleistocene-Holocene transition at 11.65 ka BP, when modern-like climatic patterns and oceanographic conditions were established (López-Correa et al. 2012; Titschack et al. 2015).

As part of the Norwegian seabed mapping program MAREANO, 595 individual coral reef mounds were surveyed with towed video on the Norwegian shelf and shelf break from 62 to 71°N at depths between 103 and 493 m. The distribution of these was also studied in relation to bottom temperature, currents, and seabed topography (Fig. 5.3). When plotted together with an additional 868 *D. pertusum* records from the literature, the bathymetric distribution of reefs displayed three peaks (90–120 m, 200–280 m, and 310–370 m depth). These depth ranges reflect three different settings: (1) fjords and coastal zone, (2) continental shelf, and (3) shelf break and upper slope, also representing common depth ranges of topographic features such as fjord sills, ridges, iceberg plough marks where hard substrates and elevated currents occur (Freiwald et al. 1999). Below 370 m depth, reefs were much less common, and the lower distribution limit fits with the depth influenced by the Arctic Intermediate Water with temperatures below 2 °C.

Reliable predictions of spatial cold-water coral reef occurrence would be a useful tool for the management of deep coastal and offshore areas (e.g., Tong et al. 2016). Several studies have indicated that the distribution of cold-water coral reefs is correlated with rough topography and slopes exceeding a certain critical angle (i.e., Thiem et al. 2006). However, this is not always the case. Two areas mapped by Norwegian seabed mapping program MAREANO illustrate this. Both areas are located within trenches crossing the continental shelf but have completely different topography. The Malangen reefs occur on a ridge crossing a trench separating Malangsgrunnen from Sveinsgrunnen, whereas the Hola reefs occur within the deeper parts of a trench on a relatively level seabed (Fig. 5.1). The reefs of the two sites differ markedly in shape: the Malangen reefs are relatively circular with summits of living corals, whereas the Hola reefs are elongated with a living up-current front (Bøe et al. 2009; Buhl-Mortensen et al. 2012). The topography of the seabed has no influence on the corals in itself but influences the environment by modifying the hydrodynamic setting. Currents accelerate over peaks and ridges and provide

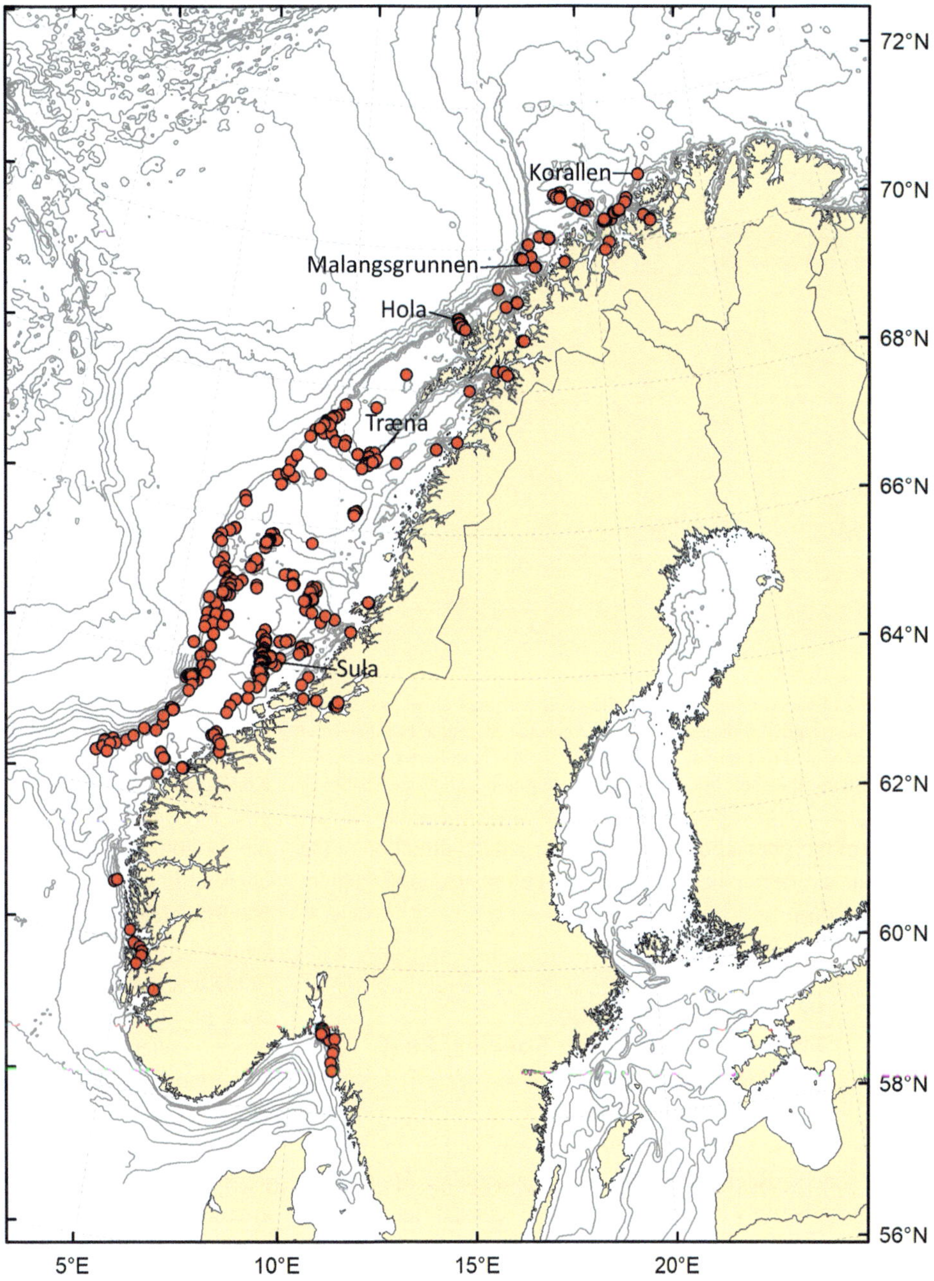

Fig. 5.1 Distribution of *Desmophyllum pertusum* reefs in Norway

environments for enhanced food encounter (Genin et al. 1986; Lim et al. 2020). At other locations the topography may induce hydrodynamic patterns concentrating food particles. In the Hola area the reefs occur at the side of the trench where the currents flow from the coast toward the shelf break. The currents are strong, and local production at the shelf may have increased the nutrient content of the water. Within the range of the coral's temperature and salinity tolerance the combination of hard-bottom substrates for coral larvae settlement and relevant food transport rates are probably more important than the topography of the seabed.

5.2.2 Oceanography

The present day oceanography is dominated by the Norwegian Current (NC) supplying warm (6–9 °C) and saline (>35‰) water from the south, as well as the Norwegian Coastal Current (NCC) which introduces runoff water from Scandinavia and the Baltic Sea as a westward thinning wedge above the NC (Blindheim 1990; Poulain et al. 1996). On a more local scale, the surface current system off mid-Norway is dominated by a northward flowing branch of the NC turning abruptly just south of Lofoten to follow the SE-NW

Fig. 5.2 Shaded relief model of the coral reef area close to the shelf break at around 300 m depth near Aktivneset at Storegga. For the scale: each individual reef mound is approximately 200 m in horizontal extension. Below the shelf break, traces of a submarine slide are visible as parallel ledges into the deep. Coral reefs are also found on these ledges down to a depth of approximately 450 m. Map produced by Terje Thorsnes (Geological survey of Norway), with bathymetry data from Norwegian Hydrographic Service

trend of the Trænadjupet Trough. The general direction of the current is reflected in the alignment of elongated reefs (Thiem et al. 2006).

5.3 Reef Structure (Major Framework-Forming Species, Reef Geomorphology)

Desmophyllum pertusum is the main reef builder in Norwegian waters. Reefs made by the scleractinian *Madrepora oculata* are only reported from one coastal location by Dons (1944), at Storegga, off mid-Norway, just below the shelf break (MAREANO), and on Sula Reef (Freiwald et al. 2002) but commonly occur together with *D. pertusum* as scattered smaller colonies (Mortensen 2000). Based on sediment coring and C^{14}-dating on one of the Traenadjupet reefs, *M. oculata* did not appear prior to 2.4 kY BP in the area (López-Correa et al. 2012).

5.3.1 Reef Types

A coral reef can be defined as an aggregation of coral skeletons completely covering the substrate underneath and thereby generating a topographic relief with time. Colonies that have grown at one site for hundreds of years transform the bottom to a cover of coral skeleton fragments via alternating growth, death, and fragmentation—the coral rubble. The area can then be termed a coral reef. Corals growing on a steep surface may not develop reefs as they lack accommodation space but are rather called coral gardens. When a coral grows large and breaks up, due to its own weight and a process called bioerosion guided by the skeleton-excavating activity of sponges, skeletal fragments will not accumulate at the site but fall deeper, outside the favorable environment (Büscher et al. 2019). The reefs may have different shapes depending on currents and seabed topography and can be circular or elliptical, elongated or droplet shaped, or occur as coral mounds.

5.3.1.1 Circular or Elliptic Reefs

On the Norwegian continental shelf, there might be tens of thousands reefs (Diesing and Thorsnes 2018) in addition to the 1462 confirmed reefs. These reefs typically have a circular or elongated outline with a maximum length of c.1000 m (Mortensen et al. 2001; Buhl-Mortensen 2017; Sundahl et al. 2020). Some of the individual reefs can be several hundred meters long and reef complexes, like the Røst Reef, can be up to 35 km long (Fosså et al. 2005). The reef structure typically has living coral at the top, surrounded by standing dead coral framework, and skeletal fragments of smaller and smaller sizes and eventually coral rubble at the base of the reef (Mortensen et al. 1995).

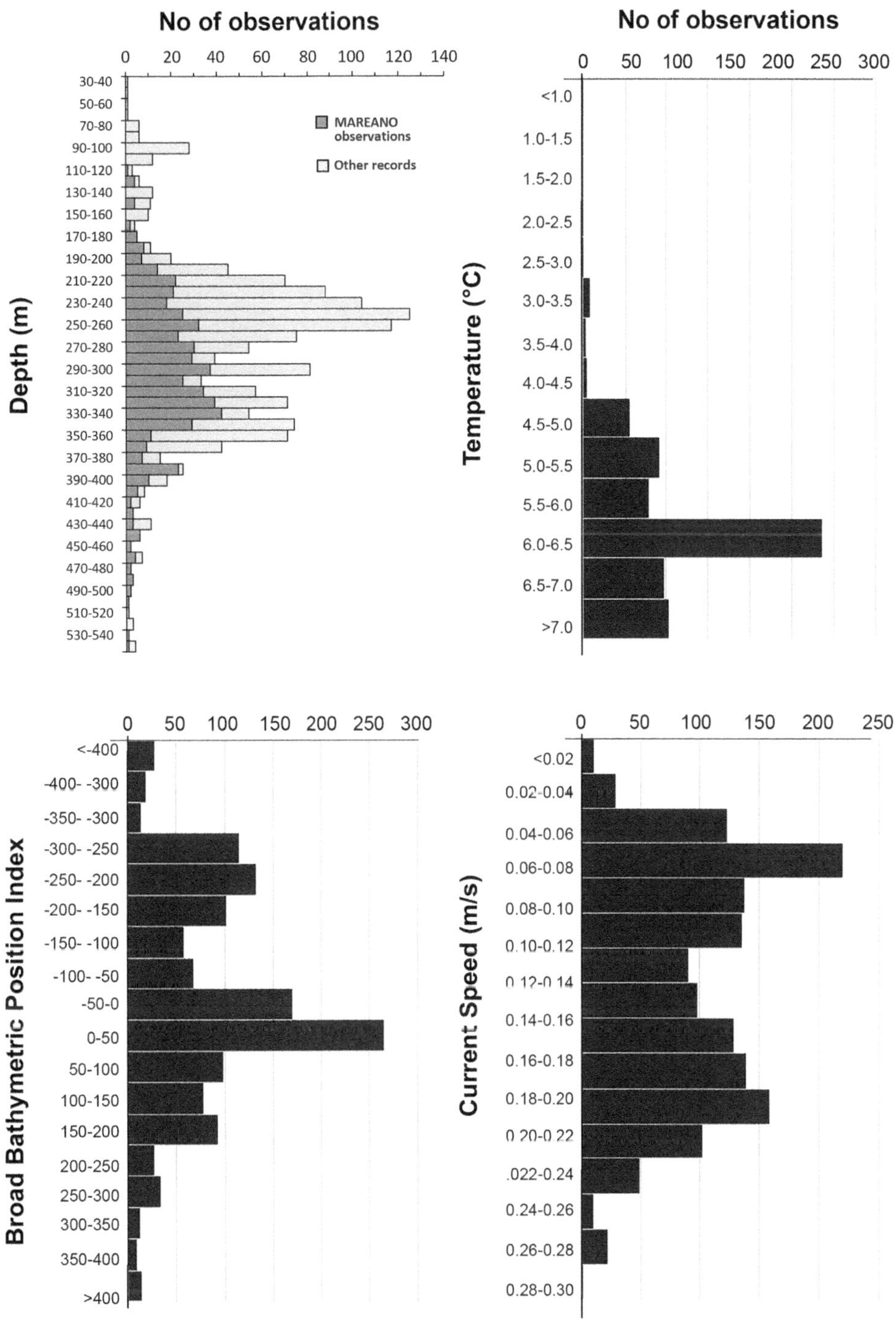

Fig. 5.3 Distribution of *Desmophyllum pertusum* reefs versus; (**a**) depth (for 1463 individual reefs part of the IMR coral database, available in OBIS), (**b**) temperature, (**c**) terrain [Broad bathymetric position index (BPI), and (**d**) current speed (for 595 reefs surveyed by the Norwegian seabed mapping program MAREANO (Buhl-Mortensen et al. 2015; Sundahl et al. 2020)]

Of the reefs surveyed by MAREANO live coral was observed on 58% of the reefs, and on average live coral habitat covered 59% of the reefs. The height and horizontal extension of reefs were positively correlated and reflected different types of linear relationships depending on the general topography. The ratio between height and linear extension was higher in steep terrain compared with level areas or small topographic structures such as morainic mounds and iceberg plough marks (Fig. 5.4c). In steep sloping terrain (i.e., below the shelf break) reef covered a depth range of up to 57 m, whereas for flat areas the reefs covered a range of up to 27 m.

The anatomy of coral reefs may vary in relation to the physical environment and local topography of the initial substrate which served as a settlement area when the reef started to develop. In Norway, Mortensen et al. (1995) described how a typical *D. pertusum* reef on the continental shelf off mid-Norway could be divided into three vertical

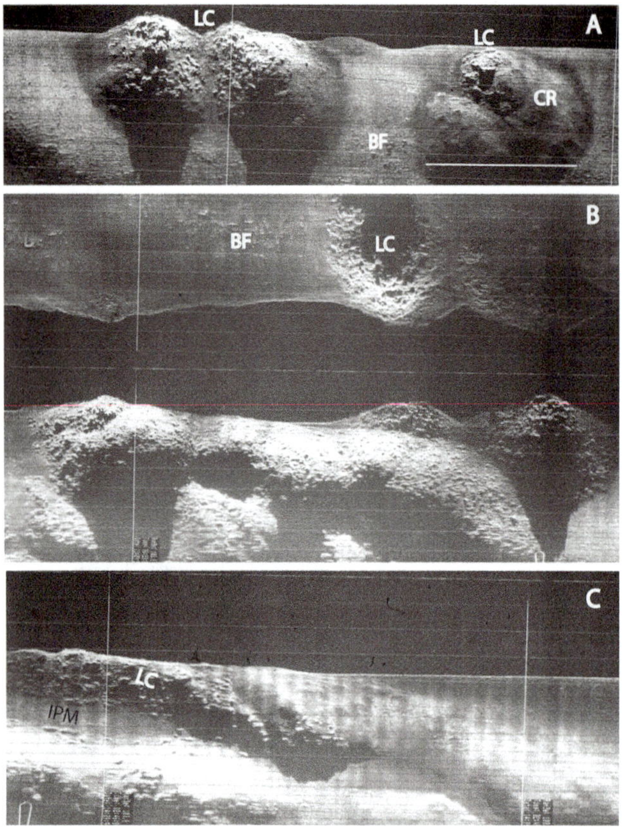

Fig. 5.4 Side scan sonar images from Sula Reef. (**a**) Circular-shaped reefs with live cauliflower-shaped *D. pertusum* colonies (LC) surrounded by coral rubble aprons (CR) and separated by boulder fields (BF). (**b**) Coalesced reefs and circular reef shapes. (**c**) Live *D. pertusum* covers the flanks of iceberg plough marks (modified from Freiwald et al. 1999, 2002). Scale bar for **a**–**c** = 50 m

Fig. 5.5 A 30-cm-high *D. pertusum* colony with live polyps and translucent tissue and/or mucus-covered skeleton and the transition to the dead colony portion below. Fouling processes of a diverse suite of sessile epifauna (foraminifers, bryozoans, serpulids amongst other) result in enhanced species richness

zones, characterized by the amount of live coral and size of dead coral fragments. The reefs used for this description were part of the Sula Reef complex where several hundreds of individual *D. pertusum* reefs occur as more or less circular mounds, of which some had coalesced to form more elongated structures (Mortensen et al. 2001; Freiwald et al. 2002; Fig. 5.4a, b).

Most of these reefs had a summit of live corals occurring above a zone of mainly dead coral blocks. Below the dead coral block zone coral rubble (fragments of dead coral skeletons <10 cm) covered the seabed. This zone was smoother and had a smaller inclination than the two zones above. Samples taken with gravity corer trough the coral rubble zone (Mortensen 2000) indicated that this area had a steady accumulation of small skeleton fragments originating from the two zones above. Dead coral fragments accumulate in the coral rubble zone, but also beneath the living summit and the dead coral block zone. However, very little information exists about the differences in accumulation rates for different parts of a coral reef. This is partly due to the differences of sampling in the complex terrain of the steep part of coral reefs, but also because of the limitations of exact georeferencing and lack of information about the local reef habitat at coring sites. All zones of a coral reef may be regarded as "graveyards." Mortensen et al. (2001) suggest that the 30 m tall reefs at the Sula Reef may consist of a 15 m thick coral layer. Coral fragments in corers from coral rubble zones at the Sula Reef occurred down to ca 2.5 m below seabed surface (Hovland and Mortensen 1999). *D. pertusum* colonies occurring in the living zone of well-developed reef are seldom higher than 2 m but may extend horizontally more than 5 m. Living coral polyps are only found in the outer parts of the live and predominantly bushy colonies. A healthy coral may have living polyps along a 30 cm growth axis, covering 4–5 polyp generations. Below these, the skeleton is dead, exposed to settlement of a diversity of invertebrates, including boring sponges and other agents of taphonomic processes (Freiwald and Wilson 1998; Fig. 5.5).

The maximum growth rates of up to 25 mm (Mortensen and Rapp 1998; Gass and Roberts 2006, 2011; Mortensen and Lepland 2007) indicate that a 2-m tall *D. pertusum* colony is around 80 years old. However, it is not clear how representative the maximum growth is compared with previously published results on average growth (Wilson 1979; Mortensen and Rapp 1998). According to the latter mentioned studies, an average growth of 7 mm indicates that a 2-m tall colony is 285 years old. Even though these two timeframes may have very different consequences for the decomposition of coral skeleton, the vertical extent of the accumulation zone underneath the summit is larger than underneath the coral rubble zone. This difference may be caused by the degree of compaction of coral skeletons. It is the small fragments that are transported furthest away from the steep bottom of two upper reef zones. However, it is not known which of these zones has the largest role as a coral "graveyard."

5.3.1.2 Elongated or Droplet Reefs

The Trænahola, just south of the Lofoten Islands is one area with elongated *D. pertusum* reefs. These reefs are up to ca 300 m long with a living zone occurring only in the up-current end of the reef. Ageing of coral fragments sampled along the "tail" of the reef showed that these reefs have been growing horizontally at a rate of 2.5 cm per year, close to the rates found in different settings by Gass and Roberts (2006, 2011) and Mortensen and Lepland (2007). Instead of a vertical zonation of live colonies, dead coral blocks, and coral rubble, these elongated reefs display a horizontal zonation with a gradual ageing and decomposition away from the live corals. One major difference between the vertical zoned and horizontally zoned reefs is that the elongated reefs grow away from the "graveyard," whereas the vertical reefs grow over the "graveyard" or the "graveyard" moves away from the living reef (Fig. 5.6).

5.3.1.3 Coral Hills

Carl Dons (1944) introduced the term "coral hills," referring to *D. pertusum* aggregations typically found on fjord sills in Norway. Such structures may occur over a greater depth range than the vertical range of circular reefs and elongate reefs. Sediment coring through several Norwegian coral reefs, including the Stjernsund Sill Reef (Freiwald et al. 1997; López Correa et al. 2012) allowed the calculation of carbonate accretion rates for the first time during the Holocene period (Titschack et al. 2015). Maximal accretion rates fall in the range of 1500 cm per 1000 years, corresponding to 15 mm per year. This value is close to the annual extensional growth rate of *D. pertusum* (see above). In detail, the high accretion rates took place during two distinct coral growth pulses in the Early and in the Late Holocene, interrupted by non- or reduced depositional phases. The maximum Norwegian coral accretion rates match the range of the tropical reef environment that have been calculated from 1000 to 3000 cm per 1000 years (summarized in Titschack et al. 2015).

5.4 Coral Biology (Physiology, Reproduction, Genetics, Microbial Associations)

Knowledge about skeletal growth rates and survival of physically damaged corals is important for assessing the potential impact of bottom fisheries and other human activities. Linear extension rates have been estimated for the cold-water corals *Desmophyllum pertusum* and *Paragorgia arborea* (Scleraxonia) by means of image analysis, aquarium experiments, and analysis of skeletal banding patterns (Mortensen and Buhl-Mortensen 2005; Gass and Roberts 2011). The results indicate highly variable growth within colonies, as well as great temporal variations within individual polyps (*D. pertusum*) and branches (*P. arborea*). Skeletal banding patterns in *D. pertusum* indicate that most young polyps generate new buds each second year, following a period with high growth [linear extension of 2–3 cm per year (Mortensen and Lepland 2007)]. Comparison of images of *Paragorgia arborea* from a Norwegian fjord revealed branch extension rates of 4–6 cm year. Visually monitoring of a *D. pertusum* reef situated off northern Norway indicates that small coral fragments may survive severe impact if the live polyps are not covered with sediments. Comparison of trawl impacted and pristine reef locations over time demonstrate that the recovery of the diversity of associated megafauna is slower than the regrowth of physically damaged *D. pertusum* colonies. Therefore, the protection of damaged

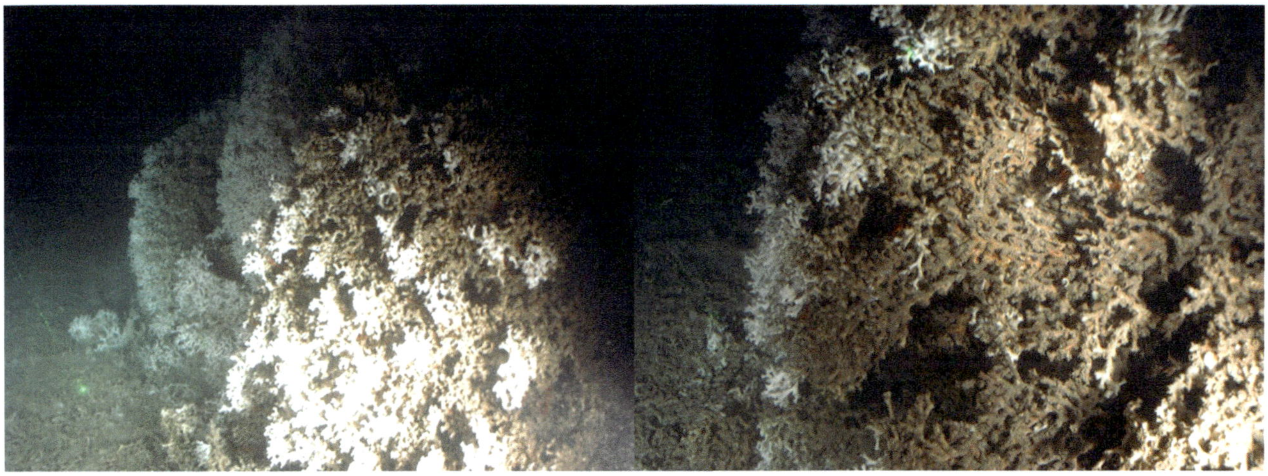

Fig. 5.6 The up-current end of an elongated reef from Traenadjupet showing a dense but thin (<10 cm thick) zone with live *D. pertusum* on top of massive still exposed dead coral framework behind © Jago Team Geomar)

cold-water coral habitats may have a faster positive effect than previously appreciated.

5.4.1 Morphology, Growth, and Feeding of *Desmophyllum pertusum*

Growth, morphology, and feeding are strongly linked in cold-water corals as in other suspension feeders. Skeletal growth patterns set the constraints of morphological variation in many species, not only for colonies, but also for the shape of reefs. The morphology serves a multitude of functions and displays a great variety of adaptations to different environmental conditions. Mechanisms of food particle capture and resistance to strong currents are probably the two most important factors for explaining the functions of azooxanthellate coral morphology (Mortensen and Buhl-Mortensen 2005; de Oliveira et al. 2021). In general, the polyp size seems to be related to the size range of utilized food particles. The colony morphology gives clues to understand the mechanisms involved in particle capture and may partly explain its wide geographical distribution.

In contrast to most shallow-water coral, the deep-water corals do not have any symbiosis with microscopic algae and rely on uptake of particulate organic matter. Recent observations from the seabed and laboratory experiments show that cold-water corals can utilize a wide range of food sources, including fine particulate matter, bacteria, phytodetritus, and large items such as ctenophores and planktonic crustaceans, thus indicating an opportunistic feeding strategy (Mueller et al. 2013; Maier et al. 2019). It is not known whether direct uptake of dissolved organic matter (e.g., amino acids) is common among the deep-water corals. In Norwegian waters crustacean plankton seems to be its main food source (Järnegren and Kutti 2014). Time-series of the behavior of the coral polyps at a seabed observatory site off Lofoten (LoVe) indicate that there is an interplay between tide cycles and the diurnal migration of copepods (Osterloff et al. 2019). When the food (copepods) are close to the bottom and the current speed is favorable for prey capture, the polyps are most active (high ratio of expanded polyps) (Fig. 5.7). However, when the currents are too strong, or the food is not present, the polyps are less active (Fig. 5.7).

Live *D. pertusum* colonies in Norway show some varieties in tissue coloration ranging from translucent (white) to orange-pigmented color morphs, often growing side by side (Strømgren 1971). According to Elde et al. (2012), the orange color morphs contain two times more astaxanthin and canthaxanthin-like carotenoids than the white morph. It was speculated that the food source, i.e., carotenoid-rich copepods, or bacteria may have an influence on the coral's tissue pigmentation (i.e., Neulinger et al. 2008). However, both color morphs often grow side by side (Fig. 5.8), thus having access to the same food source and even fusion of coral skeleton between both morphs was observed (Hennige et al. 2014). It is still an open question as to whether astaxanthin pigmentation in *D. pertusum* is genetically or environmentally controlled. In a multifactorial 1-year in-situ experiment of two contrasting reef locations, the offshore Sula Reef and the inshore Leksa Reef, Büscher et al. (2019) analyzed and compared linear extension calcification rates and mortality rates of the two color morphs. Summing up their findings, the orange color morph showed up to 30% larger linear skeletal extension rates and significantly lower mortality rates over the white morphs. Given the fact that the more robust orange morphs show significantly higher proportions in inshore and fjord *D. pertusum* reefs than the white morph-dominated offshore reefs, the supposed stronger resilience against environmental change-related stress might favor the orange morphs in this region in the future.

Fig. 5.7 Close-up of fully expanded *D. pertusum* polyps, photographed in aquarium (Pål Buhl-Mortensen)

Fig. 5.8 The orange and white (translucent tissue) color morphs of *D. pertusum* side by side in the northern Norwegian Stjernsund Reef © Jago Team, Geomar

5.4.2 Sexual Reproduction in *Desmophyllum pertusum*

The asexual reproduction, or cloning of *Desmophyllum pertusum* contributes to the local reef growth. In order to disperse to new sites, to maintain genetic diversity or re-colonize damaged areas, sexual reproduction with a planktonic larval stage is necessary. *D. pertusum* have separate sexes (Carlgren 1945; Brooke and Järnegren 2013), meaning that the colonies are either a male or a female. Very little was known about the sexual reproduction of *D. pertusum* before Brooke and Järnegren (2013) examined its reproductive periodicity in the Trondheim Fjord. By examining specimens collected at various times of the year between 2002 and 2006 they found that the coral spends around one year developing eggs. The coral has external fertilization with spawning of sperm and eggs into the water column (Brooke and Järnegren 2013). Spawning takes place over a month's time starting between January and March in Norway, depending on geographical location. The southernmost reef at Hvaler starts in mid-January while the reefs in mid-Norway start spawning mid-February (Larsson et al. 2014). The time of spawning further north is not known. Larsson et al. (2014) documented for the first time embryogenesis and larval development in *D. pertusum*. The fertilized eggs turn into embryos rather slowly. The larvae are active swimmers and likely planktotrophic (feeding during the planktonic stage) and start to settle after 3–5 weeks (Larsson et al. 2014). In laboratory they can remain planktonic for up to 8 weeks. Thus, they have great dispersal potential.

5.5 Reef Ecosystem (Community Structure, Trophic Relationships, Interactions with Surrounding Habitats) (Fig. 5.9)

Although fast advancing, the ecological significance of the deep-water coral reefs is still not fully understood. One open question is, for instance, "what quantities or proportion of different types of food does the unknown number of *D. pertusum* reefs consume at different times of the year?". During the two last decades new technology (ROV, sophisticated acoustics, benthic lander systems equipped with a broad array of probes, etc.) has greatly enhanced our ability to study the deep reefs in a way comparable to the tropical shallow-water coral reefs. In the following we briefly discuss the major ecosystem services of CWC reefs.

5.5.1 Carbon Cycling Hotspots

Experiments have clearly shown the importance of local downwelling transfer of particulate organic matter (POM) either from productive surface waters or through lateral

Fig. 5.9 Live reef habitat at 185 m depth at Lopphavet near the coast of northern Norway. Together with a pinkish orange variety of *D. pertusum* the sponge *Mycale lingua* dominates the habitat. The fish in the center is a Norway redfish (*Sebastes viviparus*)

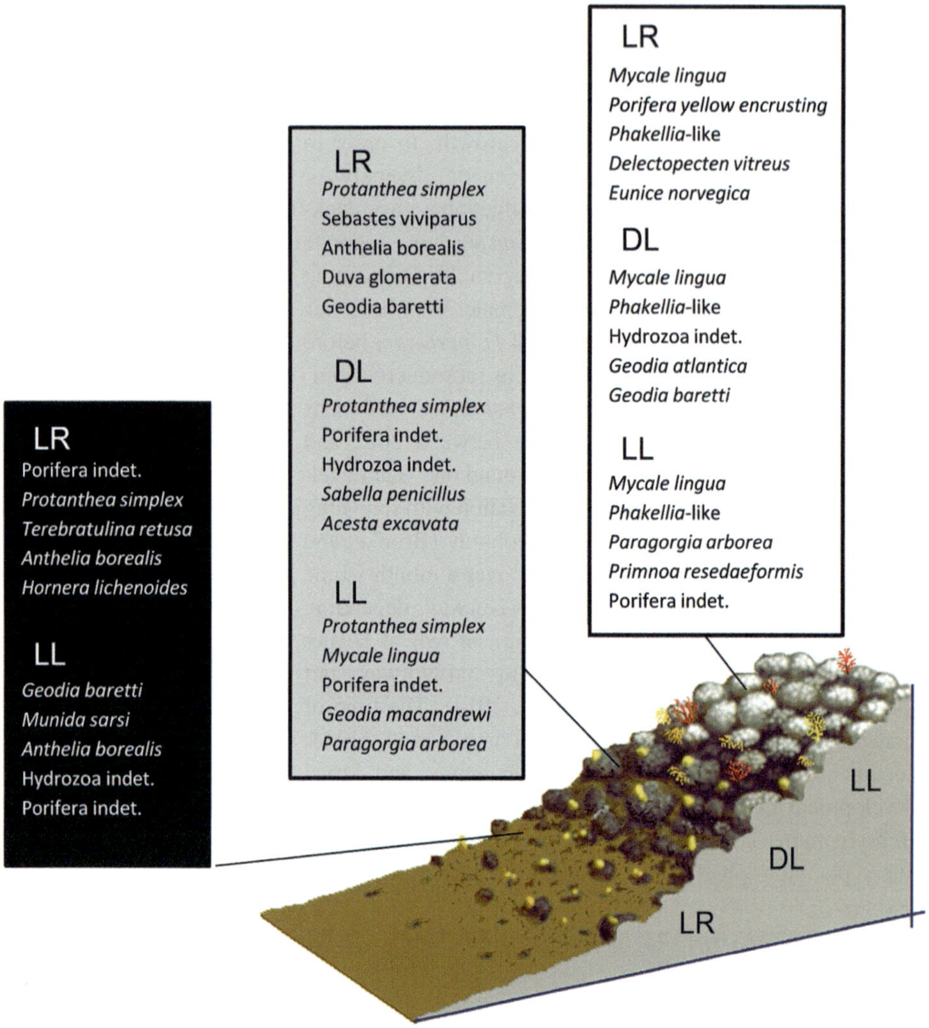

Fig. 5.10 Three successive habitats can be observed when crossing a reef: the coral rubble zone (*Lophelia* rubble—LR) consisting of small pieces of skeleton, followed by the coral block zone (Dead *Lophelia*—DL) dominated by large blocks of mainly coral skeleton, which lead to the live part of coral reef (Live *Lophelia*—LL)

advection of topographically-guided bottom currents to CWC reefs (i.e., Findlay et al. 2013; Soetaert et al. 2016; Davies et al. 2009). Compared to other communities in deeper waters, CWC reefs are generally dominated by filter- and suspension feeder, both in biomass and species diversity (e.g., Jensen and Frederiksen 1992). Van Oevelen et al. (2009) and Maier et al. (2020) have modeled the role of CWC reefs as a biofilter by significantly reducing suspended POM through digestion and subsequent deposition within the reef system. Together with coral released mucus, CWC reefs act as major hotspots of carbon cycling and deposition in bathyal depths of the seas (Wild et al. 2009; White et al. 2012). This conclusion is strongly supported by detailed measurements of oxygen consumption rates of CWC reefs, sponge grounds, and soft sediment grounds (Cathalot et al. 2015). According to their upscaled data, CWC reefs and sponge grounds take the lion share of the total benthic oxygen respiration on the Norwegian continental shelf with about 36%, thus utilizing about 5% of the total primary production (Cathalot et al. 2015). This underpins again the bioengineering role of CWC ecosystems and their importance in a wider ecosystem perspective.

5.5.2 Biodiversity Hotspots

The deep-water coral reefs represent biodiversity hotspots, and complex habitats that may function as shelter, feeding ground and nursery for a number of commercially interesting fish species (Kutti et al. 2014). Reefs represent large and spatially complex structures that significantly increase habitat diversity. The habitats surrounding the reefs are varied and present a wide range of substrates, but the complexity of the reef structures offers a range of microhabitats supporting local species diversity (Mortensen and Fosså 2006) (Fig. 5.10).

5.5.3 Associated Fauna

In a study of Norwegian reefs based on sampled *D. pertusum* Mortensen and Fosså (2006) reported a rich associated fauna, with >700 species. Mollusca, Arthropoda, and Bryozoa were the most species rich phyla contributing 48% to the total number of species. Most of these organisms are comprised of species that occur on other hard-bottom substrates and their relationship is facultative. Mortensen and Fosså (2006) found that the species diversity (H′) was highest in samples with a low proportion (1–20%) of live coral, and lowest for samples from the coral rubble zone surrounding the reefs. The number of individuals was highest in samples with a high proportion (>20%) of live coral. Deposit feeders were most common in the rubble, whereas suspension feeders dominated among live coral. Most higher taxa were represented by more species on the inshore compared to the offshore reefs. This was most evident for Cnidaria, Crustacea, Polychaeta, and Tunicata, whereas Foraminifera were more species rich offshore. The results were compared with those of three earlier studies in the Northeast Atlantic (Dons 1944; Burdon-Jones and Tambs-Lyche 1960; Jensen and Frederiksen 1992). A total of 769 species have been recorded, but only 21 were common for all four studies (Mortensen and Fosså 2006), indicating a great variation in the associated fauna at local and broad scales.

The high biodiversity associated with reef-forming, cold-water corals is primarily generated by the complex three-dimensional hard substrate rather than the living coral itself (Mortensen and Fosså 2006 and references therein). Within the reef, there are four primary microhabitats: (1) live corals, (2) dead coral skeleton, (3) microhabitats within the porous coral skeleton, and (4) spaces in the coral matrix. The live tissue of cold-water scleractinian corals seems to prevent attachment of sessile epibiotic species. The majority of the fauna on cold-water coral reefs are casual associates rather than reef-endemic species (Mortensen and Fosså 2006). Even among the few species commonly found intimately associated with living coral polyps, there are few examples of obligate relationships. However, many of these species are seldom found in other habitats.

The polychaetes *Harmothoe oculinarum* and *Eunice norvegica* (Fig. 5.11) are examples of demonstrated symbiotic relationships (see Factbox 1) on cold-water coral reefs. *E. norvegica* lives in a close relationship with *D. pertusum* and may be regarded as a mutualistic symbiont (see Factbox 1), while the polynoid polychaete *H. oculinarum* is a commensal inside the tubes of *E. norvegica*. The polychaete lays down a parchment-like tube and the coral will overgrow this with its calcium-carbonate skeleton. Eventually, *E. norvegica* will have a calcified tube with several openings near individual coral polyps (Mortensen 2001). *E. norvegica* acquires

Fig. 5.11 Close-up from the coral rubble habitat at 150 m depth, between scattered small live *D. pertusum* colonies, at the "Korallen" reef, west of Sørøya Northern Norway. In addition to the anemone *Protanthea simplex* we see cyclostome bryozoans, brittle stars, and tunicates

Fig. 5.12 The polychaete *Eunice norvegica* lives in close association with *Desmophyllum pertusum* in the Norwegian coral reefs. Photo by Pål Buhl-Mortensen (Mortensen 2001)

food from the waste associated with incomplete feeding and benefits the corals by removing inedible particles from the coral tissue or species, such as hydroids, attempting to colonize the coral skeleton (Mortensen 2001) (Fig. 5.12).

Factbox 1: Symbiosis in *Desmophyllum pertusum* Reefs

Originally symbiosis means two species living together (de Bary 1879), but a more common understanding of the term is the specified relationship known as mutualism, where both species benefit of the relationship. Commensalism is a common type of relationship where the symbiont profit from staying with the host, but the host is unaffected. Parasitism is a third type of symbiosis where the symbiont benefit, but the host suffer from its presence. All three types of symbiotic relationships can be obligatory or facultative for one or both species. Compared with shallow water scleractinians, cold-water scleractinians display few examples of mutualistic symbiosis. The best example is probably the relationship between the polychaeta *Eunice norvegica* and *Desmophyllum pertusum* (Fig. 5.11). The number of such relationships is probably higher in low latitudes than in high latitudes (Buhl-Mortensen and Mortensen 2004).

Factbox 2: *Hyrrokkin sarcophaga*, A Parasitic Obligate Foraminiferan on *Desmophyllum pertusum*

The parasitic nature and current taxonomic classification of this sclerobiont foraminifer has been depicted by Cedhagen (1994). *Hyrrokkin sarcophaga*, like other members of the family Rosalinidae live epifaunal-

Factbox 2: *Hyrrokkin sarcophaga, A Parasitic Obligate Foraminiferan on* Desmophyllum pertusum (continued)

attached, (Figs. 5.13 and 5.14) or mobile on various sorts of elevated biogenic substrates, such as on hydroids (Dobson and Haynes 1973), octocorals (Hawkes and Scott 2005), and as epiphytes on seagrass (Piazzi et al. 2016). The genus *Hyrrokkin* consists of two species, *H. carnivora* (Todd 1965) and *H. sarcophaga* Cedhagen, 1994. *Hyrrokkin* bores into its host organism thereby creating a boring trace described as *Kardopomorphos polydioryx* (Beuck et al. 2008). *H. sarcophaga* is one of the few species that successfully infest the live tissue-covered zone of *D. pertusum* and *Madrepora* corallites, mostly in larger groups (Freiwald and Schönfeld 1996; Fig. X1). However, *H. sarcophaga* also infests the cortex of larger sponges especially sponges of the order Astrophorina (Klitgaard 1995; Cárdenas and Moore 2019 and Fig. X2). Molluscs living in close association with *D. pertusum* and *Madrepora oculata* habitat are often densely colonized by *Hyrrokkin*. This has been shown for the thin-valved pectinid *Delectopecten vitreus* and the limid file clams *Acesta excavata* and *A. angolensis* (Beuck et al. 2008). In a comparative study of *Hyrrokkin* collected from *D. pertusum* and *Acesta* hosts, Schleinkofer et al. (2021) demonstrate a host-influenced geochemical signal in the tests of the parasitic foraminifer, mirroring the hosts carbonate mineralogy and isotopy through the uptake of dissolved carbonate ions from the hosts and subsequent incorporation within the foraminifer test. *Hyrrokkin* is not bound to the cold-water coral habitat as it has been found in great numbers on the polyplacophore *Leptochiton arcticus* around Iceland (Sigwart 2009). The biogeographic distribution of *H. sarcophaga* concentrates in the northern to central Northeast Atlantic, whereas *H. carnivora* is known from the Gulf of Guinea and off Angola (Gil et al. 2020). Cheng and Dai (2016) described *H. sarcophaga* on *M. oculata* and other solitary scleractinian corals from the South China Sea. *Hyrrokkin* sp. was found on *Acesta patagonica* in the Patagonian Beagle Channel (Beuck et al. 2008). The fossil record of Hyrrokkin tests and boring traces dates back at least to the Early Pleistocene of Rhodes in the eastern Mediterranean (Beuck et al. 2008) (Fig. 5.13).

5.5.4 Fish Fauna

Complex hard-bottom habitats such as corals, boulders, and steep rock walls may be essential for several fish species, and

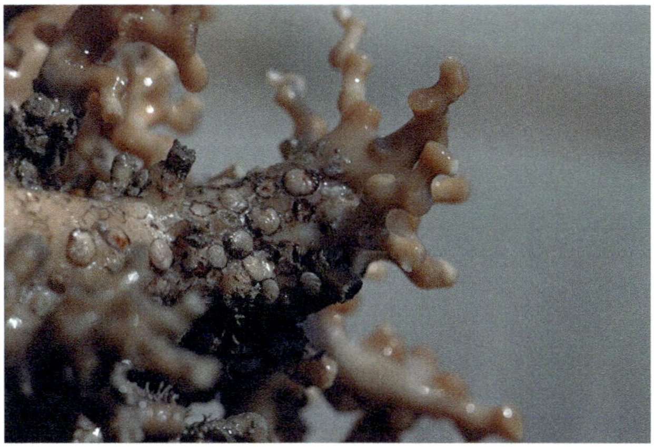

Fig. 5.13 Group of *Hyrrokkin sarcophaga* specimen measuring 3–4 mm in diameter infesting the tissue-covered area of a *Madrepora oculata* colony from the Sula Reef

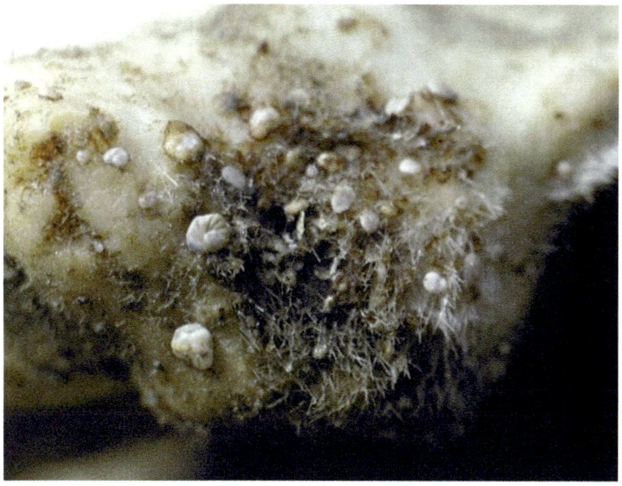

Fig. 5.14 Group of different *Hyrrokkin sarcophaga* generations on the cortex of an astrophorinid sponge

fish diversity is generally higher on the reefs than surrounding habitats in Norway (Costello et al. 2005; Mortensen et al. 2005). These habitats have different potential functions for fish during their different life history stages (shelter, food source, spawning grounds). Redfish (*Sebastes* spp.) are the most common on the reefs, but other species such as tusk (*Brosme brosme*) and ling (*Molva molva*) are also quite abundant (Mortensen 2000). The redfish share the corals food resource, plankton, but also take advantage of the complex habitat for shelter. It is not known to what degree the reefs function as nursing grounds for juvenile fish of any species, but it is reasonable to assume that the reefs can offer a rich community of potential prey for demersal fish.

5.6 Threats and Outlook

Deep-water coral reef locations have long been appreciated as good fishing grounds by local long-line fishers. Trawling on the coral reefs employing rock-hopper gear has occurred both intentionally and accidentally and has led to severe damage in coral habitats (Fosså et al. 2002) (Fig. 5.15). Based on results from recent projects, Norwegian authorities have implemented new regulations to protect *D. pertusum* reefs from destructive practices. Norwegian legislation is now prohibiting trawling on all known coral reefs, and special protection areas has been established for 18 areas offshore as well as in coastal areas.

Activities related to the petroleum industry pose a threat to *D. pertusum* reefs (and the marine ecosystem in general). These threats include physical damage (e.g., pipeline laying, anchoring, installation of seabed structures), increased particle load in the water (e.g., drill cuttings, pipeline burial), and chemical pollution (e.g., oil spill, produced water). The activities associated with normal operations have so far only been documented to have local impacts. Short-term studies (<1 year) in lab have not detected significant negative effects of exposure to particle concentrations resembling normal drilling activities (Purser 2015). In-situ studies of long-term (>1 year) effects of the exposure to enhanced particle load have not been carried out. Accidents, however, are likely to have a great negative impact.

5.6.1 Protection of Coral Reefs in Norway

There are 18 designated marine protected areas (MPAs) in Norway (Fig. 5.16). Since 1999, the Norwegian Government prohibited "all use of fishing gears that are dragged and may get in contact with the sea floor" by Norwegian and foreign vessels in the Sula Reef area as well as "all intentional destruction of coral reefs throughout Norway." Since then, new MPAs have been designated as new information about the distribution and status of *D. pertusum* reefs have been gathered (Figs. 5.15 and 5.16).

Factbox 3

The northernmost known *D. pertusum* reef, named Korallen, is found southwest of Sørøya in Finnmark county. It is a reef complex, consisting of three coral mounds separated by less than 600 m of sandy gravel. The largest of these reef mounds was approximately 1.2 km long and around 30 m high, with large parts entirely covered by live coral (Buhl-Mortensen 2017). Korallen was designated as an MPA, closed to bottom trawling in 2009 (Fig. 5.17).

Fig. 5.15 Trawling on coral reefs happens accidentally and has great negative impact on the reef and its ecosystem. The trawl doors leave deep scars easy to detect on the reef and surrounding seabed

5.7 Future Directions

As more knowledge is gained about the cold-water coral reefs, new research questions that remain to be answered arise. Also, with the development of new technology, new opportunities appear.

Sustainable management of marine resources requires information about the distribution, ecology, and health status of benthic habitats. It is essential for an ecosystem management plan to identify parameters that can provide early warnings about environmental degradation. Studies on marine benthos have traditionally focused on infauna, and indicator species for pollution have been selected from this group of organisms. However, there is little information about the response of the larger epifauna that are sensitive to destructive fisheries, oxygen deficiency, and increased

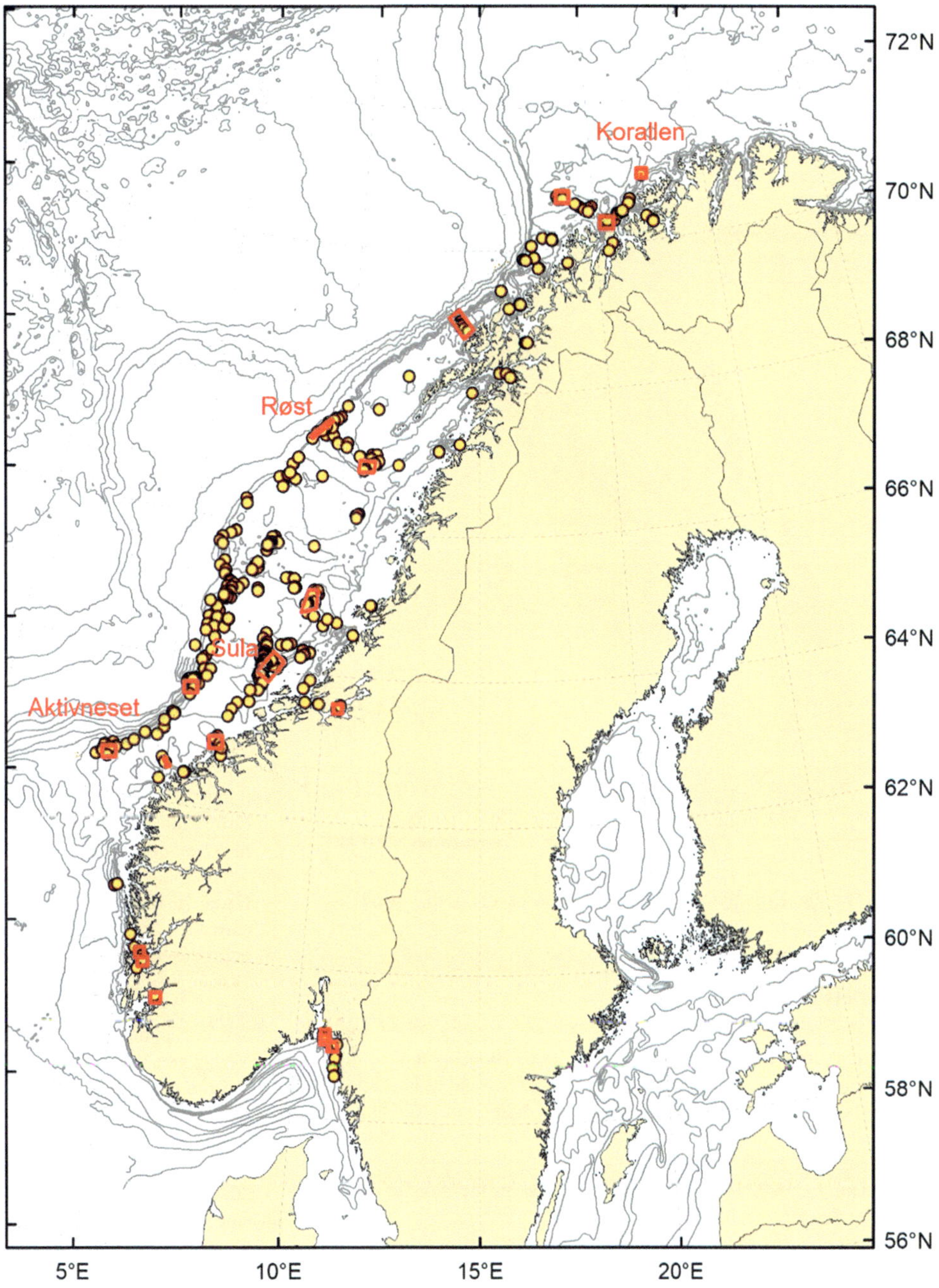

Fig. 5.16 There are 18 designated marine protected areas (marked with red borders) in Norway. Coral reefs are marked with yellow dots

particulate loads. The LoVe observatory has proven to be a good example of how remote cabled observatories can provide an on-line and real-time monitoring of the reef environment. Further studies may reveal indicators of physiological responses at a much finer time scale than is needed for measuring skeletal growth and death of whole colonies.

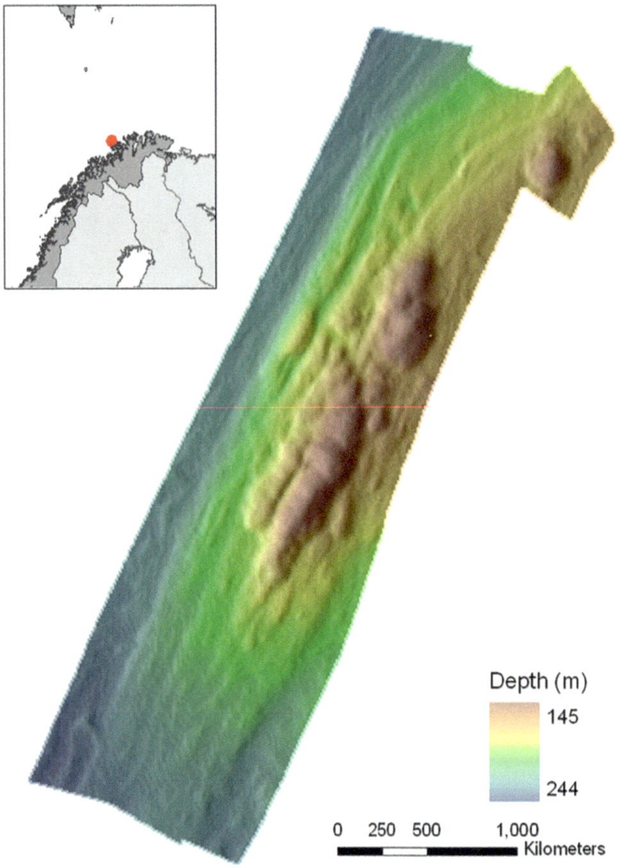

Fig. 5.17 The Korallen Reef, the northernmost reef in the world

References

Beuck L, López Correa M, Freiwald A (2008) Biogeographical distribution of *Hyrrokkin* (Rosalinidae, Foraminifera) and its host-specific morphological and textural trace variability. In: Wisshak M, Tapanila L (eds) Current developments in bioerosion. Springer, Heidelberg, pp 329–360

Blindheim J (1990) Arctic intermediate water in the Norwegian Sea. Deep-Sea Res 37:1475–1489

Bøe R, Bellec VK, Dolan MFJ, Buhl-Mortensen P, Buhl-Mortensen L, Slagstad D, Rise L (2009) Giant sandwaves in the Hola glacial trough off Vesteralen, North Norway. Mar Geol 267:36–54

Brooke S, Järnegren J (2013) Reproductive periodicity of the scleractinian coral *Lophelia pertusa* from the Trondheim Fjord, Norway. Mar Biol 160:139–153

Buhl-Mortensen L, Mortensen PB (2004) Symbiosis in deep-water corals. Symbiosis 37:33–61

Buhl-Mortensen P (2017) Coral reefs in the Southern Barents Sea: habitat description and the effects of bottom fishing. Mar Biol Res 13:1027–1040. https://doi.org/10.1080/17451000.2017.1331040

Buhl-Mortensen L, Bøe R, Dolan MFJ, Buhl-Mortensen P, Thorsnes T, Elvenes S, Hodnesdal H (2012) Banks, troughs, and canyons on the continental margin off Lofoten, Vesterålen, and Troms, Norway. In: Harris PT, Baker EK (eds) Seafloor geomorphology as benthic habitat. Elsevier, Amsterdam, pp 703–715

Buhl-Mortensen L, Buhl-Mortensen P, Dolan MFJ, Holte B (2015) The MAREANO programme – a full coverage mapping of the Norwegian off-shore benthic environment and fauna. Mar Biol Res 11:4–17

Burdon-Jones C, Tambs-Lyche H (1960) Observations on the fauna of the North Brattholmen stone-coral reef near Bergen. Årbok for Universitetet i Bergen. Mat-Naturv Serie 1960(4):1–24

Büscher JV, Wisshak M, Form AU, Titschack J, Nachtigall K, Riebesell U (2019) In situ growth and bioerosion rates of *Lophelia pertusa* in a Norwegian fjord and open shelf cold-water coral habitat. Peer J 7: e7586

Cárdenas P, Moore JA (2019) First records of *Geodia* demosponges from the New England seamounts, an opportunity to test the use of DNA mini-barcodes on museum specimens. Mar Biodivers 49:163–174

Carlgren O (1945) Polyppdyr (Coelenterata) III. Koraldyr. Danmarks fauna 51. GEC Gads, Copenhagen

Cathalot C, Van Oevelen D, Cox TJS, Kutti T, Lavaleye M, Duineveld G, Meysman FJR (2015) Cold-water coral reefs and adjacent sponge grounds: hotspots of benthic respiration and organic carbon cycling in the deep sea. Frontiers in Marine. Science 2: Article 37

Cedhagen T (1994) Taxonomy and biology of Hyrrokkin sarcophaga gen et sp. n., a parasitic foraminiferan (Rosalinidae). Sarsia 79:65–82

Cheng YR, Dai CF (2016) A bioeroding foraminifer, *Hyrrokkin sarcophaga,* on deepwater corals from the South China Sea. Coral Reefs 35:901

Costello MJ, McCrea M, Freiwald A, Lundälv T, Jonsson L, Bett BJ, van Weering TCE, de Hass H, Roberts MJ, Allen D (2005) Role of cold-water *Lophelia pertusa* coral reefs as fish habitat in the NE Atlantic. In: Freiwald A, Roberts JM (eds) Cold-water corals and ecosystems. Springer, Berlin, pp 771, 1244 pp–805

Dahlgren KIT, Vorren TO (2003) Sedimentary environment and glacial history during the last 40 ka of the Vøring continental margin, mid-Norway. Mar Geol 193:93–127

Davies AJ, Duineveld GCA, Lavaleye MSS, Bergman MJN, van Haren H, Roberts MJ (2009) Downwelling and deep-water bottom currents as food supply mechanisms to the cold-water coralLophelia pertusa(Scleractinia) at the Mingulay Reef complex. Limnol Oceanogr 54(2):620–629

de Bary A (1879) Die Erscheinung der Symbiose: Vortrag gehalten auf der Versammlung Deutscher Naturforscher und Aerzte zu Cassel. Verlag von Karl J. Trübner. Leipzig- Berlin, 30p

de Oliveira LMC, Lim A, Conti LA, Wheeler AJ (2021) 3D Classification of cold-water coral reefs: a comparison of classification techniques for 3D reconstructions of cold-water coral reefs and seabed. Front Mar Sci 8:640713. https://doi.org/10.3389/fmars.2021.640713

Diesing M, Thorsnes T (2018) Mapping of cold-water coral carbonate mounds based on geomorphometric features: an object-based approach. Geosciences 8:34. https://doi.org/10.3390/geosciences8020034

Dobson M, Haynes J (1973) Association of foraminifera with hydroids on the deep shelf. Micropaleontology 19:78–90

Dons C (1944) Norges korallrev. Det Kongelige Norske Videnskabers Selskab Forhandlinger 16:37–82

Elde AC, Pettersen R, Bruheim P, Järnegren J, Johnsen G (2012) Pigmentation and spectral absorbance signatures in deep-water corals from the Trondheimsfjord, Norway. Mar Drugs 10:1400–1411

Findlay HS, Artioli Y, Navas JM, Hennige SJ, Wicks LC, Huvenne VAI, Woodward EM, Roberts JM (2013) Tidal downwelling and implications for the carbon biogeochemistry of cold-water corals in relation to future ocean acidification and warming. Glob Chang Biol 19:2708–2719

Fosså JH, Mortensen PB, Furevik DM (2002) The deep-water coral *Lophelia pertusa* in Norwegian waters: distribution and fishery impacts. Hydrobiologia 471:1–12

Fosshagen A, Høisæter T (1992) The second Norwegian record of the deep-water coral, Desmophyllum cristagalli Milne-Edwards & Haime, 1848 (Cnidaria:Scleractinia). Sarsia 77:291–292

Fosså JH, Lindberg B, Christensen O, Lundälv T, Svellingen I, Mortensen PB, Alvsvåg J (2005) Mapping of *Lophelia* reefs in Norway: experiences and survey methods. In: Freiwald A, Roberts JM (eds) Cold-water corals and ecosystems. Springer, Berlin, pp 359–391

Freiwald A, Schönfeld J (1996) Substrate pitting and boring pattern of *Hyrrokkin sarcophaga* Cedhagen, 1994 (Foraminifera) in a modern deep-water coral reef mound. Mar Micropaleontol 28:199–207

Freiwald A, Wilson JB (1998) Taphonomy of modern deep, cold-temperate water coral reefs. Hist Biol 13:37–52

Freiwald A, Henrich R, Pätzold J (1997) Anatomy of a deep-water coral reef mound from Stjernsund, West-Finnmark, northern Norway. In: James NP, Clarke JAD (eds) Cool-water carbonates. SEPM, Special Publication, pp 141–161

Freiwald A, Wilson JB, Henrich R (1999) Grounding Pleistocene icebergs shape recent deep-water coral reefs. Sediment Geol 125: 1–8

Freiwald A, Mortensen PB (2000) The first record of the deep-water coral *Stenocyathus vermiformis* (Pourtalès, 1868) (Scleractinia: Guyniidae) from Norwegian waters. Sarsia 85:275–276

Freiwald A, Hühnerbach V, Lindberg B, Wilson JB, Campbell J (2002) The Sula Reef complex, Norwegian Shelf. Facies 47:179–200

Gass SE, Roberts JM (2006) The occurrence of the cold-water coral *Lophelia pertusa* (Scleractinia) on oil and gas platforms in the North Sea: colony growth, recruitment and environmental controls on distribution. Mar Pollut Bull 52:549–559

Gass SE, Roberts JM (2011) Growth and branching patterns of *Lophelia pertusa* (Scleractinia) from the North Sea. J Mar Biol Assoc U K 91(4):831–835

Genin A, Dayton PK, Lonsdale PF, Speiss FN (1986) Corals on seamount peaks provide evidence of current acceleration over deep-sea topography. Nature 322:59–61

Gil M, Freiwald A, Ramil F (2020) Two new species of *Rosalinda* (Cnidaria, Hydrozoa, Anthoathecata) from West African cold-water coral mounds. Deep-Sea Res 103436

Hawkes AD, Scott DB (2005) Attached benthic foraminifera as indicators of past and present distribution of the coral *Primnoa resedaeformis* on the Scotian margin. In: Freiwald A, Roberts JM (eds) Cold-water corals and ecosystems. Springer, Heidelberg, pp 881–894

Henrich R, Baumann K-H (1994) Evolution of the Norwegian Current and the Scandinavian Ice Sheets during the past 2.6 m.y,: evidence from ODP Leg 104 biogenic carbonate and terrigenous records. Palaeogeogr Palaeoclimatol Palaeoecol 108:75–94. https://doi.org/10.1016/0031-0182(94)90023-X

Hennige SJ, Morrison CL, Form AU, Büscher J, Kamenos NA, Roberts JM (2014) Self-recognition in corals facilitates deep-sea habitat engineering. Sci Rep 4:6782

Hovland MR, Farestveit R, Mortensen PB (1994) Large cold-water coral reefs off mid-Norway - a problem for pipelaying? Oceanology International, Conference Proceedings Vol. 3. Brighton, UK, 8. - 11. March 1994

Hovland M, Mortensen PB (1999) *Norske korallrev og prosesser i havbunnen* [in Norwegian]. John Grieg, Bergen, p 155

Järnegren J, Kutti T (2014) *Lophelia pertusa* in Norwegian waters. What have we learned since 2008? - NINA Report 1028. 40 pp.

Jensen A, Frederiksen R (1992) The fauna associated with the bank-forming deepwater coral *Lophelia pertusa* (Scleractinaria) on the Faroe shelf. Sarsia 77:53–69

Kenchington E, Yashayaev I, Tendal OS, Jørgensbye H (2017) Water mass characteristics and associated fauna of a recently discovered *Lophelia pertusa* (Scleractinia: Anthozoa) reef in Greenlandic waters. Polar Biol 40:321–337. https://doi.org/10.1007/s00300-016-1957-3

Klitgaard AB (1995) The fauna associated with outer shelf and upper slope sponges (Porifera, Demospongiae) at the Faroe Islands, northeastern Atlantic. Sarsia 80:1–22

Kutti T, Bergstad OA, Fosså JH, Helle K (2014) Cold-water coral mounds and sponge-beds as habitats for demersal fish on the Norwegian shelf. Deep-Sea Res II 99:122–133

Larsson AI, Järnegren J, Strömberg SM, Dahl MP, Lundälv T, Brooke S (2014) Embryogenesis and larval biology of the cold-water coral *Lophelia pertusa*. PLoS One 9(7):e102222. https://doi.org/10.1371/journal.pone.0102222

Lim A, Wheeler AJ, Price DM, O'Reilly L, Harris K, Conti L (2020) Influence of benthic currents on cold-water coral habitats: a combined benthic monitoring and 3D photogrammetric investigation. Sci Rep 10:19433. https://doi.org/10.1038/s41598-020-76446-y

López Correa M, Montagna P, Joseph N, Rüggeberg A, Fietzke J, Flögel S, Dorschel B, Goldstein SL, Wheeler A, Freiwald A (2012) Preboreal onset of cold-water coral growth beyond the Arctic Circle revealed by coupled radiocarbon and U-series dating and neodymium isotopes. Quat Sci Rev 34:24–43

Maier SR, Kutti T, Bannister RJ, van Breugel P, van Rijswijk P, van Oevelen D (2019) Survival under conditions of variable food availability: resource utilization and storage in the cold-water coral *Lophelia pertusa*. Limnol Oceanogr 64(2019):1651–1671

Maier SR, Kutti T, Bannister RJ, Fang JK-H, van Breugel P, van Rijswijk P, van Oevelen D (2020) Recycling pathways in cold-water coral reefs: use of dissolved organic matter and bacteria by key suspension feeding taxa. Sci Rep 10:9942. https://doi.org/10.1038/s41598-020-66463-2

Messing CG, Reed JK, Brooke SD, Ross SW (2008) Deep-Water coral reefs of the United States. In: Riegl BM, Dodge RE (eds) Coral reefs of the USA. Coral reefs of the world, vol 1. Springer, Dordrecht. https://doi.org/10.1007/978-1-4020-6847-8_21

Mortensen PB (2000) *Lophelia pertusa* (Scleractinia) in Norwegian waters. Distribution, growth, and associated fauna. Dr. scient. Thesis, Department of Fisheries and Marine Biology, University of Bergen, Norway

Mortensen PB (2001) Aquarium observations on the deep-water coral *Lophelia pertusa* (L., 1758) (Scleractinia) and selected associated invertebrates. Ophelia 54(2):83–104

Mortensen PB, Buhl-Mortensen L (2005) Morphology and growth of the deep-water gorgonians *Primnoa resedaeformis* and *Paragorgia arborea*. Mar Biol 147:775–788

Mortensen PB, Fosså JH (2006) Species diversity and spatial distribution of invertebrates on deep–water *Lophelia* reefs in Norway. Proceedings of the 10th International Coral Reef Symposium. Okinawa, Japan, pp 1849–1868.

Mortensen PB, Lepland A (2007) Ecological consequences of exploration drilling on coral reefs in the Træna Deep. Fisken og havet No 7-2007, 87p

Mortensen PB, Rapp HT (1998) Oxygen- and carbon isotope ratios related to growth line patterns in skeletons of Lophelia pertusa (L) (Anthozoa: Scleractinia): implications for determination of linear extension rates. Sarsia 83:433–446

Mortensen PB, Hovland M, Brattegard T, Farestveit R (1995) Deep water bioherms of the scleractinian coral *Lophelia pertusa* (L.) at 64° N on the Norwegian shelf: structure and associated megafauna. Sarsia 80:145–158

Mortensen PB, Hovland MT, Fosså JH, Furevik DM (2001) Distribution, abundance and size of *Lophelia pertusa* coral reefs in mid-Norway in relation to seabed characteristics. J Mar Biol Assoc UK 81:581–597

Mortensen PB, Buhl-Mortensen L, Gordon DC Jr, Fader GBJ, McKeown DL, Fenton DG (2005) Effects of fisheries on deep-water Gorgonian Corals in the Northeast Channel, Nova Scotia (Canada). Am Fish Soc Symp 41:369–382

Mueller CE, Larsson AI, Veuger B, Middelburg JJ, Van Oevelen D (2013) Opportunistic feeding on various organic food sources by the cold-water coral *Lophelia pertusa*. Biogeosciences 11:123–133

Neulinger SC, Järnegren J, Ludvigsen M, Lochte K, Dullo WC (2008) Phenotypic-specific bacterial communities in the cold-water coral *Lophelia pertusa* (Scleractinia) and their implications for coral's nutrition, health, and distribution. Appl Environ Microbiol 74: 7272–7285

Osterloff J, Nilssen I, Järnegren J, Engeland TV, Buhl-Mortensen P, Nattkemper TV (2019) Computer vision enables short and long-term analysis of *Lophelia pertusa* polyp behaviour and colour from an underwater observatory. Sci Rep 9:6578. https://doi.org/10.1038/s41598-019-41275-1

Ottesen D, Rise L, Knies J, Olsen L, Henriksen S (2005) The Vestfjorden-Trænadjupet palaeo-ice stream drainage system, mid-Norwegian continental shelf. Mar Geol 218:175–189

Piazzi L, Balata D, Ceccherelli G (2016) Epiphyte assemblages of the Mediterranean seagrass *Posidonia oceanica:* an overview. Mar Ecol 37:3–41

Poulain PM, Warn-Varnas A, Niiler PP (1996) Near-surface circulation of the Nordic seas as measured by Lagrangian drifters. J Geophys Res 101:18237–18258

Purser A (2015) A time series study of Lophelia pertusa and Reef Megafauna responses to drill cuttings exposure on the Norwegian margin. PLoS ONE 10(7):e0134076. https://doi.org/10.1371/journal.pone.0134076

Schleinkofer N, Evans D, Wisshak M, Büscher JV, Fiebig J, Freiwald A, Härter S, Marshall HR, Voigt S, Raddatz J (2021) Host influenced geochemical signature in the parasitic foraminifer *Hyrrokkin sarcophaga*. Biogeosciences 18:4733–4753, https://doi.org/10.5194/bg-18-4733-2021

Sigwart JD (2009) Parasitic foraminifers on a deep-sea chiton (Mollusca, Polyplacophora, Leptochinidae) from Iceland. Mar Biol Res 5:193–199

Soetaert K, Mohn C, Rengstorf A, Grehan A, Van Oevelen D (2016) Ecosystem engineering creates a direct nutritional link between 600-m deep cold-water coral mounds and surface productivity. Sci Rep 6:35057

Strømgren T (1971) Vertical and horizontal distribution of *Lophelia pertusa* (Linné) in Trondheimsfjorden on the west coast of Norway. Det Kongelige Norske Videnskabers Selskabs Skrifter 6:1–9

Sundahl H, Buhl-Mortensen P, Buhl-Mortensen L (2020) Distribution and suitable habitat of the cold-water corals *Lophelia pertusa*, *Paragorgia arborea*, and *Primnoa resedaeformis* on the Norwegian continental shelf. Front Mar Sci 7:213. https://doi.org/10.3389/fmars.2020.00213

Thiem Ø, Ravagnan E, Fosså JH, Berntsen J (2006) Food supply mechanisms for cold-water corals along a continental shelf edge. J Mar Syst 60(2006):207–219

Titschack J, Baum D, de Pol-Holz R, López Correa M, Forster N, Flögel S, Hebbeln D, Freiwald A (2015) Aggradation and carbonate accumulation of Holocene Norwegian cold-water coral reefs. Sedimentology 62:1873–1898

Todd R (1965) A new *Rosalina* (Foraminifera) parasitic on a bivalve. Deep-Sea Res 12:831–837

Tong R, Purser A, Guinan G, Unnithan V, Yu J, Zhang C (2016) Quantifying relationships between abundances of cold-water coral *Lophelia pertusa* and terrain features: a case study on the Norwegian margin. Cont Shelf Res 116:13–26. https://doi.org/10.1016/j.csr.2016.01.012

van Oevelen D, Duineveld G, Lavaleye M, Mienis F, Soetaert K, Heip CHR (2009) The cold-water coral community as a hot spot for carbon cycling on continental margins: a food-web analysis from Rockall Bank (northeast Atlantic). Limnol Oceanogr 54:1829–1844

Wheeler AJ, Beyer A, Freiwald A, de Haas H, Huvenne VAI, Kozachenko M, Olu-Le Roy K, Opderbecke J (2007) Morphology and environment of cold-water coral carbonate mounds on the NW European margin. Int J Earth Sci 96:37–56. https://doi.org/10.1007/s00531-006-0130-6

White M, Wolff GA, Lundälv T, Guihen D, Kiriakoulakis K, Lavaleye M, Duineveld G (2012) Cold-water coral ecosystem (Tisler Reef, Norwegian Shelf) may be a hotspot for carbon cycling. Mar Ecol Prog Ser 465:11–23

Wild C, Wehrmann LM, Mayr C, Schöttner S, Allers E, Lundälv T (2009) Microbial degradation of cold-water coral-derived organic matter: potential implication for organic C cycling in the water column above Tisler Reef. Aquat Biol 7:71–80

Wilson JB (1979) 'Patch' development of the deep-water coral *Lophelia pertusa* (L.) on Rockall bank. J Mar Biol Assoc U K 59:165–177

Waters of Ireland and the UK

6

Sebastian J. Hennige, Furu Mienis, Andrew Wheeler, and Veerle A. I. Huvenne

Abstract

Cold-water coral ecosystems in Irish and UK waters, particularly those created by framework-building scleractinian corals, are relatively well studied, owing in part to a rich history of geological and oceanographic research in these waters, decades before they were mapped and characterised in terms of their biodiversity. This chapter explores this history and considers the formation and diversity of coral habitats at three scales: (1) small coral ecosystems or mounds (e.g. Darwin Mounds), (2) giant carbonate mounds (e.g. Logachev Mound), and (3) vertical coral ecosystems in submarine canyons (e.g. Whittard Canyon). While corals within each of these ecosystems support a myriad of associated biodiversity and are reliant upon incoming currents delivering a consistent, rich food supply, there are significant differences in underlying oceanography and geology of these ecosystems in the areas where they are found. There is substantial connectivity driven by the dominant water masses between all three main types of coral habitat within Irish and UK waters. Hatton Bank was found to be a critical site for overall habitat connectivity, and Rosemary Bank connects Northern and Western habitats. There are significant current and future threats to all types of Irish and UK coral habitats, ranging from trawling leading to direct habitat loss, to climate pressures including ocean acidification, which could degrade the habitat quality and framework extent (and hence associated biodiversity) of the deeper habitats in particular. As we face these varied threats, effective management strategies become ever more important. While several closures have demonstrated the effectiveness of fishing reduction in preventing further habitat loss, there is now an urgent need to include projected environmental impacts into marine protected area planning to protect these vulnerable marine ecosystems in a rapidly changing ocean.

Keywords

Deep sea · Scleractinian corals · Porcupine Seabight · Marine geology · Water masses · Coral mounds · Carbonate mounds · Vertical reef · Carbon cycling · Species associations · Ocean acidification · Conservation

6.1 Introduction

This chapter will explore scleractinian cold-water corals found in the waters of Ireland and the UK. When referencing the shared margin, Ireland is referred to here first as it has a larger deep-water territory and, although both margins possess significant coral-rich habitats, Ireland's are arguably more extensive. This review includes the locations of key ecosystems, the wider oceanography of the region, the variability of cold-water coral ecosystems within these waters, and finally the threats they face. The coral ecosystems in these waters are relatively well studied in terms of being mapped, with reasonably well-defined environmental conditions and characterised associated biodiversity. Literature searches highlight that key regions in Irish and UK waters have been documented for decades, with early descriptions of coral habitats by Le Danois in 1948, followed by more focus on coral starting in 1979 (Wilson 1979a)

S. J. Hennige (✉)
School of GeoSciences, Kings Buildings, University of Edinburgh, Edinburgh, UK
e-mail: s.hennige@ed.ac.uk

F. Mienis
Department of Ocean Systems, NIOZ Royal Netherlands Institute for Sea Research, Texel, The Netherlands

A. Wheeler
School of Biological, Earth and Environmental Science/Environmental Research Institute, University College Cork, Cork, Ireland

SFI Centres for Research in Applied Geosciences/Energy, Climate and Marine, University College Cork, Cork, Ireland

V. A. I. Huvenne
National Oceanography Centre, University of Southampton Waterfront Campus, Southampton, UK

E. Cordes, F. Mienis (eds.), *Cold-Water Coral Reefs of the World*, Coral Reefs of the World 19,
https://doi.org/10.1007/978-3-031-40897-7_6

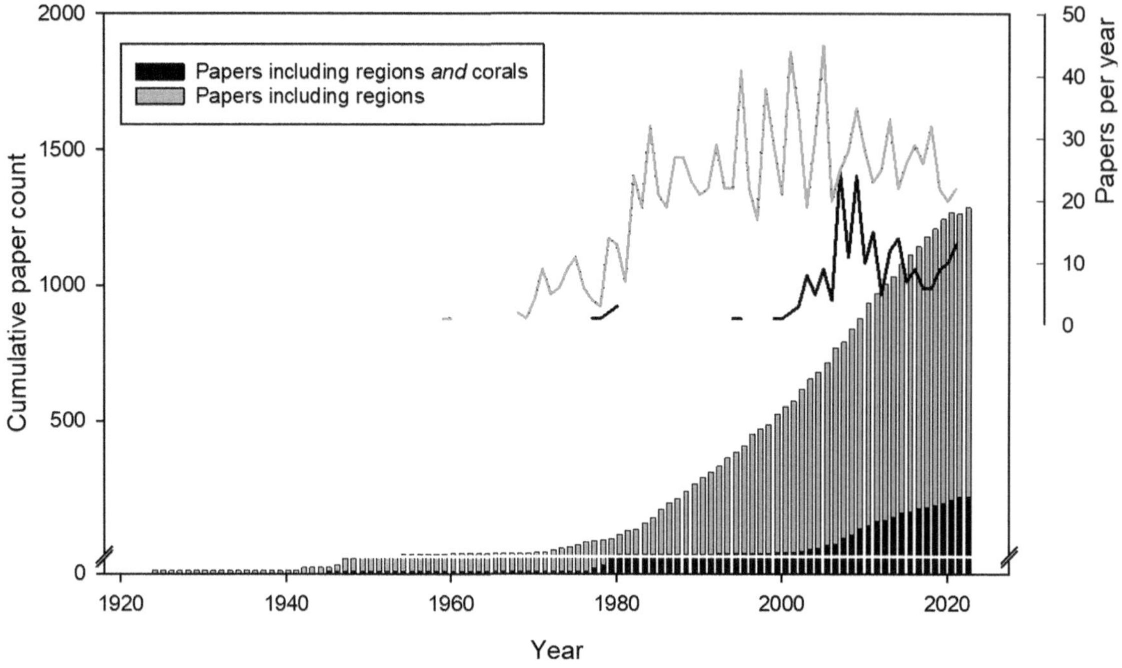

Fig. 6.1 The cumulative number of articles with a focus on a key selection of cold-water coral habitats in Irish and UK waters that are well characterised and discussed more in this chapter. The total number of papers including all sites within Irish and UK waters will be higher. This figure is broken down as articles focused on habitat regions [for example, Logachev Mounds, Porcupine Basin (Scopus citation search, May. 2022 using search criteria of 'coral' with region in journal title, abstract or keyword. Regions were Mingulay Reef Complex, Belgica Mound Province, Logachev Mounds, Whittard Canyon, Porcupine Bank, Darwin Mounds, Moira Mounds, Rockall Bank, Porcupine Seabight, Porcupine Basin, Rockall Trough, Hatton Basin, Wyville Thomson Ridge, Rosemary Bank, George Bligh Bank, Anton Dohrn Seamount, Hebrides Terrace)], and those specifically citing 'coral' in their titles, abstracts or keywords *in addition* to habitat regions. Right *y*-axis indicates annual number of papers in these two categories. Line break on left *y*-axis allows exaggeration of article numbers between 1 and 5

(Fig. 6.1). Mapping efforts of coral habitats increased in the early 2000s (e.g. Beyer et al. 2003; Huvenne et al. 2005; Mienis et al. 2006; Roberts et al. 2009a), leading to a rapid rise of articles focused on the occurrence, biology, and ecology of these vulnerable marine ecosystems. In the last few decades there have been several large-scale international consortia that have worked to this effect within Irish and UK waters, including CoralFish, HERMES, HERMIONE, MOUNDFORCE, GEOMOUND, ECOMOUND, ACES and the H2020 ATLAS and iAtlantic projects, in addition to national efforts such as through SEA7 (UK) and SeaRover (Ireland).

6.1.1 Geological Setting

The present-day morphology of the Irish-UK margin formed as a result of rift-basin formation at the opening of the Atlantic during the Mesozoic (Shannon et al. 1999). With the Atlantic eventually opening west of Hatton Bank, eastern rift grabens were preserved and now form a series of deep basins partially filled with a thick succession of Mesozoic and Cenozoic sediments (Shannon et al. 2001). Rift basins are borders for basement horst-block highs covered by thinner Mesozoic and Cenozoic sedimentary units. The post break-up history of the margin areas has been dominated by subsidence outpacing sedimentation with relatively smaller fault reactivation and uplift during the Alpine orogenic compression event.

One of the deeper bathymetric deeps is the Rockall Trough (Fig. 6.2) lying between the westerly Rockall Bank and the easterly Porcupine Bank and the continental slopes of Ireland and Scotland. In the south, the Trough deepens to >3000 m water depth and opens to the wider Atlantic, while in the north it is bound by the Wyville Thomson Ridge (~500 m water depth). The Porcupine Seabight, further east, is mostly shallower than the Rockall Trough, being infilled by thicker sedimentary units due to its proximity to the Irish landmass and location east of the Porcupine Bank. Volcanism also affected this area during the Carboniferous and later in the Cenozoic, as the Iceland Plume, further east than its present location, formed a number of volcanic seamounts especially in UK waters in addition to the Wyville Thomson Ridge bounding the northern Rockall Trough. As a consequence of this rifting and volcanism, the Irish and UK margins are very long, surrounding the horst-block banks,

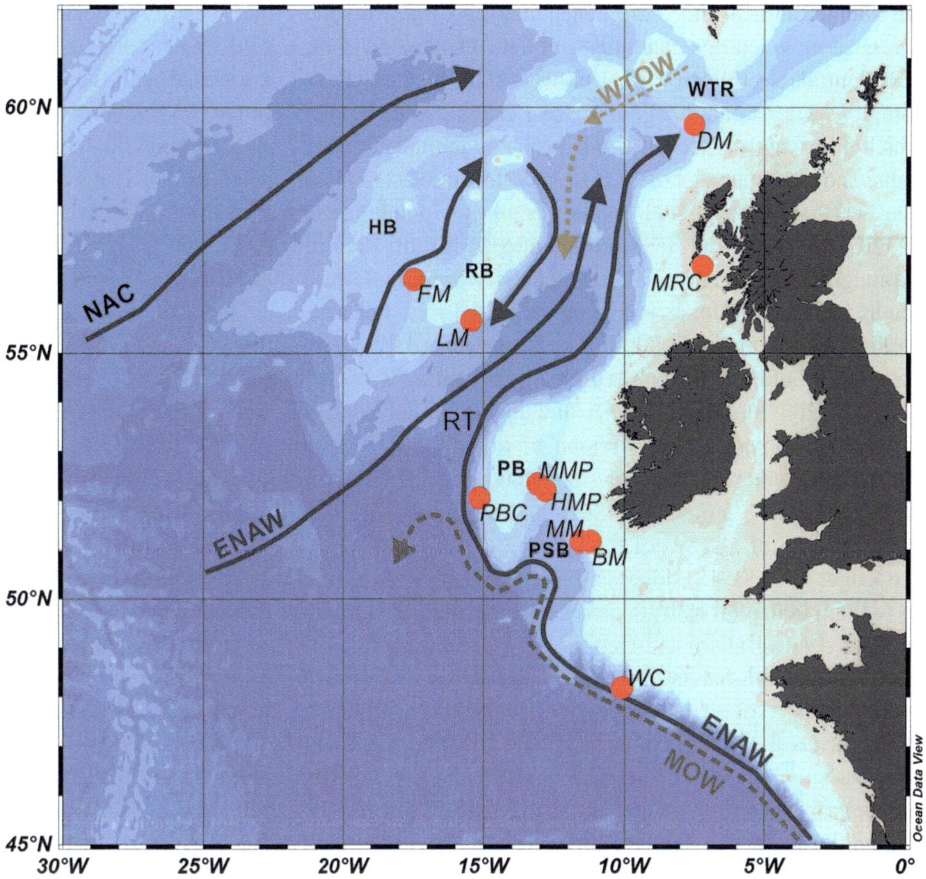

Fig. 6.2 Locations of well-characterised *Lophelia pertusa* [also referred to as *Desmophyllum pertusum* (Addamo et al. 2016)] ecosystems in the waters of the Ireland and the UK (red circles), and broad oceanography of water masses bathing the coral areas, Eastern North Atlantic Water (ENAW), Mediterranean Outflow Water (MOW), Wyville Thomson Overflow Water (WTOW), and North Atlantic Current (NAC). Indicated areas include Porcupine Bank (PB), Porcupine Bank Canyon (PCB), Porcupine Seabight (PSB), Rockall Bank (RB), Rockall Trough (RT), Hatton Bank (HB), Wyville Thomson Ridge (WTR), Moira Mounds (MM), Belgica Mounds (BM), Logachev Mounds (LM), Franken Mounds (FM), Whittard Canyon (WC), Darwin Mounds (DM), Mingulay Reef Complex (MRC), Hovland Mound Province (HMP), Magellan Mound Province (MMP)

and tracing around the embayment of the Porcupine Seabight. As cold-water corals predominate at intermediate water depths at this latitude, areas of potential colonisation are significantly increased on this margin. The margin also straddles the southern limit of European glaciation (Mienert and Weaver 2003), with submarine canyons incising the margins to the south and giving way to slope dominated margins to the north (Weaver et al. 2002). Furthermore, the Rockall and Porcupine Bank margins are isolated from direct terrestrial influence providing the Irish-UK margin with a range of sedimentary regimes influencing cold-water habitats.

Subsidence in the early Pliocene initiated the onset of a bottom current dominated regime that has created an erosional unconformity along the margin (Stoker et al. 2002). Coral mounds at the Rockall Trough margins, as well as mound areas in the Porcupine Seabight, seem to have developed on this unconformity, which offered firmgrounds for early colonisation (Mienis et al. 2006). The formation of the unconformity co-occurred with major changes in the circulation patterns in the North Atlantic. The gradual closure of the Panama seaway resulted in the onset of the North Atlantic Drift as part of the modern conveyor belt circulation pattern, introducing the flow of warmer and more saline water masses to the higher latitudes. In addition, permanent ice sheet formation started at the same time, inducing glacial-interglacial climate oscillations, which strongly influenced mound formation and growth on the Irish-UK margin (see Box 6.1, Challenger Mound). Sediment cores collected in different mound areas show that living corals were absent during the recent-most glacial times, and mounds were eroding due to the absence of a stabilising dense coral cover (Dorschel et al. 2007b; Rueggeberg et al. 2007; Mienis et al. 2009; Thierens et al. 2013; Van der Land et al. 2014).

Despite Ireland's and the UK's extensive margin, not all of the margin is colonised by cold-water corals, with some areas experiencing environmental conditions adverse to coral growth. Nevertheless, the southernmost Irish margin is dominated by submarine canyons where corals exist in a range of contexts (gardens, mounds, and vertical habitats: Morris et al. 2013), with the southern and western slopes of the Goban Spur and southern Porcupine Bank poorly explored. Further north along the Porcupine Bank on its western margin, cold-water corals are common and form carbonate mounds (van Weering et al. 2003; Wheeler et al. 2005a; Dorschel et al. 2009). However, the eastern margin of the Porcupine Bank (sloping into the Porcupine Seabight) contains limited scleractinian corals and is gently sloping. Abundant corals and giant coral mounds in the Porcupine Seabight can be found on the north-eastern Seabight margin

(Hovland, Belgica and buried Magellan and Enya mound areas) dominated by partially infilled canyon systems (Hovland et al. 1994; De Mol et al. 2002; Huvenne et al. 2007; Wheeler et al. 2007; van Rooij et al. 2008). Further west on the eastern Rockall Bank, probably the largest coral mounds on this margin exist, the Logachev Mounds (Kenyon et al. 2003) (Fig. 6.2), with smaller Franken Mounds found on the western Rockall Bank sloping into the Hatton-Rockall Basin (Wienberg et al. 2008). On the far western Hatton Bank, carbonate mounds and coral colonised rock outcrops are found (Roberts et al. 2008). Further north in UK waters notable coral mounds occur on the shelf [Mingulay Reef (see Box 6.2; Roberts et al. 2005)] and the northern Rockall Trough (Darwin Mounds: Masson et al. 2003), with the margin *per se* being less conducive to coral reef development.

6.1.2 Oceanography

Cold-water coral mounds along the Irish-UK margin have been mainly observed between 500–1000 m water depth (De Mol et al. 2002), except for the shallower Mingulay reef complex (Roberts et al. 2009a) (see Box 6.2). The oceanographic setting along the Irish margin shows the regionally differing influence of two separate southerly source regions, separated by the deep Rockall Trough (Fig. 6.2). This area is of extreme importance since the region is a conduit for transport of water from a southern origin to the north [6.6 Sv, (Houpert et al. 2020)], playing a key role in the thermohaline overturning circulation (New and Smythe-Wright 2001).

The shallowest reef in UK waters, the Mingulay reef complex is influenced by three water sources: high nutrient and high salinity Atlantic Water, Irish Sea Water and coastal water characterised by a high volume of river runoff from the Scottish mainland. A cyclonic circulation system is observed in the Sea of Hebrides with a main flow direction between the Hebrides and the Scottish mainland to the SSW-NNE (Davies et al. 2009). Water masses bathing the canyons and mounds along UK margin and the Porcupine Seabight originate in the south and are carried northward from the Bay of Biscay by the poleward shelf edge current (Huthnance 1986). The depth range of mounds with a thriving coral cover in this region (600–1000 m water depth) is marked by the boundary between the Eastern North Atlantic Water (ENAW) and the upper boundary (800–1000 m water depth) of the Mediterranean Outflow Water (MOW), characterised by a high salinity core (Van Aken and Becker 1996). The upper level of the MOW is associated with the permanent thermocline, which appears influential for the depth distribution of cold-water corals through the promotion of stronger bottom current conditions (White and Dorschel 2010). It has been suggested that the MOW depth corresponds to a peak occurrence of cold-water corals. The strong vertical density gradient associated with the permanent thermocline between ENAW and MOW supports the formation of nepheloid layers (White and Dorschel 2010), thereby enhancing the food supply and potentially playing a role in larval dispersal patterns (Dullo et al. 2008). The MOW is also present west of the Porcupine Bank, but its signal rapidly decreases above 53° north (New and Smythe-Wright 2001).

The Hatton Bank, Logachev, and Darwin mound areas (Fig. 6.2) are influenced by water masses originating in the southwest and brought to the area by a branch of the North Atlantic Current (NAC). Surface waters in this area resemble the ENAW in terms of salinity and temperature (Hansen and Osterhus 2000; New et al. 2001). Along the SW Rockall Trough margin, ENAW is relatively cool and fresh due to the interaction with the Sub-Arctic Intermediate Water (SAIW) (Hansen and Osterhus 2000), with the main pulses occurring in winter and spring (Ullgren and White 2010). An anticyclonic circulation around the Rockall Bank results in an equatorward current along the SW Rockall Trough margin (Pingree and Le Cann 1989). On several occasions, Wyville Thomson Overflow Water was observed on the SW Rockall Trough margin transporting nutrient-rich water southwards, with potential implications for particle and larvae transport (Schultz et al. 2023). Below 1000 m water depth, the Labrador Sea Water (LSW) occurs, characterised by low salinity (<34.9) and temperatures between 3.2 and 4 °C (New et al. 2001). In the northern part of the Rockall Trough, LSW occurrence is constrained by the topography of the basin, and this water mass is forced to perform a clockwise circulation in the Rockall Trough flowing to the north on the SW Rockall Trough margin and to the south on the SE Rockall Trough margin.

A specific characteristic of the upper ocean waters on the Rockall and Porcupine Bank is the deep winter mixing up to a 1000 m water depth (White et al. 2005), which is caused by weak density gradients that enhance convection and mixing of nutrients to the surface waters (Holliday et al. 2000; van Aken 2000). Increased nutrient availability over the Banks (likely kept in place by a Taylor column) results in high primary productivity in early spring (White et al. 2005), resulting in increased food supply to the cold-water corals as observed between February–June in the Logachev mound area (Duineveld et al. 2007). Two hypotheses have been proposed to explain the transfer of organic matter to the cold-water coral reefs in the Logachev mound area. In the first theory, water from the shallow banks slowly drains away through the benthic boundary layer, providing a mechanism for the downslope distribution of organic matter (Mienis et al. 2007; White et al. 2005). The second theory is derived from model simulations that show that the interaction between

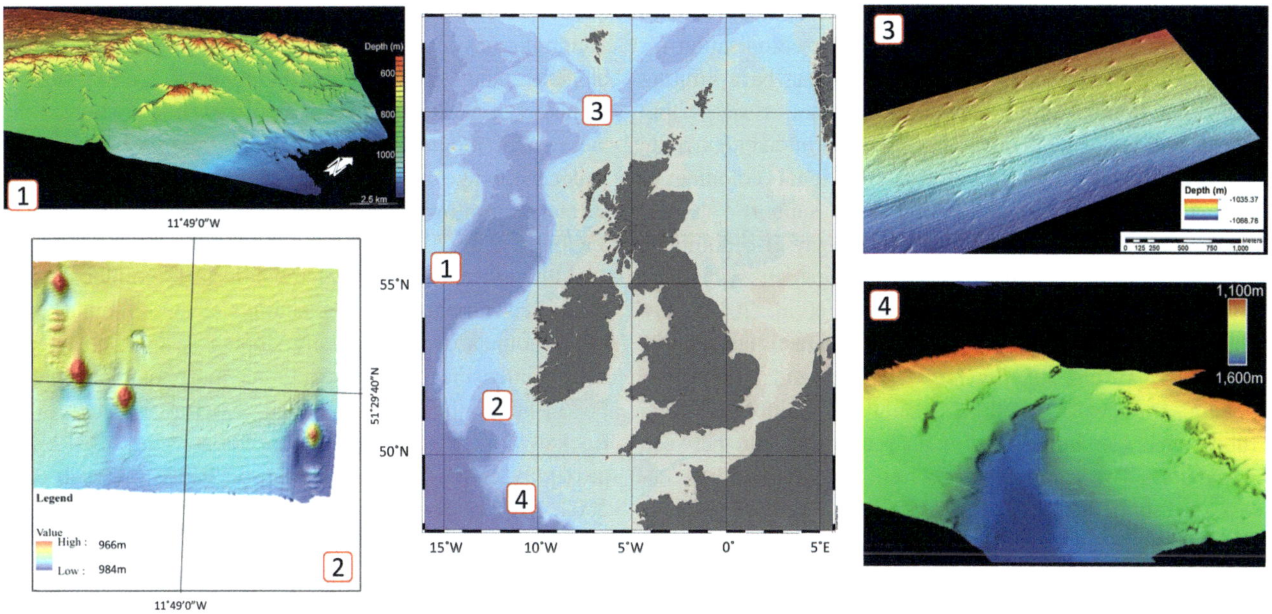

Fig. 6.3 Bathymetry examples of giant mounds (1: Logachev Mounds), small coral mounds (2: Moira Mounds; 3: Darwin Mounds (vertical scale at 2 x exaggeration)), and vertical cold-water coral ecosystems (4: Whittard Canyon) in Irish and UK waters. Imagery courtesy of NIOZ (1), copyright University College Cork (2), and National Oceanography Centre (3 & 4)

tidal currents and the coral mound topography may induce downwelling events of surface water (Soetaert et al. 2016).

6.2 Variability of Coral Ecosystems in the Region

Scleractinian cold-water coral ecosystems in the waters of Ireland and the UK range from giant mounds to small, patchy coral habitats, and even to canyon systems, where corals grow vertically along steep gradients. All of these systems support myriad biodiversity, but the geology, oceanography, and biology of each of these systems are different, and one ecosystem case study cannot represent the diversity found in these waters. Consequently, case studies here have been split into three main categories (Fig. 6.3) of giant mounds, small coral mounds, and vertical coral-water coral ecosystems. This split allows us to consider the broad differences between some major coral habitat types, and many examples of coral habitats within Irish and UK waters fall within these categories.

6.2.1 Small Coral Ecosystems/Small Mounds

Small coral ecosystems can be patchy or fragmented in nature, but may still cover quite a large area. Examples of these ecosystems are the Darwin, Moira, and Mingulay Mounds, the 'mini-mounds' on the interfluves of Whittard Canyon, and the patchy coral occurrences on Rockall Bank.

Although cold-water corals have the capacity to build large topographic reefs in Irish-UK waters, smaller occurrences are also common in areas where the environmental needs of these communities are met. These can range from individual colonies (e.g. Wilson 1979a; Wheeler et al. 2005a) to small mounds <50 m tall, with investigated examples including the Darwin Mounds (Masson et al. 2003; Wheeler et al. 2008; Huvenne et al. 2009a; Victorero et al. 2016), Moira Mounds (Foubert et al. 2011; Wheeler et al. 2011; Lim et al. 2018), Macnas Mounds (Wilson et al. 2007) and Mingulay Reefs (Roberts et al. 2005, 2009a). Initially, small coral mounds on the Irish-UK margin were indistinctly observed on regional side-scan sonar mapping surveys (e.g. Darwin Mounds on 30 kHz TOBI side-scan sonar presented in Masson et al. 2003). Closer to shore, small mounds were sometimes known anecdotally from corals being present in fishing bycatch. With the advent of higher resolution side-scan sonar surveys and ROV-mounted multibeam echosounders, smaller mounds were revealed in more detail (Wheeler et al. 2008; Foubert et al. 2011; De Clippele et al. 2017). Video mosaicking on reefs has now allowed more detailed observations down to the organism scale (Boolukos et al. 2019; Conti et al. 2019; Price et al. 2019).

Only the framework Scleractinia (stony corals), principally *Lophelia pertusa* and *Madrepora oculata*, form topographic reefs, but it is important to note that other coral orders that are not the focus of this chapter (e.g. black corals (Antipatharia) (Barrett et al. 2020) and sea pens (Pennatulacea) (Hogan et al. 2019) that are found in the

Whittard and Porcupine Bank Canyon) can also occur in dense aggregations. Available substrate for colonisation for scleractinian and non-scleractinian corals can be a limiting factor, and exposed boulder-sized dropstones supporting corals and associated organisms may form habitats quite distinct from their surroundings. Individual colonies of antipatharians and octocorals can support diverse communities (Roberts et al. 2009b) and may also dominate the seafloor as major components of coral gardens, which are complex habitats to define (Bullimore et al. 2013). For gardens and habitats that include scleractinian corals, aggregations (e.g. of *Lophelia pertusa* and *Madrepora oculata* framework) produce significant amounts of coral rubble that can induce sedimentation within and between framework. This can baffle the current, leading to sedimentation and building of topographic reefs or coral mounds (Dorschel et al. 2007a). These come in a continuum of sizes from a few metres across (Macnas Mounds) to 10 s of metres across and a few metres high (e.g. Darwin Mounds). They are typically circular to ovoid in shape becoming more elongated with stronger currents parallel to the dominant current flow directions (Wheeler et al. 2008). Small mounds are typically found in a high-current environment, characterised by sandy (contourite) sediments rather than hemipelagic muds surrounding the mounds (Moira, Darwin; Huvenne et al. 2009a). Bedforms (sandwaves, ripples) can continue onto the mounds (Foubert et al. 2011), while scour features can be found downcurrent of the mounds (Lim et al. 2018).

High-current speed benthic environments may be a prerequisite for coral mound formation, providing the bedload sediment transport which is then trapped by the coral frameworks facilitating mound growth. High current speeds also remobilise particulate organic matter (POM), although observations suggest that individual coral polyps only open for feeding during low speed current flows, e.g., in tidal slack waters. Bedforms in the surrounding seabed are often typified by sediment waves and ripples, gravel lags with exposed dropstones, and scour features caused by obstacle-induced current intensification around the mounds. On the downslope Moira Mounds chain, Lim et al. (2018) mapped scour pits around Moira Mounds using ROV-based multibeam echosounders and showed larger and deeper scours around larger mounds. Scours predominantly on the south side of the mounds possibly evidence peak flow events when most erosion would occur. New coral mounds can form within the larger scour pits in the lee of the larger mounds.

Wheeler et al. (2008) also showed a linear correlation between downstream mound elongation and current speed with high-resolution deep-tow side-scan sonar surveys over the Darwin Mounds. This demonstrates the control that current speed and bedload sediment flux have on mound morphology, with growth style strongly controlled by these two prevailing factors. Many of the coral mounds may exhibit a quite patchy coral cover surrounded by significant accumulations of sediment revealing the importance of both biogenic and geogenic processes in mound formation (Wheeler et al. 2011; Conti et al. 2019). It is postulated that most small mounds on the Irish-UK margin are probably Holocene in age and started developing when current speeds accelerated following the last glaciation (Wheeler et al. 2011; Victorero et al. 2016).

Although all small coral mounds have similar current-controlled morphologies, the settings of these mounds vary. Most mounds occur on gently sloping seabed subjected to current intensification by the regional morphological context. For instance, the Darwin Mounds exist at the head of the Rockall Trough where current speeds are intensified by the proximity of the Wyville–Thompson Ridge (Masson et al. 2003; Huvenne et al. 2009a), the Moira Mounds exist between the chain of the giant Belgica Mounds (see below) and within a submarine channel (Lim et al. 2018), and the Macnas Mounds exist within the area of strong contourite currents (Wilson et al. 2007). However, in the complex topography of submarine canyons, other contexts occur where strong currents and POM are available in abundance, such as at the interfluves of canyon branches (e.g. the Explorer and Dangaard Mounds of the Whittard Canyon complex; Stewart et al. 2014) or on the canyon rim [e.g. in the Porcupine Bank Canyon (Appah et al. 2020)].

6.2.2 Giant Carbonate Mounds

Giant Carbonate mounds, over 100 m tall, are well documented on the Irish and UK margin forming several mound provinces or areas that attest to the long term, although punctuated, suitability of the Irish-UK margin for cold-water coral reef development. Giant Mounds are almost exclusively found in the Irish waters beyond the limits of the British-Irish Ice Sheet (although it is possible that the Hovland Mounds may have been overridden by a floating ice shelf during the last glaciation). Giant carbonate mounds were first documented from the Irish margin based on industry seismic records (Hovland et al. 1994), although coral finds from fishermen had been known for many years previous, and Le Danois (1948) had hypothesised the existence of large 'massifs coralliens'. The subsequently named Hovland Mounds were groundtruthed by towed video and sampling in 1998 (Kenyon et al. 1998; De Mol et al. 2002). Other giant mounds were then mapped using hull-based multibeam echosounders and regional side-scan sonar (Beyer et al. 2003; Huvenne et al. 2005; Mienis et al. 2006), seismically imaged in higher resolution, investigated by ROVs (Huvenne et al. 2005), cored (Dorschel et al. 2007b; Rueggeberg et al. 2007), and monitored (Dorschel et al. 2007a).

Several 'provinces' of giant mounds can be found offshore of Ireland and the UK (Fig. 6.2) (Wheeler et al. 2007), including the Belgica, Hovland, and Magellan mounds in the Porcupine Seabight, the Logachev mounds on Rockall Bank, plus several clusters of mounds on Porcupine Bank (Dorschel et al. 2009). In addition, clusters of buried mounds have also been described [e.g. Viking mounds, Enya mounds (Van Rooij et al. 2009), and the majority of the Magellan mounds (Huvenne et al. 2007)]. Some of the giant mounds such as the Galway and Therese Mound (De Mol et al. 2007) in the Belgica Mound province are at present covered by dense coral assemblages and continue to trap sediment and accumulate coral rubble, therefore increasing in stature. Others lack live coral cover at present and are therefore growing more slowly or even eroding (Dorschel et al. 2009).

Studies of the internal stratigraphy of giant mounds (e.g. Dorschel et al. 2007b; Rueggeberg et al. 2007; Eisele et al. 2008; Thierens et al. 2010, see Box 6.1) reveal a climatically controlled, punctuated mound growth typified by periods of coral dominated accumulation, and periods of slower sediment accumulation or frequent erosional events. In this way giant coral mounds owe their stature to their longevity and represent sites where conditions for coral colonisation have repeatedly proved favourable. This appears to be the case for the Irish margin over the last 2.6 million years (Kano et al. 2007). The exact initiation age of most of the giant carbonate mounds is unknown, although most are found on a regional unconformity and are hence thought to have initiated broadly at the same time (Huvenne et al. 2003).

Another prerequisite for giant mound preservation is their existence beyond the limit of the European Ice sheets, which may have eroded giant mound features if they existed further north. Nevertheless, there are many examples of giant mounds that have become entombed in contourite drifts where sedimentation exceeded giant mound growth rates (e.g. the Enya Mounds—Van Rooij et al. 2009). Giant carbonate mounds rarely occur in isolation and are sometimes aligned along the continental margin to form chains. Mound morphology varies depending on current flow directions with both conical (e.g. the Therese Mound, De Mol et al. 2007) and spoked forms present (e.g. the Propeller Mound, Dorschel et al. 2005). Complex morphologies are exhibited by the Logachev Mounds on the Rockall Bank (Mienis et al. 2006) due to the interplay between along-slope contourite and up-slope tidal currents.

6.2.3 Vertical Cold-Water Coral Ecosystems: Submarine Canyons and Cliffs

In addition to the different mound types described above, cold-water corals can also form 'reef-like' ecosystems on steep deep-water topography such as vertical or overhanging walls in submarine canyons (Huvenne et al. 2011; Appah et al. 2020), or headwall scarps of submarine landslides (Huvenne et al. 2016b). The existence of such vertical cold-water corals reefs or 'hanging gardens' has only been described fairly recently, since the necessary technology to map and observe them became available (see Box 6.3). The first reports of extensive vertical cold-water coral reefs were from the Irish-UK margin, more specifically from the Whittard Canyon, where a vertical cliff up to 120 m high and >1500 m long was discovered (Huvenne et al. 2011). Since then, other deep-water vertical reefs have been registered in Irish and UK waters, and elsewhere around the world (Gori et al. 2013; Brooke and Ross 2014). This includes communities that are driven by other species (e.g. deep-sea oysters, clams), although many still host cold-water corals (Johnson et al. 2013). In addition to scleractinian cold-water corals, prominent non-scleractinian corals in Whittard Canyon, Explorer Canyon and on the Rockall Bank slide headwall escarpment include the black coral *Stichopates* sp., several morphotypes of Alcyonacea, and walls covered in *Primnoa* sp. (Robert et al. 2019). Morris et al. (2013) note that areas with high densities of *Lophelia pertusa* are characterised by lower abundances of octocorals and *vice versa*, potentially as a result of aggressive competition by *Lophelia*.

As a result of their steep setting, vertical cold-water coral reefs are often affected by intense current regimes. Internal waves and tides have been reported from the Whittard Canyon complex (Hall et al. 2017; Aslam et al. 2018) and from the Rockall Bank (van Haren et al. 2014). The flanks of Rockall Bank are also subject to strong contour currents (Stoker et al. 1998). These current regimes keep food in suspension and may focus food in specific nepheloid layers that may be washed over the vertical reefs. For example, the amplitude of the M2 internal tide measured at the main coral wall in Whittard Canyon is >80 m, washing food particles past the reefs twice a day (Hall et al. 2017), while the amplitude is even higher in the Logachev mound area (van Haren et al. 2014).

6.3 Biology and Ecology

6.3.1 Biodiversity, Communities, and Drivers

Cold-water coral ecosystems are home to a wide range of biodiversity (Roberts et al. 2006) (Fig. 6.4), which is supported by the framework of the corals and the heterogeneity created by other key organisms such as sponges. The framework that the corals create is also important as spawning ground for certain species of recreationally valuable sharks (Henry et al. 2013) (Fig. 6.4), skates (Henry et al. 2016), and commercially important fish species (Henry and

Fig. 6.4 Images of cold-water coral ecosystems in Irish and UK waters; (**a**) a greater forkbeard (*Phycis blennoides*) swimming over live and dead coral framework on the Logachev Mound Province; (**b**) dense aggregation of live *L. pertusa* coral at the Mingulay Reef Complex;

Roberts 2007; Ross and Quattrini 2007). While much focus is on the commonly occurring *Lophelia pertusa* as a cornerstone for biodiversity provision, other species contribute to the rich habitat of these ecosystems and the subsequent types and amounts of biodiversity that can be supported. This includes the other framework forming scleractinian corals in Irish and UK waters such as *Madrepora oculata* and *Solenosmilia variabilis*, which contribute to the complexity of the 3D framework, sediment accumulation, and mound growth, but also non-scleractinian corals such the Antipatharia (black corals) and Alcyonacea (octocorals) (De Clippele et al. 2019). These contribute to secondary biodiversity support in terms of shelter, nursery grounds, and as feeding platforms (Buhl-Mortensen and Mortensen 2004; De Clippele et al. 2015).

The often high biomass observed in cold-water coral reefs makes them hotspots of carbon and nitrogen cycling as compared to surrounding areas without cold-water coral cover (Van Oevelen et al. 2009; De Froe et al. 2019). Within Irish and UK waters, the distribution of scleractinian and non-scleractinian corals can be affected by a variety of factors, including local conditions such as food supply, substrate availability or type, and environmental variables highlighted in Table 6.1, including depth, bottom current speed, temperature, and salinity.

The vertical cold-water coral reef assemblages can be dominated by colonial scleractinians, such as *L. pertusa* in Whittard Canyon (Huvenne et al. 2011) or *S. variabilis,* for example on a landslide headwall escarpment on Rockall Bank (Robert et al. 2019). On other walls in Whittard and Explorer Canyon, *Primnoa* sp. is more numerous (Robert et al. 2019). Associated cold-water coral fauna include *M. oculata*, solitary cup corals including *Desmophyllum* sp., various Alcyonacea, and black corals (e.g. *Stichopates* sp.). Assemblage composition differs between walls, and between vertical and horizontal habitats, although many of the species are common between the two (Robert et al. 2019); similar to horizontal habitats, environmental drivers for the vertical assemblages include depth and substratum type. Deep-water vertical walls may be composed of different rock types, with the more competent, erosion-resistant lithologies apparently hosting more diverse assemblages. More friable rock types may not offer the right settling ground for some species, or may not be able to support coral colonies that exceed a certain size and weight. Where lithologies vary between consecutive strata, cliffs may contain a series of ledges, typically covered by sediments on top and offering hard substratum for settlement and shelter from excessive sedimentation on the underside (Robert et al. 2017). Contrary to horizontal cold-water coral habitats, no extensive accumulation of dead coral framework or coral rubble occurs in vertical cold-water coral habitats, with dead coral fragments typically accumulating in a talus at the bottom of the cliff (and potentially being washed away by strong currents).

Food supply is a consistently crucial factor for the occurrences of cold-water coral habitats and the types and amounts of biodiversity that they support (Roberts et al. 2009b). As outlined above, cold-water corals and mounds are often found where currents interact with the local topography [for example, where the bottom slope facilitates internal waves and baroclinic tidal currents (Frederiksen et al. 1992; Mohn et al. 2014), and creates favourable conditions for food supply (White et al. 2005)]. This results in pronounced bottom nepheloid layers that are important for the lateral transport of particles (Dickson and McCave 1986; Mienis et al. 2007). Such wave and tidal patterns observed on cold-water coral habitats (Dorschel et al. 2007a; Mienis et al. 2007) enhance mixing and turbulence, potentially increasing food availability in addition to supplying ecosystems with oxygen and nutrients (van Haren et al. 2014). The dynamics of food supply may change due to habitat engineering and current regimes (Mienis et al. 2006; White 2007), and is a factor to consider as habitats become large enough to significantly modify local hydrodynamics (e.g. induce tidal up- and downwelling internal waves and large intermediate nepheloid layers) (Mienis et al. 2007; Soctacrt ct al. 2016; Dc Clippclc ct al. 2019).

The quantity of food supplied from the ocean surface to the benthos has also been demonstrated to play a key role in the cold-water coral reef food webs in Irish and UK waters (Kiriakoulakis et al. 2004, 2007; Duineveld et al. 2007; Kazanidis and Witte 2016), and this can be characterised by analysing stable isotopes (e.g. carbon and nitrogen) in reef benthos. For the Mingulay Reef complex and the Logachev Mounds, this revealed that the benthos relied on fresh suspended particulate organic matter (and not decayed organic matter) as a food source, transported from the ocean surface down to the seabed (Duineveld et al. 2007, 2012).

Fig. 6.4 (Continued) (**c**) shark egg cases attached to a live *Parantipathes sp.* Sitting amongst dead coral framework; (**d**) corals (mainly 'giant' *Desmophyllum dianthus*) living on a vertical wall in the Porcupine Bank Canyon; (**e**) a *Lophelia pertusa* (pink and white polyp variants) reef on the lip on the Porcupine Bank Canyon; (**f**) a variety of cold-water corals attached to a vertical bedrock cliff in the Porcupine Bank Canyon including *Lophelia pertusa*, *Leiopathes glaberrima*, *Acanthogorgia* sp., *Stichopathes* cf. *abyssicola* and *Desmophyllum dianthus*. (**g**) and (**h**) sclerantinian communities with associated organisms (e.g. *Acesta excavata* in (**g**)) living on vertical canyon walls in Whittard Canyon. Note high water turbidity and settled sediment on the substrate from high sediment loads. Images **a**–**c** copyright Changing Oceans 2012; images d-f copyright University College Cork & Marine Institute; images g-h National Oceanography Centre

Table 6.1 Example environmental conditions of the three case study types examined here: (a) small ecosystems/small mounds, (b) giant mounds, and (c) vertical ecosystems. Well-characterised examples have been used for each ecosystem type

Case study ecosystem type	Current velocities (cm/s)	Salinity	Depth (m)	Temperature (C)	DO (µmol/l or ml/l)	Aragonite saturation	Ecosystem name	References
Small ecosystem or mounds	26–81	34.5–35.3	120–190	8.8–11.0	250–286	1.92–2.62	Mingulay reef complex	Dodds et al. (2007), Davies et al. (2009), Maier et al. (2009), Roberts et al. (2009a), Findlay et al. (2013, 2014), Hennige et al. (2014b), De Clippele et al. (2017)
Small ecosystem or mounds	33.6–25.1	35.0–35.2	900–1060	7.7–8.5	285–300		Darwin Mounds	Bett (2001), Masson et al. (2003), Kiriakoulakis et al. (2004), Huvenne et al. (2009a), Victorero et al. (2016), Cunningham (2020)
Small ecosystem or mounds	36–40	35.6–36.0	650–1150	8.4–9.6	197–286		Moira Mounds	Wheeler et al. (2011), Lim et al. (2018), Boolukos et al. (2019)
Small ecosystem or mounds	9.4–31.3 (114.2 max)	35.2–35.5	560–1050	7.0–10.8			Porcupine Bank Canyon Mounds	Appah et al. (2020), Lim et al. (2020a)
Giant mounds	17.3–60	35.1–35.4	500–1000	6.7–9.4	206–268	1.35–2.35	Logachev Mounds	Duineveld et al. (2007), Mienis et al. (2007), Wheeler et al. (2007), White et al. (2007), Findlay et al. (2014), Soetaert et al. (2016)
Giant mounds	0.2–25.5	35.4–35.6	700–1000	9–10			Belgica Mounds	Dorschel et al. (2007a), Wheeler et al. (2007, 2011), White et al. (2007), Dullo et al. (2008)
Vertical systems	20–40	35.3–35.6	1200–1400	5–8	180–190		Whittard Canyon (coral wall)	Huvenne et al. (2011), Morris et al. (2013), Pearman (2020)

Analysis such as this has demonstrated the existence of four trophic levels, with communities at the Mingulay Reef Complex characterised by elevated relative biomass of lower trophic levels compared to communities at the Logachev Mounds. Fresh suspended particulate organic matter can also support high relative biomass of suspension and filter feeders (Kazanidis and Witte 2016), and when combined with tight pelagic–benthic coupling and stronger bottom currents, can shape the structure of the benthic food webs. In habitats such as the Mingulay Reef Complex with periodic downwelling events of surface waters, this is particularly important to consider.

Variability in food supply across habitats is an important aspect in biodiversity provisioning and habitat proliferation across Irish and UK waters and would directly contribute to the variability in community composition observed in these areas (in terms of biodiversity supported by the cold-water coral framework). While important, biodiversity is not driven by food provision alone, and in a comparison between small and large mounds (using the Mingulay Reef Complex and the Logachev Mounds as an example where depth also varies, Fig. 6.2), the biodiversity variability is mainly driven by changes in depth (the Mingulay Reef Complex being far shallower than the Logachev Mounds) (Kazanidis et al. 2016). Across these two cold-water coral ecosystems, the composition of the fauna also varied due to the presence of framework forming sponges at the Mingulay Reef Complex such as *Spongosorites coralliophaga,* which can provide additional habitat/settlement surfaces for macrofauna (Kazanidis et al. 2016). Such biodiversity support by *S. coralliophaga* while common at the Mingulay Reef Complex is not typically observed at the Logachev Mounds and again highlights that depth-driven species composition is an important factor when considering the types and amounts of biodiversity that these cold-water coral systems can support. As sponges such as *S. coralliophaga* are able to process various food sources (dissolved inorganic nitrogen, dissolved organic carbon as well as microalgae and bacteria) and have the ability to synthesise *de novo* essential and non-essential hydrolysable amino acids (HAAs), their presence (also driven by depth) will contribute to the wider biodiversity that the cold-water coral ecosystems can support (Kazanidis et al. 2018). With regard to non-scleractinian corals, several species can co-occur. In a study based on the Logachev Mound Province, ten non-scleractinian coral species were identified in addition to the scleractinian corals. This diversity also serves as a potential indicator of good food supply, weak inter-species competition, and the presence of hard and heterogenous surfaces for settlement (De Clippele et al. 2019).

The high biodiversity and associated ecosystem services of cold-water corals are due to the 3-dimensional structural complexity of live or dead skeletons formed by scleractinian framework-building reef species (Mortensen et al. 1995; Price et al. 2019). While important, non-scleractinian coral species do not persist after death (although exceptions exist of large octocorals with skeletons that contribute to rubble) and do not contribute to the continued provision of stable habitat. As scleractinian cold-water coral reefs form, living coral colonies are typically observed at the top of the coral framework, with a significant proportion of the framework comprised of dead, exposed coral skeleton, which acts as a substrate for subsequent colonisation by other organisms. The longevity of the dead reef framework and cold-water coral structural complexity is key to supporting biodiversity, with decreased biodiversity seen at sites with less than 30% coral cover (Price et al. 2019). Biodiversity is highest in the transition zone of the reef, where dead framework dominates and there is only a small percentage of live colonies present (Kazanidis et al. 2016).

Within Irish and UK waters, coral colonies are observed in a variety of morphologies. Vad et al. (2017) characterised colony sizes and the proportion of live and dead *L. pertusa* skeletons using video footage from Remotely Operated Vehicle (ROV) transects at the Mingulay Reef Complex and Rockall Bank. Two distinct morphotypes were described: a 'cauliflower-shaped' morphotype which was more commonly observed at the Mingulay Reef Complex, or a more flattened 'bush-shaped' morphotype. In general, the proportion of living coral was positively correlated with whole colony size, and the live to dead ratio of coral framework never exceeded 0.27. While this ratio will vary across habitats, it further supports the importance of dead coral framework in biodiversity provision and could provide a useful metric for future consideration in monitoring and characterising cold-water coral ecosystems in Irish and UK waters (see Sect. 6.5). However, the dead coral framework is particularly vulnerable to ocean acidification and the associated shallowing of the aragonite saturation horizon (see Sect. 6.4). The clear visual contrast between white/pale living and grey/dark dead portions of the colonies gives a new way by which these habitats can be visually monitored over time, and existing data in Irish and UK waters provides a baseline to now build upon for this region.

6.3.2 Connectivity Between Cold-Water Coral Habitats

There is limited information on connectivity of cold-water coral reefs worldwide, and studies have been reliant upon genetic microsatellite analysis of populations from different locales (Morrison et al. 2011; Boavida et al. 2019). This can identify whether geographically separated cold-water coral habitats have corals that are potentially related (and how closely) or whether they are genetically distinct populations. This assessment of connectivity requires samples to be collected and analysed from different reefs using comparable techniques and genetic probes. Modelling approaches on coral larval dispersal are needed to complement genetic analysis to understand large-scale coral distribution, and how robust this may be in the face of environmental change (Ross et al. 2017).

Recent advances in experimental cold-water coral larval biology (Stromberg and Larsson 2017) have increased our understanding of the behaviour and competency time of *L. pertusa* larvae and highlighted how the dispersal of these sexually produced coral larvae and subsequent recruitment is driven by larval behaviour (e.g. their ability to migrate through the water column) in addition to local circulation patterns. Fox et al. (2016) built upon this to consider how larval dispersal can be through both subsurface and surface (wind driven Ekman Layer) drifting. This enables their distribution to be modelled with a sufficient knowledge of local currents and highlights that they could be very sensitive to future changing conditions (e.g. changes in ocean and atmospheric circulation). The modelling study of the coral larvae demonstrated dispersal consistent with the known circulation patterns (Fig. 6.2), with larvae released from the Mingulay Reef Complex reaching across the Scottish seas, the North Sea to Norway, and also towards Iceland (Fox et al. 2016). Within UK waters, certain sites were found to be critical for maintaining overall network connectivity, with Hatton Bank identified as being important for overall connectivity, and Rosemary Bank for connecting Northern and Western populations (where Northern populations include Rosemary Bank Seamount, Darwin Mounds, Wyville Thomson Ridge and the Faroe Shetlands, and Western populations include Hatton Bank and Logachev Mound sites). The ability to model the dispersal of these larvae enables us to consider whether certain populations are larvae 'donors' or 'receivers' and to assess how robust this relationship may be if environmental conditions change (Ross et al. 2017) (see Sect. 6.4).

While corals breed sexually and disperse their larvae to new locations (within their resident populations or to downstream habitats), they can also spread clonally (asexually). While this is difficult to observe in many ecosystems, due to the inability to visually differentiate between individual colonies and their ability to fuse with neighbouring, genetically distinct colonies (Hennige et al. 2014a), it can be observed clearly at Rockall Bank (Wilson 1979b). At that site, *L. pertusa* colonies were observed growing on the seabed in distinctive patches, with dead coral framework in the middle, and a 'ring' of live coral around the outside. It was reasoned that these 'Wilson Rings' are growing and repeatedly breaking, with the corals in the inner rings dying due to suboptimal conditions (likely a combination of reduced current velocities and reduced food supplies). These Wilson Rings demonstrate the process of how corals can prograde in the deep-sea, where single coral colonies can expand their range through continuous growth and breaking. These rings can go through various stages of development and this process can be seen in coral 'patches' up to 50 m across (Wilson 1979b).

6.4 Threats and Future Outlook

Threats to cold-water coral ecosystems are well documented and include both anthropogenic and environmental stressors. These stressors act on a global stage, but some of the impacts have been considered here specifically for the ecosystems around the Irish and UK waters. While not all stressors are included here, the longevity of the habitats, what these habitats will look like, and how they will interact remain critical areas to consider. These are broken down here into three main areas for discussion: (1) the habitat structure, (2) the biodiversity it can support, and (3) the sustained connectivity of these ecosystems.

6.4.1 Habitat Structure and Biodiversity Provision

The richness of species in cold-water coral habitats is directly supported by the complexity of the habitats themselves. In the previous section it was detailed how the complex 3-dimensional framework made by cold-water corals (comprising of both live and dead scleractinian coral) is crucial for biodiversity support. This habitat can be degraded either by physical damage, for example by trawling (Hall-Spencer et al. 2002), or by changes in environmental parameters (Hennige et al. 2020).

The Irish-UK margin hosts several important fishing grounds for bottom trawling fisheries, which has resulted in inevitable impacts on cold-water coral habitats in the area. Given their low relief and gentle slope, small mounds can easily be trawled over by large deep-water bottom trawling gear. The best-known example is probably that of the Darwin Mounds, where trawling impacts were reported as soon as the mounds were discovered (Wheeler et al. 2005b). This led to an emergency closure under the EU Common Fisheries Policy in 2003 (De Santo and Jones 2007), made permanent in 2004 (De Santo 2013). Unfortunately, evidence from Vessel Monitoring Systems seems to indicate that trawling activities intensified in the months before the closure, particularly in the Eastern Darwin Mounds (Davies et al. 2007). Still, the Darwin Mounds became the first deep-water Marine Protected Area (MPA) in the UK, and later also obtained the status of Special Area of Conservation under the EU Habitats Directive. Other examples of bottom trawling impacts on Irish and UK cold-water coral habitats include the so-called mini-mounds on the interfluves of the Dangaard and Explorer Canyons (part of the Whittard Canyon system) (Davies et al. 2014; Stewart et al. 2014). So far, no live framework-building corals have been discovered in that area, as the mounds are composed of coral rubble only. It is still unclear if coral demise was caused by Holocene climate and oceanographic variability, by bottom trawling, or by a combination of both (Collart 2019), but there is clear evidence of heavy bottom trawling in the area (Huvenne et al. 2011, 2016c), which will have had a destructive impact on any potential coral framework. The area is now part of the Canyons Marine Conservation Zone (UK Government (2009)). Bottom trawling impacts have also been reported from habitats such as Rockall Bank, where cold-water corals form smaller and more dispersed patches (Huvenne 2011; Howell et al. 2014a; Robert et al. 2014). Several MPAs and fisheries closures have been put in place in the area over the last decade, designated by the UK, Ireland or the North-East Atlantic Fisheries Commission. Their presence makes the evidence of trawling impacts just outside MPA boundaries even starker.

Because of their steep morphology, vertical cold-water coral reefs are *de facto* protected from direct (mechanical) bottom trawling impacts (Huvenne et al. 2011). However, they may still suffer from the excess turbidity created by trawling, particularly where the activity takes place on slopes directly above the steep cliffs (e.g. interfluves of submarine canyons, Wilson et al. 2015). Still, where cliffs are vertical or overhanging, the corals receive a certain level of protection, and the cliffs may create refugia for species sensitive to mechanical impact. It has been suggested that such vertical cold-water coral reefs may act as source populations for the potential recolonisation of areas that have been closed after impacts from bottom contact fisheries (Huvenne et al. 2011). The steep terrain is less of a deterrent to longline fisheries, though, and the high biodiversity and biomass may specifically attract longlining. Although longline fisheries are less destructive to coral habitats than bottom trawling (Pham et al. 2014a), impacts still include entangled and lost fishing gear, causing excessive littering and ghost fishing in the areas (Fig. 6.5).

As reported elsewhere around the world, recovery of cold-water coral habitats after bottom trawling impacts is slow and can take decades (Althaus et al. 2009; Williams et al. 2010). In 2011, the Darwin Mounds were re-visited for the first time since the trawling ban, and while only a few trawl marks were found, indicating good compliance, no clear recolonisation or regrowth of the main scleractinian coral species was observed. Another survey was carried out in 2019, again demonstrating that no further trawling was taking place, but

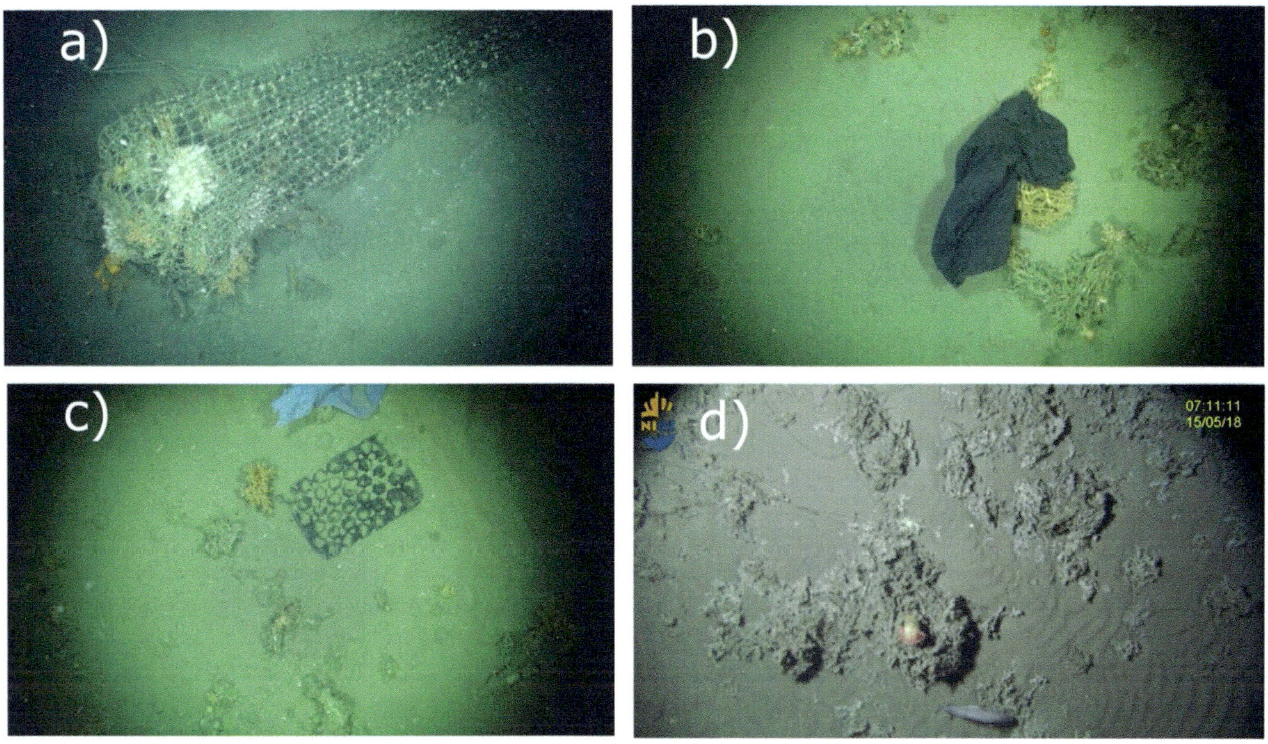

Fig. 6.5 Examples of litter in cold-water coral ecosystems. (**a**) discarded net draped around living coral, with evidence of abraded pieces in the surrounding sediment (Darwin Mounds), (**b**) a plastic bag caught on living coral (Darwin Mounds), (**c**) plastic debris next to coral (Darwin Mounds), (**d**) fishing wire observed in Whittard Canyon. Images **a**–**c** copyright National Oceanography Centre; image **d** is copyright NIOZ

equally that coral recovery was not obvious. However, some new coral recruits were found on settlement experiments left at the seabed 8 years before, raising hope that cold-water coral larvae are present in the area and that recolonisation is possible (Huvenne and Thornton 2020). The European ban on bottom trawling below 800 m water depth will further protect areas such as the Darwin Mounds, but other cold-water coral habitats such as the coral reefs on Rockall Bank are still vulnerable to the destructive impact of this fishing technique. Apart from the mechanical impact on the live coral colonies, bottom trawling also results in a reduction in the structural complexity of the habitat, which can have negative effects on the local and regional biodiversity. However, even if only limited structural complexity is left, a coral rubble habitat can still support a different community compared to the surrounding seabed (Pearman 2020). Hence, even if live coral is no longer present, formerly trawled cold-water coral habitats increase biodiversity and should be considered as part of conservation strategies.

Recent evidence has highlighted how ocean acidification can lead to increased porosity in dead coral skeletons no longer covered by coral soft tissue. Importantly, these increases in porosity occur at structurally critical places in the skeleton (Hennige et al. 2020). In instances where these cold-water coral habitats are below the aragonite saturation horizon (ASH) where aragonite becomes undersaturated, this can lead to weakening of the supporting framework structure of the habitats and physical habitat degradation from a complex system to a simpler one (Hennige et al. 2020). While this has been observed in *L. pertusa* cold-water coral habitats already living below the ASH [e.g. West coast of the USA (Hennige et al. 2020)], *L. pertusa* coral habitats in the Irish and UK waters are currently above the ASH. The potential threat of habitat degradation to these reefs then becomes linked to whether the ASH will shoal above them in the coming decades, and when. Typically, it is the deeper *L. pertusa* ecosystems that are more at risk of the ASH shoaling above them, as they are currently living closer to the ASH than shallower reefs. This means that the deeper *L. pertusa* cold-water coral ecosystems within Irish and UK waters are more at risk to a potential loss of structure. Importantly, this does not mean that these ecosystems will not persist, as live coral can continue to grow in aragonite undersaturated conditions (Hennige et al. 2020), but it will likely mean a change in the types or amounts of biodiversity it can support. A noteworthy point is that another major framework forming coral, *S. variabilis*, can be found at depths near or below the ASH in Irish and UK waters (Howell et al. 2014b) and has been documented below the ASH in other studies. The comparison of *L. pertusa* and *S. variabilis* habitat with respect to ocean acidification is therefore a key area to develop when longevity of these habitats is considered. Determining and predicting what changes in habitat structure would mean to biodiversity provision remains a priority for future work. Linked to this is further characterisation of the current habitat structure, and how much of the habitat is comprised of, and supported by, dead coral framework (e.g. Vad et al. 2017). Those habitats with a greater proportion of exposed dead coral framework could therefore be subject to a greater change in biodiversity provision than those mostly composed of live coral framework. Shallower cold-water coral ecosystems are less of risk of this as the ASH will take longer to shoal above them. The shallowest UK cold-water coral ecosystem, the Mingulay Reef Complex, is not expected to be subjected to aragonite undersaturated water (Findlay et al. 2013, 2014), and additionally has surface water tidally downwelled to it periodically (Davies et al. 2009), further ameliorating the risk of ocean acidification induced habitat loss (see Box 6.2).

Food provision, or Particulate Organic Carbon (POC) flux is crucial for the survival and growth of cold-water coral ecosystems as outlined above. Recent modelling studies (Sweetman et al. 2017) have highlighted that as the upper ocean warms more, a change in nutrient supply to phytoplankton could cause a change in the types that are exported to the seafloor and their transfer efficiency. While a general POC reduction is projected for Irish and UK benthic ecosystems (Sweetman et al. 2017), this is variable and region specific. To date, this has only been assessed on broad scales, and much higher resolution models are needed to investigate specific cold-water coral habitat areas. Quantifying potential POC loss would enable calculation of coral ecosystem accretion rates and ultimately allow assessment of potential biodiversity provision of future habitats.

6.4.2 Connectivity and Management

Recent modelling advances have highlighted how connected all the cold-water coral ecosystems are in Irish and UK waters (Fox et al. 2016; Ross et al. 2017), and how larval dispersal is strongly correlated with ocean circulation. Atmospheric driven changes in ocean circulation are projected to have an impact on larval dispersal and ultimately cold-water coral ecosystem connectivity (Fox et al. 2016). The North Atlantic Oscillation (NAO) is a key part of cold-water coral ecosystem connectivity, but this connectivity can be strong or weak, depending on variability in the NAO, driven by atmospheric sea-level pressure differences between the Arctic and subtropical Atlantic (Hurrell et al. 2003). Understanding how the NAO may change depending on climatic variability is a key consideration, as weakening of larval and habitat connectivity will increase the vulnerability of the ecosystems across the Irish and UK waters as a whole, by reducing the efficiency of larval dispersal across the wider region (Fox et al. 2016).

Linked to this is the question of how these ecosystems are managed in terms of Marine Protected Areas (MPAs) (Jackson et al. 2014). If connectivity may change in the future, then management efforts should consider whether certain MPA areas may become more important in ensuring future connectivity between regions. Fox et al. (2016) demonstrated that the Hatton Bank area is particularly important in dispersing larvae to different regions and that Rosemary Bank Seamount MPA is a vital stepping stone for MPA clusters in the North and West. As we refine further models and gain a greater understanding of the drivers and timescales of ocean circulation changes, we will be better able to predict which cold-water coral ecosystems and MPA areas may become more crucial in the future for sustained connectivity, and whether a different MPA approach is needed. Models outlining risk-based scenarios of impacts of ocean acidification to Northeast Atlantic coral reefs have highlighted that over 85% of reefs may be subjected to aragonite undersaturated water by 2060 and outline the case that marine management decisions should incorporate both managed (e.g. trawling) and unmanaged (e.g. warming and acidification) pressures (Jackson et al. 2014).

In addition to this, there is the question of whether man-made structures may provide localised populations that can contribute genetic material and larvae to the larger region (Henry et al. 2018). Cold-water corals are commonly found on oil and gas installations in the North Sea (Gass and Roberts 2006), and modelling approaches have demonstrated that the location of these installations (and cold-water coral populations) could create a highly connected anthropogenic system which would facilitate larval dispersal into the broader region, giving them a significant conservation significance. This could enhance marine connectivity in the MPAs (Boero et al. 2016) and could provide an example of how these installations could act as 'artificial reefs', supporting cold-water coral connectivity in the North Sea.

6.4.3 Pollution

Benthic samples and data from historic surveys carried out on the Mingulay Reef Complex were analysed to evaluate the presence of marine (micro) litter (La Beur et al. 2019). A total of 11% of benthic macrofauna sampled had ingested microplastics. Ingestion rates were highest in areas located in a topographic hollow along gentle sloping areas with strong currents, where it is possible that microplastic accumulation occurred. In the study by La Beur et al. (2019), most of the observed debris was identified as being related to fisheries. While this single study was not able to assign impacts of these microplastics to the organisms that ingested them, the proportion of samples that *had* ingested plastics indicates that this is an important area for further research. While microplastics may come from many sources and end up on Irish and UK reefs, macroplastic and ghost nets may also get entangled in coral habitat areas (Pham et al. 2014b; Huvenne et al. 2016a), potentially mechanically damaging or suffocating corals if flow is reduced to polyps (Fig. 6.5).

6.5 Future Directions

There are several key challenges for future conservation and management of cold-water coral habitats in Irish and UK waters, outlined below. These can be summarised into three broad categories: (1) fully mapping where these ecosystems occur within these waters, (2) understanding how robust these ecosystems are and the types of biodiversity they support, and (3) understanding how future stressors (environmental and anthropogenic) will impact this, and how we can conserve these vulnerable marine ecosystems.

6.5.1 Mapping

An ongoing challenge with mapping cold-water coral habitats is cost and time. Mapping has improved dramatically over the last decade and a half since the first cold-water coral ecosystems were characterised in Irish and UK waters, not least thanks to the Irish INFOMAR programme, that saw 100% of the Irish deep-water EEZ mapped with shipboard multibeam echosounders to a resolution of at least 25 m pixel size.

As technology improves, the resultant bathymetries have become more informative and indicative as to where corals may be growing (Dorschel et al. 2010). In addition to continuous developments in multibeam echosounder systems, new and more detailed mapping approaches are also created by the ever-expanding fleet of marine robots, including ROVs and AUVs (Huvenne et al. 2018). These robotic vehicles can be used for either acoustic mapping approaches (using multibeam echosounders, side-scan sonars, synthetic aperture sonars (e.g. Foubert et al. 2011, Huvenne et al. 2016a, Lim et al. 2018), or for optical mapping through photogrammetry and Structure-from-Motion techniques (Boolukos et al. 2019; Price et al. 2019; de Oliveira et al. 2021). The resolution of mapping efforts can therefore now be chosen ecosystem by ecosystem, although with a trade-off in area coverage. Advances in data analysis techniques, such as Object-Based Image Analysis for automated seafloor classification (Ismail et al. 2018, Conti et al. 2019) or Machine Learning approaches to automatically identify species in seabed video or photography (Purser et al. 2009), offer new possibilities for the future.

The increased use of marine autonomous survey vehicles offers an important new platform from which survey techniques could be applied to monitor deep-water marine protected areas in the future (Zelada Leon et al. 2020). As this technology develops and its use becomes more routine, this would offer an efficient way to map specific sections of the Irish and UK cold-water coral habitats in higher resolution over time. Such technology would also facilitate the characterisation of new coral areas; there are several areas in these waters where corals would be suspected to occur or are anecdotally known to occur. AUVs would provide an efficient method to survey these areas in the first instance.

Still, the extent of the Irish and UK waters is vast, and it is physically impossible to map the whole area at a high enough resolution to confirm cold-water coral presence or absence everywhere. The answer to this challenge lies in predictive coral distribution maps that can support marine spatial management, although the bathymetry resolution is critical for this, with high-resolution data recommended (Ross et al. 2015). This advances earlier work where decisions on area closures to protect habitats were based upon maps of recorded presence, and the ability to model where corals may be provides a potentially powerful tool to examine extents of listed habitats, supporting effective MPA design (Ross and Howell 2013). The classification of certain habitats in these instances is also important (Davies et al. 2015), and ensuring that the complexity of habitats (e.g. coral gardens) captured is critical, as robust maps and modelling is dependent upon agreed and reliable definitions of habitats (Bullimore et al. 2013).

6.5.2 Status of Habitats

Moving forward, the development of indicators for the assessment of cold-water coral ecosystem health status is a key challenge. There are significant anthropogenic and environmental stressors that could impact the 'health' and status of the Irish and UK cold-water coral ecosystems, both in terms of their extent and range, but also with regard to what biodiversity they can support. Recent projects have examined several of the cold-water coral ecosystems in Irish and UK waters, including the Mingulay Reef Complex, Rockall Bank, and Porcupine Seabight and suggested 24 indicators for their assessment of deep-sea environmental status in national waters and areas beyond national jurisdiction (Kazanidis et al. 2020; Orejas et al. 2020). Through the development of indicators for deep-sea health status, including identifying how associated biodiversity and biogeochemical cycles will change in the future, it will be possible to facilitate implementation of ambitious policies for the conservation of cold-water coral vulnerable marine ecosystems for the future (Kazanidis et al. 2020; Orejas et al. 2020).

6.5.3 Conserving Cold-Water Coral Ecosystems in Changing Ocean

Once we have a better understanding of where cold-water coral ecosystems are, their status, and the types of biodiversity they can support, conservation efforts could be tailored in the future. This will include understanding which cold-water coral ecosystems are at the highest risk of potential habitat degradation, and which are the most important in the connected network within Irish and UK waters (Fox et al. 2016; Ross et al. 2017). However, linked with understanding *which* ecosystems would be at the highest risk over the coming decades, improved monitoring and forecasting of environmental conditions are also needed. There are significant efforts underway with initiatives such as the Global Ocean Acidification Observation Network (GOA-ON) with the implementation of a North East Atlantic hub, which will ensure that for ocean acidification monitoring, consistent methodology is used, data is submitted to UN data centres, and that efforts are made to map and identify gaps in monitoring. Such platforms offer a way in which deep-sea monitoring (which is challenging) can be coordinated, ensuring that data is shared and can be used to increase the power of future forecasting. These examples themselves serve as a 'best practices' guide on how important environmental monitoring can be collected and shared to the communities that need them, to increase future understanding of how rapidly Irish and UK waters are changing, what risk this brings to cold-water coral ecosystems across these waters, and what this will mean for how we interact with them over the coming decades.

Box 6.1 The Challenger Mound

The Challenger Mound in the Belgica Mound Province, north-eastern Porcupine Seabight was the first and only giant carbonate mound so far to be drilled from summit to below the mound base (Ferdelman et al. 2006), producing a sedimentary record of mound initiation and growth up to the present day (Kano et al. 2007). The mound was drilled during IODP Expedition 307 to determine whether giant carbonate mounds were initiated over cold seeps and owed their stature to significant bacterial activity producing authigenic carbonates and indirectly fuelling cold-water coral growth as hypothesised by Hovland et al. (1998) and Henriet et al. (1998, 2002).

The 155 m of retrieved carbonate mound record showed the presence of cold-water corals (principally *Lophelia pertusa*) throughout the mound as bioclastic floatstones and rudstones with a micritic matrix with a

Box 6.1 (continued)

strong terrigenous input. This proved that these impressive structures were formed by successive phases of cold-water coral reef development over several million years. No significant evidence of methane-derived authigenic carbonates or gas seepage was detected (Ferdelman et al. 2006) ending the long running debate on mound development.

The mound initiated on a regional unconformity 2.6 million years ago (Kano et al. 2007) at the start of modern North Atlantic oceanographic settings and spread out over a wide area. The initial stages of mound growth saw a rapid expansion of the coral cover (both laterally and vertically, before sedimentation under the form of sandy contourites) begin to cover the surrounding terrain and the lower mound flanks (Huvenne et al. 2009b). Mound growth was then rapid until 1.5 million years ago when a major break in accumulation occurred, coinciding with the so-called Mid-Pleistocene Revolution, marking the start of major pan-European glaciations (Foubert and Henriet 2009; Thierens et al. 2010). Subsequent growth up to the present day was then much slower.

The record shows a strong climate control of reef development with coral dominated and sediment dominated intervals corresponding to interglacial and interstadial cyclicity (Thierens et al. 2013). Despite its punctuated nature, it provides one of the only records of early Pleistocene sedimentation on the Irish margin. The Challenger Mound also revealed glaciation in Ireland occurred nearly one million years earlier than previously thought, probably as mountain glaciation significantly south of the main European ice front at 2.6 million years ago (Thierens et al. 2012) (Fig. 6.6).

Box 6.2 The Mingulay Reef Complex

The nearest cold-water habitat to UK and Irish shores is the Mingulay Reef Complex (MRC). Located in the Sea of the Hebrides west of Scotland, the MRC lies south of the Little Minch in a passage between the Scottish mainland and the northern Inner and Outer Hebrides (Fig. 6.7). This complex comprises of several areas supporting reef mounds (Roberts et al. 2005) and was first mapped in 2003 (Roberts et al. 2009a). Since then, high-resolution microbathymetry of several mound areas within the complex has also been obtained (De Clippele et al. 2017) (Fig. 6.7). The MRC is the only known inshore *Lophelia pertusa*

(continued)

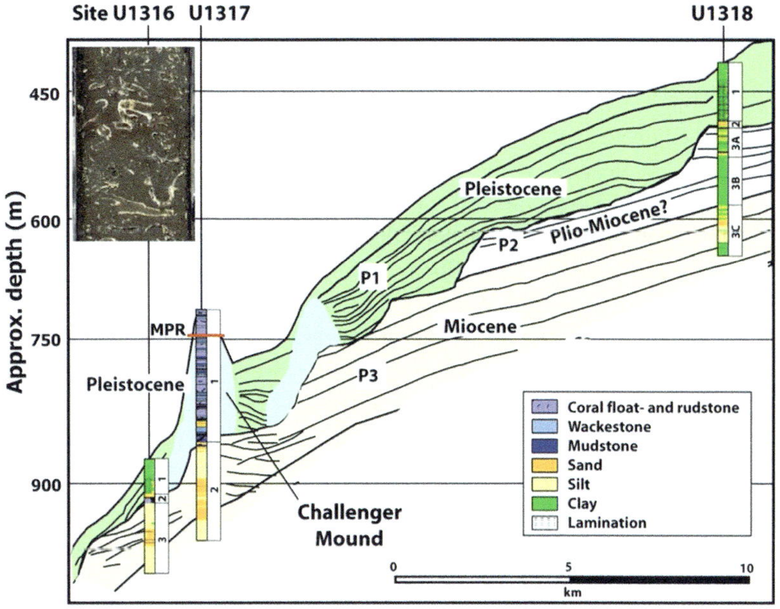

Fig. 6.6 Simplified lithostratigraphic summary of the 3 IODP EXP307 drill sites, projected on an interpreted high-resolution seismic profile across Challenger Mound (after Ferdelman et al. 2006; seismic interpretation after De Mol et al. 2002). Inset: photograph of core section containing coral rudstone

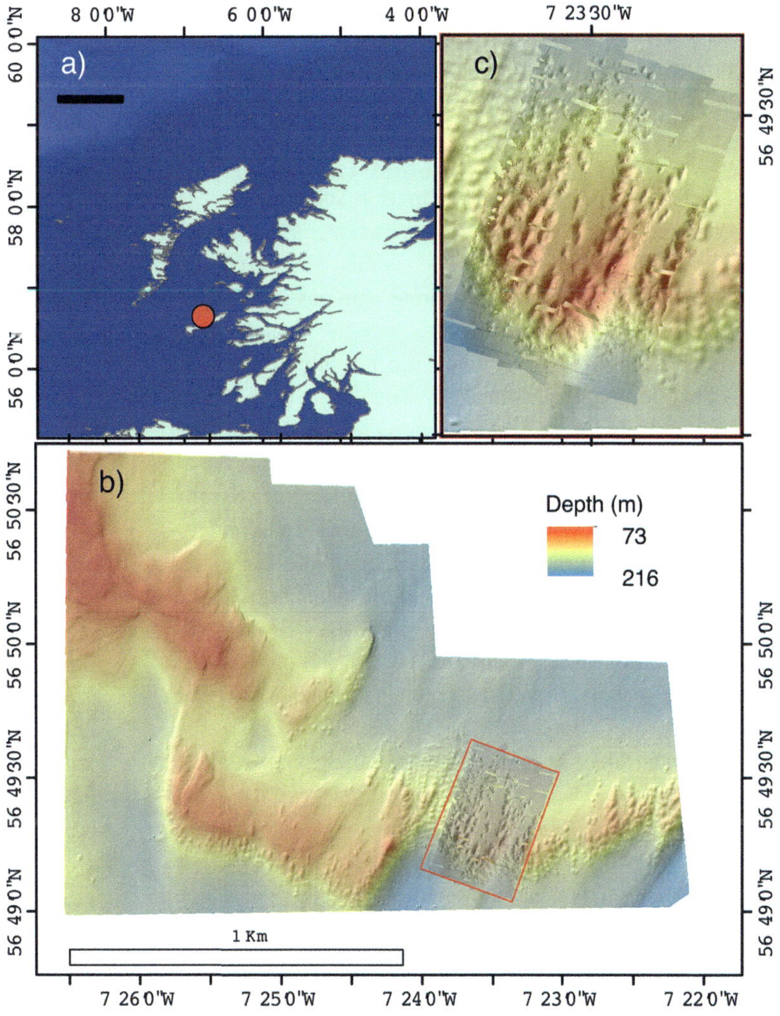

Fig. 6.7 (**a**) location of the Mingulay Reef Complex, (**b**) Mingulay Reef 01, (**c**) Microbathymetry of red box in (**b**) (based on De Clippele et al. 2017)

Box 6.2 (continued)

habitat in Irish-UK waters and is relatively shallow, starting at ~100 m and descending to ~260 m (De Clippele et al. 2017).

A key feature of the MRC, which facilitates its shallow inshore survival, is the tidally driven downwelling that periodically drives surface water down to the corals, in addition to periodic advection of high turbidity bottom waters (Davies et al. 2009). Aside from the MRC being the shallowest location of corals in Irish-UK waters, the periodic downwelling of surface waters means they experience substantial variability in environmental conditions. Research by Findlay et al. (2013) demonstrated that these downwellings resulted in significant shifts in temperature and carbonate chemistry over a 14-h period, potentially ameliorating effects of ocean acidification in the future, but also potentially increasing the risk of increased temperature impacts. Microplastics in addition to marine litter have been recorded at the MRC and provide the first evidence of ingestion of microplastics by organisms in COLD-WATER CORAL habitats in Irish-UK waters, highlighting the need to characterise this at other habitats (La Beur et al. 2019).

Box 6.3 Studying Cold-Water Corals in Submarine Canyons

The study of cold-water corals in complex environments such as submarine canyons is a major logistical challenge. The steep, occasionally overhanging walls of canyons do not allow sampling with traditional equipment lowered on a wire, while they also cause anomalous returns and high levels of noise in typically down-ward looking acoustic surveys conducted from a surface vessel. The irregular terrain makes it difficult and risky to use towed equipment, while strong currents may drag equipment off course, or may break moored equipment off its anchors (Paull et al. 2018). To overcome these challenges, researchers increasingly turn to novel technology—particularly marine robotics and autonomous systems. These include (a) Remotely Operated Vehicles or ROVs: tethered, neutrally buoyant robots that can operate within a certain radius from a ship; (b) Autonomous Underwater Vehicles or AUVs: pre-programmed, autonomous submarines; and (c) gliders: vehicles powered by a buoyancy engine that continuously profiles the water column (Hall et al. 2017; Huvenne et al. 2018). As part

Box 6.3 (continued)

of the ERC-funded CODEMAP project, the CODEMAP2015 expedition in Whittard Canyon saw the first simultaneous, coordinated operation of these three types of robotic vehicles in a single canyon, with the aim to map out the canyon environment, and its cold-water coral communities, in all their aspects and scales.

Combining a high-resolution multibeam system mounted either on an ROV or an AUV, with a ship-borne multibeam survey, enables mapping of the cold-water coral habitats in a nested fashion: high-resolution maps of limited extent can be nested within wide coverage, lower-resolution surveys conducted from the ship. However, to map out the vertical and overhanging sections of the canyon, the configuration of the multibeam system on the ROV or AUV has to be adapted. The first examples of vertical cold-water coral reef maps were obtained with an ROV, on which a multibeam echosounder was mounted in forward-looking mode rather than the usual downward-looking configuration (Huvenne et al. 2011). The ROV can then be moved parallel to the cliff at a constant distance. The set-up is illustrated in Fig. 6.8 (Huvenne et al. 2016b) and can be adapted for deployment from a hover-capable AUV as well. An alternative approach is to use a survey-class AUV, with the multibeam echosounder system rotated over a sufficient angle so it can map sideways (Robert et al. 2017). This is a good approach if the terrain is mixed, with both steep and less steep sections. The AUV's collision avoidance system will need to be adapted to track the canyon walls at a constant distance and to avoid any obstacles created by the irregular terrain. This solution typically enables more ground to be covered (thanks to the higher efficiency of survey-class AUVs), but in most cases requires the vehicle to keep a higher distance from the wall, resulting in a lower-resolution map. To obtain the finest level of detail, ROV-acquired video data can be processed with Structure-from-Motion photogrammetry resulting in 3D textured point cloud models of cold-water coral reefs and habitats in complex terrains (Robert et al. 2017; Fabri et al. 2019; Price et al. 2019). They enable quantification of live vs. dead coral coverage, and of habitat structural complexity caused by cold-water coral framework.

Acknowledgements The research reported in this chapter is based on countless research projects that studied the cold-water corals of the Irish and UK waters, and we would like to acknowledge the many researchers and authors who have dedicated many months at sea and in the lab

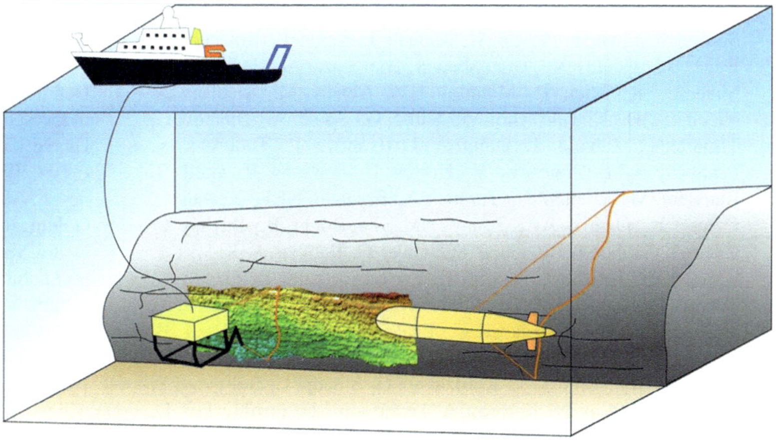

Fig. 6.8 Schematic illustrating the use of robotic technology to map steep and overhanging cold-water coral reefs, such as in submarine canyons, using multibeam echosounders mounted in either forward-looking mode on an ROV (left) or sideways-looking mode on an AUV (right)

exploring and studying the cold-water coral habitats in Irish and UK Waters. We acknowledge the generous support of international and national funding bodies who have resourced the research summarised here and would also like to thank the Irish and UK governments and the media who have worked to highlight and protect these ecosystems.

References

Addamo AM, Vertino A, Stolarski J, García-Jiménez R, Taviani M, Machordom A (2016) Merging scleractinian genera: the overwhelming genetic similarity between solitary desmophyllum and colonial lophelia. BMC Evol Biol 16. https://doi.org/10.1186/s12862-016-0654-8

Althaus F, Williams A, Schlacher TA, Kloser RJ, Green MA, Barker BA, Bax NJ, Brodie P, Schlacher-Hoenlinger MA (2009) Impacts of bottom trawling on deep-coral ecosystems of seamounts are long-lasting. Mar Ecol Prog Ser 397:279–294. https://doi.org/10.3354/meps08248

Appah JKM, Lim A, Harris K, Ramsay R, O'Reilly L, Wheeler AJ (2020) Are non-coral reef habitats as important to benthic diversity and composition as coral reef habitats in a submarine canyon? A rich analysis of organismal distribution controls in the upper Porcupine Bank Canyon, NE Atlantic. Front Mar Sci 7:571820. https://doi.org/10.3389/fmars.2020.571820

Aslam T, Hall RA, Dye SR (2018) Internal tides in a dendritic submarine canyon. Prog Oceanogr 169:20–32. https://doi.org/10.1016/j.pocean.2017.10.005

Barrett NJ, Hogan RI, Allcock AL, Molodtsova T, Hopkins K, Wheeler AJ, Yesson C (2020) Phylogenetics and mitogenome organisation in black corals (Anthozoa: Hexacorallia: Antipatharia): an order wide survey inferred from complete antipatharian mitochondrial genomes. Front Mar Sci 7:440. https://doi.org/10.3389/fmars.2020.00440

Bett BJ (2001) UK Atlantic margin environmental survey: introduction and overview of bathyal benthic ecology. Cont Shelf Res 21:917–956

Beyer A, Schenke HW, Klenke M, Niederjasper F (2003) High resolution bathymetry of the eastern slope of the Porcupine Seabight. Mar Geol 198(1–2):27–54

Boavida J, Becheler R, Choquet M, Frank N, Taviani M, Bourillet JF, Meistertzheim AL, Grehan A, Savini A, Arnaud-Haond S (2019) Out of the Mediterranean? Post-glacial colonization pathways varied among cold-water coral species. J Biogeogr 46:915–931. https://doi.org/10.1111/jbi13570

Boero F, Foglini F, Fraschetti S, Goriup P, Macpherson E, Planes S, Soukissian T, Adiloglu B, Cristens G, Delahaye C, Gregory I, Jacques S, Velkova S, Kontogianni A, Tourkolias C, Kollaras A, Damigos D, Skourtos M, Bianco L, Cesarini C, Aliani S, Angeletti L, Barbieri L, Beroldo R, Boero F, Falcieri F, Foglini F, Grande V, Griffa A, Langone L, Lazzari P, Lobato T, Miserocchi S, Palama D, Sclavo M, Solidoro C, Suaria G, Taviani M, Toncini A, Trincardi F, Vichi M, Chassanite A, Claudet J, Feral F, Marill L, Planes S, Villa E, Taquet C, Boissin E, Mangialajo L, Bottin L, Paravicini V, Baksay S, Neglia V, Legoff M, Dinoi A, Sahyoun R, Bitetto I, Lembo G, Spedicato MT, Aglieri G, Albano G, Albano O, Arena G, Asnaghi V, Bavestrello G, Belmonte G, Cecchi LB, Bevilacqua S, Bo M, Boscari E, Bulleri F, Calculli C, Capezzuto F, Capone A, Carlucci R, Carluccio A, Casagrandi R, Vietti RC, Chiantore M, Chimienti G, Ciardo L, Congiu L, Corselli C, Cutrona A, D'Onghia G, Dal Bello M, Mura ID, De Leo F, De Leo G, De Vito D, Di Bella M, Di Camillo CG, Di Mauro M, Fai S, Farella G, Forin N, Fraschetti S, Gatto M, Ghermandi A, Giangrande E, Giangrande E, Granata A, Gravili C, Guarnieri G, Guglielmo L, Ingrosso G, Laggini M, Liparoto A, Lusardi R, Maggi E, Maiorano P, Mangialajo L, Mariani P, Marino I, Hernandez LFM, Mastrototaro F, Melia P, Miglietta AM, Moscatello S, Mossa M, Nutricato R, Pansera M, Papetti C, Patarnello T, Paterno M, Petrillo A, Piraino S, Piscitelli A, Procaccini G, Rattray A, Rizzo L, Valdes LPR, Rossetto M, Savini A, Schiavina M, Scovazzi T, Sion L, Tamburello L, Tani I, Terlizzi A, Tessarolo C, Tursi A, Uttieri M, Vertino A, Zambianchi E, Zane L, Addamo AM, Andre M, Ballesteros E, Cebrian E, Coma R, Cisneros AG, Garcia-Rubies A, Guardiola M, Marquez VL, Machordom A, Macpherson E, Markandya A, Ojea E, Palacin C, Pascual M, Perez-Portela R, Ribes M, Schunter C, Serrano E, Templado J, Carreras C, Tomas F, Turon X, Uriz I, Wangensteen O, Christensen A, Mariani P, Kiorboe T, Sorensen TK, Balan S, Begun T, Briceag A, Dinu I, Dulu F, Ion G, Jipa E, Melinte M, Muresan M, Olteanu A, Opreanu G, Panin N, Popa A, Radan S, Secrieru DM, Stanica A, Teaca A, Vasiliu D, Bray L, Anastasopoulou A, Kokkali A, Assimakopoulou G, Voutsinas E, Kontoyannis H, Zervoudaki S, Pavlidou A, Axaopoloulos P, Karathanasi F, Drakopoulou P, Katsanevakis S, Krassakopoulou E, Kyriakidou H, Papadopoulou MAP, Papadopoulos A, Papadopoulos V, Papathanassiou V, Petros P, Prospathopoulos A, Reizopoulou S, Sakellariou D, Salomidi M, Soukissian T, Vassilopoulou C, Voukouvalas E, Zanou V, Berov D, Daskalov GM, Deyanova D, Dimov K, Georgieva I, Hiebaum G, Karamfilov V, Klayn S, Vassilev V, Vrabcheva Y, Macic V, Lucic D, Pestoric B, Alexandrov V, Boltacheva N, Bondareva L, Ignatyev S, Kolesnikova E, Kovardakov S, Milchakova N, Mironova N, Nevrova E, Pankeeva T, Revkov N, Sergeeva N, Shakhmatova O, Tereshchenko N, Zaika V, Deudero S, Fort J,

Vazquez-Luis M, Goni R, Aleksandrov B, Minicheva G, Ben Souissi J, Ghanem R, Rifi M, Zaouali J, Anton E, Boicenco L, Buga L, Coatu V, Golumbeanu M, Lazar L, Malciu V, Mateescu R, Micu D, Mihailov EM, Nenciu M, Nicolaev S, Niculescu D, Nita V, Oros A, Sarbu G, Spinu A, Tiganov G, Timofte F, Zaharia T, Dimitrov L, Ivanova P, Todorova V, Panayotova M, Doncheva V, Kotsev I, Prodanov B, Galil B, Ramdani M, Elkhiati N, Turan YA, Eryalcin NET, Keskin C, Ozturk B, Ozturk AA, Tonay AM, Topaloglu B, Baykal C, Cihan S, Salihoglu BF, Gucu AC, Guler G, Karahar A, Kideys A, Kideys AE, Kirezci C, Kocac M, Orek H, Ozturk D, Ozyurt G, Tezcan D, Yalciner A, Yalciner B, Zenginer A, Barabanov V, Malinovsky V, Motyzhova L, Beal S, Brown A, Creed B, Goriup P, Haynes T, Sanders N, Strike L, Arabidze M, Chitanava R, Kishinski L, Kordzakhia G, Smadja C, Bobylev LP, Davy R, Gnatyuk N, Korosov A, Pettersson LH, Pozdnyakov D, Sandven S, Gogoberidze G, Karlin L, Mamaeva M, Mikaelyan A, Shiganova T, Gagarin V, Stupnikova A, Sivkovich A, Lifanchuk A, Anohkina L, Rebetsky A, Melnikov I, Meerov M, Lukasheva T, Chasovnikov V, Podymova O, Kubrikova N, Lus V, Musaeva E, Zasko D, Pautova L, Dyakonov V, Bat L, Sezgin M, Soria PB, Rodriguez IL, Mendez MJ, Menendez M, Perez J, Rim BC, Kostyliov E, Leposhkin O, Lisovskyi R, Bratchenko O, Ivchenko E, Turina V, Kruglov A, Neprokin O, Kruschel C, Cizmek H, Kanski D, Belosevic M, Frleta D, Petricioli D, Schultz S, Beqiraj S, Nasto I, Zuna V, Hajderi E, Evans J, Schembri PJ, Terribile K, Arnold K, Braun D, Czybulka D, Homuth J, Schachtner E, Dobrev B, Staneva J, Novakova I, Dineva B, Ballerini T, Bellomo L, Berline L, Chan F, Costa A, Doglioli A, Molcard A, Petrenko A, Rammou AM, CoCoNet C (2016) CoCoNet: towards coast to coast networks of marine protected areas (from the shore to the high and deep sea), coupled with sea-based wind energy potential. Scires-It 6:1–95. https://doi.org/10.2423/i22394303v6Sp1

Boolukos C, Lim A, Wheeler AJ, Ramsay R (2019) Cold-water corals in decline - a temporal (4 year) species abundance and biodiversity appraisal of complete photomosaiced cold-water coral reef on the Irish Margin. Deep-Sea Res I 146:44–54

Brooke S, Ross SW (2014) First observations of the cold-water coral Lophelia pertusa in Mid-Atlantic canyons of the USA. Deep-Sea Res II 104:245–251. https://doi.org/10.1016/j.dsr2.2013.06.011

Buhl-Mortensen L, Mortensen PB (2004) Crustaceans associated with the deep-water gorgonian corals Paragorgia arborea (L., 1758) and Primnoa resedaeformis (Gunn., 1763). J Nat Hist 38(10):1233–1247. https://doi.org/10.1080/0022293031000155205

Bullimore RD, Foster NL, Howell KL (2013) Coral-characterized benthic assemblages of the deep Northeast Atlantic: defining "Coral Gardens" to support future habitat mapping efforts. ICES J Mar Sci 70:511–522

Collart T (2019) Cold-water corals in the Bay of Biscay: a chronicle of Holocene variability in the North Atlantic oceanic circulation. PhD, Ghent University, Ghent

Conti LA, Lim A, Wheeler AJ (2019) High resolution mappin gof a cold-water coral mound. Sci Rep 9:1016. https://doi.org/10.1038/s41598-018-37725-x

Cunningham SA (2020) Climate linked Atantic sector science & overturning in the Subpolar North Atlantic Programme Cruise Report RRS discovery cruise DY120, SAMS Internal Report 298, Scottish Association for Marine Science, Oban

Davies AJ, Roberts JM, Hall-Spencer J (2007) Preserving deep-sea natural heritage: emerging issues in offshore conservation and management. Biol Conserv 138:299–312. https://doi.org/10.1016/j.biocon.2007.05.011

Davies AJ, Duineveld GCA, Lavaleye MSS, Bergman MJN, van Haren H, Roberts JM (2009) Downwelling and deep-water bottom currents as food supply mechanisms to the cold-water coral Lophelia pertusa (Scleractinia) at the Mingulay Reef complex. Limnol Oceanogr 54(2):620–629

Davies JS, Howell KL, Stewart HA, Guinan JC, Golding N (2014) Defining biological assemblages (biotopes) of conservation interest in the submarine canyons of the South West Approaches (offshore United Kingdom) for use in marine habitat mapping. Deep-Sea Res II 104:208–229

Davies JS, Stewart HA, Narayanaswamy BE, Jacobs C, Spicer J, Golding N, Howell KL (2015) Benthic assemblages of the Anton Dohrn Seamount (NE Atlantic): defining deep-sea biotopes to support habitat mapping and management efforts with a focus on Vulnerable Marine Ecosystems. PLoS One 10(5):e0124815

De Clippele LH, Buhl-Mortensen P, Buhl-Mortensen L (2015) Fauna associated with cold water gorgonians and sea pens. Cont Shelf Res 105:67–78. https://doi.org/10.1016/j.csr.2015.06.007

De Clippele LH, Gafeira J, Robert K, Hennige S, Lavaleye MS, Duineveld GCA, Huvenne VAI, Roberts JM (2017) Using novel acoustic and visual mapping tools to predict the small-scale spatial distribution of live biogenic reef framework in cold-water coral habitats. Coral Reefs 36:255–268. https://doi.org/10.1007/s00338-016-1519-8

De Clippele LH, Huvenne VA, Molodtsova TN, Roberts JM (2019) The diversity and ecological role of non-scleractinian corals (Antipatharia and Alcyonacea) on scleractinian cold-water coral mounds. Front Mar Sci 6:184. https://doi.org/10.3389/fmars.2019.00184

De Froe E, Rovelli L, Glud RN, Maier SR, Duineveld G, Mienis F, Lavaleye M, Van Oevelen D (2019) Benthic oxygen and nitrogen exchange on a cold-water coral reef in the north-east Atlantic ocean. Front Mar Sci 6

De Mol B, Van Rensbergen P, Pillen S, Van Herreweghe K, Van Rooij D, McDonnell A, Huvenne V, Ivanov M, Swennen R, Henriet JP (2002) Large deep-water coral banks in the Porcupine Basin, southwest of Ireland. Mar Geol 188:193–231

De Mol B, Kozachenko M, Wheeler AJ, Alvares H, Henriet JP, Olu-Le Roy K (2007) Thérèse Mound: a case study of coral bank development in the Belgica Mound Province, Porcupine Seabight. Int J Earth Sci 96:103–120. https://doi.org/10.1007/s00531-005-0496-x

de Oliveira LM, Lim A, Conti LA, Wheeler AJ (2021) 3D classification of cold-water coral reefs: a comparison of classification techniques for 3D reconstructions of cold-water coral reefs and seabed. Front Mar Sci. https://doi.org/10.3389/fmars.2021.640713

De Santo EM (2013) The Darwin Mounds special area of conservation: implications for offshore marine governance. Mar Policy 41:25–32. https://doi.org/10.1016/j.marpol.2013.01.007

De Santo E, Jones PJS (2007) Offshore marine conservation policies in the North East Atlantic:emerging tensions and opportunities. Mar Policy 31:336–347. https://doi.org/10.1016/j.marpol.2006.10.001

Dickson RR, McCave IN (1986) Nepheloid layers on the continental slope west of Porcupine Bank. Deep-Sea Res 33(6):791–818

Dodds LA, Roberts JM, Taylor AC, Marubini F (2007) Metabolic tolerance of the cold-water coral Lophelia pertusa (Scleractinia) to temperature and dissolved oxygen. J Exp Mar Biol Ecol 349(2): 205–214. https://doi.org/10.1016/j.embe.2007.05.013

Dorschel B, Hebbeln D, Rueggeberg A, Dullo WC, Freiwald A (2005) Growth and erosion of a cold-water coral covered carbonate mound in the Northeast Atlantic during the Late Pleistocene and Holocene. Earth Planet Sci Lett 233:33–44

Dorschel B, Hebbeln D, Foubert A, White M, Wheeler AJ (2007a) Hydrodynamics and cold-water coral facies distribution related to recent sedimentary processes at Galway Mound west of Ireland. Mar Geol 244:184–195. https://doi.org/10.1016/j.margeo.2007.06.010

Dorschel B, Hebbeln D, Rueggeberg A, Dullo WC (2007b) Carbonate budget of a cold-water coral carbonate mound: Propeller Mound, Porcupine Seabight. Int J Earth Sci 96:73–83. https://doi.org/10.1007/s00531-005-0493-0

Dorschel B, Wheeler AJ, Huvenne VAI, de Haas H (2009) Cold-water coral mounds in an erosive environmental setting: TOBI side-scan sonar data and ROV video footage from the northwest Porcupine Bank, NE Atlantic. Mar Geol 264:218–229. https://doi.org/10.1016/j.margeo.2009.06.005

Dorschel B, Wheeler AJ, Monteys X, Verbruggen K (2010) Atlas of the deep-water seabed: Ireland. Springer, Dordrecht

Duineveld GCA, Lavaleye MSS, Bergman MJN, de Stigter HC, Mienis F (2007) Trophic structure of a cold-water coral mound community (Rockall Bank, NE Atlantic) in relation to the near-bottom particle supply and current regime. Bull Mar Sci 81(3):449–467

Duineveld GCA, Jeffreys RM, Lavaleye MSS, Davies AJ, Bergman MJN, Watmough T, Witbaard R (2012) Spatial and tidal variation in food supply to shallow cold-water coral reefs of the Mingulay Reef complex (Outer Hebrides, Scotland). Mar Ecol Prog Ser 444:97–115

Dullo WC, Flögel S, Rüggeberg A (2008) Cold-water coral growth in relation to the hydrography of the Celtic and Nordic European continental margin. Mar Ecol Prog Ser 371:165–176. https://doi.org/10.3354/meps07623

Eisele M, Hebbeln D, Wienberg C (2008) Growth history of a cold-water coral covered carbonate mound - Galway Mound, Porcupine Seabight, NE Atlantic. Mar Geol 253:160–169. https://doi.org/10.1016/j.margeo.2008.05.006

Fabri M-C, Vinha B, Allais AG, Bouhier ME, Dugornay O, Gaillot A, Arnaubec A (2019) Evaluating the ecological status of cold-water coral habitats using non-invasive methods: an example from Cassidaigne canyon, northwestern Mediterranean Sea. Prog Oceanogr 178:102172. https://doi.org/10.1016/j.pocean.2019.102172

Ferdelman TG, Kano A, Williams T, Henriet JP, the IODP Expedition 307 Scientists (2006) Proceedings of the IODP Expedition 307. Modern carbonate mounds: Porcupine drilling. IODP Management International, Washington, DC. https://doi.org/10.2204/iodp.proc.307.102.2006

Findlay HS, Arioli Y, Moreno Navas J, Hennige SJ, Wicks LC, Huvenne VAI, Woodward EMS, Roberts JM (2013) Tidal downwelling and implications for the carbon biogeochemistry of cold-water corals in relation to future ocean acidification and warming. Glob Chang Biol 19:2708–2719. https://doi.org/10.1111/gcb.12256

Findlay HS, Hennige SJ, Wicks LC, Navas JM, Woodward EMS, Roberts JM (2014) Fine-scale nutrient and carbonate system dynamics around cold-water coral reefs in the northeast Atlantic. Sci Rep 4:03671. https://doi.org/10.1038/srep.03671

Foubert A, Henriet JP (2009) Nature and significance of the recent carbonate mound record: the Mound Challenger Code, vol 126. Lecture notes in Earth Sciences. Springer, Heidelberg

Foubert A, Huvenne VAI, Wheeler AJ, Kozachenko M, Opderbecke J, Henriet JP (2011) The Moira Mounds: small cold-water coral mounds in the Porcupine Seabight, NE Atlantic: Part B - evaluating the impact of sediment dynamics through high-resolution ROV-borne bathymetric mapping. Mar Geol 282:65–78. https://doi.org/10.1016/j.margeo.2011.02.008

Fox AD, Henry LA, Corne DW, Roberts JM (2016) Sensitivity of marine protected area network connectivity to atmospheric variability. R Soc Open Sci 3(11):160494. https://doi.org/10.1098/rsos.160494

Frederiksen R, Jensen A, Westerberg H (1992) The distribution of the scleractinian coral Lophelia pertusa around the Faroe Islands and the relation to internal tidal mixing. Sarsia 77:157–171

Gass SE, Roberts JM (2006) The occurrence of the cold-water coral Lophelia pertusa (Scleractinia) on oil and gas platforms in the North Sea: colony growth, recruitment and environmental controls on distribution. Mar Pollut Bull 52:549–559

Gori A, Orejas C, Madurell T, Bramanti L, Martins M, Quintanilla E, Marti-Puig P, Lo Iacono C, Puig P, Requena S, Greenacre M, Gili J-M (2013) Bathymetrical distribution and size structure of cold-water coral populations in the Cap de Creus and Lacaze-Duthiers canyons (northwestern Mediterranean). Biogeosciences 10:2049–2060. https://doi.org/10.5194/bg-10-2049-2013

Hall RA, Aslam T, Huvenne VAI (2017) Partly standing internal tides in a dendritic submarine canyon observed by an ocean glider. Deep Sea Res I 126:73–84. https://doi.org/10.1016/j.dsr.2017.05.015

Hall-Spencer J, Allain V, Helge Fosså JH (2002) Trawling damage to Northeast Atlantic ancient coral reefs. Proc R Soc Lond B 269:507–511

Hansen B, Østerhus S (2000) North Atlantic–Nordic seas exchanges. Prog Oceanogr 45:109–208

Hennige S, Morrison CL, Form AU, Busher J, Kamenos NA, Roberts JM (2014a) Self-recognition in corals facilitates deep-sea habitat engineering. Sci Rep 4:6782. https://doi.org/10.1038/srep06782

Hennige SJ, Wicks LC, Kamenos NA, Bakker DCE, Findlay HS, Dumousseaud C, Roberts JM (2014b) Short-term metabolic and growth responses of the cold-water coral Lophelia pertusa to ocean acidification. Deep Sea Res II 99:27–35

Hennige S, Wolfram U, Wickes L, Murray F, Roberts JM, Kamenos NA, Schofield S, Groetsch A, Spiesz EM, Aubin-Tam ME, Etnoyer PJ (2020) Crumbling reefs an cold-water coral habitat loss in a future ocean: evidence of "coraloporosis" as an indicator of habitat integrity. Front Mar Sci 7:668. https://doi.org/10.3389/fmars.2020.00668

Henriet JP, Guidard S, Team the ODP "Proposal 573" Team (2002) Carbonate mounds as a possible example for microbial activity in geological processes. In: Wefer G, Billet D, Hebbeln D, Jørgensen BB, Schlüter M, van Weering T (eds) Ocean margin systems. Springer, Heidelberg, pp 439–455

Henriet JP, De Mol B, Pillen S, Vanneste M, Van Rooij D, Versteeg W, Croker PF, Shannon PM, Unnithan V, Bouriak S, Chachkine P, the Porcupine-Belgica '97 shipboard party (1998) Gas hydrate crystals may help build reefs. Nature 391:648–649

Henry LA, Roberts JM (2007) Biodiversity and ecological composition of macrobenthos on cold-water coral mounds and adjacent off-mound habitat in the bathyal Porcupine Seabight. Deep-Sea Res I 54:654–672. https://doi.org/10.1016/j.dsr.2007.01.005

Henry LA, Navas JM, Hennige SJ, Wicks LC, Vad J, Roberts JM (2013) Cold-water coral reef habitats benefit recreationally valuable sharks. Biol Conserv 161:67–70. https://doi.org/10.1016/j.biocon.2013.03.002

Henry LA, Stehmann MFW, De Clippele L, Findlay HS, Golding N, Roberts JM (2016) SEamount egg-laying grounds of the deep-water skate Bathyraja richardsoni. J Fish Biol 89(2):1473–1481. https://doi.org/10.1111/jfb.13041

Henry LA, Mayorga-Adame CG, Fox AD, Polton JA, Ferris JS, McLellan F, McCabe C, Kutti T, Roberts JM (2018) Ocean sprawl facilitates dispersal and connectivity of protected species. Sci Rep 8:11346. https://doi.org/10.1038/s41598-018-29575-4

Hogan RI, Hopkins K, Wheeler AJ, Allcock AL, Yesson C (2019) Novel diversity in mitochondrial genomes of deep-sea Pennatulacea (Cnidaria: Anthozoa: Octocorallia). Mitochondrial DNA Part A 30(6):764–777. https://doi.org/10.1080/24701394.2019.1634699

Holliday PN, Pollard RT, Read JF, Leach H (2000) Watre mass properties and fluxes in the Rockall Trough, 1975-1998. Deep-Sea Res I 47:1303–1332

Houpert L, Cunningham S, Fraser N, Johnson C, Holliday NP, Jones S et al (2020) Observed variability of the North Atlantic Current in the Rockall Trough from 4 years of mooring measurements. J Geophys Res Oceans 125:e2020JC016403. https://doi.org/10.1029/2020JC016403

Hovland M, Croker PF, Martin M (1994) Fault-associated seabed mounds (carbonate knolls?) off western Ireland and north-west Australia. Mar Pet Geol 11(2):232–246

Hovland M, Mortensen PB, Brattegard T, Strass P, Rokoengen K (1998) Ahermatypic coral banks off mid-Norway: evidence for a link with seepage of light hydrocarbons. PALAIOS 13:189–200

Howell KL, Huvenne V, Piechaud N, Robert K, Ross RE (2014a) Analysis of biological data from the JC060 survey of areas of conservation interest in deep waters off north and west Scotland. JNCC Report, vol 528. Peterborough

Howell KL, Bullimore RD, Foster NL (2014b) Qualtiy assurance in the identification of deep-sea taxa from vidfeo and image analysis: response to Henry and Roberts. Author Notes. ICES J Mar Sci 71: 899–906

Hurrell JW, Kushnir Y, Ottersen G, Visbeck M (2003) The North Atlantic Oscillation: climatic significance and environmental impact. Geophysical Monograph Series. American Geophysical Union, Washington, DC

Huthnance JM (1986) The Rockall slope current and shelf-edge processes. Proc R Soc Edinb 88B:83–101

Huvenne VAI (2011) RRS James Cook Cruise 60, 09 May-12 June 2011. Benthic habitats and the impact of human activities in Rockall Trough, on Rockall Bank and in Hatton Basin. Cruise Reports. National Oceanography Centre, Southampton, Southampton

Huvenne VAI, Thornton B (2020) RRS Discovery Cruise 108-109. CLASS - Climate-linked Atlantic System Science Darwin Mounds Marine Protected Area habitat monitoring, BioCAM first equipment trials and BLT-Recipes pilot study. Cruise Reports, vol 66. National Oceanography Centre, Southampton

Huvenne VAI, De Mol B, Henriet JP (2003) A 3D seismic study of the morphology and spatial distribution of buried coral banks in the Porcupine Basin, SW of Ireland. Mar Geol 198(1–2):5–25

Huvenne VAI, Beyer A, de Haas H, Dekindt K, Henriet JP, Kozachenko M, Olu-Le Roy K, Wheeler AJ, the TOBI/Pelagia 197 and CARACOLE cruise participants (2005) The seabed appearance of different coral bank provinces in the Porcupine Seabight, NE Atlantic: results from sidescan sonar and ROV seabed mapping. In: Freiwald A, Roberts JM (eds) Cold-water corals and ecosystems. Springer, Heidelberg, pp 535–569

Huvenne VAI, Bailey WR, Shannon P, Naeth J, di Primio R, Henriet JP, Horsfield B, de Haas H, Wheeler A, Olu-Le Roy K (2007) The Magellan mound province in the Porcupine Basin. Int J Earth Sci 96:85–101. https://doi.org/10.1007/s00531-005-0494-z

Huvenne VAI, Masson DG, Wheeler AJ (2009a) Sediment dynamics of a sandy contourite: the sedimentary context of the Darwin cold-water coral mounds, Northern Rockall Trough. Int J Earth Sci 98(4): 865–884. https://doi.org/10.1007/s00531-008-0312-5

Huvenne VAI, Van Rooij D, De Mol B, Thierens M, O'Donnell R, Foubert A (2009b) Sediment dynamics and palaeo-environmental context at key stages in the Challenger cold-water coral mound formation: clues from sediment deposits at the mound base. Deep-Sea Res I Oceanogr Res Pap 56(12):2263–2280. https://doi.org/10.1016/j.dsr.2009.08.003

Huvenne VAI, Tyler PA, Masson DG, Fisher EH, Hauton C, Hühnerbach V, Le Bas TP, Wolff GA (2011) A picture on the wall: innovative mapping reveals cold-water coral refuge in submarine canyon. PLoS One 6(12):e28755. https://doi.org/10.1371/journal.pone.0028755

Huvenne VAI, Bett BJ, Masson DG, Le Bas TP, Wheeler AJ (2016a) Effectiveness of a deep-sea cold-water coral Marine Protected Area, following eight years of fisheries closure. Biol Conserv 200:60–69. https://doi.org/10.1016/j.biocon.2016.05.030

Huvenne VAI, Georgiopoulou A, Chaumillon L, Lo Iacono C, Wynn RB (2016b) Novel method to map the morphology of submarine landslide headwall scarps using remotely operated vehicles. In: Lamarche G, Mountjoy J, Bull S et al (eds) Submarine mass movements and their consequences, 7th International Symposium, vol 41. Advances in natural and technological hazards research. Springer, Heidelberg, pp 135–144. https://doi.org/10.1007/978-3-319-20979-1_13

Huvenne VAI, Wynn RB, Gales JA (2016c) RRS James Cook Cruise 124-125-126. CODEMAP2015: habitat mapping and ROV vibrocorer trials around Whittard Canyon and Haig Fras. Cruise Reports, vol 36. National Oceanography Centre, Southampton

Huvenne VAI, Robert K, Marsh L, Lo Iacono C, Le Bas TP, Wynn RB (2018) ROVs and AUVs. In: Micallef A, Krastel S, Savini A (eds) Submarine geomorphology. Springer, Heidelberg, pp 93–108

Ismail K, Huvenne V, Robert K (2018) Quantifying spatial heterogeneity in submarine canyons. Prog Oceanogr 169:181–198. https://doi.org/10.1016/j.pocean.2018.03.006

Jackson EL, Davies AJ, Howell KL, Kershaw PJ, Hall-Spencer JM (2014) Future-proofing marine protected area networks for cold water coral reefs. ICES J Mar Sci 71(9):2621–2629

Johnson MP, White M, Wilson A, Würzberg L, Schwabe E, Folch H, Allcock L (2013) A vertical wall dominated by Acesta excavata and Neopycnodonte zibrowii, part of an undersampled group of deep-sea habitats. PLoS One 8(11):e79917. https://doi.org/10.1371/journal.pone.0079917

Kano A, Ferdelman TG, Williams T, Henriet JP, Ishikawa T, Kawagoe N, Takashima C, Kakizaki Y, Abe K, Sakai S, Browning E, Li X, the IODP Exp.307 Scientists (2007) Age constraints on the origin and growth history of a deep-water coral mound in the northeast Atlantic drilled during Integrated Ocean Drilling Program Expedition 307. Geology 35(11):1051–1054. https://doi.org/10.1130/G23917A.1

Kazanidis G, Witte UFM (2016) The trophic structure of Spongosorites coralliophaga-coral rubble communities at two northeast Atlantic cold-water coral reefs. Mar Biol Res 12:932–947

Kazanidis G, Henry LA, Roberts JM, Witte U (2016) Biodiversity of Spongosorites coralliophaga (Stephens, 1915) on coral rubble at two contrasting cold-water coral reef settings. Coral Reefs 35:193–208

Kazanidis G, van Oevelen D, Veuger B, Witte UFM (2018) Unravelling the versatile feeding and metabolic strategies of the cold-water ecosystem engineer Spongosorites coralliophaga (Stephens, 1915). Deep Sea Res I 141:71–82

Kazanidis G, Orejas C, Borja A, Kenchington E, Henry LA, Callery O, Carreiro-Silva M, Egilsdottir H, Giacomello E, Grehan A, Menot L, Morato T, Ragnarsson SA, Rueda JL, Stirling D, Stratmann T, van Oevelen D, Palialexis A, Johnson D, Roberts JM (2020) Assessming the environmental status of selected North Atlantic deep-sea ecosystems. Ecol Indic 119:106624. https://doi.org/10.1016/j.ecolind.2020.106624

Kenyon NH, Ivanov MK, Akhmetzhanov AM (eds) (1998) Cold water carbonate mounds and sediment transport on the Northeast Atlantic Margin, vol 52. IOC Technical Series. UNESCO, Paris

Kenyon NH, Akhmetzhanov AM, Wheeler AJ, van Weering TCE, de Haas H, Ivanov MK (2003) Giant carbonate mud mounds in the Southern Rockall Trough. Mar Geol 195:5–30

Kiriakoulakis K, Bett BJ, White M, Wolff GA (2004) Organic biogeochemistry of the Darwin Mounds, a deep-water coral ecosystem, of the NE Atlantic. Deep-Sea Res I 51:1937–1954

Kiriakoulakis K, Freiwald A, Fisher E, Wolff GA (2007) Organic matter quality and supply to deep-water coral/mound systems of the NW European Continental Margin. Int J Earth Sci 96(1):159–170

La Beur L, Henry LA, Kazanidis G, Hennige S, McDonald MA, Shaver MP, Roberts JM (2019) Baseline assessment of marine litter and microplastic ingestion by cold-water coral reef benthos at the East Mingulay Marine Protected Area (Sea of Hebrides, Western Scotland). Front Mar Sci 6:80. https://doi.org/10.3389/fmars.2019.00080

Le Danois E (1948) Les profondeurs de la mer. Payot, Paris

Lim A, Huvenne VAI, Vertino A, Spezzaferri S, Wheeler AJ (2018) New insights on coral mound development from groundtruthed high-resolution ROV-mounted multibeam imaging. Mar Geol 403: 225–237. https://doi.org/10.1016/j.margeo.2018.06.006

Lim A, Wheeler AJ, Price DM, O'Reilly L, Harris K, Conti L (2020a) Influence of benthic currents on cold-water coral habitats in the upper Porcupine Bank Canyon, NE Atlantic: a combined benthic monitoring an d3D photogrammetric investigation. Sci Rep 10: 19433. https://doi.org/10.1038/s41598-020-76446-y

Lim A, Wheeler AJ, Price DM, O'Reilly L, Harris K, Conti L (2020b) Ifluence of benthic currents on cold-water coral habitats in the upper Porcupine Bank Canyon, NE Atlantic: a combined benthic monitoring an d3D photogrammetric investigation. Sci Rep 10:19433. https://doi.org/10.1038/s41598-020-76446-y

Maier C, Hegeman J, Weinbauer MG, Gattuso JP (2009) Calcification of the cold-water coral Lophelia pertusa under ambient and reduced pH. Biogeosciences 6(8):1671–1680. https://doi.org/10.5194/bg-6-1671-1009

Masson DG, Bett BJ, Billett DSM, Jacobs CL, Wheeler AJ, Wynn RB (2003) The origin of deep-water, coral-topped mounds in the northern Rockall Trough, Northeast Atlantic. Mar Geol 194:159–180

Mienert J, Weaver PP (2003) European margin sediment dynamics: sidescan sonar and seismic images. Springer, Heidelberg

Mienis F, van Weering T, de Haas H, de Stigter H, Huvenne VAI, Wheeler AJ (2006) Carbonate mound development at the SW Rockall Trough margin based on high resolution TOBI and seismic recording. Mar Geol 233(1–4):1–19

Mienis F, de Stigter HC, White M, Duineveld G, de Haas H, van Weering TCE (2007) Hydrodynamic controls on cold-water coral growth and carbonate mound development at the SW and SE Rockall Trough Margin, NE Atlantic Ocean. Deep-Sea Res I 54: 1655–1674. https://doi.org/10.1016/j.dsr.2007.05.013

Mienis F, van der Land C, de Stigter HC, Van De Vorstenbosch M, De Haas H, Richter T, van Weering TCE (2009) Sediment accumulation on a cold-water carbonate mound at the Southwest Rockall Trough margin. Mar Geol 265:40–50

Mohn C, Rengstorf A, White M, Duineveld GCA, Mienis F, Soetaert K, Grehan A (2014) Linking benthic hydrodynamics and cold-water coral occurrences: a high-resolution model study at three cold-water coral provinces in the NE Atlantic. Prog Oceanogr 122:92–104. https://doi.org/10.1016/j.pocean.2013.12.003

Morris KJ, Tyler PA, Masson DG, Huvenne VAI, Rogers AD (2013) Distribution of cold-water corals in the Whittard Canyon, NE Atlantic. Deep-Sea Res II 92:136–144. https://doi.org/10.1016/j.dsr2.2013.03.036

Morrison CL, Ross SW, Nizinski MS, Brooke S, Jarnegren J, Waller RG, Johnson RL, King TL (2011) Genetic discontinuity among regional populations of Lophelia pertusa in the North Atlantic Ocean. Conserv Genet 12(3):713–729. https://doi.org/10.1007/s10592-010-0178-5

Mortensen PB, Hovland M, Brattegard T, Farestveit R (1995) Deep water bioherms of the scleractinian coral Lophelia pertusa (L.) at 64° N on the Norwegian shelf: structure and associated megafauna. Sarsia 80:145–158

New A, Smythe-Wright D (2001) Aspects of the circulation in the Rockall Trough. Cont Shelf Res 21:777–810

New AL, Barnard S, Herrmann P, Molines J-M (2001) On the origin and pathway of the saline inflow to the Nordic Seas: insights from models. Prog Oceanogr 48:255–287

Orejas C, Kenchington E, Rice J, Kazanidis G, Palialexis A, Johnson D, Gianni M, Danovaro R, Roberts JM (2020) Towards a common approach to the assessment of the environmental status of deep-sea ecosystems in areas beyond national jurisdiction. Mar Policy 121: 104182. https://doi.org/10.1016/j.marpol.2020.104182

Paull CK, Talling PJ, Maier KL, Parsons D, Xu J, Caress DW, Gwiadza R, Lundsten EM, Anderson K, Barry JP, Chaffey M, O'Reilly T, Rosenberger KJ, Gales JA, Kieft B, McGann M, Simmons SM, McCann M, Sumner EJ, Clare MA, Cartigny MJ (2018) Powerful turbidity currents driven by dense basal layers. Nat Commun 9:4114. https://doi.org/10.1038/s41467-018-06254-6

Pearman TRR (2020) Cold-water coral habitat mapping in submarine canyons. PhD, University of Southampton, Southampton

Pham CK, Diogo H, Menezes G, Porteiro F, Braga-Henriques A, Vandeperre F, Morato T (2014a) Deep-water longline fishing has reduced impact on Vulnerable Marine Ecosystems. Sci Rep 4:04837. https://doi.org/10.1038/srep04837

Pham CK, Ramirez-Llodra E, Alt CHS, Amaro T, Bergman M, Canals M, Company JB, Davies J, Duineveld G, Galgani F, Howell KL, Huvenne VAI, Isidro E, Jones DOB, Lastras G, Morato T, Gomes-Pereira JN, Purser A, Stewart H, Tojera I, Tubau X, Van Rooij D, Tyler PA (2014b) Marine litter distribution and density in European seas, from the shelves to deep basins. PLoS One 9(4): e95839. https://doi.org/10.1371/journal.pone.0095839

Pingree RD, Le Cann B (1989) Celtic and Armorican slope and shelf residual currents. Prog Oceanogr 23:303–338

Price DM, Robert K, Callaway A, Lo Iacono C, Hall RA, Huvenne VAI (2019) Using 3D photogrammetry from ROV video to quantify cold-water coral reef structural complexity and investigate its influence on biodiversity and community assemblage. Coral Reefs 38:1007–1021. https://doi.org/10.1007/s00338-019-01827-3

Purser A, Bergmann M, Lundalv T, Ontrup J, Nattkemper TW (2009) Use of machine-learning algorithms for the automated detection of cold-water coral habitats: a pilot study. Mar Ecol Prog Ser 397:241–251. https://doi.org/10.3354/meps08154

Robert K, Jones DAB, Huvenne VAI (2014) Megafaunal distribution and biodiversity in a heterogeneous landscape: the iceberg-scoured Rockall Bank, NE Atlantic. Mar Ecol Prog Ser 501:67–88. https://doi.org/10.3354/meps10677

Robert K, Huvenne VAI, Georgiopoulou A, Jones DOB, Marsh L, Carter GDO, Chaumillon L (2017) New approaches to high-resolution mapping of marine vertical structures. Nat Sci Rep 7: 9005. https://doi.org/10.1038/s41598-017-09382-z

Robert K, Jones DAB, Georgiopoulou A, Huvenne VAI (2019) Cold-water coral assemblages on vertical walls from the Northeast Atlantic. Divers Distrib 26(3):284–298. https://doi.org/10.1111/ddi.13011

Roberts JM, Brown CJ, Long D, Bates CR (2005) Acoustic mapping using a multibeam echosounder reveals cold-water coral reefs and surrounding habitats. Coral Reefs 24(4):657–669. https://doi.org/10.1007/s00338-005-0049-6

Roberts JM, Wheeler AJ, Freiwald A (2006) Reefs of the deep: the biology and geology of cold-water coral ecosystems. Science 312(5773):543–547

Roberts JM, Henry LA, Long D, Hartley JP (2008) Cold-water coral reef frameworks, megafaunal communities and evidence for coral carbonate mounds on the Hatton Bank, north east Atlantic. Facies 54: 297–316. https://doi.org/10.1007/s10347-008-0140-x

Roberts JM, Davies AJ, Henry LA, Dodds LA, Duineveld GCA, Lavaleye MSS, Maier C, van Soest RWM, Bergman MJN, Huhnerbach V, Huvenne VAI, Sinclair DJ, Watmough T, Long D, Green SL, van Haren H (2009a) Mingulay reef complex: an interdisciplinary study of cold-water coral habitat, hydrography and biodiversity. Mar Ecol Prog Ser 397:139–151

Roberts JM, Wheeler AJ, Freiwald A, Cairns SD (2009b) Cold-water corals: the biology and geology of deep-sea coral habitats. Cambridge University Press, Cambridge

Ross RE, Howell KL (2013) Use of predictive habitat modelling to assess the distribution and extent of the current protection of 'listed' deep-sea habitats'. Divers Distrib 19(4):433–445

Ross SW, Quattrini AM (2007) The fish fauna associated with deep coral banks off the southeastern United States. Deep-Sea Res I 54: 975–1007. https://doi.org/10.1016/j.dsr.2007.03.010

Ross LK, Ross RE, Stewart HA, Howell KL (2015) The influence of data resolution on predicted distribution and estimates of extent of current protection of three 'listed' deep-sea habitats. PLoS One 10(10):e0140061

Ross RE, Nimmo-Smith WAM, Howell KL (2017) Towards 'ecological coherence': assessing larval dispersal within a network of existing Marine Protected Areas. Deep Sea Res I 126:128–138. https://doi.org/10.1016/j.dsr.2017.06.004

Rueggeberg A, Dullo WC, Dorschel B, Hebbeln D (2007) Environmental changes and growth history of a cold-water carbonate mound (Propeller Mound, Porcupine Seabight). Int J Earth Sci 96:57–72. https://doi.org/10.1007/s00531-005-0504-1

Schultz K, Soetaert K, Mohn C, Korte L, Mienis F, Duineveld G, Van Oevelen D (2023) Linking large-scale circulation patterns to the distribution of cold-water corals along the eastern Rockall Bank (northeast Atlantic). J Mar Syst 212:103456

Shannon PM, Jacob AWB, O'Reilly BM, Hauser F, Readman PW, Makris J (1999) Structural setting, geological development and basin modelling in the Rockall Trough. In: Fleet AJ, Doldy SR (eds) Petroleum geology of northwest Europe. Geological Society of London, London, p 1398

Shannon PM, Corcoran DV, Haughton PDW (2001) The petroleum exploration of Ireland's offshore basins: introduction. In: Shannon PM, Haughton PDW, Corcoran DV (eds) The petroleum exploration of Ireland's offshore basins, Special Publications, vol 188. Geological Society, London, pp 1–8

Soetaert K, Mohn C, Rengstorf A, Grehan A, van Oevelen D (2016) Ecosystem engineering creates a direct nutritional link between 600-m deep cold-water coral mounds and surface productivity. Sci Rep 6:35057. https://doi.org/10.1038/srep35057

Stewart HA, Davies JS, Guinan JC, Howell KL (2014) The Dangeard and Explorer Canyons, South-West Approaches, UK: geology, sedimentology and newly discovered cold-water coral mini-mounds. Deep-Sea Res II 104:230–244. https://doi.org/10.1016/j.dsr2.2013.08.018

Stoker MS, Ackhurst MC, Howe JA, Stow DAV (1998) Sediment drifts and contourites on the continental margin off northwest Britain. Sediment Geol 115:33–51

Stoker MS, Nielsen T, van Weering TCE, Kuijpers A (2002) Towards an understanding of the Neogene tectonostratigraphic framework of the NE Atlantic margin between Ireland and the Faroe Islands. Mar Geol 188:233–248

Stromberg SM, Larsson AI (2017) Larval behavior and longevity in the cold-water coral Lophelia pertusa indicate potential for long-distance dispersal. Front Mar Sci 4:411. https://doi.org/10.3389/fmars.2017.00411

Sweetman AK, Thurber AR, Smith CR, Levin LA, Mora C, Wei CL, Gooday AJ, Jones DOB, Rex M, Yashuhara M, Ingels J, Ruhl HA, Frieder CA, Danovaro R, Wurzberg L, Baco A, Grupe BM, Pasulka A, Meyer KS, Dunlop KM, Roberts JM (2017) Major impacts of climate change on deep-sea benthic ecosystems. Elem Sci Anth 5:1–23. https://doi.org/10.1525/elementa.203

Thierens M, Titschack J, Dorschel B, Huvenne VAI, Wheeler AJ, Stuut J-B, O'Donnell R (2010) The 2.6 Ma depositional sequence from the Challenger cold-water coral carbonate mound (IODP Exp. 307): sediment contributors and hydrodynamic palaeo-environments. Mar Geol 271:260–277. https://doi.org/10.1016/j.margeo.2010.02.021

Thierens M, Pirlet H, Colin C, Latruwe K, Vanhaecke F, Lee JR, Stuut J-B, Titschack J, Huvenne VAI, Dorschel B, Wheeler AJ, Henriet JP (2012) Ice-rafting from the British-Irish ice sheet since the earliest Pleistocene (2.6 million years ago): implications for long-term mid-latitudinal ice-sheet growth in the North Atlantic region. Quat Sci Rev 44:229–240

Thierens M, Browning E, Pirlet H, Loutre MF, Dorschel B, Huvenne VAI, Titschack J, Colin C, Foubert A, Wheeler AJ (2013) Cold-water coral carbonate mounds as unique palaeo-archives: the Plio-Pleistocene Challenger Mound record (NE Atlantic). Quat Sci Rev 73:14–30. https://doi.org/10.1016/j.quascirev.2013.05.006

UK Government (2009) Marine and Coastal Access Act

Ullgren JE, White M (2010) Water mass interaction at intermediate depths in the southern Rockall Trough, northeastern North Atlantic. Deep Sea Res I 57:248–257

Vad J, Orejas C, Moreno Navas J, Findlay HS, Roberts JM (2017) Assessing the living and dead proportions of cold-water coral colonies: implications for deep-water Marine Protected Area monitoring in a changing ocean. PeerJ 5:e3705. https://doi.org/10.7717/peerj.3705

van Aken HM (2000) The hydrography of the mid-latitude Northeast Atlantic Ocean II: the intermediate water masses. Deep Sea Res I 47:789–824

Van Aken HM, Becker G (1996) Hydrography and through-flow in the north-eastern North Atlantic Ocean: The NANSEN project. Prog Oceanogr 38:297–346

van der Land C, Eisele M, Mienis F, de Haas H, Hebbeln D, Reijmer JJG, van Weering TCE (2014) Carbonate mound development in contrasting settings on the Irish margin. Deep Sea Res II 99:297–306

van Haren H, Mienis F, Duineveld GCA, Lavaleye MSS (2014) High-resolution temperature observations of a trapped nonlinear diurnal tide influencing cold-water corals on the Logachev mounds. Prog Oceanogr 125:16–25

Van Oevelen D, Duineveld G, Lavaleye M, Mienis F, Soetaert K, Heip CHR (2009) The cold-water coral community as a hot spot for carbon cycling on continental margins: A food-web analysis from Rockall Bank (northeast Atlantic). Limnol Oceanogr 54:1829–1844

Van Rooij D, Landegem K Van, Skonieczny C, Ingels J, Zaragosi S, Huvenne V, Wheeler A, Haas H De, Henriet J (2008) Neogene sedimentary processes of submarine channels, West of Ireland. American Association of Petroleum Geologists, 2008 International Conference and Exhibition, Programme and Abstract Book.

Van Rooij D, Huvenne VAI, Blamart D, Henriet JP, Wheeler AJ, de Haas H (2009) The Enya mounds: a lost mound-drift competition. Int J Earth Sci 98(4):849–863. https://doi.org/10.1007/s00531-007-0293-9

van Weering TCE, de Haas H, de Stigter H, Lykke-Andersen H, Kouvaev I (2003) Structure and development of giant carbonate mounds at the SW and SE Rockall Trough margins, NE Atlantic Ocean. Mar Geol 198(1–2):67–81

Victorero L, Blamart D, Pons-Branchu E, Mavrogordato MN, Huvenne VAI (2016) Reconstruction of the formation history of the Darwin Mounds, N Rockall Trough: how the dynamics of a sandy contourite affected cold-water coral growth. Mar Geol 378:186–195. https://doi.org/10.1016/j.margeo.2015.12.001

Weaver PPE, Wynn RB, Kenyon NH, Evans J (2002) Continental margin sedimentation, with special reference to the north-east Atlantic margin. Sedimentology 47:239–256. https://doi.org/10.1046/j.1365-3091.2000.0470s1239.x

Wheeler AJ, Beck T, Thiede J, Klages M, Grehan A, Monteys FX, Party tPAaS (2005a) Deep-water coral mounds on the Porcupine Bank, Irish Margin: preliminary results from the Polarstern ARK-XIX/3a ROV cruise. In: Freiwald A, Roberts JM (eds) Cold-water corals and ecosystems. Springer, Heidelberg, pp 393–402

Wheeler AJ, Bett BJ, Billet DSM, Masson DG, Mayor D (2005b) The impact of demersal trawling on NE Atlantic deep-water coral habitats: the case of the Darwin Mounds, UK. In: Barnes PW, Thomas JP (eds) Benthic habitats and the effects of fishing. America Fisheries Society, Bethesda, MD, pp 807–817

Wheeler AJ, Beyer A, Freiwald A, de Haas H, Huvenne VAI, Kozachenko M, Olu-Le Roy K, Opderbecke J (2007) Morphology and environment of cold-water coral carbonate mounds on the NW European margin. Int J Earth Sci 96:37–56. https://doi.org/10.1007/s00531-006-0130-6

Wheeler AJ, Kozachenko M, Masson DG, Huvenne VAI (2008) Influence of benthic sediment transport on cold-water coral bank morphology and growth: the example of the Darwin Mounds, NE Atlantic. Sedimentology 55:1875–1887. https://doi.org/10.1111/j.1365-3091.2008.00970

Wheeler AJ, Kozachenko M, Henry LA, Foubert A, de Haas H, Huvenne VAI, Masson DG, Olu K (2011) The Moira Mounds, small cold-water coral banks in the Porcupine Seabight, NE Atlantic: Part A- an early stage growth phase for future coral carbonate mounds? Mar Geol 282:53–64. https://doi.org/10.1016/j.margeo.2010.08.006

White M (2007) Benthic dynamics at the carbonate mound regions of the Porcupine Sea Bight continental margin. Int J Earth Sci 96:1–9. https://doi.org/10.1007/s00531-006-0099-1

White M, Roberts JM, Weering TV (2007) Do bottom-intensified diurnal tidal currents shape the alignment of carbonate mounds in the NE Atlantic? Geo Mar Lett 27:391–397

White M, Dorschel B (2010) The importance of the permanent thermocline to the cold water coral carbonate mound distribution in the NE Atlantic. Earth Planet Sci Lett 296(3–4):395–402. https://doi.org/10.1016/j.epsl.2010.05.025

White M, Mohn C, de Stigter H, Mottram G (2005) Deep-water coral development as a function of hydrodynamics and surface productivity around the submarine banks of the Rockall Trough, NE Atlantic. In: Freiwald A, Roberts JM (eds) Cold-water corals and ecosystems. Springer, Heidelberg, pp 503–514

Wienberg C, Beuck L, Heidikamp S, Hebbeln D, Freiwald A, Pfannkuche O, Monteys X (2008) Franken Mound: facies and biocoenoses on a newly-discovered "carbonate mound" on the western Rockall Bank, NE Atlantic. Facies 54:1–24. https://doi.org/10.1007/s10347-007-0118-0

Wilson JB (1979a) The distribution of the coral *Lophelia pertusa* (L.) [*L. prolifera* (Pallas)] in the North-East Atlantic. J Mar Biol Assoc UK 59:149–164

Wilson JB (1979b) 'Patch' development of the deep-water coral *Lophelia pertusa* (L.) on Rockall Bank. J Mar Biol Assoc UK 59:165–177

Wilson MFJ, O'Connell B, Brown C, Guinan JC, Grehan AJ (2007) Multiscale terrain analysis of multibeam bathymetry data for habitat mapping on the continental slope. Mar Geod 30(1–2):3–35. https://doi.org/10.1080/01490410701295962

Wilson AM, Kiriakoulakis K, Raine R, Gerritsen HD, Blackbird S, Allcock AL, White M (2015) Anthropogenic influence on sediment transport in the Whittard Canyon, NE Atlantic. Mar Pollut Bull 101:320–329. https://doi.org/10.1016/j.marpolbul.2015.10.067

Williams A, Schlacher TA, Rowden AA, Althaus F, Clark MR, Bowden DA, Stewart R, Bax NJ, Consalvey M, Kloser RJ (2010) Seamount megabenthic assemblages fail to recover from trawling impacts. Mar Ecol 31:183–199

Zelada Leon A, Huvenne VAI, Benoist NMA, Ferguson M, Bett BJ, Wynn RB (2020) Assessing the repeatability of automated seafloor classification algorithms, with application in Marine Protected Area monitoring. Remote Sens 12:1572. https://doi.org/10.3390/rs12101572

Life and Death of Cold-Water Corals Across the Mediterranean Sea

7

Andrea Gori, Claudia Wienberg, Jordi Grinyó, Marco Taviani, Dierk Hebbeln, Claudio Lo Iacono, André Freiwald, and Covadonga Orejas

Abstract

Deep-sea explorations performed over the last two decades have shown evidence of abundant living cold-water coral communities in the Mediterranean Sea. Many submarine canyons, escarpments, seamounts, and outer continental shelf and slope areas support communities dominated by scleractinian cold-water corals or coral gardens composed of a diverse fauna of octocoral and black coral species. Other areas across the Mediterranean Sea show evidence of past prosperity of these communities expressed by accumulations of fossil cold-water corals. Here, we review the knowledge of the present and past occurrence of cold-water corals in the Mediterranean Sea, highlighting specific features and discussing the threats that they are exposed to under high human-induced pressure in this semi-enclosed basin.

Keywords

Mediterranean Sea · Cold-water corals · Scleractinians · Gorgonians · Black corals · Submarine canyons · Escarpments · Seamounts · Continental shelves and slopes · Coral mounds · Reef outcrops · Human threats · Conservation

7.1 Flourishing Cold-Water Coral Communities in the Mediterranean Sea: "What Are the News Today?"

The development of underwater technology and the increased use of remotely operated vehicles (ROVs) and manned submersibles for the exploration of the deep sea have resulted in a recent change of paradigm about Mediterranean cold-water coral abundance and distribution (Taviani 2019). Whereas initially thought to have been almost eradicated by oceanographic changes during the last deglaciation (Zibrowius 1980; Delibrias and Taviani 1985), expeditions run in the last two decades have found evidence of abundant living cold-water coral communities in the Mediterranean, especially in its western and central basins (Taviani et al. 2017). As a result, we now know that the Mediterranean Sea offers appropriate conditions for cold-water corals in many areas, where scleractinians, gorgonians and black corals grow on both hard and soft substrata constituting emblematic ecosystems (Fig. 7.1), whereas other areas only show fossil evidence of their past prosperity (Orejas and Jiménez 2019). Here, we update the current knowledge on the present and past occurrence of cold-water corals in the Mediterranean Sea, highlighting specificities and discussing threats they are exposed to in this semi-enclosed basin under high human-induced pressure.

A. Gori (✉)
Departament de Biologia Evolutiva, Ecologia i Ciències Ambientals, Universitat de Barcelona, Barcelona, Spain

Institut de Recerca de la Biodiversitat (IRBio), Universitat de Barcelona, Barcelona, Spain
e-mail: gori@ub.edu

C. Wienberg · D. Hebbeln
Center for Marine Environmental Sciences (MARUM), University of Bremen, Bremen, Germany

J. Grinyó
Department of Ocean Systems, NIOZ Royal Netherlands Institute for Sea Research, Den Burg, The Netherland

M. Taviani
Istituto di Scienze Marine, Consiglio Nazionale della Ricerca (ISMAR-CNR), Bologna, Italy

Stazione Zoologica Anton Dohrn, Naples, Italy

C. Lo Iacono
Institut de Ciències del Mar, Consejo Superior de Investigaciones Científicas (ICM-CSIC), Barcelona, Spain

A. Freiwald
Marine Research Department, Senckenberg am Meer, Wilhelmshaven, Germany

C. Orejas
Instituto Español de Oceanografía, Centro Oceanográfico de Gijón (IEO-CSIC), Gijón, Spain

E. Cordes, F. Mienis (eds.), *Cold-Water Coral Reefs of the World*, Coral Reefs of the World 19,
https://doi.org/10.1007/978-3-031-40897-7_7

Fig. 7.1 (**a**) The scleractinian cold-water corals *Madrepora oculata* and *Lophelia pertusa* at 305 m depth in the East Cabliers Bank, Alborán Sea (courtesy SHAKE Project-Spain CGL2011–30005-C02–02). (**b**) The gorgonians *Eunicella cavolini* and *Viminella flagellum* at 200 m depth in the Menorca Channel, Balearic Sea (courtesy ICM-CSIC). (**c**) The scleractinian *M. oculata* at 405 m depth in the Corsica Channel, Tyrrhenian Sea (courtesy L. Angeletti)

7.1.1 Scleractinian Cold-Water Corals Across the Mediterranean Sea

The most frequent and widely distributed living scleractinian cold-water corals in the Mediterranean Sea are the white colonial corals *Madrepora oculata* and *Lophelia pertusa*, together with the yellow coral *Dendrophyllia cornigera* and the solitary coral *Desmophyllum dianthus* (Taviani et al. 2005a; Freiwald et al. 2009; Altuna and Poliseno 2019; Castellan et al. 2019; Chimienti et al. 2019). Based upon molecular and morphological arguments (Addamo et al. 2016), *Lophelia pertusa* has been recently reclassified as *Desmophyllum pertusum*, which is now the accepted name for the species (WoRMS Editorial Board 2021). The general requirement of these four species for hard substrates, which are exposed to bottom currents delivering sufficient food,

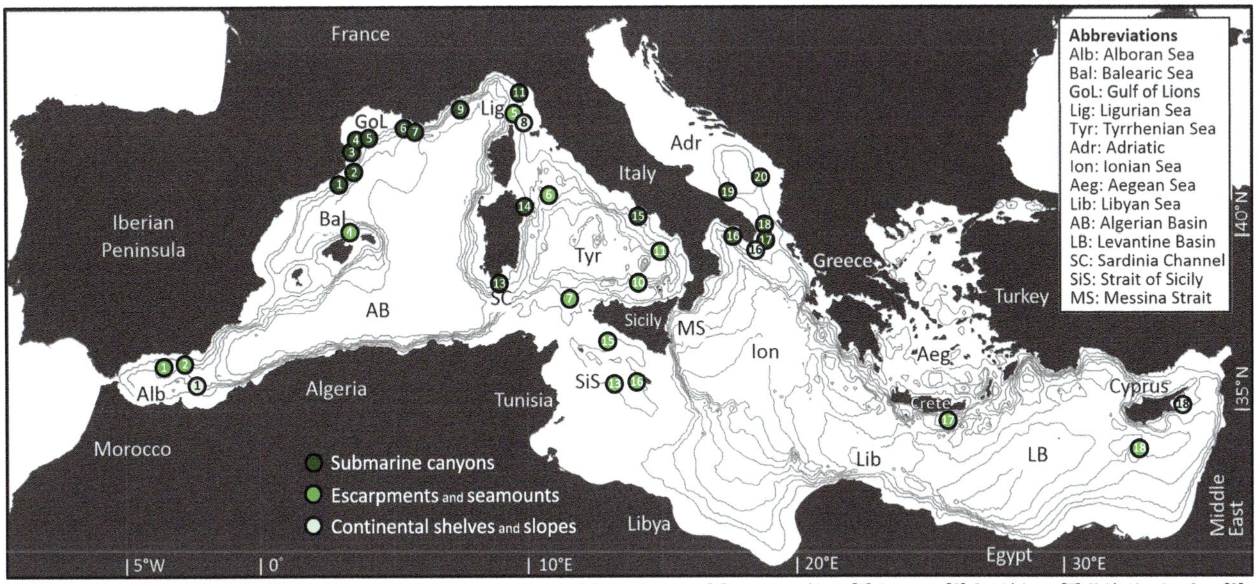

(i) Submarine canyons: Catalan margin: [1] Blanes, [2] La Fonera; Gulf of Lions: [3] Cap de Creus, [4] Lacaze-Duthiers, [5] Bourcart, [6] Cassidaigne, [7] Sicié; Ligurian Sea: [8] Stoechade, [9] Var, [10] Bordighera, [11] Levante; Balearic Sea: [12] Menorca; Sardinian Channel: [13] Nora; Tyrrhenian Sea: [14] Eastern Sardinian, [15] Dohrn; Ionian and Adriatic Sea: [16] Gallipoli, [17] Tricase, [18] Otranto, [19] Bari, [20] Montenegrian. **(ii) Escarpments and seamounts**: Alborán Sea: [1] Djibuti (El Idrissi), [2] Chella Bank (Seco de los Olivos); Algerian Basin: [3] Seco de Palos; Balearic Sea: [4] Menorca Channel; Ligurian Sea: [5] Santa Lucia Bank; Tyrrhenian Sea: [6] Vercelli Seamount, [7] Aceste Seamount, [8] Marco Bank, [9] San Vito Ridge, [10] Enarete Seamount, [11] Palinuro Seamount; Strait of Sicily: [12] Pantelleria, [13] Linosa Trough, [14] Empedocle Seamount, [15] Urania Bank, [16] Malta; Levantine Basin: [17] Danaos Seamount, [18] Eratosthenes Seamount. **(iii) Continental shelves and slopes**: Alborán Sea: [1] East Cabliers Bank, [2] Carboneras Fault; [3] Catalan Margin, [4] Cap de Creus; Balearic Sea: [5] Mallorca Channel, [6] Menorca Channel; Sardinian Channel: [7] Carloforte Shoal; Tyrrhenian Sea: [8] Corsica Channel, [9] Elba Island, [10] Montecristo Island, [11] Pontine Islands, [12] Gioia Canyon, [13] Favazina, [14] Gulf of Santa Eufemia; [15] Messina Strait; Ionian Sea: [16] Santa Maria di Leuca, [17] Cephalonia; Levantine Basin: [18] Cyprus.

Fig. 7.2 Map of the Mediterranean region showing major sites of living scleractinian cold-water coral occurrences associated with (i) submarine canyons, (ii) escarpments and seamounts and (iii) continental shelves and slopes

drives their main distribution linked to four major geomorphological features: **submarine canyons**, **escarpments**, **seamounts** and **continental shelves** and **slopes** (Fig. 7.2).

Submarine canyons

Observations of scleractinian cold-water corals thriving in **submarine canyons** by large relate to the northern margin of the Mediterranean Sea, where they have been frequently described from various canyons along the Catalan margin (north of the Balearic Sea), the Gulf of Lions, the Ligurian and Tyrrhenian Seas, as well as in the southern Adriatic (Figs. 7.2 and 7.3). Submarine canyons not just provide exposed hard substrates on their steep flanks and generally enhanced current conditions, but many of them are additionally influenced by turbulent cascading events, which periodically deliver high amounts of fresh organic matter produced in the surface waters to the deeper parts of the canyons (Canals et al. 2006).

For several submarine canyons in the Gulf of Lions, cascading flows with maximum current speeds of up to 80 cm s^{-1} have been documented (Palanques et al. 2006) with some of these canyons presenting extensive living cold-water coral communities mainly dominated by *M. oculata*. This is the case for the Cap de Creus, Lacaze-Duthiers and Cassidaigne Canyons (Fig. 7.2), where *M. oculata* occurs mainly as isolated colonies—reaching large sizes of 40 cm in height—on rocky bottoms on the canyon walls at 180–540 m depth (mainly at 180–360 m depth) (Orejas et al. 2009; Gori et al. 2013; Fabri et al. 2014, 2017; Sanchez and Fabri 2020). Long-term monitoring in the Lacaze-Duthiers Canyon showed a clear relationship of coral growth and cascading events driving fresh organic matter from the surface down to the canyon (Chapron et al. 2020).

Scattered colonies of *M. oculata* were also observed in the Bourcart and Sicié Canyons as well as in the eastern Var Canyon in the Ligurian Sea (Fig. 7.2; Fabri et al. 2014). In the Lacaze-Duthiers Canyon, *D. pertusum* has been reported at 260–540 m depth (mainly at 320–380 and 500–540 m depth) with colonies often associated with *M. oculata*. Coalescent colonies of *M. oculata* together with very large colonies of *D. pertusum* were also observed completely covering a vertical cliff at 330 m depth (Gori et al. 2013; Fabri et al. 2014). Conversely, only scattered colonies of *D. pertusum* were found in the Cap de Creus Canyon (180–320 m depth, Orejas et al. 2009; Gori et al. 2013; Lo Iacono et al. 2018). Interestingly, an opposite trend in abundance was detected for the yellow coral *D. cornigera*, which was much more abundant in the Cap de Creus Canyon (160–300 m depth) than in the Lacaze-Duthiers Canyon (220–280 m depth; Gori et al. 2013; Lo Iacono et al. 2018). *Dendrophyllia cornigera* has been also documented in the Bourcart and Cassidaigne Canyons,

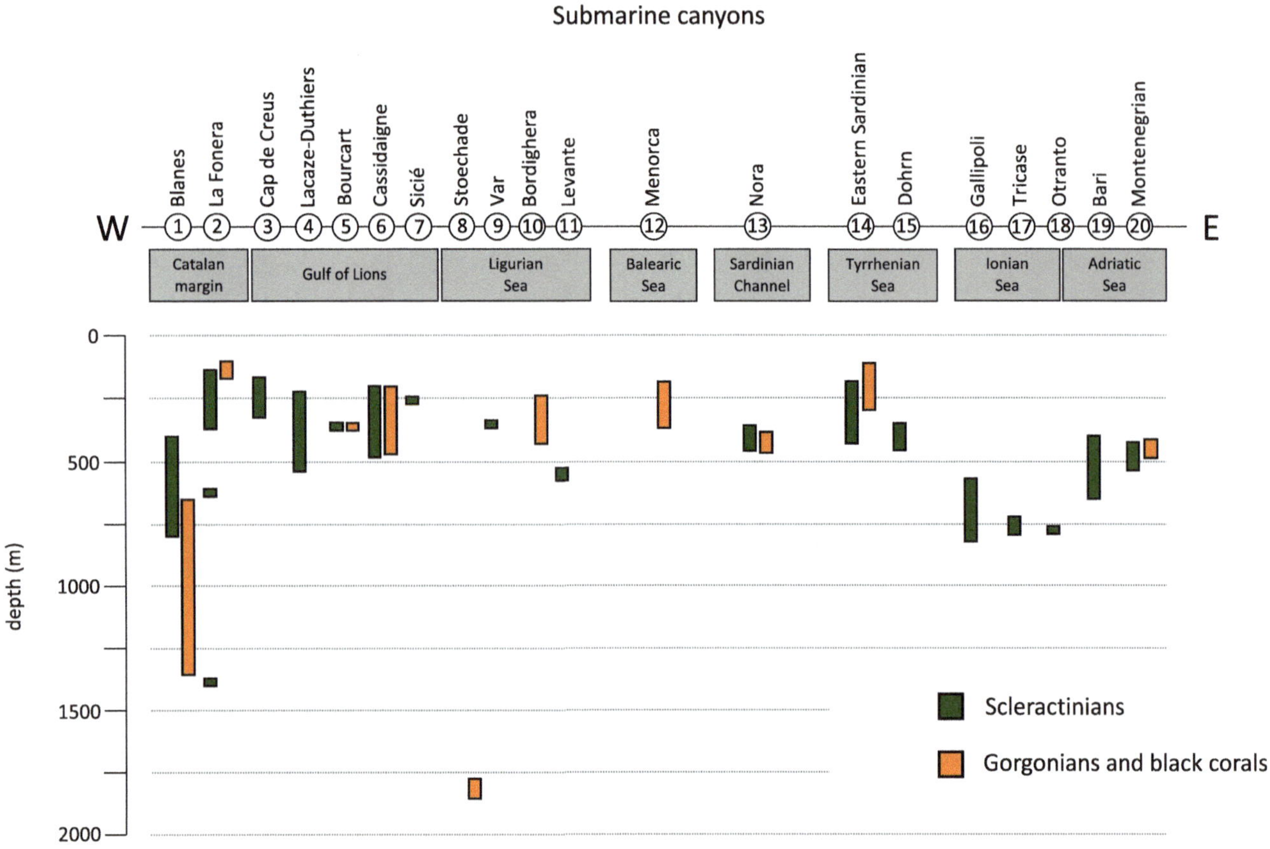

Fig. 7.3 Bathymetric distribution of known scleractinians *versus* gorgonians and black corals in Mediterranean submarine canyons (references for the individual sites in the text; numbers refer to Figs. 7.2 and 7.6)

where also the solitary coral *D. dianthus* is present (Fabri et al. 2014).

Along the Catalan margin, *M. oculata* dominates in the La Fonera and Blanes Canyons (Fig. 7.2; Lastras et al. 2016; De Leo et al. 2019; Bilan et al. 2020). In the La Fonera Canyon, *M. oculata* has been recorded on vertical and highly sloping rocky bottoms on the canyon walls at 130–370 m depth (mainly at 180–330 m depth) together with sparse colonies of *D. cornigera* at 130–330 m depth (Lastras et al. 2016), while a high abundance of *D. dianthus* has been found at 600 and 1400 m depth (Aymà et al. 2019). In the Blanes Canyon, *M. oculata* occurs together with *D. pertusum* at 400–800 m depth on vertical cliffs in the canyon head, with also high abundance of *D. dianthus* both in the canyon head and on the eastern canyon flank at 400–750 m depth (Bilan et al. 2020). In the Nora Canyon, south of Sardinia (Fig. 7.2), large colonies of *M. oculata* (up to 0.8 m height and 1.0 m width) together with *D. dianthus* and some sparse colonies of *D. pertusum* and *D. cornigera* grow on hard substrates including dead coral framework on slopping bottoms on the canyon flanks at 380–460 m depth (Taviani et al. 2017).

On the north-eastern margin of Sardinia, sparse cold-water coral communities dominated by *M. oculata* accompanied by *D. cornigera* and *D. dianthus* were documented for the Tavolara Canyon (180–270 m) and Capo Coda Cavallo Canyon (397–428 m; Eastern Sardinian; Figs. 7.2 and 7.3) (Moccia et al. 2021). These Sardinian coral assemblages are also home of a significant sponge fauna (Bertolino et al. 2019). Remarkably large *M. oculata* colonies with a height of up to 1 m were observed densely growing at 525–575 m depth on top of topographic highs and in steep areas on the southern flank of the Levante Canyon, in the Ligurian Sea (Fanelli et al. 2017).

In the Tyrrhenian Sea, the steep walls of the Dohrn Canyon cutting the shelf off Naples (Fig. 7.2) is home of small colonies of *M. oculata*, *D. pertusum* and *D. dianthus* at 350–430 m depth (Taviani et al. 2019a). These species also occur at the upper edge of the escarpment incised by a canyon south of Gallipoli, Ionian sector of the Apulian margin, at 570–820 m depth (Fig. 7.3; Freiwald et al. 2009). In the south Adriatic Sea, scleractinian cold-water corals have been widely documented (Freiwald et al. 2009; Angeletti et al. 2014; Prampolini et al. 2019), and dense water cascading events hypothesized to be a main factor controlling their distribution (Trincardi et al. 2007; Taviani et al. 2016, 2019b). Here, remarkably large colonies of *M. oculata* of up to 1 m in height occur at 400–650 m depth in the Bari Canyon (Figs. 7.2 and 7.3; Freiwald et al. 2009; Angeletti

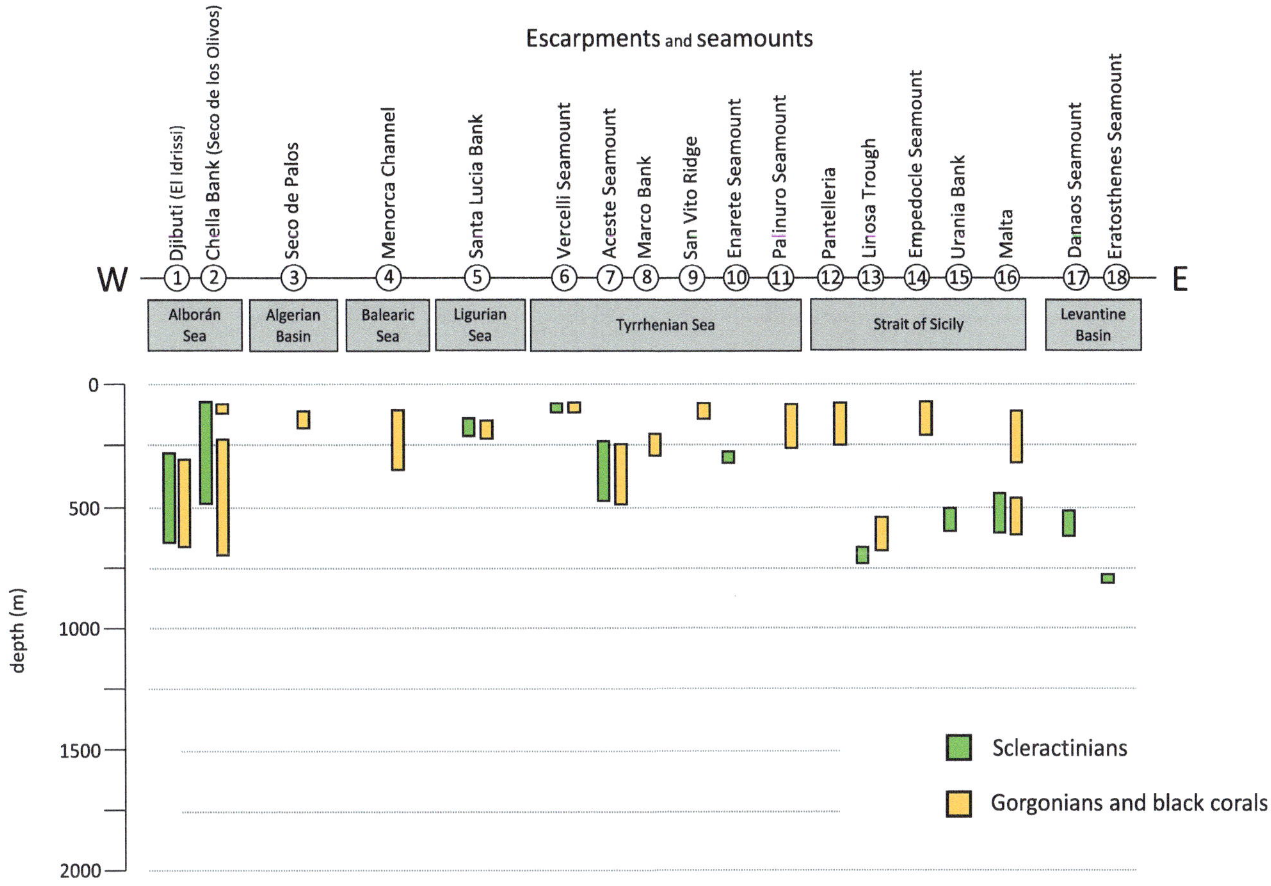

Fig. 7.4 Bathymetric distribution of known scleractinians *versus* gorgonians and black corals in Mediterranean rocky escarpments and seamounts (references for the individual sites in the text; numbers refer to Figs. 7.2 and 7.6)

et al. 2014; Prampolini et al. 2019). The coral assemblage is complemented by sparse colonies of *D. pertusum*, *D. cornigera* and frequent dense aggregations of *D. dianthus* growing on vertical cliffs as well as on the dispersed rocky substrate (Freiwald et al. 2009, 2011; Angeletti et al. 2014; D'Onghia et al. 2015). Further south along the Apulian margin, *M. oculata*, *D. pertusum* and *D. dianthus* occurred at the base of canyons in front of Otranto and Tricase at 720–780 m depth (Angeletti et al. 2014, 2019). On the opposite side, scattered colonies of *M. oculata*, *D. pertusum* and lumps of *D. dianthus* occur at 400–550 m depth in some of the small canyons incising the Montenegrin margin (Figs. 7.2 and 7.3; Angeletti et al. 2014, 2015a), where the influx of cascading processes is minimal (Taviani et al. 2016).

Escarpments

Living scleractinians have been reported for many **escarpments** flanking plateaus, banks and troughs, which similar to the submarine canyons provide habitats with rocky substrates on near-vertical walls exposed to strong currents. Especially for the Strait of Sicily remarkable examples have been reported comprising the Urania Bank, Malta and Linosa Trough (Fig. 7.2), where large fan-shaped colonies of *M. oculata* (30–70 cm in size) grow together with smaller *D. pertusum* colonies at 500–700 m depth (Fig. 7.4; Freiwald et al. 2009). An elongated ridge south of Malta shows occurrence of *M. oculata*, *D. pertusum*, *D. dianthus* and *D. cornigera* at 450–610 m depth (Figs. 7.2 and 7.4; Schembri et al. 2007; Freiwald et al. 2009; Taviani et al. 2011a, 2015; Knittweis et al. 2019). Conversely, only isolated colonies of *D. cornigera* and *M. oculata* have been reported on rocky escarpments on the margin of the Balearic Archipelago (Montagna et al. 2006; Aguilar et al. 2013; Grinyó et al. 2018a), as well as in the eastern Mediterranean basin (Vafidis et al. 1997; Taviani et al. 2016).

Seamounts

Living scleractinian cold-water corals have been documented on several Mediterranean **seamounts** (Würtz and Rovere 2015). In the Alborán Sea, *M. oculata* and *D. pertusum* are the dominant species, with frequent association with *D. dianthus* and *D. cornigera*, and less frequently with the pink coral *Dendrophyllia ramea*. This is the case of the Djibouti (El Idrissi) and the Chella Bank (Seco de los Olivos; Fig. 7.2), among others (Izquierdo et al. 1996; Gil et al. 2010;

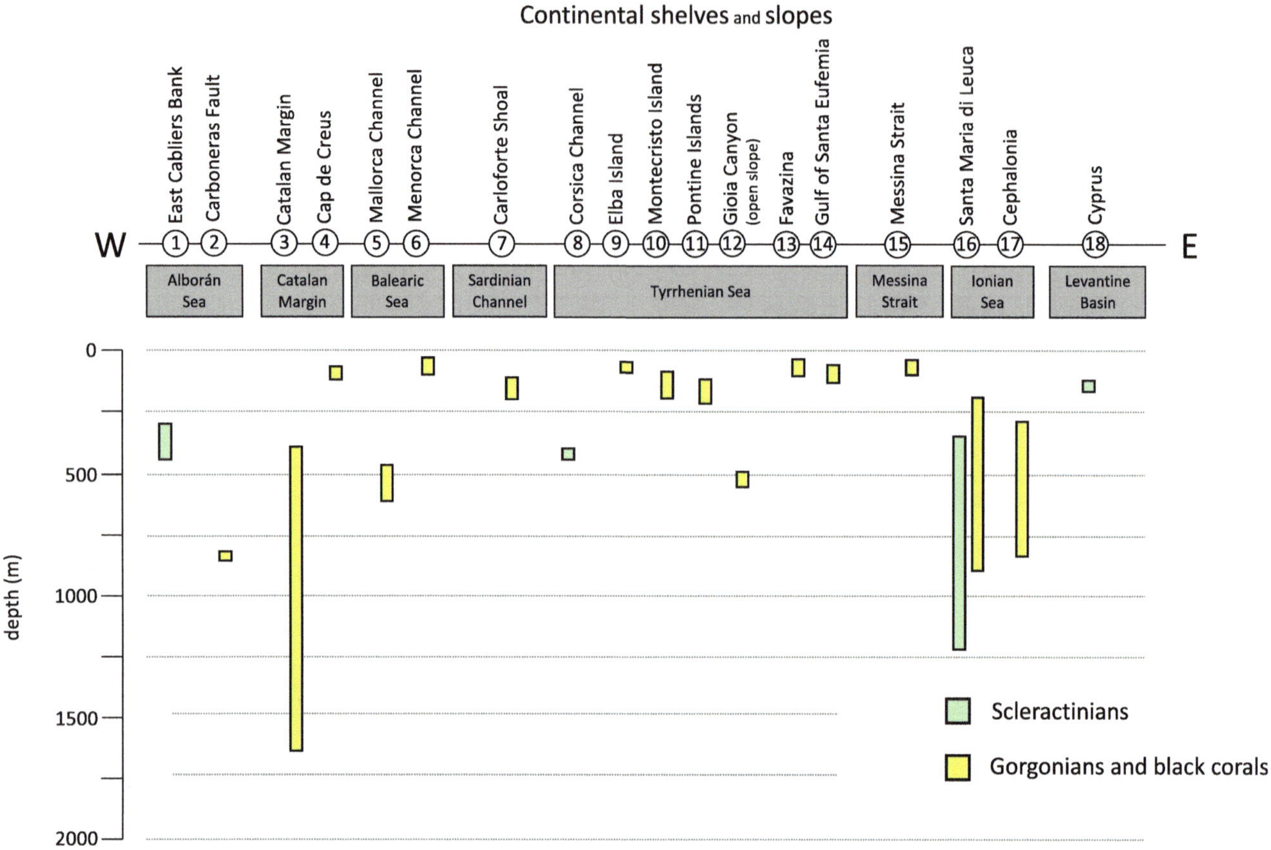

Fig. 7.5 Bathymetric distribution of known scleractinians *versus* gorgonians and black corals on Mediterranean continental shelves and slopes (references for the individual sites in the text; numbers refer to Figs. 7.2 and 7.6)

Pardo et al. 2011; Lo Iacono et al. 2012, 2019; Aguilar et al. 2013; Palomino et al. 2015). Differently from the Alborán Sea, in the central Mediterranean basin, *D. cornigera* and *D. dianthus* are the most frequent scleractinian cold-water corals on seamounts, frequently associated with dense gorgonians and black coral gardens (see Sect. 7.1.2). This is, for example, the case of the Vercelli, Aceste and Enarete Seamounts in the Tyrrhenian Sea (Bo et al. 2011a; Freiwald et al. 2011; Aguilar et al. 2013), and the Santa Lucia Bank in the Ligurian Sea (Fig. 7.2; Bo et al. 2014c). *Dendrophyllia cornigera* and *D. dianthus* are also the only scleractinian corals observed on the Danaos and Eratosthenes Seamounts in the Levantine Basin (Fig. 7.4; Galil and Zibrowius 1998; Smith et al. 2009).

Continental Shelves and Slopes

Dendrophyllia cornigera and its closely related *D. ramea* preferably live on the Mediterranean **continental shelves** (Salomidi et al. 2010; Castellan et al. 2019). *Dendrophyllia cornigera* is mainly distributed on shelves in the western basin (Castellan et al. 2019), whereas a dense population of *D. ramea* has been observed growing on a sandy bottom of the Cyprus continental shelf, in the Levantine Basin, at 125–155 m depth (Figs. 7.2 and 7.5; Orejas et al. 2019a). Conversely, cold-water coral communities dominated by living scleractinian corals have been largely documented to occur on **continental slopes** exposed to strong bottom currents. This is the case of the extensive Santa Maria di Leuca cold-water coral community, in the Ionian Sea, dominated by large fan-shaped *M. oculata* (maximum height of 90 cm) and *D. pertusum* with abundant presence of *D. cornigera* and *D. dianthus* (350–1200 m depth; Figs. 7.2 and 7.5; Tursi et al. 2004; Taviani et al. 2005a; Freiwald et al. 2009, 2011; Mastrototaro et al. 2010; Vertino et al. 2010; Savini et al. 2014; D'Onghia et al. 2016). Here, coral mounds 10–25 m high (maximum 40–50 m high) and 100–300 m wide occur on the gently inclined slope of the continental margin. Coral mounds up to 25 m high, with *M. oculata* as main frame-builder (Fig. 7.1c) and presence of *D. pertusum* and *D. dianthus*, have also been observed at 400–430 m depth on the continental slope in the Corsica Channel, in the Tyrrhenian Sea (Angeletti et al. 2020). These coral mounds have an elongated shape, probably as the result of preferential coral growth influenced by the main bottom currents (Angeletti et al. 2020). In the southern Alborán Sea several coral mounds varying in height from 10 m to 140 m

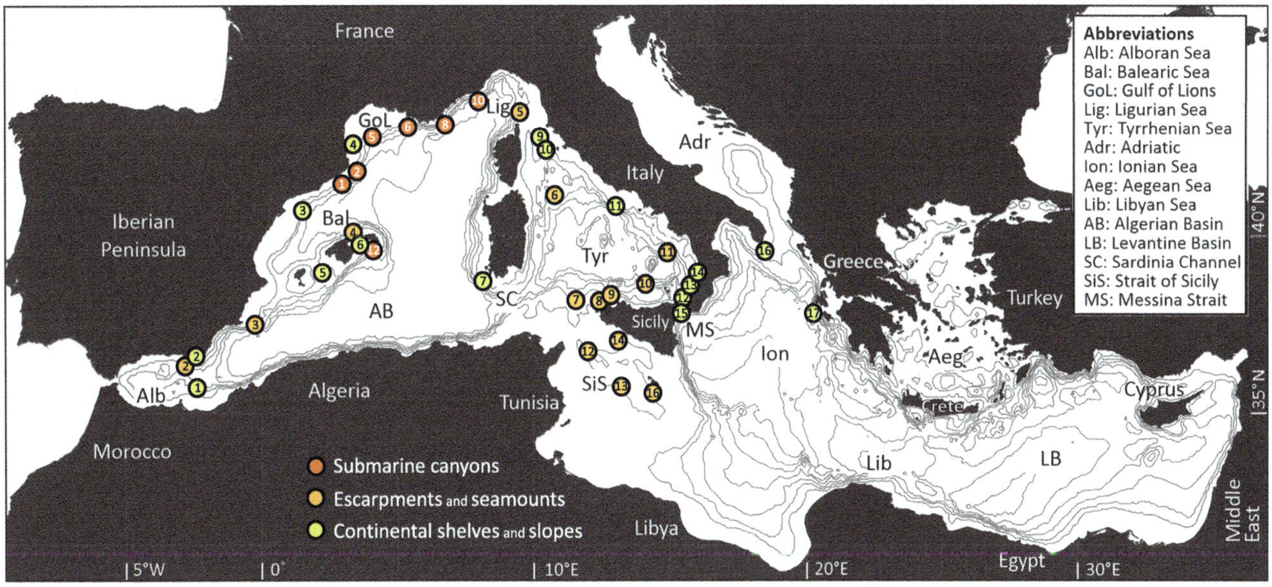

(i) Submarine canyons: Catalan margin: **[1]** Blanes, **[2]** La Fonera; Gulf of Lions: **[3]** Cap de Creus, **[4]** Lacaze-Duthiers, **[5]** Bourcart, **[6]** Cassidaigne, **[7]** Sicié; Ligurian Sea: **[8]** Stoechade, **[9]** Var, **[10]** Bordighera, **[11]** Levante; Balearic Sea: **[12]** Menorca; Sardinian Channel: **[13]** Nora; Tyrrhenian Sea: **[14]** Eastern Sardinian, **[15]** Dohrn; Ionian and Adriatic Sea: **[16]** Gallipoli, **[17]** Tricase, **[18]** Otranto, **[19]** Bari, **[20]** Montenegrian. **(ii) Escarpments and seamounts**: Alborán Sea: **[1]** Djibuti (El Idrissi), **[2]** Chella Bank (Seco de los Olivos); Algerian Basin: **[3]** Seco de Palos; Balearic Sea: **[4]** Menorca Channel; Ligurian Sea: **[5]** Santa Lucia Bank; Tyrrhenian Sea: **[6]** Vercelli Seamount, **[7]** Aceste Seamount, **[8]** Marco Bank, **[9]** San Vito Ridge, **[10]** Enarete Seamount, **[11]** Palinuro Seamount; Strait of Sicily: **[12]** Pantelleria, **[13]** Linosa Trough, **[14]** Empedocle Seamount, **[15]** Urania Bank, **[16]** Malta; Levantine Basin: **[17]** Danaos Seamount, **[18]** Eratosthenes Seamount. **(iii) Continental shelves and slopes**: Alborán Sea: **[1]** East Cabliers Bank, **[2]** Carboneras Fault; **[3]** Catalan Margin, **[4]** Cap de Creus; Balearic Sea: **[5]** Mallorca Channel, **[6]** Menorca Channel; Sardinian Channel: **[7]** Carloforte Shoal; Tyrrhenian Sea: **[8]** Corsica Channel, **[9]** Elba Island, **[10]** Montecristo Island, **[11]** Pontine Islands, **[12]** Gioia Canyon, **[13]** Favazina, **[14]** Gulf of Santa Eufemia; **[15]** Messina Strait; Ionian Sea: **[16]** Santa Maria di Leuca, **[17]** Cephalonia; Levantine Basin: **[18]** Cyprus.

Fig. 7.6 Map of the Mediterranean region showing major sites of known living cold-water gorgonian and black coral occurrences associated with (i) submarine canyons, (ii) escarpments and seamounts and (iii) continental shelves and slopes

have been identified at 200–700 m water depth (Fink et al. 2013; Lo Iacono et al. 2014; Corbera et al. 2019; Hebbeln 2019; Wienberg 2019) (see Sect. 7.2.3). Here, only sparse living colonies of *M. oculata*, *D. pertusum*, *D. dianthus* and *D. cornigera* have been observed growing on the dead coral structure of the coral mounds (Hebbeln et al. 2009; Pardo et al. 2011; Fink et al. 2013). Conversely, the East Cabliers Bank hosts the large Cabliers coral mound province (Fig. 7.2), formed by ridge-like mounds, which are up to 110 m high (Lo Iacono et al. 2014; Corbera et al. 2021). The northern portion of this bank hosts thriving reefs on its top, at 300–440 m depth (Fig. 7.5), mainly formed by *M. oculata* (with colonies up to 1 m in height, Fig. 7.1a) and some very large colonies of *D. pertusum* (up to 3 m in height) (Corbera et al. 2019).

7.1.2 Cold-Water Gorgonians and Black Corals Across the Mediterranean Sea

Compared to scleractinian cold-water corals, much more species of gorgonians and black corals grow in the deep Mediterranean Sea (Grinyó et al. 2016; Gori et al. 2017; Altuna and Poliseno 2019; Chimienti et al. 2019). Mediterranean cold-water coral gardens have been reported (associated or not to scleractinian corals) not only on rocky or soft substrates in **submarine canyons**, on **escarpments**, **seamounts,** and **continental slopes**, but also on **continental shelves** (Fig. 7.6), where they thrive under highly diverse environmental conditions encompassing a wide depth range from 40 to ~2000 m depth (Figs. 7.3, 7.4, and 7.5).

Submarine canyons

The gorgonian *Callogorgia verticillata* (with colonies up to 1 m high) occurs together with scleractinian cold-water corals in several **submarine canyons** in the Gulf of Lions, especially in the Boucart and Cassidaigne Canyons at 350 m and 200–390 m depth, respectively (Figs. 7.3 and 7.6; Fabri et al. 2014). The Cassidaigne Canyon also hosts the gorgonians *Viminella flagellum*, *Acanthogorgia hirsuta* and *Corallium rubrum*, together with the black corals *Leiopathes glaberrima*, *Antipathes dichotoma*, *Antipathella subpinnata* in co-existence with scleractinian corals (Fabri et al. 2014, 2017). The bamboo coral *Isidella elongata* has been documented on muddy bottoms on the flanks of some of the canyons in the gulf, including the Bourcart Canyon at a mean depth of 459 m (Fabri et al. 2014). *Isidella elongata* and the gorgonian *Placogorgia massiliensis* have been reported as deep as 1790–1850 m in the Stockade Canyon, where some colonies of the gorgonian *A. hirsuta* were also observed at 2180 m depth (Sartoretto and Zibrowius 2018). Only few colonies of *Eunicella cavolini* and *C. rubrum* were observed associated with scleractinian corals in the Cap de Creus Canyon (Rossi et al. 2008), whereas these species were more abundant in the La Fonera Canyon along the Catalan margin, especially at 100–160 m depth (Figs. 7.3 and 7.6; Lastras et al. 2016). Dense assemblages of the gorgonian *Muriceides lepida* were identified associated with *C. verticillata*, *Swiftia dubia*, *A. hirsuta*, *Placogorgia coronata* and the black corals *L. glaberrima*, *Parantipathes larix* and *A. dichotoma* on the rocky wall in the Blanes Canyon at 650–1350 m depth (Bilan et al. 2020). Multispecies gorgonian assemblages composed of *C. verticillata*, *Swiftia pallida*, *A. hirsuta* and *Bebryce mollis*, with sporadic presence of the black coral *L. glaberrima*, have been also observed growing on the rocky flanks of the Menorca Canyon in the Balearic Sea, at 180–360 m depth (Grinyó et al. 2016). In the northwestern Ligurian Sea, the Bordighera Canyon contains the black corals *P. larix* and *L. glaberrima* at 230–425 m depth (Figs. 7.3 and 7.6; Giusti et al. 2019), and a recent semi-quantitative study transecting the eastern Sardinian canyons identifies at intermediate depths (100–290 m; Fig. 7.3) the consistent presence of the gorgonians *V. flagellum*, *C. verticillata*, *A. hirsuta*, *E. cavolini*, *B. mollis*, *Paramuricea clavata* and *C. rubrum*, and the black corals *P. larix*, *A. subpinnata*, *A. dichotoma*, *L. glaberrima* sometimes associated with the scleractinian *D. cornigera* (Moccia et al. 2021). The scleractinian-dominated habitats of Nora Canyon, south of Sardinia, host the gorgonians *A. hirsuta* and *B. mollis* at 380–460 m (Figs. 7.3 and 7.6). In the Tyrrhenian Sea, sparse colonies of the black coral *P. larix* have been recorded on the steep wall of the northern side of the Dohrn Canyon, off Naples (Taviani et al. 2019a). In the south Adriatic, sporadic presences of the black coral *L. glaberrima* have been reported from Bari and Tricase Canyons (Angeletti et al. 2014; Prampolini et al. 2019). On the opposite side, the small canyons incising the Montenegrin margin are home to dense coral gardens at 420–490 m (Figs. 7.3 and 7.6) structured by the gorgonian *C. verticillata* with individual fans reaching up to 1 m in height, together with scattered colonies of *Paramuricea macrospina* and large colonies of the black coral *L. glaberrima* (Angeletti et al. 2014; Chimienti et al. 2020).

Escarpments

Gorgonians have been observed to form high-density multispecies assemblages on the rocky escarpments on the shelf edge of the Menorca Channel, in the Balearic Sea, at 100–180 m depth (Figs. 7.4 and 7.6). Here, coral gardens are dominated by the gorgonians *E. cavolini*, *S. pallida*, *V. flagellum* and *C. verticillata* (Fig. 7.1b), together with lower densities of *P. macrospina*, *A. hirsuta*, *B. mollis* and *C. rubrum* (Grinyó et al. 2016). The black corals *A. subpinnata*, *A. dichotoma* and *L. glaberrima* occur at low density within these assemblages, extending their distribution below 200 m depth along the escarpment on the upper continental slope, together with the gorgonians *B. mollis*, *S. pallida*, *V. flagellum* and *C. verticillata* (Grinyó et al. 2016). Rocky escarpments close to Pantelleria, in the Strait of Sicily, show populations of the gorgonian *Ellisella paraplexauroides* at 80–95 m depth (Angiolillo et al. 2012) and *V. flagellum* at 130–250 m depth (Giusti et al. 2012) (Figs. 7.4 and 7.6). The ridge south of Malta is home to coral gardens structured by *C. verticillata*, plus a variety of other gorgonians such as *P. macrospina*, *Dendrobrachia bonsai*, *C. rubrum* (the deepest known occurrence) and the black coral *L. glaberrima* at 100–300 m depth (Fig. 7.4; Taviani et al. 2010; Knittweis et al. 2016, 2019). At the boundary between the Strait of Sicily and the Ionian Sea, dense aggregations of *L. glaberrima* fringe the Malta escarpment at 313 m (Fig. 7.4; Angeletti et al. 2015b).

Seamounts

Highly diverse gorgonian assemblages also occur on several seamounts (Fig. 7.6), where Mediterranean cold-water coral gardens show their maximum development. On the top of seamounts reaching shallow waters close to the sea surface, coralligenous bottoms (Ballesteros 2006) at 60–150 m depth harbour populations of the coastal gorgonian *P. clavata* (Seco de Palos Seamount) together with *E. cavolini* (Vercelli Seamount) (Figs. 7.4 and 7.6; Bo et al. 2011a; Oceana 2014). However, it is on rocky bottoms at 100–700 m depth where maximum diversity is found in multispecies coral gardens composed of the gorgonians *C. verticillata*, *V. flagellum*, *E. cavolini*, *E. paraplexauroides*, *P. macrospina*, *A. hirsuta*, *P. coronata*, *S. pallida*, *M. lepida*, *Villogorgia bebrycoides*, *B. mollis* and *Nicella granifera*. This is, for example, the case of the Seco de Palos Seamount, among others on the northern side of the Algerian Basin, and the Aceste Seamount in the south Tyrrhenian Sea (Fig. 7.6; Aguilar et al. 2013; Oceana 2014). Some of these gorgonians also occur together with the black corals *L. glaberrima*, *A. subpinnata*, *A. dichotoma* and *P. larix* on the Chella Bank (Seco de los Olivos) in the Alborán Sea, as well as in the Santa Lucia Bank in the Ligurian Sea, and San Vito Ridge and Marco Bank in the Tyrrhenian Sea (Aguilar et al. 2013; Bo et al. 2014a, 2014b; Oceana 2014; Ingrassia et al. 2016). Muddy bottoms on seamounts also frequently support populations of the bamboo coral *I. elongata* together with sea pens, such as several seamounts close to the Balearic Islands, and the Aceste Seamount in the Tyrrhenian Sea (Fig. 7.6; Pardo et al. 2011; Aguilar et al. 2013).

Continental Shelves and Slopes

Coralligenous bottoms growing under twilight conditions on the Mediterranean continental shelf frequently host populations of the coastal gorgonians *P. clavata*, *E. singularis*, *E. cavolini* and *C. rubrum*, as observed at the Elba Island in the Tyrrhenian Sea (Priori et al. 2013; Angiolillo et al. 2016), and in the Menorca Channel in the Balearic Sea at 40–120 m depth (Figs. 7.5 and 7.6; Grinyó et al. 2016). Moreover, in the latter area, the gorgonian *P. macrospina* and the soft coral *Paralcyonium spinulosum* have been observed forming extensive high-density populations on maërl bottoms at 60–100 m depth (Fig. 7.5; Grinyó et al. 2016, 2020a). Most of the species mentioned above have been recorded growing under dark conditions (due to high turbidity in the water column) on rocky outcrops on the continental shelf of Cap de Creus and the Gulf of St. Eufemia at 70–130 m depth (Figs. 7.5 and 7.6; Bo et al. 2012; Gori et al. 2017). Gorgonians and black corals have been also frequently reported on rocky boulders, pinnacles and shoals on continental shelves and slopes (Bo et al. 2009; 2011b). *Paramuricea clavata*, *E. cavolini* and the black coral *A. subpinnata* form high-density coral gardens on rocky pinnacles at 50–100 m depth on the continental shelf near Favazzina, in the south Tyrrhenian Sea (Figs. 7.5 and 7.6; Bo et al. 2009). The black coral *P. larix* colonizes the rocky outcrops at 100–200 m depth near the island of Montecristo (Bo et al. 2014a) as well as rocky boulders interspersed within the soft sediments of the continental slope of the Pontine Islands, in the north and central Tyrrhenian Sea (Fig. 7.6), where it occurs together with the black corals *L. glaberrima* and *A. dichotoma* (Ingrassia et al. 2016). *Leiopathes glaberrima* dominates dense coral gardens at 180 200 m depth on the Carloforte shoal on the southwest continental slope of Sardinia, where the black corals *A. dichotoma* and *P. larix* and the gorgonians *C. verticillata*, *A. hirsuta*, *E. cavolini* and *B. mollis* are also present though with low abundance (Bo et al. 2015). In the same area, rocky pinnacles at 120–170 m depth on the continental margin are colonized by the gorgonian *C. verticillata*, with abundant presence of *C. rubrum*, *E. cavolini*, *A. hirsuta* and *V. flagellum* (Cau et al. 2015). Black corals are here less abundant than gorgonians, with *P. larix*, *A. subpinnata*, *A. dichotoma* and *L. glaberrima* occurring at low density in the areas more covered by mud (Cau et al. 2015).

On the continental slope, the bamboo coral *I. elongata* typically forms coral gardens on bathyal mud plains. High frequency of occurrence of the species has been recently reported at 230–700 m depth in the south Adriatic Sea (Otranto Channel) and southeastern Tyrrhenian Sea (Carbonara et al. 2020). High density populations composed of large *I. elongata* (up to 40 cm high) have been also reported in the Mallorca Channel in the Balearic Sea (480–615 m depth; Figs. 7.5 and 7.6; Mastrototaro et al. 2017), in the Carboneras fault system in the Alborán Sea (840–860 m depth; Grinyó et al. 2020b) as well as on muddy bottoms around the Carloforte rocky shoal on the continental slope of south-west Sardinia (200 m depth: Bo et al. 2015) (Figs. 7.5 and 7.6). These areas with dense *I. elongata* populations are protected from bottom trawl fishing, while considerably lower abundances have been observed on continental slopes exposed to fishing, such as along the Catalan margin at 400–1660 m depth (Maynou and Cartes 2011; Cartes et al. 2013). *Isidella elongata* has been also frequently reported from muddy bottoms around islands, such as Pontine Islands in the Tyrrhenian Sea (Ingrassia et al. 2019), or the continental slope south-west of Cephalonia in the Ionian Sea (Fig. 7.6; Mytilineou et al. 2014). *Isidella elongata* together with sea pens occur on muddy bottoms on canyon margins in the Gioia Canyon in the south Tyrrhenian Sea (Pierdomenico et al. 2016, 2018), as well as 400–500 m depth south of Malta in the border between the Sicily Strait and the Ionian Sea (Fig. 7.5; Freiwald et al. 2009).

7.1.3 Ecology of Cold-Water Corals in the Mediterranean Sea

The pivotal role played worldwide by cold-water coral not only to accommodate but also to enhance the overall diversity at intermediate depths and in the deep sea (e.g., Henry and Roberts 2017) has been largely confirmed also in the Mediterranean (Bongiorni et al. 2010; Rueda et al. 2019; D'Onghia 2019). In particular, high abundance and diversity of sponge, polychaetes, bryozoans and fish has been observed associated with the structurally complex and heterogeneous habitats generated by cold-water corals (D'Onghia et al. 2015, 2016; Santín et al. 2021).

Food availability is a paramount factor controlling cold-water coral occurrence and development. Cold-water corals present a wide trophic plasticity being capable to feed on dissolved organic matter, detritus, phytoplankton or different size of zooplankton (Kiriakoulakis et al. 2005; Gori et al. 2014b, 2018; Naumann et al. 2015; Orejas et al. 2016; Coppari et al. 2020). Isotopic analysis conducted on *M. oculata*, *D. pertusum* and *D. dianthus* in the cold-water coral province of Santa Maria di Leuca (Fig. 7.2) determined coral diet to be mainly composed of fresh zooplankton (Carlier et al. 2009). Furthermore, recent isotope and fatty acid research indicate that Mediterranean cold-water corals exploit different food resources depending on the season. In this regard, it has been observed that the black coral *A. subpinnata* mainly feeds on pico-nanoplankton in autumn and micro-zooplankton in spring (Coppari et al. 2020). In contrast, the isotopic composition of the gorgonian

P. macrospina on the continental shelf of the Menorca Channel (Fig. 7.6) remained constant throughout the year, whereas its fatty acid composition varied seasonally likely reflecting changes in gorgonian energetic demand related with gametogenesis and larval brooding (Grinyó et al. 2018b).

Variations in food resources may strongly influence cold-water coral physiology. Naumann et al. (2011) showed that zooplankton withdrawal from Mediterranean *D. dianthus* diet resulted in a 69% decline in calcification rates. The scleractinians *M. oculata, D. pertusum, D. dianthus* and *D. cornigera* from the Mediterranean have been reported to grow 0.02–0.11% day^{-1} in buoyant weight over 8 months (Orejas et al. 2011); *M. oculata* and *D. pertusum* grow 0.01–38.1 mm $year^{-1}$ in linear extension (Orejas et al. 2008, 2011; Lartaud et al. 2013, 2014, 2019). Growth rates can also vary depending on the age of the colony or the polyp (Lartaud et al. 2013; Movilla et al. 2014a). For instance, *D. pertusum* shows calcification rates being four times lower in old than in young polyps (Lartaud et al. 2013; Movilla et al. 2014a). Experimental studies have shown that cold-water coral calcification and polyp activity decrease when corals are exposed to temperatures out of their optimal thermal range (Previati et al. 2010; Gori et al. 2014a). Experimental research has also revealed a positive correlation between respiration rates and food consumption and temperature in the Mediterranean cold-water corals *D. dianthus* and *D. cornigera*, respectively (Naumann et al. 2011; Gori et al. 2014a). Conversely, under food limitation, coral respiration substantially decreased by 20% after 1 week, and by 62% after 3 weeks (Naumann et al. 2011). Similarly, *M. oculata* and *L. pertusa* have been observed to decrease their respiration rates when exposed to 3 to 4 °C lower than their 12 °C *in situ* Mediterranean temperatures (Naumann et al. 2014). Ocean acidification can also decrease calcification rates, however, species response to acidification can widely differ. In this sense, *M. oculata* and *D. pertusum* showed no variations in calcification rates after 6 months of exposure to low pH (Movilla et al. 2014a). Conversely, the calcification rate in *M. oculata* was reduced after 2 weeks under low pH (Maier et al. 2016), and the calcification rate of *D. dianthus* decreased by 70% after 314 days (Movilla et al. 2014b).

Knowledge on the reproductive biology of Mediterranean cold-water corals is very limited. No studies have been conducted up to date on the reproductive biology of scleractinian cold-water corals in the Mediterranean Sea. So far, reproductive aspects have only been studied for the black coral *A. subpinnata* and the gorgonian *P. macrospina*. A brief glimpse of the *A. subpinnata* reproductive biology was given by Gaino and Scoccia (2010), who studied colonies from the Tyrrhenian and Adriatic Seas over 5 months, revealing that this species is gonochoric, with spawning occurring during summer, and sexual maturity is likely attained when colonies reach a height of 70 cm. Furthermore, Coppari et al. (2019, 2020) recently discovered that this species can reproduce asexually when exposed to stressful conditions. Colonies of *A. subpinnata* can release branch fragments that have the capacity to reattach to substrates, which surprisingly show higher growth rates than maternal colonies. On the other hand, the reproductive biology of *P. macrospina* was studied on samples collected on the continental shelf of the Balearic Islands (Fig. 7.6). This species is gonochoric, internal brooder and presents a population sex ratio of 1:1. Oogenesis lasts ~12–14 months, whereas spermatogenesis only lasts 6 months, fertilization occurs during later summer and larvae are released in September–October (Grinyó et al. 2018b).

Another aspect of Mediterranean cold-water corals that has remained widely unexplored is their microbiome composition. Mutualist and commensal microorganisms provide corals with important nutrients (Benavides et al. 2017) and shield corals from pathogens (Rypien et al. 2010). Recently, the spatial and temporal stability of the *A. subpinnata* and *E. cavolini* microbiomes has been analysed, revealing that bacterial communities in *A. subpinnata* are highly diverse, significantly differing between seasons and locations driven by environmental variability (van de Water et al. 2020). Conversely, bacterial community of *E. cavolini* was dominated by one single core microbiome, with very limited temporal variation (van de Water et al. 2020). Differences in the microbiome of *M. oculata* and *D. pertusum* have been studied in the Mediterranean (Weinbauer et al. 2019), and recent efforts have also revealed that microbiome in *M. oculata* is less variable than that of *D. pertusum,* when coral colonies are relocated to different depths and environmental conditions or are incubated under different food availability (Galand et al. 2018, 2020; Chapron et al. 2020).

7.2 The Past Success of Cold-Water Corals in the Mediterranean: "The Old Story"

7.2.1 The Geological Significance of Scleractinian Cold-Water Corals

Even though cold-water corals encompass a large and diverse marine community (including species of the orders Octocorallia and Antipatharia), species of the order Scleractinia are of top geological relevance, providing valuable information about the "old story" of cold-water corals and their paleoceanographic significance in the Mediterranean Sea (as well as any other ocean region). This is explained by their aragonitic ("stony") skeletons, which are preserved in rocks and in deep-sea sediments and are able to store crucial paleo-climatic information. Reefs formed by scleractinian cold-water corals are commonly found in the geological record throughout Earth's history (Roberts et al. 2009). However, only few of the known scleractinian cold-

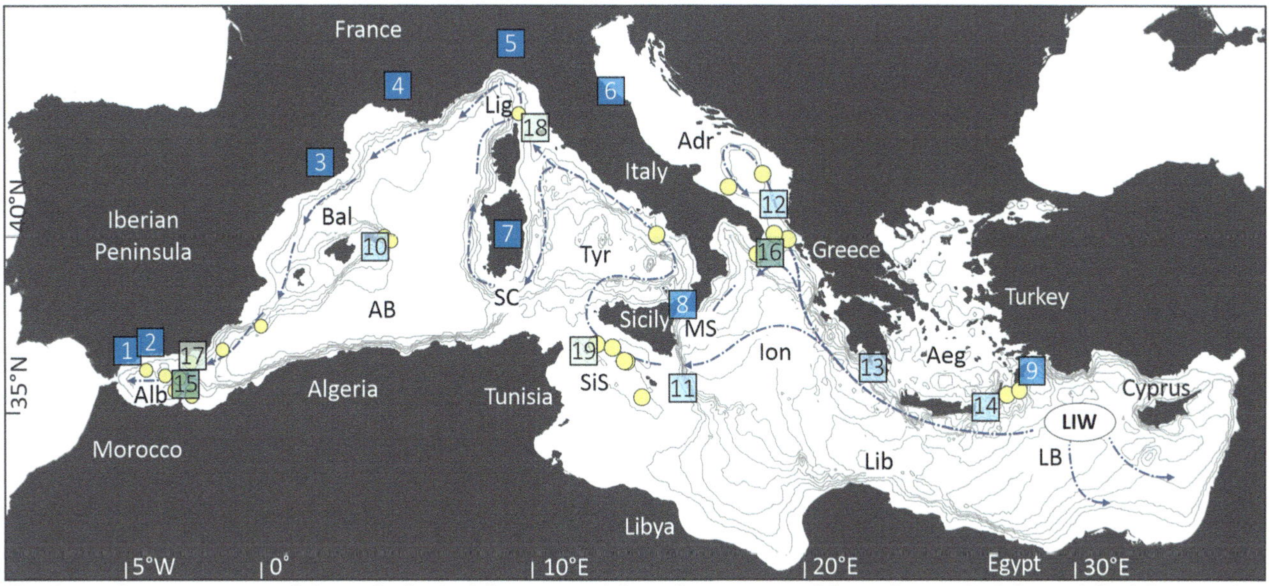

(i) Subaerial outcrops: Spain: **[1]** Manilva basin, **[2]** Estepona, **[3]** Catalonia; France: **[4]** Vaucluse; Italy: **[5]** Turin hills, **[6]** Marecchia valley, **[7]** Sardinia, **[8]** Sicily-Calabria; Greek Islands: **[9]** Rhodes-Karpathos. **(ii) Submerged outcrops**: **[10]** Balearic Islands margin, **[11]** Malta escarpment, **[12]** southern Adriatic, **[13]** Crete margin, **[14]** Hellenic trough. **(iii) Coral mounds**: **[15]** southern Alboran Sea (East & West Melilla, Cabliers), **[16]** Ionian Sea (Santa Maria di Leuca). **(iv) Potential coral mounds**: **[17]** northern Alboran Sea (Chella Bank), **[18]** NE Tyrrhenian Sea (Corsica Channel, Tuscan archipelago), **[19]** Sicily Channel (Pantellaria Island). **Abbreviations**: AB, Algerian Basin; Adr, Adriatic; Aeg, Aegean Sea; Alb, Alboran Sea; Bal, Balearic Sea; Ion, Ionian Sea; LB, Levantine Basin; Lib, Libyan Sea; Lig, Ligurian Sea; MS, Messina Strait; SC, Sardinia Channel; SiS: Strait of Sicily; Tyr, Tyrrhenian Sea.

Fig. 7.7 Map of the Mediterranean region showing major sites of known scleractinian cold-water coral occurrences. The selected sites comprise (i) subaerial (dark blue boxes) and (ii) submerged reef outcrops (light blue boxes) of Miocene to Pleistocene age, (iii) coral mounds (dark green boxes), and (iv) potential coral mounds (light green boxes). Yellow dots indicate regions, for which radiometric cold-water corals ages are available (Late Pleistocene to Holocene). A simplified scheme of the water circulation of the Levantine Intermediate Water (LIW) is shown (dashed arrows)

water corals form complex three-dimensional colonies and are considered framework-formers. The majority of species exhibit a solitary growth form (Cairns 2007). Occurring in isolated patches, solitary corals contribute to a reef structure, but they are not able to form reefs by themselves. Reef-forming species are characterised by the following key attributes: (i) they are long-lived; (ii) form topographic highs, which alter the local hydrodynamic regime and the consequent sediment transport pattern; (iii) create unique sedimentary settings by trapping supplied sediments and providing accommodation space for their deposition and (iv) provide structural habitat and refuge for a great diversity of associated marine organisms (see Roberts et al. 2009).

The prominent and cosmopolitan corals *D. pertusum* and *M. oculata* created considerable coral mound structures (>100 m high, >1 km wide) in the world's oceans, which formed through periods of successive reef development (e.g., Roberts et al. 2009; Wienberg and Titschack 2017; Lo Iacono et al. 2018). Hence, cold-water coral mounds represent sedimentary archives, which elucidate cyclic phases of regional reef proliferation and demise, controlled by climate-induced environmental changes. Coral mounds in the Mediterranean Sea document such developments predominantly for the Late Pleistocene to Holocene period (e.g., Fink et al. 2013; Stalder et al. 2015; Wang et al. 2019). In addition, outcrops unveiling fossil reefs (subaerial and submerged) in the Mediterranean region offer a glimpse on cold-water coral reef formation during much older times ranging from the Miocene to the Middle Pleistocene (Vertino et al. 2019).

7.2.2 Ancient Reef Outcrops: A Glimpse to the Pre-Middle Pleistocene Occurrence of Cold-Water Corals in the Mediterranean

In the Mediterranean region, records of fossil scleractinian cold-water corals are quite sporadic and are based on punctual information documented in rock outcrops and specimens preserved in museum collections (Corselli 2001; Taviani et al. 2005a; Vertino et al. 2014). Examples of subaerial outcrops in the Mediterranean area are best known from Italy and the Greek islands of Rhodes and Karpathos (Fig. 7.7; see reviews by Taviani et al. 2019a and Vertino et al. 2019), where deposits bearing fossil cold-water corals occur in loose sediments either attached to rocky substrate or embedded in limestone (Taviani et al. 2019a). Overall, the exposure of cold-water corals on-land is mainly related to the active tectonic setting along the convergent Eurasian-African plate boundary, leading to variable uplift rates (Taviani et al. 2005a). A prime example comes from the coral-bearing deep-sea sediments outcropping onshore eastern Sicily and southern Calabria (Fig. 7.7), related to the high uplift rates of this sector of the Mediterranean during Plio- and Pleistocene epochs (e.g., Montenat et al. 1987).

Knowledge on the age of fossil cold-water coral occurrences on-land is rather sketchy and the taxonomic classification of many described samples remains questionable as these refer to old and often poorly documented literature accounts (see also Vertino et al. 2019). Nevertheless, reef framework-forming cold-water corals are known from the Mediterranean area since at least the Miocene (Vertino et al. 2019). For example, fossilised fragments of the genera *Madrepora* and *Enallopsammia* were identified in bathyal märl deposits (~20–16 Ma) occurring in northern Italy (Turin Hills, Fig. 7.7), and a Miocene limestone of Late Tortonian age (~8–7 Ma) outcropping in southern Italy (NW Calabria) comprises a 3-m-thick coral "mound" sequence constructed by corals of the genera *Oculina* and *Dendrophyllia* (Vertino et al. 2019 and references therein). Cold-water coral occurrences of the genera *Dendrophyllia* and *Enallopsammia* were also recently reported from upper Miocene outcrops of NE Italy (valley of the Marecchia River, Rimini, Fig. 7.7; Spadini and Pizzolato 2020).

Besides these rather rare Miocene occurrences, framework-forming cold-water corals with a dominance of dendrophylliid species are known from several Pliocene (5.3–2.6 Ma) sections of northern Italy, Sardinia, the Vaucluse region (France), and Catalonia (Spain; Vertino et al. 2019 and references therein) and were also recently reported from the lower Pliocene of Estepona and Manilva Basin in southern Spain (Fig. 7.7; Aguirre et al. 2020). In addition, sites surrounding the Messina Strait (NE Sicily and SW Calabria) comprise, e.g., isolated boulders of Plio- to Pleistocene age being encrusted by *D. pertusum* and *M. oculata*, which were described to lay at the base of paleo-cliffs (Di Geronimo 1987; Barrier et al. 1996); and märl deposits displaying a highly diverse Pleistocene bathyal fauna including *D. pertusum* and *M. oculata* (e.g., Catania territory; Di Geronimo 1979).

The most remarkable rock outcrops revealing fossil corals of Pleistocene age are known from Sicily and Rhodes (Fig. 7.7). Coral limestones occurring on Sicily ("La Montagna", Messina), being dominated by *D. pertusum*, *M. oculata* and the genera *Enallopsammia*, have been assumed to represent a former cold-water coral mound due to the partly large and well-preserved coral fragments (Vertino et al. 2013). An outcrop from the island of Rhodes, comprising märl sediments and limestones containing *D. pertusum* and *M. oculata*, was interpreted as a paleo-canyon system of Middle Pleistocene age (690–760 ka; Titschack and Freiwald 2005; Titschack et al. 2013). Overall, many of the documented observations in the Mediterranean region have in common that coral assemblages developed during their lifetime on a steep and rough topography, in which submarine cliffs and canyon walls provided suitable environmental conditions for a preferential habitat, while re-deposition at the base of steep cliffs through mass-wasting processes represents a frequent fate of coral deposits within such a paleo-environmental setting (Titschack 2019).

Submerged examples of fossil cold-water corals embedded into hardgrounds are mainly found on sloping topographies undergoing the influence of bottom currents. Submerged outcrops are relatively difficult to access and fossil cold-water corals from such settings have been predominantly retrieved by geological dredges. These heavy gears are suitable to operate on structural highs with rough surfaces such as seamounts, steep canyon walls or escarpments, but suffer from a poor detail in their positioning. Nevertheless, submerged outcrops containing subfossil cold-water corals (mainly *D. pertusum* and *M. oculata*) are known from a number of sites (see Fig. 7.7) such as the margins of the Balearic Islands (Taviani and Colantoni 1979), the Tuscan Archipelago (Remia and Taviani 2005), the Strait of Sicily (Zibrowius and Taviani 2005), the Malta Escarpment (Angeletti and Taviani 2011), the Santa Maria di Leuca cold-water coral province in the Ionian Sea (Rosso et al. 2010), the southern Adriatic (Nasto et al. 2018; Taviani et al. 2019b), the Crete margin (Taviani et al. 2011b), the Hellenic Trough (Zibrowius 1981), and many seamounts in the Tyrrhenian and Ionian Seas (see also Taviani et al. 2019a and references therein). The records from all these submerged sites suggest that substantial cold-water coral growth took place in the Mediterranean since the Pliocene (after the Messinian Salinity Crisis: 7.1–5.3 Ma; Freiwald et al. 2009), and corals (i.e., *D. pertusum*, *M. oculata*) have been widespread in the basin since the Middle Pleistocene at least (Vertino et al. 2019).

7.2.3 Modern Coral Mounds: Archives of Late Pleistocene to Holocene Cold-Water Coral Reef Development in the Mediterranean

Cold-water coral mounds are formed by the interaction of (scleractinian) coral growth and sedimentary processes (supply, transport and deposition of organic-rich and terrigenous sediments) under energetic hydrodynamic conditions (Roberts et al. 2009) and (often) encompass several periods of recurring reef formation under a cyclic environmental variability. Thus, they constitute valuable archives to elucidate past periods of reef formation and demise. Coral mounds occur widespread in the entire Atlantic Ocean, where they are arranged in provinces often comprising numerous individual or merged mounds (e.g., Wheeler et al. 2007; Glogowski et al. 2015; Wienberg et al. 2018; Hebbeln et al. 2019; Tamborrino et al. 2019; Steinmann et al. 2020). In contrast, so far, only few coral mound provinces have been discovered in the Mediterranean Sea (Fig. 7.7) offering detailed insight into the most recent "old story" of the Mediterranean cold-water corals.

A "hotspot" region for coral mound formation is the southern Alborán Sea in the westernmost part of the Mediterranean, where three extended mound provinces have been discovered off Morocco (Fig. 7.7): The West Melilla, the East Melilla and the Cabliers coral mound provinces (Comas et al. 2009; Lo Iacono et al. 2014; Corbera et al. 2019). These provinces comprise individual mounds in ~200–700 m water depth, which exhibit a large variability in shape and dimension (height: few meters to ~140 m; length: hundreds of metres to several kilometres; Lo Iacono et al. 2014; Corbera et al. 2019; Hebbeln 2019). Sediment cores obtained from all the three mound provinces recorded times of coral reef proliferation and demise, which resulted in enhanced mound formation or stagnation in mound development, respectively. The post-glacial cold-water coral community of each province (dominated by *D. pertusum* and to a lesser degree by *M. oculata*) experienced a highly prolific time during the last deglaciation (starting at ~14 ka BP) until the Early Holocene, which resulted in vast mound formation with exceptionally high aggradation rates of 1–4 metres per kiloyear (Fig. 7.8; Fink et al. 2013; Stalder et al. 2015; Wang et al. 2019; Wienberg 2019; Fentimen et al. 2020; Corbera et al. 2021).

These thriving phases have been ascribed to environmental conditions of enhanced (surface and export) productivity and moderate to strong hydrodynamics supporting coral growth and sediment supply (Fink et al. 2013; Stalder et al. 2015; Fentimen et al. 2020), which in turn were most probably conditioned by the influence of internal waves acting at the interface between the Levantine Intermediate Water and the overlying Modified Atlantic Water (Wang et al. 2019; Corbera et al. 2021). However, the Mid to Late Holocene reef/mound development displays a rather contrasting pattern (Fig. 7.8). For both Melilla coral mound provinces, reef formation significantly reduced soon after the Early Holocene "booster" phase (after Henriet et al. 2002); though corals were still occasionally present on the East Melilla mounds during the entire Holocene (until today), while living coral communities vanished from the West Melilla mounds at ~4.5 ka BP (Fink et al. 2015; Wang et al. 2019). In contrast, corals found exceptional thriving conditions in the northern Cabliers province and mound formation continued until today, though at low rates (Corbera et al. 2021). Interestingly, the significant reduction in mound formation documented for the coral mounds in the southern Alborán Sea correlates with a change from a *D. pertusum*-dominated to a *M. oculata*-dominated coral community (see also Wienberg 2019; Corbera et al. 2021). In addition, even though the three known coral mound provinces in the Alborán Sea are located at the reduced distance of 50–60 km apart from each other, it is still not clear why the present-day ecological status of the reefs overlying the Alborán coral mounds varies considerably from extinct/fossil reefs (West Melilla; Lo Iacono et al. 2014) to scattered and small live coral colonies (East Melilla; Hebbeln et al. 2009), to currently thriving reefs dominated by *M. oculata* (Cabliers; Fig. 7.1a; Corbera et al. 2019), following a so far unidentified environmental pattern.

Besides the prominent mound provinces of the southern Alborán Sea, coral mounds have also been described from the northern Alborán Sea (Chella Bank *or* Seco de los Olivos) in ~300 m water depth (single up to 60-m-high mound/ridge; Lo Iacono et al. 2019), from the NE Tyrrhenian Sea (Corsica Channel, off Tuscan Archipelago) in ~350–410 m depth (>20 up to 25-m-high mounds; Angeletti et al. 2020), and from the Strait of Sicily (off Pantelleria Island) in 300–600 m (few thousand small (<10 m) mounds; Martorelli et al. 2011) (Fig. 7.7). However, even though high-resolution bathymetry mapping, seismic profiling, video observations and surface coral samples collected from these mound structures constitute initial evidence for their true coral mound nature, no sediment cores documenting their subsurface composition are available, allowing for the exploitation of these features as potential archives of past reef formation. Another special case is the Santa Maria di Leuca province in the Ionian Sea (Fig. 7.7), where cold-water corals formed small mounds on top of existing geomorphological features (slumped blocks) in 500–900 m water depth (Taviani et al. 2005b; Savini and Corselli 2010; Vertino et al. 2010; Bargain et al. 2017). However, sediment cores revealed that even these rather thin coral mound deposits serve as valuable archives, displaying, e.g., the response of the cold-water corals to the Sapropel 1 event during the Early Holocene, when decreasing oxygen contents resulted in their temporary demise (Fink et al. 2012), and consequently, in a stagnation in reef/mound formation.

7.2.4 Cold-Water Coral Presence *versus* Reef Formation During the Late Pleistocene to Holocene: How to Interpret Coral Age Data

The general good preservation of cold-water corals of Late Pleistocene to Holocene age allows for a direct absolute dating of the aragonitic skeleton using radiometric techniques (e.g., Adkins et al. 2002; Frank et al. 2004; Wefing et al.

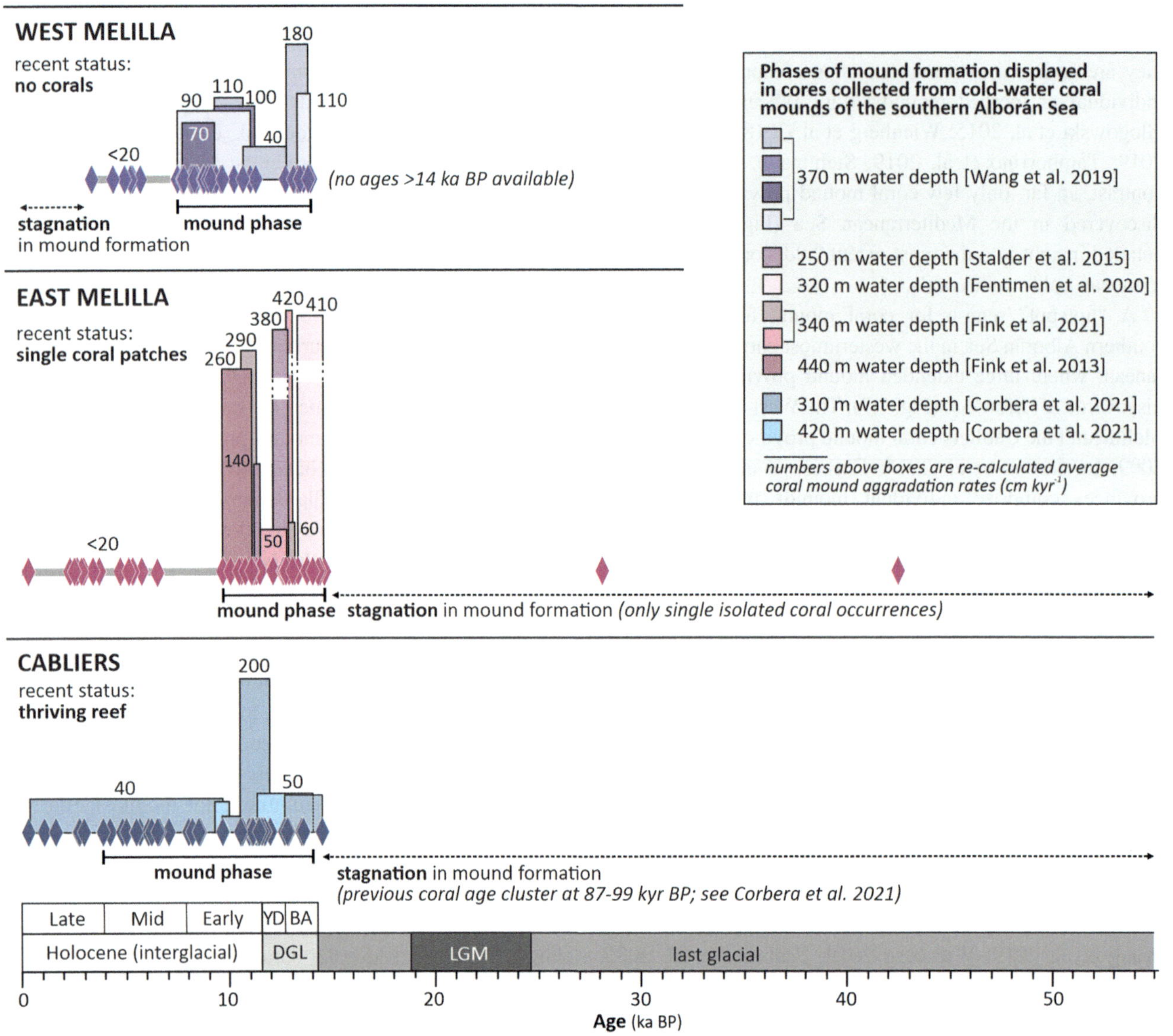

Fig. 7.8 Cold-water coral ages and mound aggradation rates (ARs) obtained from coral mounds of the southern Alborán Sea. The displayed ages (diamonds; purple: West Melilla, pink: East Melilla, blue: Cabliers) cover the last 43 kyr (note: only ages of *D. pertusum* and *M. oculata* are considered) and show that corals were largely absent from this westernmost Mediterranean region during the last glacial period (LGM: Last Glacial Maximum, DGL: deglaciation, BA: Bølling-Allerød, YD: Younger Dryas). Varying mound ARs are indicated by the coloured bars placed behind the ages and the numbers above the bars (AR given in cm kyr^{-1}). The ARs are re-calculated from coral ages obtained from core records (published in: (East Melilla) Fink et al. 2013; Stalder et al. 2015; Fentimen et al. 2020; (West Melilla) Wang et al. 2019; (Cabliers) Corbera et al. 2021). ARs of >20 cm and up to 420 cm kyr^{-1} highlight phases of pronounced mound formation, while AR of <20 cm kyr^{-1} are interpreted as a significant reduction in mound formation. The recent reef status of the three mound provinces varies from no coral occurrences (West Melilla), to coral patches (East Melilla), to presently thriving reefs (Cabliers) (Hebbeln et al. 2009; Fink et al. 2013; Lo Iacono et al. 2014; Corbera et al. 2019)

2017). Radiocarbon and Uranium-series ages obtained from reef framework-forming cold-water corals collected from the entire Mediterranean basin, including very different habitats such as canyons, steep escarpments, overhangs, seamounts and coral mounds, date back to ~43 ka BP (Schröder-Ritzrau et al. 2005; Malinverno et al. 2010; McCulloch et al. 2010; Angeletti and Taviani 2011; Frank et al. 2011; Taviani et al. 2011a; Fink et al. 2012, 2013, 2015; Stalder et al. 2015; Dubois-Dauphin et al. 2017; Taviani et al. 2019b; Wang et al. 2019; Wienberg 2019; Fentimen et al. 2020; Corbera et al. 2021). The majority of published ages (>90%) coincides with the last deglaciation and the Holocene, while last glacial ages (>14 ka BP) are comparably rare (see age compilations presented by Fink et al. 2015; Vertino et al. 2019). Fossil framework-forming corals were also reported from some mud volcanoes in the western Alborán Sea, where they have settled on clasts exhumed by mud breccia and on authigenic carbonates (Rüggeberg and Foubert 2019;

Palomino et al. 2019). Unfortunately, no corals of this very specific habitat have been dated so far, though it is suggested that they frequently colonised mud volcanoes (at least) since the last deglaciation (Margreth et al. 2011), and thus display a similar temporal pattern as coral communities of other Mediterranean habitats.

Overall, the amount of available (and reliable) coral ages is still limited and in particular coral sites from the eastern Mediterranean (Levantine Basin, Aegean Sea) and along the North African margin (e.g., Algerian Basin, southern Ionian Sea) are under-represented, preventing a basin-wide assessment of past reef formation periods. In addition, the age data often derive from single coral occurrences (often recovered through poorly positioned geological dredging with limited information on the original habitat and depth), which testifies a mere occurrence of corals and not necessarily the occurrence of large reefs, which would imply suitable environmental conditions persisting over geologic time. Therefore, even if corals have colonised a given region at a given time, this does not necessarily imply the occurrence of a reef, as it has been indicated for the coral mound provinces in the Alborán Sea (Fink et al. 2013; Wang et al. 2019; Wienberg 2019; Corbera et al. 2021), where unsuitable environmental conditions resulted in reduced mound formation or even in a stagnation of reef development (Fig. 7.8). An increase of stratigraphic records obtained from sediment cores collected on coral mounds will extend our knowledge on the exact timing of reef formation and demise in the Mediterranean Sea (Fink et al. 2012, 2013; Stalder et al. 2015; Wang et al. 2019; Fentimen et al. 2020; Corbera et al. 2021).

The several oceanographic expeditions targeting coral mounds in the Mediterranean Sea in recent years revealed the progress in successfully applying conventional coring (e.g., gravity and piston coring) and drilling techniques (such as the seafloor driling rig MeBo; Freudenthal and Wefer 2013), supported by improved mapping tools and video-guided equipment, to obtain valuable records to study coral reefs in an unprecedented stratigraphic resolution going far back in time (e.g., Hebbeln et al. 2015; Corbera et al. 2021). A further exploration in the so far poorly explored regions of the Mediterranean should benefit of a multidisciplinary and integrated approach in the study of cold-water coral reefs, providing crucial information for a more comprehensive and better understanding of coral reef development in this basin.

7.2.5 Outlook: Increasing Discoveries of Coral Mound Provinces to Be Expected

The distribution of cold-water coral mounds in the Mediterranean Sea and the knowledge of their past development is still limited and largely concentrates on the coral mound provinces of the Alborán Sea (Fig. 7.7). However, further discoveries of coral mound provinces are likely to be expected in the coming years. Strong indicators for the potential presence of coral mounds are contourites, as both deposits often occur side-by-side due to their dependence on strong bottom currents and sediment transport processes (Hebbeln et al. 2016). In the Mediterranean Sea, several observations of contourite drifts describe these deposits as mostly being located at ~200–600 m formed by the action of the Levantine Intermediate Water (see Rebesco and Taviani 2019 and references therein). Interestingly, the known present-day coral mound provinces are located in the same depth range, and their formation is also assumed to be linked to enhanced turbulence (probably generated by internal waves) associated with the flow of the Levantine Intermediate Water (e.g., Wang et al. 2019; Corbera et al. 2021). The co-occurrence of coral mounds and contourites has already been described for the southern Alborán Sea, the Corsica Channel and the Strait of Sicily (Martorelli et al. 2011; Hebbeln et al. 2015, 2016; Rebesco and Taviani 2019; Angeletti et al. 2020), while for other contourite deposits, such as in the southern Tyrrhenian Sea and in the eastern Levantine basin (Schattner et al. 2015; Cattaneo et al. 2017), no indications of coral mounds are available so far. An exception exists for the outer shelf off Israel, where mound-like structures have been described (tentatively interpreted as "carbonate" mounds; Lazar et al. 2016); however, this assumption is based on seismic data only and additional data (videos, seabed samples) are needed to verify the true nature of these seabed morphologies.

Finally, even though the Moroccan margin in the southern Alborán Sea exhibits numerous coral mounds, the remaining North African margin (off Algeria, Tunisia, Libya, and Egypt) is still a white spot on the distribution map of Mediterranean cold-water corals, and consequently of related reef and mound structures, as it is largely underexplored. The observation of the large coral mound provinces on the northern African margin off Morocco might imply that the remainder of the African Mediterranean margin could also host cold-water coral reefs and mounds (either in the present-day or recent past), making this region a prime target area for future oceanographic explorations.

7.3 Mediterranean Cold-Water Corals in the Global Context: "Beyond the Local Story"

The present Mediterranean basin is aptly described as a *Madrepora* sea with respect to its cold-water coral population. The dominance up to pre-dominance of this colonial scleractinian over all other frame-builders is unquestionable. In fact, all recent Mediterranean largest cold-water coral "reefs" and coral mounds from the western and central basins have a *M. oculata* core, as documented in the previous pages. The presence of *D. pertusum* is generally ancillary when not null. This ecological trait differs from most situations we know in the ocean today, where the mutual role of these two scleractinian species seems reversed. In the Atlantic Ocean, at least on both sides of its northern range, *D. pertusum* is reportedly considered as the most relevant frame-builder (Roberts et al. 2009). As admirably depicted by the comprehensive study by Arnaud-Haond et al. (2017), the presence of the other "pillar", *M. oculata*, has been somehow underestimated since it consistently co-occurs with *D. pertusum* in the NE Atlantic. Yet, quantitatively the massive presence of *D. pertusum* in the ocean is factual, as it is its minor role in the Mediterranean at present. A similar but as yet unexplained *M. oculata* dominance (sometimes with *Enallopsammia rostrata*) occurs on the north-western Australian shelf edge, as well as in the Philippine Archipelagoes and off New Caledonia (Freiwald et al. 2017). Recent studies also highlight a remarkable presence of *M. oculata* in the Southeast Atlantic *D. pertusum*-dominated coral reefs, with colonies of the largest sizes ever documented until now (Orejas et al. 2021).

What about the past? Because of widespread collisional tectonics, the Mediterranean basin is marked by uplifted marine deposits contributing to its younger mountain chains, such as the Apennines, and islands (Taviani et al. 2005a). Early-Middle Pleistocene deep-water sedimentary units contain at places abundant fossil remains of previous cold-water coral habitats with extant taxa (Taviani et al. 2019c; Vertino et al. 2019, with references therein). Furthermore, later Pleistocene submarine records complete our knowledge about temporal situations of cold-water coral life in the Mediterranean (Wienberg 2019; Taviani et al. 2019c). This palaeontological heritage indicates the presence of *D. pertusum* at places seemingly prevailing over *M. oculata* (Vertino et al. 2019), or at least matching its presence. Furthermore, it also provides evidence that another colonial scleractinian, *Enallopsammia*, contributed to the cold-water coral frame-builders (Vertino et al. 2019). This scleractinian, known to form reefs and bioconstructions in the Atlantic and Indo-Pacific Oceans (Roberts et al. 2009; Hebbeln et al. 2014; Baco et al. 2017) is (temporarily?) absent from the present Mediterranean Sea. The lesson that we can derive is that the present *Madrepora* domination is only but one of the many cold-water coral configurations experienced by the Mediterranean Sea since the time it achieved its status of semi-enclosed basin and after its post-Messinian fauna renewal. We can safely anticipate that the cold-water coral fauna of the Mediterranean will undergo again major qualitative and quantitative ecological changes in the future at the shorter (decadal/centennial) up to geological temporal scales.

This leads us to consider the potential role of the Mediterranean Sea as *refugium* for cold-water corals at times of environmental crises (De Mol et al. 2005; Henry et al. 2014). In other terms, did the Mediterranean play a pivotal role in safeguarding cold-water coral stocks in the Atlantic Ocean? Answering this question would not be trivial, but could also have relevance to future climatic scenarios. *Desmophyllum pertusum* and *M. oculata* have been documented since the Eocene and the Jurassic, respectively (Squires 1957; Stolarski and Vertino 2007). This long-lasting presence in the ocean combined with their habit to thrive in the relatively more stable deep-sea habitats, secured their almost universal presence in the ocean, except truly polar environments (Roberts et al. 2009). Therefore, we can consider the open ocean like an immense reservoir with numerous options to preserve and accommodate cold-water coral populations whilst some areas get exposed to deleterious situations. Even intuitively, this does not seem to be the case of the peripheral Mediterranean basin, given its exposition to a number of relevant stressors of its inhabiting fauna since the Messinian (Taviani 2002).

If we then restrict further our observations to the last glacial-postglacial times, with the support of precise dating, a global picture emerges about the dynamic response of cold-water corals to oceanographic perturbations. It clearly documents important turnovers at local to regional scales in all the oceans, with periods of complete or partial cold-water coral demise and times of re-occupancy (Frank et al. 2011; Thresher et al. 2011; Wienberg and Titschack 2017; Wienberg et al. 2018; Trotter et al. 2019). The marginal Mediterranean is obviously not immune from such climatically-driven oceanographic changes (McCulloch et al. 2010; Fink et al. 2015; Taviani et al. 2019b), and because of its semi-closeness and latitudinal placement, the consequences for its ecosystems are amplified. This reasoning argues against the hypothesis of erecting the Mediterranean as an ideal *refugium* for the cold-water corals. To lend support to this view, it is sufficient to recall that eastern Mediterranean cold-water coral ecosystems, still flourishing during the late Pleistocene (Taviani et al. 2011a), have yet to recover from the regional extinction they underwent for an intra-Mediterranean short-level event of oxygen depletion (Fink et al. 2012). At best, the Mediterranean could have contributed intermittently as one (but probably not *the* one) regional *refugium* during the Pleistocene (see Boavida et al. 2019a, b) but no conclusive data exist to validate this possibility.

Over larger temporal scales, the Mediterranean could have played a significant role in the dispersal and connectivity of cold-water corals on a global scale. As seen from fossils, many extant cold-water corals inhabited the waters of Mediterranean precursors in the Miocene and perhaps earlier, when open seaways connected the Atlantic and Indo-Pacific oceans. Since the middle Miocene onwards, eastern connections no longer existed, making the Atlantic the sole mother basin. The relationships among the various cold-water coral stocks with the Mediterranean, however, cannot be evaluated directly requiring the assistance of molecular studies on modern cold-water corals at a global scale. The few molecular attempts, based upon cosmopolitan cold-water coral species (also occurring in the Mediterranean), to unravel phylogeographic links proved promising but still insufficient to extract unquestionable evolutionary and biogeographic pathways to date (Addamo et al. 2012, 2020; Boavida et al. 2019a, b).

7.4 The Fate of Mediterranean Cold-Water Coral Communities: "How Will the Story Continue?"

7.4.1 Major Threats to Mediterranean Cold-Water Corals

The Mediterranean Sea only comprises 0.8% of the Earth's surface and 0.3% of the oceans' volume, but it is a biodiversity hotspot as it hosts 7.5% of all known marine species, including deep-sea organisms (Coll et al. 2010). At the same time, the Mediterranean Sea suffers from pressure from human activities over thousands of years, endangering its biodiversity. Total human population of Mediterranean countries has grown from 420 million in 2000 to 470 million in 2010 and is predicted to surpass 570 million by 2030, with more than one third of the population living along the Mediterranean coast (Mediterranean Quality Status Report 2017). As a consequence, impacts from human activities are high and densely concentrated (Claudet and Fraschetti 2010; Ramírez et al. 2018). Among human-induced pressures, cold-water corals have been shown to be mainly threatened by bottom-contact fishing (longlines, trawling and entangling nets), marine litter disposal, pollution and regional consequences of global climate change (Ramirez-Llodra et al. 2011; Otero and Marin 2019).

The direct and mechanical effects of fisheries cause undoubtedly the most important pressure currently affecting cold-water corals in the Mediterranean Sea. Due to the overfishing of shallow coastal waters, the fishing industry developed new equipment and strategies allowing operations not only in deep waters but also in rough terrain such as canyons, and consequently impact common cold-water coral habitats (Otero and Marin 2019 and references therein). At the same time, coral habitats play an important role as a nursery, feeding ground or refuge for many invertebrates and fish, and therefore the increasing fishing pressure poses a severe threat to entire ecosystems (D'Onghia 2019). In the Mediterranean Sea, evidence of direct impact of longline fishing on cold-water corals has been reported (Mytilineou et al. 2014; Angiolillo et al. 2021), and lost fishing gears (longlines and nets) have been frequently observed entangling cold-water corals in submarine canyons (Orejas et al. 2009; Angeletti et al. 2014; Fabri et al. 2014; Giusti et al. 2019; Chimienti et al. 2020), as well as on continental slopes (Taviani et al. 2011b; D'Onghia et al. 2016) and seamounts (Hebbeln et al. 2009; Bo et al. 2014c, 2020). Bottom trawl fishing has been shown to strongly impact populations of the bamboo coral *I. elongata* typically distributed on bathyal mud plains (Maynou and Cartes 2011; Cartes et al. 2013; Mastrototaro et al. 2017; Pierdomenico et al. 2018; Grinyó et al. 2020b).

Moreover, bottom trawling increases the concentration of suspended sediments in the water column as well as the deposition of sediments adjacent to the fished areas (Martín et al. 2014). Experimental work has shown that exposure to sediments caused polyp mortality and decreased growth rates in the scleractinian *D. pertusum*, whose polyps did not fully extend, possibly precluding efficient feeding (Larsson and Purser 2011; Larsson et al. 2013). Deep-sea ecosystems do not escape from chemical pollution (Ramirez-Llodra et al. 2010), and abundant litter from land and river discharge additionally concentrates in several Mediterranean canyons (Fabri et al. 2014; Taviani et al. 2019a; Dominguez-Carrió et al. 2020). Interactions with macro- and micro-litter have been shown to be detrimental for the scleractinian *D. pertusum*, impacting its food capture and growth rates (Chapron et al. 2018; Mouchi et al. 2019).

Compounding all of these direct impacts are the pressures driven by the ongoing global climate change (ocean warming, ocean acidification, deoxygenation, reduction in food supply) (Levin and Le Bris 2015; Sweetman et al. 2017). In the Mediterranean Sea, the low surface productivity (especially in the eastern basin) and the relatively high temperatures of intermediate and deep waters (~13 °C) make ecosystems far more vulnerable to climate change effects than in other oceanic regions (Danovaro 2018). Ocean warming may become life-threatening for several Mediterranean cold-water coral species as they have a limited thermal tolerance (Brooke et al. 2013) and are already today living close to their upper temperature threshold (Ramirez-Llodra et al. 2010). Among scleractinians, *D. pertusum* and *D. dianthus* were shown to significantly reduce growth and energy storage under increased temperatures, whereas *M. oculata* and especially *D. cornigera* can tolerate higher temperatures (Naumann et al. 2013; Gori et al. 2014a, 2015, 2016). Species-specific response may consequently result in future major changes in the depth distribution of Mediterranean species, and in the overall composition of cold-water coral communities (Castellan et al. 2019).

The recent appreciation that *M. oculata* owns a remarkable tolerance to low oxygen levels (Orejas et al. 2021) providing an enhanced resilience to oceanographic stresses than other cold-water corals might explain its comparative success in the Mediterranean basin. Although most Mediterranean scleractinian cold-water corals have shown to acclimate to reduced pH (Maier et al. 2012; Movilla et al. 2014a, b), possible synergistic effects with increased temperatures cannot be excluded (Gori et al. 2016). Moreover, since the observed acclimation is supposed to be sustained by extra energy obtained from feeding (Hennige et al. 2015), the predicted reduction in food supply to the deep-sea may pose an additional severe threat for cold-water corals (but see Maier et al. 2016) in the overall oligotrophic Mediterranean Sea (Danovaro 2018).

At the same time, the Mediterranean is also getting saltier related to a steadily increasing evaporation driven by the rapid surface warming, which is further enhanced by a long-term decrease of river freshwater discharge since the early 1960s, reflecting the effect of river damming as well as recent climate change. Consequently, the salt content of the basin strongly increased over the last decades with this salinification signal rapidly proceeding from the surface into deep waters (Skliris 2019). Indeed, an increase in salinity (~0.1 psu) has been reported for deep waters (250–2500 m) together with a clear rise in temperatures (~0.12 °C) observed over the last 30 years (Bethoux et al. 1990; Danovaro 2018). Testing the response of Mediterranean cold-water corals to the stressors mentioned above has been, and will further be in the future, strongly favoured by the implementation of ex-situ experiments under controlled conditions, based on the successful maintenance of several cold-water corals in aquaria (Orejas et al. 2019b).

7.4.2 Conservation of Mediterranean Cold-Water Corals

Management measures to protect cold-water corals against damage caused by anthropogenic activities are still rather limited in the Mediterranean Sea. Among worldwide conservation policy frameworks, Sustainable Development Goal 14 ("*Conserve and sustainably use the oceans, seas and marine resources for sustainable development*"), defined by the United Nations, is the global framework to develop policies and actions in the marine realm. This goal has been also recently integrated in the Mediterranean regional strategies (GFCM Mid-Term Strategy 2017–2020; UNEP-MAP Mediterranean Strategy for Sustainable Development 2016–2025). Further, the Aichi Target 11 from the Strategic Plan for Biodiversity 2011–2020 (Convention on Biological Diversity (CBD)), called for commitment to preserve at least 10% of coastal and marine areas by 2020 (Rees et al. 2018), emphasizing the protection of ecosystems such as cold-water coral reefs (CBD COP11 Decision XI/16). Within the CBD context, the Ecologically and Biologically Significant Marine Areas (EBSAs) process in the Mediterranean culminated at the CBD 12th COP with the endorsement of 15 EBSAs, including off-shore areas. This was an important milestone highlighting the biological and ecological significance of deep-sea habitats in the Mediterranean Sea to ensure their conservation.

Besides these overarching frameworks, the regional instrument for the conservation of biodiversity in the Mediterranean Sea is the Protocol concerning Specially Protected Areas and Biological Diversity (SPA/BD Protocol) adopted in 1995 within the framework of the Barcelona Convention (Convention for the Protection of the Marine Environment and the Coastal Region of the Mediterranean). This Protocol also serves as main tool for implementing the CBD regarding the sustainable management of coastal and marine biodiversity. Currently, 13 cnidarian species are listed as endangered or threatened in Annex II (see Otero and Marin 2019 for details), whereas only *C. rubrum* is considered an exploited species in need of regulation. Thanks to the IUCN regional Red List assessment (2020), several anthozoan species with deep-water distribution were additionally classified as "threatened", including the bamboo coral *I. elongata* which is described as "critically endangered", mainly as a result of deep-water bottom trawling (Otero et al. 2017).

Despite the existence of these international and regional frameworks, the implementation of these policies into management and conservation measures in the Mediterranean Sea is still limited, and frequently inefficient for the effective conservation of cold-water corals, because no specific protection measures are enforced (Otero and Marin 2019). The scarce management measures currently implemented in the Mediterranean Sea to protect cold-water corals, especially from fishing damage, become evident when compared with the increasing number of Marine Protected Areas (MPAs) having been declared for the Northeast Atlantic Ocean (e.g., Norway, Azores) and the still scarce number of offshore MPAs in the Mediterranean. Indeed, the presence of the framework-forming species *D. pertusum* and *M. oculata* occurs in a few designated MPAs from the western Mediterranean Sea such as the National Park of Calanques in France or the Chella Bank (Seco de los Olivos) in Spain, this last one declared in 2014 as Site of Community Importance (SCI). In addition, several submarine canyons in the Gulf of Lions inhabited by cold-water coral communities (Figs. 7.2 and 7.6) are part of the Gulf of Lyons Natural Park established by France in 2011, and some areas characterized by the presence of cold-water corals have been recently included in the Natura 2000 sites network by the European Union, but specific management measures still need to be defined and enforced. It is worthy to mention that while MPAs and Other Effective area-based Conservation Measures (OECMs) coverage increased from 4.56% (2012) to 7.14% (2016) in the

Mediterranean basin, more than 55% of this increment just covers depth ranges between 0 and 200 m leaving deep-sea areas unprotected.

Besides MPAs, the General Fisheries Commission for the Mediterranean (GFCM) has declared the Santa Maria di Leuca cold-water coral province in the Ionian Sea, and the Eratosthenes Seamounts south of Cyprus (Fig. 7.2), as Fishery Restricted Areas (FRAs) for the protection of Vulnerable Marine Ecosystems (VME, FAO 2009), and five additional FRAs including sites with cold-water coral communities have been established for the governance of Essential Fish Habitats (EFHs). Discussions at the Mediterranean level for the implementation of management measures to enhance conservation of VMEs (including cold-water corals) are currently ongoing within the GFCM Working Group of Vulnerable Marine Ecosystems, with special focus at present on the Bari Canyon cold-water coral community (Angeletti et al. 2021). Following a precautionary approach, bottom trawl fishing has been restricted below 1000 m depth in the entire Mediterranean Sea since 2016. However, this measure leaves many cold-water coral habitats unprotected, as the known bathymetric distribution for a large number of cold-water coral species lies mostly between 100 and 600 m depth (see also Figs. 7.3 and 7.5).

Looking forward, the application of the Marine Strategy Framework Directive (MSFD 2008/56/EC), representing the EU's Integrated Maritime Policy tool to achieve Good Environmental Status (GES) of European marine waters, including the deep sea, could provide a management framework to define a sound ecosystem-based approach for the management and protection of Mediterranean deep-sea ecosystems (Orejas et al. 2020; Danovaro et al. 2020; Fanelli et al. 2021). Given the lack of a complete census of the Mediterranean cold-water coral occurrences, the implementation of refined suitability models would prove an important tool for future conservation actions (Bargain et al. 2017; Lauria et al. 2021; Matos et al. 2021).

7.4.3 The Way Forward

Knowledge on the current cold-water corals distribution in the Mediterranean Sea is still imperfect, a bias to their effective conservation trough marine spatial planning and protection areas. Exploration in areas that have still no or have just been little observed is required, with a special need for increasing depths. Indeed, most of the explorations have been so far performed in relatively shallow waters (100–600 m depth), with scant information available for deepest depths. At the same time, there is a strong need for more biological and ecological studies to assess the vulnerability and adaptation capacities of cold-water coral communities in this rapidly changing marine region under multiple anthropogenic stressors, including climate change.

Acknowledgements We are grateful to Furu Mienis and Erik Cordes for the kind invitation to contribute to this book with our point of view about cold-water corals in the Mediterranean Sea. We are also grateful to Nadine Le Bris and Andres Rüggeberg for their constructive and helpful reviews, which significantly enhanced the content and form of this chapter. Information provided in this chapter by and large originates from a number of European (Moundforce, Hermes, Hermione, Coconet, Coralfish, Idem, Life-Indemares) and national (Aplabes) research projects. This study was supported by the German Science Foundation (DFG) and the Cluster of Excellence "The Ocean Floor—Earth's Uncharted Interface", and by the European Union's Horizon 2020 Research and Innovation Program, under the MSC Project Habiss (Grant Agreement No. GA890815), Atlas project (Grant Agreement No. 678760) and iAtlantic project (Grant agreement No. 818123). This output reflects only the author's view, and the European Union cannot be held responsible for any use that may be made of the information contained therein. This is Ismar-CNR bologna scientific contribution n 2042.

References

Addamo AM, Reimer JD, Taviani M, Freiwald A, Machordom A (2012) *Desmophyllum dianthus* (Esper, 1794) in the scleractinian phylogeny and its intraspecific diversity. PLoS One 7(11):e50215

Addamo AM, Vertino A, Stolarski J, García-Jiménez R, Taviani M, Machordom A (2016) Merging scleractinian genera: the overwhelming genetic similarity between solitary *Desmophyllum* and colonial *Lophelia*. BMC Evol Biol 16:108

Addamo AM, Miller KJ, Taviani M, Machordom A (2020) Global-scale genetic structure of a cosmopolitan cold-water coral species. Aquat Conserv Mar Freshwat Ecosyst 31:1–14

Adkins J, Griffin S, Kashgarian M, Cheng H, Druffel ERM, Boyle EA, Edwards RL, Shen CC (2002) Radiocarbon dating of deep-sea corals. Radiocarbon 44:567–580

Aguilar R, Pastor X, García S, Marín P (2013) Importance of seamount-like feature for conserving Mediterranean marine habitats and threatened species. www.oceana.org

Aguirre J, Ocaña O, Pérez-Asensio JN, Domènech R, Martinell J, Mayoral E, Santos A (2020) Mesophotic azooxanthellate coral communities and submarine seascape during the early Pliocene in Manilva Basin (S Spain). Coral Reefs 39:1739–1752

Altuna A, Poliseno A (2019) Taxonomy, genetics and biodiversity of Mediterranean deep-sea corals and cold-water corals. In: Saco O, del Valle C, Jiménez C (eds) Mediterranean cold-water corals: past, present and future, Springer series: coral reefs of the world. Springer, pp 121–156

Angeletti L, Taviani M (2011) Entrapment, preservation and incipient fossilization of benthic predatory molluscs within deep-water coral frames in the Mediterranean Sea. Geobios 44:543–548

Angeletti L, Taviani M, Canese S, Foglini F, Mastrototaro F, Argnani A, Trincardi F, Bakran-Petricioli T, Ceregato A, Chimienti G, Mačić V, Poliseno A (2014) New deep-water cnidarian sites in the southern Adriatic Sea. Mediterranean Mar Sci 15(2):263–273

Angeletti L, Canese S, Franchi F, Montagna P, Reitner J, Walliser EO, Taviani M (2015a) The "chimney forest" of the deep Montenegrin margin, south-eastern Adriatic Sea. Mar Pet Geol 66:542–554

Angeletti L, Mecho A, Doya C, Micallef A, Huvenne V, Georgiopoulou A, Taviani M (2015b) First report of live deep-water cnidarian assemblages from the Malta Escarpment. Ital J Zool 82:291–297

Angeletti L, Prampolini M, Foglini F, Grande V, Taviani M (2019) Cold-water coral habitat in the Bari Canyon System, Southern Adriatic Sea (Mediterranean Sea) In: Harris P, Baker E (eds) Seafloor geomorphology as benthic habitat, Chapter 49 - Geohab Atlas and seafloor geomorphic features and benthic habitats (2nd edn), pp 811–824

Angeletti L, Castellan G, Montagna P, Remia A, Taviani M (2020) The "Corsica Channel Cold-Water Coral Province" (Mediterranean Sea). Front Mar Sci 7:1–15

Angeletti L, D'Onghia G, Otero MM, Settanni Taviani M (2021) A perspective for best governance of the bari canyon deep-sea ecosystems. Water 13(12):1646. https://doi.org/10.3390/w13121646

Angiolillo M, Bo M, Bavestrello G, Giusti M, Salvati E, Canese S (2012) Record of *Ellisella paraplexauroides* (Anthozoa: Alcyonacea: Ellisellidae) in Italian waters (Mediterranean Sea). Mar Biodivers Rec 5:e4

Angiolillo M, Gori A, Canese S, Bo M, Priori C, Bavestrello G, Salvati E, Erra F, Greenacre M, Santangelo G (2016) Distribution and population structure of deep-dwelling red coral in the Northwest Mediterranean. Mar Ecol 37:294–310

Angiolillo M, La Mesa G, Giusti M, Salvati E, Di Lorenzo B, Rossi L, Canese S, Tunesi L (2021) New records of scleractinian cold-water coral (CWC) assemblages in the southern Tyrrhenian Sea (western Mediterranean Sea): human impacts and conservation prospects. Prog Oceanogr 197. https://doi.org/10.1016/j.pocean.2021.102656

Arnaud-Haond S, Van den Beld IMJ, Becheler R, Orejas C, Menot L, Frank N, Grehan A, Bourillet JF (2017) Two "pillars" of cold-water coral reefs along Atlantic European margins: prevalent association of *Madrepora oculata* with *Lophelia pertusa*, from reef to colony scale. Deep Sea Res Part II 145:110–119

Aymà A, Aguzzi J, Canals M, Company JB, Lastras G, Mecho A, Lo Iacono C (2019) Occurrence of living cold-water corals at large depths within submarine canyons of the northwestern Mediterranean Sea. In: Orejas Saco del Valle C, Jiménez C (eds) Mediterranean cold-water corals: past, present and future, springer series: coral reefs of the world. Springer, pp 271–284

Baco AR, Morgan N, Roark EB, Silva M, Shamberger KEF, Miller K (2017) Defying dissolution: discovery of deep-sea scleractinian coral reefs in the north Pacific. Sci Rep 7:5436

Ballesteros E (2006) Mediterranean coralligenous assemblages: a synthesis of present knowledge. Oceanogr Mar Biol Annu Rev 44:123–195

Bargain A, Marchese F, Savini A, Taviani M, Fabri MC (2017) Santa Maria di Leuca Province (Mediterranean Sea): identification of suitable mounds for cold-water coral settlement using geomorphometric proxies and maxent methods. Front Mar Sci 4:338

Barrier P, Di Geronimo I, La Perna R, et al. (1996) Taphonomy of deep-sea hard and soft bottom communities: the Pleistocene of Lazzàro (Southern Italy). In: Meléndez Hevia G, Blasco Sancho M, Pérez Urresti I (eds) II Meet Taphonomy Fossilization, Institución "Fernando el Católico", Zaragoza, pp 39–45

Benavides M, Bednarz VN, Ferrier-Pagès C (2017) Diazotrophs: overlooked key players within the coral symbiosis and tropical reef ecosystems? Front Mar Sci 4:10

Bertolino M, Ricci S, Canese S, Cau A, Bavestrello G, Pansini M, Bo M (2019) Diversity of the sponge fauna associated with white coral banks from two Sardinian canyons (Mediterranean Sea). J Mar Biol Assoc U K 99:1735–1751

Bethoux JP, Gentili B, Raunet J, Tailliez D (1990) Warming trend in the western Mediterranean deep water. Nature 347:660–662

Bilan M, Grinyó J, Ambroso S, Lo Iacono C, Huvenne V, Fabri MC, Duran R, Paradis S, Arjona-Camas M, Palanques A, Santín A, Gori A, Piraino S, Rossi S, Gili JM, Puig P (2020) eDeep sea biology symposium 2020

Bo M, Bavestrello G, Canese S, Giusti M, Salvati E, Angiolillo M, Greco S (2009) Characteristics of a black coral meadow in the twilight zone of the central Mediterranean Sea. Mar Ecol Prog Ser 397:53–61

Bo M, Bertolino M, Borghini M, Castellano M, Harriague C, Di Camillo CG, Gasparini GP, Misic C, Povero P, Schroeder K, Bavestrello G (2011a) Characteristics of the mesophotic megabenthic assemblage of the Vercelli Seamount (North Tyrrhenian Sea). PLoS One 6: e16357

Bo M, Bavestrello G, Canese S, Giusti M, Angiolillo M, Cerrano C, Salvati E, Greco S (2011b) Coral assemblages off the Calabrian Coast (South Italy) with new observations on living colonies of *Antipathes dichotoma*. Ital J Zool 78:231–242

Bo M, Canese S, Spaggiari C, Pusceddu A, Bertolino M, Angiolillo M, Giusti M, Loreto MF, Salvati E, Greco S, Bavestrello G (2012) Deep coral oases in the South Tyrrhenian Sea. PLoS One 7:e49870

Bo M, Canese S, Bavestrello G (2014a) Discovering Mediterranean black coral forests: *Parantipathes larix* (Anthozoa: Hexacorallia) in the Tuscan Archipelago. Ital J Zool 81:112–125

Bo M, Cerrano C, Canese S, Salvati E, Angiolillo M, Santangelo G, Bavestrello G (2014b) The coral assemblages of an off-shore deep Mediterranean rocky bank (NW Sicily, Italy). Mar Ecol 35:332–342

Bo M, Bava S, Canese S, Angiolillo M, Cattaneo-Vietti R, Bavestrello G (2014c) Fishing impact on deep Mediterranean rocky habitats as revealed by ROV investigation. Biol Conserv 171:167–176

Bo M, Bavestrello G, Angiolillo A, Calcagnile L, Canese S, Cannas R, Cau A, D'Elia M, D'Oriano F, Follesa MC, Quarta G, Cau A (2015) Persistence of pristine deep-sea coral gardens in the Mediterranean Sea (SW Sardinia). PLoS One 10:e0119393

Bo M, Coppari M, Betti F, Enrichetti F, Bertolino M, Massa F, Bava S, Gay G, Cattaneo-Vietti R, Bavestrello G (2020) The high biodiversity and vulnerability of two Mediterranean bathyal seamounts supports the need for creating offshore protected areas. Aquat Conserv Mar Freshwat Ecosyst 31:543–566

Boavida J, Becheler R, Addamo AM, Sylvestre F, Arnaud-Haond S (2019a) Past, present and future connectivity of Mediterranean cold-water corals: Patterns, drivers and fate in a technically and environmentally changing world. In: Orejas Saco del Valle C, Jiménez C (eds) Mediterranean cold-water corals: past, present and future, Springer series: coral reefs of the world. Springer, pp 357–372

Boavida J, Becheler R, Choquet M, Frank N, Taviani M, Bourillet J-F, Meistertzheim AL, Grehan A, Savini A, Arnaud-Haond S (2019b) Out of the Mediterranean? Post-glacial colonization pathways varied among cold-water coral species. J Biogeogr 46:915–931

Bongiorni L, Mea M, Gambi C, Pusceddu A, Taviani M, Danovaro R (2010) Deep-water scleractinian corals promote higher biodiversity in deep-sea meiofaunal assemblages along continental margins. Biol Conserv 143:1687–1700

Brooke S, Ross SW, Bane JM, Seim HE, Young CM (2013) Temperature tolerance of the deep-sea coral *Lophelia pertusa* from the southeastern United States. Deep Sea Res II 92:240–248

Cairns SD (2007) Deep-water corals: an overview with special reference to diversity and distribution of deep-water scleractinian corals. Bull Mar Sci 81:311–322

Canals M, Puig P, de Madron X, Heusner S, Palanques A, Fabres J (2006) Flushing submarine canyons. Nature 444:354–357

Carbonara P, Zupa W, Follesa MC, Cau A, Capezzuto F, Chimienti G, D'Onghia G, Lembo G, Pesci P, Porcu C, Bitetto I, Spedicato MT, Maiorano P (2020) Exploring a deep-sea vulnerable marine ecosystem: *Isidella elongata* (Esper, 1788) species assemblages in the Western and Central Mediterranean. Deep Sea Res I 166:103406

Carlier A, Le Guilloux E, Olu K, Sarrazin J, Mastrototaro F, Taviani M, Clavier J (2009) Trophic relationships in a deep Mediterranean cold-water coral bank (Santa Maria di Leuca, Ionian Sea). Mar Ecol Prog Ser 397:125–137

Cartes JE, Lo Iacono C, Mamouridis V, López-Pérez C, Rodríguez P (2013) Geomorphological, trophic and human influences on the bamboo coral *Isidella elongata* assemblages in the deep

Mediterranean: to what extent does Isidella form habitat for fish and invertebrates? Deep Sea Res I 76:52–65

Castellan G, Angeletti L, Taviani M, Montagna P (2019) The yellow coral *Dendrophyllia cornigera* in a warming ocean. Sci Rep 6:692

Cattaneo A, Miramontes E, Samalens K, Garreau P, Caillaud M, Marsset B, Corradi N, Migeon S (2017) Contourite identification along Italian margins: the case of the Portofino drift (Ligurian Sea). Mar Pet Geol 87:137–147

Cau A, Follesa MC, Moccia D et al (2015) Deepwater corals biodiversity along roche du large ecosystems with different habitat complexity along the south Sardinia continental margin (CW Mediterranean Sea). Mar Biol 162:1865–1878

Chapron L, Peru E, Engler A, Ghiglione JF, Meistertzheim AL, Pruski AM, Purser A, Vétion G, Galand PE, Lartaud F (2018) Macro- and microplastics affect cold-water corals growth, feeding and behaviour. Sci Rep 8:15299

Chapron L, Lartaud F, Le Bris N, Peru E, Galand PE (2020) Local variability in microbiome composition and growth suggests habitat preferences for two reef-building cold-water coral species. Front Microbiol 11:275

Chimienti G, Bo M, Taviani M, Mastrototaro F (2019) Occurrence and biogeography of Mediterranean cold-water corals. In: Saco O, del Valle C, Jiménez C (eds) Mediterranean cold-water corals: past, present and future, Springer series: coral reefs of the world. Springer, pp 213–243

Chimienti G, Angeletti L, Furfaro G, Canese S, Taviani M (2020) Habitat, morphology and trophism of *Tritonia callogorgiae* sp. nov., a large nudibranch inhabiting *Callogorgia verticillata* forests in the Mediterranean Sea. Deep Sea Res Part I 165

Claudet J, Fraschetti S (2010) Human-driven impacts on marine habitats: a regional meta-analysis in the Mediterranean Sea. Biol Conserv 143:2195–2206

Coll M, Piroddi C, Steenbeek J, Kaschner K, Ben Rais Lasram F, Aguzzi J, Ballesteros E, Bianchi CN, Corbera J, Dailianis T, Danovaro R, Estrada M, Froglia C, Galil BS, Gasol JM, Gertwagen R, Gil J, Guilhaumon F, Kesner-Reyes K, Kitsos MS, Koukouras A, Lampadariou L, Laxamana E, López-Fé de la Cuadra CM, Lotze HK, Martin D, Mouillot D, Oro D, Raicevich S, Rius Barile J, Saiz-Salinas JI, San Vicente C, Somot S, Templado J, Turon X, Vafidis D, Villanueva R, Voultsiadou E (2010) The biodiversity of the Mediterranean Sea: estimates, patterns, and threats. PLoS One 5:e11842

Comas M, Pinheiro LM, Ivanov M, and TTR-17 scientific party (2009) Deep-water coral mounds in the Alboran Sea: the Melilla mound field revisited. Geo-Marine Research on the Mediterranean and European-Atlantic Margins International Conference and TTR-17 post-cruise meeting of the training-through-research programme:8–9

Coppari M, Mestice F, Betti F, Bavestrello G, Castellano L, Bo M (2019) Fragmentation, re-attachment ability and growth rate of the Mediterranean black coral *Antipathella subpinnata*. Coral Reefs 38: 1–14

Coppari M, Fumarola L, Bramanti L, Romans P, Pillot R, Bavestrello G, Bo M (2020) Unveiling asexual reproductive traits in black corals: polyp bail-out in *Antipathella subpinnata*. Coral Reefs 39:1517–1523

Corbera G, Lo Iacono C, Gràcia E, Grinyó J, Pierdomenico M, Huvenne VAI, Aguilar R, Gili JM (2019) Ecological characterisation of a Mediterranean cold-water coral reef: Cabliers Coral Mound Province (Alboran Sea, western Mediterranean). Prog Oceanogr 175: 245–262

Corbera G, Lo Iacono C, Standish CD, Anagnostou E, Titschack J, Katsamenis O, Cacho I, Van Rooij D, Huvenne VAI, Foster GL (2021) Glacio-eustatic variations and Sapropel events as main controls on the Middle Pleistocene-Holocene evolution of the Cabliers Coral Mound Province (W Mediterranean). Quat Sci Rev

Corselli C (2001) Change and diversity: the Mediterranean deep corals from the Miocene to the present. In: Faranda FM, Guglielmo L, Spezie G (eds) Mediterranean ecosystems: structures and processes. Springer, Milano, pp 361–366

D'Onghia G (2019) Cold-water corals as shelter, feeding and life-history critical habitats for fish species: Ecological interactions and fishing impact. In: Orejas Saco del Valle C, Jiménez C (eds) Mediterranean cold-water corals: past, present and future, springer series: coral reefs of the world. Springer, pp 335–356

D'Onghia G, Capezzuto F, Cardone F, Carlucci R, Carluccio A, Chimienti G, Corriero G, Longo C, Maiorano P, Mastrototaro F, Panetta P, Rosso A, Sanfilippo R, Sion L, Tursi A (2015) Macro- and mega- fauna recorded in the submarine Bari Canyon (southern Adriatic Sea, Mediterranean Sea) using different tools. Mediterr Mar Sci 16:180–196

D'Onghia G, Calculli E, Capezzuto F, Carlucci R, Carluccio A, Maiorano P, Pollice A, Ricci P, Sion L, Tursi A (2016) New records of cold-water coral sites and fish fauna characterization of a potential network existing in the Mediterranean Sea. Mar Ecol 37:1398–1422

Danovaro R (2018) Climate change impacts on the biota and on vulnerable habitats of the deep Mediterranean Sea. Rendiconti Lincei. Scienze Fisiche e Naturali 29:525–541

Danovaro R, Fanelli E, Canals M, Ciuffardi T, Fabri MC, Taviani M, Argyrou M, Azzurro E, Bianchelli S, Cantafaro A, Carugati L, Corinaldesi C, de Haan WP, Dell'Anno A, Evans J, Foglini F, Galil B, Gianni M, Soldevila E (2020) Towards a marine strategy for the deep Mediterranean Sea: analysis of current ecological status. Mar Policy 112:103781

De Leo FC, Puig P, Lo Iacono C, Durán R, Grinyó J, Ambroso S, Arjona-Camas M, Paradis S, Palanques S, ABIDES cruise team (2019) First observations of living cold-water corals surrounded by fishing grounds in Blanes Canyon (NW Mediterranean). International symposium on deep-sea corals 2019, Cartagena

De Mol B, Henriet JP, Canals M (2005) Development of coral banks in Porcupine Seabight: Do they have Mediterranean ancestors? In: Freiwald A, Roberts JM (eds) Cold-water corals and ecosystems. Springer, Heidelberg, pp 515–533

Delibrias G, Taviani M (1985) Dating the death of Mediterranean deep-sea scleractinian corals. Mar Geol 62:175–180

Di Geronimo I (1979) Il Pleistocene in facies batiale di Valle Palione (Grammichele, Catania). Bollettino Malacologico 15:85–156

Di Geronimo I (1987) Bionomie des peuplements benthiques des substrats meubles et rocheux plioquaternaires du Détroit de Messine. Documents et Travaux IGAL 11:153–169

Dominguez-Carrió C, Sanchez-Vidal A, Estournel C, Corbera G, Riera JL, Orejas C, Canals M, Gili JM (2020) Seafloor litter sorting in different domains of Cap de Creus continental shelf and submarine canyon (NW Mediterranean Sea). Mar Pollut Bull 161:111744

Dubois-Dauphin Q, Montagna P, Siani G, Douville E, Wienberg C, Hebbeln D, Liu Z, Kallel N, Dapoigny A, Revel M, Pons-Branchu E, Taviani M, Colin C (2017) Hydrological variations of the intermediate water masses of the western Mediterranean Sea during the past 20 ka inferred from neodymium isotopic composition in foraminifera and cold-water corals. Clim Past 13:17–37

Fabri MC, Pedel L, Beuck L, Galgani F, Hebbeln D, Freiwald A (2014) Megafauna of vulnerable marine ecosystems in French Mediterranean submarine canyons: spatial distribution and anthropogenic impacts. Deep-Sea Res II 104:184–207

Fabri MC, Bargain A, Pairaud I, Pedel L, Taupier-Letage I (2017) Cold-water coral ecosystems in Cassidaigne Canyon: an assessment of their environmental living conditions. Deep-Sea Resh II 137:436–453

Fanelli E, Delbono I, Cocito S, Peirano A (2017) Cold-water coral *Madrepora oculata* in the eastern Ligurian Sea (NW Mediterranean): historical and recent findings. Aquat Conserv Mar Freshwat Ecosyst 27:965–975

Fanelli E, Bianchelli S, Foglini F, Canals M, Castellan G, Güell-Bujons Q, Galil B, Goren M, Evans J, Fabri MC, Vaz S, Ciuffardi T, Schembri PJ, Angeletti L, Taviani M, Danovaro R (2021) Identifying priorities for the protection of deep Mediterranean Sea ecosystems through an integrated approach. Front Mar Sci 8. https://doi.org/10.3389/fmars.2021.698890

FAO (2009) International guidelines for the management of deep-sea fisheries in the high seas. FAO, Rome, p 73

Fentimen R, Feenstra E, Rüggeberg A, Vennemann T, Hajdas I, Adatte T, Van Rooij D, Foubert A (2020) Cold-water coral mound archive provides unique insights into intermediate water mass dynamics in the Alboran Sea during the last deglaciation. Front Mar Sci 7:354

Fink HG, Wienberg C, Hebbeln D, McGregor HV, Schmiedl G, Taviani M, Freiwald A (2012) Oxygen control on Holocene cold-water coral development in the eastern Mediterranean Sea. Deep Sea Res Part I 62:89–96

Fink HG, Wienberg C, De Pol-Holz R, Wintersteller P, Hebbeln D (2013) Cold-water coral growth in the Alboran Sea related to high productivity during the Late Pleistocene and Holocene. Mar Geol 339:71–82

Fink HG, Wienberg C, De Pol-Holz R, Hebbeln D (2015) Spatio-temporal distribution patterns of Mediterranean cold-water corals (*Lophelia pertusa* and *Madrepora oculata*) during the past 14,000 years. Deep-Sea Res I 103:37–48

Frank N, Paterne M, Ayliffe L, van Weering TCE, Henriet J-P, Blamart D (2004) Eastern North Atlantic deep-sea corals: tracing upper intermediate water $\Delta^{14}C$ during the Holocene. Earth Planet Sci Lett 219:297–309

Frank N, Freiwald A, López Correa M, Wienberg C, Eisele M, Hebbeln D, Van Rooij D, Henriet JP, Colin C, van Weering T, de Haas H, Buhl-Mortensen P, Roberts JM, De Mol B, Douville E, Blamart D, Hatte C (2011) Northeastern Atlantic cold-water coral reefs and climate. Geology 39:743–746

Freiwald A, Beuck L, Rüggeberg A, Taviani M, Hebbeln D, R/V Meteor Cruise M70-1 participants (2009) The white coral community in the central Mediterranean Sea revealed by ROV surveys. Oceanography 22:58–74

Freiwald A, Boetius A, Bohrmann G (2011) Deep water ecosystems of the Eastern Mediterranean, Cruise No. 70, Leg 1-3, September 24–December 8, 2006, La Valetta (Malta) - Heraklion (Greece). METEOR-Berichte, 11-5. Leitstelle Deutsche Forschun- gsschiffe Institut für Meereskunde der Universität Hamburg

Freiwald A, Rogers A, Hall-Spencer J, Guinotte JM, Davies AJ, Yesson C, Martin CS, Weatherdon LV (2017) Global distribution of cold-water corals (version 5.0). Fifth update to the dataset in Freiwald et al. (2004) by UNEP-WCMC, in collaboration with Andre Freiwald and John Guinotte. In: Centre UEWCM (ed), Cambridge

Freudenthal T, Wefer G (2013) Drilling cores on the sea floor with the remote-controlled sea floor drilling rig MeBo. Geosci Instrum Methods Data Syst 2:329–337

Gaino E, Scoccia F (2010) Gamete spawning in Antipathella subpinnata (Anthozoa, Antipatharia): a structural and ultrastructural investigation. Zoomorphology 129:213–219

Galand PE, Chapron L, Meistertzheim AL, Peru E, Lartaud F (2018) The effect of captivity on the dynamics of active bacterial communities differs between two deep-sea coral species. Front Microbiol 9:2565

Galand PE, Remize M, Meistertzheim AL, Pruski AM, Peru E, Suhrhoff TJ, Le Bris N, Vétion G, Lartaud F (2020) Diet shapes cold-water corals bacterial communities. Environ Microbiol 22:354–368

Galil B, Zibrowius H (1998) First benthos samples from Eratosthenes Seamount, eastern Mediterranean. Senckenberg Marit 28:111–121

Gil J, Rueda JL, Baro J, Canoura J, Palomino D, DEEPER Team (2010) Composición y estructura de la comunidad bentóni- co-demersal de montes submarinos del Mar de Alborán. Libro de Resúmenes. XVI Simposio Ibérico de Estudios de Biología Marina. Alicante (España), 6 al 10 Septiembre 2010

Giusti M, Bo M, Bavestrello G, Angiolillo M, Salvati E, Canese S (2012) Record of *Viminella flagellum* (Alcyonacea: Ellisellidae) in Italian waters (Mediterranean Sea). Mar Biodivers Rec 5:e34

Giusti M, Canese S, Fourt M, Bo M, Innocenti C, Goujard A, Daniel B, Angeletti L, Taviani M, Aquilina L, Tunesi L (2019) Coral forests and derelict fishing gears in submarine canyon systems of the Ligurian Sea. Prog Oceanogr 178:102186

Glogowski S, Dullo WC, Feldens P, Liebetrau V, von Reumont J, Hühnerbach V, Krastel S, Wynn RB, Flögel S (2015) The Eugen Seibold coral mounds offshore western Morocco: oceanographic and bathymetric boundary conditions of a newly discovered cold-water coral province. Geo-Mar Lett 35:257–269

Gori A, Orejas C, Madurell T, Bramanti L, Martins M, Quintanilla E, Marti-Puig P, Lo Iacono C, Puig P, Requena S, Greenacre M, Gili JM (2013) Bathymetrical distribution and size structure of cold-water coral populations in the Cap de Creus and Lacaze-Duthiers canyons (northwestern Mediterranean). Biogeosciences 10:2049–2060

Gori A, Reynaud S, Orejas C, Gili JM, Ferrier-Pagès C (2014a) Physiological performance of the cold-water coral *Dendrophyllia cornigera* reveals its preference for temperate environments. Coral Reefs 33: 665–674

Gori A, Grover R, Orejas C, Sikorski S, Ferrier-Pagès C (2014b) Uptake of dissolved free amino acids by four cold-water coral species from the Mediterranean Sea. Deep-Sea Res II 99:42–50

Gori A, Reynaud S, Orejas C, Ferrier-Pagès C (2015) The influence of flow velocity and temperature on zooplankton capture rates by the cold-water coral *Dendrophyllia cornigera*. J Exp Mar Biol Ecol 466: 92–97

Gori A, Ferrier-Pagès C, Hennige SJ, Murray F, Rottier C, Wicks LC, Roberts JM (2016) Physiological response of the cold-water coral *Desmophyllum dianthus* to thermal stress and ocean acidification. PeerJ 4:e1606

Gori A, Bavestrello G, Grinyó J, Dominguez-Carrió C, Ambroso S, Bo M (2017) Animal forests in deep coastal bottoms and continental shelf of the Mediterranean Sea. In: Rossi S, Bramanti L, Gori A, Orejas C (eds) Marine animal forests. Springer, pp 207–234

Gori A, Tolosa I, Orejas C, Rueda L, Viladrich N, Grinyó J, Flögel S, Gili JM, Grover R, Ferrier-Pagès C (2018) Biochemical composition of the cold-water coral *Dendrophyllia cornigera* under contrasting productivity regimes: insights from lipid biomarkers and compound-specific isotopes. Deep-Sea Res I 141:106–117

Grinyó J, Gori A, Ambroso S, Purroy A, Calatayud C, Dominguez-Carrió C, Coppari M, Lo Iacono C, López-González PJ, Gili JM (2016) Diversity distribution and population size structure of deep Mediterranean gorgonian assemblages (Menorca Channel, Western Mediterranean Sea). Prog Oceanogr 145:42–56

Grinyó J, Gori A, Greenacre M, Requena S, Canepa A, Lo Iacono C, Ambroso S, Purroy A, Gili JM (2018a) Meagebentic assemables in the continental shelf edgea nd upper slope of the Menorca Channel. Western Mediterranean Sea Prog Oceanogr 162:40–51

Grinyó J, Viladrich N, Díaz D, Muñoz A, Mallol S, Slaazar J, Castillo R, Gili JM, Gori A (2018b) Reprodcution, energy sotrage and metabolic requirement sin a mesophotic population of the gorgonian *Paramuricea macrospina*. PLoS One 13:e0203308

Grinyó J, Garriga A, Soler-Membrives A, Santín A, Ambroso S, López-González PJ, Díaz D (2020a) Soft corals assemblages in deep enviornments of the Menorca Channel (Western Mediterranean Sea). Prog Oceanogr 188:102435

Grinyó J, Lo Iacono C, Pierdomenico M, Conlon S, Corbera G, Gràcia E (2020b) Evidences of human impact on megabenthic assemblages of bathyal sediments in the Alboran Sea (western Mediterranean). Deep Sea Res Part I 165:103369

Hebbeln D (2019) Highly variable submarine landscapes in the Alboran Sea created by cold-water corals. In: Saco O, del Valle C, Jiménez C (eds) Mediterranean cold-water corals: past, present and future, Springer series: coral reefs of the world. Springer, pp 61–65

Hebbeln D, Wienberg C, Beuck L, Freiwald A, Wintersteller P, Cruise Participants (2009) Report and preliminary results of RV POSEIDON Cruise POS 385 "Cold-Water Corals of the Alboran Sea (western Mediterranean Sea)", Faro - Toulon, May 29–June 16 2009 Reports of the Department of Geosciences (GeoB), University of Bremen, No 273 79

Hebbeln D, Wienberg C, Wintersteller P, Freiwald A, Becker M, Beuck L, Dullo C, Eberli GP, Glogowski S, Matos L, Forster N, Reyes-Bonilla H, Taviani M, MSM 20-4 shipboard scientific party (2014) Environmental forcing of the Campeche cold-water coral province, southern Gulf of Mexico. Biogeosciences 11:1799–1815

Hebbeln D, Wienberg C, Bartels M, Bergenthal M, Frank N, Gaide S, Henriet JP, Kaszemeik K, Klar S, Klein T, Krengel T, Kuhnert M, Meyer-Schack B, Noorlander C, Reuter M, Rosiak U, Schmidt W, Seeba H, Seiter C, Strange N, Terhzaz L, Van Rooij D (2015) MoccoMeBo: climate-driven development of Moroccan cold-water coral mounds revealed by MeBo-drilling: Atlantic vs. Mediterranean settings - Cruise MSM36 - February 18- March 14, 2014 - Malaga (Spain) - Las Palmas (Spain) MARIA S MERIAN-Reports, MSM36. DFG-Senatskommission für Ozeanographie 47

Hebbeln D, Van Rooij D, Wienberg C (2016) Good neighbours shaped by vigorous currents: cold-water coral mounds and contourites in the North Atlantic. Mar Geol 378:171–185

Hebbeln D, Bender M, Gaide S, Titschack J, Vandorpe T, Van Rooij D, Wintersteller P, Wienberg C (2019) Thousands of cold-water coral mounds along the Moroccan Atlantic continental margin: distribution and morphometry. Mar Geol 411:51–61

Hennige SJ, Wicks LC, Kamenos NA, Findlay HS, Roberts JM (2015) Hidden impacts of coral acclimation to ocean acidification. Proc R Soc B Biol Sci 282:20150990

Henriet J-P, Guidard S, and the ODP "Proposal 573" Team (2002) Carbonate mounds as a possible example for microbial activity in geological processes. In: Wefer G, Billett D, Hebbeln D, Jørgensen BB, Schlüter M, van Weering T, Ocean margin systems, Springer Berlin, pp. 437–455

Henry LA, Roberts JM (2017) Global biodiversity in cold-water coral reef ecosystems. In: Rossi S, Bramanti L, Gori A, Orejas C (eds) Marine animal forests. Springer, pp 235–256

Henry LA, Frank N, Hebbeln D, Wienberg C, Robinson L, van de Flierdt T, Dahl M, Douarin M, Morrison CL, Lopez Correa M, Rogers AD, Ruckelshausen M, Roberts JM (2014) Global ocean conveyor lowers extinction risk in the deep sea. Deep-Sea Res I 88: 8–16

Ingrassia M, Macelloni L, Bosman A, Chiocci FL, Cerrano C, Martorelli E (2016) Black coral (Anthozoa, Antipatharia) forest near the western Pontine Islands (Tyrrhenian Sea). Mar Biodivers 46:285–290

Ingrassia M, Martorelli E, Bosman A, Chiocci FL (2019) Isidella elongata (Cnidaria: Alcyonacea): first report in the Ventotene Basin (Pontine Islands, western Mediterranean Sea). Reg Stud Mar Sci 25:100494

IUCN (2020) The IUCN red list of threatened species. Version 2020-3. www.iucnredlist.org

Izquierdo FJ, Esteras M, Sandoval NG (1996) Lithified coral mounds in the Gibraltar Strait. Geogaceta 20:401404

Kiriakoulakis K, Fisher E, Wolff GA, Freiwald A, Grehan A, Roberts JM (2005) Lipids and nitrogen isotopes of two deep-water corals from the North-east Atlantic: initial results and implications for their nutrition. In: Freiwald A, Roberts JM (eds) Cold-water corals and ecosystems. Springer, Heidelberg, pp 715–729

Knittweis L, Aguilar R, Alvarez H, Borg JA, Evans J, Garcia S, Schembri PJ (2016) New depth record of the precious red coral Corallium rubrum for the Mediterranean. Rapport Commission Internationale pour l'Exploration Scientifique de la Mer Méditerranée 41:467

Knittweis L, Evans J, Aguilar R, Alvarez H, Borg JA, Garcia S, Schembri PJ (2019) Recent discoveries of extensive cold-water coral assemblages in Maltese waters. In: Orejas Saco del Valle C, Jiménez C (eds) Mediterranean cold-water corals: past, present and future, springer series: coral reefs of the world. Springer, pp 253–255

Larsson AI, Purser A (2011) Sedimentation on the cold-water coral *Lophelia pertusa*: cleaning efficiency from natural sediments and drill cuttings. Mar Pollut Bull 62:1159–1168

Larsson AI, van Oevelen D, Purser A, Thomsen L (2013) Tolerance to long-term exposure of suspended benthic sediments and drill cuttings in the cold-water coral *Lophelia pertusa*. Mar Pollut Bull 70:176–188

Lartaud F, Pareige S, de Rafelis M, Feuillassier L, Bideau M, Peru E, Romans P, Alcala F, Le Bris N (2013) A new approach for assessing cold-water coral growth in situ using fluorescent calcein staining. Aquat Living Resour 26:187–196

Lartaud F, Pareige S, de Rafelis M, Feuillassier L, Bideau M, Peru E, De la Vega E, Nedoncelle K, Romans P, Le Bris N (2014) Temporal changes in the growth of two Mediterranean cold-water coral species, in situ and in aquaria. Deep-Sea Res II 99:64–70

Lartaud F, Mouchi V, Chapron L, Meistertzheim AL, Le Bris N (2019) Growth patterns of mediterranean calcifying cold-water corals. In: Saco O, del Valle C, Jiménez C (eds) Mediterranean cold-water corals: past, present and future, Springer series: Coral Reefs of the World. Springer, pp 405–422

Lastras G, Canals M, Ballesteros E, Gili JM, Sanchez-Vidal A (2016) Cold-water corals and anthropogenic impacts in La Fonera submarine canyon head, Northwestern Mediterranean Sea. PLoS One 11: e0155729

Lauria V, Massi D, Fiorentino F, Milisenda G, Cillari T (2021) Habitat suitability mapping of the black coral *Leiopathes glaberrima* to support conservation of vulnerable marine ecosystems. Sci Rep 11: 15661

Lazar M, Lang G, Schattner U (2016) Coincidence or not? Interconnected gas/fluid migration and ocean–atmosphere oscillations in the Levant Basin. Geo Mar Lett 36:293–306

Levin LA, Le Bris N (2015) The deep ocean under climate change. Science 350:766–768

Lo Iacono C, Orejas C, Gori A, Gili JM, Requena S, Puig P, Ribó M (2012) Habitats of the Cap de Creus continental shelf and Cap de Creus canyon, northwestern Mediterranean. In: Harris PH, Baker EK (eds) Seafloor geomorphology as benthic habitat. Elsevier, pp 457–469

Lo Iacono C, Gràcia E, Ranero CR, Emelianov M, Huvenne VAI, Bartolomé R, Booth-Rea G, Prades J (2014) The West Melilla cold water coral mounds, Eastern Alboran Sea: morphological characterization and environmental context. Deep Sea Res Part II 99:316–326

Lo Iacono C, Savini A, Basso D (2018) Cold-water carbonate bioconstructions. In: Micallef A, Krastel S, Savini A (eds) Submarine geomorphology. Springer International Publishing, Cham, pp 425–455

Lo Iacono C, Savini A, Huvenne VAI, Gràcia E (2019) Habitat mapping of cold-water corals in the Mediterranean Sea. In: Saco O, del Valle C, Jiménez C (eds) Mediterranean cold-water corals: past, present and future, Springer series: coral reefs of the world. Springer, pp 157–171

Maier C, Watremez P, Taviani M, Gattuso JP (2012) Calcification rates and the effect of ocean acidification on Mediterranean cold-water corals. Proc R Soc B 279:1716–1723

Maier C, Popp P, Sollfrank N, Winbauer MG, Wild C, Gattuso JP (2016) Effects of elevated pCO_2 and feeding on net calcification and energy budget of the Mediterranean cold-water coral *Madrepora oculata*. J Exp Biol 219:3208–3217

Malinverno E, Taviani M, Rosso A, Violanti D, Villa I, Savini A, Vertino A, Remia A, Corselli C (2010) Stratigraphic framework of the Apulian deep-water coral province, Ionian Sea. Deep Sea Res Part II 57:345–359

Margreth S, Gennari G, Rüggeberg A, Comas MC, Pinheiro LM, Spezzaferri S (2011) Growth and demise of cold-water coral ecosystems on mud volcanoes in the West Alboran Sea: the messages from the planktonic and benthic foraminifera. Mar Geol 282:26–39

Martín J, Puig P, Masqué P, Palanques A, Sánchez-Gómez A (2014) Impact of bottom trawling on deep-sea sediment properties along the flanks of a submarine canyon. PLoS One 9:e104536

Martorelli E, Petroni G, Chiocci FL, and the Pantelleria Scientific P (2011) Contourites offshore Pantelleria Island (Sicily Channel, Mediterranean Sea): depositional, erosional and biogenic elements. Geo-Mar Lett 31:481–493

Mastrototaro F, D'Onghia G, Corriero G, Matarrese A, Maiorano P, Panetta P, Gherardi M, Longo C, Rosso A, Sciuto F, Sanfilippo R, Gravilli C, Boero F, Taviani M, Tursi A (2010) Biodiversity of the white coral bank off Cape Santa Maria di Leuca (Mediterranean Sea): An update. Deep Sea Res Part II 57:412–430

Mastrototaro F, Chimienti G, Acosta J, Blanco J, Garcia S, Rivera J, Aguilar R (2017) *Isidella elongata* (Cnidaria: Alcyonacea) facies in the western Mediterranean Sea: visual surveys and descriptions of its ecological role. Eur Zool J 84:209–225

Matos FL, Company JB, Cunha M (2021) Mediterranean seascape sutability for *Lophelia pertusa*: living on the edge. Deep Sea Res Part I 170:103496

Maynou F, Cartes JE (2011) Effects of trawling on fish and invertebrates from deep-sea coral facies of *Isidella elongata* in the western Mediterranean. J Mar Biol Assoc UK 92:1501–1507

McCulloch M, Taviani M, Montagna P, López Correa M, Remia A, Mortimer G (2010) Proliferation and demise of deep-sea corals in the Mediterranean during the Younger Dryas. Earth Planet Sci Lett 298: 143–152

Mediterranean Quality Status Report (2017). www.medqsr.org. Downloaded on 24 January 2021

Moccia D, Cau A, Bramanti L, Carugati L, Canese S, Follesa MC, Cannas R (2021) Spatial distribution and habitat characterization of marine animal forest assemblages along nine submarine canyons of Eastern Sardinia (central Mediterranean Sea). Deep-Sea Res I 167: 103422

Montagna P, McCulloch M, Taviani M, Mazzoli C, Vendrell B (2006) Phosphorus in cold-water corals as a proxy for seawater nutrient chemistry. Science 312:1788–1791

Montenat C, Barrier P, Di Geronimo I (1987) The strait of Messina, past and present: a review. Documents et Travaux IGAL 11:7–13

Mouchi V, Chapron L, Peru E, Pruski AM, Meistertzheim AL, Vétion G, Galand PE, Lartaud F (2019) Long-term aquaria study suggests species-specific responses of two cold-water corals to macro-and microplastics exposure. Environ Pollut 253:322–329

Movilla J, Gori A, Calvo E, Orejas C, López-Sanz À, Domínguez-Carrió C, Grinyó J, Pelejero C (2014a) Resistance of two Mediterranean cold-water coral species to low-pH conditions. Water 6:59–67

Movilla J, Orejas C, Calvo E, Gori A, López-Sanz À, Grinyó J, Domínguez-Carrió C, Pelejero C (2014b) Differential response of two Mediterranean cold-water coral species to ocean acidification. Coral Reefs 33:675–686

Mytilineou C, Smith CJ, Anastasopoulou A, Papadopoulou KN, Christidis G, Bekas P, Kavadas S, Dokos J (2014) New cold-water coral occurrences in the Eastern Ionian Sea: results from experimental long line fishing. Deep Sea Res Part II 99:146–157

Nasto I, Cardone F, Mastrototaro F, Panetta P, Rosso A, Sanfilippo R, Taviani M, Tursi A (2018) Benthic invertebrates associated with subfossil cold-water coral frames and hardgrounds in the Albanian deep waters (Adriatic Sea). Turk J Zool 42:360–371

Naumann MS, Orejas C, Wild C, Ferrier-Pagès C (2011) First evidence for zooplankton feeding sustaining key physiological processes in a scleractinian cold-water coral. J Exp Biol 214:3570–3576

Naumann MS, Orejas C, Ferrier-Pagès C (2013) High thermal tolerance of two Mediterranean cold-water coral species maintained in aquaria. Coral Reefs 32:749–754

Naumann MS, Orejas C, Ferrier-Pagès C (2014) Species-specific physiological response by the cold-water corals *Lophelia pertusa* and *Madrepora oculata* to variations within their natural temperature range. Deep Sea Res Part II 99:36–41

Naumann MS, Tolosa I, Taviani M, Grover R, Ferrier-Pagès C (2015) Trophic ecology of two cold-water coral species from the Mediterranean Sea revealed by lipid biomarkers and compound-specific isotope analyses. Coral Reefs 34:1165–1175

OCEANA (2014) Scientific information to describe areas meeting scientific criteria for Mediterranean EBSAs. OCEANA, Madrid

Orejas C, Jiménez C (2019) Mediterranean cold-water corals: past, present and future, springer series: coral reefs of the world. Springer

Orejas C, Gori A, Gili JM (2008) Growth rates of live *Lophelia pertusa* and *Madrepora oculata* from the Mediterranean Sea maintained in aquaria. Coral Reefs 27:255

Orejas C, Gori A, Lo Iacono C, Puig P, Gili JM, Dale MRT (2009) Cold-water corals in the Cap de Creus canyon, northwestern Mediterranean: spatial distribution, density and anthropogenic impact. Mar Ecol Prog Ser 397:37–51

Orejas C, Ferrier-Pagès C, Reynaud S, Gori A, Beraud E, Tsounis G, Allemand D, Gili JM (2011) Long-term growth rate measurements of four Mediterranean cold water coral species (*Madrepora oculata*, *Lophelia pertusa*, *Desmophyllum cristagalli* and *Dendrophyllia cornigera*) maintained in aquaria. Mar Ecol Prog Ser 429:57–65

Orejas C, Gori A, Rad-Menéndez C, Last KS, Davies AJ, Beveridge CM, Sadd D, Kiriakoulakis K, Witte U, Roberts JM (2016) The effect of flow speed and food size on the capture efficiency and feeding behaviour of the cold-water coral *Lophelia pertusa*. J Exp Mar Biol Ecol 481:34–40

Orejas C, Gori A, Jiménez C, Rivera J, Kamidis N, Alhaija RA, Lo Iacono C (2019a) Occurrence and distribution of the coral *Dendrophyllia ramea* in Cyprus insular shelf: Environmental setting and anthropogenic impacts. Deep Sea Res Part II 164:190–205

Orejas C, Taviani M, Ambroso S, Andreou V, Bilan M, Bo M, Brooke S, Buhl-Mortensen P, Cordes E, Dominguez-Carrió C, Ferrier-Pagès C, Godinho A, Gori A, Grinyó J, Gutiérrez Zárate C, Hennige S, Jiménez C, Larsson AI, Lartaud F, Lunden J, Maier C, Maier SR, Movilla J, Murray F, Peru E, Purser A, Rakka M, Reynaud S, Roberts JM, Siles P, Strömberg SM, Thomsen M, van Oevelen D, Veiga A, Carreiro-Silva M (2019b) Cold-water coral in aquaria: advances and challenges. A focus on the Mediterranean. In: Orejas Saco del Valle C, Jiménez C (eds) Mediterranean cold-water corals: past, present and future, springer series: coral reefs of the world. Springer, pp 435–471

Orejas C, Kenchington E, Rice J, Kazanidis G, Palialexis A, Johnson D, Gianni M, Danovaro R, Roberts JM (2020) Towards a common approach to the assessment of the environmental status of deep-sea ecosystems in areas beyond national jurisdiction. Mar Policy 121: 104182

Orejas C, Wienberg C, Titschack J, Tamborrino L, Freiwald A, Hebbeln D (2021) *Madrepora oculata* forms large frameworks in hypoxic waters off Angola (SE Atlantic). Sci Rep 11:14170

Otero MDM, Marin P (2019) Conservation of cold-water corals in the Mediterranean: current status and future prospects for improvement. In: Saco O, del Valle C, Jiménez C (eds) Mediterranean cold-water corals: past, present and future, Springer series: coral reefs of the world. Springer, pp 535–545

Otero MDM, Numa C, Bo M, Orejas C, Garrabou J, Cerrano C, Kruzic P, Antoniadou C, Aguilar R, Kipson S, Linares C, Terrón-Sigler A, Brossard J, Kersting D, Casado-Amezúa P, García S,

Goffredo S, Ocaña O, Caroselli E, Maldonado M, Bavestrello G, Cattaneo-Vietti R, Özalp B (2017) Overview of the conservation status of Mediterranean Anthozoa. International Union for Conservation of Nature. IUCN, Malaga

Palanques A, Durrieu de Madron X, Puig P, Fabres J, Guillén J, Calafat A, Canals M, Heussner S, Bonnin J (2006) Suspended sediment fluxes and transport processes in the Gulf of Lions submarine canyons. The role of storms and dense water cascading. Mar Geol 234:43–61

Palomino D, Alonso B, Lo Iacono CL, Casas D, d'Acremont E, Ercilla G, Gorini C, Vazquez JT (2015) Seamounts and seamount-like structures of the Alborán Sea. In: Würtz M, Rovere M (eds) Atlas of the Mediterranean seamounts and seamount-like structures. IUCN, Gland, pp 21–58

Palomino D, Rueda JL, Vázquez JT, Urra J, Sánchez-Guillamón O, González-García E, López-González N, Fernández-Salas LM (2019) Cold-water corals in fluid venting submarine structures. In: Orejas Saco del Valle C, Jiménez C (eds) Mediterranean cold-water corals: past, present and future, Springer series: coral reefs of the world. Springer, pp 261–264

Pardo E, Rubio RA, García S, Ubero J (2011) Documentación de arrecifes de corales de agua fría en el Mediterráneo occidental (Mar de Alborán). Chronica Naturae 1:20–34

Pierdomenico M, Martorelli E, Dominguez-Carrió C, Gili JM, Chiocci FL (2016) Seafloor characterization and benthic megafaunal distribution of an active submarine canyon and surrounding sectors: the case of Gioia Canyon (Southern Tyrrhenian Sea). J Mar Syst 157: 101–117

Pierdomenico M, Russo T, Ambroso S, Gori A, Martorelli E, D'Andrea L, Gili JM, Chiocci FL (2018) Effects of trawling activity on the bamboo-coral *Isidella elongata* and the sea pen *Funiculina quadrangularis* along the Gioia Canyon (Western Mediterranean, southern Tyrrhenian Sea). Prog Oceanogr 169:214–226

Prampolini M, Angeletti L, Grande V, Taviani M, Foglini F (2019) Tricase Submarine Canyon: cold-water coral habitats in the south-westernmost Apulian margin (Mediterranean Sea). In: Harris P, Baker E (eds) Seafloor geomorphology as benthic habitat. Elsevier, pp 793–810

Previati M, Scinto A, Cerrano C, Osinga R (2010) Oxygen consumption in Mediterranean octocorals under different temperatures. J Exp Mar Biol Ecol 390:39–48

Priori C, Mastascusa V, Erra F, Angiolillo M, Canese S, Santangelo G (2013) Demography of deep-dwelling red coral populations: age and reproductive structure of a highly valued marine species. Estuar Coast Shelf Sci 118:43–49

Ramírez F, Coll M, Navarro J, Bustamante J, Green AJ (2018) Spatial congruence between multiple stressors in the Mediterranean Sea may reduce its resilience to climate impacts. Sci Rep 8:1–8

Ramirez-Llodra E, Brandt A, Danovaro R, De Mol BD, Escobar E, German CR, Levin LA, Martinez Arbizu P, Menot L, Buhl-Mortensen P, Narayanaswamy BE, Smith CR, Tittensor DP, Tyler PA, Vanreusel A, Vecchione M (2010) Deep, diverse and definitely different: unique attributes of the world's largest ecosystem. Biogeosciences 7:2851–2899

Ramirez-Llodra E, Tyler PA, Baker MC, Bergstad OA, Clark MR, Escobar E, Levin LA, Menot L, Rowden AA, Smith CR, Van Dover CL (2011) Man and the last great wilderness: human impact on the deep sea. PLoS One 6(8):e22588

Rebesco M, Taviani M (2019) A turbulent story: Mediterranean contourites and cold-water corals. In: Saco O, del Valle C, Jiménez C (eds) Mediterranean cold-water corals: past, present and future, Springer series: Coral Reefs of the World. Springer, pp 34–46

Rees SE, Foster NL, Langmead O, Pittman S, Johnson DE (2018) Defining the qualitative elements of Aichi Biodiversity Target 11 with regard to the marine and coastal environment in order to strengthen global efforts for marine biodiversity conservation outlined in the United Nations Sustainable Development Goal 14. Mar Policy 93:241–250

Remia A, Taviani M (2005) Shallow-buried Pleistocene *Madrepora*-dominated coral mounds on a muddy continental slope, Tuscan Archipelago, NE Tyrrhenian Sea. Facies 50:419–425

Roberts JM, Wheeler AJ, Freiwald A, Cairns SD (2009) Cold-water corals. The biology and geology of deep-sea coral habitats. Cambridge University Press

Rossi S, Tsounis G, Orejas C, Padrón T, Gili JM, Bramanti L, Teixidó N, Gutt J (2008) Survey of deep-dwelling red coral (*Corallium rubrum*) populations at Cap de Creus (NW Mediterranean). Mar Biol 154:533–545

Rosso A, Vertino A, Di Geronimo I, Sanfilippo R, Sciuto F, Di Geronimo R, Violanti D, Corselli C, Taviani M, Mastrototaro F, Tursi A (2010) Hard- and soft-bottom thanatofacies from the Santa Maria di Leuca deep-water coral province, Mediterranean. Deep Sea Res II 57:360–379

Rueda JM, Urra J, Aguilar R, Angeletti L, Bo M, García-Ruiz C, González-Duarte MM, López E, Madurell T, Maldonado M, Ramírez Á M, Megina C, Moreira J, Moya F, Ramalho LV, Rosso A, Sitjà C, Taviani M (2019) In: Orejas Saco del Valle C, Jiménez C (eds) Mediterranean cold-water corals: past, present and future, Springer series: coral reefs of the world. Springer, pp 295–333

Rüggeberg A, Foubert A (2019) Cold-water corals and mud volcanoes: Life on a dynamic substrate. In: Saco O, del Valle C, Jiménez C (eds) Mediterranean cold-water corals: past, present and future, Springer series: Coral Reefs of the World. Springer, pp 265–269

Rypien KL, Ward JR, Azam F (2010) Antagonistic interactions among coral-associated bacteria. Environ Microbiol 12:28–39

Salomidi M, Zibrowius H, Issaris Y, Milionis K (2010) *Dendrophyllia* in Greek waters, Mediterranean Sea, with the first record of *D. ramea* (Cnidaria, Scleractinia) from the area. Mediterr Mar Sci 11:189–194

Sanchez P, Fabri M-C (2020) Ecologie et cartographie des coraux d'eau froide du canyon de Lacaze-Duthiers en Méditerranée; Données de la campagne CALADU 2019. Rapport de stage Ifremer, University of Montpellier, 63 p

Santín A, Grinyó J, Uriz MJ, Lo Iacono C, Gili JM, Puig P (2021) Mediterranean coral provinces as a sponge diversity reservoir: is there a mediterranean cold-water coral sponge fauna? Front Mar Sci 8:662899. https://doi.org/10.3389/fmars.2021.662899

Sartoretto S, Zibrowius H (2018) Note on new records of living Scleractinia and Gorgonaria between 1700 and 2200 m depth in the western Mediterranean Sea. Mar Biodivers 48:689–694

Savini A, Corselli C (2010) High-resolution bathymetry and acoustic geophysical data from Santa Maria di Leuca Cold Water Coral province (Northern Ionian Sea-Apulian continental slope). Deep Sea Res II 57:326–344

Savini A, Vertino A, Marchese F, Beuck L, Freiwald A (2014) Mapping cold-water coral habitats at different scales within the Northern Ionian Sea (Central Mediterranean): an assessment of coral coverage and associated vulnerability. PLoS One 9:e87108

Schattner U, Gurevich M, Kanari M, Lazar M (2015) Levant jet system-effect of post LGM seafloor currents on Nile sediment transport in the eastern Mediterranean. Sediment Geol 329:28–39

Schembri PJ, Dimech M, Camilleri M, Page R (2007) Living deep-water *Lophelia* and *Madrepora* corals in Maltese waters (Strait of Sicily, Mediterranean Sea). Cah Biol Mar 48:77–83

Schröder-Ritzrau A, Freiwald A, Mangini A (2005) U/Th-dating of deep-water corals from the eastern North Atlantic and the western Mediterranean Sea. In: Freiwald A, Roberts JM (eds) Cold-water corals and ecosystems. Springer, Heidelberg, pp 691–700

Skliris N (2019) The Mediterranean is getting saltier: From the past to the future. In: Saco O, del Valle C, Jiménez C (eds) Mediterranean cold-water corals: past, present and future, Springer series: coral reefs of the world. Springer, pp 506–512

Smith C, Sakellariou D, McCoy F, Wachsmann S (2009) Deep coral environments south of Crete. 9th Symposium on oceanography and fisheries, 2009 - Proceedings, Volume I

Spadini V, Pizzolato F (2020) Gli antozoi di Torriana (Val Marecchia, RN). Quaderno di Studi e Notizie di Storia Naturale della Romagna 51:1–18

Squires FD (1957) New species of caryophylliid corals from the Gulf Coast Tertiary. J Paleontol 31:992–996

Stalder C, Vertino A, Rosso A, Rüggeberg A, Pirkenseer C, Spangenberg JE, Spezzaferri S, Camozzi O, Rappo S, Hajdas I (2015) Microfossils, a key to unravel cold-water carbonate mound evolution through time: evidence from the eastern Alboran Sea. PLoS One 10:e0140223

Steinmann L, Baques M, Wenau S, Schwenk T, Spiess V, Piola A, Bozzano G, Violante R, Kasten S (2020) Discovery of a giant cold-water coral mound province along the northern Argentine margin and its link to the regional Contourite Depositional System and oceanographic setting. Mar Geol 106223

Stolarski J, Vertino A (2007) First Mesozoic record of the scleractinian *Madrepora* from the Maastrichtian siliceous limestones of Poland. Facies 53:67–78

Sweetman AK, Thurber AR, Smith CR, Levin LA, Mora C, Wei CL, Gooday AJ, Jones DOB, Rex M, Yasuhara M, Ingels J, Ruhl HA, Frieder CA, Danovaro R, Würzberg L, Baco A, Grupe BM, Pasulka A, Meyer KS, Dunlop KM, Henry LA, Roberts JM (2017) Major impacts of climate change on deep-sea benthic ecosystems. Elem Sci Anth 5:4

Tamborrino L, Wienberg C, Titschack J, Wintersteller P, Mienis F, Schröder-Ritzrau A, Freiwald A, Orejas C, Dullo W-C, Haberkern J, Hebbeln D (2019) Mid-Holocene extinction of cold-water corals on the Namibian shelf steered by the Benguela oxygen minimum zone. Geology 47:1185–1188

Taviani M (2002) The Mediterranean benthos from late Miocene up to present: ten million years of dramatic climatic and geologic vicissitudes. Biol Mar Mediterr 9:445–463

Taviani M (2019) Changing views about Mediterranean cold-water corals. In: Saco O, del Valle C, Jiménez C (eds) Mediterranean cold-water corals: past, present and future, Springer series: Coral Reefs of the World. Springer, pp 115–118

Taviani M, Colantoni P (1979) Thanatocoenoses wurmiennes associées aux coraux blancs. Rapport Commission Internationale pour l'Exploration Scientifique de la Mer Méditerranée 25(26):141–142

Taviani M, Freiwald A, Zibrowius H (2005a) Deep coral growth in the Mediterranean Sea: an overview. In: Freiwald A, Roberts JM (eds) Cold-water corals and ecosystems. Springer, Heidelberg, pp 137–156

Taviani M, Remia A, Corselli C, Freiwald A, Malinverno E, Mastrototaro F, Savini A, Tursi A (2005b) First geo-marine survey of living cold-water *Lophelia* reefs in the Ionian Sea (Mediterranean basin). Facies 50:409–417

Taviani M, Freiwald A, Beuck L, Angeletti L, Remia A, Vertino A, Dimech M, Schembri PJ (2010) The deepest known occurrence of the precious red coral *Corallium rubrum* (L. 1758) in the Mediterranean Sea. In: Bussoletti E, Cottingham D, Bruckner A, Roberts G, Sandulli R (eds) Proceedings of the international workshop on red coral science, management, and trade: lessons from the mediterranean, NOAA technical memorandum CRCP-13. Silver Spring, MD, pp 87–93

Taviani M, Angeletti L, Antolini B, Ceregato A, Froglia C, Lopez Correa M, Montagna P, Remia A, Trincardi F, Vertino A (2011a) Geo-biology of Mediterranean deep-water coral ecosystems. In: Brugnoli E, Cavarretta G, Mazzola S, Trincardi F, Ravaioli M, Santoleri R (eds) Marine research at CNR, Volume DTA, vol 6, pp 809–822

Taviani M, Vertino A, López Correa M, Savini A, De Mol B, Remia A, Montagna P, Angeletti L, Zibrowius H, Alves T, Salomidi M, Ritt B, Henry P (2011b) Pleistocene to recent scleractinian deep-water corals and coral facies in the Eastern Mediterranean. Facies 57: 579–603

Taviani M, Franchi F, Angeletti CA, López Correa M, Maselli V, Mazzoli C, Peckmann J (2015) Biodetrital carbonates on the Adriatic continental shelf imprinted by oxidation of seeping hydrocarbons. Mar Pet Geol 66:511–531

Taviani M, Angeletti L, Beuck L, Campiani E, Canese S, Foglini F, Freiwald A, Montagna P, Trincardi F (2016) Reprint of 'On and off the beaten track: megafaunal sessile life and Adriatic cascading processes'. Mar Geol 375:146–160

Taviani M, Angeletti L, Canese S, Cannas R, Cardone F, Cau A, Cau AB, Follesa MC, Marchese F, Montagna P, Tessarolo C (2017) The "Sardinian cold-water coral province" in the context of the Mediterranean coral ecosystems. Deep Sea Res Part II 145:61–78

Taviani M, Angeletti L, Cardone F, Montagna P, Danovaro R (2019a) A cold-water coral-bivalve biotope new to the Mediterranean offshore the Naples megalopolis: just discovered and under threat. Sci Rep 9: 3411

Taviani M, Angeletti L, Foglini F, Corselli C, Nasto I, Pons-Branchu E, Montagna P (2019b) U/Th dating records of cold-water coral colonization in submarine canyons and adjacent sectors of the southern Adriatic Sea since the Last Glacial Maximum. Prog Oceanogr 175: 300–308

Taviani M, Vertino A, Angeletti L, Montagna P, Remia A (2019c) Paleoecology of Mediterranean cold-water corals. In: Saco O, del Valle C, Jiménez C (eds) Mediterranean cold-water corals: past, present and future, Springer series: Coral Reefs of the World. Springer, pp 15–30

Thresher RE, Tilbrook B, Fallon S, Wilson NC, Adkins J (2011) Effects of chronic low carbonate saturation levels on the distribution, growth and skeletal chemistry of deep-sea corals and other seamount megabenthos. Mar Ecol Prog Ser 442:87–99

Titschack J (2019) Bathyal corals within the Aegean Sea and the adjacent Hellenic Trench. In: Saco O, del Valle C, Jiménez C (eds) Mediterranean cold-water corals: past, present and future, Springer series: Coral Reefs of the World. Springer, pp 85–94

Titschack J, Freiwald A (2005) Growth, deposition, and facies of Pleistocene bathyal coral communities from Rhodes, Greece. In: Freiwald A, Roberts JM (eds) Cold-water corals and ecosystems. Springer, Heidelberg, pp 41–59

Titschack J, Joseph N, Fietzke J, Freiwald A, Bromley RG (2013) Record of a tectonically-controlled regression captured by changes in carbonate skeletal associations on a structured island shelf (mid-Pleistocene, Rhodes, Greece). Sediment Geol 283:15–33

Trincardi F, Foglini F, Verdicchio G, Asioli A, Correggiari A, Minisini D, Piva A, Remia A, Ridente D, Taviani M (2007) The impact of cascading currents on the Bari Canyon System, SW-Adriatic Margin (Central Mediterranean). Mar Geol 246:208–230

Trotter JA, Pattiaratchi C, Montagna P, Taviani M, Falter J, Thresher R, Hosie A, Haig D, Foglini F, Hua Q, McCulloch MT (2019) First ROV exploration of the Perth Canyon: Canyon setting, faunal observations, and anthropogenic impacts. Front Mar Sci 6:173

Tursi A, Mastrototaro F, Matarrese A, Maiorano P, D'Onghia G (2004) Biodiversity of the white coral reefs in the Ionian Sea (Central Mediterranean). Chem Ecol 20:107–116

Vafidis D, Koukouras A, Voultsiadou-Koukoura E (1997) Actiniaria, Corallimorpharja, and Scleractinia (Hexacorallia, Anthozoa) of the Aegean Sea, with a checklist of the eastern Mediterranean and Black Sea species. Isr J Ecol Evol 43:55–70

Van de Water JA, Coppari M, Enrichetti F, Ferrier-Pagès C, Bo M (2020) Local conditions influence the prokaryotic communities associated with the mesophotic black coral *Antipathella subpinnata*. Front Microbiol 11:2423

Vertino A, Savini A, Rosso A, Di Geronimo I, Mastrototaro F, Sanfilippo R, Gay G, Etiope G (2010) Benthic habitat characterization and distribution from two representative sites of the deep-water SML Coral Province (Mediterranean). Deep Sea Res Part II 57:380–396

Vertino A, Titschack J, Rosso A, Di Geronimo I, Pino P (2013) Messina Strait: cold-water coral Pleistocene deposits from "La Montagna" (Messina). Bridging off-shore and on-land research on carbonate mounds: common concepts and techniques Buildups from Sicily: Triassic to Quaternary examples, COCARDE Workshop and Field Seminar 2013, September 23–27th:4–11

Vertino A, Stolarski J, Bosellini FR, Taviani M (2014) Mediterranean corals through time: from Miocene to present. In: Goffredo S, Dubinsky Z (eds) The Mediterranean Sea: its history and present challenges. Springer, Dordrecht, pp 257–274

Vertino A, Taviani M, Corselli C (2019) Spatio-temporal distribution of Mediterranean cold-water corals. In: Saco O, del Valle C, Jiménez C (eds) Mediterranean cold-water corals: past, present and future. Coral Reefs of the World. Springer, Springer series, pp 67–83

Wang H, Lo Iacono C, Wienberg C, Titschack J, Hebbeln D (2019) Cold-water coral mounds in the southern Alboran Sea (western Mediterranean Sea): Internal waves as an important driver for mound formation since the last deglaciation. Mar Geol 412:1–18

Wefing A-M, Arps J, Blaser P, Wienberg C, Hebbeln D, Frank N (2017) High precision U-series dating of scleractinian cold-water corals using an automated chromatographic U and Th extraction. Chem Geol 475:140–148

Weinbauer MG, Oregioni D, Großkurth A, Kerros ME, Harder T, DuBow M, Gattuso JP, Maier C (2019) Diversity of bacteria associated with the cold-water corals *Lophelia pertusa* and *Madrepora oculata*. In: Orejas Saco del Valle C, Jiménez C (eds) Mediterranean cold-water corals: past, present and future, Springer series: coral reefs of the world. Springer, pp 377–386

Wheeler AJ, Beyer A, Freiwald A, de Haas H, Huvenne VAI, Kozachenko M, Olu-Le Roy K, Opderbecke J (2007) Morphology and environment of cold-water coral carbonate mounds on the NW European margin. Int J Earth Sci 96:37–56

Wienberg C (2019) A deglacial cold-water coral boom in the Alboran Sea: from coral mounds and species dominance. In: Saco O, del Valle C, Jiménez C (eds) Mediterranean cold-water corals: past, present and future, Springer series: Coral Reefs of the World. Springer, pp 57–60

Wienberg C, Titschack J (2017) Framework-forming scleractinian cold-water corals through space and time: a late Quaternary North Atlantic perspective. In: Rossi S, Bramanti L, Gori A, Saco O, del Valle C (eds) Marine animal forests: the ecology of benthic biodiversity hotspots. Springer, Cham, pp 699–732

Wienberg C, Titschack J, Freiwald A, Frank N, Lundälv T, Taviani M, Beuck L, Schröder-Ritzrau A, Krengel T, Hebbeln D (2018) The giant Mauritanian cold-water coral mound province: Oxygen control on coral mound formation. Quat Sci Rev 185:135–152

WoRMS Editorial Board (2021) World register of marine species. www.marinespecies.org

Würtz M, Rovere M (2015) Atlas of the Mediterranean seamounts and seamount-like structures. IUCN, Gland

Zibrowius H (1980) Les Scléractiniaires de la Méditerranée et de l'Atlantique nord-oriental. Mémoires de l'Institut océanographique, Monaco

Zibrowius H (1981) Thanatocoenose plèistocène profonde à spongiaires et scléractiniaires dans la Fosse Hellénique. In: Journées d'études sur la systèmatique évolutive et la biogéographie en Méditerranée Cagliari, 13 et 14 octobre 1980, Commission Internationale pour l'Exploration Scientifique de la Mer Méditerranée [CIESM], Monaco, pp 133–136

Zibrowius H, Taviani M (2005) Remarkable sessile fauna associated with deep coral and other calcareous substrates in the Strait of Sicily, Mediterranean Sea. In: Freiwald A, Roberts JM (eds) Cold-water corals and ecosystems. Springer, Heidelberg, pp 807–819

Cold-Water Coral Reefs in the Oxygen Minimum Zones Off West Africa

8

Claudia Wienberg, André Freiwald, Norbert Frank, Furu Mienis, Jürgen Titschack, Covadonga Orejas, and Dierk Hebbeln

Abstract

The discoveries of large reefs within cold-water coral mound provinces revealed that the West African margin is a coral hotspot area in the Atlantic Ocean. The most striking observation is that cold-water corals thrive in extensive oxygen minimum zones under extreme conditions. This points to a wide tolerance of cold-water corals in these regions to low oxygen concentrations. The coral mound provinces off Mauritania, Angola, and Namibia, which are located in the centre of the local oxygen minimum zones, were selected as key study areas, and their regional oceanographic, bio-ecological, and geo-morphological settings are described in detail. Even though all three provinces are characterised by highly productive, oxygen-depleted, and relatively warm environmental conditions, they differ considerably with respect to the present-day reef status and the timing of mound formation during the last glacial-interglacial cycle. Today's bottom-trawl fishing and oil and gas exploration pose severe threats to the coral communities, and together with predicted ocean warming and deoxygenation, these areas may not continue to support living coral reefs. To fully understand the ecology of the West African cold-water corals and the regional environmental control mechanisms, research strategies following a multidisciplinary and integrated approach are needed.

Keywords

Scleractinian framework-building cold-water corals · Coral mound province · Oxygen minimum zone · Upwelling · Threats · West African continental margin

8.1 Introduction

Until the beginning of this century, our knowledge of the West African cold-water corals (CWCs) was mainly based on dredged material obtained during historical oceanographic expeditions circumnavigating the world (such as the Challenger, Valdivia, Galathea, and Prince of Monaco expeditions; Gravier 1920; Zibrowius 1980; Zibrowius and Gili 1990), comprising samples of *Lophelia pertusa* (recently assigned to the genus *Desmophyllum*; Addamo et al. 2016) collected from the Congo canyon off Angola and from the Valdivia seamount on the Walvis Ridge. Thus, for a long time, little was known about the distribution, ecological status, and community structure of the West African CWC populations. During the last two decades, research along the West African margin increased considerably and revealed that CWCs occur as an almost continuous belt, which extends from the Strait of Gibraltar to the southern tip of South Africa (Fig. 8.1). Despite the fact that this belt still exhibits some distinct gaps, the numerous coral findings discovered in recent years highlight that the West African margin is an important 'hotspot' for CWCs in the world's oceans.

The most remarkable finding unquestionably was the discovery of CWCs occurring in the extended oxygen minimum zones (OMZs) beneath the upwelling regimes off Mauritania, Angola, and Namibia. These—at first glance—offer a rather

C. Wienberg (✉) · D. Hebbeln
Center for Marine Environmental Sciences (MARUM), University of Bremen, Bremen, Germany
e-mail: cwberg@marum.de

A. Freiwald · J. Titschack
Center for Marine Environmental Sciences (MARUM), University of Bremen, Bremen, Germany

Marine Research Department, Senckenberg am Meer, Wilhelmshaven, Germany

N. Frank
Institut für Umweltphysik & Institut für Geowissenschaften (IUP), Universität Heidelberg, Heidelberg, Germany

F. Mienis
Royal Netherlands Institute for Sea Research (NIOZ), 't Horntje, Texel, The Netherlands

C. Orejas
Centro Oceanográfico de Gijón, Spanish Institute of Oceanography (IEO-CSIC), Gijón, Spain

E. Cordes, F. Mienis (eds.), *Cold-Water Coral Reefs of the World*, Coral Reefs of the World 19,
https://doi.org/10.1007/978-3-031-40897-7_8

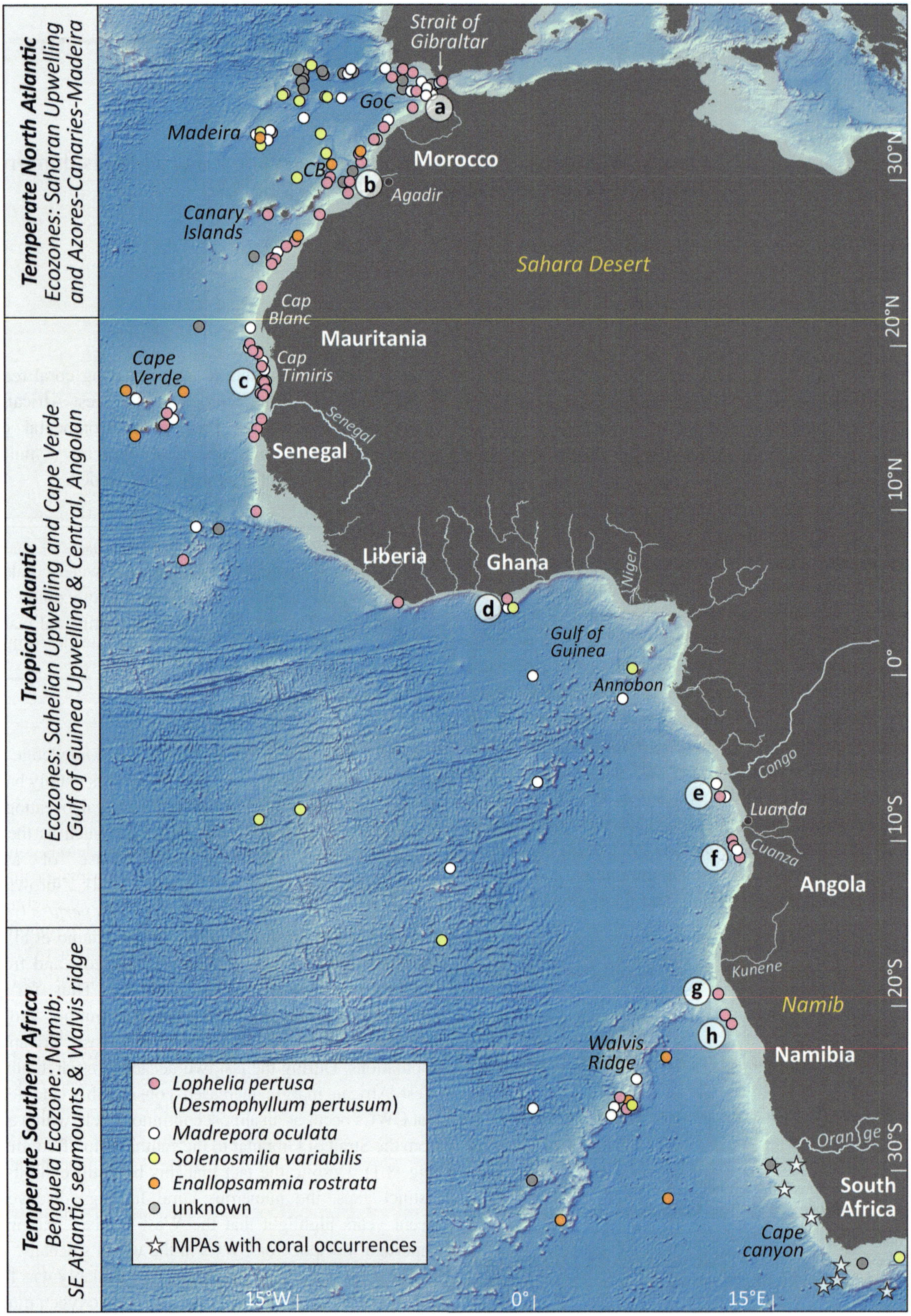

Fig. 8.1 Cold-water corals off West Africa. Overview map of the West African margin showing the distribution of the reported main framework-building cold-water corals (coloured dots; source: UNEP; complemented by reports from various cruise reports; for references, see text). Coral mounds are known from the Strait of Gibraltar, off Morocco (**a**: Gulf of Cádiz, GoC; **b**: Agadir), on the Concepcion Bank (CB; NE

extreme environment due to the prevailing extremely low dissolved oxygen (DO) concentrations of less than 1.5 mL L^{-1} (Le Guilloux et al. 2009; Ramos et al. 2017a; Hanz et al. 2019). Prior to their discovery, habitat suitability models predicted a rather low potential for the occurrence of CWCs (i.e., *L. pertusa* and *Madrepora oculata*) in these hypoxic Atlantic regions (Tittensor et al. 2009; Davies and Guinotte 2011). However, CWCs were found in extended coral mound provinces, which together make up the largest coherent mound provinces known worldwide (Hebbeln et al. 2017; Ramos et al. 2017a; Wienberg et al. 2018; Tamborrino et al. 2019, 2022). Living CWCs forming dispersed reef frameworks on the Mauritanian coral mounds and—even more impressive—thriving *Lophelia*-reefs covering the Angolan mounds demonstrate that at least some CWC species tolerate much lower DO concentrations than previously assumed (e.g., Dodds et al. 2007; Davies and Guinotte 2011; Hebbeln et al. 2020; Orejas et al. 2021). This might indicate that the corals have developed certain metabolic strategies to regionally adapt to extreme, oxygen-depleted environmental conditions (Hebbeln et al. 2020). Although today living corals are absent from the Namibian mound province, the >1200 mounds formed by *L. pertusa* (Tamborrino et al. 2022) clearly display the past proliferation of this species, before it became regionally extinct ~4.5 kyr ago, possibly caused by dropping oxygen levels (Tamborrino et al. 2019). However, we are just at the beginning of understanding the ecology of the West African coral reefs and starting to identify the key environmental parameters (and potential stressors) controlling their proliferation or their decline.

This review chapter aims at providing a comprehensive overview of the known CWC occurrence off West Africa following the margin from north to south (Sect. 8.2). The compilation focusses on reports of scleractinian framework-building corals, as this group has the capacity to form large reefs, which are highly diverse ecosystems in the deep sea. In addition, CWC reefs can develop into extended coral mound provinces over geologic timescales. Coral mounds represent important sedimentary archives, which record successive phases of regional reef proliferation and hiatuses in growth controlled by past environmental changes, and provide valuable information to predict their fate in the future (e.g., Wienberg and Titschack 2017; Hebbeln et al. 2019a). The remarkably large coral mound provinces off Mauritania, Angola, and Namibia are the key study areas, and their regional oceanographic (Sect. 8.3), bio-ecological (Sect. 8.4), and geo-morphological (Sect. 8.5) settings are described and compared. All three provinces are characterised by highly productive, oxygen-depleted, and relatively warm conditions, but they differ considerably with respect to their present-day reef status (ranging from prosperous reefs to extinct coral populations) and the timing of mound formation during the last glacial-interglacial cycle (Sect. 8.6).

Although the controls behind these patterns are still not completely deciphered, we already face a race against time. In particular, the Mauritanian and Angolan coral reefs are threatened by oil and gas exploration and by intense deep-sea bottom trawling (Colman et al. 2005; Le Guilloux et al. 2009) and will likely suffer from the predicted deoxygenation and ocean warming induced by global climate change (e.g., Schmidtko et al. 2017) (Sect. 8.7). Finally, some regions along the West African margin are poorly explored at this point and still await the discovery of potentially important CWC sites, which would fill the existing knowledge gaps of the West African coral belt. Further explorations should benefit from a multidisciplinary and integrated approach to gain oceanographic, bio-ecological, and geo-morphological data in high spatial and temporal resolution (Sect. 8.8). Such an approach will provide vital information for a more comprehensive and better understanding of the development of reefs and coral mounds along the West African margin.

8.2 Cold-Water Corals Off West Africa

8.2.1 The Main Framework-Building Cold-Water Corals Off West Africa

The major ecosystem-engineering CWC along the West African margin is *L. pertusa* (Table 8.1; Fig. 8.1). This species dominates the (present-day and fossil) coral reef communities known from off Morocco, Mauritania, Ghana, and Angola (e.g., Le Guilloux et al. 2009; Glogowski et al. 2015; Buhl-Mortensen et al. 2017a; Wienberg et al. 2018) and was the principal contributor to the fossil Holocene coral mound assemblage off Namibia (Tamborrino et al. 2019). *Madrepora oculata* is much less common, but the species dominates in certain Mauritanian slope locations and was also abundant in the last glacial Moroccan coral mound assemblages (Wienberg et al. 2009). On coral mounds off

Fig. 8.1 (Continued) Canary Islands), off Mauritania (**c**: south of Cap Timiris), off Ghana (**d**: Gulf of Guinea), off Angola (**e**: south of the Congo fan; **f**: south of Luanda), and off Namibia (**g**: south of Kunene river; **h**: southeast of Walvis Ridge); for references, see text. Marine Protected Areas (MPAs) off South Africa with documented cold-water corals (white stars) are compiled from Sink et al. (2019). Marine Ecozones of the Temperate North Atlantic, Tropical Atlantic, and Temperate Southern Africa are indicated according to Spalding et al. (2007)

Table 8.1 Presence of live, dead, or fossil framework-building cold-water corals along West African Marine Ecozones (after Spalding et al. 2007), and from the Strait of Gibraltar, the Walvis Ridge and Southeast Atlantic seamounts. X: living coral occurrences, X(f): dead or fossil coral occurrences. References are given in the text

			Framework-building cold-water coral			
Realm	Province	Ecozone (numbers refer to ecoregions defined by Spalding et al. 2007)	*Lophelia pertusa (Desmophyllum pertusum)*	*Madrepora oculata*	*Enallopsammia rostrata*	*Solenosmilia variabilis*
Temperate North Atlantic	Lusitanian	Strait of Gibraltar	X	X		
		Saharan Upwelling (28) (e.g., Morocco)	X	X	X	X
		Azores, Canaries, Madeira (29)	X	X	X	X
Tropical Atlantic	West African Transition	Cape Verde (79)	X	X		
		Sahelian Upwelling (80) (e.g., Mauritania)	X	X	X(f)	X(f)
	Gulf of Guinea	Gulf of Guinea Upwelling (82) (e.g., Ghana)	X	X		X
		Gulf of Guinea Central (83) (e.g., Nigeria)	X		X	
		Angolan (86)	X	X		
Temperate Southern Africa	Benguela	Namib (190)	X(f)			
		Walvis Ridge & Southeast Atlantic seamounts	X(f)		X	X

Angola, *M. oculata* displays some dense patches with exceptionally large colonies (Orejas et al. 2021). Live populations of *Enallopsammia rostrata* are known from the temperate Northeast Atlantic (off Morocco and around the volcanic islands of the Azores, Canaries, Madeira, and Cape Verde; e.g., Hansteen et al. 2014) and were also frequently encountered on several Southeast Atlantic seamounts off Namibia (e.g., on Walvis Ridge) and South Africa (Table 8.1; Bergstad et al. 2019). Off Mauritania, the presence of *E. rostrata* as well as *Solenosmilia variabilis* is documented only for the fossil mound record (Colman et al. 2005), while the rare presence of live *S. variabilis* is also reported from the volcanic islands of the Azores, Canaries, and Madeira (Zibrowius 1980) and from the *Lophelia*-dominated coral mound off Ghana (Table 8.1; Buhl-Mortensen et al. 2017a).

8.2.2 A North to South Compilation of the Occurrence of West African Cold-Water Corals

Here, the main records and observations of framework-forming CWCs are compiled, following the West African continental margin from north to south (Fig. 8.1). In addition, the occurrence of the West African CWCs is described according to the marine biogeographic classification scheme (Marine Ecoregions of the World) proposed by Spalding et al. (2007). Based on taxonomic configurations influenced by evolutionary history, patterns of dispersal, and isolation, this global system is hierarchically structured comprising realms, provinces, and ecoregions. A complete overview of ecozones defined for the West African margin is listed in Table 8.1.

8.2.2.1 Cold-Water Corals of the Strait of Gibraltar and the Saharan Upwelling Ecozone in the Temperate North Atlantic

The northernmost CWC distribution along the West African margin is documented in the Strait of Gibraltar (36°N; Fig. 8.1) that hosts a scleractinian coral community dominated by *L. pertusa* and to a lesser degree by *M. oculata*. On the Spanish side of the strait, CWCs occur in a water depth range of 30–460 m, while living corals appear to be limited to 180–300 m depth on mound structures that reach heights of 10–40 m (Álvarez-Pérez et al. 2005; De Mol et al. 2012). Corals were rarely dredged from the Moroccan side between 570 and 720 m depth, but high-resolution bathymetry and geophysical data predict a wide distribution of CWCs also on this side of the strait (De Mol et al. 2012).

Following the Moroccan margin to the south (~35°N; mainly encompassing the Gulf of Cádiz) and entering the Saharan Upwelling Ecozone, fossil accumulations of CWCs are widespread, comprising *L. pertusa* and *M. oculata*, while living corals are the exception (Foubert et al. 2008; Wienberg et al. 2009). The first discoveries of corals were made on top of mud volcanoes in the region (see Box 8.1; Pinheiro et al. 2003; Somoza et al. 2003) and since then have been identified to be a common feature of most mud volcanoes in water depths <1200 m (León et al. 2012; Urra et al. 2021). However, even more impressive is the presence of numerous coral mounds, which range in height from <10 m to ~60 m (indicated as 'a' in Fig. 8.1; Foubert et al. 2008; Van Rooij et al. 2011; Hebbeln et al. 2019b; Rincón-Tomás et al. 2019). The mounds are grouped to provinces, which often occur in close vicinity to large mud volcanoes and tectonic ridges (e.g., the Pen Duick or Pompeia mound provinces; Van Rooij et al. 2011; Rincón-Tomás et al. 2019). The largest coral mound province comprises >3400 mounds covering an area of ~1450 km^2, which are arranged in two slope-parallel belts that centre in water depths between 720 and 870 m and between 890 and 980 m, respectively (Hebbeln et al. 2019b). Overall, the coral mounds of the entire region largely encompass a water depth interval between 500 and 1200 m, while only few mounds occur also in shallower depths of 200–350 m (Al Idrisi mound province; see Foubert et al. 2008). Radiometric dating of fossil CWCs revealed their regional preference for glacial conditions, when higher productivity and stronger hydrodynamics probably provided a suitable habitat, while their occurrence significantly declined since the Early Holocene (Wienberg et al. 2010; Frank et al. 2011; Hebbeln et al. 2019b; Vandorpe et al. 2023).

Close to the Agadir canyon (~31°N), a further large Moroccan coral mound province (Eugen Seibold mound province; indicated as 'b' in Fig. 8.1) comprising numerous rather small mounds (height: <12 m) covers an area of 410 km^2. These mounds occur in water depths of 680–860 m, while a living *Lophelia*-dominated coral community was only observed at depths as deep as 720 m (Glogowski et al. 2015). Between this province and Cap Blanc, records relate mainly to dredged samples of live and dead *L. pertusa* (Fig. 8.1).

8.2.2.2 Cold-Water Corals of the Sahelian Upwelling, Gulf of Guinea and Angolan Ecozones in the Tropical Atlantic

Cap Blanc at the border between Morocco and Mauritania marks the transition to the Sahelian Ecozone of the eastern Tropical Atlantic. From almost the entire Mauritanian margin (21°N-16°N), CWC communities are reported, which are dominated by *L. pertusa* with contributions from *M. oculata* (see Sect. 8.4). Most spectacular are up to 100-m-high coral mounds at 400–550 m water depth (see Sect. 8.5), merged to almost continuous slope-parallel chains extending over ~400 km between Cap Timiris and the Senegalese border (indicated as 'c' in Fig. 8.1; Colman et al. 2005; Ramos et al. 2017a). The Mauritanian mounds are today positioned in the extensive OMZ beneath the Mauritanian upwelling regime (see Sect. 8.3) and are exposed to DO concentrations of 1–1.3 mL L^{-1} (Ramos et al. 2017a), which might explain the rather sparse occurrence of living CWCs. However, living corals are more common in and close to the numerous canyons dissecting the entire Mauritanian margin, which probably serve as a refuge for the corals (Wienberg et al. 2018).

Leaving the coral hotspot area off Mauritania, CWC records are rare and individual findings of *L. pertusa* are reported from the northern Senegalese margin (Zibrowius 1980) and off Liberia (Rogers 1999). For the Gulf of Guinea Upwelling Ecozone, documentations of corals remain low with one remarkable exception. Off Ghana (~5°N), one large CWC mound of 70 m height has been discovered in a water depth of 400 m (indicated as 'd' in Fig. 8.1; Buhl-Mortensen et al. 2017a). This mound has an arcuate shape and extends over 1400 m perpendicular to the main current direction and is probably coalesced from several smaller mounds. The mound is topped by a thriving *Lophelia*-dominated reef accompanied by *M. oculata* and *S. variabilis* (Buhl-Mortensen et al. 2017a).

In contrast to this rather poorly documented section of the West African margin, several records exist for the adjacent Angolan Marine Ecozone (Fig. 8.1). On the northern side of the Congo canyon (~6°S), two mound-like structures have been observed in ~400 m water depth, which are ~30 m high and covered by living CWCs (Gay et al. 2007), while south of this canyon (~6.6°S), isolated colonies of *L. pertusa* and *M. oculata* are reported (Zibrowius 1980). The first real evidence for a coral mound province exists for the northern Angolan margin (~7.3°S), where several up to 30-m-high and 1.5-km-long coral mounds have been detected in ~340–410 m water depth covering (at least) 12 km^2 (indicated as 'e' in Fig. 8.1; Le Guilloux et al. 2009). The summits of the mounds are covered with living reefs dominated by *L. pertusa* with minor contributions of *M. oculata* (Le Guilloux et al. 2009). An even larger mound province was recently detected south of the Cuanza river (9.5°S-10°S) in water depths of 250–500 m (indicated as 'f' in Fig. 8.1; Hebbeln et al. 2017). Here, numerous coral mounds and ridges reaching heights of up to 100 m extend over (at least) 60 km along the margin (see Sect. 8.5). Thriving reefs exist at 330–470 m water depth, although at these depths they are exposed to hypoxic conditions (DO: 0.5–1.5 mL L^{-1}) in the centre of the local OMZ (see Sect. 8.3) (Hanz et al. 2019). The coral community is dominated by *L. pertusa* (see Sect. 8.4); though *M. oculata* is comparably rare, some relatively dense spots were documented and the species forms partly exceptionally large colonies of >1 m height (Orejas et al. 2021).

8.2.2.3 Cold-Water Corals in the Namib Ecozone and Offshore South Africa

For the Namib Ecozone, there are two coral mound provinces that are also exposed to hypoxic conditions in the local OMZ (see Sect. 8.3). A first province south of the Kunene river

(~17.5°) comprises rather small mounds (height: <15 m) covering an area of ~110 km^2 at 230–290 m water depth (indicated as 'g' in Fig. 8.1; Rush et al. 2019). These mounds have circular to ellipsoidal footprints and are arranged to elongated clusters (up to ~4 km in length) or follow a conspicuous sickle-shaped line. They are covered by coral rubble dominated by *L. pertusa* with few living corals (Rush et al. 2019). A second province, located SE of the Walvis Ridge (20.2°S-21°S), extends over (at least) 80 km and consists of >2000 coral mounds in 160–270 m water depth (indicated as 'h' in Fig. 8.1; Tamborrino et al. 2019). The up to 20-m-high mounds have formed during the Early to Mid-Holocene (9.5–4.5 kyr; Tamborrino et al. 2019) and are covered by coral rubble composed of *L. pertusa* (see Sects. 8.5 and 8.6). Today, no live corals have been observed and the benthic fauna on the mounds is overall impoverished, which probably reflects the extremely low oxygen conditions in this region (DO: <0.15 mL L^{-1}; Hanz et al. 2019).

Offshore South Africa (~29.5°S–37°S), CWCs have been indicated for several Marine Protected Areas (MPAs; Fig. 8.1) such as the Childs Bank (~200–300 m), the Cape Canyon, and the Browns Bank, and on some of the deep (>800 m) Southeast Atlantic seamounts, the latter providing habitat for *E. rostrata* (Bergstad et al. 2019; Sink et al. 2019). However, further investigations are needed for the South African margin to document the composition and distribution of the CWC community.

8.2.2.4 Open Ocean Habitats

The Southeast Atlantic seamounts (including the Walvis Ridge) are open ocean habitats for CWCs (see Box 8.1), which also encompass seamounts and volcanic islands of the Ecozones of the Azores-Canaries-Madeira and Cape Verde in the northern hemisphere (e.g., Zibrowius 1980; Zibrowius and Gili 1990; Rogers 1999). At these settings, CWCs occur widespread (Table 8.1), but the generally steep topography and the lack of substantial near-by sediment sources hamper the formation of coral mounds. An exceptional situation has been found on Concepcion Bank (Canary Islands; Fig. 8.1), where small coral mounds have developed on the extensive summit plateau of this seamount, which descends slowly to the west (Rivera et al. 2016). These mounds are 20 m in height and are concentrated at 600 m water depth at the slowly descending western part of the summit plateau. This morphological summit configuration probably provided a suitable habitat for mound initiation, while mound formation was likely supported by its location close to the West African margin allowing the delivery of large amounts of aeolian dust from the Sahara (Holz et al. 2004).

8.2.3 Cold-Water Corals Occurring Under Extreme Environmental Conditions off Mauritania, Angola, and Namibia: The MAN Key Study Areas

As summarised above, CWCs are widespread in West African waters and occupy different habitats, but in particular the coral reefs and mounds formed by them on the outer shelf and upper slope of the continental margin are of peculiar interest. These sites provide detailed insight into the present-day and past ecosystem structure governed by the regional oceanographic setting. For this study, the coral mound provinces in the upwelling regions off Mauritania, Angola, and Namibia (in the following referred to as MAN) were selected as key study areas, because of their large extension, but above all because of their occurrence in the centre of the local OMZs. Although the prevailing hypoxic conditions should pose severe stress for the CWCs, the presence of live coral populations, some even forming prosperous reefs off Angola, points to their capability to adapt to such an extreme environment (Hebbeln et al. 2020). By directly comparing the MAN coral mound provinces, the following three sections describe the specific local oceanographic, bio-ecological, and geo-morphological characteristics in order to highlight any differences and commonalities between these regions.

8.3 MAN: General Oceanographic Pathways and Regional Hydrographic Settings

Cold-water corals are sensitive to the regional and local hydrographic environment in which they grow. Current strength, temperature, salinity, oxygen levels, downward labile carbon fluxes as well as the supply of sediments and larval transport pathways are driven by surface and thermocline water circulations, density boundaries between water masses, residual flow, and tidal movements (e.g., Frederiksen et al. 1992; White and Dorschel 2010; Mohn et al. 2014; Juva et al. 2020; Puerta et al. 2020; Schulz et al. 2020). Thus, CWC reefs along the West African margin are likely to respond to basin-scale advection patterns as well as to local near-bed water-mass dynamics. Regarding the growth of framework-building CWCs, these general parameters have been recognised to control coral proliferation in various modern CWC reefs in the North Atlantic (e.g., Dorschel et al. 2007; Mienis et al. 2007, 2014; Davies et al. 2010; Rüggeberg et al. 2011; van Haren et al. 2014).

The MAN coral mound provinces on the West African continental margin reveal significant differences with respect to their present-day and past reef status. For instance, some

25 kyr ago during the last glacial period, the Mauritanian and Angolan margins in the sub-tropical eastern Atlantic have been the home for coral reefs resulting in vast coral mound formation. However, only off Angola, CWCs thrive and mound formation persists until today, while off Mauritania, the presence of CWCs significantly declined with the onset of the Holocene resulting in a stagnation in mound formation (see Sect. 8.5; Wefing et al. 2017; Wienberg et al. 2018). Hence, during the last major global warming, inter-hemispheric differences in the coral environments must have been created that steered the waning of CWCs in the northern hemisphere (Mauritania) and their resilience in the southern hemisphere (Angola). However, the Namibian margin does not fit to this pattern, as CWCs seem to have vanished from coral mounds on the outer shelf after a short and intense mound formation 'pulse' during the Early to Mid-Holocene (see Sect. 8.5; Tamborrino et al. 2019).

While fresh labile carbon (and hence food) flux is one major factor governing the proliferation of CWCs, the regional and local flow and the properties of water masses beneath the seasonal thermocline are also important (e.g., Davies et al. 2009; White and Dorschel 2010). The cross-equatorial Atlantic flow of water between 30°N and 30°S is very complex and stems, for example, from the low Coriolis parameter and the generally northward displaced near-equatorial pressure trough (Hastenrath and Lamb 2004). In the following, the wind-driven surface circulation across the equator and the underlying eddy-driven thermocline advection patterns are described, before local to regional aspects intimately connected with the large-scale flow regime and directly affecting the MAN coral mound provinces along the West African margin are emphasised.

8.3.1 Surface Circulation Patterns

The surface waters of the eastern equatorial to sub-tropical Atlantic are predominantly influenced by the atmospheric wind stress variability resulting from the seasonal migration of the Inter-Tropical Convergence Zone (ITCZ), and thus, from the strength and position of the Hadley circulation in both hemispheres. The northerly position of the near-equatorial pressure trough over North Africa in the boreal summer drives a net cross-equatorial flow of waters from the southern to the northern hemisphere (Joyce 1988; Stramma et al. 2003). The southerly wind component in the southern hemisphere produces westward upper-ocean Ekman transport, oceanic upwelling, and a tongue of cold surface waters centred to the south of the equator. Upon crossing the equator, the surface winds are subject to rightward-directed acceleration resulting in water-mass divergence north of the equator (Molinari et al. 2003). Further into the northern hemisphere, wind stress is reduced through the competing north-easterly and south-westerly winds that converge, defining the seasonally variable position of the ITCZ. Over the central and western Atlantic, winds in the equatorial band are predominantly from the east, and thus the upper-ocean Ekman transport is directed away from the equator, and the upwelling and resulting cold tongue are centred along the eastern basin boundary, which is most pronounced during boreal winter.

Vigorous zonal currents carry heat and mass in the eastern equatorial to the sub-tropical Atlantic Ocean. Starting in the southern hemisphere, the Benguela Current (BC) carries about 20 Sverdrup of water from the southern tip of Africa and the polar Southern Ocean north-westward contributing to the southern hemisphere sub-tropical gyre (Stramma and Peterson 1989; Garzoli et al. 1996; Boyer and Hampton 2001; Veitch et al. 2010). In the surface layer, the cold water receives strong contributions of warm and saline Indian Ocean water through the Agulhas leakage (Biastoch et al. 2008; Weijer and van Sebille 2014; Biastoch et al. 2015). The BC meets the southward moving Angolan Current (AC) creating the Angola-Benguela Front (ABF; Fig. 8.2), which separates the warm North and South Equatorial Counter Current (NECC and SECC) waters from the colder BC waters at a seasonally varying latitude of 15 ± 3°S (Meeuwis and Lutjeharms 1990). The cyclonic gyre in the Gulf of Guinea causes a particular oceanic feature known as the (cold) Angola Dome (AD), which in fact represents several upwelling domes with seasonally different positions (Doi et al. 2007). The BC, however, turns westward towards the equator becoming the South Equatorial Current (SEC; Fig. 8.2) that eventually feeds the Brazil Current and the Gulf Stream contributing to the Atlantic cross-equatorial flow. Slightly north of the equator, the NECC returns parts of the warm and saline water to the eastern Atlantic. In the northern hemisphere (>10°N), the Mauritania Current (MC) develops as a coastal component of the NECC and flows northward along the Mauritanian margin (Fig. 8.2). At the Cape Verde Frontal Zone (CVFZ), the MC turns westward contributing to the tropical anti-clockwise gyre circulation. The southward-flowing Canary Current (CC) is the eastern basin expression of the northern hemisphere sub-tropical gyre. It turns westward at the CVFZ and becomes the North Equatorial Current (NEC; Fig. 8.2). Hence, the CVFZ, which extends from 21°N in the east (Cap Blanc) to 16°N in the west (Cape Verde), is a highly dynamic frontal system separating northern and southern Atlantic water masses.

The massive heat transport from the southern to the northern hemisphere is indicated by a strong cross-equatorial latitudinal temperature gradient that spans from roughly 17 °C at

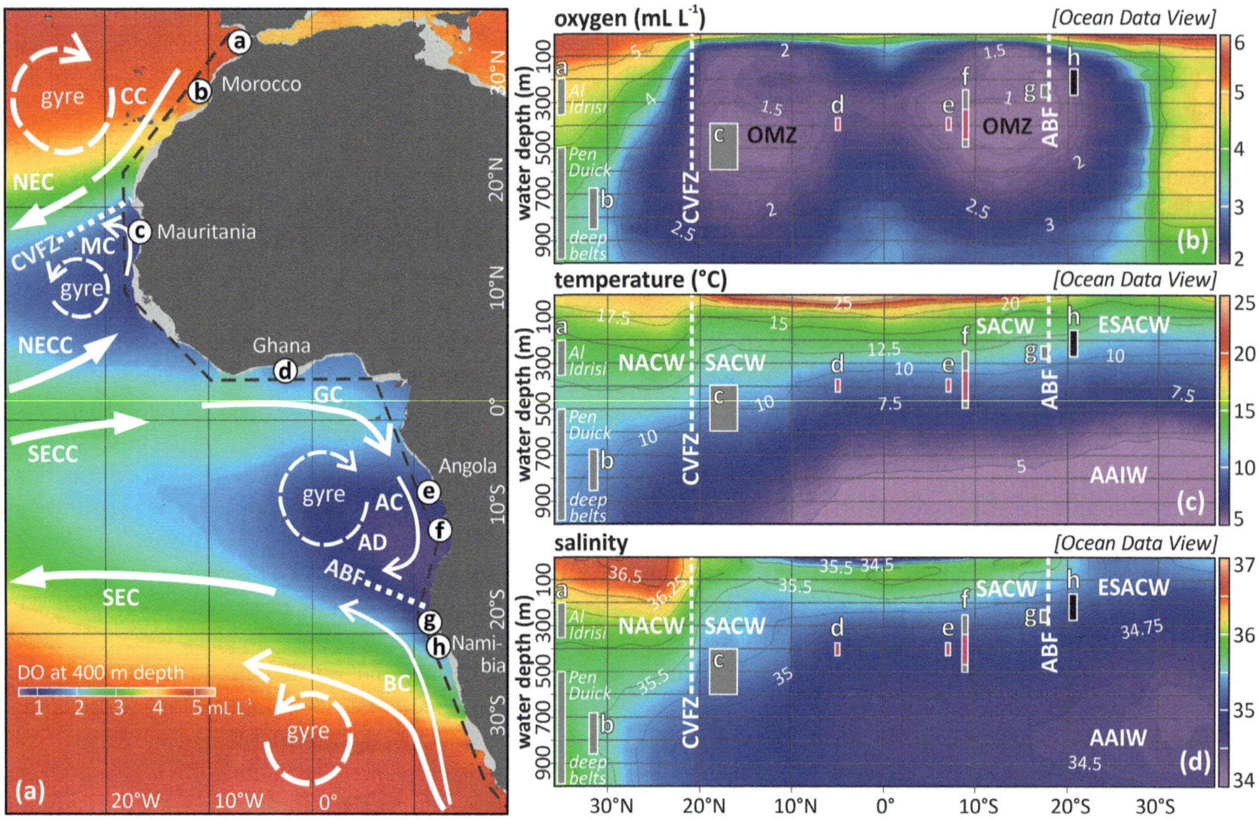

Oceanographic features - AC: Angola Current, AD: Angola Dome, BC: Benguela Current, CC: Canary Current, GC: Guinea Current, MC: Mauritania Current; NEC/SEC: North/South Equatorial Current, NECC/SECC: North/South Equatorial Counter Current, OMZ: oxygen minimum zone; **Frontal zones -** ABF: Angola-Benguela Front, CVFZ: Cape Verde Frontal Zone; **Water masses -** NACW: North Atlantic Central Water, SACW: South Atlantic Central Water, ESACW: Eastern South Atlantic Central Water, AAIW: Antarctic Intermediate Water; **Coral mound areas - a**: Morocco (Gulf of Cádiz), **b**: Morocco (off Agadir), **c**: Mauritania, **d**: Ghana, **e**: Angola (off Congo river), **f**: Angola (S of Luanda), **g**: Namibia (S of Kunene), **h**: Namibia (SE Walvis Ridge).

Fig. 8.2 The oceanography off West Africa. (**a**) Overview map showing the general circulation pattern (white bold arrows: surface currents; white dashed arrows: anti-clockwise and clockwise gyre circulation) and frontal zones (white dotted lines) off West Africa (a to h indicate coral mound provinces off West Africa; see Fig. 8.1 for details). Dissolved oxygen concentrations (DO) at 400 m water depth are displayed. The black dashed line shows the position of the N-S cross profiles shown in **b–d**. Depth-oxygen (**b**), depth-temperature (**c**), and depth-salinity (**d**) profiles along the West African margin extending from ~35°N to ~35°S (data source WOA13; Garcia et al. 2013; Locarnini et al. 2013; Zweng et al. 2013; generated with ODV v5.1.5, R. Schlitzer 2018). Frontal zones and intermediate water masses are indicated. Note: Values given here—representing the conditions down to 1000 m water depth—might slightly deviate from *in situ* measurements mentioned in the text and listed in Table 8.2, which are partly obtained in much shallower waters closer to the coast. Coral mound areas are shown (labelled as a to h) highlighting their present-day reef status (pink box: thriving reefs, grey box: sparse live colonies, black box: fossil reefs; see also Table 8.2)

the ABF in the south to more than 25 °C at the CVFZ in the north (Fig. 8.2). In addition, the thermal structure of the underlying thermocline waters differs significantly between the two hemispheres, nicely illustrated by the cross-equatorial deepening of the 10 °C isothermal from ~300 m depth at 20°S to >500 m depth at 20°N (Fig. 8.2). Hence, the eastern Atlantic in the northern hemisphere receives larger heat fluxes compared to the southern hemisphere governed by the large-scale surface circulation patterns. The salinity strongly follows the cross-equatorial trends marked by a significant northward increase in salinity by >1‰, which is most prominent near the CVFZ (Fig. 8.2). Overall, the identification and position of the CVFZ are not just achieved by the salinity gradient but also through the nutrient content as the southern water masses contain higher contents of silicate, phosphate, and nitrate compared to the northern water masses (Zenk et al. 1991; Liu and Tanhua 2021). In contrast, the position and dynamics of the ABF are indicated by the temperature gradient (Fig. 8.2).

8.3.2 Circulation Patterns at Thermocline Depths

The West African margin is characterised by two major eastern boundary coastal upwelling systems: The Canary upwelling system (including the Mauritanian upwelling region being part of the Sahelian Upwelling Ecozone; see Sect. 8.2) in the sub-tropical North Atlantic (e.g., Arístegui et al. 2009) and the Benguela upwelling system (off Angola and Namibia) in the sub-tropical South Atlantic (e.g., Mohrholz et al. 2014). Overall, the upwelling of nutrient-rich cold waters results in high primary productivity (e.g., Carr and Kearns 2003), while the remineralisation of high fluxes of organic particles in the water column results in severe mid-depth oxygen depletion and pronounced OMZs. Consequently, the most obvious and significant feature of the thermocline waters is the generally low oxygen level. The OMZs off West Africa extend between ~20°N and 25°S and the MAN coral mound provinces are situated in the centre of these OMZs (Fig. 8.2). The extension of the local OMZs is highly variable, not only due to the seasonal variability in upwelling intensity triggered by the local wind forcing, but also due to the poleward advection of oxygen-depleted thermocline waters.

The Mauritanian and Angolan coral mound provinces positioned on the West African upper slope are today bathed by the oxygen-depleted and nutrient-rich South Atlantic Central Water (SACW; Fig. 8.2). Off Mauritania, the SACW is carried northwards, more and more competing with the southward-flowing, and comparatively nutrient-poor and oxygen-rich, North Atlantic Central Water (NACW). Both upper thermocline water masses are separated by the CVFZ down to ~600 m water depth (Fig. 8.2; Zenk et al. 1991; Peña-Izquierdo et al. 2015). Off Angola, the SACW is transported southwards and converges at the ABF with the equatorward-flowing cooler and oxygen-enriched Eastern SACW (ESACW; Mohrholz et al. 2008), which is the main water mass influencing the extinct Namibian coral mounds on the outer shelf (Fig. 8.2). The position and dynamics of both frontal zones, and hence the poleward propagation of oxygen-depleted SACW, are seasonally variable due to the changes in flow intensity of the respective eastern boundary currents (MC and AC). In addition, distinct poleward undercurrents play an important role in the upwelling regions in the northern and southern hemispheres (Veitch et al. 2010; Pelegrí et al. 2017; Kounta et al. 2018). These narrow and vigorous slope currents allow parts of the poleward-flowing thermocline waters to cross the frontal zones. For instance, for the North Atlantic, it has been found that the poleward undercurrent is continuous as far north as Iberia (Arístegui et al. 2009).

In the Mauritanian upwelling system, elevated poleward flow of the SACW is observed from May to July, in comparison with the main upwelling season lasting from December to April (Klenz et al. 2018). In the Benguela upwelling system off Angola and Namibia, the enhanced poleward advection of oxygen-depleted and nutrient-rich SACW from the Angola gyre results in an extended OMZ from October to April (austral summer), whereas during the austral winter season the oxygen-rich ESACW spreads further northward (Mohrholz et al. 2008). The advection of the oxygen-poor SACW has particularly severe impacts on the environmental conditions on the Namibian margin as it causes hypoxic to anoxic conditions and even may support sulphur outbreaks on the shelf (Ohde and Mohrholz 2011; Junker et al. 2017).

The key mechanism driving the residual northward flow is the eddy kinetic energy provided by the wind-driven geostrophic circulation (Ollitrault and Colin de Verdière 2014), though the circulation below the thermocline is far more sluggish than the wind-driven surface circulation. However, the basin-scale circulation at thermocline depths benefits from the production of mid-depth waters, i.e. Antarctic Intermediate Water (AAIW), in the subpolar South Atlantic. Those cold waters are subducted underneath the tropical warm waters due to their lower temperature (Fig. 8.2), forcing the geostrophic flow northward (e.g., Talley et al. 2011).

8.3.3 Local to Regional Hydrographic Settings

The most striking feature of the MAN coral mound provinces is that living CWCs are exposed to hypoxic conditions at the lowest range of DO concentrations (<1.5 mL L^{-1}; Ramos et al. 2017a; Hanz et al. 2019) so far recorded for CWC reefs worldwide. However, except for the very low DO concentrations, local near-bed environmental conditions like temperature and current speed (for an overview, see Table 8.2) fall within the ranges that allow for active CWC growth (Davies et al. 2008; Davies and Guinotte 2011). Interestingly, the distribution of CWC reefs and mounds occurring along the entire West African margin follows the overall thermal structure and are found in much shallower depths in the cooler southern hemisphere environments (mainly above 500 m depth) as compared to the warmer northern hemisphere environments (mainly below 500 m depth; Fig. 8.2).

Extremely low DO concentrations (<0.15 mL L^{-1}) and relatively high bottom-water temperatures (>11.8 °C; Table 8.2; Hanz et al. 2019), which are near the currently known upper tolerable range of *L. pertusa*, likely explain the absence of living coral populations on the Namibian shelf (SE of Walvis Ridge; see Sect. 8.5). Along the Mauritanian

Table 8.2 Near-bed water mass characteristics of cold-water coral mound provinces along the West African margin (see Figs. 8.1 and 8.2 for location). Values for temperature (*T*), salinity (*S*), dissolved oxygen (*DO*), current speed (*CS*) and pH derive from *in situ* measurements (benthic lander systems, CTD, ADCP) and from the WOA13 database (numbers in brackets; Garcia et al. 2013; Zweng et al. 2013). Data for net primary productivity (*NPP*; annual mean) according to Behrenfeld and Falkowski (1997; http://data.guillaumemaze.org/ocean_productivity). Realm nomenclature according to Spalding et al. 2007. *NACW* North Atlantic Central Water, *AAIW* Antarctic Intermediate Water, *MOW* Mediterranean Outflow Water, *SACW* South Atlantic Central Water, *ESACW* Eastern South Atlantic Central Water

Realm	Coral region	Depth (m)	Reef status; coral cover	Water mass	T (°C)	S	DO (mL L^{-1})	NPP (g C m^{-2} day^{-1})	CS (m s^{-1})	pH	Refs.
Temperate North Atlantic	Morocco (Pen Duick) 35.3°N/6.8°W	500–550	fossil; sparse live corals	NACW	10.7–11.8	35.6–35.7	(>4)	370	mean: 0.09 max: 0.3	7.98	Mienis et al. (2012), Flögel et al. (2014)
	Morocco (deep belts) 35.3°N/7°W	700–950	fossil; sparse live corals	AAIW MOW	9.9–11.2	35.7–35.8	3.5–3.7	380	./.	7.92	Flögel et al. (2014), Hebbeln et al. (2015)
	Morocco (Agadir) 31°N/11°W	680–860	unknown; sparse live corals >720 m	NACW	8.9–10.3	35.5–35.7	3.1–3.3	1660	./.	./.	Glogowski et al. (2015)
Tropical Atlantic	Mauritania 19-16°N/17°W	400–550	fossil; sparse live corals	SACW	9.4–11.7	35.2–36.6	1.1–1.4	19°N: 5990 16°N: 750	./.	7.83–7.86	Flögel et al. (2014), Ramos et al. 2017a
	Ghana 4.8°N/3°W	~400	thriving	SACW	8.5–9	34.8–34.9	(~1.5–2)	1180	./.	./.	Buhl-Mortensen et al. (2017a)
	Angola (Congo) 7°S;12°E	340–410	thriving	SACW	7.7–8.8	(~34.8–35)	(~1–1.5)	7050	./.	./.	Le Guilloux et al. (2009)
	Angola (Luanda) 9°S/13°E	250–500	thriving; live corals at 330–470 m	SACW	6.7–14.2	34.6–35.8	0.5–1.5	6240	mean: 0.1 max: 0.3	8.12	Hanz et al. (2019), Hebbeln et al. (2020)
Temperate Southern Africa	Namibia (Kunene) 17°S/11°E	230–290	fossil; sparse live corals	SACW	10.6–12.9	35–35.6	0.5–0.6	1690	mean: 0.06 max: 0.15	./.	Rush et al. (2019)
	Namibia (SE Walvis Ridge) 20°S/13°E	160–270	fossil; no live corals	ESACW	11.8–13.2	34.5–35.4	0–0.15	2260	mean: 0.09 max: 0.21	8.01	Hanz et al. (2019)

and Angolan margins, DO concentrations are still low (0.5 to 1.5 mL L^{-1}) and bottom-water temperatures vary around 10.5 °C (Table 8.2; Flögel et al. 2014; Ramos et al. 2017a; Hanz et al. 2019), but here coral growth is not hampered, though their proliferation varies from sparse live colonies to thriving reefs (see Sect. 8.4). A recent short-term aquarium experiment with *L. pertusa* specimens from the Angolan margin showed that hypoxic conditions had no effect on their respiration rate, indicating that low-oxygen conditions are not a general limiting factor for the overall distribution of this species (Gori et al. 2023).

As observed for many other CWC reefs in the Atlantic (e.g., Frederiksen et al. 1992; Mohn et al. 2014; Mosquera Giménez et al. 2019), internal tides, resulting in vertical movements of 70–100 m (Hanz et al. 2019; Juva et al. 2020) also play an important role in the MAN coral mound provinces. On the Namibian margin, a negative correlation of DO and bottom-water temperatures was observed during the internal tidal cycle, acting as a combined stressor (low DO and high temperatures; Hanz et al. 2019). This is in contrast to the Angolan margin, where CWC growth is confined to 250–500 m water depth, and thriving reefs are constrained to 330–470 m (Hebbeln et al. 2020). Here, prolific CWC growth is likely related to the fact that at the shallower depths the highest bottom-water temperatures coincide with the highest DO concentrations, while at greater depths, the opposite pattern was observed (Hanz et al. 2019); hence, only one stressor is active at the respective depths. In addition, the increase of one stressor is likely compensated by another positive factor such as food supply. The observations from the MAN coral mound provinces constrain the compound effect of ecosystem failure (tipping point) to oceanographic properties somewhere between 0.5 and 0.15 mL L^{-1} oxygen and temperatures between 10 °C and 11.5 °C, notwithstanding the possibility that many other local influences and processes contribute to the consequence of reef failure.

The presence of living colonies and even thriving reefs demonstrates that *L. pertusa* and *M. oculata* are capable of adapting to extreme conditions (low oxygen, high temperatures), when food supply is adequate (Hanz et al. 2019; Hebbeln et al. 2020; Orejas et al. 2021) as facilitated by high net primary productivity (NPP) in the Mauritanian and Angolan regions (NPP: 750–7050 g C m^{-2} day^{-1}; Table 8.2; Behrenfeld and Falkowski 1997). On the contrary, even though the Namibian shelf is also marked by high NPP (<380 g C m^{-2} day^{-1}; Table 8.2; Behrenfeld and Falkowski 1997), the prevailing anoxic conditions are apparently too extreme to allow coral proliferation (Hanz et al. 2019). In addition, and to emphasise how complex and sensitive the various environmental parameters interact, the present-day absence of thriving reefs on the northern Moroccan margin is rather ascribed to the lack of sufficient food supply (Wienberg et al. 2010; Mienis et al. 2012), while DO concentrations of >3.5 mL L^{-1} (Table 8.2; Mienis et al. 2012) would fit the metabolic needs of CWCs (Dodds et al. 2007). Moreover, stressors like high temperatures can affect the calcification, respiration, and growth rates of benthic species, increasing their energy demand (Coma 2002; Gori et al. 2016). Experiments have shown that indeed multiple stressors can be counteracted by sufficient food supply (Büscher et al. 2017). However, future ocean warming and further oxygen loss will certainly decrease the chance of survival of the live coral populations off Mauritania and Angola (see Sect. 8.7).

8.4 MAN: Cold-Water Coral Ecosystems and Reef Habitats

8.4.1 The Cold-Water Coral Ecosystem Structure

The CWC sites off Mauritania and Angola belong to those Marine Ecozones of the eastern Tropical Atlantic, which are bordered by oceanic frontal systems and are influenced by coastal upwelling of nutrient-rich and cold waters (e.g., Carr and Kearns 2003). Even though these high-productivity conditions further result in severe oxygen depletion at mid-depths, the West African upwelling regions show much higher faunal richness than the typical tropical regions with warm waters and reduced salinities (Le Lœuff and von Cosel 1998). In addition, CWC habitats are hotspots of biomass as they play an important role in carbon cycling (van Oevelen et al. 2009; Cathalot et al. 2015) and have a major influence on species diversity as they provide a multitude of micro habitats (Buhl-Mortensen et al. 2010, 2017b). Some of these habitats are even unique for CWCs, like the tissue-covered coral skeleton and the cavities within the coral skeleton made by excavating bioeroders and subsequent chasmoliths (e.g., Beuck and Freiwald 2005; Beuck et al. 2007), but also the diverse micro-habitats related to the hard substrate provided by dead coral skeletons facilitate a high biodiversity (Mortensen and Fosså 2006; Buhl-Mortensen et al. 2017b; Korpanty et al. 2023). However, the regional environmental conditions (temperature, salinity, DO, food supply, current speed; see Table 8.2) exert a strong control on overall ecosystem health. Hence, the local hydrographic variations between the MAN coral mound provinces might explain the large regional differences of the West African coral ecosystems revealed by (i) the principal framework constructing architectures and colony shapes and (ii) the biological interactions and faunal composition, which are described in detail in the following.

8.4.2 Colony Framework Architectures

Along the Mauritanian margin, two major geoform types with living CWC frameworks are present: submarine canyons and open slope coral mounds (for a broad overview on the Mauritanian slope biodiversity and biomass distribution, see the seminal book of Ramos et al. 2017b). For submarine canyons along the northern margin (~20.2° N-18.6°N), only recently settled colonies of *L. pertusa* (Tanoûdêrt canyon) or mixed assemblages of *L. pertusa* and *M. oculata* (Nouamghar, Inchiri and Tioulit canyons) attached to the steep (partly near vertical) canyon flanks were observed (for a detailed geo-morphological description of the Mauritanian canyons, see Sanz et al. 2017). Focussing on the northernmost Tanoûdêrt canyon as an example, the majority of *L. pertusa*-colonies thriving on canyon flanks have arborescent growth forms, meaning that the individual main branches grow in any direction upwardly and are open spaced probably due to high particle load in the water column (Fig. 8.3a; see also Sanna and Freiwald 2021); this is similar to morphotypes previously observed in other canyon systems in the Atlantic (e.g., Brooke et al. 2009). The colony heights vary from 3 to about 10 cm with 3 to 8 polyp generations. These small colonies are entirely alive, from the base to the upper parts, and the ageing process initiating the separation of an upper live zone and a basal dead zone has not commenced yet. The situation changes on the upper canyon plain area, where locally up to 1-m-high *L. pertusa*-frameworks grouped in dense coral thickets on rock outcrops prevail. Interestingly, this is the only known location in the area where environmental conditions are today favourable to support the growth and longevity of such large frameworks. Most of the colony portions are dead, upward growing like individual towers, and are speckled either with surviving or, and more common, with recently settled *L. pertusa*-colonies, which show the same growth attributes as described before (Fig. 8.3b). Higher up and along an isolated coral mound that has developed on the upper canyon plain (see Sect. 8.5), live *L. pertusa*-colonies become less dense. Individual colony shapes change to strictly upward growing forms with short and 'crowded' branch arrangements. Overall, it appears that the majority of the live *L. pertusa*-colonies result from a recent settlement event within the past 10 years, which is valid for all observed canyons.

The majority of the Mauritanian coral mound complexes along the open slope host sparse live coral occurrences. Older and *in situ* coral framework was rarely encountered (Fig. 8.3c), rather there are mostly spatially dispersed and overturned *L. pertusa* 'towers' that continue to grow with few surviving branches. Individual overturned towers measure up to 50 cm in length and the live portions face against the prevailing bottom currents. Both the dead and live colony portions exhibit a crowded and short-branched pattern compared to the long-branched colonies from the canyons. Most active coral growth takes place on some mound summits shielded with rocks made of lithified hemipelagic sediments and coral limestone. These rocks are biodiversity hotspots as they attract recently settled *L. pertusa*- and *M. oculata*-colonies next to other sessile and mobile invertebrates and fish. The rock-attached colonies measure between 5 and 10 cm in height and also show a crowded branching pattern.

The major framework constructors of the Angolan coral mounds are *L. pertusa* and, to a much lesser extent, *M. oculata* (Orejas et al. 2021). The Angolan sites yield the most vivid and flourishing coral growth, albeit thriving under the lowest oxygen contents ever measured for both species (Table 8.2; Hanz et al. 2019). The construction architecture differs largely from the Mauritanian growth pattern by arcuate coral galleries, or thickets, up to 3-m wide and up to ~80 cm in height. The colonies are composed of 20 to 30-cm-long zig-zagging branches comprising 15–25 polyp generations (Fig. 8.3d, e). The formation of such large coral galleries follows at least four evolutionary steps: (1) Aligned settlement of several colonies most likely on dead framework slabs or coral rubble of a former growth pulse—not a stable foundation at all. (2) With time, the neighboured colonies develop into a closed 'thicket' and architectural stability is provided by intermingling of the branches, partly through fusion (Hennige et al. 2014). This induces a reduction in lateral colony growth. Due to the unstable settling ground, the thickets start to collapse in place with the basal and dead colony stems pointing inward and the surviving colony parts pointing outward. A side effect is the enlargement of the thicket perimeter and an increased provision of hard substrate and micro-niches for associated fauna. (3) Continued colony growth, however, with re-oriented growth direction to close the gaps in the thicket after the collapsing process with renewed open-spaced branch growth. This results in the cauliflower-like appearance of thriving thickets (Fig. 8.3d). (4) With time, the thickets begin to sink and slide down the steep mound slopes contributing to the coral rubble accumulations, triggered by negative buoyancy and the unstable foundation. The age estimation of a mature *M. oculata*-thicket points to a lifetime of ~100 years (Orejas et al. 2021), further some of the few existing *M. oculata*-colonies have remarkable heights of up to 1.2 m and represent some of the largest ever discovered (Orejas et al. 2021).

The extinct coral mounds off Namibia (SE of Walvis Ridge; Fig. 8.1) were built by *L. pertusa*. A large colony sample recovered from one of the Namibian mounds (belonging to the Coral Belt mound cluster; see Tamborrino et al. 2022) indicates that branch growth direction may change after overturning of the colony (Fig. 8.3f) as also described for the Angolan coral mounds. North of these extinct mounds, small living *L. pertusa*-colonies have been observed

Fig. 8.3 Colony framework architectures. (**a**) Several recently settled, scattered *Lophelia pertusa* (a.k.a. *Desmophyllum pertusum*)-colonies on rock with open-spaced growth (Mauritania, Tanoûdêrt canyon, flank). Scale bar: 10 cm. (**b**) Larger *L. pertusa*-framework with some living colony portions (Mauritania, Tanoûdêrt canyon, shoulder). Scale bar: 10 cm. (**c**) Isolated *L. pertusa*-'tower' with dead framework portion

in the Kunene coral mound province (see Sect. 8.2; Fig. 8.1). The living colonies were growing on exposed *L. pertusa* rubble on the upper slope and summit of the mounds. The presence of live corals is likely related to slightly higher DO concentrations in this area (0.5–0.6 mL L^{-1}; Table 8.2; Rush et al. 2019) compared to the extinct southern Namibian coral mounds (Table 8.2).

8.4.3 Biological Interactions and Faunal Composition

The MAN coral mound provinces exhibit a variety of hard substrate habitats provided by (i) live corals, (ii) dead coral framework, (iii) coral rubble, and (iv) the hard substrate fauna. In the following, evident biological interactions and faunal groups that characterise these different habitats are described for the Mauritanian and Angolan areas.

8.4.3.1 The Live Coral Zone

In the Mauritanian and Angolan coral mound provinces, live *L. pertusa*- and *M. oculata*-colonies suffer from grazing pressure of echinoderms (mainly *Gracilechinus acutus, G. elegans,* and *Peltaster placenta*) revealed by echinoid gut contents collected from live corals yielding scraped-off coral skeleton pieces (see also Stevenson and Rocha 2013). An even more efficient coral predator is the gastropod *Calliostoma bullatum* (a 'living fossil'; see Box 8.2). Provided with an extractable proboscis and radula, this relatively large gastropod was frequently observed on Mauritanian coral colonies (*L. pertusa* and *M. oculata*) sucking off live coral polyps and tissue (Fig. 8.4a), leaving a tissue-barren skeleton behind (Hoffman et al. 2019c). Further common on live Mauritanian corals are the corallivore gastropod *Coralliaphila richardi* and nudibranchs.

The well-known symbiotic relationship between *L. pertusa* and *M. oculata* with tube and gall-forming *Eunice* polychaete species (i.e., Mueller et al. 2013) appears prominent off Mauritania and Angola but has not been seen in Holocene coral remains off Namibia. The majority of freshly settled coral colonies are associated with calcified tubes of *Eunice* at their bases. *Eunice* tubes at later growth stages of the coral colonies were hardly detected, thus underpinning the functional role of *Eunice* as an ecosystem engineer (Mueller et al. 2013). A unique coral-sponge association is known from live Mauritanian corals (Göcke et al. 2016; see Box 8.2). Unlike other known *Cladorhiza* species, the carnivorous sponge *Cladorhiza coralliophila* is confined to coral colonies (*L. pertusa* and *M. oculata*) for attachment and is considered an obligate symbiont of its coral host (Fig. 8.4b).

8.4.3.2 The Tissue-barren Coral Framework

As previously described for various CWC sites (e.g., Jensen and Frederiksen 1992; Mortensen and Fosså 2006; Buhl-Mortensen et al. 2017b; Henry and Roberts 2017), also for Mauritania and Angola, the faunal richness in the microhabitat provided by the dead coral framework is the richest. This is revealed by a plethora of filter- and suspension feeders, such as encrusting and erect demosponges, solitary scleractinians, antipatharians, acanthogorgiid, clavulariid and plexaurid octocorals, serpulids, actiniarians, epizoanthids, bryozoans, and hydroids (Fig. 8.4c). The most appealing findings in the Mauritanian dead coral framework are aggregations of byssus-attached *Acesta* cf. *excavata* (Fig. 8.4d). This species is known for extreme high filtration and respiration rates (Järnegren and Altin 2006). Hexactinellid sponges are rarely present in the Mauritanian coral habitat but dominant off Angola with large *Aphrocallistes beatrix, Farrea* sp., and *Sympagella* sp. aggregations (Fig. 8.4c). It is tempting to assume that the abundant presence of hexactinellid sponges in Angolan coral systems might be facilitated by the silica excess provided from pelagic siliceous algae (diatoms; Van Bennekom and Berger 1984; Maldonado et al. 2005, 2020). Another source of silicate presumably derived from decaying seagrass, which occur frequently in river deltas. Seagrass shoots and seeds are seasonally torn-off by waves, drift out to the ocean and subsequently sink to the seabed, thereby exporting silica (and carbon) from the shallow to deep waters (Duarte and Krause-Jensen 2017). In addition, the Angolan mounds are bathed by southern-sourced waters, which are generally enriched in silica (Kawase and Sarmiento 1985).

8.4.3.3 The Coral Rubble Plains

The largest spatial coverage of coral habitat is on the coral rubble plains draping Mauritanian coral mounds. These plains primarily represent lag deposits as indicated by dating

Fig. 8.3 (Continued) underneath and squat lobsters (*Eumunida bella*) (Mauritania, Timiris mound complex). The colony shows short and crowded branches. Scale bar: 10 cm. (**d**) Live *L. pertusa*-colony with open-spaced zig-zagging growth pattern, characteristic for Angolan coral mounds. Scale bar: 10 cm. (**e**) Large *L. pertusa* 'hedges' aligned on the upper flanks of Angolan coral mounds. The entire framework construction lies loose on the seabed and the initial parts of the hedges are pointing upward (X) as a matter of a slow collapse process. Scale bar: 30 cm. (**f**) A huge *L. pertusa*-colony sampled from the Holocene Namibian coral mounds. Two different main growth axes document a former change in colony orientation. Presumably after the older colony (yellow arrow) has fallen aside, the surviving colony parts re-orientated their main growth direction (white arrow). Scale bar: 10 cm. Credits ROV images **a**–**c**: Tomas Lundälv, Tjärnö Marine Laboratory, University of Gothenburg, Sweden; ROV images **d**–**e**: MARUM ROV SQUID, Bremen, Germany

Fig. 8.4 Biological interactions and faunal groups found for cold-water coral habitats off West Africa. (**a**) Partial mortality of a *Madrepora oculata*-colony (whitish branches) induced by *Calliostoma bullatum*, a very common corallivore gastropod in Mauritanian coral habitats. Scale bar: 2 cm. (**b**) The presumed obligate symbiosis of the white carnivore sponge *Cladorhiza corallophila* and live colonies of *Lophelia pertusa* (a.k.a. *Desmophyllum pertusum;* Mauritania). Scale bar: 6 cm. (**c**) Live and dead assemblages of the glass sponges *Aphrocallistes beatrix* (A) and *Sympagella* sp. (S) display a very abundant component of the Angolan coral mound fauna. Scale bar: ~10 cm in the foreground. (**d**) Aggregation of *Acesta* c.f. *excavata* byssus attached to coral framework (Mauritania). Scale bar: 10 cm. (**e**) Intriguing view on the Holocene coral mound surfaces off Namibia. The yellow spots represent masses of the bryozoan *Metroperiella* sp. encrusting fossil coral fragments. Scale bar: 10 cm. (**f**) The other very common bryozoan on the Namibian mound surfaces is the erect-bushy *Klugeflustra* sp. Scale bar: ~10 cm in the foreground. Credits ROV images **a–c**: Tomas Lundälv, Tjärnö Marine Laboratory University of Gothenburg, Sweden; ROV images **d–f**: MARUM ROV SQUID, Bremen, Germany

of coral rubble yielding ages as old as ~100 kyr (Wienberg et al. 2018). In current-exposed locations on the coral mound surfaces, octocoral gardens occur comprising *Acanthogorgia armata, A. granulata, Clavularia borealis, Paramuricea placomus*, *Swiftia* sp.n., and *Thesea talismani* (De Matos-Pita and Ramil 2015b; Buschewski 2016; Sampaio et al. 2019). Coral rubble on the Angolan coral mound flanks is not as purely developed as off Mauritania as it forms thin veneers mixed with sandy mud deposits that are sparsely colonised.

8.4.3.4 The Hard Substrate Fauna

The submarine canyons off Mauritania have cut deeply into the shelf (Sanz et al. 2017), and the transition from the canyons to the shallower shelf exhibits rocky substrates with little loose sediment. The aforementioned megafauna groups are basically present as hard substrate colonisers on the rocks, but in different proportions than in the deeper canyon area. The dominance of *L. pertusa* gives way to the dominance of *M. oculata*, as their population densities decrease significantly towards the shallower areas. In addition, *Dendrophyllia cornigera* is more common, and large astrophorid sponges up to one metre high are frequent. Most conspicuous was the discovery of live giant deep-sea oyster populations, *Neopycnodonte zibrowii,* which attached themselves to the lower flanks of individual rocks in the Nouamghar canyon. This finding represents the first evidence for the occurrence of this bathyal oyster species south of the temperate North Atlantic (see Box 8.2). However, the southernmost known occurrence of this bathyal oyster species was found in a rocky area off Angola (Anna ridge). At both sites, CWCs were still present albeit in low numbers, while the oysters become common.

8.4.4 The Faunal Composition on Extinct Namibian Coral Mounds

The extinct Namibian coral mounds (SE of Walvis Ridge; Fig. 8.1) are spectacular in several ways. First, they provide a unique example of an ecosystem that crossed an ecological tipping point, in this case extreme deoxygenation, and experienced an evolution from prosperous reef communities to extinct coral populations (Tamborrino et al. 2019). Second, the role of preserved dead coral mounds as a habitat for the recent community, adapted to near anoxic conditions (Hanz et al. 2019), provides clues for deep time extinction events in Earth's history, where flourishing ecosystems vanished by anoxia and were subsequently recolonised by a completely different community—known as 'disaster fauna' by palaeontologists.

The most intriguing aspect found for the fossil Namibian coral mounds was the extremely dense colonisation of a yellow-coloured coral skeleton-encrusting bryozoan species that is assigned to the genus *Metroperiella* (K. Matsuyama, pers. comm.; Fig. 8.4e). This noticeable bryozoan also colonises ascidian, soft sponge bodies, and outcropping rocks. Triggered with UV light, *Metroperiella* emits a bright green fluorescence light—not reported before from bryozoans. It is yet not known, if the fluorescence is genetically confined to the bryozoan, or if *Metroperiella* has ingested luminous bacteria from the fertile waters in the Benguela upwelling system. Argument for the latter hypothesis is that highly fluorescing water caused by luminous bacteria and dinoflagellates coincides with the upwelling zones in the world (i.e., Haddock et al. 2010). Green fluorescence has also been documented in some erect sponge species collected from the Namibian mounds (Hebbeln et al. 2017). Other common sessile organism groups are erect-foliose bryozoans, assigned to the genus *Klugeflustra* (K. Matsuyama, pers. comm.; Fig. 8.4f), ascidians and white zoanthids. Compared to the diverse crustacean fauna in deeper waters on the Namibian slope (>400 m; Macpherson 1991), on the shallow (<270 m) Namibian mounds only *Bathynectes piperitus, Macropipus australis,* and the endemic stomatopod *Pterygosquilla armata* were observed (E. Macpherson, pers. comm.).

8.5 MAN: Geo-morphology and Past Formation of Cold-Water Coral Mounds

The MAN coral mound province contains impressive seabed features on the West African margin extending several 10 s to 100 s of kilometres. *L. pertusa* is the dominant mound constructor in all settings (see Sect. 8.4), yet they reveal distinct differences with respect to their morphological appearance and arrangement (Table 8.3). Moreover, the apparent differences in present-day reef status, varying from sparse live corals (Mauritania) to prosperous reefs (Angola) to extinct coral populations (Namibia), suggest that the past proliferation of *L. pertusa*, and hence, the formation history of the West African coral mounds also reveals distinct variations. Here, we describe the most recent mound formation period coinciding with the last glacial and the Holocene.

8.5.1 Morphological Mound Characteristics and Geological Features of the Continental Margin

The largest coherent coral mound province known from the West African margin exists along the north-south-trending Mauritanian slope (indicated as 'c' in Fig. 8.1; Colman et al. 2005; Ramos et al. 2017a). Coral mounds occur along the entire margin in water depths between 400 and 550 m, but

Table 8.3 Comparison of geological features of the margin, mound morphology and configuration, reef status and timing of mound formation, between the three major West African coral mound provinces off Mauritania, Angola, and Namibia (for references see text)

		Mauritania	Angola	Namibia
Province	Extent	• Max. 580 km (20.2–16°N) • Continuous chain of ~400 km (19–16°N)	• >60 km (9.5–10°S)	• >80 km (20.2–21°S)
	Position	• Upper to mid slope	• Upper to mid slope	• Outer shelf
	Water depth	• 400–550 m	• 250–500 m	• 160–270 m
	Configuration	• Mounds merged to almost continuous chain, partly two parallel chains • Slope-parallel orientation • Deeply buried coral mounds	• Mounds arranged as and partly merged to chains • Orientation parallel or in acute angle to slope • Superficially buried mounds	• >1200 individual mounds grouped to clusters • Superficially buried coral mounds
Mound	Max. Height	• 100 m	• 100 m	• 20 m
	Shape	• Elongated & straight, arcuate • Mostly coalesced to reticular mound complexes	• Elongated & straight, arcuate • Partly coalesced to reticular mound complexes	• Round, elongated • Randomly mounds merged to small patches
	Mound-forming coral	• *Lophelia pertusa* • Minor: *Madrepora oculata*	• *Lophelia pertusa* • Minor: *Madrepora oculata*	• *Lophelia pertusa*
	Reef status	• Sparse live colonies	• Thriving reefs	• No living corals
	Coral age range	• Recent to (at least) 193 kyr	• Recent to (at least) 53 kyr	• 4.5–9.5 kyr
	Initiation time	• Unknown	• Unknown	• Early Holocene
	Mound status	• Dormant: since ~11 kyr,	• Active, but slowed-down	• Dormant: since 4.5 kyr
	Formation periods (or pulses)	• Last interglacial (MIS 5e) • Last glacial (MIS2-4) • Last deglaciation	• Last glacial • Last deglaciation • Recent interglacial (Early and Mid-Holocene)	• Recent interglacial (Early to Mid-Holocene)
Margin	Present sediment input from land	• Sahara desert • Senegal river (reduced input since dam constructions)	• Cuanza & Congo rivers	• Namib desert • Orange river
	Erosional features	• Numerous inactive canyons (related to palaeo-river systems) • Contourites, moats	• Few inactive canyons • Contourites, moats	• Straight escarpment: 63 km long, 45 m high • Low-relief channels, ridges
	Gravitational feature	• Many large submarine slides	• Submarine slides	• Small-scale slides at the base of the escarpment
	Tectonic features	• Faults/pockmarks • Gas/oil occurrence	• Active salt tectonics • Faults/pockmarks • Gas/oil occurrence	• None

only between Cap Timiris (19°N) and the Senegalese border (16°N), the mounds are merged to an almost continuous slope-parallel chain, which extends over ~400 km (Fig. 8.5a; Ramos et al. 2017a). In places, even two (sub-) parallel mound chains are present with a shallower chain arising to 400–450 m water depth, while the summits of the deeper mounds reach depths of ~500–550 m (Fig. 8.5d; Wienberg et al. 2018). The Mauritanian mounds are up to 100 m high and show highly variable shapes that vary from straight and elongated to arcuate shapes, with arcuate mounds often coalescing to form complex reticular structures (Fig. 8.5g; e.g., Tamxat mound complex; Wienberg et al. 2018). Deep moats (~20–50 m) are developed at the eastern and western base of the mound chain(s) (Fig. 8.5a) pointing to strong and erosive bottom currents (Colman et al. 2005; Ramos et al. 2017a). Locally, the mound chain is interrupted by canyons, which are common erosional features along the entire Mauritanian margin (e.g., Wynn et al. 2000; Henrich et al. 2010; Sanz et al. 2017).

While for the coral mounds on the open slope only sparse live corals were encountered, recently settled colonies are frequently reported from most Mauritanian canyons (see Sect. 8.4). Moreover, isolated coral mounds, up to 60-m-high occur where the upper steep canyon flank morphology changes into almost even plains (e.g., found for the Tanoûdêrt and Tioulit canyons). The canyons have no connection to any modern river system, but were likely associated with palaeo-river systems, which had been active during past African humid periods ('Green Sahara'; Demenocal and Tierney 2012; Pausata et al. 2020) contributing to the delivery of sediments to the (sub-)tropical Atlantic margin (e.g., Skonieczny et al. 2015). Today, fluvial sediments only derive from the Senegal river in the south (Fig. 8.1; Table 8.3), though the construction of large dams—

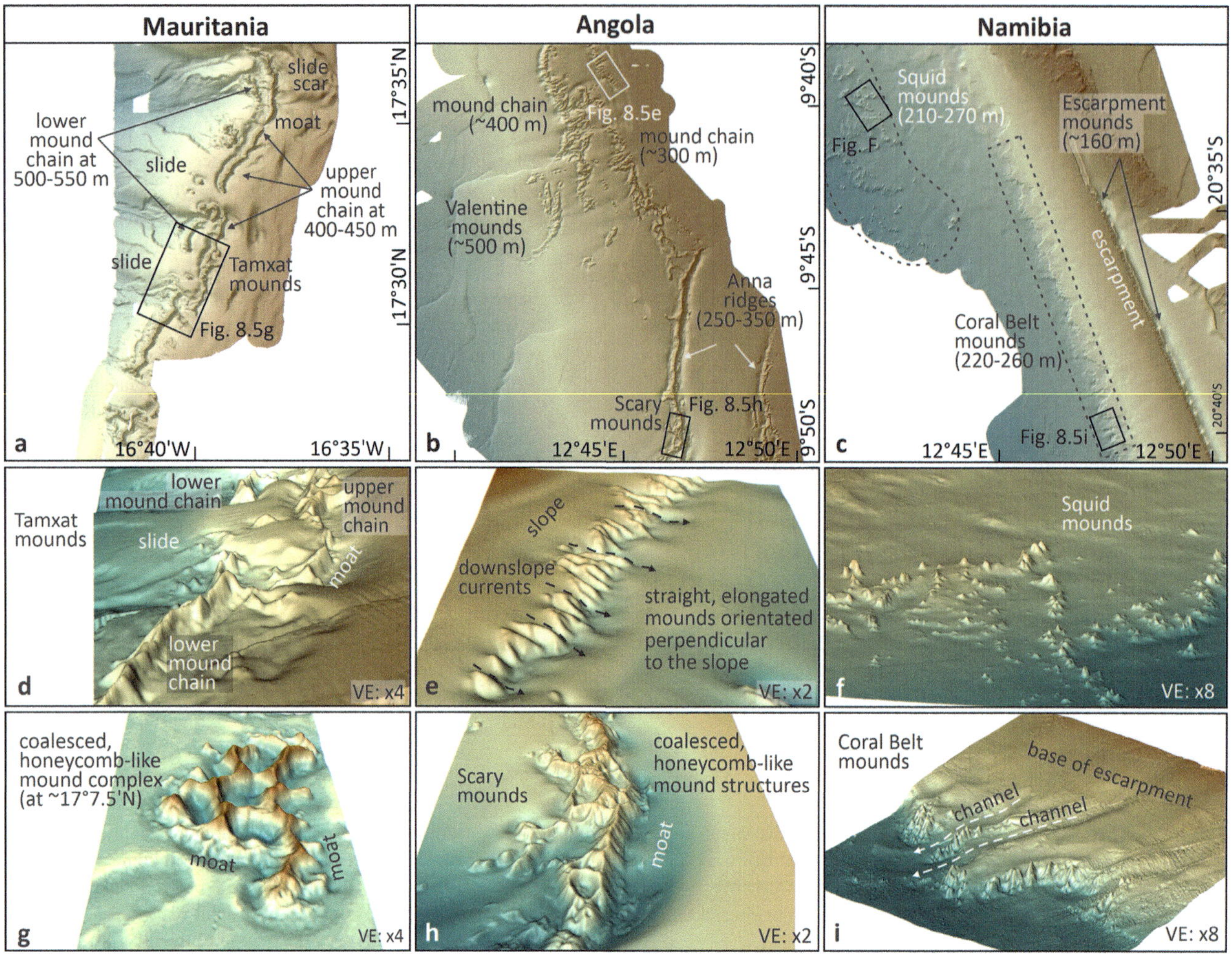

Fig. 8.5 Overview and detailed maps showing the three MAN cold-water coral mound provinces off West Africa. (**a**) Mauritania: coral mounds in 400–550 m water depth merged to two slope-parallel mound chains, (**b**) Angola: coral mounds in 250–500 m water depth arranged in chains following underlying faults (Anna ridges) or being orientated parallel to the slope following distinct depth levels, (**c**) Namibia: more than 1200 small coral mounds on the outer shelf in 160–270 m water depth arranged in three main clusters. (**d–i**) Detailed views of specific mound shapes and configurations observed for the MAN coral mound provinces as discussed in the text (VE: vertical exaggeration)

in response to the recurrent series of droughts during the 1970s—considerably reduced the sediment load discharged by the river (Niang et al. 2019). However, large quantities of aeolian sediments are transported from the hyperarid Sahara desert to the deep-sea environment (Harrison et al. 2001). Associated with the canyons, several large submarine slides are present (Fig. 8.5a) such as the prominent Mauritania slide. This slide complex resulted from a series of retrogressive failure events, triggered by earthquakes and diapiric uplift, and is one of the largest slope failures in the entire NE Atlantic (Antobreh and Krastel 2007). Some coral mounds seem to have had a stabilising effect on the slope, as they stopped the retrogressive upslope movement of the slide headwalls (De Mol et al. 2009).

The recently discovered Angolan coral mound province (~9.5–10°S; Hebbeln et al. 2017) stretches over ~60 km starting approximately 10 km south of the Cuanza river (Luanda; indicated as 'f' in Fig. 8.1), although its actual extent might be even larger. About 300 km to the north, close to the Congo canyon, additional coral mounds were mapped along a 6-km-long transect (indicated as 'e' in Fig. 8.1; see Sect. 8.2; Le Guilloux et al. 2009), and it is likely that coral mounds are also present between both sites, although this still needs to be verified by future mapping campaigns. The coral mounds and ridges south of Luanda are organised in distinct chains between 250 and 500 m water depth, which in contrast to the Mauritanian mound chains are only partly orientated parallel to the slope. For instance, two straight and parallel ridges (Anna Ridges), stretching over 12–14 km and being ~5 km apart, follow a N-S direction, hence, they are orientated in an acute angle (~20°) to the slope that overall follows a NW-SE direction (Fig. 8.5b). The ridges are confined to rather shallow water depths of ~250–350 m and rise from an erosional surface 75–100 m above the surrounding seafloor. Their orientation is explained by their relation to major faults, which were created by salt diapirism (Fort et al. 2004; Kukla et al. 2018) long before mound formation commenced (Haberkern 2017). Numerous faults are present in the Angolan mound province, but not all of them are associated with coral mounds (Haberkern 2017), and in turn, a number of coral mounds and ridges without underlying faults are observed. Thus, salt tectonics might have exerted an important role as a pre-conditioning factor for some mounds, but thick contouritic deposits and deep moats that developed at the base of most Angolan coral mounds hint to a strong interaction with the local bottom current regime (Hebbeln et al. 2017).

North of the fault-related ridges, two mound chains follow the 300 and 400 m isobaths (Fig. 8.5b) and fuse at their northern edges. The chain at 300 m depth consists of several straight elongated mounds with their long axis (~300 m) being orientated perpendicular to the slope (Fig. 8.5e; Haberkern 2017). Their particular shape and orientation might be the result of up- and downslope currents induced by internal waves, a relationship that was already assumed for coral mounds along the Moroccan margin (Hebbeln et al. 2019b). The mounds at 400 m depth are highly variable in shape and are merged to reticular complexes. The same configuration has also been observed for the deepest mounds (Valentine mounds), which occur further to the west in water depths down to ~500 m. Few canyons are developed at the southern end of the two fault-related ridges. They end below the shelf (at ~300 m water depth) and are not connected to any currently active river system (Hebbeln et al. 2017). While for the Mauritanian canyons only single conical mounds were observed, the Angolan canyons are flanked by several mounds with heights of up to 50 m. These mounds have an elongated to arcuate shape and are coalesced to complex honeycomb-like structures (Fig. 8.5h; Scary mounds), which very much resemble the Mauritanian mound complexes.

The Namibian coral mound province (SE of Walvis Ridge; 20.2–21°S; indicated as 'h' in Fig. 8.1) is bordered to the east by the Namib desert, which directly touches the coastline, while significant fluvial sediment input derives mainly from the large Orange river situated at the border between Namibia and South Africa (Rogers and Rau 2006). With respect to the marginal setting and the mound characteristics, the Namibian coral mound province displays a completely different morphological appearance compared to the Mauritanian and Angolan sites. First of all, the Namibian coral mounds developed on the northern Namibian shelf and are constrained to shallow water depths of 160–270 m, and hence, represent one of the very few mound provinces known from shelf environments (see also Fosså et al. 2005; Douarin et al. 2013). Secondly, the Namibian mounds are rather small not exceeding heights of 20 m, but they occur in high numbers with more than 1200 individual mounds that extend over >80 km parallel to the coast (Tamborrino et al. 2019, 2022); and probably beyond, as further small mounds were recently found to the north (~17.5°S) in 230–290 m water depth (Kunene mounds; indicated as 'g' in Fig. 8.1; see Sect. 8.2; Rush et al. 2019). Thirdly, the Namibian mounds are not arranged in coherent chains, instead numerous individual mounds are grouped in clusters, which occur west and on top of a remarkably large (height: 45 m, length: 63 km) and straight escarpment trending from NNW to SSE (Fig. 8.5c). The coral mounds mostly have round to slightly elongated shapes, with few mounds being merged to irregular small patches. While only few mounds occur on top of the escarpment at 160 m water depth (Escarpment mounds), the majority of mounds are arranged in two main clusters (Squid and Coral Belt mounds; Figs. 8.5c,f,i) west of the escarpment in water depths between 210 and 270 m. Their distribution is linked to various erosional features (e.g., small ridges, channel walls; Fig. 8.5i), which mark the lower flank of the escarpment (Tamborrino et al. 2019, 2022). Overall, the geo-morphology of the Namibian mound province is not affected by tectonics, including the prominent escarpment, whose origin is still enigmatic. Many of the coral mounds seem to be related to outcrops of older strata (of unknown age, but probably comprising Neogene shelf clastics; Light et al. 1993), which provided a suitable substrate that once enabled the settling of coral larvae and initiated mound formation. Moreover, the mounds root more or less on the depth level of the seafloor, meaning they have no sub-seafloor extension and the measured heights of 4–20 m are their true heights (Tamborrino et al. 2022). In contrast, the large Mauritanian and Angolan mounds root on deeper horizons below the recent seafloor, hence have a significant subsurface extension. For Mauritania as well as for Morocco, buried mounds (see Box 8.3), covered by several (10 s of) metres of sediment and not reaching the seafloor anymore, have been detected (Colman et al. 2005; Vandorpe et al. 2017), while for the Namibian and Angolan regions some mounds are superficially buried by hemipelagic sediments, but are still detectable as an elevation on the seabed (Haberkern 2017; Hebbeln et al. 2017).

8.5.2 Cold-Water Coral Mound Formation Across the Last Glacial-Interglacial Cycle

The West African coral mounds recorded periods of reef proliferation terminated by regional coral extinction events followed by periods of live CWC absence across glacial-interglacial cycles (as revealed by high precision Th/U ages

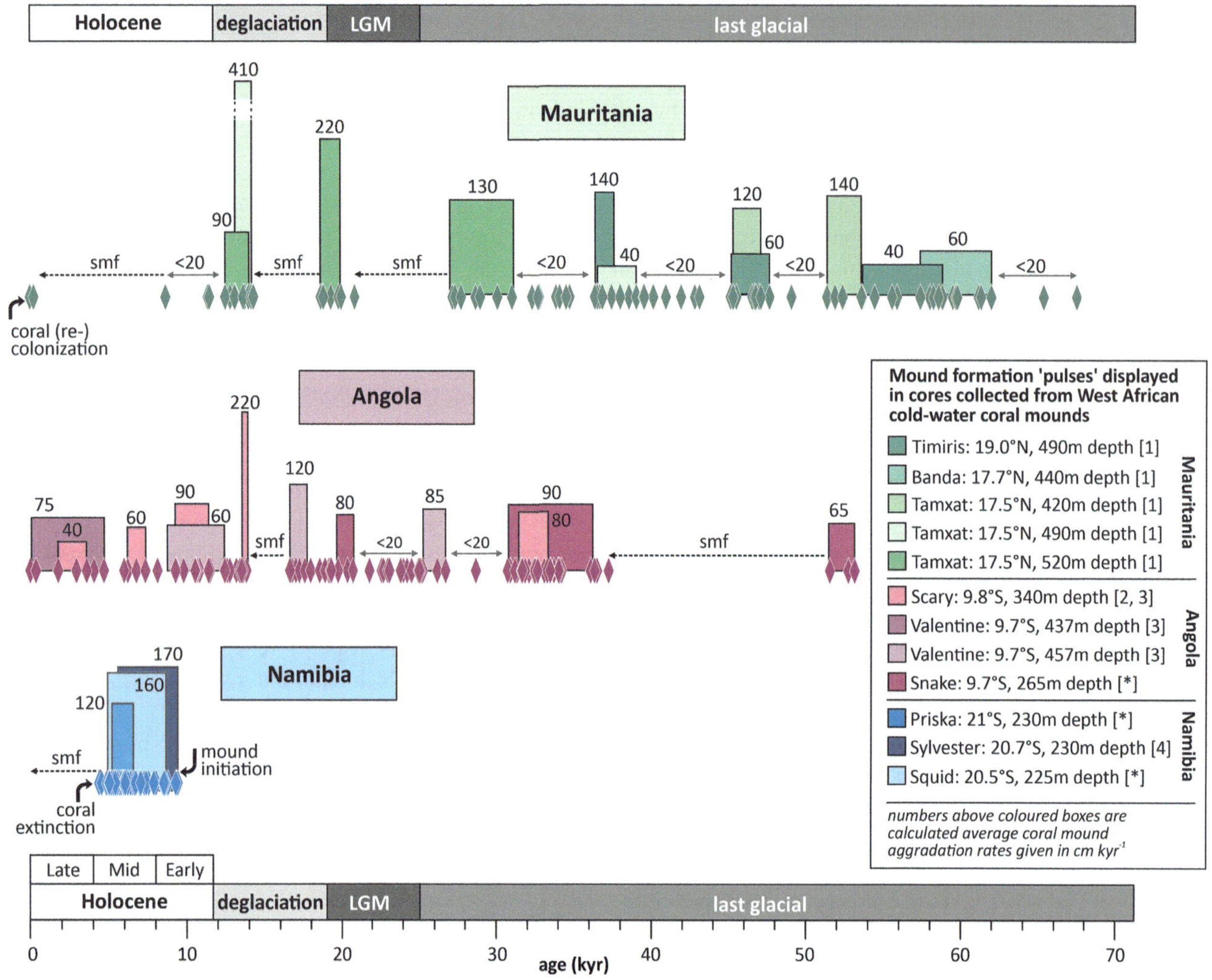

Fig. 8.6 Cold-water coral ages and mound aggradation rates (ARs) obtained from coral mounds of the MAN coral mound provinces off West Africa. The displayed coral ages (diamonds) have been obtained from various sediment cores and surface coral samples collected off Mauritania, Angola, and Namibia (published in Wienberg et al. 2018 [1]; Wefing et al. 2017 [2]; Beisel et al. 2023 [3]; Tamborrino et al. 2019 [4]; and unpublished data MARUM[*]). The ages cover the last 71 kyr, which comprise the last glacial, the Last Glacial Maximum (LGM), the last deglaciation (associated with a major global warming event marking the transition from 'cold' to 'warm' climate conditions), and the present interglacial (Holocene). Vertical coloured bars indicate duration and intensity of mound formation 'pulses'; numbers above bars represent re-calculated ARs (given in cm kyr^{-1}). Mound formation pulses with high ARs of up to 410 cm kyr^{-1} lasted for only a few hundreds or thousands of years, while in between mound formation slowed down (AR: <20 cm kyr^{-1}; bounded by grey arrows) or temporarily stagnated (dashed black arrows; *smf* stagnation in mound formation) due to a reduced occurrence or absence of corals

obtained from fossil CWC; see Box 8.4) that vary considerably between the three provinces. The Namibian coral mounds experienced only one rather short but intense period of continuous mound formation (average mound aggradation rate (øAR): ~120–170 cm kyr^{-1}) during the recent 'warm' interglacial period (Holocene; Fig. 8.6; Tamborrino et al. 2019; unpublished data MARUM). Mound development initiated during the Early Holocene (~9.5 kyr; indicated by one core record penetrating the base of a mound), but had already ceased 5000 years later during the Mid-Holocene (~4.5 kyr), and was never re-activated again (Tamborrino et al. 2019). This explains their rather small dimension compared to the large Mauritanian and Angolan coral mounds, which developed during several recurring periods of coral proliferation also prior to the Holocene. For the Angolan coral mound province, available coral ages range from >53 kyr to recent ages (though only encompassing the upper 10 m of the up to 100-m-high CWC mounds), hence coral mound formation took place (at least) since the last 'cold' glacial and continued until today (Fig. 8.6; Wefing et al. 2017; Beisel et al. 2023; unpublished data MARUM). However, the record shows some variations in mound development with recurring periods of increased mound formation (øAR: ~40–220 cm kyr^{-1}) coinciding with rather short-time intervals lasting less than 35 kyr, while in between mound formation significantly slowed down (øAR: <20 cm kyr^{-1}; e.g., during the Last Glacial Maximum), or even temporarily stagnated (Fig. 8.6).

In contrast to the Angolan province, coral mound formation along the Mauritanian margin already ceased with the onset of the Holocene at ~11 kyr and did not resume until recently (Eisele et al. 2011; Wienberg et al. 2018). Nevertheless, the observation of live corals, though in rather low numbers, supplemented by sub-recent to Late Holocene ages of less than 1000 years obtained for corals collected from canyon walls and mound tops, indicates that the corals started to re-colonise the area in recent times (see Sect. 8.4; Wienberg et al. 2018). The pre-Holocene formation history of the Mauritanian mound province shows that corals were almost continuously present in the area during the last glacial (~68–19 kyr) and the last deglaciation (~14–11.4 kyr; Fig. 8.6), but also at certain times during the two last 'warm' interglacial periods (~123–108 kyr, ~193 kyr; Wienberg et al. 2018), thus, contradicting the initially supposed strict glacial-interglacial contrast of coral growth off Mauritania (Eisele et al. 2011). In addition, and as already observed for the Angolan province, the Mauritanian mounds experienced short-time formation 'pulses' during the last glacial and deglaciation with enhanced mound formation (øAR: ~40–410 cm kyr^{-1}) lasting for only a few thousands or even just a few hundreds of years (Fig. 8.6; Wienberg et al. 2018). Finally, the Mauritanian mound province exhibits a conspicuous inter-mound variability expressed by distinct temporal differences in coral presence and mound development from shallow to deeper waters and from north to south (Fig. 8.6; Wienberg et al. 2018).

In general, coral mounds provide unique archives, which allow to assess the key environmental parameters and oceanographic processes controlling reef/mound formation. Even though the last glacial to Holocene mound formation histories of the MAN coral mound provinces differ to some extent, it is supposed that their development was triggered by very similar processes influencing the regional environmental conditions. All provinces occur in highly productive eastern boundary upwelling regions characterised by extensive OMZs providing hypoxic conditions for the corals. Because minor differences in the oxygen content might explain the present-day variation in coral vitality, past variations in the oxygen content might also be responsible for the observed pattern in mound formation, though the importance of further environmental parameters (e.g., fluctuations in food supply) cannot be excluded (see Sect. 8.3). Past variations in DO were the consequence of vertical (related to water-mass boundaries) and/or lateral (related to the frontal zones, CVFZ and ABF) changes in the intermediate water-mass structure that triggered the intensification or weakening of the OMZs and/or placed the mounds near or out of oxygen-depleted water masses (Wienberg et al. 2018; Tamborrino et al. 2019). However, we are just at the beginning of understanding the implications of past oceanographic changes on the proliferation or decline of the West African CWCs. In addition, effects of past changes of the hydroclimate in the Saharan-Sahelian region—today characterised by hyper- to semiarid conditions—may need to be considered as well. Milankovitch-driven shifts of the ITCZ not just interacted with the Atlantic surface and thermocline circulations (e.g., Portilho-Ramos et al. 2017), they also governed variations of the West African monsoon strength. This resulted in dramatic precipitation and vegetation shifts, as expressed by the recurring greening of the Sahara (Demenocal and Tierney 2012; Singarayer and Burrough 2015), and might have imposed important land-atmosphere-ocean feedback processes (Pausata et al. 2020), which still need to be fully explored.

8.6 Differences and Commonalities of the MAN Coral Mound Provinces

The MAN coral mound provinces occur in upwelling regions, which are bordered by dynamic oceanic frontal systems: The Cape Verde Frontal Zone (CVFZ) in the northern and the Angola-Benguela Front (ABF) in the southern hemispheres (Fig. 8.2), both separating poleward-flowing oxygen-depleted thermocline water masses (South Atlantic Central Waters) from equator-ward flowing oxygen-enriched waters (Mohrholz et al. 2008; Peña-Izquierdo et al. 2015). These fronts are so distinct that they coincide with major biogeographic limits as they are expressed by the definition of marine ecozones (Spalding et al. 2007). Nevertheless, the oceanic fronts experience some seasonal variability in position and dynamics, which together with the seasonally variable upwelling intensity steers the expansion of the local OMZs directly impacting the CWCs of the MAN regions. In addition, in both hemispheres, vigorous poleward undercurrents have the potential to cross these frontal systems at least seasonally (Arístegui et al. 2009; Veitch et al. 2010) at the depth of the MAN coral mounds. Also, in the past, the frontal systems changed their latitudinal positions in line with Milankovitch cyclicity, leaving a clear imprint on the development of the coral reefs and the entire marine ecosystems within the MAN regions across (at least) the last glacial-interglacial cycle (Wefing et al. 2017; Wienberg et al. 2018; Tamborrino et al. 2019).

The great environmental commonality in the MAN coral mound provinces is not just their connection to hypoxic and warm conditions in the given highly productive upwelling zones, but also the pivotal role of *L. pertusa* as the major framework constructor. However, despite these distinct commonalities, a closer look on the ecological health of the coral habitats shows a different picture. The absence of living CWCs on the Namibian shelf (<270 m water depth) indeed might be related to very low DO (<0.5 mL L^{-1}; Hanz et al. 2019), which belong to the lowest oxygen levels measured for CWC reefs along the West African margin. Their local extinction at around 4.5 kyr was also likely triggered by oxygen depletion, which imposed a deadly stress for the Holocene coral communities and the associated faunal assemblages (Tamborrino et al. 2019). However, off Angola, vivid CWC growth happens under the lowest DO (>0.5 mL L^{-1}; Hanz et al. 2019) ever observed for live occurrences of *L. pertusa* and *M. oculata*, and even thriving reefs with a characteristic hedge growth pattern developed at depths of ~330–470 m (Hebbeln et al. 2020). Although DO concentrations off Mauritania are considerably higher (>1 mL L^{-1}; Ramos et al. 2017a), in contrast to Angola, only sparse living corals are observed. Living coral populations are here mostly confined to localised spots, such as rocky hardground outcrops on the summits of some coral mounds (~400–550 m depth), but are clearly more common along the upper flanks (430–640 m depth) of submarine canyons, which obviously provide more suitable environmental conditions in comparison with the open slope, and therefore, might act as a refuge for the CWCs (Wienberg et al. 2018). The age structure and general appearance of the Mauritanian CWCs indicate a very recent re-colonisation event with healthy looking colonies mixed with older colonies consisting of a higher proportion of standing dead coral skeletons. However, the recently settled CWCs already suffer from the grazing pressure of echinoderms and gastropods (Hoffman et al. 2019c).

Overall, the reef status pattern observed for the MAN coral mound provinces indicates that beside oxygen other environmental parameters play an important role in controlling the distribution of the CWCs. For all three regions, the ambient bottom-water temperatures are comparably high with maximum values of 11–14 °C (Table 8.2), which is close to the highest temperatures ever observed for *L. pertusa* (14–15 °C; Freiwald et al. 2009; Mienis et al. 2014). Therefore, since *L. pertusa* increases its respiration rate with increasing temperatures (based on laboratory experiments on specimens collected from the NE Atlantic; Dodds et al. 2007), temperature may act as a further stressor in the MAN coral mound provinces. However, the CWCs seem to compensate to some extent the increasing demand in energy, and hence the environmental stress, by high food availability (Hanz et al. 2019; Hebbeln et al. 2020; Vinha et al. 2023). As the MAN regions are marked by a very high primary productivity due to the upwelling of nutrient-rich subsurface waters, it is probably the high food supply that enables coral life in these hypoxic and rather warm waters (Hebbeln et al. 2020). Thus, the distribution of living CWCs off Mauritanian and Angola are likely controlled by the interaction of multiple factors such as DO, temperature, and food supply. However, based on the existing knowledge, an important role of further environmental factors cannot be excluded. In addition, it will be vital to identify to which extent individual parameters fit the metabolic needs of CWCs, which are adapted to specific or even extreme regional environments, and at which (upper and lower) threshold these become a life-threatening stressor. In addition, our understanding of an amplification or compensation of multiple stressors/factors is still very limited, but will be indispensable to decode the past as well as to assess the future development of CWC reefs worldwide.

8.7 Threats and Outlook

The main present-day threats for CWC reefs worldwide are mechanical damage by deep-sea bottom trawling, pollution by waste (e.g., plastic) being disposed to the oceans, disturbance by human installations (e.g., cables, pipelines), and contamination by activities around the offshore oil and gas exploration (e.g., Foley et al. 2010; Purser and Thomsen 2012; Aguilar et al. 2017; Mouchi et al. 2019). All these human-induced threats are also affecting to a minor or major degree the CWCs off West Africa, which are briefly summarised with a focus on the Mauritanian and Angolan coral mound provinces as these sites host living CWCs and even thriving reefs today. Finally, threats induced by global climate change are discussed, with the predicted ocean warming and deoxygenation (e.g., Sweetman et al. 2017) likely going to exert the most severe pressure on the West African corals, as already today, they live in rather warm and hypoxic waters.

8.7.1 Deep-Sea Bottom Trawling and Waste

Induced by the intense coastal upwelling, the waters off Mauritanian, Angola, and Namibia belong to the most productive regions in the world's oceans and are rich in pelagic and demersal fish populations. CWC reefs, forming structurally complex habitats, play an important role in the life cycle of many fish species by providing shelter, food, and mating grounds (Biber et al. 2014), and therefore, are preferred target areas for the local and international fisheries off West Africa (e.g., Colman et al. 2005). The development of new deep-sea fishing technologies has enabled the commercial fishing industry to exploit bottom-dwelling species down to 2000 m water depth, hence covering the full depth range of coral reefs on the continental slope, seamounts, and volcanic ridges. In particular, bottom-trawl fishing causes substantial damage to CWC reefs (Hall-Spencer et al. 2002; IMROP

Fig. 8.7 Threats to cold-water corals off West Africa. (**a**) Map showing the heavy traffic of fishing vessels (orange symbols) off Mauritania at one given moment (source: www.marinetraffic.com; date: 24/09/2020, 14:00 UTC). (**b**) Trawler fleet in the fishing harbour of Nouadhibou, Mauritania, and (**c**) local fishermen, Nouakchott, Mauritania (photos: A. Freiwald). ROV images documenting fishing activities and waste disposal on the Mauritanian cold-water coral mounds: (**d**) Deep trawl mark incising the surface of a coral mound covered with coral rubble. Scale bar: 20 cm; (**e**) longline that got entangled with a colony of *Madrepora oculata*. Scale bar: 10 cm; (**f**) lost fishing equipment with *Helicolenus dactylopterus* resting on top; (**g**) a plastic bag covering a live coral. Scale bar: 10 cm. Credits ROV images: Tomas Lundälv, Tjärnö Marine Laboratory, University of Gothenburg, Sweden

2013; Hinz 2017; Victorero et al. 2018) that is effectively non-renewable due to the slow growth rates of the corals (e.g., Roberts et al. 2009; Orejas et al. 2011).

Demersal and pelagic trawling fleets operate in Mauritanian waters and the intensity of fishing operations is impressively shown by the heavy traffic of a large number of fishing vessels along the entire Mauritanian margin (Fig. 8.7). The consequences of the intense fishing off Mauritania are displayed by deep trawl marks incising the mound surfaces and frequent lost fishing equipment (Fig. 8.7; Colman et al. 2005; Westphal et al. 2012). In contrast to the situation on the open slope, the canyons along the Mauritanian margin with their rugged surfaces seem to provide some natural protection against trawling; however, recent technological advances (e.g., canyonbusters, rockhoppers) allow bottom trawlers to access those areas that they once avoided because of the risk to damage their nets. Even though the Namibian and Angolan waters are also influenced by intense fishing (Boyer and Hampton 2001), visible human impact in the area of the coral mounds seems to be low (Le Guilloux et al. 2009; Hebbeln et al. 2017). During extended video observations in 2016 only one lost net with a corresponding trawl mark was detected in the Namibian province, and one lost longline, but no signs of trawl marks were observed in the Angolan area (Hebbeln et al. 2017). An emerging topic in this context is the impact of bottom trawling in dysoxic to anoxic sediments that triggers unwanted releases of CO_2 and CH_4 to ambient water masses, and finally, to the atmosphere (Sala et al. 2021). This is of particular interest for the oxygen-poor sediments prevailing near coral mounds off Mauritania and Angola, where industrial fishing efforts are at greatest.

For all three provinces occasional findings of pollution have been reported (e.g., glass bottles and plastic debris; Fig. 8.7), but no significant amounts of waste were detected. However, during an expedition in January 2016 (RV Meteor M122 'ANNA'), a large carpet of floating plastic waste was observed at the sea surface off Angola (Hebbeln et al. 2017), which might be related to uncontrolled on-land waste disposal, being transported via wind or rivers (e.g., Cuanza river) towards the open ocean, and then being distributed over large distances by currents. Plastic waste can become a severe problem for CWCs (Mouchi et al. 2019). For example, plastic bags get stuck on the branched structure of the coral framework and cover the polyps (Fig. 8.7) that can neither breathe nor feed correctly anymore (Ramos et al. 2019).

8.7.2 Oil and Gas Exploration off Mauritania and Angola

Oil and gas exploration pose another serious threat to the West African coral reefs. During the last 20 years, a large number of significant oil and gas fields have been discovered off Mauritania and Angola (Gulf of Guinea), with the exploration and development activities being accompanied by increased ship traffic and by the installation of infrastructure on the seafloor (platforms, pipelines etc.). Technological advances allowed these activities to progress towards increasing depths, and hence, are reaching the depth level of the sensitive coral reefs located on the slope. For example, the Chinguetti oil field, discovered in 2001, is located at ~800 m water depth just below the Mauritanian coral mound chain (Colman et al. 2005). Overall, as oil exploration moves farther offshore into deeper waters, the risk of disastrous accidents increases, and contingency planning for oil spills is still a matter of discussion (Cordes et al. 2016). This is most dramatically expressed in the consequences of the explosion of the Deepwater Horizon drill platform in the Gulf of Mexico in 2010. The release of >500 million litres of crude oil and the subsequent extensive use of oil dispersants caused irreversible damage to many coral communities (and numerous other organisms) living in the deep (Fisher et al. 2014; Etnoyer et al. 2016; Weinnig et al. 2020).

The awareness of the importance of CWC reefs providing various ecosystem services for human society (e.g., fish stock security, carbon storage, medical/biochemical resources; Armstrong et al. 2014), but also of their vulnerability against human threats has increased in recent years. As a consequence, European countries such as Norway, the UK, and Portugal (Azores) have declared more than 20 regions as MPAs as a management measure to protect CWC ecosystems against bottom trawling and oil/gas exploration. An international expert panel, commissioned by the Mauritanian authorities, recommended the establishment of five MPAs (total area: >1100 km^2) along the Mauritanian continental slope, comprising canyons densely colonised by corals as well as parts of the coral mound chain, where deep-sea fishing should be strictly forbidden and oil and gas development should be subject to thorough environmental assessments (Ramos et al. 2019). Despite these recommendations, in 2018 British Petroleum gained approval from the Mauritania and Senegal governments to commence drilling on the maritime border of Mauritania and Senegal. This new exploration project (Greater Tortue Ahmeyim Project) encompasses an area of 33,000 km^2 involving drilling to a water depth of 2000 m, which is directly impacting CWCs in multiple ways, including the routing of underwater pipelines through the fragile reefs.

8.7.3 Consequences of Global Climate Change

Even though MPAs may protect vulnerable CWCs against damage and pollution, they offer no protection against threats induced by global climate and ocean change. Pressures derived from ocean warming, food depletion, deoxygenation, and acidification are hard to manage (Jackson et al. 2014), and it is expected that CWCs will suffer dramatically from these future environmental changes in the oceans (e.g., Guinotte et al. 2006; Turley et al. 2007; Tittensor et al. 2010; Lunden et al. 2014; Sweetman et al. 2017; Hennige et al. 2020). For the West African coral reefs, deoxygenation might pose the most serious threat and will likely become a life-threatening stressor (Laffoley and Baxter 2019). CWCs thriving within the extensive OMZs off Mauritania and Angola are today already exposed to hypoxic conditions with very low DO of <1.5 mL L^{-1} (Ramos et al. 2017a; Hanz et al. 2019; Hebbeln et al. 2020; Orejas et al. 2021) and will be most vulnerable for the expected decrease in oxygen (~2% until 2100; Sweetman et al. 2017). The predicted scenario could even cause their regional extinction, as it has been observed for the Namibian coral mound province, where corals went extinct during the Mid-Holocene due to extremely low oxygen levels (Tamborrino et al. 2019). A further important stressor might be the increase in water temperatures and enhanced stratification, as already today the Mauritanian (max. 11.7 °C; Flögel et al. 2014; Ramos et al. 2017a) and Angolan (max. 14.2 °C; Hanz et al. 2019; Hebbeln et al. 2020) corals live at temperatures that are close to their upper thermal tolerance (~15 °C; Mienis et al. 2014). In turn, this may have also implications for the corals' respiration rate as their need for oxygen increases with increasing temperatures (as observed for *L. pertusa* in the NE Atlantic; Dodds et al. 2007). Finally, the gradual decrease in seawater pH as a consequence of the increasing uptake of atmospheric CO_2 by the world's oceans will have a severe influence on calcification, respiration, and prey capture rates of CWCs as it has been shown in several experimental studies conducted on *L. pertusa* (e.g., Maier et al. 2009; Form and Riebesell 2012; Georgian et al. 2016; Büscher et al. 2017). Ocean acidification will certainly also have an impact on the MAN coral areas, where the Angolan CWC communities still live in saturated waters with respect to aragonite (average pH of ~8.12 measured in the direct vicinity of thriving reefs by benthic landers; Table 8.2; Hanz et al. 2019), while much lower pH of 7.83 and an aragonite saturation of 1.305 mmol kg^{-1} are already today measured for the dispersed coral communities of the Mauritanian coral mound province (Table 8.2; Flögel et al. 2014).

It is assumed that enhanced availability of food in the highly productive upwelling regions off Mauritanian and Angola allows the corals to compensate metabolic stress already caused by the prevailing low oxygen and high temperature conditions (Hanz et al. 2019), and the high food availability might potentially also allow them to acclimate and adapt to the increasing seawater acidity in the future (e.g., Georgian et al. 2016). However, coupled with the effects of global warming, deoxygenation, and acidification, many regions in the deep sea may become severely food limited in future oceans, for instance due to an expected decline in particulate organic carbon flux (~30% until 2100; Sweetman et al. 2017). This will lead to a multi-stressor environment that might become too extreme to cope with for the CWCs in the MAN areas, and it is likely that the corals' rate of adaptation cannot occur rapidly enough to keep up with the pace of global change (Hoegh-Guldberg 2014).

8.8 Future Directions

For a long time, CWCs along the West African margin remained poorly explored and our knowledge regarding their distribution and ecological status was mainly based on scattered dredge samples obtained during historical expeditions (Zibrowius 1980; Zibrowius and Gili 1990). Only in the past two decades has coral research gained momentum due to the discovery of large reefs and extended coral mound provinces (e.g., Colman et al. 2005; Le Guilloux et al. 2009; Wienberg et al. 2009; Buhl-Mortensen et al. 2017a; Tamborrino et al. 2019; Hebbeln et al. 2020), steadily increasing our knowledge on the ecology and geology of the West African CWCs and their oceanographic controls. However, even though research has made tremendous progress, there are still many open questions and uncertainties, precluding a comprehensive understanding of the (past, present, and future) functioning of these important deep-sea ecosystems.

For some regions along the West African margin, the potential to provide suitable habitat for CWCs is still not fully explored, leaving them as blank spots on the coral distribution map (Fig. 8.1). In particular, the continental margins between the Senegal and the Congo river are promising target areas for future oceanographic explorations, considering the occasional coral findings reported so far, including the large *Lophelia*-dominated reef discovered off Ghana (Buhl-Mortensen et al. 2017a). Overall, large mapping campaigns to obtain high-resolution bathymetry and seismic data are required to explore the full extent and variety of coral habitats off West Africa. In addition, innovative mapping tools such as remotely and autonomous operated vehicles (ROV, H-ROV, AUV) equipped with acoustic devices, sensors, and cameras will allow the exploration of areas that are currently difficult to access, such as canyons (e.g., Huvenne et al. 2011).

Although technological advances have improved our understanding of CWC habitats, these are very dynamic environments, and many of the obtained data only provide snapshots in time. They do not inform on changes of the coral ecosystem structure or any seasonal to inter-annual variability of their environment. To overcome the temporal sampling bias, continuous monitoring programmes are urgently needed. For example, year-long measurements of physical and chemical water mass properties (temperature, salinity, DO, pH, current velocity, sediment flux, etc.), either recorded at the seafloor by benthic lander systems and cabled observatories (e.g., Van Engeland et al. 2019) or measured through the water column by moorings and floats, are required to examine any seasonal environmental changes and to record episodic or extreme events (such as storms). For example, knowing the extent of seasonal fluctuations of DO concentrations is essential to fully understand the tolerance range of the corals thriving within OMZs. Moreover, the installation of deep-sea seafloor observatories (e.g., permanent platforms equipped with sensors and time-lapse cameras) would allow an examination of temporal variations in structure, composition, and diversity of the West African coral populations and their associated fauna. Continuous monitoring of coral activities (e.g., growth, feeding, spawning, polyp activity) as well as faunal interaction (with fishes, molluscs, sponges, crustaceans, etc.) would improve our understanding of coral physiology and overall ecosystem functioning (e.g., Chapron et al. 2020). In addition, repetitive non-invasive investigations (such as video mapping, mosaicking, photogrammetry, 3D models) would provide data to evaluate the ecological status of CWC habitats and temporal changes in biological composition by measuring the abundance, spatial distribution, density, and size structure of the coral populations (e.g., Orejas et al. 2009; Gori et al. 2013; Boolukos et al. 2019; Fabri et al. 2019; Girard et al. 2019; Orejas et al. 2021).

However, it is important to understand not just the present-day ecology of the West African CWCs and any regional environmental controls, but also their development over geologic timescales that has been controlled by climate-driven environmental changes. The correlation of geological coral records with palaeoceanographic data provides valuable information about times when ecological tipping points had been crossed resulting in the demise of entire reef ecosystems (Hebbeln et al. 2019a; Portilho-Ramos et al. 2022), as observed for the Namibian coral site during the Holocene (Tamborrino et al. 2019). Due to the lack of present-day observations of such regional extinction events, this geological approach is of great importance to identify the key environmental parameters and to unravel their complex interplay that may directly or indirectly control coral vitality (Wienberg and Titschack 2017; Hebbeln et al. 2019a). With respect to the large dimension of individual mound provinces and the temporal variability in coral mound development between individual mounds of one province (e.g., Wienberg et al. 2018), future campaigns should sample mounds located at different water depths and, for extended mound provinces, at different locations to improve our understanding of the influence of past changes in water-mass circulation and water column structure (e.g., displacements of water-mass boundaries or frontal zones; e.g., Wienberg et al. 2020). To evaluate the entire coral mound palaeo-archive and to understand the initiation of coral reef formation, coral mounds need to be sampled down to their base using efficient drilling techniques (e.g., Kano et al. 2007; Wienberg et al. 2022), which would allow for an understanding of the history of coral reef initiation, proliferation, and decline off West Africa going far beyond the last glacial-interglacial cycle.

Totally unexplored is the integration of CWC mound studies as paleo-archives into broad-scale biosphere-hydrosphere-coupled Earth's system models. In addition, the rise and demise of modern CWC populations might be early warning indicators of the onset of sudden environmental change. For instance, the proposed decline in DO and the ongoing expansion of OMZs will not just impact CWC ecosystems in the deep sea, but will increasingly affect shallower waters (Stramma et al. 2010). This might also result in the loss of habitable space for pelagic ecosystems causing accelerated shifts in animal distributions and changes in ecosystem structure, which consequently may culminate in the collapse of whole ecosystems and economies (such as fisheries; e.g., Worm et al. 2005).

Therefore, further explorations along the West African margin would clearly benefit from a multidisciplinary and integrated approach. In particular, the combination of geological-palaeoceanographic studies documenting the past reef ecology with innovative biological and oceanographic techniques monitoring the present-day reef ecology will provide crucial information for a more comprehensive and better understanding of the West African CWC reefs and mounds. In addition, past continental climate shifts, in particular variations of the West African monsoon strength that caused the recurring on–/offset of African humid periods and the de–/re-activation of large African river systems (Skonieczny et al. 2015; Pausata et al. 2020), should be integrated in future strategic studies to better predict sea and landscape changes. Such an approach will have a great potential to assess the fate of the West African CWCs in times of accelerating global change and to improve the sustainable management of these important deep-sea ecosystems. Due to the increasing anthropogenic pressures on CWC reefs impacting their sustainability and provision of ecosystem services, the assessment of the status of the West African CWC reefs is of utmost importance to inform policy makers and managers to establish sustainable measures for CWC conservation and protection in this region and beyond.

Box 8.1: Cold-Water Coral Habitats and Their Geological Legacy

by Dierk Hebbeln

Along the West African margin, CWCs occupy a variety of habitats demonstrating their opportunistic behaviour. Open ocean habitats, such as oceanic islands (e.g., Cape Verde), seamounts, and volcanic ridges (e.g., Walvis Ridge), offer an elevated position, exposed rocks for coral larvae settlement, and are characterised by increased turbulence and productivity (e.g., White et al. 2007). Due to their remote position, open ocean habitats are rather sediment-starved. Furthermore, their steep topography favours sediment bypass and mass wasting. Consequently, dead corals remain exposed without any protecting sediment cover and are prone to bioerosion (Freiwald and Wilson 1998) and redeposition along the steep topography. Hence, fossil coral accumulations are likely small-scaled, very localised, and difficult to sample so that their potential as paleoenvironmental archives for reconstructing the development of their habitat through time is limited. Suitable areas for coral mound development are extremely limited at open ocean habitats,

(continued)

Box 8.1 (continued)
but not excluded (Rivera et al. 2016; Somoza et al. 2014).

The numerous canyons along the Mauritanian margin are a further common CWC habitat (Wienberg et al. 2018). Canyons provide suitable living conditions for the corals, because they are episodically influenced by processes such as cascading of dense shelf waters and turbidity currents that enhance the downslope transport of fresh (well-ventilated) surface waters and particulate organic matter (food) (e.g., Orejas et al. 2009; Taviani et al. 2016). As corals preferentially grow on the rocky steep canyon flanks (e.g., Freiwald et al. 2009; Huvenne et al. 2011), they tend to fall down after they died to accumulate at the canyon thalweg, where they are prone to further downslope transport. Consequently, the localised and small-scaled coral deposits in canyons are difficult to sample and make the reconstruction of the temporal development of the canyon habitat a challenge.

Another peculiar coral habitat are the numerous mud volcanoes off NW Morocco (e.g., Somoza et al. 2003; León et al. 2012). Exhumed methane-derived authigenic carbonates provide a suitable hard substrate for corals (e.g., Rüggeberg and Foubert 2019). In addition, the corals seem to benefit from the elevated position of the mud volcanoes causing a topography-induced intensification of the hydrodynamics associated with an enhanced food supply. Once corals settle on a mud volcano and as long as they do not become buried by mud breccia extruded from the volcano, they can continue to grow for millennia triggering the deposition of >1-m-thick coral-bearing sediments (Pinheiro et al. 2003; Wienberg et al. 2009), which represent a valuable geological archive.

Coral mounds are the most common habitat for CWCs along the entire West African margin. They range from a few metres to >100 m in height and occur in large numbers arranged in provinces (e.g., Le Guilloux et al. 2009; Glogowski et al. 2015; Wienberg et al. 2018; Hebbeln et al. 2019b; Tamborrino et al. 2022). Coral mounds form through the interaction of the coral framework and bypassing bottom waters from which the corals baffle suspended sediments (Hebbeln et al. 2016), and hence, are mainly composed of coral fragments and hemipelagic sediments (Titschack et al. 2009; Wang et al. 2021). With respect to the large dimension of coral mounds, the mound deposits provide a high-resolution geological record of the coral development as well as of their habitat through time (Hebbeln et al. 2019a).

Box 8.2: Discovery of New Species Associated with West African Coral Habitats
by André Freiwald

The German and Spanish cruises, dedicated to West African CWC habitats since 2006, have increased our knowledge on species that live on, from or with the CWCs. A yet not published overview on species assessments from CWC habitats off Mauritania sums to about 800 valid species that are spread over the phyla Foraminifera (#176), Porifera (#29), Cnidaria (#66), Polychaeta (#3), Mollusca (#335), Arthropoda (#39), Echinodermata (#41), Bryozoa (#62), and 'Fish' (#48). This snapshot indicates clearly understudied groups including their function. Moreover, more than twenty species new to science were described from Mauritanian CWC habitats comprising the potentially obligate symbiotic carnivorous sponge, *Cladorhiza corallophila* (see Fig. 8.4b; Göcke et al. 2016), twelve molluscs (Gastropoda: Buccinidae, Calliotropidae, Costellariidae, Colloniidae, Skeneidae, Columbellidae; Bivalvia: Vesicomyiidae, Cuspidariidae; Castillo and Ramil 2018; Hoffman and Freiwald 2018a, 2019, 2020; Hoffman et al. 2018a, b, 2019a, b), four decapods (Diogenidae, Axiidae, Callianopsidae, Mathildellidae; De Matos-Pita and Ramil 2015a, b, 2016; Sakai et al. 2015), and two bryozoan species (Matsuyama et al. 2015). Finally, four new species of hydroids described for Mauritanian and Angolan CWC habitats complete the new species gallery off West Africa (Gil et al. 2020a, b) and will likely further increase as taxonomic studies from the Angolan and Namibian CWC mounds are ongoing.

Noteworthy is also the range extension of the giant deep-sea oyster, *Neopycnodonte zibrowii*, which was previously known from the temperate North Atlantic (north of ~30°N: Bay of Biscay, the Azores, Gulf of Cádiz, off NW Africa) and the Mediterranean Sea (Beuck et al. 2016), but living specimens have now also been documented for the tropical Atlantic off Mauritania (~19°N) and even in the southern hemisphere off Angola (~9°S).

In addition, 'living fossils' have been detected off Mauritania: at least two living gastropod species, *Calliostoma bullatum* (Fig. 8.4a) and *Fissurisepta papillosa*, have been found to predate coral polyps and probably sponges (Hoffman and Freiwald 2018b; Hoffman et al. 2019c). These gastropods were previously only known from Early- to Mid-Pleistocene outcrops in southern Italy and on Rhodes, yielding bathyal assemblages with CWCs (Titschack et al.

(continued)

Box 8.2 (continued)

2005; La Perna and D'Abramo 2010). It might be speculated if these living Mauritanian gastropod populations are the last representatives of a former widespread and connected East Atlantic and Mediterranean distribution that went extinct in the Mediterranean during the course of drastic environmental and oceanographic changes. Further examples for such a connectivity derive from various taxa, such as for cold-water Scleractinia and for Mollusca (Di Geronimo and La Perna 1997; Vertino et al. 2019).

Finally, a new type of a vulnerable marine ecosystem, the *Swiftia* spp. octocoral garden, is suggested by Sampaio et al. (2022). This new species of a plexaurid octocoral, thriving on the coral rubble surfaces of some Mauritanian coral mounds, lived so far unrecognised as a widespread habitat-forming species. These findings highlight that not only the discovery of species new to science makes faunistic studies in West African CWC habitats challenging. Overall, the examples touched here imply that along the West African margin many species still await their discovery underpinning our great knowledge gaps but also the great potential to intensify joint research activities on these African deep-sea ecosystems.

Box 8.3: Buried Cold-Water Coral Mounds

by Jürgen Titschack

Cold-water coral mounds are prominent and common seabed obstacles along the continental margins of the Atlantic Ocean. Less well-known are their buried counterparts as geophysical data of the sub-seafloor is needed for their detection (e.g., Beyer et al. 2003; Huvenne et al. 2003). Along the West African margin, buried coral mounds were described off Morocco (Foubert et al. 2008; Vandorpe et al. 2017), Mauritania (Colman et al. 2005), Angola and Namibia (Haberkern 2017; Hebbeln et al. 2017). Such 'fossil' mounds are either superficially buried by hemipelagic sediments still recognisable as an elevation on the seafloor (Lo Iacono et al. 2014) or are buried several metres below the seabed leaving no imprint on the present-day seafloor (Huvenne et al. 2003; Colman et al. 2005).

In general, the development of a coral mound as well as its burial is controlled by the interplay of coral mound formation processes and the sedimentary processes on the adjacent (off-mound) seafloor (Hebbeln et al. 2016). This results in two scenarios: (i) if

Box 8.3 (continued)

off-mound sediment deposition < coral mound formation, the mound gains seafloor relief; and (ii) if off-mound sediment deposition > coral mound formation, the mound loses relief (De Mol et al. 2005; Hebbeln et al. 2016). While the coral reef framework growth depends exclusively on environmental conditions, mound formation as well as the deposition of contourites needs sediment input. The continuous rain of particles from surface waters provides a sediment flow to the deep sea, where the local hydrodynamic regime governs their distribution and defines areas of erosion (formation of moats), sediment bypass and (contourite) deposition. Additionally, bottom currents effectively deliver food and sediment to the coral reef framework fostering mound formation. As long as a coral mound is covered by a thriving reef, the reef framework traps the suspension load of bypassing waters, which results in fast mound formation easily outpacing off-mound sediment deposition (e.g., Huvenne et al. 2009; Titschack et al. 2015; Wang et al. 2021). When coral reef framework growth is impeded due to unfavourable environmental conditions, mound formation ceases, and the topographically-steered acceleration of currents often hinders sediment deposition on the mound, while off-mound deposition continues. While the off-mound deposition along continental margins is generally subject to only minor variations, mound formation is highly variable and discontinuous (e.g., Wienberg and Titschack 2017; Wienberg et al. 2018). Thus, the fate of a coral mound on geological timescales depends predominantly on the mound formation duration and rate. The co-occurrence of buried and exposed mounds within the same mound province (e.g., De Mol et al. 2005; Vandorpe et al. 2017; Hebbeln et al. 2019b) clearly suggests that already small differences in local environmental conditions and sediment input can decide on their survival or burial.

Box 8.4: Cold-Water Corals: A Unique Geochemical Toolbox

by Norbert Frank

The aragonite skeletons of scleractinian CWCs (e.g., *Lophelia pertusa*, *Madrepora oculata*, *Desmophyllum dianthus*) are unique sources of information because they record time, local environmental conditions, nutrient cycling, and calcification

(continued)

Box 8.4 (continued)
processes. During calcification, the corals take up minor and trace elements from seawater, which are either replacing Calcium or just entering the skeleton as an alien atom (e.g., Robinson et al. 2014). For instance, the incorporation of Uranium isotopes starts a radioactive clock that provides high precision Th/U ages spanning from years to >600 millennia (Cheng et al. 2000; Frank et al. 2011; Wefing et al. 2017; Kerber et al. 2023). The solid skeleton withstands chemical and physical weathering and is very little open to the exchange of Uranium with the coral's environment, and hence, provides even accurate measures of the $^{234}U/^{238}U$ isotopic composition of seawater serving as a proxy for continental weathering and discharge (Chen et al. 2016). The first geochemical skeleton studies focused on stable oxygen and carbon isotopes. The calcifying process leads to correlated isotope fractionation, which has been thought to serve as a temperature tracer (Smith et al. 1997). Nevertheless, this was neglecting the complexity of bio-mineralisation (Adkins et al. 2003; Lutringer et al. 2005). In contrast to the stable carbon isotopes, the radiocarbon does provide quantitative measures of the seawater ventilation age (Adkins et al. 1998; Mangini et al. 1998), as the independent absolute Th/U age allows re-calculating the original seawater ^{14}C-content at the time of calcification (Frank et al. 2004; Robinson et al. 2005; Burke and Robinson 2012; Chen et al. 2015; Beisel et al. 2023). In addition, it has been shown that the so-called clumped isotopes Δ47 and the Li/Mg element ratio provide valuable measures of ambient water temperatures (Raddatz et al. 2013; Montagna et al. 2014; Spooner et al. 2016; Cuny-Guirriec et al. 2019). Corals further store information on the surface ocean nutrient cycle and productivity, since the skeleton's Ba-isotope composition and its P/Ca and Ba/Ca ratios as well as the aragonite bound $^{15}N/^{14}N$ ratio are directly linked to nutrient availability and consumption (Anagnostou et al. 2012; Wang et al. 2014, 2017; Hemsing et al. 2018). The provenance of water masses can be retrieved from Neodymium isotopes trapped in the skeleton (e.g., Colin et al. 2010; van de Flierdt et al. 2010; Montero-Serrano et al. 2013; Dubois-Dauphin et al. 2017; Struve et al. 2017), while Boron isotopes yield measures of ocean acidification considering that corals upregulate the pH in their calcifying fluids (Anagnostou et al. 2012; McCulloch et al. 2012; Jurikova et al. 2019). This geochemical toolbox greatly serves palaeoceanography

Box 8.4 (continued)
on precise timescales, and it can be applied largely independent of the aragonite-forming species, which make the retrieval of information only dependent on the occurrence of corals itself.

Acknowledgements The authors would like to thank all nautical and scientific crews for their on-board assistance during expeditions exploring the West African cold-water corals (e.g., RV Meteor expedition M122 '*ANNA*', RV Maria S. Merian expedition MSM16-3 '*PHAETON*'). In particular, the tremendous efforts of the ROV-*Squid* team (Nicolas Nowald, Christian Seiter, Vincent Vittori and Tom Leymann; MARUM, Bremen, Germany) and the fabulous ROV-pilot Tomas Lundälv (University of Gothenburg, Sweden) are greatly acknowledged. This research was supported by (i) the German Science Foundation (DFG) and the Cluster of Excellence '*The Ocean Floor-Earth's Uncharted Interface*' (CW, AF, JT, DH); (ii) the GIZ-project '*West African Biodiversity under Pressure*' (WASP; Contract 81248171) (AF); (iii) the Innovational Research Incentives Scheme of the Netherlands Organisation for Scientific Research (NWO-VIDI grant 016.161.360) (FM); (iv) the DFG-project '*Cold-water coral mound development in a tropical upwelling cell—The great wall of (f) Mauritania*' (Ti 706/3-1) (JT); (v) the German Academic Exchange Service (DAAD; Fellowship No. 91723955 within the Re-invitation Programme for Former Scholarship Holders, 2019, No. 57440916) (CO); and (vi) the European Union's Horizon 2020 research and innovation programme under grant agreement No. 679849 ('*SponGES*') (FM) and No. 818123 ('*iAtlantic*') (DH, CO). This output reflects only the authors' views and the European Union cannot be held responsible for any use that may be made of the information contained therein.

References

Addamo AM, Vertino A, Stolarski J, García-Jiménez R, Taviani M, Machordom A (2016) Merging scleractinian genera: the overwhelming genetic similarity between solitary *Desmophyllum* and colonial *Lophelia*. BMC Evol Biol 16:108. https://doi.org/10.1186/s12862-12016-10654-12868

Adkins JF, Cheng H, Boyle EA, Druffel ERM, Edwards RL (1998) Deep-sea coral evidence for rapid change in ventilation of the Deep North Atlantic 15,400 years ago. Science 280:725–728

Adkins JF, Boyle EA, Curry WB, Lutringer A (2003) Stable isotopes in deep-sea corals and a new mechanism for "vital effects". Geochim Cosmochim Acta 67:1129–1143

Aguilar R, Perry AL, López J (2017) Conservation and management of vulnerable marine benthic ecosystems. In: Rossi S, Bramanti L, Gori A, Orejas C (eds) Marine animal forests: the ecology of benthic biodiversity hotspots. Springer, Cham, pp 1165–1207

Álvarez-Pérez G, Busquets P, De Mol B, Sandoval NG, Canals M, Casomar JL (2005) Deep-water coral occurrences in the Strait of Gibraltar. In: Freiwald A, Roberts JM (eds) Cold-water corals and ecosystems. Springer, Heidelberg, pp 207–221

Anagnostou E, Huang KF, You CF, Sikes EL, Sherrell RM (2012) Evaluation of boron isotope ratio as a pH proxy in the deep sea coral *Desmophyllum dianthus*: evidence of physiological pH adjustment. Earth Planet Sci Lett 349–350:251–260

Antobreh AA, Krastel S (2007) Mauritania Slide Complex: morphology, seismic characterisation and processes of formation. Int J Earth Sci 96:451–472

Arístegui J, Barton ED, Álvarez-Salgado XA, Santos AMP, Figueiras FG, Kifani S, Hernández-León S, Mason E, Machú E, Demarcq H (2009) Sub-regional ecosystem variability in the Canary Current upwelling. Prog Oceanogr 83:33–48

Armstrong CW, Foley NS, Kahui V, Grehan A (2014) Cold-water coral reef management from an ecosystem service perspective. Mar Policy 50:126–134

Behrenfeld MJ, Falkowski PG (1997) Photosynthetic rates derived from satellite-based chlorophyll concentration. Limnol Oceanogr 42:1–20

Beisel E, Frank N, Robinson LF, Lausecker M, Friedrich R, Therre S, Schröder-Ritzrau A, Butzin M (2023) Climate induced thermocline aging and ventilation in the Eastern Atlantic over the last 32,000 years. Paleoceanogr Paleoclimatol 38:e2023PA004662

Bergstad OA, Gil M, Høines Å, Sarralde R, Maletzky E, Mostarda E, Singh L, António MA, Ramil F, Clerkin P, Campanis G (2019) Megabenthos and benthopelagic fishes on Southeast Atlantic seamounts. Afr J Mar Sci 41:29–50

Beuck L, Freiwald A (2005) Bioerosion patterns in a deep-water *Lophelia pertusa* (Scleractinia) thicket (Propeller Mound, northern Porcupine Seabight). In: Freiwald A, Roberts JM (eds) Cold-water corals and ecosystems. Springer, Heidelberg, pp 915–936

Beuck L, Vertino A, Stepina E, Karolczak M, Pfannkuche O (2007) Skeletal response of *Lophelia pertusa* (Scleractinia) to bioeroding sponge infestion visualised with micro-computed tomography. Facies 53:157–173

Beuck L, Aguilar R, Fabri M, Freiwald A, Gofas S, Hebbeln D, López Correa M, Ramos Matos A, Ramil F, Sánchez Delgado F, Taviani M, Wienberg C, Wisshak M, Zibrowius H (2016) Biotope characterisation and compiled geographical distribution of the deep-water oyster *Neopycnodonte zibrowii* in the Atlantic Ocean and Mediterranean Sea. Rapport Commission International Mer Méditerranée 41:462

Beyer A, Schenke HW, Klenke M, Niederjasper F (2003) High resolution bathymetry of the eastern slope of the Porcupine Seabight. Mar Geol 198:27–54

Biastoch A, Böning CW, Lutjeharms JRE (2008) Agulhas leakage dynamics affects decadal variability in Atlantic overturning circulation. Nature 456:489–492

Biastoch A, Durgadoo JV, Morrison AK, van Sebille E, Weijer W, Griffies SM (2015) Atlantic multi-decadal oscillation covaries with Agulhas leakage. Nat Commun 6:10082. https://doi.org/10.11038/ncomms10082

Biber MF, Duineveld GCA, Lavaleye MSS, Davies AJ, Bergman MJN, van den Beld IMJ (2014) Investigating the association of fish abundance and biomass with cold-water corals in the deep Northeast Atlantic Ocean using a generalised linear modelling approach. Deep-Sea Res II 99:134–145

Boolukos CM, Lim A, O'Riordan RM, Wheeler AJ (2019) Cold-water corals in decline – a temporal (4 year) species abundance and biodiversity appraisal of complete photomosaiced cold-water coral reef on the Irish Margin. Deep-Sea Res I 146:44–54

Boyer DC, Hampton I (2001) An overview of the living marine resources of Namibia. S Afr J Mar Sci 23:5–35

Brooke SD, Holmes MW, Young CM (2009) Sediment tolerance of two different morphotypes of the deep-sea coral *Lophelia pertusa* from the Gulf of Mexico. Mar Ecol Prog Ser 390:137–144

Buhl-Mortensen L, Vanreusel A, Gooday AJ, Levin LA, Priede IG, Buhl-Mortensen P, Gheerardyn H, King NJ, Raes M (2010) Biological structures as a source of habitat heterogeneity and biodiversity on the deep ocean margins. Mar Ecol 31:21–50

Buhl-Mortensen L, Serigstad B, Buhl-Mortensen P, Olsen MN, Ostrowski M, Błażewicz-Paszkowycz M, Appoh E (2017a) First observations of the structure and megafaunal community of a large *Lophelia* reef on the Ghanaian shelf (the Gulf of Guinea). Deep-Sea Res II 137:148–156

Buhl-Mortensen P, Buhl-Mortensen L, Purser A (2017b) Trophic ecology and habitat provision in cold-water coral ecosystems. In: Rossi S, Bramanti L, Gori A, Orejas C (eds) Marine animal forests: the ecology of benthic biodiversity hotspots. Springer, Cham, pp 919–944

Burke A, Robinson LF (2012) The Southern Ocean's role in carbon exchange during the last deglaciation. Science 335:557–561

Büscher JV, Form AU, Riebesell U (2017) Interactive effects of ocean acidification and warming on growth, fitness and survival of the cold-water coral *Lophelia pertusa* under different food availabilities. Front Mar Sci 4:101. https://doi.org/10.3389/fmars.2017.00101

Buschewski U (2016) Die Gattung *Acanthogorgia* (Anthozoa: Octocorallia) in Kaltwasserriffen des europäischen Kontinentalrandes – Charakterisierung der Arten anhand der funktionalen Ordnung des Skleroms mittels verschiedener morphologischer Methoden. Greifswald University, p 53

Carr M-E, Kearns EJ (2003) Production regimes in four Eastern Boundary Current systems. Deep-Sea Res II 50:3199–3221

Castillo S, Ramil F (2018) A new species of *Cuspidaria* (Mollusca: Bivalvia: Anomalodesmata) from the Mauritanian deep continental margin (North-Eastern Atlantic). J Conchol 43:135–140

Cathalot C, Van Oevelen D, Cox TJS, Kutti T, Lavaleye M, Duineveld G, Meysman FJR (2015) Cold-water coral reefs and adjacent sponge grounds: hotspots of benthic respiration and organic carbon cycling in the deep sea. Front Mar Sci 2:37. https://doi.org/10.3389/fmars.2015.00037

Chapron L, Le Bris N, Durrieu de Madron X, Peru E, Galand PE, Lartaud F (2020) Long term monitoring of cold-water coral growth shows response to episodic meteorological events in the NW Mediterranean. Deep-Sea Res I 160:103255. https://doi.org/10.1016/j.dsr.2020.103255

Chen T, Robinson LF, Burke A, Southon J, Spooner P, Morris PJ, Ng HC (2015) Synchronous centennial abrupt events in the ocean and atmosphere during the last deglaciation. Science 349:1537–1541

Chen T, Robinson LF, Beasley MP, Claxton LM, Andersen MB, Gregoire LJ, Wadham J, Fornari DJ, Harpp KS (2016) Ocean mixing and ice-sheet control of seawater $^{234}U/^{238}U$ during the last deglaciation. Science 354:626–629

Cheng H, Adkins J, Edwards RL, Boyle EA (2000) U-Th dating of deep-sea corals. Geochim Cosmochim Acta 64:2401–2416

Colin C, Frank N, Copard K, Douville E (2010) Neodymium isotopic composition of deep-sea corals from the NE Atlantic: implications for past hydrological changes during the Holocene. Quat Sci Rev 29:2509–2517

Colman JG, Gordon DM, Lane AP, Forde MJ, Fitzpatrick J (2005) Carbonate mounds off Mauritania, Northwest Africa: status of deep-water corals and implications for management of fishing and oil exploration activities. In: Freiwald A, Roberts JM (eds) Cold-water corals and ecosystems. Springer, Heidelberg, pp 417–441

Coma R (2002) Seasonality of in situ respiration rate in three temperate benthic suspension feeders. Limnol Oceanogr 47:324–331

Cordes EE, Jones DOB, Schlacher TA, Amon DJ, Bernardino AF, Brooke S, Carney R, DeLeo DM, Dunlop KM, Escobar-Briones EG, Gates AR, Génio L, Gobin J, Henry L-A, Herrera S, Hoyt S, Joye M, Kark S, Mestre NC, Metaxas A, Pfeifer S, Sink K, Sweetman AK, Witte U (2016) Environmental impacts of the deep-water oil and gas industry: a review to guide management strategies. Front Environ Sci 4:58. https://doi.org/10.3389/fenvs.2016.00058

Cuny-Guirriec K, Douville E, Reynaud S, Allemand D, Bordier L, Canesi M, Mazzoli C, Taviani M, Canese S, McCulloch M, Trotter J, Rico-Esenaro SD, Sanchez-Cabeza J-A, Ruiz-Fernández AC, Carricart-Ganivet JP, Scott PM, Sadekov A, Montagna P (2019) Coral Li/Mg thermometry: caveats and constraints. Chem Geol 523:162–178

Davies AJ, Guinotte JM (2011) Global habitat suitability for framework-forming cold-water corals. PLoS One 6(4):e18483. https://doi.org/10.1371/journal.pone.0018483
Davies AJ, Wisshak M, Orr JC, Roberts JM (2008) Predicting suitable habitat for the cold-water coral *Lophelia pertusa* (Scleractinia). Deep-Sea Res I 55:1048–1062
Davies AJ, Duineveld G, Lavaleye M, Bergman MJ, van Haren H, Roberts JM (2009) Downwelling and deep-water bottom currents as food supply mechanisms to the cold-water coral *Lophelia pertusa* (Scleractinia) at the Mingulay Reef Complex. Limnol Oceanogr 54: 620–629
Davies AJ, Duineveld GCA, van Weering TCE, Mienis F, Quattrini AM, Seim HE, Bane JM, Ross SW (2010) Short-term environmental variability in cold-water coral habitat at Viosca Knoll, Gulf of Mexico. Deep-Sea Res I 57:199–212
De Matos-Pita SS, Ramil F (2015a) Hermit crabs (Decapoda: Crustacea) from deep Mauritanian waters (NW Africa) with the description of a new species. Zootaxa 3926:151–190
De Matos-Pita SS, Ramil F (2015b) Additions to thalassinidean fauna (Crustacea: Decapoda) off Mauritania (NW Africa) with the description of a new genus and a new species. Zootaxa 4020:571–587
De Matos-Pita SS, Ramil F (2016) New species of *Neopilumnoplax* Serène in Guinot, 1969 (Decapoda, Brachyura, Mathildellidae) from Northwest Africa with a key to the genus. Mar Biodivers 46:253–260
De Mol B, Henriet J-P, Canals M (2005) Development of coral banks in the Porcupine Seabight: do they have Mediterranean ancestors? In: Freiwald A, Roberts JM (eds) Cold-water corals and ecosystems. Springer, Heidelberg, pp 515–533
De Mol B, Huvenne VAI, Canals M (2009) Cold-water coral banks and submarine landslides: a review. Int J Earth Sci 98:885–899
De Mol B, Amblas D, Alvarez G, Busquets P, Calafat A, Canals M, Duran R, Lavoie C, Acosta J, Muñoz A (2012) Cold-water coral distribution in an erosional environment: The Strait of Gibraltar gateway. In: Harris PT, Baker EK (eds) Seafloor geomorphology as benthic habitat. Elsevier, London, pp 635–643
Demenocal P, Tierney EJ (2012) Green Sahara: African humid periods paced by Earth's orbital changes. Nat Educ Knowl 3:12
Di Geronimo I, La Perna R (1997) Pleistocene bathyal molluscan assemblages from southern Italy. Riv Ital Paleontol Stratigr 103: 389–426
Dodds LA, Roberts JM, Taylor AC, Marubini F (2007) Metabolic tolerance of the cold-water coral *Lophelia pertusa* (Scleractinia) to temperature and dissolved oxgen change. J Exp Mar Biol Ecol 349: 205–214
Doi T, Tozuka T, Sasaki H, Masumoto Y, Yamagata T (2007) Seasonal and interannual variations of oceanic conditions in the Angola Dome. J Phys Oceanogr 37:2698–2713
Dorschel B, Hebbeln D, Foubert A, White M, Wheeler AJ (2007) Hydrodynamics and cold-water coral facies distribution related to recent sedimentary processes at Galway Mound west of Ireland. Mar Geol 244:184–195
Douarin M, Elliot M, Noble SR, Sinclair D, Henry L-A, Long D, Moreton SG, Murray Roberts J (2013) Growth of north-east Atlantic cold-water coral reefs and mounds during the Holocene: a high resolution U-series and ^{14}C chronology. Earth Planet Sci Lett 375: 176–187
Duarte CM, Krause-Jensen D (2017) Export from seagrass meadows contributes to marine carbon sequestration. Front Mar Sci 4:13. https://doi.org/10.3389/fmars.2017.00013
Dubois-Dauphin Q, Montagna P, Siani G, Douville E, Wienberg C, Hebbeln D, Liu Z, Kallel N, Dapoigny A, Revel M, Pons-Branchu E, Taviani M, Colin C (2017) Hydrological variations of the intermediate water masses of the western Mediterranean Sea during the past 20 ka inferred from neodymium isotopic composition in foraminifera and cold-water corals. Clim Past 13:17–37
Eisele M, Frank N, Wienberg C, Hebbeln D, López Correa M, Douville E, Freiwald A (2011) Productivity controlled cold-water coral growth periods during the last glacial off Mauritania. Mar Geol 280:143–149
Etnoyer PJ, Wickes LN, Silva M, Dubick JD, Balthis L, Salgado E, MacDonald IR (2016) Decline in condition of gorgonian octocorals on mesophotic reefs in the northern Gulf of Mexico: before and after the Deepwater Horizon oil spill. Coral Reefs 35:77–90
Fabri M-C, Vinha B, Allais A-G, Bouhier M-E, Dugornay O, Gaillot A, Arnaubec A (2019) Evaluating the ecological status of cold-water coral habitats using non-invasive methods: an example from Cassidaigne canyon, northwestern Mediterranean Sea. Prog Oceanogr 178:102172
Fisher CR, Hsing P-Y, Kaiser CL, Yoerger DR, Roberts HH, Shedd WW, Cordes EE, Shank TM, Berlet SP, Saunders MG, Larcom EA, Brooks JM (2014) Footprint of Deepwater Horizon blowout impact to deep-water coral communities. Proc Natl Acad Sci U S A 111: 11744–11749
Flögel S, Dullo WC, Pfannkuche O, Kiriakoulakis K, Rüggeberg A (2014) Geochemical and physical constraints for the occurrence of living cold-water corals. Deep-Sea Res II 99:19–26
Foley N, van Rensburg T, Armstrong CW (2010) The ecological and economic value of cold-water coral ecosystems. Ocean Coast Manag 53:313–326
Form AU, Riebesell U (2012) Acclimation to ocean acidification during long-term CO_2 exposure in the cold-water coral *Lophelia pertusa*. Glob Chang Biol 18:843–853
Fort X, Brun J-P, Chauvel F (2004) Salt tectonics on the Angolan margin, synsedimentary deformation processes. AAPG Bull 88: 1523–1544
Fosså JH, Lindberg B, Christensen O, Lundälv T, Svellingen I, Mortensen PB, Alsvag J (2005) Mapping of *Lophelia* reefs in Norway: experiences and survey methods. In: Freiwald A, Roberts JM (eds) Cold-water corals and ecosystems. Springer, Heidelberg, pp 359–391
Foubert A, Depreiter D, Beck T, Maignien L, Pannemans B, Frank N, Blamart D, Henriet J-P (2008) Carbonate mounds in a mud volcano province off north-west Morocco: key to processes and controls. Mar Geol 248:74–96
Frank N, Paterne M, Ayliffe L, van Weering TCE, Henriet J-P, Blamart D (2004) Eastern North Atlantic deep sea corals: tracing upper intermediate water $\Delta^{14}C$ during the Holocene. Earth Planet Sci Lett 219:297–309
Frank N, Freiwald A, Lopez Correa M, Wienberg C, Eisele M, Hebbeln D, Van Rooij D, Henriet J-P, Colin C, van Weering T, de Haas H, Buhl-Mortensen P, Roberts JM, De Mol B, Douville E, Blamart D, Hatte C (2011) Northeastern Atlantic cold-water coral reefs and climate. Geology 39:743–746
Frederiksen R, Jensen A, Westerberg H (1992) The distribution of the scleractinian coral *Lophelia pertusa* around the Faroe islands and the relation to internal tidal mixing. Sarsia 77:157–171
Freiwald A, Wilson JB (1998) Taphonomy of modern deep, cold-temperate water coral reefs. Hist Biol 13:37–52
Freiwald A, Beuck L, Rüggeberg A, Taviani M, Hebbeln D, R/V Meteor Cruise M70-1 participants (2009) The white coral community in the central Mediterranean Sea revealed by ROV surveys. Oceanography 22:58–74
Garcia HE, Locarnini RA, Boyer TP, Antonov JI, Mishonov AV, Baranova OK, Zweng MM, Reagan JR, Johnson DR (2013) World Ocean Atlas 2013. Volume 3: dissolved oxygen, apparent oxygen utilization, and oxygen saturation. In: Levitus S, Mishonov A (eds) NOAA Atlas NESDIS 75. National Oceanographic Data Center, Ocean Climate Laboratory United States, National Environmental Satellite Data Information Service, p 27

Garzoli S, Gordon AL, Kamenkovich V, Pillsbury D, Duncombe-Rae C (1996) Variability and sources of the southeastern Atlantic circulation. J Mar Res 54:1039–1071

Gay A, Lopez M, Berndt C, Séranne M (2007) Geological controls on focused fluid flow associated with seafloor seeps in the Lower Congo Basin. Mar Geol 244:68–92

Georgian SE, Dupont S, Kurman M, Butler A, Strömberg SM, Larsson AI, Cordes EE (2016) Biogeographic variability in the physiological response of the cold-water coral Lophelia pertusa to ocean acidification. Mar Ecol 37:1345–1359

Gil M, Ramil F, Ansín Agís J (2020a) Hydroids (Cnidaria, Hydrozoa) from Mauritanian coral mounds. Zootaxa 4878:412–466

Gil M, Freiwald A, Ramil F (2020b) Two new species of *Rosalinda* (Cnidaria, Hydrozoa, Anthoathecata) from West African cold-water coral mounds. Deep-Sea Res I 168:103436. https://doi.org/10.1016/j.dsr.2020.103436

Girard F, Cruz R, Glickman O, Harpster T, Fisher CR (2019) In situ growth of deep-sea octocorals after the Deepwater Horizon oil spill. Elem Sci Anth 7:12. https://doi.org/10.1525/elementa.1349

Glogowski S, Dullo WC, Feldens P, Liebetrau V, von Reumont J, Hühnerbach V, Krastel S, Wynn RB, Flögel S (2015) The Eugen Seibold coral mounds offshore western Morocco: oceanographic and bathymetric boundary conditions of a newly discovered cold-water coral province. Geo-Mar Lett 35:257–269

Göcke C, Hestetun JT, Uhlir C, Freiwald A, Beuck L, Janussen D (2016) *Cladorhiza corallophila* sp. nov., a new carnivorous sponge (Cladorhizidae, Demospongiae) living in close association with *Lophelia pertusa* and *Madrepora oculata* (Scleractinia). Zootaxa 4168:512–524

Gori A, Orejas C, Madurell T, Bramanti L, Martins M, Quintanilla E, Marti-Puig P, Lo Iacono C, Puig P, Requena S, Greenacre M, Gili JM (2013) Bathymetrical distribution and size structure of cold-water coral populations in the Cap de Creus and Lacaze-Duthiers canyons (northwestern Mediterranean). Biogeosciences 10:2049

Gori A, Ferrier-Pagès C, Hennige SJ, Murray F, Rottier C, Wicks LC, Roberts JM (2016) Physiological response of the cold-water coral *Desmophyllum dianthus* to thermal stress and ocean acidification. PeerJ 4:e1606. https://doi.org/10.7717/peerj.1606

Gori A, Orejas C, Mienis F, Ferrier-Pagès C, Bilan M, Flöter S, Reynaud S, Sweetman AK, Roberts JM, Wienberg C, Hebbeln D (2023) Natural hypoxic conditions do not affect the respiration rates of the cold-water coral Lophelia pertusa living in the Angola margin (Southeastern Atlantic Ocean). Deep Sea Res Part I 197:104052. https://doi.org/10.1016/j.dsr.2023.104052

Gravier C (1920) Madréporaires provenant des campagnes des yachts Princesse-Alice et Hirondelle II (1893-1913). Résultats des campagnes Scientifiques accomplies sur son yacht par Albert Ier Prince Souverain de Monaco 55:1–123

Guinotte JM, Orr J, Cairns S, Freiwald A, Morgan L, George R (2006) Will human-induced changes in seawater chemistry alter the distribution of deep-sea scleractinian corals? Front Ecol Environ 4:141–146

Haberkern J (2017) Contouritic depositional systems influenced by complex seafloor topography – Late Cenozoic seismoacoustic reconstructions from the Galicia and Angola Continental Margins. Department of Geosciences, University of Bremen, PhD-Thesis, p 119. http://nbn-resolving.de/urn:nbn:de:gbv:46-00106386-17

Haddock SHD, Moline MA, Case JF (2010) Bioluminescence in the sea. Annu Rev Mar Sci 2:443–493

Hall-Spencer JH, Allain V, Fosså JH (2002) Trawling damage to NE Atlantic ancient coral reefs. Proc R Soc Lond 269:507–511

Hansteen T, Kwasnitschka T, Klügel A (2014) Cape Verde Seamounts - Cruise No. M80/3 - December 29, 2009 - February 1, 2010 - Dakar (Senegal) - Las Palmas de Gran Canaria (Spain). METEOR-Reports, M80/3, DFG-Senatskommission für Ozeanographie, p 42. https://doi.org/10.2312/cr_m80_3

Hanz U, Wienberg C, Hebbeln D, Duineveld G, Lavaleye M, Juva K, Dullo W-C, Freiwald A, Tamborrino L, Reichart G-J, Flögel S, Mienis F (2019) Environmental factors influencing cold-water coral ecosystems in the oxygen minimum zones on the Angolan and Namibian margins. Biogeosciences 16:4337–4356

Harrison SP, Kohfeld KE, Roelandt C, Claquin T (2001) The role of dust in climate changes today, at the Last Glacial Maximum and in the future. Earth Sci Rev 54:43–80

Hastenrath S, Lamb PJ (2004) Climate dynamics of atmosphere and ocean in the equatorial zone: a synthesis. Int J Climatol 24:1601–1612

Hebbeln D, Wienberg C, Bartels M, Bergenthal M, Frank N, Gaide S, Henriet JP, Kaszemeik K, Klar S, Klein T, Krengel T, Kuhnert M, Meyer-Schack B, Noorlander C, Reuter M, Rosiak U, Schmidt W, Seeba H, Seiter C, Strange N, Terhzaz L, Van Rooij D (2015) MoccoMeBo: climate-driven development of Moroccan cold-water coral mounds revealed by MeBo-drilling: Atlantic vs. Mediterranean settings - Cruise MSM36 - February 18 - March 14, 2014 - Malaga (Spain) - Las Palmas (Spain). MARIA S. MERIAN-Reports, MSM36, DFG-Senatskommission für Ozeanographie, p 47. https://doi.org/10.2312/cr_msm36

Hebbeln D, Van Rooij D, Wienberg C (2016) Good neighbours shaped by vigorous currents: Cold-water coral mounds and contourites in the North Atlantic. Mar Geol 378:171–185

Hebbeln D, Wienberg C, Bender M, Bergmann F, Dehning K, Dullo W-C, Eichstädter R, Flöter S, Freiwald A, Gori A, Haberkern J, Hoffmann L, João FM, Lavaleye M, Leymann T, Matsuyama K, Meyer-Schack B, Mienis F, Moçambique IB, Nowald N, Orejas Saco del Valle C, Ramos Cordova C, Saturov D, Seiter C, Titschack J, Vittori V, Wefing A-M, Wilsenack M, Wintersteller P (2017) ANNA cold-water coral ecosystems off angola and namibia - Cruise No. M122 - December 30, 2015 - January 31, 2016 - Walvis Bay (Namibia) - Walvis Bay (Namibia). METEOR-Reports, M122, DFG-Senatskommission für Ozeanographie, p 74. https://doi.org/10.2312/cr_m122

Hebbeln D, Portilho-Ramos R, Wienberg C, Titschack J (2019a) The fate of cold-water corals in a changing world: a geological perspective. Front Mar Sci 6. https://doi.org/10.3389/fmars.2019.00119

Hebbeln D, Bender M, Gaide S, Titschack J, Vandorpe T, Van Rooij D, Wintersteller P, Wienberg C (2019b) Thousands of cold-water coral mounds along the Moroccan Atlantic continental margin: distribution and morphometry. Mar Geol 411:51–61

Hebbeln D, Wienberg C, Dullo W-C, Freiwald A, Mienis F, Orejas C, Titschack J (2020) Cold-water coral reefs thriving under hypoxia. Coral Reefs 39:853–859

Hemsing F, Hsieh Y-T, Bridgestock L, Spooner PT, Robinson LF, Frank N, Henderson GM (2018) Barium isotopes in cold-water corals. Earth Planet Sci Lett 491:183–192

Hennige SJ, Morrison CL, Form AU, Büscher J, Kamenos NA, Roberts JM (2014) Self-recognition in corals facilitates deep-sea habitat engineering. Sci Rep 4:6782

Hennige SJ, Wolfram U, Wickes L, Murray F, Roberts JM, Kamenos NA, Schofield S, Groetsch A, Spiesz EM, Aubin-Tam M-E, Etnoyer PJ (2020) Crumbling reefs and cold-water coral habitat loss in a Future Ocean: evidence of "coralporosis" as an indicator of habitat integrity. Front Mar Sci 7

Henrich R, Cherubini Y, Meggers H (2010) Climate and sea level induced turbidite activity in a canyon system offshore the hyperarid Western Sahara (Mauritania): The Timiris Canyon. Mar Geol 275:178–198

Henry L-A, Roberts JM (2017) Global biodiversity in cold-water coral reef ecosystems. In: Rossi S, Bramanti L, Gori A, Orejas C (eds) Marine animal forests: the ecology of benthic biodiversity hotspots. Springer, Cham, pp 235–256

Hinz H (2017) Impact of bottom fishing on animal forests: Science, conservation, and fisheries management. In: Rossi S, Bramanti L,

Gori A, Orejas C (eds) Marine animal forests: the ecology of benthic biodiversity hotspots. Springer, Cham, pp 1041–1059

Hoegh-Guldberg O (2014) Coral reef sustainability through adaptation: glimmer of hope or persistent mirage? Curr Opin Environ Sustain 7: 127–133

Hoffman L, Freiwald A (2018a) A new genus and two new species in Calliotropidae (Mollusca: Gastropoda) from NE Atlantic cold-water coral habitats. Misc Malacol 7:89–96

Hoffman L, Freiwald A (2018b) Last snails standing: a tale of Fissurellidae (Gastropoda) from deep-water coral habitats off Mauritania since the Pleistocene. Misc Malacol 7:115–126

Hoffman L, Freiwald A (2019) *Vexillum monscorallum* n. sp. (Gastropoda: Costellariidae) from deep-water coral habitats off Mauritania. Misc Malacol 8:37–40

Hoffman L, Freiwald A (2020) *Cantrainea mauritanea* n. sp. (Gastropoda: Colloniidae) from deep-water coral habitats off Mauritania. Misc Malacol 8:69–73

Hoffman L, Fraussen K, Freiwald A (2018a) The genus *Chauvetia* (Gastropoda, Buccinoidea) in deep-water coral habitats on the continental slope of Mauritania. Misc Malacol 7:127–138

Hoffman L, Freiwald A, van Heugten B, Lavaleye M (2018b) New information on *Ganesa*, *Lopheliella* and *Trochaclis* (Gastropoda) in cold-water coral habitats in the temperate and tropical North Atlantic. Misc Malacol 8:27–34

Hoffman L, Monsecour K, Freiwald A (2019a) Columbellidae (Gastropoda) from deep-water coral habitats off Mauritania. Misc Malacol 8:41–49

Hoffman L, Von Cosel R, Freiwald A (2019b) New and little-known Pliocardiinae (Bivalvia, Vesicomyidae) from the continental slope off Mauritania. Basteria 83:151–157

Hoffman L, Beuck L, Van Heugten B, Lavaleye M, Freiwald A (2019c) Last snails standing since the Pleistocene, a tale of Calliostomatidae (Gastropoda) living in deep-water coral habitats in the Northeast Atlantic. Zootaxa 4613:93–110

Holz C, Stuut JBW, Henrich R (2004) Terrigenous sedimentation processes along the continental margin off NW Africa: implications from grain-size analysis of seabed sediments. Sedimentology 51: 1145–1154

Huvenne VAI, De Mol B, Henriet J-P (2003) A 3D seismic study of the morphology and spatial distribution of buried coral banks in the Porcupine Basin, SW of Ireland. Mar Geol 198:5–25

Huvenne VAI, Van Rooij D, De Mol B, Thierens M, O'Donnell R, Foubert A (2009) Sediment dynamics and palaeo-environmental context at key stages in the challenger cold-water coral mound formation: clues from sediment deposits at the mound base. Deep-Sea Res I 56:2263–2280

Huvenne VAI, Tyler PA, Masson DG, Fisher EH, Hauton C, Hühnerbach V, Le Bas TP, Wolff GA (2011) A picture on the wall: Innovative mapping reveals cold-water coral refuge in submarine canyon. PLoS One 6(12):e28755. https://doi.org/10.21371/journal.pone.0028755

IMROP (2013) Atlas maritime des zones vulnérables en Mauritanie. l'Institut Mauritanien de Recherches Océanographiques et des Pêches, Nouadhibou, p 151

Jackson EL, Davies AJ, Howell KL, Kershaw PJ, Hall-Spencer JM (2014) Future-proofing marine protected area networks for cold-water coral reefs. ICES J Mar Sci 71:2621–2629

Järnegren J, Altin D (2006) Filtration and respiration of the deep living bivalve *Acesta excavata* (J.C. Fabricius, 1779) (Bivalvia; Limidae). J Exp Mar Biol Ecol 334:122–129

Jensen A, Frederiksen R (1992) The fauna associated with the bank-forming deepwater coral *Lophelia pertusa* (Scleractinia) on the Faroe shelf. Sarsia 77:53–69

Joyce TM (1988) On the wind-rriven cross-equatorial flow in the Atlantic Ocean. J Phys Oceanogr 18:793–799

Junker T, Mohrholz V, Siegfried L, van der Plas A (2017) Seasonal to interannual variability of water mass characteristics and currents on the Namibian shelf. J Mar Syst 165:36–46

Jurikova H, Liebetrau V, Raddatz J, Fietzke J, Trotter J, Rocholl A, Krause S, McCulloch M, Rüggeberg A, Eisenhauer A (2019) Boron isotope composition of the cold-water coral *Lophelia pertusa* along the Norwegian margin: Zooming into a potential pH-proxy by combining bulk and high-resolution approaches. Chem Geol 513:143–152

Juva K, Flögel S, Karstensen J, Linke P, Dullo W-C (2020) Tidal dynamics control on cold-water coral growth: a high-resolution multivariable study on eastern Atlantic cold-water coral sites. Front Mar Sci 7:132. https://doi.org/10.3389/fmars.2020.00132

Kano A, Ferdelman TG, Williams T, Henriet J-P, Ishikawa T, Kawagoe N, Takashima C, Kakizaki Y, Abe K, Sakai S, Browning EL, Li X, Integrated Ocean Drilling Program Expedition 307 Scientists (2007) Age constraints on the origin and growth history of a deep-water coral mound in the northeast Atlantic drilled during Integrated Ocean Drilling Program Expedition 307. Geology 35:1051–1054

Kawase M, Sarmiento JL (1985) Nutrients in the Atlantic thermocline. J Geophys Res Oceans 90:8961–8979

Kerber IK, Arps J, Eichstädter R, Kontor F, Dornick C, Schröder-Ritzrau A, Babu A, Warken S, Frank N (2023) Simultaneous U and Th isotope measurements for U-series dating using MCICPMS. Nucl Instrum Methods Phys Res Sect B: Beam Interactions Materials Atoms 539:169–178

Klenz T, Dengler M, Brandt P (2018) Seasonal variability of the Mauritania current and hydrography at 18°N. J Geophys Res Oceans 123:8122–8137

Korpanty C, Hoffman L, Portilho-Ramos R, Titschack J, Wienberg C, Hebbeln D (2023) Decline in cold-water coral growth promotes molluscan diversity: a paleon to logical perspective from a cold-water coral mound in the western Mediterranean Sea. Front Mar Sci 9:895946. https://doi.org/10.3389/fmars.2022.895946

Kounta L, Capet X, Jouanno J, Kolodziejczyk N, Sow B, Gaye AT (2018) A model perspective on the dynamics of the shadow zone of the eastern tropical North Atlantic – Part 1: the poleward slope currents along West Africa. Ocean Sci 14:971–997

Kukla PA, Strozyk F, Mohriak WU (2018) South Atlantic salt basins – witnesses of complex passive margin evolution. Gondwana Res 53: 41–57

La Perna R, D'Abramo M (2010) Una collezione di G. Seguenza conservata presso l'Istituto Tecnico Commerciale "O.G. Costa" di Lecce. Boll Malacol 46:29–35

Laffoley D, Baxter JM (2019) Ocean deoxygenation: everyone's problem - Causes, impacts, consequences and solutions. IUCN, International Union for Conservation of Nature and Natural Resources Gland, Switzerland:580

Le Guilloux E, Olu-Le Roy K, Bourillet JF, Savoye B, Iglésias SP, Sibuet M (2009) First observations of deep-sea coral reefs along the Angola margin. Deep-Sea Res II 56:2394–2403

Le Lœuff P, von Cosel R (1998) Biodiversity patterns of the marine benthic fauna on the Atlantic coast of tropical Africa in relation to hydroclimatic conditions and paleogeographic events. Acta Oecol 19:309–321

León R, Somoza L, Medialdea T, Vázquez JT, González FJ, López-González N, Casas D, del Pilar MM, del Fernández-Puga MC, Giménez-Moreno CJ, Díaz-del-Río V (2012) New discoveries of mud volcanoes on the Moroccan Atlantic continental margin (Gulf of Cádiz): morpho-structural characterization. Geo-Mar Lett 32: 473–488

Light MPR, Maslanyj MP, Greenwood RJ, Banks NL (1993) Seismic sequence stratigraphy and tectonics offshore Namibia. Geol Soc Lond, Spec Publ 71:163–191

Liu M, Tanhua T (2021) Water masses in the Atlantic Ocean: characteristics and distributions. Ocean Sci 17:463–486

Lo Iacono C, Gràcia E, Ranero CR, Emelianov M, Huvenne VAI, Bartolomé R, Booth-Rea G, Prades J (2014) The West Melilla cold water coral mounds, Eastern Alboran Sea: morphological characterization and environmental context. Deep-Sea Res II 99:316–326

Locarnini RA, Mishonov AV, Antonov JI, Boyer TP, Garcia HE, Baranova OK, Zweng MM, Paver CR, Reagan JR, Johnson DR, Hamilton M, Seidov D, Levitus S (2013) World Ocean Atlas 2013. Volume 1: Temperature. In: Levitus S, Mishonov A (eds) NOAA Atlas NESDIS 73. National Oceanographic Data Center, Ocean Climate Laboratory United States, National Environmental Satellite Data Information Service, p 40

Lunden JJ, McNicholl CG, Sears CR, Morrison CL, Cordes EE (2014) Acute survivorship of the deep-sea coral *Lophelia pertusa* from the Gulf of Mexico under acidification, warming, and deoxygenation. Front Mar Sci 1:78. https://doi.org/10.3389/fmars.2014.00078

Lutringer A, Blamart D, Frank N, Labeyrie L (2005) Paleotemperatures from deep-sea corals: scale effects. In: Freiwald A, Roberts JM (eds) Cold-water corals and ecosystems. Springer, Heidelberg, pp 1081–1096

Macpherson E (1991) Biogeography and community structure of the decapod crustacean fauna off Namibia (Southeast Atlantic). J Crustac Biol 11:401–415

Maier C, Hegeman J, Weinbauer MG, Gattuso JP (2009) Calcification of the cold-water coral *Lophelia pertusa* under ambient and reduced pH. Biogeosciences 6:1671–1680

Maldonado M, Carmona MC, Velásquez Z, Puig A, Cruzado A, López A, Young CM (2005) Siliceous sponges as a silicon sink: An overlooked aspect of benthopelagic coupling in the marine silicon cycle. Limnol Oceanogr 50:799–809

Maldonado M, Beazley L, López-Acosta M, Kenchington E, Casault B, Hanz U, Mienis F (2020) Massive silicon utilization facilitated by a benthic-pelagic coupled feedback sustains deep-sea sponge aggregations. Limnol Oceanogr 66:366–391

Mangini A, Lomitschka M, Eichstädter R, Frank N, Vogler S, Bonani G, Hajdas I, Patzold J (1998) Coral provides way to age deep water. Nature 392:347–348

Matsuyama K, Titschack J, Baum D, Freiwald A (2015) Two new species of erect Bryozoa (Gymnolaemata: Cheilostomata) and the application of non-destructive imaging methods for quantitative taxonomy. Zootaxa 4020:81–100

McCulloch M, Falter J, Trotter J, Montagna P (2012) Coral resilience to ocean acidification and global warming through pH up-regulation. Nat Clim Chang 2:623–627

Meeuwis JM, Lutjeharms JRE (1990) Surface thermal characteristics of the Angola-Benguela front. S Afr J Mar Sci 9:261–279

Mienis F, de Stigter HC, White M, Duineveld G, de Haas H, van Weering TCE (2007) Hydrodynamic controls on cold-water coral growth and carbonate-mound development at the SW and SE Rockall Trough Margin, NE Atlantic Ocean. Deep-Sea Res I 54: 1655–1674

Mienis F, De Stigter HC, De Haas H, Van der Land C, Van Weering TCE (2012) Hydrodynamic conditions in a cold-water coral mound area on the Renard Ridge, southern Gulf of Cadiz. J Mar Syst 96–97: 61–71

Mienis F, Duineveld GCA, Davies AJ, Lavaleye MMS, Ross SW, Seim H, Bane J, van Haren H, Bergman MJN, de Haas H, Brooke S, van Weering TCE (2014) Cold-water coral growth under extreme environmental conditions, the Cape Lookout area, NW Atlantic. Biogeosciences 11:2543–2560

Mohn C, Rengstorf A, White M, Duineveld G, Mienis F, Soetaert K, Grehan A (2014) Linking benthic hydrodynamics and cold-water coral occurrences: a high-resolution model study at three cold-water coral provinces in the NE Atlantic. Prog Oceanogr 122:92–104

Mohrholz V, Bartholomae CH, van der Plas AK, Lass HU (2008) The seasonal variability of the northern Benguela undercurrent and its relation to the oxygen budget on the shelf. Cont Shelf Res 28:424–441

Mohrholz V, Eggert A, Junker T, Nausch G, Ohde T, Schmidt M (2014) Cross shelf hydrographic and hydrochemical conditions and their short term variability at the northern Benguela during a normal upwelling season. J Mar Syst 140:92–110

Molinari RL, Bauer S, Snowden D, Johnson GC, Bourles B, Gouriou Y, Mercier H (2003) A comparison of kinematic evidence for tropical cells in the Atlantic and Pacific oceans. In: Goni GJ, Malanotte-Rizzoli P (eds) Elsevier oceanography series. Elsevier, pp 269–286

Montagna P, McCulloch M, Douville E, López Correa M, Trotter J, Rodolfo-Metalpa R, Dissard D, Ferrier-Pagès C, Frank N, Freiwald A, Goldstein S, Mazzoli C, Reynaud S, Rüggeberg A, Russo S, Taviani M (2014) Li/Mg systematics in scleractinian corals: calibration of the thermometer. Geochim Cosmochim Acta 132:288–310

Montero-Serrano J-C, Frank N, Tisnérat-Laborde N, Colin C, Wu C-C, Lin K, Shen C-C, Copard K, Orejas C, Gori A, De Mol L, Van Rooij D, Reverdin G, Douville E (2013) Decadal changes in the mid-depth water mass dynamic of the Northeastern Atlantic margin (Bay of Biscay). Earth Planet Sci Lett 364:134–144

Mortensen P, Fosså J (2006) Species diversity and spatial distribution of invertebrates on *Lophelia* reefs in Norway. Proceedings of the 10th international coral reef symposium

Mosquera Giménez Á, Vélez-Belchí P, Rivera J, Piñeiro S, Fajar N, Caínzos V, Balbín R, Jiménez Aparicio JA, Dominguez-Carrió C, Blasco-Ferre J, Carreiro-Silva M, Morato T, Puerta P, Orejas C (2019) Ocean circulation over North Atlantic underwater features in the path of the Mediterranean Outflow Water: The Ormonde and Formigas seamounts, and the Gazul mud volcano. Front Mar Sci 6: 702. https://doi.org/10.3389/fmars.2019.00702

Mouchi V, Chapron L, Peru E, Pruski AM, Meistertzheim A-L, Vétion G, Galand PE, Lartaud F (2019) Long-term aquaria study suggests species-specific responses of two cold-water corals to macro-and microplastics exposure. Environ Pollut 253:322–329

Mueller CE, Lundälv T, Middelburg JJ, Van Oevelen D (2013) The symbiosis between *Lophelia pertusa* and *Eunice norvegica* stimulates coral calcification and worm assimilation. PLoS One 8(3):e58660. https://doi.org/10.51371/journal.pone.0058660

Niang A, Scheren P, Diop S, Kane C, Koulibaly CT (2019) The Senegal and Pangani rivers: examples of over-used river systems within water-stressed environments in Africa. In: Wolanski E, Day JW, Elliott M, Ramachandran R (eds) Coasts and Estuaries. Elsevier, pp 311–320

Ohde T, Mohrholz V (2011) Interannual variability of sulphur plumes off the Namibian coast. Int J Remote Sens 32:9327–9342

Ollitrault M, Colin de Verdière A (2014) The ocean general circulation near 1000-m depth. J Phys Oceanogr 44:384–409

Orejas C, Gori A, Lo Iacono C, Puig P, Gili JM, Dale MR (2009) Cold-water corals in the Cap de Creus canyon (north-western Mediterranean): spatial distribution, density and anthropogenic impact. Mar Ecol Prog Ser 397:37–51

Orejas C, Ferrier-Pagès C, Reynaud S, Gori A, Beraud E, Tsounis G, Allemand D, Gili JM (2011) Long-term growth rates of four Mediterranean cold-water coral species maintained in aquaria. Mar Ecol Prog Ser 429:57–65

Orejas C, Wienberg C, Titschack J, Tamborrino L, Freiwald A, Hebbeln D (2021) *Madrepora oculata* forms large frameworks in hypoxic waters off Angola (SE Atlantic). Sci Rep 11:15170. https://doi.org/10.11038/s41598-15021-94579-15176

Pausata FSR, Gaetani M, Messori G, Berg A, Maia de Souza D, Sage RF, deMenocal PB (2020) The greening of the Sahara: past changes and future implications. One Earth 2:235–250

Pelegrí JL, Peña-Izquierdo J, Machín F, Meiners C, Presas-Navarro C (2017) Oceanography of the Cape Verde Basin and Mauritanian Slope Waters. In: Ramos A, Ramil F, Sanz JL (eds) Deep-sea ecosystems off Mauritania: research of marine biodiversity and habitats in the Northwest African Margin. Springer, Dordrecht, pp 119–153

Peña-Izquierdo J, van Sebille E, Pelegrí JL, Sprintall J, Mason E, Llanillo PJ, Machín F (2015) Water mass pathways to the North Atlantic oxygen minimum zone. J Geophys Res Oceans 120:3350–3372

Pinheiro LM, Ivanov MK, Sautkin A, Akhmanov G, Magalhaes VH, Volkonskaya A, Monteiro JH, Somoza L, Gardner J, Hamouni N, Cunha MR (2003) Mud volcanism in the Gulf of Cádiz: results from the TTR-10 cruise. Mar Geol 195:131–151

Portilho-Ramos R, Chiessi CM, Zhang Y, Mulitza S, Kucera M, Siccha M, Prange M, Paul A (2017) Coupling of equatorial Atlantic surface stratification to glacial shifts in the tropical rainbelt. Sci Rep 7:1561. https://doi.org/10.1038/s41598-41017-01629-z

Portilho-Ramos R, Titschack J, Wienberg C, Siccha M, Yokohama Y, Hebbeln D (2022) Major environmental drivers determining life and death of cold-water corals through time. PLoS Biol 20(5):e3001628. https://doi.org/10.1371/journal.pbio.3001628

Puerta P, Johnson C, Carreiro-Silva M, Henry L-A, Kenchington E, Morato T, Kazanidis G, Rueda JL, Urra J, Ross S, Wei C-L, González-Irusta JM, Arnaud-Haond S, Orejas C (2020) Influence of water masses on the biodiversity and biogeography of deep-sea benthic ecosystems in the North Atlantic. Front Mar Sci 7:239. https://doi.org/10.3389/fmars.2020.00239

Purser A, Thomsen L (2012) Monitoring strategies for drill cutting discharge in the vicinity of cold-water coral ecosystems. Mar Pollut Bull 64:2309–2316

Raddatz J, Liebetrau V, Rüggeberg A, Hathorne E, Krabbenhoft A, Eisenhauer A, Böhm F, Vollstaedt H, Fietzke J, López Correa M, Freiwald A, Dullo WC (2013) Stable Sr-isotope, Sr/Ca, Mg/Ca, Li/Ca and Mg/Li ratios in the scleractinian cold-water coral *Lophelia pertusa*. Chem Geol 352:143–152

Ramos A, Sanz JL, Ramil F, Agudo LM, Presas-Navarro C (2017a) The Giant Cold-Water Coral Mounds Barrier Off Mauritania. In: Ramos A, Ramil F, Sanz JL (eds) Deep-sea ecosystems off Mauritania: research of marine biodiversity and habitats in the Northwest African Margin. Springer, Dordrecht, pp 481–525

Ramos A, Ramil F, Sanz JL (2017b) Deep-sea ecosystems off Mauritania: research of marine biodiversity and habitats in the Northwest African Margin. Springer, Dordrecht

Ramos A, Ramil F, Freiwald A, Beuck L, Bouzouma ME, Khallahi B, Kloff S (2019) A second life for the Chinguetti fishing exclusion zone: a marine protected area network on the Mauritanian slope for prosperous fisheries and improved understanding of marine biodiversity. Report for the Ministry of Fisheries and Maritime Economy, Ministry of Environment and Sustainable Development, Ministry of Education, Nouakchot, p 93

Rincón-Tomás B, Duda JP, Somoza L, González FJ, Schneider D, Medialdea T, Santofimia E, López-Pamo E, Madureira P, Hoppert M, Reitner J (2019) Cold-water corals and hydrocarbon-rich seepage in Pompeia Province (Gulf of Cádiz) – living on the edge. Biogeosciences 16:1607–1627

Rivera J, Canals M, Lastras G, Hermida N, Amblas D, Arrese B, Martín-Sosa P, Acosta Yepes J (2016) Morphometry of Concepcion Bank: evidence of geological and biological processes on a large volcanic seamount of the Canary Islands Seamount Province. PLoS One 11(5):e0156337. https://doi.org/10.1371/journal.pone.0156337

Roberts JM, Wheeler AJ, Freiwald A, Cairns SD (2009) Cold-water corals. The biology and geology of deep-sea coral habitats. Cambridge University Press

Robinson LF, Adkins J, Keigwin LD, Southon J, Fernandez DP, Wang S-L, Scheirer DS (2005) Radiocarbon variability in the western North Atlantic during the last deglaciation. Science 310:1469–1473

Robinson LF, Adkins JF, Frank N, Gagnon AC, Prouty NG, Brendan Roark E, de Flierdt TV (2014) The geochemistry of deep-sea coral skeletons: a review of vital effects and applications for palaeoceanography. Deep-Sea Res II 99:184–198

Rogers AD (1999) The biology of *Lophelia pertusa* (Linnaeus 1758) and other deep-water reef-forming corals and impacts from human activities. Int Rev Hydrobiol 84:315–406

Rogers J, Rau AJ (2006) Surficial sediments of the wave-dominated Orange River Delta and the adjacent continental margin off south-western Africa. Afr J Mar Sci 28:511–524

Rüggeberg A, Foubert A (2019) Cold-water corals and mud volcanoes: life on a dynamic substrate. In: Orejas C, Jiménez C (eds) Mediterranean cold-water corals: past, present and future, Springer series: Coral Reefs of the World. Springer, pp 265–269

Rüggeberg A, Flögel S, Dullo W-C, Hissmann K, Freiwald A (2011) Water mass characteristics and sill dynamics in a subpolar cold-water coral reef setting at Stjernsund, northern Norway. Mar Geol 282:5–12

Rush D, Lavaleye M, and cruise participants (2019) RV Pelagia cruise report 64PE449 KUNOXI - Kunene Coral Mound Provinces and Namibian Oxygen Depleted Biogeochemistry - February 15th-March 10th, 2019. Royal NIOZ, Texel, p 114

Sakai K, Türkay M, Beuck L, Freiwald A (2015) A collection of the Infraorder Callianassidea (Decapoda, Pleocyemata) with one new genus and five new species from the Eastern Atlantic off Mauritania (R/V Maria S. Merian cruise MSM 16/3 "PHAETON"). Mar Biodivers 45:113–133

Sala E, Mayorga J, Bradley D, Cabral RB, Atwood TB, Auber A, Cheung W, Costello C, Ferretti F, Friedlander AM, Gaines SD, Garilao C, Goodell W, Halpern BS, Hinson A, Kaschner K, Kesner-Reyes K, Leprieur F, McGowan J, Morgan LE, Mouillot D, Palacios-Abrantes J, Possingham HP, Rechberger KD, Worm B, Lubchenco J (2021) Protecting the global ocean for biodiversity, food and climate. Nature 592:397–402

Sampaio Í, Carreiro-Silva M, Freiwald A, Menezes G, Grasshoff M (2019) Natural history collections as a basis for sound biodiversity assessments: Plexauridae (Octocorallia, Holaxonia) of the Naturalis CANCAP and Tyro Mauritania II expeditions. ZooKeys 870:1–32

Sampaio Í, Beuck L, Freiwald A (2022) A new octocoral species of *Swiftia* (Holaxonia: Plexauridae) from the upper bathyal off Mauritanian Slope (NE Atlantic). ZooKeys 1106:121–140

Sanna G, Freiwald A (2021) Deciphering the composite morphological diversity of Lophelia pertusa, a cosmopolitan deep-water ecosystem engineer. Ecosphere 12:e03802

Sanz JL, Maestro A, Agudo LM (2017) The Mauritanian Margin. Bathymetric and geomorphological characteristics. In: Ramos A, Ramil F, Sanz JL (eds) Deep-sea ecosystems off Mauritania: research of marine biodiversity and habitats in the Northwest African Margin. Springer, Dordrecht, pp 53–117

Schmidtko S, Stramma L, Visbeck M (2017) Decline in global oceanic oxygen content during the past five decades. Nature 542:335–339

Schulz K, Soetaert K, Mohn C, Korte L, Mienis F, Duineveld G, van Oevelen D (2020) Linking large-scale circulation patterns to the distribution of cold water corals along the eastern Rockall Bank (northeast Atlantic). J Mar Syst 212:103456

Singarayer JS, Burrough SL (2015) Interhemispheric dynamics of the African rainbelt during the late Quaternary. Quat Sci Rev 124:48–67

Sink K, Adams R, Mann J, Whitehead O, Franken M, Maze K (2019) South Africa's New Marine Protected Areas. South African National Biodiversity Institute, an entity of the Department of Environmental Affairs, Pretoria

Skonieczny C, Paillou P, Bory A, Bayon G, Biscara L, Crosta X, Eynaud F, Malaizé B, Revel M, Aleman N, Barusseau J-P,

Vernet R, Lopez S, Grousset F (2015) African humid periods triggered the reactivation of a large river system in Western Sahara. Nat Commun 6:8751. https://doi.org/10.1038/ncomms9751

Smith JE, Risk MJ, Schwarcz HP, McConnaughey TA (1997) Rapid climate change in the North Atlantic during the Younger Dryas recorded by deep-sea corals. Nature 386:818–820

Somoza L, Diaz-del-Rio V, Leon R, Ivanov M, Fernandez-Puga MC, Gardner JM, Hernandez-Molina FJ, Pinheiro LM, Rodero J, Lobato A (2003) Seabed morphology and hydrocarbon seepage in the Gulf of Cádiz mud volcano area: Acoustic imagery, multibeam and ultra-high resolution seismic data. Mar Geol 195:153–176

Somoza L, Ercilla G, Urgorri V, León R, Medialdea T, Paredes M, Gonzalez FJ, Nombela MA (2014) Detection and mapping of cold-water coral mounds and living Lophelia reefs in the Galicia Bank, Atlantic NW Iberia margin. Mar Geol 349:73–90

Spalding M, Fox H, Allen G, Davidson N, Ferdaña Z, Finlayson M, Halpern B, Jorge M, Lombana A, Lourie S, Martin K, McManus E, Molnar J, Recchia C, Robertson J (2007) Marine ecoregions of the world: a bioregionalization of coastal and shelf areas. Bioscience 57: 573–583

Spooner PT, Guo W, Robinson LF, Thiagarajan N, Hendry KR, Rosenheim BE, Leng MJ (2016) Clumped isotope composition of cold-water corals: a role for vital effects? Geochim Cosmochim Acta 179:123–141

Stevenson A, Rocha C (2013) Evidence for the bioerosion of deep-water corals by echinoids in the Northeast Atlantic. Deep-Sea Res I 71:73–78

Stramma L, Peterson RG (1989) Geostrophic transport in the Benguela Current region. J Phys Oceanogr 19:1440–1448

Stramma L, Fischer J, Brandt P, Schott F (2003) Circulation, variability and near-equatorial meridional flow in the central tropical Atlantic. In: Goni GJ, Malanotte-Rizzoli P (eds) Elsevier oceanography series. Elsevier, pp 1–22

Stramma L, Schmidtko S, Levin LA, Johnson GC (2010) Ocean oxygen minima expansions and their biological impacts. Deep-Sea Res I 57: 587–595

Struve T, van de Flierdt T, Burke A, Robinson LF, Hammond SJ, Crocket KC, Bradtmiller LI, Auro ME, Mohamed KJ, White NJ (2017) Neodymium isotopes and concentrations in aragonitic scleractinian cold-water coral skeletons - Modern calibration and evaluation of palaeo-applications. Chem Geol 453:146–168

Sweetman AK, Thurber AR, Smith CR, Levin LA, Mora C, Wei C-L, Gooday AJ, Jones DOB, Rex M, Yasuhara M, Ingels J, Ruhl HA, Frieder CA, Danovaro R, Würzberg L, Baco A, Grupe BM, Pasulka A, Meyer KS, Dunlop KM, Henry L-A, Roberts JM (2017) Major impacts of climate change on deep-sea benthic ecosystems. Elem Sci Anth 5:4. https://doi.org/10.1525/elementa.1203

Talley LD, Pickard GL, Emery WJ, Swift JH (2011) Atlantic Ocean. In: Talley LD, Pickard GL, Emery WJ, Swift JH (eds) Descriptive physical oceanography, 6th edn. Academic, Boston, pp 245–301

Tamborrino L, Wienberg C, Titschack J, Wintersteller P, Mienis F, Schröder-Ritzrau A, Freiwald A, Orejas C, Dullo W-C, Haberkern J, Hebbeln D (2019) Mid-Holocene extinction of cold-water corals on the Namibian shelf steered by the Benguela oxygen minimum zone. Geology 47:1185–1188

Tamborrino L, Titschack J, Wienberg C, Purkis S, Eberli GP, Hebbeln D (2022) Spatial distribution and morphometry of the Namibian coral mounds controlled by the hydrodynamic regime and outer-shelf topography. Front Mar Sci 9:877616. https://doi.org/10.3389/fmars.2022.877616

Taviani M, Angeletti L, Beuck L, Campiani E, Canese S, Foglini F, Freiwald A, Montagna P, Trincardi F (2016) Reprint of 'On and off the beaten track: megafaunal sessile life and Adriatic cascading processes'. Mar Geol 375:146–160

Titschack J, Bromley RG, Freiwald A (2005) Plio-Pleistocene cliff-bound, wedge-shaped, warm-temperate carbonate deposits from Rhodes (Greece): sedimentology and facies. Sediment Geol 180: 29–56

Titschack J, Thierens M, Dorschel B, Schulbert C, Freiwald A, Kano A, Takashima C, Kawagoe N, Li X (2009) Carbonate budget of a cold-water coral mound (Challenger Mound, IODP Exp. 307). Mar Geol 259:36–46

Titschack J, Baum D, De Pol HR, López Correa M, Forster N, Flögel S, Hebbeln D, Freiwald A (2015) Aggradation and carbonate accumulation of Holocene Norwegian cold-water coral reefs. Sedimentology 62:1873–1898

Tittensor DP, Baco AR, Brewin PE, Clark MR, Consalvey M, Hall-Spencer J, Rowden AA, Schlacher T, Stocks KI, Rogers AD (2009) Predicting global habitat suitability for stony corals on seamounts. J Biogeogr 36:1111–1128

Tittensor DP, Baco AR, Hall-Spencer JM, Orr JC, Rogers AD (2010) Seamounts as refugia from ocean acidification for cold-water stony corals. Mar Ecol 31:212–225

Turley CM, Roberts JM, Guinotte JM (2007) Corals in deep-water: will the unseen hand of ocean acidification destroy cold-water ecosystems? Coral Reefs 26:445–448

Urra J, Palomino D, Lozano P, González-García E, Farias C, Mateo-Ramírez Á, Fernández-Salas LM, López-González N, Vila Y, Orejas C, Puerta P, Rivera J, Henry L-A, Rueda JL (2021) Deep-sea habitat characterization using acoustic data and underwater imagery in Gazul mud volcano (Gulf of Cádiz, NE Atlantic). Deep Sea Res Part I 169:103458. https://doi.org/10.1016/j.dsr.2020.103458

Van Bennekom AJ, Berger GW (1984) Hydrography and silica budget of the Angola Basin. Neth J Sea Res 17:149–200

van de Flierdt T, Robinson LF, Adkins JF (2010) Deep-sea coral aragonite as a recorder for the neodymium isotopic composition of seawater. Geochim Cosmochim Acta 74:6014–6032

Vandorpe T, Delivet S, Blamart D, Wienberg C, Bassinot F, Mienis F, Stuut J-BW, Van Rooij D (2023) Palaeoceanographic and hydrodynamic variability for the last 47 kyr in the southern Gulf of Cádiz (Atlantic Moroccan margin): sedimentary and climatic implications. Depos Rec 9(1):30–51. https://doi.org/10.1002/dep2.212

Van Engeland T, Godø OR, Johnsen E, Duineveld GCA, van Oevelen D (2019) Cabled ocean observatory data reveal food supply mechanisms to a cold-water coral reef. Prog Oceanogr 172:51–64

van Haren H, Mienis F, Duineveld GCA, Lavaleye MSS (2014) High-resolution temperature observations of a trapped nonlinear diurnal tide influencing cold-water corals on the Logachev mounds. Prog Oceanogr 125:16–25

van Oevelen D, Duineveld G, Lavaleye M, Mienis F, Soetaert K, Heip CHR (2009) The cold-water coral community as hotspot of carbon cycling on continental margins: a food-web analysis from Rockall Bank (northeast Atlantic). Limnol Oceanogr 54:1829–1844

Van Rooij D, Blamart D, De Mol L, Mienis F, Pirlet H, Wehrmann LM, Barbieri R, Maignien L, Templer SP, de Haas H, Hebbeln D, Frank N, Larmagnat S, Stadnitskaia A, Stivaletta N, van Weering T, Zhang Y, Hamoumi N, Cnudde V, Duyck P, Henriet JP (2011) Cold-water coral mounds on the Pen Duick Escarpment, Gulf of Cadiz: the MiCROSYSTEMS project approach. Mar Geol 282:102–117

Vandorpe T, Wienberg C, Hebbeln D, Van den Berghe M, Gaide S, Wintersteller P, Van Rooij D (2017) Multiple generations of buried cold-water coral mounds since the Early-Middle Pleistocene Transition in the Atlantic Moroccan Coral Province, southern Gulf of Cádiz. Palaeogeogr Palaeoclimatol Palaeoecol 485:293–304

Veitch J, Penven P, Shillington F (2010) Modeling Equilibrium Dynamics of the Benguela Current System. J Phys Oceanogr 40:1942–1964

Vertino A, Taviani M, Corselli C (2019) Spatio-temporal distribution of Mediterranean cold-water corals. In: Saco O, del Valle C, Jiménez C (eds) Mediterranean cold-water corals: past, present and future. Coral Reefs of the World. Springer, Springer series, pp 67–83

Victorero L, Watling L, Deng Palomares ML, Nouvian C (2018) Out of sight, but within reach: A global history of bottom-trawled deep-sea fisheries from >400 m depth. Front Mar Sci 5:98. https://doi.org/10.3389/fmars.2018.00098

Vinha B, Rossi S, Gori A, Hanz U, Pennetta A, Mienis F, Huvenne VAI, Hebbeln D, Wienberg C, Titschack J, Freiwald A, Piraino S, Orejas C (2023) Trophic ecology of Angolan cold-water coral mounds reefs (SE Atlantic) based on stable isotope analyses. Sci Rep 13:9933. https://doi.org/10.1038/s41598-023-37035-x

Wang XT, Prokopenko MG, Sigman DM, Adkins JF, Robinson LF, Ren H, Oleynik S, Williams B, Haug GH (2014) Isotopic composition of carbonate-bound organic nitrogen in deep-sea scleractinian corals: a new window into past biogeochemical change. Earth Planet Sci Lett 400:243–250

Wang XT, Sigman DM, Prokopenko MG, Adkins JF, Robinson LF, Hines SK, Chai J, Studer AS, Martínez-García A, Chen T, Haug GH (2017) Deep-sea coral evidence for lower Southern Ocean surface nitrate concentrations during the last ice age. Proc Natl Acad Sci 114:3352–3357

Wang H, Titschack J, Wienberg C, Korpanty C, Hebbeln D (2021) The importance of ecological accommodation space and sediment supply for cold-water coral mound formation, a case study from the Western Mediterranean Sea. Front Mar Sci 8:760909

Wefing A-M, Arps J, Blaser P, Wienberg C, Hebbeln D, Frank N (2017) High precision U-series dating of scleractinian cold-water corals using an automated chromatographic U and Th extraction. Chem Geol 475:140–148. https://doi.org/10.1016/j.chemgeo.2017.10.036

Weijer W, van Sebille E (2014) Impact of agulhas leakage on the atlantic overturning circulation in the CCSM4. J Clim 27:101–110

Weinnig AM, Gómez CE, Hallaj A, Cordes EE (2020) Cold-water coral (*Lophelia pertusa*) response to multiple stressors: High temperature affects recovery from short-term pollution exposure. Sci Rep 10: 1768. https://doi.org/10.1038/s41598-41020-58556-41599

Westphal H, Beuck L, Braun S, Freiwald A, Hanebuth T, Hetzinger S, Klicpera A, Kudrass H, Lantzsch H, Lundälv T, Vicens GM, Preto N, von Reumont J, Schilling S, Taviani M, Wienberg C (2012) Phaeton - Paleoceanographic and paleo-climatic record on the Mauritanian Shelf – Cruise No. MSM16/3 - Oct 13 – Nov 20, 2010 - Bremerhaven (Germany) – Mindelo (Cap Verde) MARIA S. MERIAN-Berichte, MSM16/3, DFG-Senatskommission für Ozeanographie, p 57. https://doi.org/10.2312/cr_msm16_3

White M, Dorschel B (2010) The importance of the permanent thermocline to the cold water coral carbonate mound distribution in the NE Atlantic. Earth Planet Sci Lett 296:395–402

White M, Bashmachnikov I, Aristegui J, Martins A (2007) Physical processes and seamount productivity. In: Pitcher TJ, Morato T, Hart PJB, Clark MR, Haggan N, Santos RS (eds) Seamounts: ecology, fisheries & conservation. Blackwell Publishing, Oxford, pp 65–84

Wienberg C, Titschack J (2017) Framework-forming scleractinian cold-water corals through space and time: a late Quaternary North Atlantic perspective. In: Rossi S, Bramanti L, Gori A, Orejas C (eds) Marine animal forests: the ecology of benthic biodiversity hotspots. Springer, Cham, pp 699–732

Wienberg C, Hebbeln D, Fink HG, Mienis F, Dorschel B, Vertino A, Lopez Correa M, Freiwald A (2009) Scleractinian cold-water corals in the Gulf of Cádiz - first clues about their spatial and temporal distribution. Deep-Sea Res I 56:1873–1893

Wienberg C, Frank N, Mertens KN, Stuut J-B, Marchant M, Fietzke J, Mienis F, Hebbeln D (2010) Glacial cold-water coral growth in the Gulf of Cádiz: implications of increased palaeo-productivity. Earth Planet Sci Lett 298:405–416

Wienberg C, Titschack J, Freiwald A, Frank N, Lundälv T, Taviani M, Beuck L, Schröder-Ritzrau A, Krengel T, Hebbeln D (2018) The giant Mauritanian cold-water coral mound province: oxygen control on coral mound formation. Quat Sci Rev 185:135–152. https://doi.org/10.1016/j.quascirev.2018.02.012

Wienberg C, Titschack J, Frank N, De Pol-Holz R, Fietzke J, Eisele M, Kremer A, Hebbeln D (2020) Deglacial upslope shift of NE Atlantic intermediate waters controlled slope erosion and cold-water coral mound formation (Porcupine Seabight, Irish margin). Quat Sci Rev 237:106310. https://doi.org/10.1016/j.quascirev.102020.106310

Wienberg C, Krengel T, Frank N, Wang H, Van Rooij D, Hebbeln D (2022) Cold-water coral mounds in the western Mediterranean Sea: new insights into their initiation and development since the Mid-Pleistocene in response to changes of African hydroclimate. Quat Sci Rev 293:107723. https://doi.org/10.1016/j.quascirev.2022.107723

Worm B, Sandow M, Oschlies A, Lotze HK, Myers RA (2005) Global patterns of predator diversity in the Open Oceans. Science 309: 1365–1369

Wynn RB, Masson DG, Stow DA, Weaver PP (2000) The Northwest African slope apron: a modern analogue for deep-water systems with complex seafloor topography. Mar Pet Geol 17:253–265

Zenk W, Klein B, Schroder M (1991) Cape verde frontal zone. Deep-Sea Res Part A 38(Suppl 1):S505–S530

Zibrowius H (1980) Les Sclératiniaires de la Méditerranée et de l'Atlantique nord-oriental. Mémoires de l'Institut Océanographique, Monaco 11:284

Zibrowius H, Gili JM (1990) Deep-water Scleractinia (Cnidaria: Anthozoa) from Namibia, South Africa, and Walvis Ridge, southeast Atlantic. Sci Mar 54:19–46

Zweng MM, Reagan JR, Antonov JI, Locarnini RA, Mishonov AV, Boyer TP, Garcia HE, Baranova OK, Johnson DR, Seidov D, Biddle MM, Levitus S (2013) World Ocean Atlas 2013. Volume 2: Salinity. In: Levitus S, Mishonov A (eds) NOAA Atlas NESDIS 74. National Oceanographic Data Center, Ocean Climate Laboratory United States, National Environmental Satellite Data Information Service, p 39

New Zealand: Southwest Pacific Region

9

Dianne M. Tracey, Malcolm R. Clark, and Ashley A. Rowden

Abstract

The New Zealand region, in the South Pacific, supports a diverse and abundant deep-sea coral fauna. In particular, framework-forming scleractinians that can form reefs can support diverse benthic assemblages. In this chapter, we first set the broader environmental scene by describing the geological and bathymetric setting of the region and its complex oceanography at both the surface and in the deep sea. We then describe current knowledge of the biology and ecology of the reef-forming species, before discussing the various threats to them from the exploitation of natural resources for commercial use, as well as a changing regime of temperature and ocean acidification due to climate change. Key knowledge gaps for the region include coral distribution data in areas without commercial fishing, information on coral age and growth, and an evaluation of the efficacy of existing protections for cold-water corals.

Keywords

Deep sea · Marine geology · Seabed mapping · Scleractinian corals · Species distribution · Predictive habitat modelling · Coral age and growth · Reproduction · Genetic connectivity · Fishing impacts

9.1 Introduction

The New Zealand region (Fig. 9.1) supports a diverse and abundant deep-sea coral fauna (Cairns 1995, 2007; Gordon 2009), where corals are important components of deep-sea ecosystems, providing habitat and refuge for a variety of invertebrate and fish species. In particular, the calcifying, framework-forming scleractinians (branching stony corals) that can form reefs (referred to as reef-corals here, and the subject group of this chapter) are a widespread and abundant group in New Zealand waters (Tracey et al. 2011) that can support diverse benthic assemblages (Clark and Rowden 2009).

In this chapter, we first set the broader environmental scene by describing the geomorphology and bathymetry of the region and its complex oceanography at both the surface and in the deep sea—thus focusing on the key elements that define the environmental characteristics of deep-sea coral-reef habitat. We then describe current knowledge of the ecology of the reef-forming species, before discussing the various threats to them from the exploitation of natural resources for commercial use, as well as a changing regime of temperature and ocean acidification (OA) due to climate change. Key knowledge gaps are then highlighted.

9.2 Geology and Bathymetric Setting

The geological history of New Zealand and its tectonic activity have resulted in a complex topography composed of a number of distinct bathymetric features including seamounts, rises, volcanic ridges, plateaus, submarine canyons, deep troughs, and abyssal plains (Fig. 9.1). Many of these features have rocky hard substrata which can support reef-coral species, especially seamount features which are widespread throughout the region as a result of subduction processes along the tectonic plate boundary, or intraplate volcanism which was common in the Cenozoic and has ages of 10–40 million years (Rowden et al. 2005; Timm et al. 2010).

The age of the reef-corals, both live and dead components of the reef matrix, is described in Sect. 9.5.2. However, linked with the active geological history of the New Zealand region are occurrences on land of fossil coral thickets in the southern part of the North Island (see Squires 1964). Records include a thicket of *Goniocorella dumosa* in mudstones, with foraminifera data indicating an age of Late

D. M. Tracey (✉) · M. R. Clark · A. A. Rowden
National Institute of Water and Atmospheric Research (NIWA), Wellington, New Zealand
e-mail: Di.Tracey@niwa.co.nz

E. Cordes, F. Mienis (eds.), *Cold-Water Coral Reefs of the World*, Coral Reefs of the World 19,
https://doi.org/10.1007/978-3-031-40897-7_9

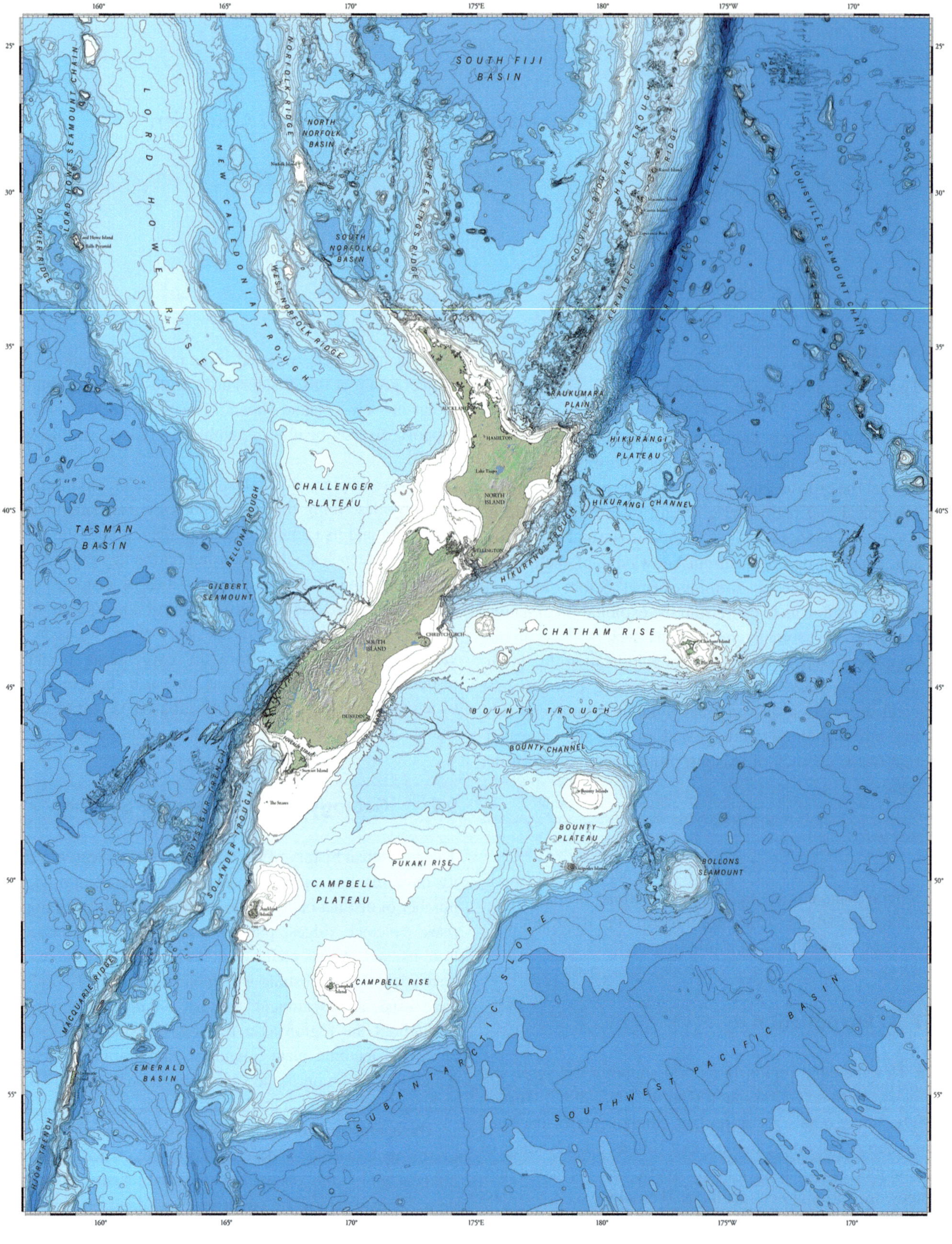

Fig. 9.1 Bathymetry map of the New Zealand region (Mitchell et al. 2012)

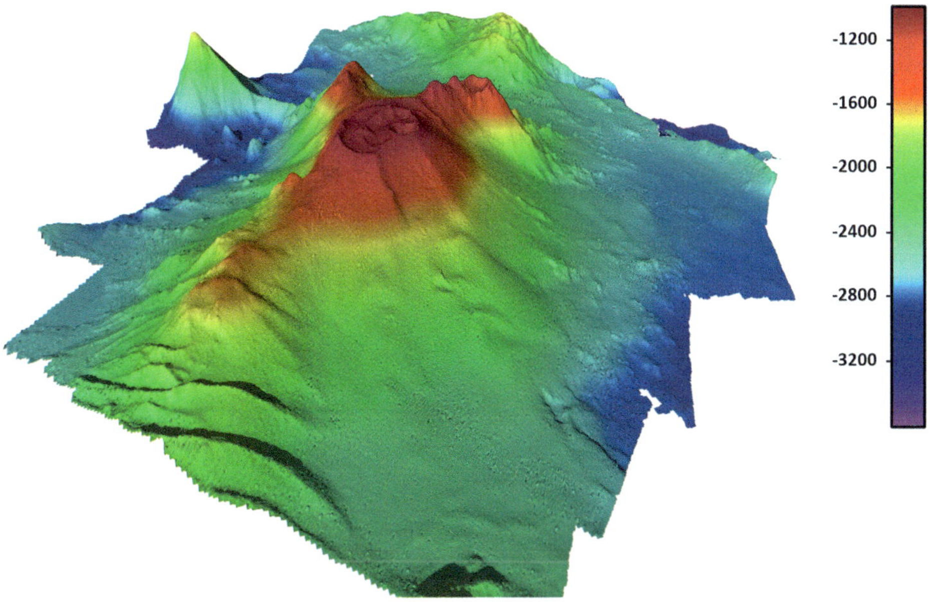

Fig. 9.2 Digital terrain model derived from multi-beam echosounder bathymetric data showing the 'Sonne' Seamount feature, located in the Kermadec (Rangitahua) region, northern New Zealand

Miocene and deposition in around 300 m of water (Wells 1986).

Multi-beam echosounder (MBES) data are synthesised into a bathymetric map of the ocean floor, which reveals the geomorphology at regional scales. The use of these instruments since the start of the millenium has significantly increased our knowledge of the distribution of seafloor features around New Zealand (Rowden et al. 2005). MBES data have been used to produce detailed maps of deep-sea coral-reef habitats on topographical features such as hills, knolls, seamounts, guyots, calderas, and ridges (Fig. 9.2).

Most reef-corals in the region are found associated with seamount summits and flanks and thrive between depths of 400 and 1500 m (e.g., Clark and Rowden 2009; Tracey et al. 2011), and results of habitat suitability modelling studies show that factors of seamount feature and seafloor slope are key drivers of reef-coral distributions (Anderson et al. 2014). However, reef-corals can often have a patchy distribution on a seamount feature and highly detailed and accurate bathymetric and backscatter data are required to produce more resolved estimates of their distribution (Anderson et al. 2016a). The use of such data in habitat modelling studies indicates that relatively high abundances of reef-coral are generally predicted to be restricted to elevated ridge- and cone-like features on seamounts (Rowden et al. 2017).

9.3 Oceanography

The New Zealand landmass sits at the crossroads between warm, nutrient-poor, subtropical surface waters flowing in from the north, and cool, nutrient-rich, subantarctic surface waters flowing in from the south. The complex bathymetry exerts controls over many of the oceanographic currents (Chiswell et al. 2015). Figure 9.3 shows the present understanding of the water masses and key ocean currents in the region as summarised by Chiswell et al. (2015) and Stevens et al. (2021). The region is fed with Subtropical Water from the north by the East Australia Current and the Tasman Front, which becomes the East Auckland Current (EAUC). This current merges with highly modified Subtropical Water (STW) from the south in flows associated with the Subtropical Front that forms the Southland Current, which is the local name for the Subtropical Front (SAF) as it runs up the New Zealand South Island east coast. The front separates modified STW inshore and Subantarctic Water (SAW) offshore (Sutton 2003). While the current is formed by Pacific Ocean regional processes, fluctuations are predominately wind-driven (Chiswell 1996). There is also a much smaller contribution directly from the Tasman Sea to the west, although this is relatively weak in terms of energetics (Chiswell et al. 2015). How these currents affect the shelf region varies in both space and time and depends on a confluence of drivers, one being tides (Stanton et al. 2001).

The variability in regional currents influences the distribution of deep-sea corals through their influence on dispersal of coral larvae, food delivery, and sediment redistribution (Miller and Gunasekera 2017).Deep-sea reef-corals are also affected by surface currents and eddies that influence and enhance primary productivity, which is the ultimate source of food at the seafloor (Stevens et al. 2021). For example, surface currents form the Subtropical Front (STF), the position of which is influenced by the Chatham Rise, a plateau between Te Waipounamu (the South Island) and the Chatham

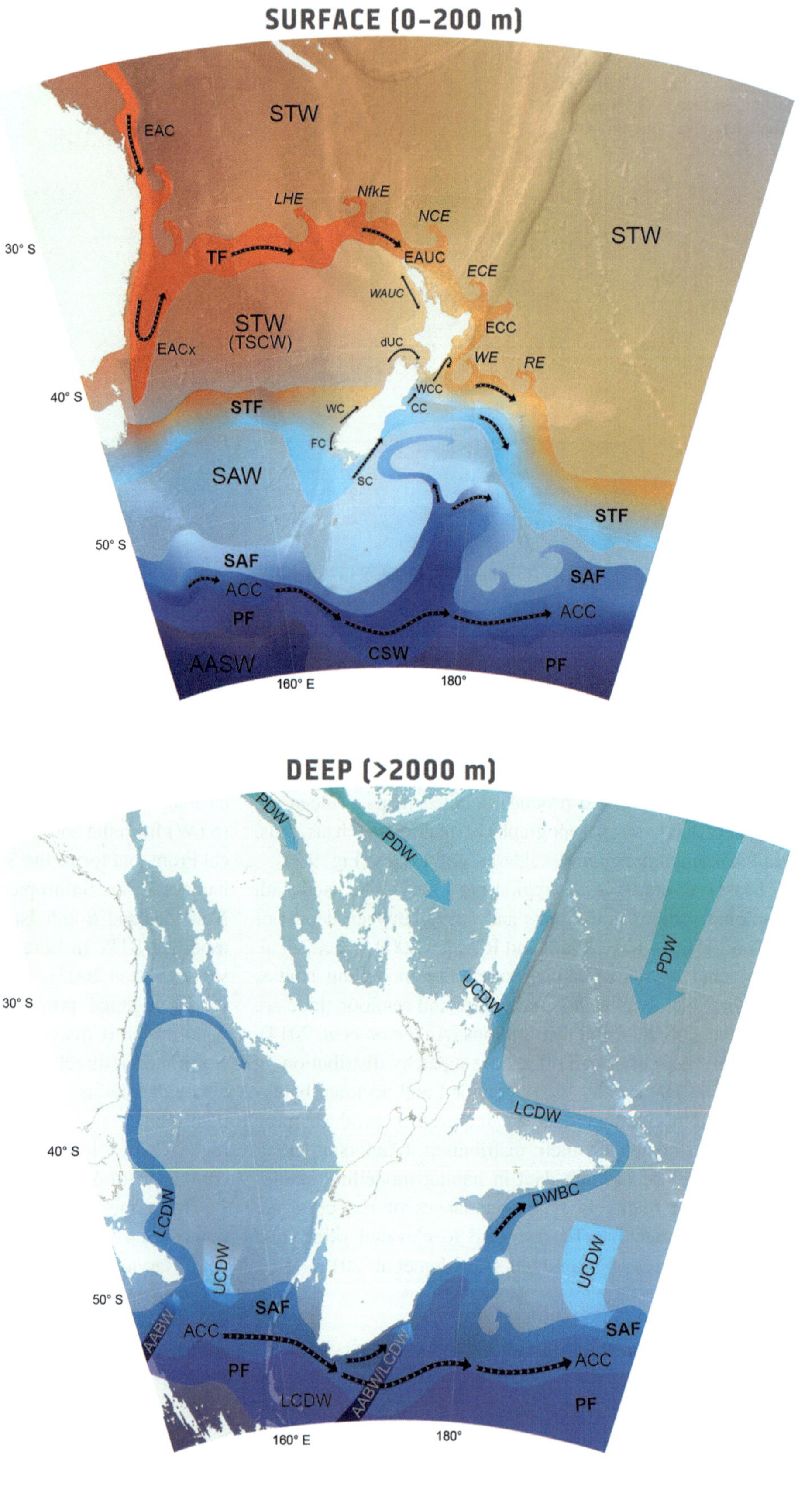

Fig. 9.3 Description of the oceanography around New Zealand (from Chiswell et al. 2015). Key ocean currents in the top image are the Subtropical Surface Waters (STW), Subantarctic Waters (SAW), Subantarctic Front (SAF), East Auckland Current (EAUC), and the Subtropical Front (STF). In the bottom image Pacific Deep Water (PDW) are old waters with low oxygen and high nutrients and they flow south from the North Pacific. Circumpolar Deep Water (CDW) flows in directly from the Southern Ocean in the Deep Western Boundary Current, the main flow of water into the South Pacific. SAF is the northern edge of the Antarctic Circumpolar Current (ACC). Polar Front (PF) is another ACC front

Islands (also called Rekohu/Wharekauri) (Fig. 9.1). The front is the meeting point of Subtropical surface waters and Subantarctic waters sourced from the Southern Ocean, which results in a region of high nutrients and therefore high primary productivity. The reef-coral *Goniocorella dumosa* is abundant on the Chatham Rise, where it is associated with the position of the STF (Chiswell et al. 2015).

The environmental characteristics of the water masses at depth, including temperature, salinity, dissolved oxygen, nitrate and phosphate, and carbonate content, along with variations in these variables, specifically the aragonite saturation state, all influence the distribution of corals (see section below).

9.4 Reef-Corals

9.4.1 Description

There are six key species of scleractinian reef-corals in New Zealand waters (Cairns 1995; Tracey et al. 2011): *Solenosmilia variabilis* (Family: Caryophylliidae); *Madrepora oculata* (Family: Oculinidae); *Enallopsammia rostrata* and *E. pusilla* (*E. marenzelleri*) (Family: Dendrophylliidae), which are also found globally (Roberts et al. 2009); *Goniocorella dumosa* (Family: Caryophylliidae) which is found only in the Indo-Pacific and New Zealand regions; and *Oculina virgosa* (Family: Oculinidae) which is located in New Zealand and New Caledonian waters (Kitahara and Cairns in press).

The living and dead frameworks of most of these species form reef-like structures (Fig. 9.4) that provide habitat for other fauna (see Sect. 9.7). In the literature, the structures are variously described as 3D matrices, coppices, bioherms, and thickets.

The continuous and semi-continuous patches of reef-coral habitat are often topographically constrained, and their size can vary considerably from square metres to square kilometres. The species *S. variabilis* and *M. oculata,* for example, typically form reef-like 3D matrix structures, on the Graveyard Seamount complex on Chatham Rise that can range from 100 m^2 to 0.05 km^2 (derived from Clark et al. 2021) and be up to 2–3 m high on unfished seamounts (Fig. 9.4). On the large guyots of the Louisville Seamount Chain, reefs of *S. variabilis* range from 600 m^2 to 0.04 km^2 (Rowden et al. 2017). These reefs on the Louisville and Graveyard seamounts are smaller than those reported on seamounts off Tasmania, Australia (*see* location in Box insert Fig. 9.15), (0.02–1.16 km^2, Williams et al. 2020a, b).

In contrast to the reef structures formed by *S. variabilis,* the two *Enallopsammia* species and *G. dumosa* form thickets or dense clumps of individual colonies less than 0.5 m high and these can be located on seamounts but also intermittently scattered over distances of 50 km on the continental slope (Rowden et al. 2014a, b). On the crest of the Chatham Rise concentrations of glauconite and phosphate nodules (Chatham Rock Phosphate Ltd 2014) provide suitable hard substrate for *G. dumosa* (Fig. 9.5), in a typically muddy-sand sediment environment (Rowden et al. 2014a, b).

Oculina virgosa is a more sparsely branched species which can also form thicket-like structures that are sometimes continuous when along a ridge. The species has a more northern distribution than the other species and can be found quite shallow, often observed by divers on cliff overhangs (Tracey and Hjorvarsdottir 2019).

As well as reefs and/or thickets made up of a mixture of live and dead coral framework, extensive areas of dead matrix can occasionally be found at the periphery of some reefs, such as on some seamounts of the Louisville Seamount Chain where they occur at greater depths than live coral (Clark et al. 2015). The dead matrix occurs on both fished and unfished seamounts and their present distribution is possibly due to a major oceanographic event in the past, changes in sea-level, or a gradual deepening of the seamounts with age through tectonic movement (Clark et al. 2015). The dead matrix can also form habitat for other invertebrates (see below) and also off Tasmania (Thresher et al. 2014). The reef-corals provide habitat for other fauna not only at the spatial scales described above, but also at the scale of their individual colony structures.

While the focus of this chapter is on the reef-corals, other coral groups found in abundance throughout the region also provide habitat e.g., large tree-like gorgonian octocoral taxa *Paragorgia* and *Primnoa* (Sánchez 2005; Cairns 2016; Thresher et al. 2014), and bushy black corals such as *Dendrobathypathes* and *Telopathes* (Opresko et al. 2014).

9.4.2 Distribution

The distribution of reef-corals collected in the New Zealand region is shown in Fig. 9.6.

Records indicate that most of the reef-coral species are widely distributed through the region in temperate water masses. They occur primarily at depths of 200 to over 2000 m, with elevated abundance at 800 to 1000 m (Tracey et al. 2011). The exception is *Oculina virgosa,* which is found primarily in warmer northern latitudes, and unlike the other species it extends into a shallow water environment (Tracey et al. 2011).

Areas with high incidences of seamounts and ridges and where reef species have been frequently observed include the Kermadec, Colville, Three Kings, Norfolk, and Macquarie ridges, the East Coast North Island, and seamount features in the Chatham Rise region such as the Graveyard and Andes complexes (Tracey et al. 2011).

Fig. 9.4 Key habitat-forming coral-reef species found in the New Zealand region: *Solenosmilia variabilis* (**a**) (visible are abundant live coral heads and dead coral matrix), *Enallopsammia rostrata* (**b**), *Madrepora oculata* (**c**) *Goniocorella dumosa* (**d**), and *Oculina virgosa* (**e**). Live and dead reef is clearly visible in images a-c. The more clump-like thicket structures are shown for *G. dumosa* and *O. virgosa*. (Images NIWA)

9.4.3 Predicting Reef-Coral Habitats

Physical sampling efforts in the deep sea around New Zealand have been extensive, but there are still many gaps in our knowledge about the distribution of benthic species. A range of modelling methods to predict habitat suitability and species distributions for corals, including scleractinians, in unsampled or poorly sampled areas have been developed and applied to the New Zealand region and the wider Southwest Pacific Ocean (Tracey et al. 2011; Baird

Fig. 9.5 In situ image of *Goniocorella dumosa* attached to phosphate nodules (Left), and on a hard substrate, seamount flank, Chatham Rise, eastern New Zealand waters (Right). (Images NIWA)

et al. 2013; Anderson et al. 2014, 2015, 2016a, b; Rowden et al. 2015, 2017; Stephenson et al. 2018; Georgian et al. 2019). Some of these studies have utilised predictions of future oceanographic conditions from Earth System Models to identify deep coral habitats at risk from commercial fishing activities and potential future climactic refugia (Anderson et al. 2015). Other studies have focussed on specific sites and individual seamounts, using detailed drop-camera surveys to provide coral presence data and local oceano graphic information along with terrain variables derived from MBES surveys (Rowden et al. 2017). Results from two different model approaches are shown in Fig. 9.7.

Results from the various regional-scale modelling studies generally predict faunal distributions across similar seabed areas. Anderson et al. (2016a) used two habitat suitability modelling techniques, boosted regression trees, and maximum entropy, to describe the distribution for Vulnerable Marine Ecosystem indicator taxa in the New Zealand area and adjacent seas, including four of the reef-corals discussed in this chapter. For *G. dumosa* the highest habitat suitability indices were over the Chatham Rise, parts of the Challenger and Campbell plateaus, some regions north of New Zealand, and much of the Louisville Seamount Chain. In contrast, the pattern for the three deeper living species *S. variabilis, E. rostrata,* and *M. oculata* showed areas of greatest suitability situated primarily in deeper parts of the Campbell Plateau as well as the fringes of the Chatham Rise and other features in the west and north of the region, at latitudes of 30 to 35 degrees S. A similar pattern was shown for *S. variabilis* in the Georgian et al. (2019) study (Fig. 9.7).

These modelling studies indicate that the distributions of reef-coral habitats are dependent primarily on the predicator variables dynamic topography (a proxy for large-scale current patterns) and temperature (Fig. 9.8), with seabed slope, seamount, ocean chemistry (specifically aragonite), and dissolved organic matter representing further important variables (e.g., Anderson et al. 2015, 2016a).

9.5 Coral Biology

9.5.1 Physiology

Deep sea corals form calcium-carbonate skeletons consisting of the minerals aragonite, calcite, and high magnesium calcite. As with other deep-sea reef-corals, the scleractinians in New Zealand waters are all made of aragonite (Guinotte et al. 2006; Thresher et al. 2011a, b; Bostock et al. 2015).

In the New Zealand region 85% of the reef-corals (*S. variabilis, E. rostrata, G. dumosa, M. oculata,* and *O. virgosa*) are found above the Aragonite Saturation Horizon (ASH), which sits between depths of 1050 and 1250 m (shallower south of Chatham Rise and deeper north of Chatham Rise; Bostock et al. 2015). However, part of the distribution of *S. variabilis* occurs below the ASH. This species appears to be more resilient to undersaturation of aragonite than other reef-coral species (Bostock et al. 2015). Nevertheless, because aragonite is relatively more soluble than some of the other minerals in coral skeletons, these New Zealand species can be susceptible to dissolution with climate change. Because of their physiology, coral mortality will also occur if conditions such as oxygen and temperature are sub-optimal.

9.5.2 Age

From radiocarbon dating (^{14}C), colony ages of various reef-coral species have been obtained for the New Zealand region (Fig. 9.9). The age of *S. variabilis* species has been assessed

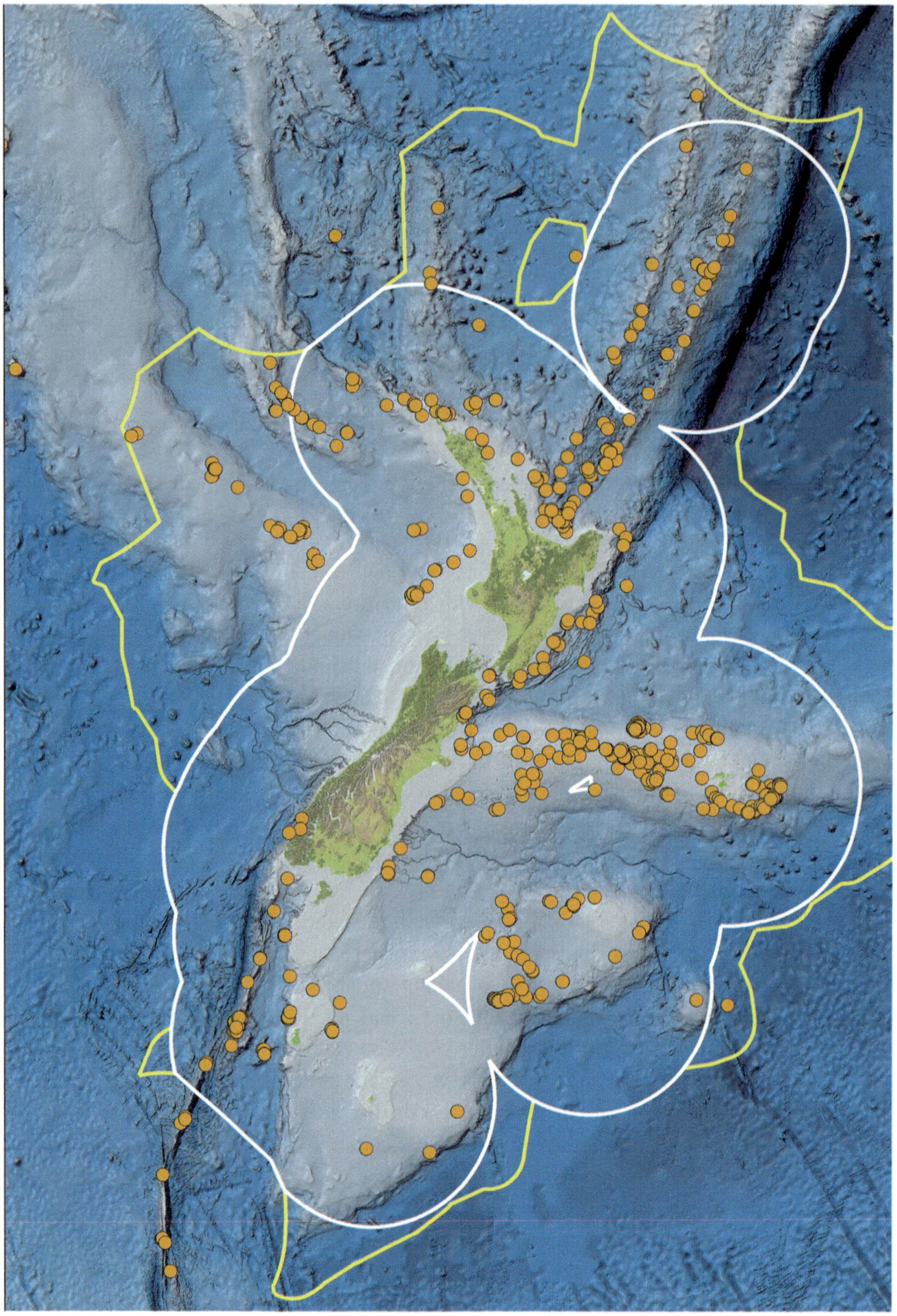

Fig. 9.6 Distribution of reef-coral records in the New Zealand region. The white line shows the Exclusive Economic Zone (EEZ) and the yellow line the Extended Continental Shelf (ECS)

for 'long dead' (old and discoloured manganese coated matrix), 'recently dead' (no polyps but white-grey skeleton), and live branch sections of coral sampled from seamounts on the Chatham Rise and Louisville Seamount Chain. Overall, linear growth rates for *S. variabilis* ranged from 0.25 to 1.3 mm year^{-1}. Using a conservative estimate of matrix height of ~20 cm, it could take hundreds of years (~150–660 years) for a colony to attain this height, or ~2000 years, to build a colony diameter of 1 m (NIWA unpub. data; Tracey et al. 2019). Similar growth rates were reported for *S. variabilis* on seamounts off Tasmania, Australia, ranging from 0.84 to 1.25 mm year^{-1} linear extension (Fallon et al. 2014).

Other New Zealand reef-coral species show similar growth rates to *S. variabilis* but colony ages differ. For the Chatham Rise and Bay of Plenty areas, *M. oculata* colony ages determined by ^{14}C ranged from 752 to 1123 years, with overall growth rates of 0.39–0.8 mm year^{-1}. For *G. dumosa* from the Chatham Rise, ^{14}C colony ages were

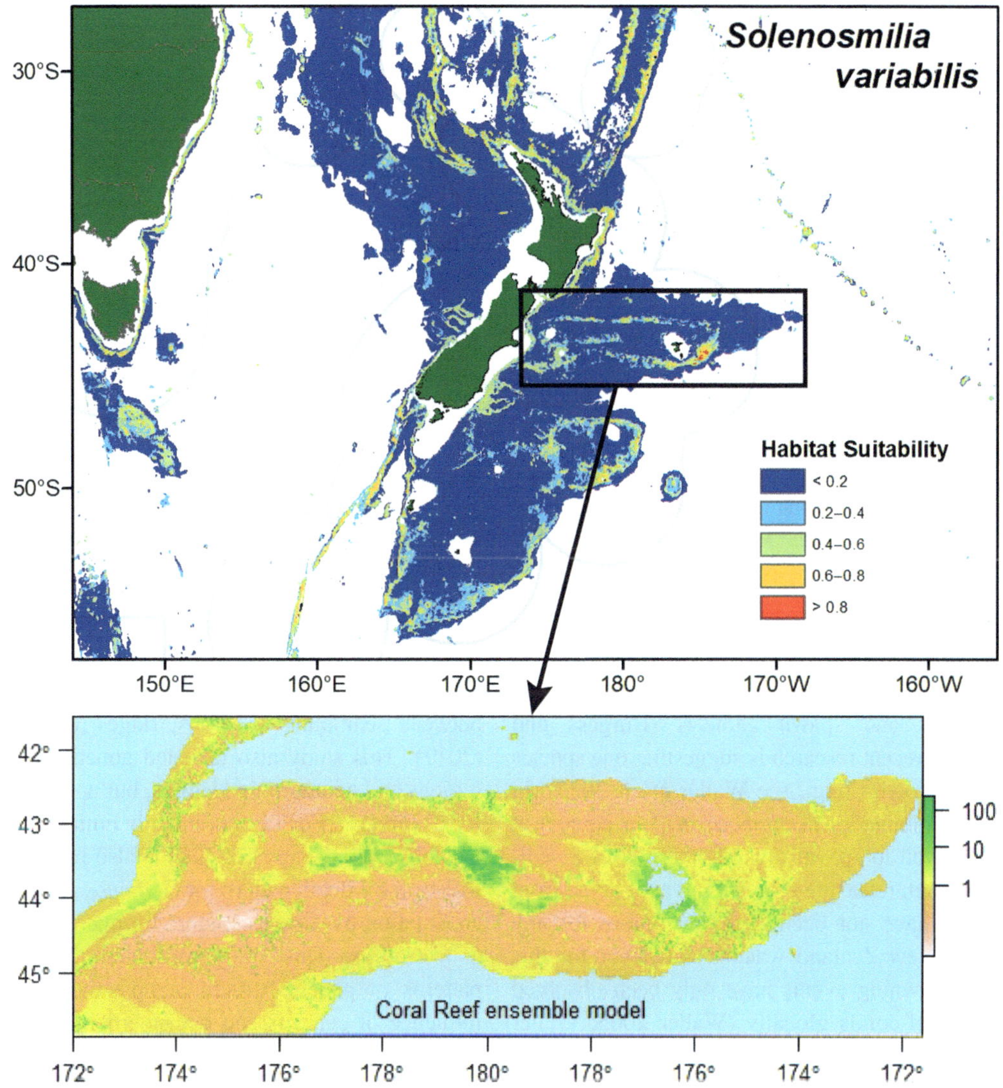

Fig. 9.7 Habitat suitability models: Top: Ensemble model habitat suitability map for the stony coral *Solenosmilia variabilis* in the southwest Pacific Ocean (modified from Georgian et al. 2019). Bottom: Ensemble model predictions of abundance (individuals 1000 m $^{-2}$) on the Chatham Rise for four coral-reef species as a group: *Enallopsammia rostrata*, *Madrepora oculata*, *Solenosmilia variabilis*, *and Goniocorella dumosa* (from Bowden et al. 2019)

372–422 years, and linear growth was estimated to be 0.48 mm year^{-1} (Tracey et al. 2019).

As noted in Sect. 9.4.1, the flanks of several Louisville Ridge seamounts have extensive areas of dead coral matrix (Clark et al. 2015). The age of the apparent coral die-off events that have occurred was estimated at over 10,000 radiocarbon (^{14}C) years, which coincides with a time of significant oceanographic change in the New Zealand region (Bostock et al. 2013). The oldest reef matrix age in Tasmanian waters, located in a different water mass (*see* location in Box insert Fig. 9.15), was radiocarbon dated to be a lot older, 47,400 radiocarbon years (Fallon et al. 2014). This age estimate, while treated with caution by the authors as the measured results were at the limit of radiocarbon dating, suggested the Tasmanian seamounts have been occupied by *S. variabilis* since before the Last Glacial Maximum.

9.5.3 Reproduction

In New Zealand waters the reproduction of *S. variabilis*, *M. oculata*, *G. dumosa*, and *E. rostrata* are thought to exhibit seasonality, with fertilisation occurring after the austral summer in April/May in synchrony with increased food availability (Burgess and Babcock 2005). No studies have been carried out on the reproduction of *O. virgosa*, and so it is uncertain whether this species exhibits a similar reproductive cycle to *O. varicosa,* which occurs off the east coast of Florida. For the latter species, the gametogenic cycle appears

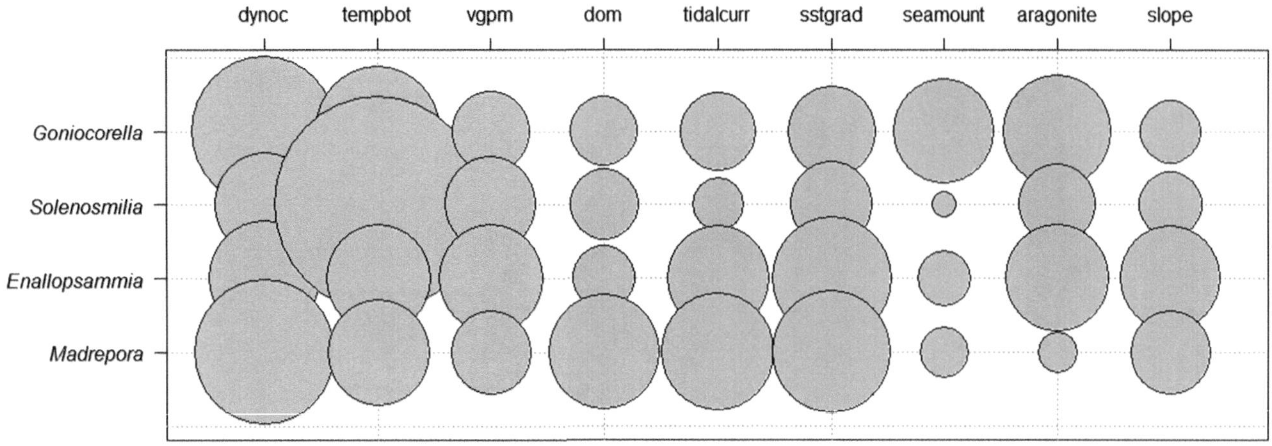

Fig. 9.8 The relative influence of explanatory variables on the predictive models for four reef-forming coral taxa (revised from Anderson et al. 2014). Variable descriptions are: Dynoc: dynamic topography; tempbot: bottom temperature; vgpm: surface water primary productivity; dom: dissolved organic matter; tidalcurr: tidal current velocity; sstgrad: sea surface temperature gradient; seamount: on or off a seamount; aragonite: seafloor aragonite saturation state; calcite: seafloor calcite saturation state; slope: seafloor slope

to begin in early boreal summer, with spawning occurring during late boreal summer/fall (Brooke and Young 2003).

Previously, New Zealand reef-corals were believed to be broadcast spawners and spawn gametes (Burgess and Babcock 2005), but recent research is suggesting one species may instead be a brooder (e.g., see Waller et al. 2023). In September 2020, colonies of *G. dumosa* held in aquaria at NIWA were observed to spawn towards the end of a sedimentation experiment, releasing brooded larvae. Larvae from a spawning event have not been observed before for any deep-sea coral in New Zealand waters, and never for this species. Indeed, spawning events have only been observed for a handful of reef-corals globally (Waller 2005; Waller et al. 2023) even for very well-studied species like *Lophelia pertusa* (Larsson et al. 2014). Some of the larvae settled after only 24 h (Fig. 9.10). Subsequent research from this serendipitous observation of *G. dumosa* describing aspects of brooding, larval behaviour, settlement medium, fecundity, length of larval phase prior to settlement, and possible settlement cues is now completed and the paper describing the science is in review).

9.5.4 Genetics

There have been a few genetic studies of the reef-coral species off New Zealand, but they indicate that connectivity patterns are not uniform across the species and in some cases have varied between studies, molecular markers applied, or geographical areas sampled. For example, low connectivity was observed for *S. variabilis* with local structure noted between seamounts off New Zealand and Tasmania (Miller and Gunasekera 2017). In contrast, genetic differentiation was observed for the same species at the province and regional scales, as well as the geomorphic feature scale (i.e., between seamounts, plateaus, ridges, etc.) by Zeng et al. (2017). This study also detected genetic structure amongst regions and geomorphic features, but not between provinces, for *G. dumosa*. For *M. oculata* only limited regional structure was observed. Zeng et al. (2017) also recorded a significant isolation-by-depth pattern for *G. dumosa* and *M. oculata*, and an isolation-by-distance pattern for microsatellite variation in *S. variabilis*. This study attributed the observed spatial patterns of genetic structure/connectivity primarily to the influence of currents and oceanographic fronts. Zeng et al. (2020) in a subsequent seascape analysis study found that dynamic topography (*G. dumosa*), sea surface temperature and surface water primary productivity (*M. oculata*), and tidal current speed (*S. variabilis*) were the most important variables explaining patterns of genetic structure observed in these reef-coral species. At the population level, factors related to current and food source explained the genetic structure in all species, whilst at the geomorphic features level, factors related to food source and topography were most important. Environmental variation in these parameters may be acting as a barrier to gene flow at different scales. The study of Zeng et al. (2020) highlights the utility of seascape genetic studies to better understand the processes shaping the genetic structure of organisms and to identify environmental factors that can be used to locate sites for the protection of deep-sea corals (see Sect. 9.8).

There are limited genetic data available for *E. rostrata*, with analysis findings being somewhat ambiguous and highlighting the need for further genetic data (Holland et al.

Fig. 9.9 NIWA staff Peter Marriott and Di Tracey examining a *Madrepora oculata* specimen selected for transverse sectioning and subsequent radiocarbon dating (Top). *M. oculata* branch showing the various sample sites where section cuts were made for sample milling along the colony to estimate linear and step-wise growth (Bottom). (NIWA images). Note: *M. oculata* displays three different morphologies in New Zealand waters. One (pictured above) has calices that alternate sympodially on each side of thick branches in a flute-like orientation (Cairns 1995)

2020). No studies on the genetic structure of *O. virgosa* populations have been carried out to date.

9.6 Reef-Corals as Habitat

Deep-sea reef-corals in general are known to provide habitat for other fauna, including a range of invertebrates and fish (Rogers 1999; Krieger and Wing 2002; Stone 2006; Mortensen et al. 2008; D'Onghia et al. 2010; Soffker et al. 2011; Rooper et al. 2019; Henderson et al. 2020; Schnabel and Mills 2019). These associations can be highly dependent (i.e., associated fauna not found outside of coral habitat) or spatially and temporally variable in their strength (e.g., used as nursery grounds for limited periods of time).

9.6.1 Invertebrates

One of the most commonly observed associations is between eunicid polychaetes and scleractinian corals. Eunicids form parchment-like tubes on the matrix of reef-corals, and the corals commonly calcify over the tubes (Mueller et al. 2013). Three species in this family have been observed to form their tubes on *S. variabilis* in the New Zealand region (Schnabel and Mills 2019). However, nothing is known about the potential enhancement of calcification and reef growth for New Zealand coral species.

Ophiuroid assemblages sampled from *S. variabilis* coral reefs off Australia and New Zealand are distinct from those of seamount habitats and from continental slope (O'Hara et al. 2008). The coral-associated ophiuroids were generally in the coral matrix rather than on the top of the reef. The species found in the matrix lack the specialised hook-like arm spines that species living on arborescent corals usually possess. Instead, the ophiuroids of the coral reef typically had long flexible arms used for filter- or deposit-feeding.

Other echinoderms such as brisingid seastars and echinoids are regularly seen on live reef-corals. Stevenson et al. (2018) examined echinoid distributions off France, Australia, and New Zealand. The New Zealand records were primarily from seamounts, and 43% of observations

Fig. 9.10 Left: Sectioned *G. dumosa* corallite showing several planulae larvae ready to be released. Image shows 4 of the 10 larvae in this single polyp. Right: Recently settled larva on a dead branch area of *G. dumosa* showing a newly formed corallite with tentacles extending from the calice area and coenenchyme forming at the base region of the pedicel. (NIWA images)

were associated with live coral reef. *Dermechinus horridus*, a common echinoid on seamounts off New Zealand and Tasmania, was found primarily distributed amongst dense live coral, potentially because of a likely suspension-feeding mode where living on elevated coral matrix is an advantage (Stevenson et al. 2018).

Galatheoid squat lobsters, including the globally distributed *Munida* species, are frequently encountered throughout the region, but some species are distinctly 'seamount associated' species (Rowden et al. 2010). For example, *M. isos* is often observed in high abundance in the living and dead-standing reef-coral matrix (Schnabel 2020) as well as in accumulations of coral rubble at the base of the reef (Fig. 9.11).

An analysis of megabenthos assemblages off Tasmania between ~700 and 4000 m on the rocky continental margin (Thresher et al. 2014) showed species richness was highest at intermediate depths (1000–1300 m), as a result of an extensive coral-reef habitat comprising both live and dead *Solenosmilia variabilis*. The community associated with this habitat was similar to that off New Zealand. They included brisingid seastars, *Dermechinus horridus*, octocorals, sponges, and crinoids on the surface of the coral, while common interstitial organisms included galatheid squat lobsters, shrimps, crabs, worms, and ophiuroids. Overall, the observed depth zonation in this region appeared to be based in part on changing species assemblages with depth, but also an underlying reef-coral biota that was common across all depths.

9.6.2 Fish

Compared to elsewhere, there has been limited research looking at the nature of associations between fish and reef-corals in New Zealand. Seamounts with high densities of reef-corals often host large populations of commercial fish species, including orange roughy (*Hoplostethus atlanticus*), alfonsino (*Beryx splendens*, and black cardinalfish (*Epigonus telescopus*)). A study of research trawl catches from the Graveyard Seamounts complex showed that the composition and abundance of fish populations can vary between seamounts on the kilometre scale (Tracey et al. 2012), in contrast to observed differences in fish communities and abundance between widely distributed seamounts at ocean-basin scales and influenced by different water masses (Clark et al. 2010c). However, the specificity of the associations among fishes and corals are uncertain, including the extent to which these corals might provide biogenic structure for shelter, feeding, spawning, or as nursery sites (Morrison et al. 2014; Clark and Dunn 2012; Jones et al. 2016). A recent study used video surveys from six seamounts on the Chatham Rise to examine the habitat associations of demersal fish including those with reef-corals. Over 4000 individual fish and their associations with a range of habitats were recorded and showed significant and consistent associations between certain fish species and the deep-sea coral-reef matrix (Simmons 2016; Tracey and Hjorvarsdottir 2019).

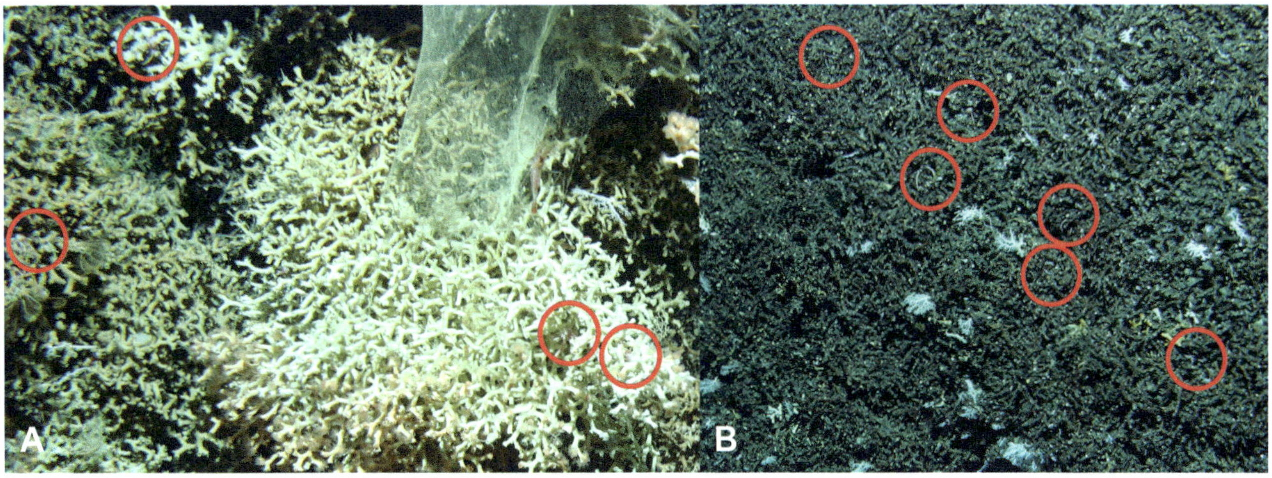

Fig. 9.11 Squat lobsters (red circle) identified in live and dead matrix (**a**) as well as on dead coral rubble at the base of the reef (**b**). (NIWA image)

9.6.3 Microbes

Microbial communities associated with deep-sea corals are beginning to be studied worldwide, and it appears they can be important contributors to the coral host's function (e.g., see Vega Thurber et al. 2008). As an indication of their functional importance, microbes may help the coral host adjust to prevailing environmental conditions (e.g., see research by Röthig et al. (2017) on the deep-sea coral *Eguchipsammia fistula*, a cup-like form of stony coral, and a species abundant in the New Zealand region). Using 16S rRNA gene amplicon sequencing, the microbiome for *G. dumosa* colonies collected from two contrasting habitats of the Chatham Rise, a sedimented and a hard-bottom environment, is currently being researched to see if the microbiome adjusts and supports the coral holobiont under changing environmental conditions. It is anticipated that this work will be able to assess how stress from different sedimentation levels might manifest itself as a change in the coral microbiome community, thereby acting as a proxy of host health.

9.7 Threats and Outlook

The reef-coral species focused on here are protected species within the New Zealand EEZ (New Zealand Wildlife Act 1953) and are also considered vulnerable marine ecosystem (VME) indicator taxa in the South Pacific Regional Fishery Management Organisation (SPRFMO) region (Parker et al. 2009). S. *variabilis, M. oculata, G. dumosa*, and *E. rostrata* are all considered 'At Risk' and 'Declining' according to the New Zealand Threat Classification (Freeman et al. 2013).

Due to their fragile skeletal composition, erect growth form and habitat preferences (e.g., seamounts) deep-sea reef-coral species are vulnerable to various physical disturbances, such as trawling and potential seabed mining (Clark et al. 2016). In the following sections, threats to reef-coral species from various sources in the region are described.

9.7.1 Fishing Impact

A number of studies have investigated the effects of bottom trawling on deep-sea macro-invertebrate assemblages (including reef-corals) around New Zealand and Australia (e.g., Koslow et al. 2001; Clark & O'Driscoll 2003; Anderson & Clark 2003; Clark and Rowden 2009; Althaus et al. 2009; Clark et al. 2010a, b, 2019a; Williams et al. 2020a, b). These studies have demonstrated appreciable impacts from trawling, with changes in the extent of the reef matrix on fished seamounts, as well as changes in community composition and diversity associated with the reef-corals and reductions in relative abundance of certain taxa. The matrix of *S. variabilis* on seamounts can have a vertical extent of several metres, and combined with its brittle skeletal structure, heavy ground gear from bottom trawling can have a severe impact after only a few passes with a trawl (Fig. 9.12). The reduction from 20 to 50% coral cover on unfished seamounts to levels of 1–5% on fished features can occur after less than 10 trawls.

Over the period of 20 years where time series of surveys have been carried out, there is little or no evidence of recovery of the reef-coral species (Althaus et al. 2009; Williams et al. 2010; Clark et al. 2019a; Goode et al. 2020) (***see* Box on seamount recovery**) although remnant patches occur on most fished seamounts which may aid localised recovery of them (Clark et al. 2021).

More detailed analyses of trawl effort in certain areas and on individual seamounts have also been carried out for the

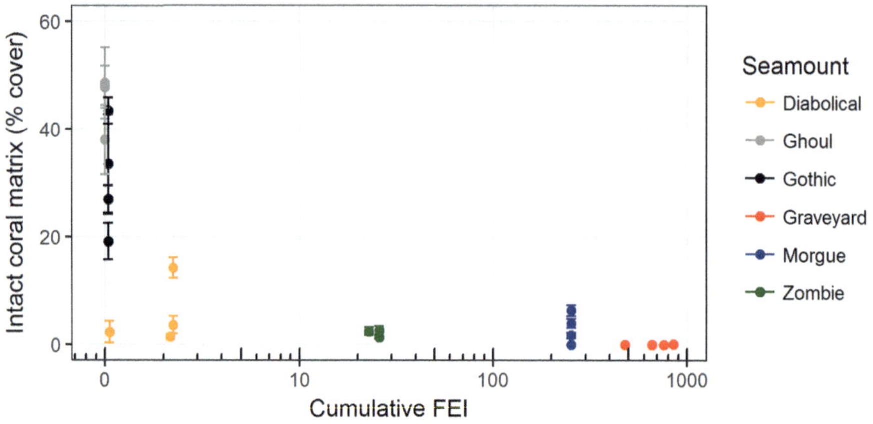

Fig. 9.12 Plot of decreasing reef-coral matrix (Y axis; mean per cent cover measured in individual seabed photographs, from six seamounts on the Chatham Rise over time), in relation to increasing fishing effort (Fishing Effects Index; FEI) (X axis) (from Clark et al. 2019a)

New Zealand EEZ, to ascertain how much fishing occurs in such habitat, and how its distribution compares with the likely or known distribution of corals. This work combined data on the trawl footprint of commercial deep-sea fisheries and records of reef-corals (Clark et al. 2014a, b). The trawl footprint shows a clear overlap between reef-like corals and targeted tows for orange roughy on various seamount features of the Chatham Rise in depths of 800 to 1000 m (Fig. 9.13).

The spatial extent of the corals or suitable habitat (such as seamount features or from habitat suitability models) in relation to the distribution of commercial trawling (trawl footprint) is regularly quantified (e.g., Anderson et al. 2014; Baird and Wood 2018) as well as analyses of bycatch in commercial fisheries (e.g., Anderson et al. 2017). Similar overlap-type analyses of predicted changes in distributions due to climate change are also evaluated (e.g., Anderson et al. 2020).

9.7.2 Sedimentation

The direct physical effects of human activities in the deep sea, such as bottom contact fisheries, are relatively well documented globally and around New Zealand (see Clark et al. 2016). Indirect effects, however, are not, and a major concern for the sustainability of reef-corals is the possible impact of sedimentation from plumes generated by trawling or potential seabed mining that can extend over a wide area. Current studies focused on the Chatham Rise are using a combination of field survey experimentation with in situ field observations, and laboratory-based experiments to determine the effects of various levels of suspended sediment

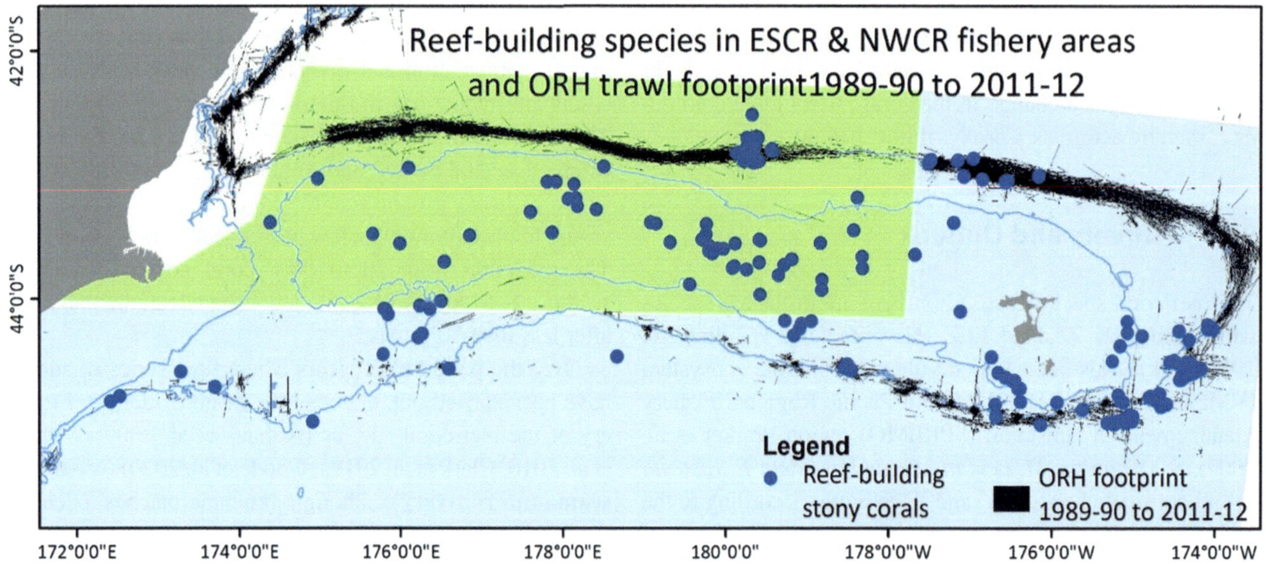

Fig. 9.13 Plot of the location of reef-corals on the Chatham Rise overlain is the trawl footprint for orange roughy (ORH) for the fishing years 1989 to 2012 (from Clark et al. 2014a, b). East South Chatham Rise (ESCR) = blue rectangle, Northwest Chatham Rise (NWCR) = Green rectangle

on benthic corals and sponges, including *G. dumosa* (Clark et al. 2018).

The laboratory programme has involved holding live *G. dumosa* in tanks and exposing them to various levels and duration of suspended sediment concentrations from 0 to 500 mg l^{-1} of particles. Data from experiments including pulses of sediment and continuous exposure for up to 4 weeks indicate that *G. dumosa* is resilient to levels of 100 mg l^{-1} or less for periods of several weeks, and also to 500 mg l^{-1} if exposure was not constant (i.e., pulses were also tested). With constant exposure effects included death of some coral polyps and tissue atrophy, and changes in respiration rates for periods of time (Mobilia 2021, NIWA unpublished data).

9.7.3 Ocean Climate

Cumulative impacts are an important consideration for the management of deep-sea corals. Along with impacts from fishing and potential seabed mining, other threats faced by these corals in the New Zealand region include those resulting from, or associated with, climate change. Numerous global studies describe how climate change could significantly shift and reduce the availability of suitable habitats for deep-sea species by 2100 (see Sweetman et al. 2017 for an overview). While there has been limited research in the New Zealand region to investigate impacts of climate change on deep-sea fauna, work by Thresher et al. (2015), Anderson et al. (2015), Bostock et al. (2015), Gammon et al. (2018), Mikaloff-Fletcher et al. (2017), and Law et al. (2018) predict a rapid shoaling of the ASH in the region that represents a significant threat to deep-sea corals in construction and maintenance of their skeletons. Deep-sea coral geochemistry archives are being used to reconstruct New Zealand's baseline ocean dynamics to help trace the changing health of the marine environment (Davis et al. 2021; Hitt et al. 2020; Hitt 2021).

9.7.3.1 Temperature

A general trend in recent decades of increasing surface water temperatures has been exacerbated by marine heatwaves, notably off the south-east coast of New Zealand's North Island in 2017 and 2018, where surface temperatures were up to 4 °C above average, and warmer water reached deeper (down to 200 m) than the mixed layer (Sutton and Bowen 2019). Ocean warming research is showing less phytoplankton production in the Tasman Sea associated with a warming environment as deep as 850 m, but conversely there is more in the mixed waters of the subtropical front east of New Zealand (Chiswell and Sutton 2020). The surface water changes will likely impact *O. virgosa*, but the deeper warming may also have effects on the other New Zealand reef-corals.

Habitat suitability modelling indicates that the distribution of reef-coral habitat is only weakly dependent on surface water primary productivity (phytoplankton) and dissolved organic matter (see Fig. 9.8). Even so, the aforementioned changes in phytoplankton production, as a result of future ocean warming, could impact strongly upon reef-coral distribution in New Zealand waters as temperature is a strong predictor variable. It is worth noting here that productivity is likely driven by several local dynamics such as phytoplankton community structure, sea surface temperature, upwelling, and gyre circulation and that there is natural variation in these variables on a variety of timescales (Hitt 2021).

9.7.3.2 Ocean Chemistry

With their delicate aragonitic skeleton, scleractinian reef-corals will be under threat from ocean acidification (Law et al. 2017). Few studies have taken place on ocean acidification impacts on deep-sea corals, and laboratory aquaria experiments, while they can be problematic in terms of cost or knowledge of food availability (Büscher et al. 2017), are often the only option. Gammon et al. (2018) investigated impacts from a change in pH from current conditions (7.88) and an end-of-century scenario (7.65), on the skeletons of *S. variabilis* from New Zealand waters. This in-aquaria research showed loss of coenenchyme tissue, but no significant changes in growth rates or respiration with the reduced pH. The loss of coenenchyme indicated a reallocation of energy with physiological processes (e.g., growth and respiration) being maintained at the expense of coenenchyme production (Gammon et al. 2018).

Water chemistry and carbonate mineralogy have a strong influence on the distribution of deep-sea reef-corals (Feely et al. 2012), and like elsewhere, New Zealand corals are at risk from future changes in these parameters (Law et al. 2018). The depth of ASH has already shallowed since pre-industrial times by 50–100 m over much of the New Zealand region (Mikaloff-Fletcher et al. 2017). This shallowing has demonstrable implications for the habitat range of marine organisms with aragonite shells or exoskeletons and will contribute to a decline in suitable habitat for reef-corals at depths where aragonite levels in the water column will decrease, and thereby impact the deep-sea ecosystems and biodiversity they support (Anderson et al. 2015; Bostock et al. 2015). The shallower regions of the Chatham Rise may provide refugia for reef-corals from the impact of changes in ocean acidification (Bostock et al. 2015).

9.8 Future Directions

While research on reef-corals in the New Zealand region has grown substantially in the last 10–15 years, there still remain significant limitations in knowledge. A recent gap analysis by Hjorvarsdottir and Tracey (2017) and research needs identified by Anderson et al. (2018) covered six gap categories: Biological, Environmental, Spatial, Modelling, Threat and Pressure, and Data, Management and Communication. These gap analyses formed the basis of a Draft Medium-Term Research Plan for Protected Coral Species which will guide future research efforts for New Zealand waters, when finalised (DOC 2023). Below we highlight some issues that relate to some of the threats posed to reef-corals.

Deep-sea reef-coral species have protected status in the New Zealand EEZ but are nonetheless subjected to impacts from human activities, in particular bottom trawling (see Sects. 9.7.1 and 9.7.2). Thus, the status of being a protected species does not by itself guarantee that these species, and the functions they perform, are conserved. The key for effective conservation is to ensure that species, communities, and populations are maintained above a level where adverse effects (from human impacts but also natural factors) do not risk the sustainability of these biological entities (Freeman and Cryer 2019). Inside the New Zealand EEZ there are Benthic Protection Areas (BPAs) (Helson et al. 2010) and Seamount Closure Areas (SCAs) (Brodie and Clark 2003) where bottom trawling and dredging are prohibited, which will protect reef-corals in those areas. However, it is uncertain if these closed areas together constitute an adequate connected network of suitably representative areas that will ensure that reef-coral populations are sustained across the region. Studies to date suggest that these BPAs and SCAs are potentially inadequate for maintaining the genetic structure and diversity of *S. variabilis*, *M. oculata*, and *G. dumosa* populations (Zeng et al. 2017, 2020), but clearly further work is required to establish the efficacy of the current protection measures (and see Clark et al. 2019b). For example, while monitoring of reef-coral populations has taken place for some seamounts in the Graveyard Seamount Complex, there is currently no widespread deep-sea monitoring programme to track the stability or otherwise of coral populations.

While there are many records for reef-corals in the New Zealand region (Fig. 9.6), most of these have come from fishery by-catch surveys or research surveys that focused on particular areas/habitats, making the records spatially biased. This results in distributional gaps in sampling, particularly for depths greater than 2000 m, where the information on coral species composition is sparse or lacking entirely. Additional surveys (both direct physical sampling and less-destructive seafloor imagery) would provide a better understanding of the distribution of reef-corals and their population structure, as well as improve knowledge of population connectivity (through more genetic studies). Data from such additional sampling will also help test and improve the predictive habitat suitability modelling that has already been carried out for reef-corals in the New Zealand region. These data would also help to reveal the response of corals to changes in carbonate saturation horizon depth. Furthermore, future surveys should include quantitative density and biomass data to more accurately reflect the relative importance of coral species in different areas. Achieving this will require a shift towards sampling with submersibles, remotely-operated vehicles (ROVs), and autonomous underwater vehicles (AUVs) to support these quantitative analyses.

A preliminary ecological risk assessment of bottom trawl fishing impacts on deep-sea corals on the Chatham Rise (Clark et al. 2014) concluded that, although vulnerability criteria could be assessed reasonably well, data on coral productivity was limited. This highlighted the need for research to improve knowledge of age and growth characteristics of many deep-sea coral species. Similarly, there is little information on reproductive biology of these corals. Although obtaining this information is often challenging for deep-sea corals, research in this area is increasing (e.g., Feehan and Waller 2015; Tracey et al. 2021; Waller et al. 2023) to determine spawning patterns, and hence how coral communities may grow or recolonise impacted areas.

The above knowledge gap issues are just some of those detailed in the draft research plan published by the New Zealand Department of Conservation. The extent of the research needed ensures the sustainability of deep-sea reef-coral groups is challenging. There is an urgent need to better understand the ecology of deep-sea corals in the region and evaluate the nature and extent of natural and human impacts. It will require more sampling to address distributional gaps, improve data on abundance (both on coral-reef density and biomass), data on coral productivity, and long-term monitoring programmes.

Recovery Box/Spotlight

A number of the deep-sea reef-corals that are found around New Zealand are broadly distributed in the South Pacific Ocean (Cairns 1995; Tracey et al. 2011; Anderson et al. 2016a, b). Off the Australian island of Tasmania, in particular, seamount and knoll features have very similar coral assemblages to those described in this chapter for New Zealand, where benthic communities are dominated by the reef-corals *Solenosmilia variabilis* and *Madrepora oculata* (Thresher et al. 2014; Williams et al. 2010; Fig. 9.14).

(continued)

Fig. 9.14 *Solenosmilia variabilis* reef-coral on the summit of seamounts off southern Tasmania (Images courtesy of Alan Williams, CSIRO, Hobart)

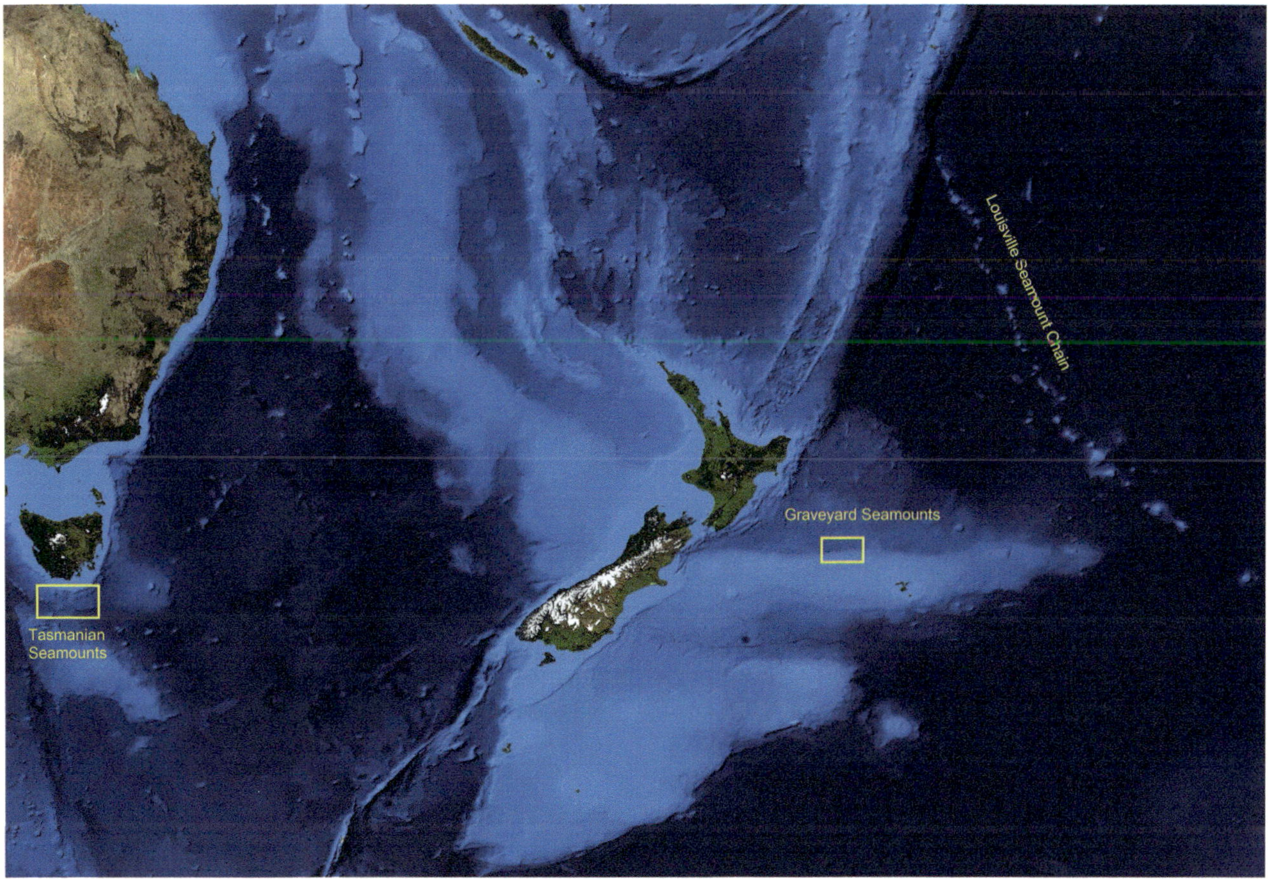

Fig. 9.15 The general location of the Tasmanian seamounts off southern Australia, and of the Graveyard Seamounts off New Zealand

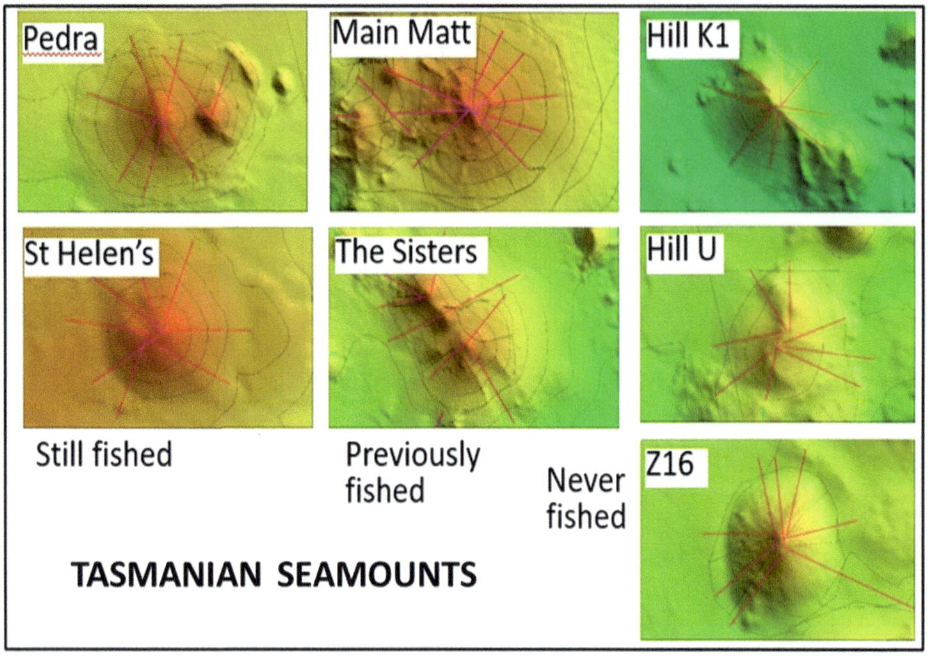

Fig. 9.16 Core set of seamounts surveyed over time off southern Tasmania. Depths range from 600 m (red) through 1200 m (yellow) to 1700 m (green). The DTM is 3× vertical exaggeration. North is to the top of the page. The red lines show photographic transect lines from a survey in 2018

Another feature in common between Australia and New Zealand is the occurrence of bottom trawl fisheries for orange roughy, and both countries have undertaken research into the effects of trawling on corals and other benthic invertebrates on seamounts. A key objective of this research has been to examine the resilience of the corals to trawling pressure and to understand the dynamics of recovery post-trawling when a number of seamounts off New Zealand and southern Tasmania (Fig. 9.15) were closed to trawling in the 1990s and early 2000s (Brodie and Clark 2003; Koslow et al. 2001). Time series of surveys have been carried out in both areas to monitor changes over time.

Both areas have seamount/knoll features in close proximity (within km-10 s km), of broadly similar size (1–10 km^2) and depth ranges (700–1200 m summit depths) and contrasting fishing states. This situation has enabled 'compare and contrast' photographic studies of features that have continually been fished, some previously fished and then closed, and others that have remained unfished (Figs. 9.16 and 9.17).

Comparison of benthic communities on fished and unfished seamounts has shown strong contrast and demonstrated clearly that bottom trawling can damage the coral-reef structure, and repeated trawling can leave a seamount almost without any live intact reef (Koslow et al. 2001; Clark and Rowden 2009; Althaus et al. 2009; Clark et al. 2016). The results are not unexpected, given the fragility of these reef-coral structures (Clark et al. 2010b), but a key management issue arising from this research is whether such communities can recover if management action closes an area to trawling, and if so, over what sort of time period?

Repeat photographic surveys have been carried out on the Graveyard seamounts in 2001, 2006, 2009, 2015, and 2020. Repeat surveys of Tasmanian seamounts have occurred in 1998, 2007, and 2018. The 5-survey series off New Zealand and 3-survey comparison off Tasmania each spanning about 20 years, respectively, reveal the same patterns: (1) there are consistent and strong differences in faunal diversity and abundance between fished and unfished seamounts; (2) various faunal groups exhibit different responses over time, but the benthic communities of fished and then closed seamounts show little indication of steps towards recovery; and (3) the reef-corals only persist in areas where trawling was not possible because of rough bottom, or in remnant patches in-between trawling regions (Clark et al. 2019a; Williams et al. 2020a, b; Goode et al. 2020).

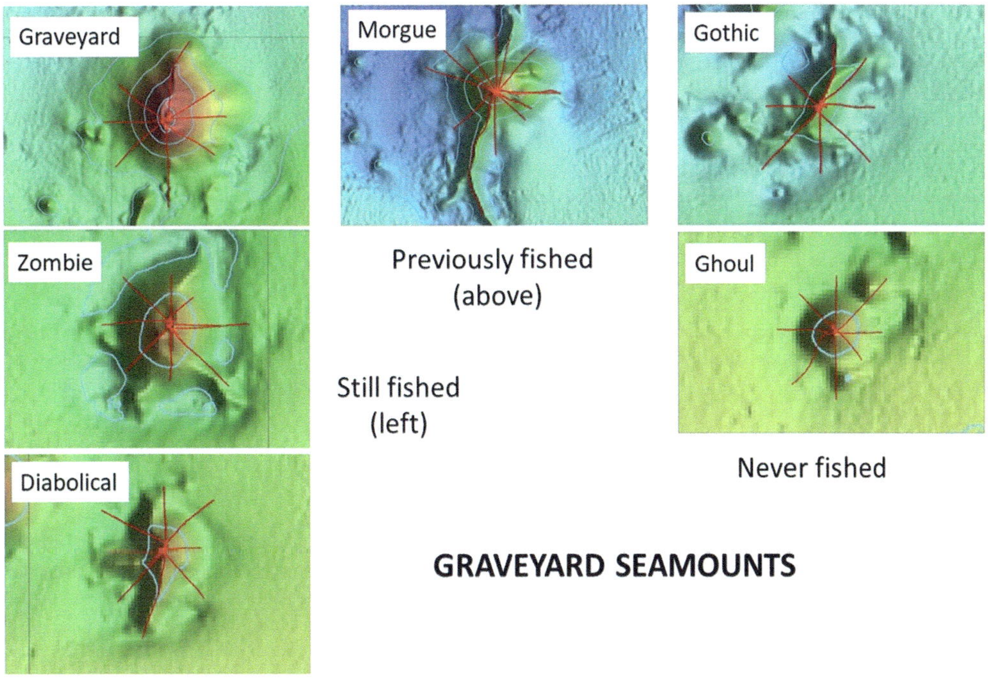

Fig. 9.17 Core set of seamounts surveyed over time off New Zealand. Depths range from 750 m (reddish) through 1000 m (green) to 1200 m (blue). The DTM is 3× vertical exaggeration. North is to the top of the page. The red lines show photographic transect lines from a survey in 2015

Acknowledgements A key source of information used for this chapter is a recent report on the 'State of Knowledge of Deep-sea Corals in the New Zealand Region' compiled by Tracey and Hjorvarsdottir (2019). We acknowledge the many contributors of that report, Erika Mackay, Kareen Schnabel, Rob Stewart, Owen Anderson, Jennifer Beaumont, Kevin Mackay (all NIWA) for the provision of figures and images used in this chapter, Stephen Cairns for his dedication to the taxonomic descriptions of the region's scleractinian reef-corals, as well as more recent research efforts by NIWA scientists that continue to contribute to an improved understanding of deep sea reef corals in New Zealand waters. Various New Zealand Government agencies have funded deep-sea coral research over the years, these include the Department of Conservation, Fisheries New Zealand, and Ministry for Business Innovation and Employment. Finally, our thanks to Erik Cordes and Furu Mienis for their significant effort in leading this Cold-Water Corals volume in the 'Coral Reefs of the World' series.

References

Althaus F, Williams A, Schlacher TA, Kloser RJ, Green MA, Barker BA, Bax NJ, Brodie P, Schlacher-Hoenlinger MA (2009) Impacts of bottom trawling on deep-coral ecosystems of seamounts are long-lasting. Mar Ecol Prog Ser 397:279–294

Anderson OF (2014) Fish and invertebrate bycatch in New Zealand deepwater fisheries from 1990–91 until 2011–12. New Zealand Aquatic Environment and Biodiversity Report 139: 60

Anderson OF, Clark MR (2003) Analysis of bycatch in the fishery for orange roughy, Hoplostethus atlanticus, on the South Tasman Rise. Marine and Freshwater Research 54: 643–652

Anderson O, Tracey D, Bostock H, Williams M, Clark M (2014) Refined habitat suitability modelling for protected coral species in the New Zealand EEZ. NIWA client report prepared for Department of Conservation WLG2014-69: 46

Anderson O, Mikaloff Fletcher S, Bostock H (2015) Development of models for predicting future distributions of protected coral species in the New Zealand region. NIWA client report to Department of Conservation WLG2015-65: 28

Anderson O, Guinotte J, Rowden A, Tracey D, Mackay K, Clark M (2016a) Habitat suitability models for predicting the occurrence of vulnerable marine ecosystems in the seas around New Zealand. Deep-Sea Res I Oceanogr Res Pap 115:265–292. https://doi.org/10.1016/j.dsr.2016.07.006

Anderson O, Guinotte J, Rowden A, Clark M, Mormede S, Davies A, Bowden D (2016b) Field validation of habitat suitability models for vulnerable marine ecosystems in the South Pacific Ocean: implications for the use of broad-scale models in fisheries management. Ocean Coast Manag 120:110–126. https://doi.org/10.1016/j.ocecoaman.2015.11.025

Anderson OF, Ballara SL, Edwards CTT (2017) Fish and invertebrate bycatch and discards in New Zealand orange roughy and oreo trawl fisheries from 2001–02 until 2014–15. New Zealand Aquatic Environment and Biodiversity Report 190: 216

Anderson T, Morrison M, MacDiarmid A, D'Archino R, Nelson W, Tracey D, Clark M, Gordon D, Read G, Morrisey D, Kettles H, Wood A, Anderson O, Smith A, Page M, Paul-Burke K, Schnabel K, Wadhwa S (2018) Review of New Zealand's key biogenic habitats. NIWA client report prepared for the Ministry for the Environment Project MFE18301: 184

Anderson OA, Stevensen F, Behrens E (2020). Updated habitat suitability modelling for protected corals in New Zealand waters. NIWA client report no. 2020174WN prepared for Department of Conservation. [https://www.doc.govt.nz/our-work/conservation-services-programme/csp-reports/201920/improved-habitat-suitability-modelling-for-protected-corals-in-new-zealand-waters/]

Baird SJ, Wood BA (2018) Extent of bottom contact by New Zealand commercial trawl fishing for deepwater tier 1 and tier 2 target fishstocks, 1989–90 to 2015–16. New Zealand Aquatic Environment and Biodiversity Report No 193: 102

Baird SJ, Tracey D, Mormede S, Clark M (2013) The distribution of protected corals in New Zealand waters. Research report for the Department of Conservation. Available for download from http://www.doc.govt.nz/publications/conservation/marine-and-coastal/conservationservices-programme/csp-reports/distribution-of-protected-corals/

Bostock HC, Sutton PJ, Williams MJM, Opdyke BN (2013) Reviewing the circulation and mixing of Antarctic intermediate water in the South Pacific using evidence from geochemical tracers and Argo float trajectories. Deep Sea Res I Oceanogr Res Pap 73:84–98

Bostock H, Tracey D, Currie K, Dunbar G, Handler M, Mikaloff-Fletcher S, Smith A, Williams M (2015) The carbonate mineralogy and distribution of habitat-forming deep-sea corals in the Southwest Pacific region. Deep Sea Res Part 1 Oceanogr Res Pap 100:88–104. https://doi.org/10.1016/j.dsr.2015.02.008

Bowden DA, Anderson OA, Escobar-Flores P, Rowden AA, Clark MR (2019) Quantifying benthic biodiversity: using seafloor image data to build single-taxon and community distribution models for Chatham rise, New Zealand. New Zealand Aquatic Environment and Biodiversity Report No 235: 65

Brodie S, Clark MR (2003) The New Zealand seamount management strategy – steps towards conserving offshore marine habitat. In: Beumer JP, Grant A, Smith DC (eds) Aquatic protected areas: what works best and how do we know? Proceedings of the world congress on aquatic protected areas, Cairns, Australia, August 2002: 664–673. Australian Society of Fish Biology, Cairns

Brooke S, Young CM (2003) Reproductive ecology of a deep-water scleractinian coral, Oculina varicosa, from the Southeast Florida shelf. Cont Shelf Res 23:847–858. https://doi.org/10.1016/S0278-4343(03)00080-3

Burgess S, Babcock R (eds) (2005) Reproductive ecology of three reef-forming, deep- sea corals in the New Zealand region. Cold-water corals and ecosystems. Springer, Berlin, pp 701–713

Büscher JV, Form AU, Riebesell U (2017) Interactive effects of ocean acidification and warming on growth, fitness and survival of the cold-water coral *Lophelia pertusa* under different food availabilities. Front Mar Sci 4:101. https://doi.org/10.3389/fmars.2017.00101

Cairns S (1995) The marine fauna of New Zealand: Scleractinia (Cnidaria: Anthozoa). NZ Oceanogr Inst Mem 103:210

Cairns S (2007) Deep-water corals: an overview with special reference to diversity and distribution of deep-water scleractinian corals. Bull Mar Sci 81:311–322

Cairns SD (2016) The Marine fauna of New Zealand: Primnoid octocorals (Anthozoa, Alcyonacea) -- Part 2. *Primnoella*, *Callozostron*, *Metafannyella*, *Callogorgia*, *Fanellia* and other genera. NIWA Biodivers Mem 129:1–131

Chatham Rock Phosphate Ltd (2014) Review of sediment chemistry and effects of mining (Golder 2014a). Appendix 11. No. Chatham Rock Phosphate (CRP) Ltd, p 95

Chiswell SM (1996) Variability in the southland current, New Zealand. N Z J Mar Freshw Res 30(1):1–17

Chiswell SM, Sutton PJH (2020) Relationships between long-term ocean warming, marine heat waves and primary production in the New Zealand region. Null 54(4):614–635. https://doi.org/10.1080/00288330.2020.1713181

Chiswell SM, Bostock HC, Sutton PJH, Williams MJM (2015) Physical oceanography of the deep seas around New Zealand: a review, New Zealand. J Mar Freshw Res 49:286. https://doi.org/10.1080/00288330.2014.992918

Clark M, O'Driscoll R (2003) Deepwater fisheries and aspects of their impact on seamount habitat in New Zealand. Journal of Northwest Atlantic Fishery Science 31: 441–458

Clark MR, Dunn MR (2012) Spatial management of deep-sea seamount fisheries: balancing sustainable exploitation and habitat conservation. Environ Conserv 39:204–214

Clark M, Rowden A (2009) Effects of Deepwater trawling on the macro-invertebrate assemblages of seamounts on the Chatham Rise, New Zealand. Deep-Sea Res I 56:1540–1554. https://doi.org/10.1016/j.dsr.2009.04.015

Clark M, Bowden D, Baird S, Stewart R (2010a) Effects of fishing on the benthic biodiversity of seamounts of the "Graveyard" complex, northern Chatham Rise. New Zealand Aquatic Environment and Biodiversity Report No 46: 40

Clark MR, Rowden AA, Schlacher T, Williams A, Consalvey M, Stocks KI, Rogers AD, O'Hara TD, White M, Shank TM, Hall-Spencer JM (2010b) The ecology of seamounts: structure, function, and human impacts. Annu Rev Mar Sci 2:253–278

Clark MR, Althaus F, Williams A, Niklitschek E, Menezes G, Hareide N-R, Sutton P, O'Donnell C (2010c) Are deep-sea fish assemblages globally homogenous? Insights from seamounts. Mar Ecol 31(Suppl 1):39–51

Clark M, Anderson O, Mackay K, Notman P, Roux MJ, Tracey D (2014a) Determination or orange roughy trawl footprint in relation to protected coral species distribution. NIWA Client Report WLG2014-56: 57

Clark M, Tracey D, Anderson O, Parker S (2014b) Pilot ecological risk assessment for protected corals. Report prepared by the National Institute of Water and Atmospheric Research for the New Zealand Department of Conservation, Wellington: 32. http://www.doc.govt.nz/Documents/conservation/marine-and-coastal/marine-conservation-services/reports/pilot-ecological-risk-assessment-for-protected-corals-final-report.pdf

Clark MR, Anderson O, Bowden D, Chin C, George S, Glasgow D, Guinotte J, Hererra S, Osterhage D, Pallentin A, Parker S, Rowden AA, Rowley S, Stewart R, Tracey D, Wood S, Zeng C (2015) Vulnerable marine ecosystems of the Louisville seamount chain: voyage report of a survey to evaluate the efficacy of preliminary habitat suitability models. New Zealand Aquatic Environment and Biodiversity Report No 149. 86 p

Clark M, Althaus F, Schlacher T, Williams A, Bowden D, Rowden A (2016) The impacts of deep-sea fisheries on benthic communities: a review. ICES J Mar Sci 73:51–69

Clark M, Rauhina-August L, Tracey D (2018) The resilience of deep-sea benthic communities to the effects of sedimentation. Information flyer 1. The Programme. [https://niwa.co.nz/coasts-and-oceans/research-projects/resilience-of-deep-sea-benthic-fauna-to-sedimentation-from-seabed-mining]

Clark MR, Bowden DA, Rowden AA, Stewart R (2019a) Little evidence of benthic community resilience to bottom trawling on seamounts after 15 years. Front Mar Sci 6(63):16p

Clark MR, Mills S, Leduc D, Anderson OF, Rowden AA (2019b) Biodiversity of benthic protection areas and seamount closure areas: a description of available benthic invertebrate data, and a preliminary evaluation of the effectiveness of BPAs for biodiversity protection. New Zealand Aquatic Environment and Biodiversity Report No. 227. 270 p

Clark MR, Bowden DA, Stewart R, Schnabel K, Quinn W, Lennard B, Goode SL, Davis A (2021) Seamount recovery: factual voyage report of a survey of seamounts on the Northwest and Southeast Chatham Rise (TAN2009) New Zealand Aquatic Environment and Biodiversity Report No. 262. 117 p

D'Onghia G, Maiorano P, Sion L, Giove A, Capezzuto F, Carlucci R, Tursi A (2010) Effects of deep-water coral banks on the abundance and size structure of the megafauna in the Mediterranean Sea. Deep-Sea Res II Top Stud Oceanogr 57(5–6):397–411. ISSN 0967-0645. https://doi.org/10.1016/j.dsr2.2009.08.022

Davis AN, Hitt NT, Sinclair DJ, Tracey DM, Fallon SJ, Behrens E, Komugabe-Dixson AF, Hellstrom JC (2021) Three centuries of Southwest Pacific Gyre biogeochemistry reconstructed from a New Zealand Black Coral. Presentation to American Geophysical Union (AGU)

Department of Conservation (2023). Conservation Services Programme Annual Plan 2023/24. 98 p. https://www.doc.govt.nz/globalassets/documents/conservation/marine-and-coastal/marine-conservationservices/plans-and-submissions/202324/final-csp-annual-plan-2023-24.pdf

Fallon S, Thresher R, Adkins J (2014) Age and growth of the cold-water scleractinian *Solenosmilia variabilis* and its reef on SW Pacific seamounts. Coral Reefs 33:31–38

Feehan KA, Waller RG (2015) Notes on reproduction of eight species of Eastern Pacific cold-water octocorals. J Mar Biol Assoc United Kingdom 95:691–696. https://doi.org/10.1017/S0025315415000053

Feely RA, Sabine CL, Byrne RH, Millero FJ, Dickson AG, Wanninkhof R, Murata A, Miller LA, Greeley D (2012) Decadal changes in the aragonite and calcite saturation state of the Pacific Ocean. Glob Biogeochem Cycles 26:15. GB3001. https://doi.org/10.1029/2011GB004157

Freeman D, Cryer M (2019) Current management measures and threats, Chapter 9. in: Tracey DM, Hjorvarsdottir F (eds) The state of knowledge of deep-sea corals in the New Zealand region. NIWA science and technology series number 84, 140 p

Freeman D, Schnabel K, Marshall B, Gordon D, Wing S, Tracey D, Hitchmough R (2013) Conservation status of New Zealand marine invertebrates, New Zealand threat classification series 9. Department of Conservation, Wellington, 20 p

Gammon MJ, Tracey DM, Marriott PM, Cummings VJ, Davy SK (2018) The physiological response of the deep-sea coral *Solenosmilia variabilis* to ocean acidification. PeerJ 6:e5236. https://doi.org/10.7717/peerj.5236

Georgian SE, Anderson OF, Rowden AA (2019) Ensemble habitat suitability modelling of vulnerable marine ecosystem indicator taxa to inform deep-sea fisheries management in the South Pacific Ocean. Fish Res 211:256–274

Goode S, Rowden AA, Bowden DA, Clark MR (2020) Resilience of seamount benthic communities to trawling disturbance. Mar Environ Res 161:1–15. https://doi.org/10.1016/j.marenvres.2020.105086

Gordon DP (ed.) (2009) New Zealand inventory of biodiversity, vol 1: Kingdom Animalia. Canterbury University Press, pp 59–101

Guinotte JM, Orr J, Cairns S, Freiwald A, Morgan L, George R (2006) Will human-induced changes in sea water chemistry alter the distribution of scleractinian bioherms? Front Ecol Environ 4:141–146

Helson J, Leslie S, Clement G, Wells R, Wood R (2010) Private rights, public benefits: Industrydriven seabed protection. Marine Policy 34(3): 557–566.

Henderson MJ, Huff DD, Yoklavich MM (2020) Deep-sea coral and sponge taxa increase demersal fish diversity and the probability of fish presence. Front Mar Sci 7:593844. https://doi.org/10.3389/fmars.2020.593384

Hitt N (2021) Southwest Pacific Paleoceanography during the Late Holocene 3000 years of ocean circulation and marine biogeochemistry reconstructed from New Zealand deep-sea black corals. A thesis submitted to the Victoria University of Wellington in fulfilment of the requirements for the degree of Doctor of Philosophy in Geology Victoria University of Wellington, 502 p

Hitt NT, Sinclair DJ, Fallon SJ, Neil HL, Tracey DM, Komugabe-Dixson A, Marriott P (2020) Growth and longevity of New Zealand black corals. Deep-Sea Res I Oceanogr Res Pap 162: 103298

Hjorvarsdottir F, Tracey D (2017) Workshop on research needs for protected corals in New Zealand waters: Workshop minutes: 18. https://www.doc.govt.nz/pagefiles/174857/doc-coral-workshop-minutes-and-gaps-nov-2017-final.pdf

Holland LP, Rowden AA, Hamilton JZ, Clark MR, Chiswell SM, Gardner JPA (2020) Genetic connectivity of deep-sea corals in the New Zealand region. New Zealand Aquatic Environment and Biodiversity Report No. 245

Jones EG, Morrison MA, Davey N, Hartill BW, Sutton C (2016) Biogenic habitats on New Zealand's continental shelf. Part I: local ecological knowledge. New Zealand Aquatic Environment and Biodiversity Report 174: 95

Kitahara MV, Cairns SD (in press) Azooxanthellate Scleractinia (Cnidaria, Anthozoa) from New Caledonia, in Tropical deep- Sea benthos

Koslow JA, Gowlett-Holmes K, Lowry JK, O'Hara T, Poore GCB, Williams A (2001) Seamount benthic macrofauna off southern Tasmania: community structure and impacts of trawling. Mar Ecol Prog Ser 213:111–125. https://doi.org/10.3354/meps213111

Krieger KJ, Wing BL (2002) Megafauna associations with Deepwater corals (*Primnoa* spp.) in the Gulf of Alaska. Hydrobiologia 471:83–90

Larsson AI, Järnegren J, Strömberg SM, Dahl MP, Lundälv T, Brooke S (2014) Embryogenesis and larval biology of the cold-water coral *Lophelia pertusa*. PLoS One 9:e102222. https://doi.org/10.1371/journal.pone.0102222

Law CS, Bell JJ, Bostock HC, Cornwall CE, Cummings VJ, Currie K, Davy SK, Gammon M, Hepburn CD, Hurd CL, Lamare M, Mikaloff-Fletcher SE, Nelson WA, Parsons DM, Ragg NLC, Sewell MA, Smith AM, Tracey DM (2017) Ocean acidification in New Zealand waters: trends and impacts. N Z J Mar Freshw Res 52(2):155–195. https://doi.org/10.1080/00288330.2017.1374983

Law CS, Rickard GJ, Mikaloff-Fletcher SE, Pinkerton MH, Behrens E, Chiswell SM, Currie K (2018) Climate change projections for the surface ocean around New Zealand. N Z J Mar Freshw Res 52(3): 309–335. https://doi.org/10.1080/00288330.2017.1390772

Mikaloff-Fletcher SE, Bostock HC, Williams MJ, Forcen A (2017) Modelling the effects of ocean acidification in New Zealand. New Zealand Aquatic Environment and Biodiversity Report 187: 13. Modelling the Effects of Ocean Acidification in New Zealand (fish.govt.nz)

Miller KJ, Gunasekera RM (2017) A comparison of genetic connectivity in two deep sea corals to examine whether seamounts are isolated islands or stepping stones for dispersal. Sci Rep 7:46103

Mitchell JS, Mackay KA, Neil HL, Mackay EJ, Pallentin A, Notman P (2012) Undersea New Zealand, 1:5,000,000. NIWA Chart, Miscellaneous Series 92

Mobilia V (2021) Sediment impacts on sponges and a deep-sea coral in New Zealand. A Thesis submitted to the Victoria University of Wellington, New Zealand. 178 p

Morrison MA, Jones E, Consalvey M, Berkenbusch K (2014) Linking marine fisheries species to biogenic habitats in New Zealand: a review and synthesis of knowledge. New Zealand Aquatic Environment and Biodiversity Report No 130. 156 p

Mortensen PB, Buhl-Mortensen L, Gebruk AV, Krylova EM (2008) Occurrence of deep-water corals on the Mid-Atlantic Ridge based on MAR-ECO data. Deep-Sea Res II 55(2008):142–152

Mueller CE, Lundälv T, Middelburg JJ, van Oevelen D (2013) The symbiosis between *Lophelia pertusa* and *Eunice norvegica* stimulates coral calcification and worm assimilation. PLoS One 8(3):e58660. https://doi.org/10.1371/journal.pone.0058660

O'Hara TD, Rowden AA, Williams A (2008) Cold-water coral habitats on seamounts: do they have a specialist fauna? Divers Distrib 14(6): 925–934. https://doi.org/10.1111/j.1472-4642.2008.00495.x

Opresko D, Tracey D, Mackay E (2014) Antipatharia (black corals) for the New Zealand region. A field guide of commonly sampled New Zealand black corals including illustrations highlighting technical terms and black coral morphology. NIWA Client Report for the New Zealand Aquatic Environment and Biodiversity Report No. 131. 20 p

Parker SJ, Penney AJ, Clark MR (2009) Detection criteria for managing trawl impacts on vulnerable marine ecosystems in high seas fisheries of the South Pacific Ocean. Mar Ecol Prog Ser 397:309–317. https://doi.org/10.3354/meps08115

Roberts J, Wheeler A, Freiwald A, Cairns S (2009) Cold water corals: the biology and geology of Deep-Sea Coral habitats. Cold-Water Corals: The Biology and Geology of Deep-Sea Coral Habitats: 1–350. https://doi.org/10.1017/CBO9780511581588

Rogers AD (1999) The biology of *Lophelia pertusa* (Linnaeus 1758) and other Deepwater reef-forming corals and impacts from human activities. Int Rev Hydrobiol 84:315–406

Rooper CN, Goddard P, Wilborn R (2019) Are fish associations with corals and sponges more than an affinity to structure? Evidence across two widely divergent ecosystems. Can J Fish Aquat Sci 76: 2184–2198

Röthig T, Roik A, Yum LK, Voolstra CR (2017) Distinct bacterial microbiomes associate with the Deep-Sea Coral *Eguchipsammia fistula* from the Red Sea and from aquaria settings. Front Mar Sci 4:259. https://doi.org/10.3389/fmars.2017.00259

Rowden A, Clark M (2010) Benthic biodiversity of seven seamounts on the southern end of the Kermadec volcanic arc, Northeast New Zealand. New Zealand Aquatic Environment and Biodiversity Report 62:31

Rowden AA, Clark MR, Wright IC (2005) Physical characterisation and a biologically focused classification of "seamounts" in the New Zealand region. N Z J Mar Freshw Res 39(5):1039–1059

Rowden AA, Schnabel KE, Schlacher TA, Macpherson E, Ahyong ST, Richer de Forges B (2010) Species composition of benthic assemblages on seamounts is distinct from some but not all deep-sea habitats. Mar Ecol 31:63–83

Rowden A, Leduc D, Torres L, Bowden D, Hart A, Chin C, Davey N, Nodder S, Pallentin A, Mackay K, Northcote L, Sturman J (2014a) Benthic epifauna communities of the central Chatham Rise crest. NIWA Client Report WLG2014–9 prepared for Chatham Rock Phosphate Ltd, 121p

Rowden A, Leduc D, Torres L, Bowden D, Hart A, Chin C, Davey N, Nodder SD, Pallentin A, Mackay KA, Northcote L, Sturman J (2014b) Benthic epifauna communities of the central Chatham Rise crest. NIWA Client Report to Chatham Rock Phosphate Ltd. WLG2014-9: 116

Rowden AA, Clark MR, Lundquist CJ, Guinotte JM, Anderson OF, Julian KA, Mackay KA, Tracey DM, Gerring PK (2015) Developing spatial management options for the protection of vulnerable marine ecosystems in the South Pacific Ocean region. New Zealand Aquatic Environment and Biodiversity Report 155:76

Rowden A, Anderson OF, Georgian SE, Bowden DA, Clark MR, Pallentin A, Miller A (2017) High-resolution habitat suitability models for the conservation and management of vulnerable marine ecosystems on the Louisville Seamount Chain, South Pacific Ocean. Front Mar Sci 4:1–19

Sánchez J (2005) Systematics of the bubblegum corals (Cnidaria: Octocorallia: Paragorgiidae) with description of new species from New Zealand and the Eastern Pacific. Zootaxa 1014:1–72

Schnabel KE (2020) The marine fauna of New Zealand. Squat lobsters (Crustacea, Decapoda, Chirostyloidea). NIWA biodiversity memoir, ISSN 1174-0043; 132

Schnabel K, Mills S (2019) Species associations with deep coral communities, Chapter 4. In: Tracey DM & Hjorvarsdottir F (eds) The state of knowledge of deep-sea corals in the New Zealand region. NIWA Science and Technology Series Number 84: 140

Simmons JM (2016) Diving into the deep. Unpublished placement report by Jessica Marie Simmons (Bangor University, Wales), Deepsea community project DSCA163: 44

Soffker M, Slomam KA, Hall-Spencer JM (2011) In situ observations of fish associated with coral reefs off Ireland. Deep-Sea Res I 58:818–825

Squires DF (1964) Fossil coral thickets in Wairarapa, New Zealand. J Paleontol 38:904–915

Stanton BR, Goring DG, Bell RG (2001) Observed and modelled tidal currents in the New Zealand region. N Z J Mar Freshw Res 35:397–415

Stephenson F, Leathwick JR, Geange SW, Bulmer RH, Hewitt JE, Anderson OF, Rowden AA, Lundquist CJ (2018) Using gradient forests to summarize patterns in species turnover across large spatial scales and inform conservation planning. Divers Distrib 24(11): 1641–1656. https://doi.org/10.1111/ddi.12787

Stevens CL, O'Callaghan JM, Chiswell SM, Hadfield MG (2021) Physical oceanography of New Zealand/Aotearoa shelf seas – a review. N Z J Mar Freshw Res 55(1):6–45. https://doi.org/10.1080/00288330.2019.1588746

Stevenson A, Davies JS, Williams A, Althaus F, Rowden AA, Bowden DA, Clark MR, Mitchell FJG (2018) Echinoid associations with coral habitats differ with taxon in the deep sea and the influence of other echinoids, depth, and fishing history on their distribution. Deep-Sea Res Part I-Oceanogr Res Pap 133:27–38. https://doi.org/10.1016/j.dsr.2018.01.007

Stone RP (2006) Coral habitat in the Aleutian Islands of Alaska: depth distribution, fine scale species association, and fisheries interactions. Coral Reefs 25:229–238

Sutton PJ (2003) The Southland current: a subantarctic current. N Z J Mar Freshw Res 37(3):645–652

Sutton PJH, Bowen M (2019) Ocean temperature change around New Zealand over the last 36 years. N Z J Mar. Freshw Res 1–22. https://doi.org/10.1080/00288330.2018.1562945

Sweetman AK, Thurber AR, Smith CR, Levin LA, Mora C et al (2017) Major impacts of climate change on deep-sea benthic ecosystems. Elem Sci Anth 5:4. https://doi.org/10.1525/elementa.203

Thresher R, Adkins J, Fallon S, Gowlett-Holmes K, Althaus F, Williams A (2011a) Extraordinarily high biomass benthic community on Southern Ocean seamounts. Sci Rep 1:119

Thresher RE, Tilbrook B, Fallon S, Wilson NC, Adkins J (2011b) Effects of chronic low carbonate saturation levels on the distribution, growth and skeletal chemistry of deep-sea corals and other seamount megabenthos. Mar Ecol Prog Ser 442:87–99. https://doi.org/10.3354/meps09400

Thresher R, Althaus F, Adkins J, Gowlett-Holmes K, Alderslade P, Dowdney J, Cho W, Gagnon A, Staples D, McEnnulty F, Williams A (2014) Strong depth-related zonation of megabenthos on a rocky continental margin (700-4000 m) off southern Tasmania, Australia. PLoS One 9(1):e85872. https://doi.org/10.1371/journal.pone.0085872

Thresher RE, Guinotte JM, Matear RJ, Hobday AJ (2015) Options for managing impacts of climate change on a deep-sea community. Nat Clim Chang 5:635–639

Timm C, Hoernle K, Werner R, Hauff F, van den Bogaard P et al (2010) Temporal and geochemical evolution of the Cenozoic intraplate volcanism of Zealandia. Earth Sci Rev 98:38–64

Tracey DM, Hjorvarsdottir F (eds, comps) (2019) The state of knowledge of deep-sea corals in the New Zealand Region. NIWA Science and Technology Series Number 84: 140. HYPERLINK "https://niwa.co.nz/sites/niwa.co.nz/files/Deepsea-corals-NZ-2019-NIWA-SciTechSeries-84.pdf" Deepsea-corals-NZ-2019-NIWA-SciTechSeries-84.pdf

Tracey D, Rowden A, Mackay K, Compton T (2011) Habitat-forming cold-water corals show affinity for seamounts in the New Zealand region. Mar Ecol Prog Ser 430:1–22

Tracey DM, Clark MR, Anderson OF, Kim S (2012) Deep-sea fish distribution varies between seamounts: results from a seamount complex off New Zealand. PLoS One 7(6):e36897. https://doi.org/10.1371/journal.pone.0036897

Tracey D, Marriott P, Bostock H, Bilewitch J, Komugabee-Dixson A, Holland L, Hitt N (2019) Coral biology, Chapter 2. In: Tracey DM, Hjorvarsdottir F (eds) The state of knowledge of deep-sea corals in the New Zealand Region. NIWA Science and Technology Series Number 84: 140

Tracey DM, Goode S, Marriott P, Waller R (2021) Protected coral reproduction. Report prepared for Department of Conservation. Project protected coral reproduction. Milestone 1. Literature review and sample analysis: 29 p

Vega Thurber RL, Barott KL, Hall D, Liu H, Rodriguez-Mueller B, Desnues C, Edwards RA, Haynes M, Angly FE, Wegley L, Rohwer FL (2008) Metagenomic analysis indicates that stressors induce production of herpes-like viruses in the coral Porites compressa. Proceedings of the National Academy of Sciences 105(47): 18413–18418

Waller RG (ed) (2005) Deep-water Scleractinia (Cnidaria: Anthozoa): current knowledge of reproductive processes. Cold-water corals and ecosystems. Springer, Berlin, pp 691–700

Waller RG, Goode S, Tracey D, Johnstone J, Mercier A (2023) A review of current knowledge on reproductive and larval processes of deep-sea corals. Mar Biol 170:58. https://doi.org/10.1007/s00227-023-04182-8

Wells PE (1986) Record of an Upper Miocene fossil *Goniocorella* coral thicket, Mt. Bruce, Wairarapa, New Zealand. J R Soc N Z 16:139–144

Williams A, Schlacher TA, Rowden AA, Althaus F, Clark MR, Bowden DA, Stewart R, Bax NJ, Consalvey M, Kloser RJ (2010) Seamount megabenthos assemblages fail to recover from trawling impacts. Mar Ecol 31(Suppl 1):183–199

Williams A, Althaus F, Green M, Maguire K, Untiedt C, Mortimer N, Clark MR, Bax N, Pitcher R, Schlacher T (2020a) True size matters for conservation: deep-sea coral reefs are typically small and estimates of their size are remarkably robust to a method used to define them. Front Mar Sci 7. https://doi.org/10.3389/fmars.2020.00187

Williams A, Althaus F, Maguire K, Green M, Untiedt C, Alderslade P, Clark MR, Bax N, Schlacher TA (2020b) The fate of deep-sea coral reefs on seamounts in a fishery-seascape: what are the impacts, what remains, and what is protected? Front Mar Sci 7(798). https://doi.org/10.3389/fmars.2020.567002

Zeng C, Rowden AA, Clark MR, Gardner JP (2017) Population genetic structure and connectivity of deep-sea stony corals (order Scleractinia) in the New Zealand region: implications for the conservation and management of vulnerable marine ecosystems. Evol Appl 10(10):1040–1054

Zeng C, Rowden AA, Clark MR, Gardner JP (2020) Species-specific genetic variation in response to deep-sea environmental variation amongst vulnerable marine ecosystem indicator taxa. Sci Rep 10: 2844. https://doi.org/10.1038/s41598-020-59210-0

Deep-Sea Corals of the North and Central Pacific Seamounts

10

Amy R. Baco, Frank A. Parrish, Steven Auscavitch, Stephen Cairns, Beatriz E. Mejia-Mercado, Virginia Biede, Nicole Morgan, E. Brendan Roark, and W. Ben Brantley

Abstract

The North and Central Pacific are home to a number of major archipelagos including Hawaii, the Marianas, Samoa, Line Islands, Phoenix Islands, and Marshall Islands, and numerous seamounts associated with each of these island groups. These islands span from 14°S, the latitude of American Samoa in the South Pacific, to Kure Atoll at 28°N latitude in the Central North Pacific, with seamounts continuing along the Hawaiian Ridge northwest into the Emperor Seamount Chain. Although all are generally remote, some of these island and seamount chains are so isolated (Hawaii and the Northwestern Hawaiian Islands) that they have some of the world's highest levels of marine endemism. Differences across the Central Pacific have been documented in the extensive shallow tropical coral reef communities through two decades of directed assessment and monitoring efforts for the subgroup of sites that include the US Pacific Islands. It is likely that deeper coral communities also differ across this region, however, broader explorations of mesophotic and deep-sea communities are needed to document these patterns. Extensive explorations have been undertaken in the Hawaiian Archipelago and started to expand to the rest of the Central Pacific in 2015. Within the North and Central Pacific, deep-sea scleractinian reefs are so far only known to occur from the far end of the Northwestern Hawaiian Islands (NWHI) and lower end of the Emperor Seamount Chain (ESC), and from the Phoenix Islands, thus this chapter will focus on those regions. In all areas of the North and Central Pacific so far explored, dense beds of octocorals and antipatharian corals are instead generally the dominant benthic megafaunal taxa on deep-sea hard substrates, occupying the same functional roles and niche as scleractinian reefs, but far more common and extensively distributed, thus they will also be frequently mentioned.

Keywords

Hawaii Archipclago · Scamounts · Equatorial Pacific · North Pacific Subtropical Gyre · Oceanography · Ocean acidification · Deep-water reefs · Coral phylogeny · Phoenix Islands · Coral reproduction · Reef ecosystem · Trophic structure

10.1 Oceanography

10.1.1 Major Currents

Surface currents in the Pacific are dominated by two large gyres, one north and one south, between which the trade winds drive a set of equatorial currents (Fig. 10.1). The South Equatorial Current (SEC) moves westerly between 15°S and 3°N latitude. Within this current the Cromwell Current, or Pacific Equatorial Undercurrent (EUC), flows in the opposite direction at depths of 70–200 m, creating a belt 450 km wide for a distance of up to 13,000 km (Tchernia 1980; Thurman 1981). On the north side of the equator the North Equatorial Current (NEC) flows westward at 1 km/h between 8° and 20°N latitude. The North Equatorial Counter Current (NECC) is located between the North and South

A. R. Baco (✉) · B. E. Mejia-Mercado · V. Biede · N. Morgan · W. B. Brantley
Department°f Earth, Ocean, and Atmospheric Sciences, Florida State University, Tallahassee, FL, USA
e-mail: abacotaylor@fsu.edu

F. A. Parrish
Pacific Islands Fisheries Science Center, National Oceanic and Atmospheric Administration, Honolulu, HI, USA

S. Auscavitch
Department of Biology, Temple University, Philadelphia, PA, USA

S. Cairns
Department of Invertebrate Zoology, National Museum of Natural History, Smithsonian Institution, Washington, DC, USA

E. B. Roark
Department of Geography, Texas A&M University, College Station, TX, USA

E. Cordes, F. Mienis (eds.), *Cold-Water Coral Reefs of the World*, Coral Reefs of the World 19,
https://doi.org/10.1007/978-3-031-40897-7_10

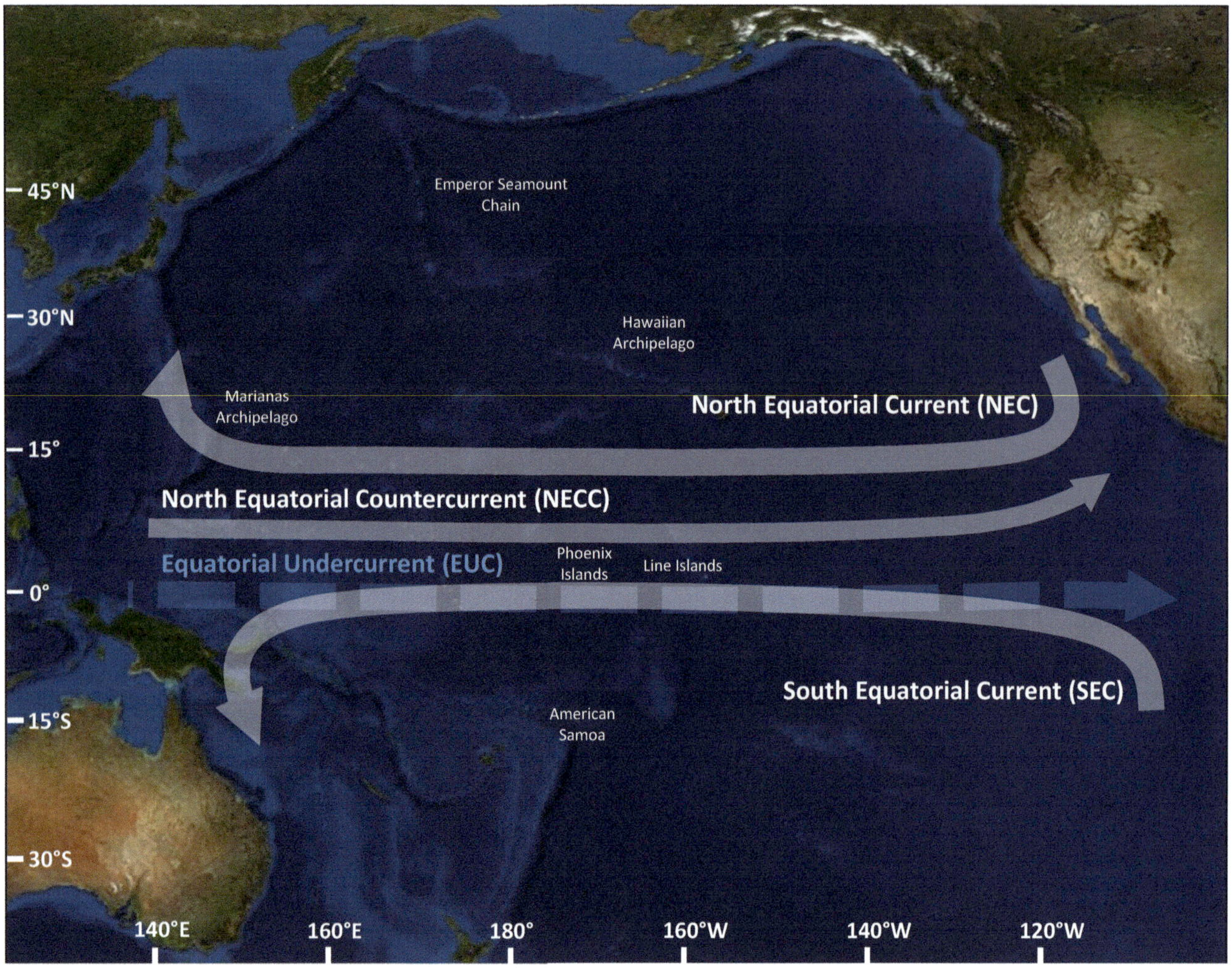

Fig. 10.1 Map of the Pacific Basin with the primary Island groups labeled and arrows to indicate the intensity and direction of equatorial current flow. Figure adapted from Brainard et al. (2019)

Equatorial Currents and flows eastward at 2 km/h (Thurman 1981). The overlap in the boundaries of these currents is an important driver of the long-distance dispersal of pelagic larvae in the region.

Deep-water seafloor features (seamounts, pinnacles, and other structures) obstruct flow and generate both cold- and warm-water eddies of variable size and intensity (Genin et al. 1986; Mullineaux and Mills 1997; Coutis and Middleton 2002). Cold-water eddies (anti-cyclonic) can be formed and stay above a seamount whereas when a warm-water eddy (cyclonic) is formed, it wanders downstream behind the seamount (Kamenkovich et al. 1986).

Satellite data are currently our best means to identify and monitor environmental differences and oceanographic features. The patterns in ocean circulation vary considerably by Pacific region and are known to drive the ecological processes of shallow coral reef ecosystems (Gove et al. 2006, 2013; Brainard et al. 2018; Venegas et al. 2019) but are increasingly being studied in relation to mesophotic (Rooney et al. 2010; Pyle et al. 2016) and deep-sea reef ecosystems (Bauer et al. 2016; Auscavitch et al. 2020a; Parrish and Oliver 2020).

The Northwestern Hawaiian Islands (NWHI) are located in the nutrient-poor surface waters of the North Pacific Subtropical Gyre, but their proximity to the nutrient-rich surface waters of the North Pacific Subpolar Gyre influences their physico-chemical characteristics with particular physical processes throughout the year (Leonard et al. 2001). The front created by these two gyres presents seasonal changes (Polovina et al. 2001) that allow for the input of colder and nutrient-rich waters mainly toward the west of the Islands from November through May (Desch et al. 2009). This front shows interannual and decadal changes (Nakamura and Kazmin 2003) that affect the productivity of the whole region (Polovina and Haight 1999; Polovina 2005). During the summer months the front moves northward and during the winter months it moves southward (Seki et al. 2002). High chlorophyll waters with high nutrient concentrations arrive at the northern portions of the NWHI during southward winter migrations of the subtropical front (Friedlander et al. 2005).

On a smaller scale, the productivity in the NWHI is affected by local and regional factors (Firing and Brainard 2006). Phytoplankton distribution along the Hawaiian Islands is influenced by the North Hawaiian Ridge Current (NHRC), which flows through the northwest along the north side of the islands from the main Hawaiian Islands, the Hawaiian Lee Countercurrent (HLCC), which flows through the east along the south side, and the Subtropical Countercurrent (SCC), which flows from the Kuroshio current through the east.

Further south in the U.S. waters around Howland and Baker Island and the Phoenix Islands of Kiribati, two major currents dominate in the surface and subsurface waters (Obura et al. 2011). At the surface, the warmer SEC moves westward at velocities around 0.3 m/s. In contrast, the deeper (70–200 m), rapid (~1 m/s), eastward flowing EUC brings cool, nutrient-rich waters toward the eastern Pacific, impinging on Howland and Baker Island as well as the northern reaches of the Phoenix Islands Protected Area in the vicinity of Winslow Reef. The result of the impinging of the EUC on these features is known to be a regulator of sea surface temperature (Gove et al. 2006), productivity (Maragos et al. 2008), and shallow-water coral health in the region across multi-year timescales (Vargas-Ángel et al. 2011; Obura and Mangubhai 2011).

10.1.2 Water Masses

There are several water masses that bathe the islands and seamounts of the North and Central Pacific. At depths greater than ~1500 m, south of the Hawaiian Ridge, the main water mass is Antarctic Bottom Water (AABW). From the Hawaiian Ridge northward, Circumpolar Deep Water (CDW) (also called Pacific Bottom Water), with temperatures between 5.0 and 12°C, is the result of deep-water flow into the Pacific Basin from the Atlantic, entering from the southwest between New Zealand and Antarctica spreading northward at 180° to 160°W at depths below 2500 m and southward above 2500 m (Knauss 1996). After crossing the equator, branches of the CDW spread northwest and northeast along basins into the North Pacific. North of the equator, Pacific Deep Water is the dominant water mass below ~800 m. Pacific Deep Water (or Pacific Common Water) flows in a clockwise pattern to the eastern North Pacific at its farthest reaches. Water reaching the North Pacific must ultimately rise into the mid-depth layers above 2000 m (Schmitz Jr 1996).

At intermediate depths water masses are more complicated, with North Pacific Intermediate Water (NPIW) extending across the Emperor Seamount Chain (ESC) and the northern half of the NWHI to 15–20°N within the 400–800 m depth range. This water mass is characterized by a salinity minimum (33.6–34.4 ‰), relatively high oxygen levels ranging from 2.5 to 4 ml/l, and has a narrow density range centered around 26.7–26.9 σ_θ (Talley 1993). NPIW is deepest at ~800 m in the west rising to 300 m in the east. NPIW is formed by sinking in the mixed water region of the Oyashio and Kuroshio currents and is ventilated in the Sea of Okhotsk and to some extent the Alaskan Gyre. The remainder of the Hawaiian Archipelago at intermediate depths falls into a mixing zone between NPIW and the Equatorial Pacific Intermediate Waters (EqPIW). Formed by mixing of Antarctic Intermediate Water and upwelling of Pacific Deep Waters, the EqPIW are further subdivided into North Equatorial (NEqPIW) and South Equatorial Intermediate Waters (SEqPIW) at about 2°N. One of the biggest differences between these water masses is in oxygen concentration, with the NEqPIW generally having much lower oxygen than the SEqPIW (Bostock et al. 2010). The Phoenix Islands at intermediate depths fall entirely within the SEqPIW, which extends to about 15°S.

There are five surface water masses (0–500 m) across the North and Central Pacific. Western North Pacific Central Water (WNPCW) and the Eastern North Pacific Central Water (ENPCW) split the NWHI, and surface waters in the South Pacific are split into Western and Eastern South Pacific Central Water. Along the equator, the main surface water mass is the Pacific Equatorial Water (PEW) (Emery 2001).

10.1.3 Temperature, Climate, and Other Oceanographic Data

Cooler waters surround the Hawaiian Archipelago (sea surface temperatures 23–27°C), compared to the warmer water of the Marianas and American Samoa Archipelagos (28–29°C). Mean sea surface temperatures (SSTs) at Wake and Johnston Atolls were between these extremes (Fig. 10.2a). The atolls of Wake, Johnston, and the Pacific Remote Islands (including the US Pacific Remote Islands Marine National Monument (PRIMNM) and the Phoenix Islands) are widely dispersed, resulting in high variability in their mean temperatures. For example, the equatorial locations of Jarvis and the Howland and Baker Islands experienced noticeably cooler mean SSTs due to the locally-intense topographic upwelling of the eastward flowing subsurface Equatorial Undercurrent (Gove et al. 2006). In contrast, Palmyra Atoll and Kingman Reef are further to the north and are less influenced by the equatorial current and associated equatorial upwelling.

Patterns in the long-term mean chlorophyll-*a* concentrations (Fig. 10.2b) show primary productivity is

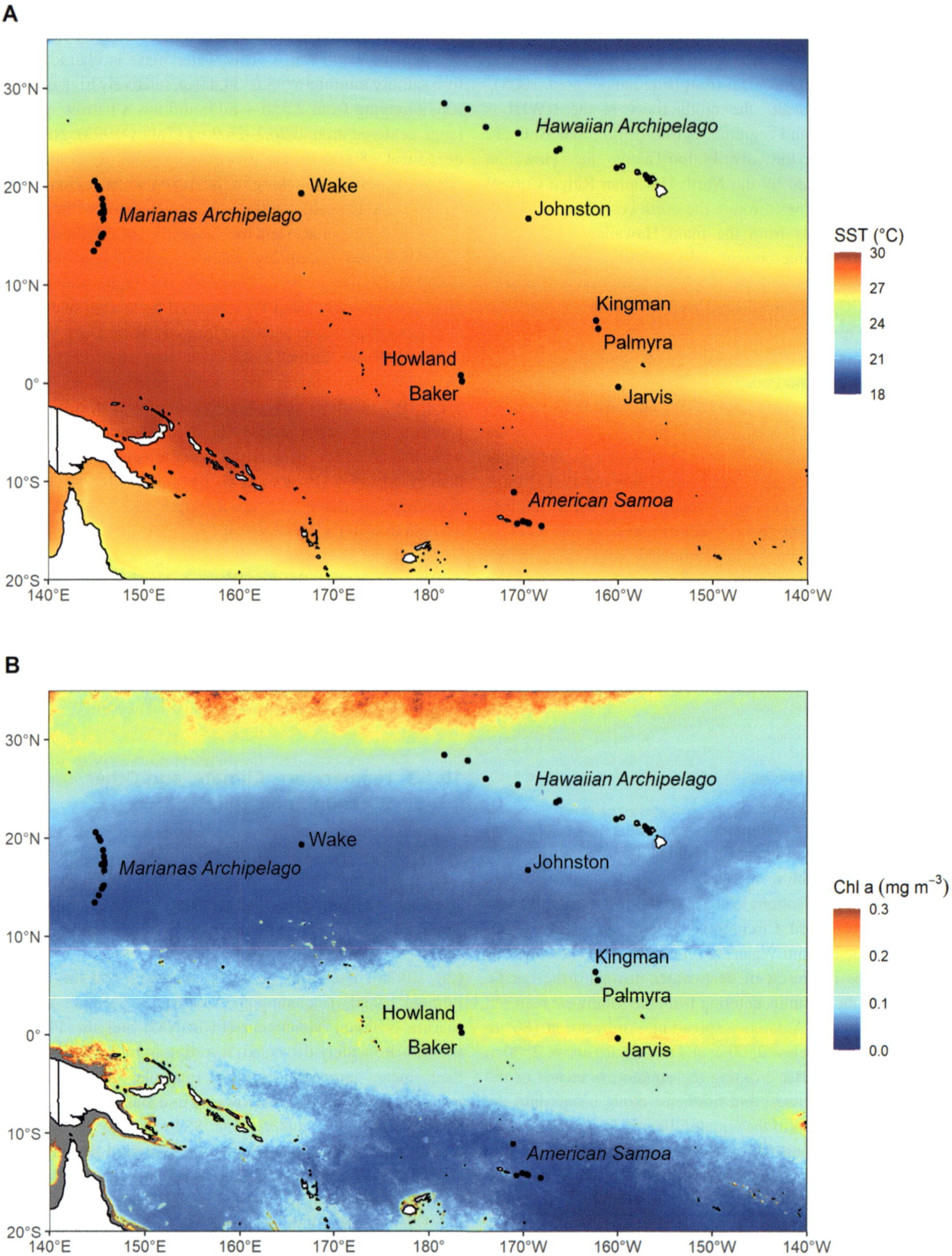

Fig. 10.2 Panel A shows the Climatological sea surface temperature averaged across 1981–2018. Panel B shows the chl-a levels for the same area and time period. Figure adapted from Brainard et al. (2019)

highest (0.2 mg/m^3) in the water around Jarvis, Howland, and Baker Islands (Gove et al. 2006). They are the most productive because they sit in the path of the Pacific Equatorial Undercurrent that extends to the west. Topographic upwelling by seafloor features forces cool, nutrient-rich water up toward the surface and enhances productivity (Maragos et al. 2008). In contrast, locations away from the equator have lower biological productivity as indicated by lower chlorophyll-*a* concentrations at Wake Atoll, Johnston Atoll, and the Archipelagos of the Marianas and Samoa. More moderate levels of productivity are seen around Palmyra Atoll and Kingman Reef because they are situated in the gradient between the productive equatorial belt and the lower productivity in the oligotrophic gyres.

Observations of key climate variables such as sea surface temperature, sea level pressure, wind vectors, and precipitation have established the existence of coupled atmosphere-ocean modes of variability in the Pacific. On interannual timescales, the most important of these coupled modes is the El Niño-Southern Oscillation (ENSO) which has its genesis in the tropical Pacific. Longer frequency companion modes to ENSO exist of which the most well-known is the Pacific Decadal Oscillation, or PDO (e.g., Zhang and Levitus 1997). The pattern of the SST anomaly field associated with longer frequency oscillations (decadal-multi-decadal) including the PDO has led researchers to describe them as a long-lived or persistent warm phase of the ENSO with warm SST anomalies in the tropics and cool anomalies at the high latitudes during the warm phase and the converse during the cool phase (Zhang and Levitus 1997). Evaluation of ENSO indices and the PDO have noted correspondence wherein ENSO appears to be modulated by the background state of the climate system and the PDO. When the PDO is in the positive phase more frequent El Niño's tend to occur and when the PDO is in the negative phase, the converse is true.

Interannual patterns in SST and chlorophyll that change with the shifting cycle of ENSO can cause changes throughout the region but the most dramatic interannual differences are seen in the equatorial waters (Brainard et al. 2018). The La Niña events are periods of strong trade winds that push cool water and increased productivity across the central equatorial Pacific. During El Niño, the trade winds abate, which slows the movement of surface water and equatorial upwelling. Temperatures rise and chlorophyll concentrations drop severely, impacting what are considered to be normal conditions. These impacts are most extreme for sites near the equator (Jarvis, Howland, and Baker Islands), somewhat less for sites of the outer equatorial transition zone including Kingman Reef and Palmyra Atoll (Gove et al. 2006). Johnston and Wake Atolls and the rest of the archipelagos are far enough removed to not feel sustained impacts from intense El Niño warming.

The underlying cause of the decadal modulation of ENSO and its interaction with decadal and interdecadal climate variability is unknown. One simple view is that the shallow circulation of the Pacific is the equivalent of an atmospheric Hadley Cell, wherein much of the water that upwells along the equator has subducted in the subtropical gyres of the extra-tropical South and North Pacific (also referred to as the Pacific Subtropical Cell (STC)) (*c.f.* Schott et al. 2004). The upwelled water returns to the subducting regions of the subtropics through an Ekman-driven surface flow. Pacific STC varies on both interannual and decadal time scales (Mcphaden and Zhang 2004; Hong et al. 2014) with STC convergence increasing during episodic La Nina events on interannual time scales (Mcphaden and Zhang 2004; Schott et al. 2004). On decadal timescales STC convergence decreased from the mid-1950s to the mid-1990s (Mcphaden and Zhang 2004; Schott et al. 2004) followed by an increase of STC convergence from the mid-1990s to 1998–2003 along with cooling of upwelled equatorial waters (Mcphaden and Zhang 2004).

While the exact mechanisms controlling STC variability on decadal time scales are not clear, there are clear links between STC variability and the PDO. For example, Mcphaden and Zhang (2004) have shown that the shift from the 1976/1977 negative to positive phases of the PDO coincide with a slowdown of the STC. In one scenario, an oceanic teleconnection is evoked where the PDO can affect the equatorial zone as a result of variations in the STC transport (Kleeman et al. 1999) and transport of temperature anomalies originating at the sea surface in the mid-latitudes advect into the equatorial thermocline via the subsurface, thus changing the thermal structure of the equatorial pycnocline (Deser et al. 1996; Gu and Philander 1997, Zhang et al. 1998). In another scenario it is argued that decadal changes in STC can play a role in modulating climate changes like PDO (Gu and Philander 1997). More recent work based on the reanalysis of 138 years of ocean data shows that decadal variability of STC is associated with the PDO by two processes involving change in the northeast trade winds and density changes driven by subduction (Hong et al. 2014) suggesting aspects of both scenarios are at play. The one thing that is clear is that current datasets are not sufficient to fully understand decadal variability in the Pacific.

The Pacific Basin also has gradients in pH and the saturation state of aragonite. A higher saturation state means more carbonate ions, which reduces the energy demand for growth in corals and other marine calcifiers. Net carbonate accretion rates for shallow-water reef systems have been reported to range from 0.016 g $CaCO_3$/cm^2/year in the Northwestern Hawaiian Islands to 0.128 g $CaCO_3$/cm^2/year in American Samoa; the rest of the monitored sites fall between these two extremes (Brainard et al. 2019). In deeper waters of the North

Pacific the aragonite saturation state is generally low, with the Aragonite Saturation Horizon (ASH) occurring at depths generally less than 500 m. The carbonate dissolution rate is also exceptionally high in the North Pacific, peaking at depths of 400–600 m (Feely et al. 2004). These factors are thought to contribute to the low occurrence of reef-forming corals in deep waters of the North Pacific (Guinotte et al. 2006; Cairns 2007). Further south in the equatorial Central Pacific (Phoenix Islands), the depth of the ASH is slightly deeper, typically between 500 and 700 m (Fig. 10.3).

Deep-water oxygen levels in the Pacific are low because of the hundreds of years that the water takes to move in and across the Pacific Basin. Oxygen averages 0.5–4.5 ml as opposed to the significantly higher (as much as 6 times) Atlantic values (Thurman 1981). The influence of oxygen level patterns on deep-sea corals, sponges, and other benthic organisms is poorly understood but has been identified as an important consideration (Wishner et al. 1990; Hanz et al. 2019). The potential change in these environmental parameters with anthropogenic impacts is discussed in Sect. 10.6.3.

10.2 Geological Setting

The oceanic islands and submarine ridges of the Pacific are volcanic pinnacles that rise steeply from abyssal depths of 4–7 km. Subjected to plate drift, subsidence, isostasy, and sea level rise, many of these features have drowned and now exist as submerged banks and seamounts. The northwesterly drift of the Pacific Plate has formed linear chains of islands and seamounts as it passes over stationary volcanic hotspots (Kennett 1982). The youngest feature in each chain is typically closest to the volcanically active spot (e.g. Island of Hawai'i) and features age moving away from the hotspot, with the furthest sites representing the oldest volcanism (e.g. Oahu—3 Ma, Kauai—5 Ma, Midway—27 ma) (Grigg 1988). Given this plate movement and the volcanic history in the Pacific Basin, it has some of the youngest (Main Hawaiian Islands) and oldest (Line Islands) volcanic archipelagos in the world. Many of the other Pacific features occur as widely spaced groups that are diverse in geologic age, spread across a large geographic area (i.e., Phoenix Islands, Line Islands, Mid-Pacific Mountains oceanic plateau) rather than belonging to a single geologic archipelago (e.g. Hawaiian or Mariana Islands). Wake Atoll is isolated on its own and is one of the oldest seafloor features in the region estimated to have formed 85–90 Ma (Hein et al. 2005). Younger features include Johnston Atoll as the northernmost surface feature of the submerged north-south Line Islands Ridge that includes Kingman Reef, Palmyra Atoll, and Jarvis Island (Koppers et al. 2003).

The seabed in the vicinity of the Phoenix Islands (including Howland and Baker Islands) has been aged at between 120 and 132 Ma, while the islands and seamounts themselves are substantially younger at 60–72 Ma (Koppers and Staudigel 2005). Together with the Howland and Baker Islands to the west, these features are estimated at between 40 and 72 Ma (Hein et al. 2005; Maragos et al. 2008). Howland and Baker were formed as part of the Tokelau submarine ridge and are the northern extent of the Phoenix Islands group. Seafloor features in this area are thought to have originated from the Macdonald hotspot (Finlayson et al. 2018).

Younger seafloor, 111–113 Ma, can be found further east in the vicinity of the Line Islands, including Kingman Reef, Palmyra Atoll, and Jarvis Island (Maragos et al. 2008; Müller et al. 1997). The Line Islands, seamounts, and undersea ridges are suspected to be of multiple volcanic origins resulting from lithospheric extension (Davis et al. 2002). Along the Line Islands, several subsided seamounts and atolls have been identified as having significant shallow tropical reef carbonate caps dating back to the mid to late Cretaceous (Maragos et al. 2008). Recent observations from seamounts near Jarvis Island have indicated that the lack of more developed carbonate cliffs suggests that some of these deeper seamounts may not have been subaerially exposed during their geological history (Auscavitch et al. 2020b).

10.3 The Reef Structure

Although the term "reef" is a well-known concept for shallow-water structures, both recent and fossil, the word becomes less accurate in describing deep-water coral structures (Roberts et al. 2009). From the purist point of view, the word "reef" derives from the Old Norse seafaring term "*rif*"—a submerged structure rising from the surrounding seafloor shallow enough to endanger sailing ships. Clearly a coral structure occurring in deep water, e.g., 500 m, might endanger a submarine but not a sailing vessel. Thus, a new term had to be coined to apply to deep-water coral structures, and many have been suggested, e.g., knoll, bank (Dons 1944), construction, hill, coral garden, lithoherm (Neumann et al. 1977), bioherm (Mullins et al. 1981), thicket (Squires 1964), coppice (Squires 1964), and mound (Henriet et al. 2001). The growing consensus was to use the word bank, although mound is used by many paleontologists (Roberts et al. 2009), and many, including government regulatory agencies, still use the word reef for deep-water structures (Roberts et al. 2009). For the purposes of consistency with the rest of the book, here we will use the term "reef." Reef mound development is usually understood to occur over millennial or even million-year time scales (Roberts et al. 2009). The occurrence of living deep-water

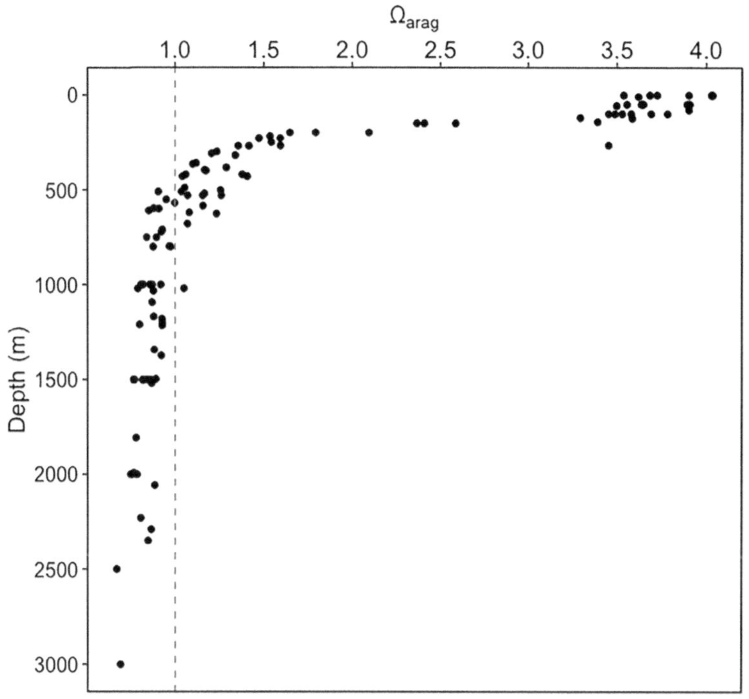

Fig. 10.3 Aragonite saturation state profile (Ω_{arag}) for the Phoenix Islands Protected Area collected from downcast CTD rosette and ROV *SuBastian* Niskin bottles during cruise FK171005 between the surface to 3000 m. A red dashed vertical line at $\Omega_{arag} = 1.0$ indicates the depth of the aragonite saturation horizon, occurring between 500 and 700 m. Data from Auscavitch et al. (2020a)

reefs is governed by: (1) location on a hard substrate with sufficient submarine topography below the depth of zooxanthellate reef-building activities, such as a seamount (2) association with a vigorous current and abundance of organics and nutrients carried in the current, such as an area of upwelling, the mouth of a fjord, or the flank of a seamount (Cairns and Stanley 1981), and (3) presence of framework-forming scleractinian corals (see below). As implied above, seamounts present an excellent opportunity for deep-water coral reefs; there are estimated to be at least 38,000 (Yesson et al. 2021) and may be as many as 100,000 (Rogers et al. 2007) in the world oceans, many of which probably support coral reefs. The first deep-water reef to be discovered, composed of *Desmophyllum pertusum* (formerly *Lophelia pertusa*), was reported by Sars (1865) off Norway. Subsequent reviews of the locations and characteristics of recent and fossil deep-water coral reefs are given by Teichert (1958), Cairns and Stanley (1981), and Stanley and Cairns (1988). They have been found to be abundant worldwide, occurring from 50 to 1500 m and from 70°N to 55°S.

Most deep-water reefs consist of a single framework-forming colonial coral species that gives structure to the reef, traps sediment, and provides niches for other animals in which to live. *Desmophyllum* (formerly *Lophelia*) *pertusum* (Linnaeus 1758) is the quintessential deep-water framework coral, which forms reefs in the northeastern and northwestern Atlantic. *Goniocorella dumosa* (Alcock 1902) is the favored species off New Zealand. *Oculina varicosa* (Lesueur 1821) forms relatively shallow (45–100 m) reefs off Florida, whereas *Enallopsammia profunda* (Pourtalès 1867) and *Madrepora oculata* (Linnaeus 1758) form the reefs along with *D. pertusum* father to the north and in deeper water on the Blake Plateau from Florida to off North Carolina. Pseudocolonial *Desmophyllum dianthus* (Ehrenberg 1832) forms the basis for the shallow reefs in the Chilean fjords. *Solenosmilia variabilis* (Duncan 1873) is the framework former off Tasmania and the south Pacific. Roberts et al. (2009) list 17 scleractinian species considered to be framework-forming species, to which can be added: *Bathelia candida* (Moseley 1881) (off Argentina), *Eguchipsammia fistula* (Alcock 1902) (Red Sea), and *Dendrophyllia alcocki* (Wells 1954) (New Caledonia). Thus, only about 2.8% of recent deep-water scleractinian species (20 of approximately 700, see Cairns 2007) are considered to be framework forming species. In addition to scleractinian corals, other Cnidaria, such as Antipatharia, Octocorallia (e.g., *Paragorgia, Corallium, Primnoa* and *Isidella*), and even stylasterids can contribute substantially to reef structure (Roberts et al. 2009).

Deep-water fossil reefs are less commonly reported and less well known, the oldest suspected reef being of Triassic age from western North America, composed of the scleractinian coral *Thecosmilia* (Stanley 1979). A Late Cretaceous reef, also from the western USA composed of *Archohelia*, was reported by Coates and Kauffman (1973). Relatively shallow reefs from the early Tertiary of Greenland and Denmark are reported by Floris (1972), constructed of *Oculina* and *Dendrophyllia*.

Given their longevity in geological time and their wide distribution, deep-water coral reefs are significant paleo-environmental archives, that can be used to estimate past seawater temperatures as well as for studies that follow the ventilation history of the ocean (Roberts et al. 2006; Rogers et al. 2007). They are thus important to the study of climate change as reflected in the ocean, as well as serving as important centers of speciation in the deep sea (Roberts et al. 2016).

10.3.1 Taxonomy

The Scleractinia of the Hawaiian Islands were first revised by Vaughan (1907), who reported 32 deep-water species and 41 zooxanthellate shallow-water species. He considered this fauna, both the shallow and deep-water components, to be a depauperate, impoverished Indo-West Pacific fauna, but having a high percentage (i.e., 70%) endemic component. Subsequent revisions of the Hawaiian Scleractinia by Cairns (1984, 2006), based on substantially more specimens, increased the number of deep-water species to 56. These species have been meticulously enumerated by Parrish and Baco (2007) and Parrish et al. (2017, 2022). Cairns (2006) found a substantial decrease of endemic species (i.e. 21%) as it was discovered that the Hawaiian fauna had much more in common with the South Pacific realm than previously thought. Forty-four of the 56 Hawaiian deep-water species are solitary in growth form and thus rather small, and therefore cannot be considered as framework-building corals. Of the remaining 12 colonial species, only three, all of which are cosmopolitan in distribution, are large and densely populated enough to be considered as a framework-building coral. The most abundant is *Enallopsammia rostrata* (Pourtalès 1878), which can form colonies several meters in height and breadth, and is quite common throughout the Archipelago. It comes in a yellow and purple form, and although only slight morphological differences can be found between these forms, it remains for molecular confirmation whether these color forms are different species or simply polymorphic color variants. *Solenosmilia variabilis* (Duncan 1873) also forms deep-water reefs in the NWHI, as described by Baco et al. (2017) (Fig. 10.4a), but is not common in this region. Finally, *Madrepora oculata* (Linnaeus 1758), which forms large framework structures in the Atlantic, occurs in the Hawaiian Islands but is quite rare (Cairns 1984). However, recent observations of mixed *Enallopsammia rostrata* and *Madrepora oculata* reef structures from Jarvis Island in the Line Islands have suggested that significant coral rubble and debris fields on island flanks may act as a foundation for reef development (Auscavitch et al. 2020b, Fig. 10.4b)

Parrish et al. (2017) list 32 taxa of structure-forming deep-water corals *sensu lato* (Cairns 2007) for the Hawaiian Archipelago and Johnston Atoll, including five scleractinian corals: the three mentioned above, plus *Madracis kauaiensis* Vaughan 1907 and *Eguchipsammia* sp., neither of which produce large structures. Their list also includes five species of octocorals in the family Coralliidae (*Pleurocorallium* and *Hemicorallium*), which, due to their dense axes formed from closely fused sclerites, form robust and large (up to 2 m height) coralla that can be considered as framework structures. Their list also includes three species of Isididae, six Primnoidae, two Chrysogorgiidae, and three Antipatharian species, all of which can produce large and hard skeletons that may be considered as framework-worthy. Although not on their list, and for good reason, there are six species of stylasterid corals (Hydrozoa: Stylasteridae) that occur in low abundance throughout the Archipelago (Cairns 2005, 2017); however, their colonies are much too small to qualify as framework species.

A survey of the annotation database for submersible and ROV dives completed by the Hawaii Undersea Research Laboratory indicates a few additional potential reef locations (HURL 2020). A single dive on Lanikai, off the Island of Oahu found a small area with a significant accumulation of *Madrepora oculata* at a depth of ~430 m. Another dive on the north side of French Frigate Shoals found an area of *Solenosmilia* reef at a depth of 701 m; however, there were no live colonies observed within the reef (C. Kelley pers. comm.). While there are numerous additional records of *Enallopsammia rostrata* and other *Enallopsammia* sp. throughout the NWHI, and Necker Ridge, there has not been an effort to compile the full distribution of this/these species. Formation of a reef by this species has only been noted at the sites indicated above, the rest of the records are of individual colonies (Fig 10.5a). *Eguchipsammia* is another colonial scleractinian that occurs commonly on some NWHI sites, as well as being recorded from Jarvis Island in the Line Islands (C. Kelley pers. comm.), but again its distribution has not been compiled and it has not been reported to form reef structures.

Outside the Hawaiian Archipelago and Phoenix Islands, in the Central Pacific, Cairns (1999) lists 134 azooxanthellate scleractinians from the Wallis and Futuna Islands and Vanuatu in the tropical Central Pacific, from depths greater than 100 m, but only 14 of these are colonial, and of those only three could be considered structure-forming: *Madrepora oculata*, *Desmophyllum pertusum*, and *Enallopsammia rostrata*. Cairns (1999) notes that these Islands had more species overlap with New Zealand than with the Hawaiian Archipelago.

10.3.2 Phylogenetics

The most abundant framework-forming deep-sea scleractinians that occur in the Central Pacific fall into three families, *Solenosmilia* is in the Caryophylliidae, *Enallopsammia* is in the Dendrophylliidae, and *Madrepora*

Fig. 10.4 (**a**) Close-up of a patch of *Solenosmilia variabilis* reef with live polyps on Southeast Hancock Seamount in the NWHI. For scale the leading edge of the biobox is ~30 cm. Photo Credit: A. Baco-Taylor FSU, and E.B. Roark TAMU, AUV Sentry, Pisces submersible Pilots T. Kerby and M. Cremer, and NSF. (**b**) An accumulation of colonial scleractinian corals composed of *Madrepora oculata* and *Enallopsammia rostrata* on the southeastern ridge of Jarvis Island between 600 and 1000 m. This image was taken at 874 m during a 2019 cruise (NA110) on the E/V *Nautilus* using ROV *Hercules*, Science Lead: S. Auscavitch, R. Pockalny. Fish from the Oreosomatidae (*Neocyttus* sp.) and Lophiidae (*Sladenia* sp.) and echinoderms (brisingid sea stars and crinoids) were most commonly associated with these reef features. Scale bar indicates 10 cm distance. Photo credit: Ocean Exploration Trust

falls into the Oculinidae. A number of phylogenetic studies have contributed to our understanding of the relationships among these families. A mitochondrial 16S phylogenetic tree shows that Scleractinia can be split into "complex" and

Fig. 10.5 (**a**) An example image of the arborescent morphology taken on by many colonial scleractinians in the Hawaiian Archipelago. (**b**) An example image of the extensive reef structure on Colahan Seamount. Seamount. Photo Credits: A. Baco-Taylor FSU, and E.B. Roark TAMU, AUV Sentry, Pisces submersible Pilots T. Kerby and M. Cremer, and NSF. Additional reef images can be found in Baco et al. (2017, 2019, 2020)

"robust" clades based on gross colony morphology, with many families paraphyletic among these clades. The families Caryophylliidae and Oculinidae are paraphyletic with representatives in both clades, while dendrophylliids fall entirely within the complex clade (Le Goff-Vitry et al. 2004). In this phylogeny, *M. oculata* does not group with any other Oculinidae and is likely its own family.

Phylogenetic analysis from COI data also suggests many scleractinian families are paraphyletic, including Caryophylliidae, Oculinidae, and Anthemiphyliidae. This phylogeny also suggests that the *Madrepora* genus should be raised to the family level. Other colonial and solitary families common to the North Pacific, Dendrophylliidae, Flabellidae, and Fungiacyanthidae, were monophyletic (Kitahara et al. 2010). Even with 12 DNA markers (Kitahara et al. 2016), Caryophylliidae and Oculinidae are paraphyletic. *Madrepora* still falls on its own branch, closest to the families Pocilloporidae and Astrocoeniidae (which is also paraphyletic). The genera *Solenosmilia, Desmophyllum,* and *Caryophyllia* all fall within the "robust" branch of Caryophylliidae and seem to be well sorted phylogenetically. Of the "complex" deep-sea families, Dendrophylliidae and Flabellidae are monophyletic and also appear to be well defined by molecular markers.

10.3.3 Deep-Sea Coral Distributions Within the Hawaiian Archipelago

Hard substrate areas on the deep slopes of the seamounts of the Hawaiian Ridge and Emperor Seamount Chain (ESC) tend to be dominated by dense octocoral and antipatharian beds as well as a diversity of sponges, rather than by reef-forming scleractinians (Parrish and Baco 2007; Baco 2007; Parrish et al. 2015; Long and Baco 2014; Morgan et al. 2015, 2019). In fact, there has been a general paucity of records of deep-sea reef-forming scleractinians in the North Pacific. This has been attributed to the shallow aragonite saturation and high carbonate dissolution rates in this region which were thought to prevent reef formation (reviewed in Sect. 10.1.3,

Feely et al. 2004, Guinotte et al. 2006, Cairns 2007). Until 2017, all scleractinians reported in the Hawaiian Archipelago were solitary cup corals, or colonial forms that generally take on a two-dimensional arborescent morphology (Fig. 10.5a) and are easily confused with an octocorallian sea fan (see Sect. 10.3.1 for a review).

Unexpectedly, Baco et al. (2017) reported the discovery of deep-sea scleractinian reefs on six seamounts of the NWHI and ESC, with subsequent observations on a seventh seamount, Colahan, by submersible and fisheries surveys (Fig. 10.5b) (Hayashibara and Nishida 2017; Miyamoto et al. 2017; Baco et al. 2019, 2020). The initial discoveries of the reefs were made during Autonomous Underwater Vehicle (AUV) imaging surveys. Sampling was not possible during these surveys, but based on images, the species comprising the reefs were tentatively identified as *Solenosmilia variabilis* and *Enallopsammia rostrata* (Baco et al. 2017). Subsequent collections on later cruises allowed for morphological examinations that confirmed these identifications. As the NWHI and ESC populations are significantly isolated from other populations of these species, genetic analyses are underway to determine if these may be divergent enough to constitute new species.

The most well-surveyed depths within the NWHI and ESC are from ~300-2000 m. Reefs were found to occur from 535 to 732 m. This was the maximum surveyed depth of the study that first observed the reefs, so they may extend deeper than that range, and in fact, a recent fisheries survey found reefs occurred to a depth of at least 850 m on Colahan (Hayashibara and Nishida 2017). The reefs showed a pattern of occurring at shallower depths moving to the northwest. Of all the features that have been surveyed, reefs have only been observed in this depth range at Koko, Yuryaku, Kammu, Colahan, NW Hancock, SE Hancock, and Academician Seamounts (Baco et al. 2017, 2019; Hayashibara and Nishida 2017).

10.3.4 Phoenix Islands Reefs

The Phoenix Islands lie along the equator between 170 and 180°W within the waters of the Republic of Kiribati, but also include the U.S. islands of Howland and Baker. The majority of the Phoenix Islands lie within the Kiribati Phoenix Islands Protected Area (Rotjan et al. 2014). To the north, Howland and Baker Islands compose a part of the U.S. Pacific Remote Islands Marine National Monument (PRINM, Howland and Baker unit). Features in these regions consist of sharp-peaked seamounts and guyots, submerged shallow-water reefs (<10 m summit depth), atolls, and islands. The Tokelau ridge runs along the western boundary of the Phoenix Islands Protected Area (PIPA) and into the Howland and Baker unit of PRINM. East of the Tokelau ridge lies a cluster of seamounts and islands, including Kanton Atoll, Rawaki Island, and Orona Atoll. Two large submerged reef complexes, Winslow Reef and Carondelet Reef lie at the northwest and southwest corners of PIPA, respectively. Prior to 2017, the deep seafloor in much of this area had not been explored leading to a dearth of knowledge of coral species inventories and gaps in biogeographic knowledge in the Pacific (Rotjan et al. 2014; Demopoulos et al. 2018).

Since 2017, a series of explorations by the NOAA Ship *Okeanos Explorer* (EX-1703), Schmidt Ocean Institute R/V *Falkor* (FK171005), and E/V *Nautilus* (NA114) have yielded substantial insights into the abundance and diversity of the Pacific bathyal coral fauna in this region. Within PIPA, framework-forming corals were largely dominated by the cosmopolitan scleractinians *Enallopsammia rostrata* (360–1522 m) and *Madrepora oculata* (644–1578 m) (Auscavitch et al. 2020a, b). The largest and most extensive stands of *Enallopsammia* were observed at Nikumaroro between 500 and 650 m where they were occasionally observed to form reef-like structures in excess of 1 m in height (Fig. 10.6). Colonies greater than 20–30 cm in height were often observed with live polyps and tissue visible in the upper third of the colony, with varying levels of fouling on the exposed skeleton below and on the underlying rubble. At the shallower end of the depth range where *Enallopsammia* and *Madrepora* corals were observed, substantial coral rubble and debris fields were often observed on these features as material fell downslope.

Colony size, combined with estimates of the growth rates of *Enallopsammia* colonies from the Central Pacific indicate these reef structures could be many hundreds of years old (see growth rates in Houlbreque et al. 2010). A second species of *Enallopsammia*, *E. pusilla*, was also observed overlapping the depth range of its congener, but only between 425 and 692 m and colonies were typically smaller in size. Framework-forming species were not observed deeper than 1600 m in the Phoenix Islands, suggesting a hard bathymetric boundary for these corals (Auscavitch et al. 2020a, b). While the number of solitary coral species outnumbered the colonial taxa, the solitary scleractinian coral *Crispatotrochus rubsescens* was frequently observed to grow in dense pseudocolonial clusters (>20 live individuals) between 519 and 636 m (Auscavitch et al. 2020a).

In the Howland and Baker Island unit of PRIMNM, structure-forming corals *Enallopsammia rostrata* and *Madrepora oculata* were observed in 2017 on the flanks off Howland Island in the 357–584 m depth range (Demopoulos et al. 2018), but were rarely observed to accumulate sufficient undergrowth or rubble debris to be identified as reefs, possibly owing to the relatively steep slope of the seafloor. However, a more recent exploration of the ridge extending northwest of Howland Island in 2019 (see Petruncio et al. 2020) revealed the existence of massive living *Madrepora*

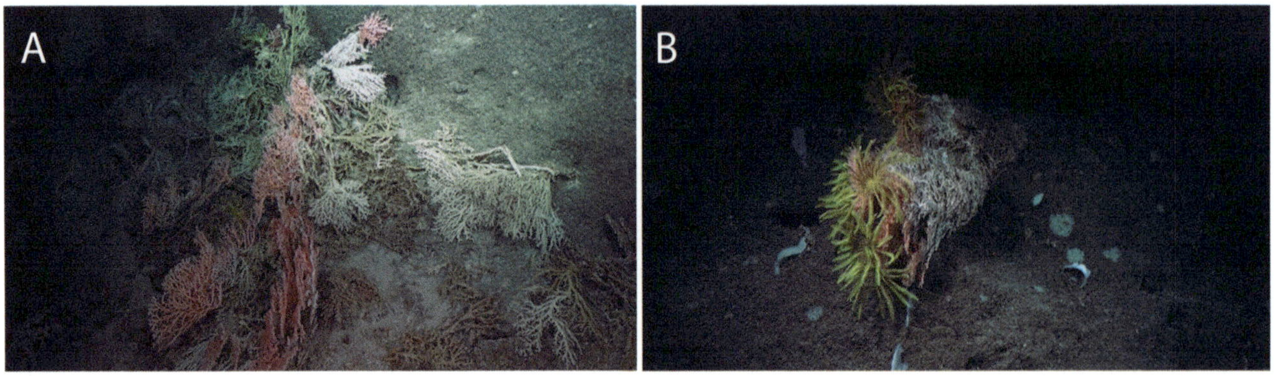

Fig. 10.6 *Enallopsammia rostrata* reef structures from Nikumaroro in the Phoenix Islands Protected Area. (**a**) Live colonies growing on an accumulation of coral rubble in sloped areas were observed to form reef structures around 620 m. (**b**) On steeper terrain, colonies were found to form reefs that projected outward into the water column. Photo credit: Schmidt Ocean Institute ROV *SuBastian*, R/V Falkor cruise FK171005, Chief Scientist: E. Cordes, Temple University

oculata reefs at 950 m depth (Fig. 10.7). These reefs were often associated with oreosomatid fishes and other associated invertebrate fauna like crinoids, brisingid sea stars, and anemones. Throughout visual surveys of the northern and southern Phoenix Islands (Demopoulos et al. 2018; Auscavitch et al. 2020a), infrequently observed calcified hydrocorals in the family Stylasteridae were not found to form the significant structures that have been observed at higher latitudes in the North and South Pacific.

10.3.5 Coral Assemblage Structure

Although the discovery of deep-sea scleractinian reefs in the NWHI and ESC was significant and unexpected due to the carbonate chemistry of the North Pacific, the reefs cover a very small area of the deep-sea slopes of these seamounts and are currently a minor component of their known coral diversity. Early work examining the distribution and community structure of deep-sea corals in the NWHI indicated that there were at least two distinct deep-sea coral communities, with an abrupt change in species composition at depths of about 600–700 m (Baco 2007; Parrish and Baco 2007). Above that depth, the economically important precious corals in the octocoral family Coralliidae and in the zoanthidean genus *Kulumanamana* were among the dominant benthic megafaunal taxa on hard substrates and were the focus of most studies. However, a high diversity of other octocorals and antipatharians are also found in these depths. Below 600–700 m, octocorals of the families Primnoidae, Isididae, Chrysogorgiidae, and Coralliidae are the dominant taxa.

Clark et al. (2009) further found three depth assemblages for the benthic megafauna of the Hawaiian Archipelago based on a meta-analysis of species annotations from videos taken with the Pisces submersibles from the Hawaii Undersea Research Lab's video annotations database. These were from 200–350 m, 360–600 m, and 600–1800 m. This work showed that cnidarians were the most diverse benthic megafaunal group in the Hawaiian Archipelago across the studied depth ranges, representing about 30% of the observed species. Further analyses of the same dataset by Schlacher et al. (2014) also found significant variation with depth and geographically within and among seamounts.

Continued explorations in the Archipelago have allowed for a more detailed understanding of the distributions of species and the variability in community structure among and within deep-sea coral beds of the Hawaiian Ridge. Long and Baco (2014) focused on the largest known precious coral bed in the Archipelago at Makapu'u off the island of O'ahu. They found that within a single continuous coral bed, there was a significant changeover in community structure, with six distinct assemblages across a depth range of only 320–534 m. Among the groups, there was a 93% dissimilarity in community structure. Primnoid and coralliid octocorals were the dominant families within the bed; 9 total octocoral families occurred along with abundant anitpatharians and sponges. Colonial scleractinians did occur at this site, with *Enallopsammia rostrata* dominating one of the 6 coral assemblages and occurring from depths of 473–534 m. However, at this site, all of the occurrences were of the arborescent form rather than the framework-building form, similar to Fig. 10.5a.

Morgan et al. (2015) looked at the distribution of deep-sea megafauna on hard substrates at deeper depths of 1400–1800 m along the Necker Ridge. They found that community structure varied less with depth, but more among sites. Northern sites were dominated by octocorals, principally the families Isididae and Chrysogorgiidae, and southern sites were dominated by crinoids. The colonial scleractinian *Enallopsammia rostrata* also occurred at a subset of the sites of this study, but again only as an arborescent form.

Fig. 10.7 A massive *Madrepora* reef structure approximately 2 m in height northwest of Howland Island at 950 m depth. This habitat was associated with Oreosomatid fishes (*Neocyttus* sp.), cutthroat eels (Synaphobranchidae), attached anemones, and echinoderms (crinoids, brisingid stars). Imagery was acquired by the Ocean Exploration Trust E/V *Nautilus* and ROV *Hercules* during cruise NA114 in 2019, Science Leads: E. Petruncio, A. Weinnig.

In the most extensive survey of a single large seamount so far published, Morgan et al. (2019) surveyed depths of 200–700 m on 3 sides of Mokumanamana using an AUV. This seamount had a very high proportion of soft sediment substrate, so the results are not as informative for the distribution of deep-sea corals, but did show an abundance of octocorals and antipatharians in the few areas of hard substrate. What is significant about this study in terms of the distribution of megafauna on seamounts though is that significant variation in benthic megafaunal communities was observed, not only with depth, but also among the different sides of the same seamount. Analyses of community structure showed nine distinct assemblages, with only one occurring on all three sides. The others were a subset of sides and depth combinations. Similar to the Long and Baco (2014) study, this work also found that small depth differences can make a large difference in assemblage structure and dominant species. This study covered a larger geographic area than the 2014 study and found that horizontal distances of only 30–80 km along a depth contour can also result in substantially different benthic communities. Similar results were found for fish assemblages from the same transects (Mejía-Mercado et al. 2019). Together these results indicate that seamounts can have strong variability in habitats, communities, and dominant species over relatively short spatial and depth scales. They also indicate that surveying small areas of a seamount will result in missing a large proportion of the total species diversity and many assemblage types. This fact was exemplified by the very recent discovery of scleractinian reefs (Baco et al. 2017, outlined in previous sections) that remained undetected after nearly two decades of exploration in the Hawaiian Archipelago.

Recent explorations of deep-water features in the Phoenix Islands Protected Area have also revealed fine-scale oceanographic and bathymetric changes in community structure across the bathyal depth range. Using a combined dataset from *Okeanos Explorer* cruise EX1703 (Demopoulos et al. 2018) and R/V *Falkor* expedition FK171005, Auscavitch et al. (2020a, b) examined occurrences of deep-water corals from the Octocorallia, Scleractinia, Antipatharia, and Zoantharia among 17 unique features across 24 dives between 152 and 2439 m in PIPA. Octocorals were found to be the most diverse and abundant taxa in the region with 78% of all coral observations and 113 identified morphospecies. High species turnover was observed with increasing depth across the upper 600 m of the water column, indicating significant changes in species distribution primarily along rapidly changing temperature and dissolved oxygen gradients. Below 600 m, coral communities trended toward greater similarity in composition, but revealed smaller-scale assemblage differences. For example, a multivariate analysis of all morphospecies in increments of 100 m depth bins revealed 8 distinct bathymetric assemblages: 100–300 m, 500–700 m, 800–1000 m, 1100–1300 m, 1300–1500 m, 1600–1800 m, 1800–2000 m, and 2000–2300 m (Auscavitch et al. 2020a, b). When comparing assemblages among feature types (e.g., seamounts, atolls, islands, and shallow submerged reefs) and controlling for depth, few significant differences were identified indicating that feature types generally hosted similar coral communities at similar depths.

10.3.6 Deep-Sea Coral Distributions in Relation to Environment

Without question the strongest factor determining benthic species distributions and community structure within the Hawaiian Archipelago hard substrate habitats is depth (Long and Baco 2014; Schlacher et al. 2014; Morgan et al. 2019; Mejía-Mercado et al. 2019). However, depth itself may not be as much the influencing factor as the many environmental parameters that are strongly correlated with depth such as temperature, pressure, particulate organic carbon flux, etc. Depth appears to have a greater influence at the shallower end of the depth range than at the deeper end, as species turnover with depth has been observed to be much greater above 600 m (Long and Baco 2014; Morgan et al. 2019; Mejía-Mercado et al. 2019) than it is below 1500 m (Morgan et al. 2015; Schlacher et al. 2014). This pattern of beta diversity is consistent with large-scale depth patterns observed in other taxonomic groups (e.g., Howell et al. 2002).

As many factors are correlated with depth, deep-ocean water masses can serve as important factors to summarize the effects of multiple oceanographic variables on deep-sea coral assemblages. Overlying water mass variables have been shown to be important factors in community assembly in the equatorial Central Pacific (Auscavitch et al. 2020a), as well as exhibiting strong effects on beta diversity (species replacement) along interfaces between water masses (Victorero et al. 2018). Further examination of Phoenix Islands coral community structure among bathyal water masses in the equatorial Central Pacific found significant effects of overlying water mass on coral community structure. In the shallowest water mass, Pacific Equatorial Water (PEW, 200–500 m), assemblage similarities were driven by the cosmopolitan scleractinian *Enallopsammia rostrata*, plexaurids, and the Pacific primnoid *Paracalyptrophora hawaiiensis*. Underlying this, in Equatorial Pacific Intermediate Water (EqPIW, 500–1500 m) similarities were likewise driven by abundances of *Enallopsammia rostrata*, plexaurids, but also the widely distributed chrysogorgiid *Iridogorgia magnispiralis*. In Pacific Deep Water (PDW, 1500–2439 m), widespread Pacific chrysogorgiid taxa dominated including *Iridogorgia magnispiralis*, *Chrysogorgia* sp., and *Pleurogorgia* cf. *militaris*. Given the large spatial extent and coverage of these water masses and associated currents at Pacific bathyal depths, they are likely to contribute to connectivity among isolated seamounts and islands throughout the basin (Watling et al. 2013).

For scleractinian reefs specifically, other factors appear to have an influence on their distributions. Because they are carbonate structures, the ASH would be expected to have the strongest influence on the distribution of scleractinian reefs (e.g. Tittensor et al. 2009, 2010, Davies and Guinotte 2011); however, the distribution pattern of the reefs does not match the current depth of the ASH, with the reefs occurring at Ω aragonite levels as low as 0.71. Additionally, the distribution of reefs gets shallower as the ASH becomes deeper (Baco et al. 2017). Baco et al. (2017) explored other environmental variables that might instead explain the locations where the reefs occur. They found that there is higher chlorophyll-α and a change in surface current direction at the sites where the reefs occur. However, neither of these factors can explain the depth distribution of the reefs.

For deep-sea coral assemblages in the Hawaiian Archipelago in general, besides depth, seafloor rugosity is another common factor related to assemblage distributions (Long and Baco 2014; Morgan et al. 2015). Additional factors correlated with deep-sea coral assemblage distributions include slope and relief (Long and Baco 2014), as well as latitude, percent sediment cover, temperature, oxygen, and salinity (Morgan et al. 2015). Morgan et al. (2019) found that for the overall benthic community assemblages on Mokumanamana, sediment, oxygen, substrate variability and rugosity, POC, and surface currents were most strongly related to assemblage distributions.

An open question of interest is the effect of substrate type (carbonate, manganese, lava, sunken vessels) on the colonization of corals. Community development is driven by the environment, substrate type, and elapsed time, so the importance of substrate type is hard to discern until we better understand the corals' environmental requirements. Available studies in the Pacific do describe seafloor differences among coral community assemblages between islands and reefs spanning a broad range of geologic ages (e.g. Dolan et al. 2008; Parrish 2015). One study by Putts et al. (2019) compared geologically dated lava substrates on the flanks of Hawaii's Mauna Loa volcano, to examine the effect of elapsed time (decades to millennia) on community development where the substrate and the passing water mass were consistent. They found an increase in diversity measures with flow age within historic flows (Putts et al. 2019).

Another aspect of substrate that is important in this region due to potential mining are manganese crusts (see Sect. 10.6.2). Schlacher et al. (2014) did not make an overall comparison of assemblages to environmental data, but instead compared the community structure of cobalt-rich manganese crust sites to that of non-cobalt rich sites and found that they were significantly different. They also found that species richness did not differ between the two crust types. Future survey work should look for further opportunities for similar comparative work paired with environmental monitoring.

An environmental factor for deep-sea corals that is likely to be important, but far more challenging to study, is fluid flow. Environmental forces operate at different scales throughout the Pacific Islands causing wide variations in the level of productivity and rate of food delivery to seafloor suspension feeding corals. These include basin-level scales associated with the latitudinal productivity gradient and other forces that are more regional-level scales associated with feature-related upwelling (e.g. currents and seafloor forcing). Hydrography is also an important factor in shaping Atlantic cold-water reefs, bioherms, and mounds. These cold-water coral mounds are more likely found in areas of intensified near-bottom currents, compared to investigated locations without corals present (Mohn et al. 2014). Currents are an important factor in models of *M. oculata*, *S. variabilis*, and *E. rostrata* habitats on New Zealand seamounts as slow tidal and orbital velocity flows were more indicative of finding coral communities compared to regional flow regimes (Tracey et al. 2011). Submarine canyon *D. pertusum* and *M. oculata* studied *in situ* over an 8-year period experienced growth that was correlated to annual and interannual variability in hydrodynamics, with high budding rates and high linear extension relating to a medium intensity of water plume events and increased organic material supply, versus the high intensity and low intensity events (Chapron et al. 2020).

Seamount communities are especially influenced by current but there is a high degree of variation in flow regimes on different seamounts. Tide has been reported as the primary flow signal (Matsuyama et al. 1993; Musgrave et al. 2016) with semidiurnal (M_2 period) tides dominant in Hawaii (Merrifield et al. 2001; Merrifield and Holloway 2002; Rudnick et al. 2003). The required flow rate for coral growth is dependent on the organic component suspended in a passing water mass (Emery 2001). Few studies have tried to identify threshold environmental values for coral growth. The most recent effort was some comparative observations made between three known coral patches in Hawaii (Parrish and Oliver 2020). A coral patch in the upwelling lee of the island of Hawaii (Bidigare et al. 2003; Seki et al. 2002) showed a 1/3 greater backscatter strength than other patches in the main islands suggesting more food is carried by passing water (Iida et al. 1996) and explained the presence of deep corals at a lower delivery rates (Orejas et al. 2011; Carlier et al. 2009; Naumann et al. 2011). The observations were focused on gorgonian octocorals and zoanthids but a cluster of *Eguchipsammia* sp. was seen at the shallowest site (311 m) on the high flow peak (mean of 9.6–13.5 cm/s) of a pinnacle feature.

10.3.7 Habitat Suitability and Species Distribution Modeling

Data on framework forming scleractinians in the North and Central Pacific have only recently become available, so their records have not been incorporated into deep-sea coral or habitat suitability models or species distribution modeling to date (e.g., Costa et al. 2015; Bauer et al. 2016). However, efforts to create these models are currently underway by at least 2 groups (unpubl data, M. Poti pers comm.). Until these models are published, we can refer to global level habitat suitability models to gain insights into the potential habitat areas for framework-forming scleractinians in the North and Central Pacific. The first such global modeling effort included all Scleractinia on seamounts (Tittensor et al. 2009). It showed that there was no suitable habitat for scleractinians in the Central Pacific below 1000 m and only a small amount of habitat suitability (0.3 on a 0 to 1 scale) in the Northwest Pacific north of about 20°N and west of about 160°W. This study also found that the environmental factors of aragonite saturation state, and oxygen saturation and concentration were most strongly correlated with the distribution of deep-sea scleractinians, along with low levels of dissolved inorganic carbon, nitrate, phosphate, and silicate.

Davies and Guinotte (2011) subsequently focused specifically on five species of framework forming scleractinians including *Enallopsammia rostrata, Madrepora oculata,* and *Solenosmilia variabilis*. They conclude that there is generally very low habitat suitability in the Central Pacific for all species, however, a closer examination of the figures from this paper shows very high suitability for *Solenosmilia* on Kammu and Koko Seamounts in the far NWHI and ESC, but not at any other sites north of about 10°S. *Enallopsammia rostrata* had much higher habitat suitability in general across the North and Central Pacific, with extremely high habitat suitability on many of the seamounts near the bend of the ESC and far NWHI as well as throughout the entire Hawaiian Archipelago. The Phoenix Islands also showed high suitability for this species on several seamounts. It is important to note though that this study did not distinguish between the individual colony form of *E. rostrata* and the reef-forming occurrences in its model building. *Madrepora oculata* showed high habitat suitability on some sites within the Phoenix Islands and on the northern slopes of many of the seamounts of the Hawaiian Archipelago. It also showed very high suitability on many of the seamounts near the intersection of the NWHI and ESC. Intriguingly, *Desmophyllum*

(Lophelia) pertusum and *Goniocorella dumosa* also show moderate to extremely high habitat suitability, respectively, on Koko and Kammu seamounts.

Overall for all five species in the Davies and Guinotte (2011) study, occurrences were most strongly correlated with the environmental factors of depth, temperature, aragonite saturation state, and salinity. Based on environmental data, the occurrences of both *Solenosmilia variabilis* and *Enallopsammia rostrata* were concentrated at lower temperatures, lower aragonite saturation state, lower POC flux, lower oxygen, and higher phosphate than the other 3 species studied, even though the depth range and slope parameters were comparable.

These early studies included fairly low-resolution data for many environmental parameters including bathymetry and in some areas may have overestimated suitable habitat (Davies and Guinotte 2011). Subsequently higher resolution data have been used to examine the distribution for *S. variabilis* in the South Pacific. Although the area does not overlap with our geographic areas of interest, this work is highly relevant as it showed that *S. variabilis* habitat occurs most commonly at the summit/slope break and along the ridges and pinnacles on the slopes of seamounts in that region (Rowden et al. 2017).

Global level habitat suitability studies have also been undertaken for octocorals and show extremely high habitat suitability for octocorals throughout the Central Pacific (Yesson et al. 2012). Parameters most correlated with octocoral distributions included temperature, salinity, broad-scale slope, productivity, oxygen, and calcite saturation state. Antipatharian corals also show consistently high habitat suitability across all seamounts of the North and Central Pacific (Yesson et al. 2017). In this model, temperature was the strongest correlate with occurrence data, along with topography, surface productivity, and oxygen. These models are consistent with observations that show octocorals and antipatharians as abundant taxa on all sites so far studied in the North and Central Pacific (e.g. Parrish and Baco 2007; Parrish et al. 2015; Baco 2007; Auscavitch et al. 2020a, b)

10.3.8 Reef Formation and Growth

The ages and growth rates of individual deep-sea corals prevalent in North and Central Pacific Seamounts have been well documented. The proteinaceous colonial zoanthid *Kulumanamana haumeaae* (Gold coral) and the deep-water black coral *Leiopathes* sp. were found to be among the longest-lived organisms with lifespans in excess of 2700 years (radial growth rate of 41 ± 20 μm/year) and 4000 years, respectively (radial growth rate of less than 10 μm/year) (Roark et al. 2006, 2009; Parrish and Roark 2009). *Corallium* sp. (pink and red coral) colony age estimates range between 50 and 180 years, with radial growth rates of 100s of μm/year (Druffel et al. 1990; Marschal et al. 2004; Andrews et al. 2005; Roark et al. 2006) while bamboo coral (Family Isididae) have lifespans ranging from 50 to 420 years and radial growth rates ranging 50 to 100 μm/year (Roark et al. 2005; Hill et al. 2011; Thresher et al. 2011).

Most of the research on reef-building scleractinian corals has focused on the Atlantic Ocean. *Desmophyllum pertusum* and *Oculina varicosa* are two of the most important constituents of deep-water coral reefs as they form bioherms or mounds up to tens of kilometers long and up to 30 m high (Reed 2004). They are often found in association with *Enallopsammia profunda*, also a major structure-forming species (Cairns 1979; Rogers 1999) and *Madrepora oculata* and *Solenosmilia variabilis* (Reed 2002; Reed et al. 2006) as secondary structure-forming species. Uranium/thorium (U/Th) dating has shown that *D. pertusum* reefs from lower latitudes like those off of NW-Africa, the Mid-Atlantic Ridge and the Western Mediterranean Sea have been continuously growing for the last 50,000 years (Schröder-Ritzrau et al. 2005). Higher latitude carbonate mounds appear to have more ephemeral growth based on changing environmental conditions associated with glacial-interglacial cycles (Raddatz et al. 2020).

In contrast to Atlantic Ocean reefs that form mounds, North and Central Pacific deep-sea reefs associated with seamounts have not been found to accumulate sediments and thus the accumulation rate of those reefs likely depends on the growth rates of the most important constituent of the reefs; *S. variabilis* and *Enallopsammia rostrata* (Baco et al. 2017). Fallon et al. (2014) estimated linear extension rates of 0.84–1.25 mm for *S. variabilis* from Tasmania. *Enallopsammia rostrata* collected in the Line Islands and dated by the U/Th method have lifespans ranging from 209 ±8 to 605 ±7 years (radial extension rate 12–70 μm/year) (Houlbreque et al. 2010) while a single *E. rostrata* specimen from the North Atlantic had a lifespan of ~100 years (radial extension rate 70 μm/year) and a linear extension rate of 5 mm/year based on lead isotope (^{210}Pb) measurements (Adkins et al. 2004). The growth rates of *S. variabilis* and *E. rostrata* are similar to *Desmophyllum pertusum* and *Oculina varicosa*, the dominant corals forming bioherms in the Atlantic. *D. pertusum* in the northeastern Atlantic have axial growth rates of 5–26 mm/year based on aquaria observations, isotopic analyses, and estimates of age inferred from growth on man-made structures (Mortensen et al. 1998, 2001; Gass and Roberts 2006). Transplanted *D. pertusum* from the Gulf of Mexico had a range of axial growth rates with new polyps exhibiting faster growth rates (16 mm/year) than more mature polyps (≤5 mm/year) (Brooke and Young 2009). *Oculina varicosa* living at 80 m depth had an axial branch growth rate of 16.1 mm/year (Reed 1981).

With the exception of the *S. variabilis* reefs accumulating at a rate of 0.27 mm/year on Tasmanian Seamounts over the last 47,000 years (Fallon et al. 2014), no other estimates exist for reef accumulation rates in the Pacific Ocean. Raddatz et al. (2020) measured accumulation rates of a mound structure off the Brazilian Margin using ^{230}Th/U age constraints where *S. variabilis* was the dominant framework-building deep-sea coral. In this setting there were six clusters of distinct aggradation periods of *S. variabilis* lasting between ~0.5 to 4.0 ka (0.47, 0.52, 0.69, 1.76, 2.04, and 4.0 ka years) between 13.65 to 158.41 ka years (Raddatz et al. 2020). Accumulation rates during these periods ranged from 54 to 80 cm/ka years (~0.7 mm/year) (Raddatz et al. 2020), very similar to the *S. variabilis* reefs accumulation rate (0.27 mm/year) on Tasmanian Seamounts. Clearly more work is needed to determine what the accumulation rates of deep-sea reefs are in the North and Central Pacific, whether that accumulation has been continuous or sporadic, and if accumulation has been sporadic, what are the environmental controls on accumulation rates. Luckily both *S. variabilis* and *E. rostrata* can be precisely dated using both radiocarbon (up to 50,000 years ago) and U/Th over the time periods of interest (0–200,000 years).

10.4 Coral Biology

Because of their very recent discoveries, publications on the growth, biology, connectivity, and reproduction of the North and Central Pacific reef corals are not yet available. However we can infer important information on these topics for the key North and Central Pacific species from studies of the same species in other geographic locations.

10.4.1 Reproduction

Cold-water scleractinian corals are typically gonochoric with separate male and female polyps. Only three species of *Caryophyllia* in the North Atlantic are known to have hermaphroditic individuals (Waller 2005; Waller et al. 2005; Pires et al. 2014). Of those three, *C. ambrosia* (Alcock 1898) is also listed as a North Pacific species (Parrish et al. 2017). Evidence of hermaphroditism was also found in *Madrepora oculata*, though the presence of both gametes in a polyp was only found in one sampled colony; the other colonies were gonochoric (Pires et al. 2014).

Gonochoric colonies typically have two modes of timing of gamete release, continuous or seasonal. *Enallopsammia rostrata* is a known continuous spawner (Burgess and Babcock 2005; Pires et al. 2014) while *Solenosmilia variabilis* has seasonal release, typically associated with the Austral fall plankton bloom in New Zealand (Burgess and Babcock 2005). However, evidence of continual gametogenesis was also found in the South Atlantic, suggesting different modes for different ocean basins (Pires et al. 2014). *M. oculata* showed seasonal spawning in the North Atlantic (Waller and Tyler 2005), but showed continuous spawning in the South Atlantic (Pires et al. 2014).

Nearly all cold-water scleractinian corals studied release large oocytes and have lecithotrophic larvae (reviewed in Waller 2005, Pires et al. 2014, Feehan et al. 2019). Brooding is rare among deep-sea scleractinians, with only three Antarctic species of *Flabellum* solitary corals known to be brooders (reviewed in Waller 2005), though in shallow waters off the California coast *Balanophyllia elegans* also broods its larvae (Fadlallah and Pearse 1982). No studies have yet shown if species of *Balanophyllia* or *Flabellum* in the North Pacific are also brooders. As with brooding, asexual reproduction also appears to be far rarer in cold-water scleractinians than is found in shallow-water species, especially in hermatypic corals (Harrison and Wallace 1990). Only *Solenosmilia variabilis* in the South Atlantic and *Eguchipsammia fistula* in the Red Sea showed evidence of asexual reproduction (Mughal 2012; Pires et al. 2014). It is highly likely that many cold-water corals could reproduce through fragmentation, but few deep-water ecosystems experience the same wave-based disturbance that fragments shallow-water corals. Fragmentation can come from human impacts however, and areas recovering from heavy trawling in the NWHI showed evidence of regrowth from fragments for both *S. variabilis* and the octocoral *Hemicorallium lauuense* (see Baco et al. 2019).

10.4.2 Connectivity

Population genetics studies for scleractinian species in the North and Central Pacific have not been undertaken, limiting our understanding of the connectivity of these corals in this area. However, studies for four species have been done within the South Pacific, with three of those species also commonly found in the North and Central Pacific.

The common solitary coral *Desmophyllum dianthus* has been assessed for connectivity using mitochondrial markers (16S, MtC), nuclear markers (ITS) (Miller et al. 2011), and microsatellite markers (Miller and Gunasekera 2017). Populations throughout the South Pacific showed strong differentiation by depth when comparing the mitochondrial and nuclear markers between large geographic distances. The mitochondrial and nuclear markers had numerous haplotypes distinct to deeper waters off Australia's shelf while shallower samples shared more haplotypes, even across the ocean basin between New Zealand and Chile (Miller et al. 2011). At a smaller scale of within Tasmanian seamounts, microsatellite markers showed similar depth stratification for *D. dianthus*, with different populations occurring between ~600 m and 1000 m depth, yet showing moderate dispersal capabilities between populations within the same depth stratum (Miller and Gunasekera 2017). *D. dianthus* also shows high genetic diversity in the microsatellite markers, with only two pairs of shared multi-locus genotypes from 326 samples (Miller and Gunasekera 2017), which also indicated very low levels of asexual reproduction.

For the reef-forming coral *Solenosmilia variabilis* on Tasmanian seamounts, microsatellite markers showed low genetic diversity and low gene flow between seamounts, indicating populations were likely isolated (Miller and Gunasekera 2017). High numbers of clonal individuals also indicated high levels of asexual reproduction (Miller and Gunasekera 2017). For *S. variabilis* populations on New Zealand seamounts over a wider geographic area, population structure was not found between seamounts using ITS or D-loop haplotypes, but 27 microsatellite loci were able to determine population structure at fine scales between seamounts and at large scales between regions (Zeng et al. 2017). The same project assessed population structuring for *Goniocorella dumosa* and *M. oculata* and found genetic structuring for *M. oculata* at the regional scale for ITS, but no structuring for *G. dumosa* for either mitochondrial marker. Microsatellite markers for *M. oculata* showed population differentiation again only at the larger regional scale and not between features. *G. dumosa* interestingly showed differentiation between regions and between seamounts, but not within regions between provinces (Zeng et al. 2017). Estimated population sizes were low for all species, suggesting only 20-60 individuals contribute to the current recruitment stock, potentially due to late maturity and long lives of coral species (Zeng et al. 2017). If the framework-forming populations of the Central Pacific follow these patterns, we can predict low effective population sizes and strong structuring among populations. Both of these characteristics make species extremely vulnerable to human impacts.

10.5 Reef Ecosystem

Structure-forming species are recognized as critical biodiversity hotspots that offer a range of microhabitats that are likely to be used for other species to feed and reproduce (Buhl-Mortensen et al. 2010; Watling et al. 2011). Nevertheless, numerous anthropogenic threats, such as trawling and mining, are destroying and degrading these species and, consequently, species associated with these structures (Koslow et al. 2001; Baillon et al. 2012). For instance, 30–50% of *Desmophyllum pertusum* reefs have been damaged by the use of trawls off Norway (Fosså et al. 2002) and above 90% of *Oculina* present on a reserve off the east coast of Florida have been diminished to piles of rubble (Koenig et al. 2000). In the Northwestern Hawaiian Ridge (NHR) and lower Emperor Seamount Chain (ESC), the seamounts' ecosystems have been affected by extensive fishing between the 1960s and 1980s, focused on two pelagic fish species; armourhead (*Pseudopentaceros wheeleri*) and alfonsino (*Beryx splendens*) and on deep-sea precious corals (Grigg 2002; Clark et al. 2007; Baco et al. 2019, 2020). Because of the potential interdependence of a variety of invertebrates and fishes with deep-sea corals, and the extensive impacts to deep-sea corals, it is critical to gain a better understanding of the diversity, species compositions, and nature of relationships of coral-associated fauna. This information can help to predict the responses of the species involved when there is a loss or disturbance to these habitats (Matias et al. 2011).

10.5.1 Invertebrate Associations

A direct study of the invertebrates associated with North and Central Pacific deep-sea reefs has not yet been undertaken; however, we have observed crinoids, anemones, urchins, brisingids, and ophiuroids through general processing of reef images from the ESC and NWHI sites for other studies (unpubl data). These observations are consistent with observations on deep-sea reefs in other locations where associated taxa include other coral species and suspension feeders that utilize the corals for protection and for additional height to rise above the benthic boundary layer into faster current regimes (O'Hara et al. 2008; Buhl-Mortensen et al. 2016). Ophiuroid assemblages collected from living, and dead *Solenosmilia* and rubble from New Zealand seamounts and the continental slope were found to be distinct from adjacent habitats without *Solenosmilia* thickets, evidence of an association that is dependent on the physical structure provided by the coral (O'Hara et al. 2008). Besides providing structure, corals may be sloppy feeders that provide undigested food that can be utilized by shrimps and amphipods and other fauna, which may contribute to the abundance of diversity of species found in these communities (Buhl-Mortensen et al. 2017).

Although the species diversity of the framework-forming corals themselves is relatively low on any particular deep-water reef, the total associated faunistic diversity can be quite high, just as it is in shallow-water tropical coral reefs. For instance, Burdon-Jones and Tambs-Lyche (1960) reported 300 species associated with a Norwegian coral reef. Koslow et al. (2001) list 242 benthic invertebrates on Tasmanian seamount *Solenosmilia* reefs. Rogers et al. (2007) recorded 526 species of coral species alone on reefs and seamounts worldwide (although not all seamounts harbor coral reefs). Approximately 32 benthic groups, representing 13 phyla, were collected from a Subantarctic South Pacific seamount at 550 m, along with literally tons of the framework-forming coral *Solenosmilia variabilis*. And finally, over 1300 species have been found on a *Desmophyllum pertusum* reef in the northeast Atlantic (Roberts et al. 2006).

Specialization of associates to the coral ecosystem is demonstrated by one of the many associated species to *S. variabilis* reefs in the South Pacific, a stylasterid coral (*Adelopora pseudothyron* Cairns 1982). This species has a moveable, hinged operculum (0.5 mm in width) that completely seals off its gastrozooid and seemed to have evolved the perfect adaptation in this highly competitive environment, but several opercula were found to have a circular drill hole (0.1 mm in diameter) probably made by a molluscan radula that penetrated the center of the operculum, showing that the evolutionary arms race is alive and well on deep-water coral reefs (Cairns 1982; Cairns and Stanley 1981).

10.5.2 Fish Associations

Associations between deep-sea fishes and structure-forming deep-sea corals have been increasingly studied worldwide (Mortensen et al. 1995; Fosså et al. 2002; Heifetz 2002; Husebø et al. 2002; Reed 2002; Auster et al. 2005, Costello et al. 2005; Morgan et al. 2005; Parrish 2006; Ross and Quattrini 2007; Buhl-Mortensen et al. 2010; D'Onghia et al. 2012). Two different types of associations have been identified between deep-sea fishes and structure-forming deep-sea corals: obligate and facultative. Known obligate relationships include the use of specific deep-sea coral species for reproduction (e.g., egg nurseries) or for shelter. For example, Mortensen et al. (1995), Husebø et al. (2002), and Costello et al. (2005) observed the importance of reefs formed by *Desmophyllum pertusum* for some fishes in the northeastern Atlantic. In the same area, Foley et al. (2010) noted that cold-water corals are fundamental habitats for the redfish *Sebastes* spp.

Facultative associations occur when deep-sea fishes indiscriminately use different species of deep-sea corals, sponges or other attributes of the landscape as a refuge, to feed, rest, and/or reproduce (Auster et al. 2005; Costello et al. 2005), without drastic changes in abundance when these features are absent (Auster et al. 1995). Auster et al. (2005) and Tissot et al. (2006) implied that deep-sea corals are not as important as other reef structures for fishes in the northwestern Atlantic and off Southern California, respectively. However, regardless of the type of association that may occur, studies of correlation and predictive models have demonstrated higher densities and sizes of adult fish close to deep-sea corals than in non-coral areas (Husebø et al. 2002; Auster et al. 2005), arguing that the occurrence of high abundances of structure-forming deep-sea species favors high abundances of fishes, whether or not there is a specific association (Auster et al. 2005).

Studies conducted to understand these two types of associations in the deep-sea have mainly focused on the stony corals belonging to the subclass Hexacorallia, order Scleractinia, genera *Desmophyllum* and *Oculina* (Auster et al. 2005; Buhl-Mortensen et al. 2010). On stony corals, Fosså et al. (2002) observed that reefs of *Desmophyllum pertusum* at depths of 300–500 m off the coast of Norway harbored significant densities of *Sebastes viviparus* in a gravid stage. Husebø et al. (2002) carried out a fisheries study in the same area to make comparisons inside and outside of *Desmophyllum pertusum* reefs and observed inside of the reef an increase in the catch and in the size of redfish (*Sebastes marinus*), ling (*Molva molva*), and tusk (*Brosme brosme*). Gilmore and Jones (1992) and Koenig et al. (2000) found a large number of individuals of gag grouper (*Mycteroperca microlepis*) and scamp (*M. phenax*) aggregated for reproduction in the stony coral *Oculina varicosa*, while juveniles of speckled hind (*Epinephelus drummondhayi*) were using the same species of coral as a refuge.

Although scleractinian corals are not the most abundant type of coral in the NWHI (Baco 2007), previously unpublished observations of AUV and ROV images and videos captured at depths of 250–650 m on seven seamounts in this area show associations of these structure-forming species with some species of fishes. From images that also contained scleractinian reef, 28 fish species were taxonomically identified in association with the corals. Yuryaku, Hancock, and Colahan seamounts had the highest number of individuals and species compared to the Academician, Bank 11, Kammu, and Koko seamounts.

On Yuryaku, species of fishes associated with scleractinians included chimeras, sharks, and bony fishes (Fig. 10.8 a, b, and c, respectively). There was also an observation of a deep-water ray egg pouch. Some of the species were observed resting on the coral, while others were observed swimming near or above the coral.

On the Hancock Seamounts many fish species were observed resting on the scleractinians. *Hoplostethus* sp., *Lophiodes* cf. *bruchius*, *Pentaceros wheeleri*, and *Physiculus grinnelli* (Fig. 10.8 d, e, f, and g, respectively) were examples of this association at NW Hancock, while *Neocyttus* cf. *acanthorhynchus* and *Physiculus nigripinnis* (Fig. 10.8 h and i, respectively) were examples on SE Hancock.

A high number of benthic and demersal fish species associated with scleractinians were identified on Colahan seamount, in potentially obligate and facultative associations. Based on what has been observed in other areas with rockfish species, in a possible obligate association with scleractinians, individual rockfish were observed resting on the coral

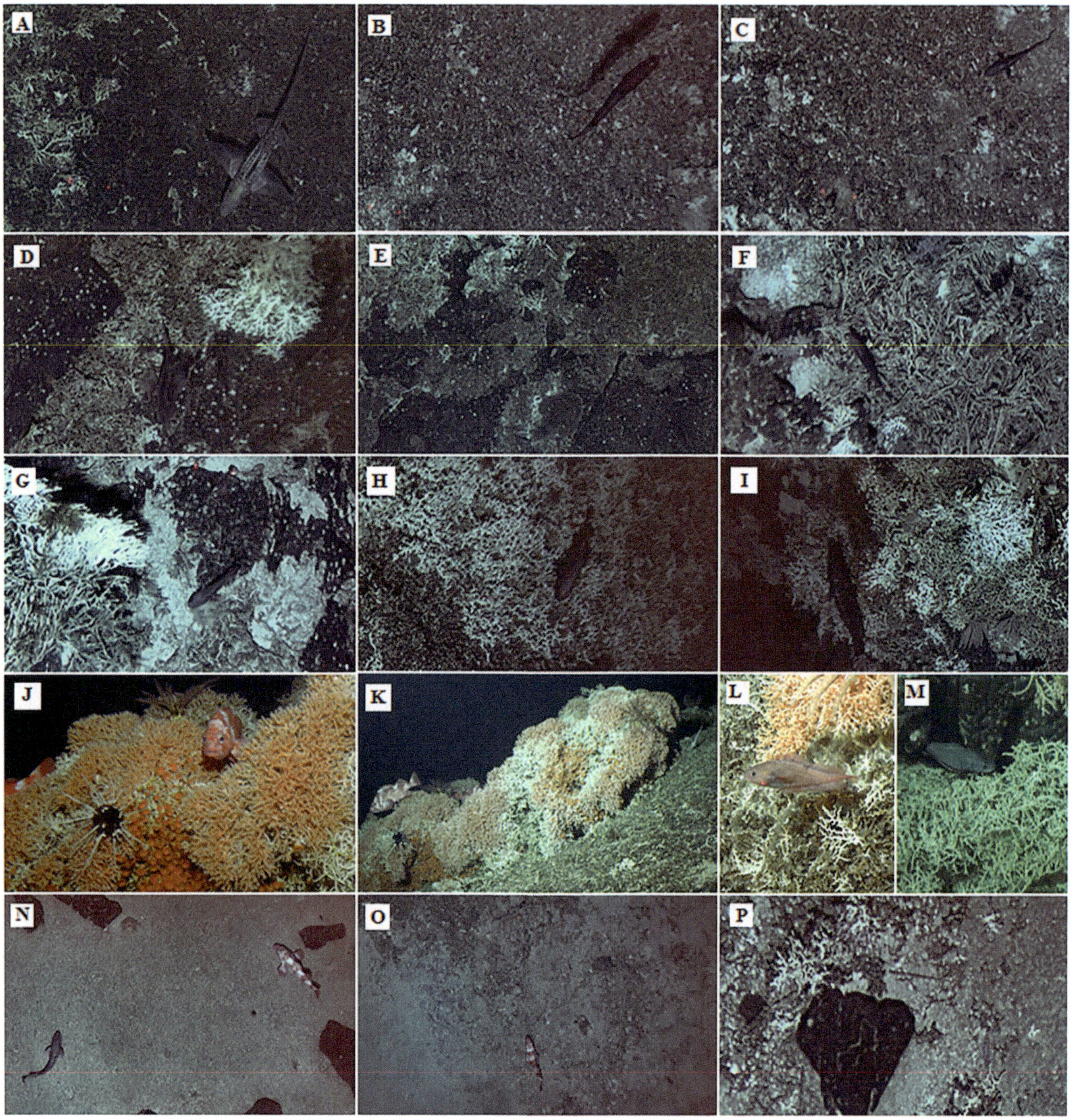

Fig. 10.8 Examples of fish associations on different seamounts located in the Northwestern Hawaiian Ridge and lower Emperor Seamount Chain. Source of the photos: A. Baco-Taylor FSU, and E.B. Roark TAMU, AUV Sentry, Pisces submersible Pilots T. Kerby and M. Cremer, and NSF. (**a**) *Hydrolagus owstoni* (**b**) *Etmopterus pusillus* (**c**) *Coelorinchus aratrum* (**d**) *Hoplostethus* sp. (**e**) *Lophiodes* cf. *bruchius* (**f**) *Pentaceros wheeleri* (**g**) *Physiculus grinneli* (**h**) *Neocyttus* cf. *acanthorhynchus* (**i**) *Physiculus grinneli* (**j**) Scorpaenidae (**k**) (**l**) *Gadella* sp. (**m**) *Physiculus grinnelli* (**n**) Scorpaenidae and morid (**o**) Scorpaenidae (**p**) *Symphurus* sp

(Fig. 10.8 j), using it as a shelter or for feeding purposes. This last hypothetical reason may be due to the observation of small fishes and invertebrates on the same reef (Fig. 10.8 k). On the other hand, in a possible facultative association, some species of morids were observed swimming through the scleractinian reef (*Gadella* sp., Fig. 10.8 l) and resting near the coral, but at the same time other individuals of the same species were observed hiding in a cave in rocky substrate (*Physiculus grinnelli*, Fig. 10.8 m).

On Kammu, one species of rockfish and one species of morid were observed on sandy sediments close to the scleractinians, but not in a direct association (Fig. 10.8 n and o). In contrast, a flatfish was observed in close proximity to the coral (*Symphurus* sp., Fig. 10.8 p).

Based on these observations multiple fish species associate with the reefs of the NWHI and ESC. More research will be needed to determine whether these associations are facultative or obligate and to determine the importance of these and other deep-sea coral structures to fish assemblages and stocks.

10.5.3 Trophic Information

The interactions between associated reef organisms and the habitat-forming stony corals in the North and Central Pacific have not been addressed, and there are also few studies globally of stony coral food webs for deep-sea scleractinians. Recent studies of habitat-forming cold-water corals indicate *D. pertusum* and *Madrepora oculata* on NE Atlantic reefs are supported mainly by zooplankton (Kiriakoulakis et al. 2006). Within the Mediterranean, lipid markers showed a preference for omnivorous or carnivorous zooplankton (Naumann et al. 2015) but both were also found to actively absorb dissolved free amino acids (Gori et al. 2014).

On North Atlantic reefs, carbon is also recycled through a detrital food web (van Oevelen et al. 2009). Associated taxa and the corals themselves can also contribute to the detritus on the reef through sloppy feeding that releases undigested food available for shrimp and amphipods (Buhl-Mortensen et al. 2017). The diverse array of invertebrate and fish fauna associated with these structures include other suspension feeders atop the structure, various ophiuroids within the structure, as well as crustaceans and fishes throughout, all of which may contribute to carbon flow and recycling through the reef ecosystem (O'Hara et al. 2008; Buhl-Mortensen et al. 2010, 2016, 2017).

Sponges are frequently found adjacent to corals in this region and may also be an important factor in the recycling of nutrients on seamounts. The sponge-loop has been theorized as an ubiquitous recycler of coral mucus in deep and shallow corals that is used in sponge filtration cells that are quickly sloughed and recycled into particulate detritus ready for uptake (Rix et al. 2016; De Goeij et al. 2013).

Central Pacific seamount hydrography is not well understood but is also likely to be very important to the trophic dynamics of the habitat-forming stony coral communities. Deep-water masses are structural influences on the communities of corals found on seamounts within the Pacific (Auscavitch et al. 2020a) and currents are also an important factor influencing community structure and occurrence of reef communities (Baco et al. 2017). Reef communities are likely dependent on primary production, zooplankton and particulate organic material retained by the seamounts and local recycling (Rogers 2018). To understand the trophic dynamics occurring on specific reefs of seamounts in the North and Central Pacific, future studies should include hydrodynamics around reefs as well as further investigations into associated species and trophic connections.

10.6 Threats and Outlook

10.6.1 Effects of Bottom-Contact Fishing

The expansion of fishing beyond continental shelves into the deep sea has targeted fish stocks and benthic communities that are neither well understood nor expected to be resilient to frequent disturbance (Morato et al. 2006; Clark et al. 2016; Victorero et al. 2018). Bottom-contact fisheries directly disturb large swaths of sessile organisms on the seafloor, leaving only the barest of substrate where there once had been a thriving benthic community supporting cnidarians, echinoderms, crustaceans, and many important fish species (Koslow et al. 2001; Hall-Spencer et al. 2002; Heifetz et al. 2009; Clark et al. 2019; Baco et al. 2020). Seamounts in particular have experienced exceptionally high effort, but few have been well studied (Clark et al. 2010; Victorero et al. 2018). North and Central Pacific deep fisheries include bottom-contact fisheries that have targeted seamounts, particularly for precious corals, alfonsino, and pelagic armorhead, with fishing effort continuing on many seamounts today (Clark and Koslow 2007; Victorero et al. 2018). From the NWHI and ESC between the years of 1968–1984, a reported 899,658 metric tons of the target fish species were extracted, a high effort and pressure for such a short period of time (Clark and Koslow 2007). As much as 200,000 kg of precious corals were harvested per year in the same region and time period (reviewed in Grigg 2002). This exploitation occurred in the same area as scleractinian reefs and extensive octocoral beds (Baco et al. 2017, 2020).

The North and Central Pacific seamounts support many heterogeneous communities with patchy distributions (Parrish and Baco 2007; Baco 2007; Long and Baco 2014; Schlacher et al. 2014; Morgan et al. 2015; Parrish et al. 2015; Auscavitch et al. 2020a, b). Reefs are only a portion of the entire seamount community and are often formed by *Enallopsammia rostrata*, *Madrepora oculata*, *Solenosmilia variabilis*, *Enallopsammia pusilla*, or *Eguchipsammia fistula* (Baco et al. 2017; Auscavitch et al. 2020a, b). These reef-building stony corals are slow growing, on the order of μm-mm per year (e.g. Houlbreque et al. 2010). This slow growth and fragility make these reefs highly vulnerable to bottom-contact fishing pressure (Clark et al. 2016). The disturbance and removal of reefs leads to a removal of reef species and loss of biodiversity (e.g. Althaus et al. 2009). Not only are the sessile organisms affected but also those that rely on these reefs for shelter, refuge, and food also experience a loss of habitat (Clark and Rowden 2009). Several decades after high fishing effort there may be recovery of seamount deep-water coral communities in the Pacific (Baco et al. 2019), but occurrence of scleractinian reefs remains rare (Baco et al. 2017; Clark et al. 2019). Koko seamount in the ESC has experienced high levels of fishing effort historically and is still actively trawled. Yet, here there are some recovering scleractinian patches and remnant coral rubble with living polyps that remain unprotected (Baco et al. 2020). Colahan also saw some of the highest removal of alfonsino and armorhead globally (Clark and Koslow 2007), yet due to topography and the inability of nets to disturb some locations, there are some patch reefs on the seamount that have not been disturbed (Baco et al. 2020). With one of the key species on these seamounts, *Solenosmilia variabilis*, having been shown to maintain isolated populations that appear dependent on local reproduction (Miller and Gunasekera 2017; Zeng et al. 2017), these remnant local communities untouched by direct trawling impact may be vital sources of propagules for recovery on Central Pacific seamounts (Baco et al. 2020).

In addition to the impact of direct contact with scleractinians and the substrate found in a region targeted for exploitation, fishing efforts can also lead to the deposition of marine debris in the ecosystem. Derelict fishing gear has been found along seamounts, including in areas identified as possible scleractinian reef habitat, in the Western and Central Pacific Ocean (Baco et al. 2019, 2020; Amon et al. 2020). Fishing debris includes plastics, metals, fibrous netting, cables, and overboard debris, most of which have the potential to persist in the deep-sea environment for centuries (Schlining et al. 2013). The impact of abandoned fishing gear is dependent on both the size and material of the debris. Larger types of fishing equipment, such as gill nets and longlines, have been observed entangled in or even completely covering deep-sea coral (e.g. Roberts and Hirshfield 2004; Baco et al. 2019). The complete covering of coral can result in mortality, while abrasions and increased suspended sediment can affect coral health. However, there is evidence of colonization or regrowth from fragments on fishing debris by two scleractinian species, *S. variabilis* has been observed regrowing from fragments on and around derelict fishing gear in the NWHI and ESC (Baco et al. 2019), and *M. oculata* has been observed colonizing submerged floats within the Mediterranean (Battaglia et al. 2019). Non-scleractinian corals as well as hydroids and ophiuroids have similarly been observed on fishing debris within the Central Pacific (unpubl data, Baco et al. 2019, Amon et al. 2020).

Because of their fragile nature, role as habitat, and slow recovery, deep-sea coral habitats and seamounts in general are considered vulnerable marine ecosystems (VMEs) by the FAO (UNGA 61/105, 2006). Fisheries expressly targeting seamounts endanger these VMEs. There is evidence of high disturbance by bottom-contact fishing in this region that can take decades to show signs of coral recovery (Baco et al. 2019). There are calls to protect high seas regions of the Central Pacific still experiencing fishing effort where known reefs are present, such as the northwestern Hawaiian Ridge Seamounts and the ESC (Baco et al. 2020). Explorations of seamounts throughout the NWHI, ESC, and other parts of the Central Pacific have all found large areas of deep-sea coral communities, suggesting most seamounts in the Central Pacific are highly likely to host deep-sea coral VMEs that can be impacted by any historical or active bottom-contact fishing. Therefore, impacts of bottom-contact fishing may be more widespread than currently documented and efforts to mitigate the effects of trawling should be undertaken.

10.6.2 Cobalt-Rich Manganese Crust Mining

Global demand for rare earth minerals and metals has accelerated the interest in exploiting deep-sea mineral deposits (Miller et al. 2018). The areas with the highest probability of mining for both cobalt-rich ferromanganese crusts and nodules fall within the Central Pacific and include the Prime Crust Zone (PCZ) for crusts and the Clarion-Clipperton Fracture zone for nodules (Fig. 10.9). The International Seabed Authority has already granted exploratory permits for crust mining within the Western Central Pacific as well as multiple exploratory and reserved leases for polymetallic nodule mining in the Clarion-Clipperton Fracture Zone and the western part of the Prime Crust Zone (Fig. 10.9) (Petersen et al. 2016). Crust mining targets seamount ridges, flanks, and summit crusts as areas of increased rare earth element concentrations and crust thickness, generally thickest between depths of 800 and 2500 m (Hein et al. 2013). Polymetallic nodule mining, on the other hand, targets the abyssal plains and areas between seamounts where polymetallic nodules have formed in greater densities (Hein et al. 2013).

Mining of crusts will disrupt and remove corals and sponges on seamounts along with the habitat they create. Seamounts within the exploratory crust regions have not been examined for benthic fauna but are expected to support vulnerable marine ecosystems including extensive deep-sea coral beds and scleractinian reefs, as has been observed in all other explored areas of the North and Central Pacific (Fig. 10.9) (Parrish and Baco 2007; Baco 2007; Parrish et al. 2015; Morgan et al. 2015; Auscavitch et al. 2020a, b). This expectation is supported by habitat suitability modeling as outlined in Sect. 10.3.7. Deep-sea corals and sponges have strongly overlapping habitat requirements with areas where cobalt-rich crusts are likely to form on seamounts. For example, for *S. variabilis* reefs in the South Pacific, seamount slopes and ridges were found to be the most suitable habitat at depths between 860 and 2000 m (the model's maximum

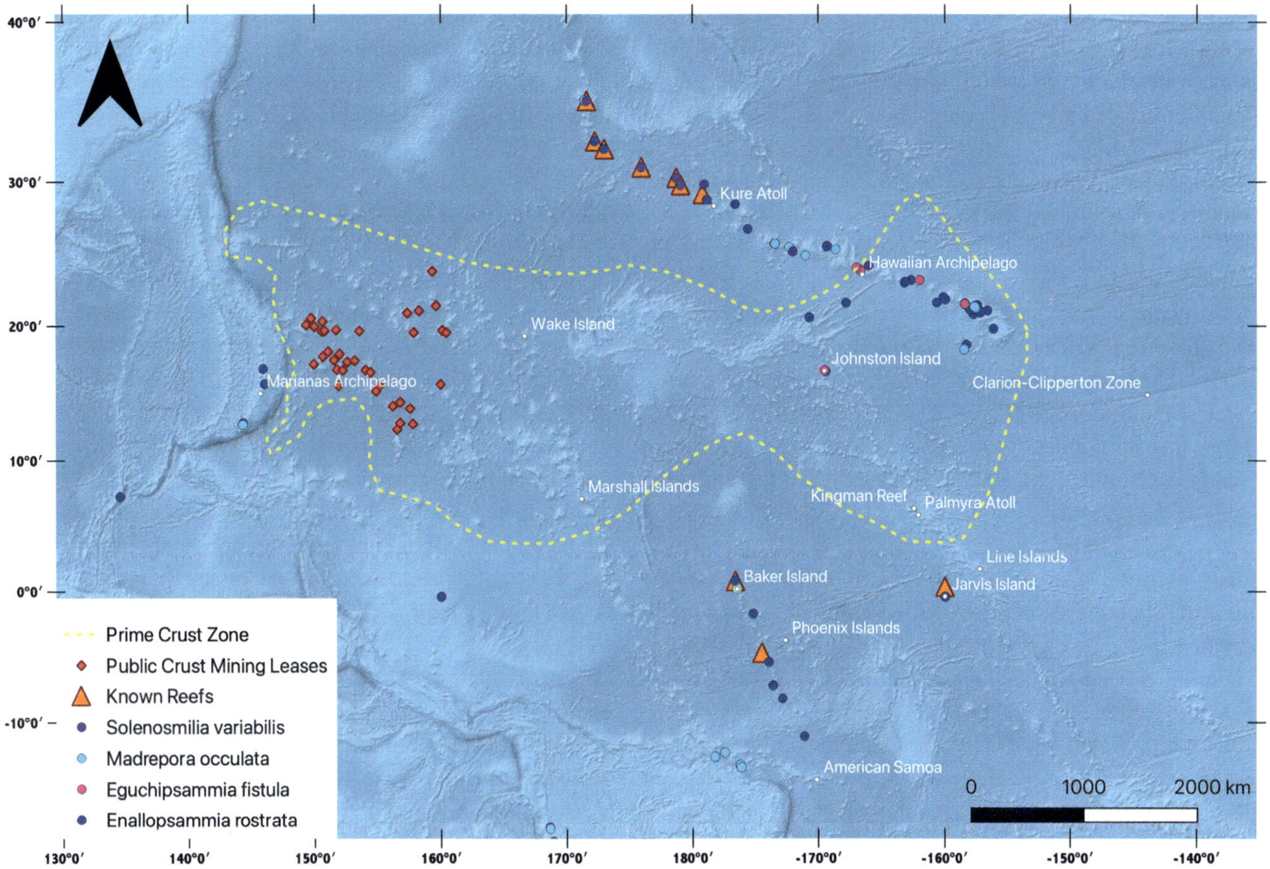

Fig. 10.9 Presence of known scleractinian reef-forming species in the North and Central Pacific denoted by orange triangles as well as observations of four reef-forming species from Smithsonian samples and OER imagery in the region (NMNH 2020; NOAA 2020). The Prime Crust Zone (adapted from Hein et al. 2009), area of intense mining interest, is outlined in yellow and the current publicly available leased mining areas are indicated by red diamonds (Petersen et al. 2016)

depth) (Rowden et al. 2017). Modeled reef-forming coral habitat suitability in the Central Pacific shows similar patterns of potential suitable habitat on seamount peaks and ridges (Davies and Guinotte 2011). The physical features which make these sites most suitable for reef-forming corals *E. rostrata* and *S. variabilis* included hard substrate, specific depth zones, and increased flow velocity. These same areas on seamounts are also likely to attract increased mining efforts, as relatively shallow areas with low sedimentation due to high degree of current provide optimum conditions for mining processes (Petersen et al. 2016; Rowden et al. 2017).

The direct impact of mining efforts on deep-sea benthos can be both severe and long-lasting (Levin et al. 2016). Seabed and seamount mining processes not only remove existing organisms but also change the physical structure of the benthic environment by removing hard substrate, possibly rendering the area uninhabitable by the former residents (Boschen et al. 2013). The removal of corals can be especially deleterious, as deep-sea corals are slow growing, long lived, and may take hundreds or even thousands of years to reform their complex biogenic structures (e.g., Fallon et al. 2014), assuming that the habitat is not altered past the point of suitability for these organisms. The impact of mining efforts would be greater than that of other anthropogenic disturbances, such as trawling, due to the complete removal of the substrate (Miller et al. 2018). Crusts and nodules form astoundingly slowly, on the order of 10–20 mm of deposition per million years (Hein 2016). Therefore, disruption of this magnitude could lead to a permanent shift in community structure and significantly decrease biodiversity for a given seamount (Gollner et al. 2017).

Mining seamount crusts will also have indirect impacts such as the generation of sediment plumes, pumping sediment and water below the surface, and the potential for tailings containing toxic metals and pollutants. Cutting of seafloor crusts would lead to creation of particulates and resuspension of sediment (Christiansen et al. 2020). The downstream particulate deposition of crust mining on a seamount in the Atlantic has been estimated up to 1.4 km from the crust mining site, with a 100 m radius of substantially greater deposition (Spearman et al. 2020). Similarly, modeling of resuspended sediment from polymetallic nodule mining in the German lease of the CCZ led to a regional increase in deposition that would blanket the seafloor 4–9 km away from the mining site (Gillard et al. 2019). Although the exact impact area has not been studied on the Central Pacific seamounts, coral communities directly next to and downstream of the mined substrate would be susceptible to the potential effects of higher than usual sedimentation, burial and smothering, as well as the stress of light and noise pollution from the mining operations.

Sediment plumes are also generated after the mined substrate has been processed at the surface and processing waste products, including non-target metals, are returned into the water column along with surface water (Levin et al. 2020). Depending on the depth of release, this sediment and warm surface water will disperse to differing extents. The shallower in the water column wastes are released, the further the area of dispersal of fine particles. At abyssal depths, a plume will have a narrower dispersal and affect only local communities, but at mesophotic depths the sediment and pollutants could spread across a much wider range (Christiansen et al. 2020). Exposure of deep-sea coral communities to surface water and increased sediment could lead to heightened stress due to temperature changes, pollution, the energetic cost of consuming inorganic materials (Larsson et al. 2013; Christiansen et al. 2020), as well as potential suffocation and burial. *M. oculata* is found to experience significant changes in physiological responses to temperature changes of only 3°C outside of the ambient temperature in the Mediterranean (Naumann et al. 2014). Depending on the location of the release of the partially processed sediment and surface water slurry, corals above the already impacted region may become disturbed by the warm surface waters and increased sedimentation.

The impact of mining efforts, both direct and indirect, will play a role in shaping the Pacific deep-sea coral habitat in years to come. As global demand for metals found in this area continues to rise, the potential degradation of coral habitat will also increase. Future planning of Pacific seamount mining must address these issues in order to mitigate serious harm to the resident vulnerable marine ecosystems. Seamounts in the Central Pacific contain many distinct communities, with community composition and abundance varying among seamounts, and even within the scale of a single seamount among different depths, substrate types, and sides of the seamount (Clark et al. 2009; Long and Baco 2014; Schlacher et al. 2014; Morgan et al. 2015, 2019; Mejía-Mercado et al. 2019; Auscavitch et al. 2020a). This can lead to variability in the response of these features to mining impacts; therefore, each seamount community should be managed separately. The patchy nature of heterogeneous seamount communities adds to the difficulty of planning and managing the shock to this sensitive ecosystem; however, harm reduction must be considered when planning mining operations to ensure the continuation of vulnerable deep-sea coral communities. Deep-sea corals are exceptionally vulnerable to disturbance and the loss of these habitat-forming corals would lead to high losses of biodiversity.

10.6.3 Global Change

As humans continue to modify the atmosphere and oceans, significant changes to environmental conditions are occurring and are expected to continue to occur throughout the oceans, even in the deep sea (Mora et al. 2013; Sweetman et al. 2017). Deep-sea organisms have evolved in an environment with a low range of variability, thus many are expected to lack the ability to adapt to environmental change. At bathyal depths in the Pacific, by 2100, it has been projected that temperatures will increase by up to 3.6°C, oxygen will decrease by 0.05 ml/l, pH will decrease by up to 0.3 units, and POC flux may decrease by as much as 5.2 $mgC/m^2/d$ (Sweetman et al. 2017). There are not published studies of the effects of changes of this magnitude. However, habitat suitability modeling in the Atlantic has been used to model current suitable habitat for *Desmophyllum pertusum*, *Madrepora oculat*a, and *Desmophyllum dianthus* as well as to predict future effects of climate on the suitable habitat for these three species of scleractinian corals, along with three octocorals and six fishes (Morato et al. 2020). After incorporating 2100 climate projections from the business as usual RCP8.5 trajectory, *D. pertusum* decreased in habitat by over 79%. *M. oculata* and *D. dianthus* suitable habitat decreased by 30–55% as well, due to changes in aragonite saturation, food availability, and deep-water warming (Morato et al. 2020). The habitat loss of *M. oculata* and *D. dianthus* may be buffered by the northward extension of both species between 1.9° and 4° of latitude (Morato et al. 2020), but would depend on the recruitment ability of the species and rates of colonization of the newly habitable areas. These drastic decreases in suitable habitat for corals would lead to substantial losses in the habitat they create and could also lead to considerable decreases in connectivity.

Besides colonizing higher latitudes to adjust to changing temperatures, species may also colonize deeper depths to maintain their distributions with respect to temperature. However, currently in the Pacific, the aragonite saturation horizon (ASH), a factor that is thought to determine the lower limit of occurrence for most reef-forming corals (e.g. Guinotte et al. 2006), is already much shallower in the Pacific than in the Atlantic (Feely et al. 2004). Additionally, the ASH has been found to be shoaling in the Pacific due to ocean acidification (Feely et al. 2012; Carter et al. 2017). The combination of a shoaling ASH and increasing temperatures will reduce or eliminate the range of the depth band of suitable habitat for reef-forming species.

Along with a shoaling ASH, another important component of ocean acidification is pH. A decrease in pH has been shown to lead to increased porosity of the reef framework that could lead to decreased reef stability and decreased three-dimensional structure (Hennige et al. 2020). Some species may be buffered from physiological impacts, as living *D. pertusum* and *M. oculata* exposed to decreased pH or increased CO_2 concentrations have not exhibited altered respiration or responses during short-term experiments in aquaria (Maier et al. 2013a, b, 2016; Movilla et al. 2014; Büscher et al. 2017). In contrast, significant effects were found in Lunden et al. (2014), Georgian et al. (2016), and Kurman et al. (2017). Also see Gómez et al. (2018) for an example from the Pacific populations of *Desmophyllum pertusum* in California. Similarly, in a study that included coral health, a significant change in the coral tissue color of *Solenosmilia variabilis* indicated a loss of the coenenchyme with decreasing pH and increased temperature (Gammon et al. 2018). *S. variabilis* is one of the previously discussed key reef species in the Pacific and a loss of the protective tissue could leave these reefs even more vulnerable. Therefore, changing climate in the Pacific with increased temperatures, a shallower ASH, and decreased pH could lead to a decrease in *S. variabilis* health as well as the loss of the three-dimensional reef structure.

Finally, in the already oligotrophic waters of the North Pacific, any change to POC flux is likely to have dramatic impacts on benthic food webs. A decrease in suitable habitat for cold-water corals and fish species was correlated with decreased POC in the Atlantic (Morato et al. 2020). The potential decrease of POC to the Pacific (−44.5%, Jones et al. 2014) is almost twice that of the Atlantic. Combined with temperature and other changing factors, the end result is likely significantly reduced habitat for deep-sea corals. Not only would the corals be affected, but the other fauna within the community would be as well. All of these expected changes will synergistically impact the long-lived sessile deep-sea coral communities, which are unlikely to be able to migrate to different depths or higher latitude to find more favorable conditions over the time scales at which these changes are occurring.

10.7 Future Directions

As can easily be discerned from the text of this chapter, because of their recent discovery, beyond a few location points, there is a paucity of data on most aspects of the biology and ecology of framework-forming scleractinians in the North and Central Pacific. With the current threats to seamount coral communities including trawl fisheries, climate change, and ocean acidification, as well as the rising likelihood of cobalt-rich manganese crust mining in the North and Central Pacific, it is a critical time to gain a better understanding of these ecosystems. The basic level at which to start is to undertake further explorations would be to determine the size of individual reefs and their full geographic extent in this region. As knowledge about where reefs occur increases, we can then move on to more process-oriented and hypothesis-driven questions to understand the biology and ecology of these species, including life history strategies, growth rates, genetic connectivity, microbial associations, species interactions, resilience, and trophic relationships. For the documented sites in the NWHI and ESC, further studies currently underway include phylogenetics and species distribution modeling. A large-scale study is also underway focused specifically on the question: How is it that these reefs can form and persist in waters undersaturated with respect to aragonite?

In the Phoenix Islands, planned future studies will further examine the factors that influence large stony coral reef structures, primarily composed of *Enallopsammia* spp. and *Madrepora* sp., and which environmental and seafloor terrain variables drive their distribution, longevity, growth rates, and resilience. Additionally, the diversity of fish and invertebrate fauna hosted by these reef structures is poorly understood in the region and remains to be quantified through visual and direct sampling.

With the rising anthropogenic impacts to deep-sea ecosystems, and given the fact that framework-forming scleractinians are a minor component of the deep-sea fauna of this region, one could also argue that an even greater imperative for the North and Central Pacific is to gain a better understanding of the dense octocoral and antipatharian communities that are far more common on the seamounts of this region. Knowledge of their distributions, biology, and ecology is not at a much more advanced state than the current knowledge of the scleractinians. Octocorals beds in the North and Central Pacific occupy an essential niche that is the same that would otherwise be occupied by framework-forming scleractinians, playing a comparable role to reefs in their support of invertebrate and fish communities, including for species of commercial importance (e.g., Husebø et al. 2002; Krieger and Wing 2002; Parrish and Baco 2007; Baco 2007; Mosher and Watling 2009; Buhl-Mortensen et al. 2010; Baillon et al. 2012, 2014). As the dominant taxa playing these roles for the Central and North Pacific, octocorals, antipatharians, and sponges should be recognized for these roles and awarded the same level of importance as are reefs in the South Pacific and North Atlantic. The bias of attention and priority toward stony corals as more important than octocorals is a relic of study of shallow-water coral reef studies and ignores the critical role and importance of octocorals in the vast majority of deep Pacific waters.

Acknowledgments We would like to thank the editors for the invitation to participate in this chapter. Our thanks to Chris Kelley who reviewed the Hawaii Undersea Research Laboratory (HURL) database and provided unpublished HURL data records to help determine potential additional location of scleractinian reefs. We also thank Hannah Barkley and Ari Halperin who provided Figs. 10.1 and 10.2. Unpublished data on fish associations were collected through NSF grants OCE-1334652 to ARB and OCE-1334675 to EBR.

References

Adkins JF, Henderson GM, Wang SL et al (2004) Growth rates of the deep-sea scleractinia *Desmophyllum cristagalli* and *Enallopsammia rostrata*. Earth Planet Sci Lett 227:481–490. https://doi.org/10.1016/j.epsl.2004.08.022

Alcock A (1898) An account of the deep-sea Madreporaria collected by the Royal Indian Marine Survey Ship "Investigator". Trustees of the Indian Museum, Calcutta. pp. 1–29, pls. 1–3, available online at https://doi.org/10.5962/bhl.title.27614

Alcock A (1902) Diagnoses and descriptions of new species of corals from the "Siboga-Expedition". Tijdschrift der Nederlandse Dierkundige Vereeniging (Ser. 2) 7:89–115

Althaus F, Williams A, Schlacher TA et al (2009) Impacts of bottom trawling on deep-coral ecosystems of seamounts are long-lasting. Mar Ecol Prog Ser 397:279–294. https://doi.org/10.3354/meps08248

Amon DJ, Kennedy BRC, Cantwell K et al (2020) Deep-sea debris in the Central and Western Pacific Ocean. Front Mar Sci 7:1–16. https://doi.org/10.3389/fmars.2020.00369

Andrews AH, Cailliet GM, Kerr LA et al (2005) In: Friewald A, Roberts JM (eds) Investigations of age and growth for three deep-sea corals from the Davidson Seamount off central California. Springer-Verlag, pp 1021–1038

Auscavitch SR, Deere MC, Keller AG et al (2020a) Oceanographic drivers of deep-sea coral species distribution and community assembly on seamounts, islands, atolls, and reefs within the Phoenix Islands protected area. Front Mar Sci 7:1–15. https://doi.org/10.3389/fmars.2020.00042

Auscavitch S, Pockalny R, Konrad K et al (2020b) Deepwater exploration of Kingman Reef, Palmyra Atoll, and Jarvis Island: geological and biological discoveries from the US Line Islands. Oceanography 33:38–39. https://doi.org/10.5670/oceanog.2020.supplement.01

Auster PJ, Malatesta RJ, LaRosa SC (1995) Patterns of microhabitat utilization by mobile megafauna on the southern New England (USA) continental shelf and slope. Mar Ecol Progr Ser 127:77–85

Auster PJ, Moore J, Heinonen KB, Watling L (2005) A habitat classification scheme for seamount landscapes: assessing the functional role of deep-water corals as fish habitat. In: Friewald A, Roberts JM (eds) . Springer, New York, pp 761–769

Baco AR (2007) Exploration for deep-sea corals on North Pacific seamounts and islands. Oceanography 20:108–117. https://doi.org/10.5670/oceanog.2007.11

Baco AR, Morgan N, Roark EB et al (2017) Defying dissolution: discovery of deep-sea scleractinian coral reefs in the North Pacific. Sci Rep 7. https://doi.org/10.1038/s41598-017-05492-w

Baco AR, Roark EB, Morgan NB (2019) Amid fields of rubble, scars, and lost gear, signs of recovery observed on seamounts on 30-40 year time scales. Sci Adv 5:eaaw 4513

Baco AR, Morgan NB, Roark EB (2020) Observations of vulnerable marine ecosystems and significant adverse impacts on high seas seamounts of the Northwestern Hawaiian Ridge and Emperor Seamount Chain. Mar Policy 115. https://doi.org/10.1016/j.marpol.2020.103834

Baillon S, Hamel J-F, Wareham VE, Mercier A (2012) Deep cold-water corals as nurseries for fish larvae. Front Ecol Environ 10:351–356. https://doi.org/10.1890/120022

Baillon S, Hamel J, Mercier A (2014) Diversity, distribution and nature of faunal associations with deep-sea pennatulacean corals in the Northwest Atlantic. PLoS One 9:e111519

Battaglia P, Consoli P, Ammendolia G et al (2019) Colonization of floats from submerged derelict fishing gears by four protected species of deep-sea corals and barnacles in the Strait of Messina (central Mediterranean Sea). Mar Pollut Bull 148:61–65. https://doi.org/10.1016/j.marpolbul.2019.07.073

Bauer LM, M Poti, BM Costa, et al (2016) Benthic habitats and corals. In: BM Costa and MS Kendall (eds.) Marine biogeographic assessment of the Main Hawaiian Islands. Bureau of ocean energy management and national oceanic and atmospheric administration. OCS study BOEM 2016-035 and NOAA Technical Memorandum NOS NCCOS 214, pp 57–136

Bidigare RR, Benitez-Nelson C, Leonard CL et al (2003) Influence of a cyclonic eddy on microheterotroph biomass and carbon export in the lee of Hawaii. Geophys Res Lett 30. https://doi.org/10.1029/2002GL016393

Boschen RE, Rowden, AA, Clark, MR, Gardner JPA (2013) Mining of deep-sea seafloor massive sulfides: a review of the deposits, their benthic communities, impacts from mining, regulatory frameworks and management strategies. Ocean Coast Manag 84:54–67. https://doi.org/10.1016/j.ocecoaman.2013.07.005

Bostock HC, Opdyke BN, Williams MJM (2010) Characterising the intermediate depth waters of the Pacific Ocean using δ13C and other geochemical tracers. Deep Res Part I Oceanogr Res Pap 57:847–859. https://doi.org/10.1016/j.dsr.2010.04.005

Brainard RE, Oliver T, McPhaden MJ, Cohen A, Venegas R, Heenan A, Vargas-Ángel B, Rotjan R, Mangubhai S, Flint E, Hunter S (2018) Ecological impacts of the 2015/16 El Niño in the Central Equatorial Pacific. Am Meteorol Soc 99:521–526

Brainard RE, Acoba T, Asher MAM, et al (2019) Chapter 9: Pacific Remote Islands Marine National Monument in the Pacific-wide context. In: Coral reef ecosystem monitoring report for the Pacific Remote Islands Marine National Monument 2000–2017. Pacific Islands Fisheries Science Center, PIFSC Special Publication, SP-19-006i. 72 p. doi:https://doi.org/10.25923/rwd2-2118

Brooke S, Young C (2009) In situ measurement of survival and growth of *Lophelia pertusa* in the northern Gulf of Mexico. Mar Ecol Prog Ser 397:153–161. https://doi.org/10.3354/meps08344

Buhl-Mortensen L, Vanreusel A, Gooday AJ et al (2010) Biological structures as a source of habitat heterogeneity and biodiversity on the deep ocean margins. Mar Ecol 31:21–50. https://doi.org/10.1111/j.1439-0485.2010.00359.x

Buhl-Mortensen P, Buhl-Mortensen L, Purser A (2016) Trophic ecology and habitat provision in cold-water coral ecosystems. Mar Anim For. https://doi.org/10.1007/978-3-319-17001-5

Buhl-Mortensen P, Buhl-Mortensen L, Purser A (2017) Marine animal forests. Mar Anim For. https://doi.org/10.1007/978-3-319-17001-5

Burgess SN, Babcock RC (2005) Reproductive ecology of three reef-forming, deep-sea corals in the New Zealand region. In: Freiwald A, Roberts JM (eds) Cold-water corals and ecosystems. Springer, Berlin, pp 701–713

Büscher JV, Form AU, Riebesell U (2017) Interactive effects of ocean acidification and warming on growth, fitness and survival of the cold-water coral *Lophelia pertusa* under different food availabilities. Front Mar Sci 4(Apr):1–14. https://doi.org/10.3389/fmars.2017.00101

Cairns SD (1979) The deep-water Scleractinia of the Caribbean sea and adjacent waters. Stud Fauna Curacao other Caribb Islands

Cairns SD (1982) Antarctic and Subantarctic Scleractinia. Biol Antarct Seas XI 34:1–74. https://doi.org/10.1029/ar034p0001
Cairns SD (1984) New records of ahermatypic corals (Scleractinia) from the Hawaiian and Line Islands. Bish Museum Occas Pap XXV:1–30
Cairns SD (1999) Species richness of recent Scleractinia. National Museum of Natural History, Smithsonian Institution
Cairns SD (2005) Revision of the Hawaiian stylasteridae (Cnidaria: Hydrozoa: Athecata). Pacific Sci 59:439–451. https://doi.org/10.1353/psc.2005.0034
Cairns SD (2006) New records of Azooxanthellate Scleractinia from the Hawaiian Islands. Bish Museum Occas Pap 87:45–53
Cairns SD (2007) Deep-water corals: an overview with special reference to diversity and distribution of deep-water scleractinian corals. Bull Mar Sci 81:311–322
Cairns SD (2017) New species of stylasterid (Cnidaria: Hydrozoa: Anthoathecata: Stylasteridae) from the Northwestern Hawaiian Islands. Pacific Sci 71:77–81. https://doi.org/10.2984/71.1.7
Cairns SD, Stanley GD (1981) Ahermatypic coral banks: living and fossil counterparts. Proc Fourth Int Coral Reef Symp Manila 1(1):611–618
Carlier A, Le Guilloux E, Olu K et al (2009) Trophic relationships in a deep Mediterranean cold-water coral bank (Santa Maria di Leuca Ionian Sea). Mar Ecol Prog Ser 397:125–137. https://doi.org/10.3354/meps08361
Carter BR, Feely RA, Mecking S et al (2017) Two decades of Pacific anthropogenic carbon storage and ocean acidification along Global Ocean Ship-based Hydrographic Investigations Program sections P16 and P02. Glob Biogeochem Cycles 31:306–327. https://doi.org/10.1002/2016GB005485
Chapron L, Le Bris N, Durrieu de Madron X et al (2020) Long term monitoring of cold-water coral growth shows response to episodic meteorological events in the NW Mediterranean. Deep Res Part I Oceanogr Res Pap 160. https://doi.org/10.1016/j.dsr.2020.103255
Christiansen B, Denda A, Christiansen S (2020) Potential effects of deep seabed mining on pelagic and benthopelagic biota. Mar Policy 114: 103442. https://doi.org/10.1016/j.marpol.2019.02.014
Clark MR, Koslow JA (2007) Impacts of fisheries on seamounts. In: Pitcher T, Morato T, Hart PJB, et al. (eds). Wiley Online Library, pp 413–441
Clark MR, Rowden AA (2009) Effect of deepwater trawling on the macro invertebrate assemblages of seamounts on the Chatham Rise, New Zealand. Deep Sea Res Part I Oceanogr Res Pap 56:1540–1554. https://doi.org/10.1016/j.dsr.2009.04.015
Clark MR, Vinnichenko VI, Gordon JD, Beck-Bulat GZ, Kukharev NN, Kakora AF (2007) Large-scale distant water trawl fisheries on seamounts. In: Pitcher TJ, Morato T, Hart PJ, Clark MR, Haggan N, Santos RS (eds) Seamounts: ecology, fisheries, and conservation, Blackwell fisheries and aquatic resources series, pp 361–399
Clark MR, Kelley C, Baco A, Rowden AA (2009) Fauna of cobalt-rich crust seamounts. WLG2009 63:1–61
Clark MR, Shank TM, Hall-Spencer JM et al (2010) The ecology of seamounts: structure, function, and human impacts. Ann Rev Mar Sci 2:253–278. https://doi.org/10.1146/annurev-marine-120308-081109
Clark MR, Althaus F, Schlacher TA et al (2016) The impacts of deep-sea fisheries on benthic communities: a review. ICES J Mar Sci 73: i51–i69. https://doi.org/10.1093/icesjms/fsv123
Clark MR, Bowden DA, Rowden AA, Stewart R (2019) Little evidence of benthic community resilience to bottom trawling on seamounts after 15 years. Front Mar Sci 6:1–16. https://doi.org/10.3389/fmars.2019.00063
Coates AG, Kauffman EG (1973) Stratigraphy, paleontology and paleoenvironment of a Cretaceous coral thicket, Lamy, New Mexico. J Paleontol 47:953–968
Costa B, Kendall MS, Parrish FA, et al (2015) Identifying suitable locations for mesophotic hard corals offshore of Maui, Hawaii. PLoS One. doi:https://doi.org/10.1371/journal.pone.0130285
Costello MJ, McCrea M, Freiwald A et al (2005) Role of cold-water *Lophelia pertusa* coral reefs as fish habitat in the NE Atlantic. In: Cold-water corals and ecosystems. Springer, Berlin, pp 771–805
Coutis PF, Middleton JH (2002) The physical and biological impact of a small island wake in the deep ocean. Deep Res Part I Oceanogr Res Pap 49:1341–1361
D'Onghia G, Maiorano P, Carlucci R et al (2012) Comparing deep-sea fish fauna between coral and non-coral "megahabitats" in the Santa Maria di Leuca cold-water coral province (Mediterranean Sea). PLoS One 7(9):e44509
Davies AJ, Guinotte JM (2011) Global habitat suitability for framework-forming cold-water *Lophelia pertusa* coral reefs as fish habitat in the NE Atlantic. In: Cold-water corals and ecosystems. Springer, Berlin, pp 771–805
Davis AS, Gray LB, Clague DA, Hein JR (2002) The Line Islands revisited: New $^{40}Ar/^{39}Ar$ geochronologic evidence for episodes of volcanism due to lithospheric extension. Geochem Geophys Geosystems 3:1–28. https://doi.org/10.1029/2001gc000190
De Goeij JM, Van Oevelen D, Vermeij MJA et al (2013) Surviving in a marine desert: the sponge loop retains resources within coral reefs. Science (80) 342:108–110. 6:e18483. https://doi.org/10.1126/science.1241981
Demopoulos A, Auscavitch S, Sowers D et al (2018) Discovering the deep: exploring remote Pacific marine protected areas. Oceanog 31:76–77. https://10.5670/oceanog.2018.supplement.01
Desch A, Wynne T, Brainard R, Friedlander A, et al (2009) Oceanographic and physical setting. A marine biogeographic assessment of the Northwestern Hawaiian Islands. Silver Spring: NOAA, 17–63.
Deser C, Alexander MA, Timlin MS (1996) Upper-ocean thermal variations in the North Pacific during 1970-1991. J Clim 9:1840–1855
Dolan MFJ, Grehan AJ, Guinan JC, Brown C (2008) Modelling the local distribution of cold-water corals in relation to bathymetric variables: adding spatial context to deep-sea video data. Deep Res Part I Oceanogr Res Pap 55:1564–1579. https://doi.org/10.1016/j.dsr.2008.06.0101371/journal.pone.0018483
Dons C (1944) Norges korallrev. Det kongelige norske videnskabers selskabs forhandlinger. 16:37–82
Druffel ERM, King LL, Belastock RA, Buesseler KO (1990) Growth rate of a deep-sea coral using 210Pb and other isotopes. Geochim Cosmochim Acta 54:1493–1499. https://doi.org/10.1016/0016-7037(90)90174-J
Duncan M (1873) A description of the Madreporaria dredged up during the expeditions of H.M.S. 'Porcupine' in 1869 and 1870. Trans Zool Soc London 8:303–343
Ehrenberg CG (1832) Beiträge zur physiologischen Kenntniss der Corallenthiere im allgemeinen, und besonders des Rothen Meeres, nebst einem Versuch zur physiologischen Systematik derselben. Physikalische-Mathematische Abhandlungen der Königlichen Akademie der Wissenschaften zu Berlin 1:225–380
Emery WJ (2001) Water types and water masses. https://doi.org/10.1006/rwos.2001.0108
Fadlallah YH, Pearse JS (1982) Sexual reproduction in solitary corals: overlapping oogenic and brooding cycles, and benthic planulas in *Balanophyllia elegans*. Mar Biol 71:223–231. https://doi.org/10.1007/BF00397039
Fallon SJ, Thresher RE, Adkins J (2014) Age and growth of the cold-water scleractinian *Solenosmilia variabilis* and its reef on SW Pacific seamounts. Coral Reefs 33:31–38. https://doi.org/10.1007/s00338-013-1097-y
Feehan KA, Waller RG, Häussermann V (2019) Highly seasonal reproduction in deep-water emergent *Desmophyllum dianthus*

(Scleractinia: Caryophylliidae) from the Northern Patagonian Fjords. Mar Biol 166:1–13. https://doi.org/10.1007/s00227-019-3495-3
Feely RA, Sabine CL, Lee K et al (2004) Impact of anthropogenic CO2 on the CaCO3 system in the oceans. Science 305:362–366. https://doi.org/10.1126/science.1097329
Feely RA, Sabine CL, Byrne RH et al (2012) Decadal changes in the aragonite and calcite saturation state of the Pacific Ocean. Glob Biogeochem Cycles 26. https://doi.org/10.1029/2011GB004157
Finlayson VA, Konter JG, Konrad K et al (2018) Sr–Pb–Nd–Hf isotopes and $^{40}Ar/^{39}Ar$ ages reveal a Hawaii–Emperor-style bend in the Rurutu hotspot. Earth Planet Sci Lett 500:168–179
Firing J, Brainard RE (2006) Ten years of shipboard ADCP measurements along the Northwestern Hawaiian Islands. Atoll Res Bull 543:347–363
Floris S (1972) Scleractinian corals from the Upper Cretaceous and Lower Tertiary of Nugssuaq, West Greenland
Foley NS, Kahui V, Armstrong CW, Van Rensburg TM (2010) Estimating linkages between redfish and cold-water coral on the Norwegian coast. Mar Resour Econ 25:105–120
Fosså JH, Mortensen PB, Furevik DM (2002) The deep-water coral *Lophelia pertusa* in Norwegian waters: distribution and fishery impacts. Hydrobiologia 471(1–3):1–12
Friedlander A, Aeby G, Brainard, R, et al (2005) The state of the coral reef ecosystem of the Northwestern Hawaiian Islands. In: Waddell JE (ed) The State of Coral Reef Ecosystems of the United States and Pacific Freely Associated States, pp 270–311
Gammon MJ, Tracey DM, Marriott PM et al (2018) The physiological response of the deep- sea coral *Solenosmilia variabilis* to ocean acidification. PeerJ 2018:1–24. https://doi.org/10.7717/peerj.5236
Gass SE, Roberts JM (2006) The occurrence of the cold-water coral *Lophelia pertusa* (Scleractinia) on oil and gas platforms in the North Sea: Colony growth, recruitment and environmental controls on distribution. Mar Pollut Bull 52:549–559. https://doi.org/10.1016/j.marpolbul.2005.10.002
Genin A, Dayton PK, Lonsdale PF, Spiess FN (1986) Corals on seamount peaks provide evidence of current acceleration over deep-sea topography. Nature 322:59–61. https://doi.org/10.1038/322059a0
Georgian SE, Dupont S, Kurman M, Butler A, Stromberg SM, Larsson AI, Cordes EE (2016) Biogeographic variability in the physiological response of the cold-water coral *Lophelia pertusa* to ocean acidification. Mar Ecol 37(6):1345–1359
Gillard B, Purkiani K, Chatzievangelou D et al (2019) Physical and hydrodynamic properties of deep sea mining-generated, abyssal sediment plumes in the Clarion Clipperton Fracture Zone (eastern-central Pacific). Elem Sci Anth 7. https://doi.org/10.1525/elementa.343
Gilmore RG, Jones RS (1992) Color variation and associated behavior in the epihepheline groupers, *Mycteroperca microlepis* (Goode and Bean) and *M. phenax* Jordan and Swain. Bull Mar Sci 51:83–103
Gollner S, Kaiser S, Menzel L et al (2017) Resilience of benthic deep-sea fauna to mining activities. Mar Environ Res 129:76–101. https://doi.org/10.1016/j.marenvres.2017.04.010
Gómez CE, Wickes L, Deegan D et al (2018) Growth and feeding of deep-sea coral *Lophelia pertusa* from the California margin under simulated ocean acidification conditions. PeerJ. https://doi.org/10.7717/peerj.5671
Gori A, Grover R, Orejas C et al (2014) Uptake of dissolved free amino acids by four cold-water coral species from the Mediterranean Sea. Deep Sea Res Part II Top Stud Oceanogr 99:42–50. https://doi.org/10.1016/j.dsr2.2013.06.007
Gove JM, Merrifield MA, Brainard RE (2006) Temporal variability of current-driven upwelling at Jarvis Island. J Geophys Res 111: C12011. https://doi.org/10.1029/2005JC003161
Gove JM, Williams GJ, McManus MA et al (2013) Quantifying climatological ranges and anomalies for Pacific coral reef ecosystems. PLoS One 8:e61974. https://doi.org/10.1371/journal.pone.0061974
Grigg RW (1988) Paleoceanography of coral reefs in the Hawaiian-Emperor Chain. Science 240:1737–1743. https://doi.org/10.1126/science.240.4860.1737
Grigg RW (2002) Precious corals in Hawaii: discovery of a new bed and revised management measures for existing beds. Mar Fish Res Rev 64:13–20
Gu D, Philander SGH (1997) Interdecadal climate fluctuations that depend on exchanges between the tropics and extratropics. Science 275:805–807. https://doi.org/10.1126/science.275.5301.805
Guinotte JM, Orr J, Cairns S et al (2006) Will human-induced changes in seawater chemistry alter the distribution of deep-sea scleractinian corals? Front Ecol Environ 4:141–146
Hall-Spencer J, Allain V, Fosså JH (2002) Trawling damage to Northeast Atlantic ancient coral reefs. Proc R Soc Lond B Biol Sci 269: 507–511. https://doi.org/10.1098/rspb.2001.1910
Hanz U, Wienberg C, Hebbeln D, Duineveld G, Lavaleye M, Juva K et al (2019) Environmental factors influencing benthic communities in the oxygen minimum zones on the Angolan and Namibian margins. Biogeosciences 16:4337–4356. https://doi.org/10.5194/bg-16-4337-2019
Harrison PL, Wallace CC (1990) Reproduction, dispersal and recruitment of scleractinian corals. In: Dubinsky Z (ed) Ecosystems of the world: coral reefs. Elsevier, Amsterdam, pp 133–207
Hayashibara T, Nishida K (2017) Results of the bottom environmental survey of the Emperor Seamounts Chain trawl fishing grounds in 2016: exploration for spatial extent of known coral assemblages and distribution of bycatch corals collected by a trawl operation. NPFC-2017-SSC VME02-WP4. 13p
Heifetz J (2002) Coral in Alaska: distribution, abundance and species associations. Hydrobiologia 471(1-3):19–28
Heifetz J, Stone R, Shotwell S (2009) Damage and disturbance to coral and sponge habitat of the Aleutian Archipelago. Mar Ecol Prog Ser 397:295–303. https://doi.org/10.3354/meps08304
Hein JR (2016) Manganese nodules. In: Harff J (ed) Encyclopedia of marine geosciences. Encyclopedia of earth sciences series. Springer, Dordrecht, pp 408–412
Hein JR, McIntyre BR, Piper DZ (2005) Marine mineral resources of Pacific Islands - a review of the exclusive economic zones of Islands of U.S. affiliation, excluding the State of Hawaii. US Geol Surv Circ, pp 1–68
Hein JR, Conrad TA, Dunham RE (2009) Seamount characteristics and mine-site model applied to exploration- and mining-lease-block selection for cobalt-rich ferromanganese crusts. Mar Georesour Geotechnol 27:160–176. https://doi.org/10.1080/10641190902852485
Hein JR, Mizell K, Koschinsky A, Conrad TA (2013) Deep-ocean mineral deposits as a source of critical metals for high- and green-technology applications: comparison with land-based resources. Ore Geol Rev 51:1–14. https://doi.org/10.1016/j.oregeorev.2012.12.001
Hennige S, Wolfram U, Wickes L et al (2020) Crumbling reefs and cold-water coral habitat loss in a future ocean: evidence of "coralporosis" as an indicator of habitat integrity. Front Mar Sci 7:1–16. https://doi.org/10.3389/fmars/2020/00668
Henriet JP, De Mol B, Vanneste M et al (2001) Carbonate mounds and slope failures in the Porcupine Basin: a development model involving fluid venting. Geol Soc Spec Publ 188:375–383. https://doi.org/10.1144/GSL.SP.2001.188.01.22
Hill TM, Spero HJ, Guilderson T et al (2011) Temperature and vital effect controls on bamboo coral (*Isididae*) isotope geochemistry: a test of the "lines method". Geochem Geophys Geosyst 12. https://doi.org/10.1029/2010GC003443
Hong L, Zhang L, Chen Z, Wu L (2014) Linkage between the Pacific Decadal Oscillation and the low frequency variability of the Pacific Subtropical Cell. J Geophys Res Oceans 119(6):3464–3477
Houlbreque F, McCulloch M, Roark B et al (2010) Uranium-series dating and growth characteristics of the deep-sea scleractinian

coral: *Enallopsammia rostrata* from the Equatorial Pacific. Geochim Cosmochim Acta 74:2380–2395

Howell KL, Billett DSM, Tyler PA (2002) Depth-related distribution and abundance of seastars (Echinodermata: Asteroidea) in the Porcupine Seabight and Porcupine Abyssal Plain, N.E. Atlantic. Deep Sea Res Part I Oceanogr Res Pap 49:1901–1920. https://doi.org/10.1016/S0967-0637(02)00090-0

Husebø Å, Nøttestad L, Fosså JH et al (2002) Distribution and abundance of fish in deep-sea coral habitats. Hydrobiologia 471(1-3):91–99

Iida K et al (1996) Relationship between acoustic backscattering strength and density of zooplankton in the sound-scattering layer. ICES J Mar Sci 53:507–512. https://doi.org/10.1006/jmsc.1996.0073

Jones DO, Yool A, Wei C et al (2014) Global reductions in seafloor biomass in response to climate change. Glob Chang Biol 20:1861–1872. https://doi.org/10.1111/gcb.12480

Kahng SE, Garcia-Sais JR, Spalding HL et al (2010) Community ecology of mesophotic coral reef ecosystems. Coral Reefs 29:255–275. https://doi.org/10.1007/s00338-010-0593-6

Kamenkovich VM, Koshlyakov MN, Monin AS (1986) Eddies in the open ocean. In: Synoptic eddies in the ocean. Springer, Netherlands, pp 265–376

Kane C, Kosaki RK, Wagner D (2014) High levels of mesophotic reef fish endemism in the Northwestern Hawaiian Islands. Bull Mar Sci 90(2):693–703

Kennett JP (1982) The geologic record of bottom currents. In: Marine Geology. Prentice Hall, Englewood Cliffs, NJ, pp 505–534

Kiriakoulakis K, Fisher E, Wolff GA et al (2006) Lipids and nitrogen isotopes of two deep-water corals from the North-East Atlantic: initial results and implications for their nutrition. Cold-Water Corals Ecosyst 715–729. https://doi.org/10.1007/3-540-27673-4_37

Kitahara MV, Cairns SD, Stolarski J, Blair D, Miller DJ (2010) A comprehensive phylogenetic analysis of the Scleractinia (Cnidaria, Anthozoa) based on mitochondrial CO1 sequence data. PLoS One 5(7):e11490

Kitahara MV, Fukami H, Benzoni F, Huang D (2016) The new systematics of scleractinia: integrating molecular and morphological evidence. In: Goffredo S, Dubinsky Z (eds) The cnidaria, past, present and future. Springer International, pp 41–59

Kleeman R, McCreary JP Jr, Klinger BA (1999) A mechanism for generating ENSO decadal variability. Geophys Res Lett 26(12): 1743–1746

Knauss JA (1996) Introduction to physical oceanography. Prentice Hall, Englewood Cliffs, NJ, p 338

Koenig CC, Coleman FC, Grimes CB et al (2000) Protection of fish spawning habitat for the conservation of warm temperate reef fish fisheries of shelf-edge reefs of Florida. Bull Mar Sci 66:593–616

Koppers AAP, Staudigel H (2005) Asynchronous bends in Pacific seamount trails: a case for extensional volcanism? Science 307: 904–907. https://doi.org/10.1126/science.1107260

Koppers AAP, Staudigel H, Pringle MS, Wijbrans JR (2003) Short-lived and discontinuous intraplate volcanism in the South Pacific: Hot spots or extensional volcanism? Geochem Geophys Geosyst 4. https://doi.org/10.1029/2003GC000533

Koslow JA, Gowlett-Holmes K, Lowry JK et al (2001) Seamount benthic macrofauna off southern Tasmania: community structure and impacts of trawling. Mar Ecol Progr Ser Mar Ecol Prog Ser 213:111–125. https://doi.org/10.3354/meps213111

Krieger KJ, Wing B (2002) Megafauna associated with deepwater (*Primnoa* spp.) in the Gulf of Alaska. In Watling L, Risk M (eds.) Biology of cold water corals. Hydrobiologia 471: 83–90.

Kurman MD, Gomez CE, Georgian SE, Lunden JJ, Cordes EE (2017) Intra-specific variation reveals potential for adaptation to ocean acidification in a cold-water coral from the Gulf of Mexico. Front Mar Sci 4:111. https://doi.org/10.3389/fmars.2017.00111

Larsson AI, van Oevelen D, Purser A, Thomsen L (2013) Tolerance to long-term exposure of suspended benthic sediments and drill cuttings in the cold-water coral *Lophelia pertusa*. Mar Pollut Bull 70:176–188. https://doi.org/10.1016/j.marpolbul.2013.02.033

Le Goff-Vitry MC, Rogers AD, Baglow D (2004) A deep-sea slant on the molecular phylogeny of the Scleractinia. Mol Phylogenet Evol 30:167–177. https://doi.org/10.1016/S1055-7903(03)00162-3

Leonard CL, Bidigare RR, Seki MP et al (2001) Interannual mesoscale physical and biological variability in the North Pacific Central Gyre. Prog Oceanogr 49(1–4):227–244

LeSueur CP (1821) Description de plusieurs animaux appartement aux polypiers lamellifères de M le Chev. de Lamarck. Mém Mus Hist Nat Paris 6:271–298

Levin LA, Mengerink K, Gjerde KM et al (2016) Defining "serious harm" to the marine environment in the context of deep-seabed mining. Mar Policy 74:245–259. https://doi.org/10.1016/j.marpol.2016.09.032

Levin LA, Wei CL, Dunn DC et al (2020) Climate change considerations are fundamental to management of deep-sea resource extraction. Glob Chang Biol 26:4664–4678. https://doi.org/10.1111/gcb.15223

Linnaeus C (1758) Tomus I. Systems naturae per regna tria naturae, secundum classes, ordines, genera, species, cum characteribus, differentiis, synonymis, locis. Ed decima, reformata Holmiae (Laurentii Salvii) 1:1–824

Long DJ, Baco AR (2014) Rapid change with depth in megabenthic structure-forming communities of the Makapu'u deep-sea coral bed. Deep Res Part II Top Stud Oceanogr 99:158–168. https://doi.org/10.1016/j.dsr2.2013.05.032

Lunden JJ, McNicholl CG, Sears CR, Morrison CL, Cordes EE (2014) Acute survivorship of the deep-sea coral *Lophelia pertusa* from the Gulf of Mexico under acidification, warming, and deoxygenation. Front Mar Sci 1:78. https://doi.org/10.3389/fmars.2014.00078

Maier C, Bils F, Weinbauer MG et al (2013a) Respiration of Mediterranean cold-water corals is not affected by ocean acidification as projected for the end of the century. Biogeosciences 10(8):5671–5680. https://doi.org/10.5194/bg-10-5671-2013

Maier C, Schubert A, Berzunza Sànchez MM et al (2013b) End of the century pCO2 levels do not impact calcification in mediterranean cold-water corals. PLoS One 8(4). https://doi.org/10.1371/journal.pone.0062655

Maier C, Popp P, Sollfrank N et al (2016) Effects of elevated pCO2 and feeding on net calcification and energy budget of the Mediterranean cold-water coral *Madrepora oculata*. J Exp Biol 219:3208–3217. https://doi.org/10.1242/jeb.127159

Maragos J, Miller J, Gove J et al (2008) US coral reefs in the Line and Phoenix Islands, Central Pacific Ocean: history, geology, oceanography, and biology. In: Coral reefs of the USA. Springer, Dordrecht, pp 595–641

Marschal C, Garrabou J, Harmelin JG, Pichon M (2004) A new method for measuring growth and age in the precious red coral *Corallium rubrum* (L.). Coral Reefs 23:423–432. https://doi.org/10.1007/s00338-004-0398-6

Matias MG, Underwood AJ, Hochuli DF et al (2011) Habitat identity influences species–area relationships in heterogeneous habitats. Mar Ecol Prog Ser 437:135–145

Matsuyama M, Ohta S, Hibiya T, Yamada H (1993) Strong tidal currents observed near the bottom in the Suruga Trough, central Japan. J Oceanogr 49:683–696. https://doi.org/10.1007/BF02276752

McPhaden MJ, Zhang D (2004) Pacific Ocean circulation rebounds. Geophys Res Lett 31(18)

Mejía-Mercado BE, Mundy B, Baco AR (2019) Variation in the structure of the deep-sea fish assemblages on Necker Island, Northwestern Hawaiian Islands. Deep Sea Res Part I Oceanogr Res Pap 152: 103086

Merrifield MA, Holloway PE (2002) Model estimates of M2 internal tide energetics at the Hawaiian Ridge. J Geophys Res Ocean 107:5–1. https://doi.org/10.1029/2001jc000996

Merrifield MA, Holloway PE, Johnston TMS (2001) The generation of internal tides at the Hawaiian Ridge. Geophys Res Lett 28:559–562. https://doi.org/10.1029/2000GL011749

Miller KJ, Gunasekera RM (2017) A comparison of genetic connectivity in two deep sea corals to examine whether seamounts are isolated islands or stepping stones for dispersal. Sci Rep 7:1–14. https://doi.org/10.1038/srep46103

Miller KJ, Rowden AA, Williams A, Häussermann V (2011) Out of their depth? isolated deep populations of the cosmopolitan coral *Desmophyllum dianthus* may be highly vulnerable to environmental change. PLoS One 6. https://doi.org/10.1371/journal.pone.0019004

Miller KA, Thompson KF, Johnston P, Santillo D (2018) An overview of seabed mining including the current state of development, environmental impacts, and knowledge gaps. Front Mar Sci 4. https://doi.org/10.3389/fmars.2017.00418

Miyamoto M, Kiyota M, Hayashibara T et al (2017) Megafaunal composition of cold-water corals and other deep-sea benthos in the southern Emperor Seamounts area, North Pacific Ocean. Galaxea J Coral Reef Stud 19:19–30. https://doi.org/10.3755/galaxea.19.1_19

Mohn C, Rengstorf A, White M et al (2014) Linking benthic hydrodynamics and cold-water coral occurrences: a high-resolution model study at three cold-water coral provinces in the NE Atlantic. Prog Oceanogr 122:92–104. https://doi.org/10.1016/j.pocean.2013.12.003

Mora C, Wei CL, Rollo A et al (2013) Biotic and human vulnerability to projected changes in ocean biogeochemistry over the 21st century. PLoS Biol 11(10). https://doi.org/10.1371/journal.pbio.1001682

Morato T, Watson R, Pitcher TJ, Pauly D (2006) Fishing down the deep. Fish Fish 7:24–34. https://doi.org/10.1111/j.1467-2979.2006.00205.x

Morato T, González-Irusta JM, Dominguez-Carrió C et al (2020) Climate-induced changes in the suitable habitat of cold-water corals and commercially important deep-sea fishes in the North Atlantic. Glob Chang Biol *26*(4):2181–2202. https://doi.org/10.1111/gcb.14996

Morgan LE, Etnoyer P, Scholz AJ et al (2005) Conservation and management implications of deep-sea coral and fishing effort distributions in the Northeast Pacific Ocean. In: Freiwald A, Roberts JM (eds) Cold-water corals and ecosystems. Springer, Berlin, pp 1171–1187

Morgan NB, Cairns S, Reiswig H, Baco AR (2015) Benthic megafaunal community structure of cobalt-rich manganese crusts on Necker Ridge. Deep Res I Oceanogr Res Pap 104. https://doi.org/10.1016/j.dsr.2015.07.003

Morgan NB, Goode S, Roark EB, Baco AR (2019) Fine scale assemblage structure of benthic invertebrate megafauna on the North Pacific Seamount Mokumanamana. Front Mar Sci 6. https://doi.org/10.3389/fmars.2019.00715

Mortensen PB, Hovland M, Brattegard T et al (1995) Deep water bioherms of the scleractinian coral *Lophelia pertusa* (L.) at 641N on the Norwegian shelf: structure and associated megafauna. Sarsia 80:145–158

Mortensen PB, Rapp HT, Båmstedt U (1998) Oxygen and carbon isotope ratios related to growth line patterns in skeletons of *Lophelia pertusa* (L) (Anthozoa, Scleractinia): Implications for determination of linear extension rate. Sarsia 83(5):433–446

Mortensen PB, Hovland T, Fossa JH, Furevik DM (2001) Distribution, abundance and size of *Lophelia pertusa* coral reefs in mid-Norway in relation to seabed characteristics. J Mar Biolog Assoc UK 81(4):581–597

Moseley HN (1881) Report on certain Hydroid, Alcyonarian and Madreporarian Corals procured during the voyage of HMS Challenger, vol 2

Mosher CV, Watling L (2009) Partners for life: a brittle star and its octocoral host. Mar Ecol Progr Ser 397:81–88. https://doi.org/10.3354/meps08113

Movilla J, Gori A, Calvo E et al (2014) Resistance of two mediterranean cold-water coral species to low-pH conditions. Water 6(1):59–67. https://doi.org/10.3390/w6010059

Mughal M (2012) *Eguchipsammia fistula* microsatellite development and population analysis. King Abdullah University of Science and Technology

Müller RD, Roest WR, Royer JY et al (1997) Digital isochrons of the world's ocean floor. J Geophys Res Solid Earth 102(B2):3211–3214

Mullineaux LS, Mills SW (1997) A test of the larval retention hypothesis in seamount-generated flows. Deep Sea Res Part I Oceanogr Res Pap 44:745–770

Mullins HT, Newton CR, Heath K, Van Buren HM (1981) Modern deep-water coral mounds north of Little Bahama Bank; criteria for recognition of deep-water coral bioherms in the rock record. J Sediment Res 51(3):999–1013

Musgrave RC, Pinkel R, MacKinnon JA et al (2016) Stratified tidal flow over a tall ridge above and below Mullineaux LS, Mills SW (1997) A test of the turning latitude. J Fluid Mech 793:933–957. https://doi.org/10.1017/jfm.2016.150

Nakamura H, Kazmin AS (2003) Decadal changes in the North Pacific oceanic frontal zones as revealed in ship and satellite observations. J. Geophys Res 108(C3):3078

National Museum of Natural History (NMNH), Smithsonian Institution, 10th and Constitution Ave. N.W., Washington, DC 20560-0193 (https://collections.nmnh.si.edu/search/) Accessed October 5, 2020.

Naumann MS, Orejas C, Wild C, Ferrier-Pagès C (2011) First evidence for zooplankton feeding sustaining key physiological processes in a scleractinian cold-water coral. J Exp Biol 214:3570–3576. https://doi.org/10.1242/jeb.061390

Naumann MS, Orejas C, Ferrier-Pagès C (2014) Species-specific physiological response by the cold-water corals *Lophelia pertusa* and *Madrepora oculata* to variations within their natural temperature range. Deep Sea Res Part II Top Stud Oceanogr 99:36–41

Naumann MS, Tolosa I, Taviani M et al (2015) Trophic ecology of two cold-water coral species from the Mediterranean Sea revealed by lipid biomarkers and compound-specific isotope analyses. Coral Reefs 34:1165–1175. https://doi.org/10.1007/s00338-015-1325-8

Neumann AC, Kofoed JW, Keller GH (1977) Lithoherms in the straits of Florida. Geology 5(1):4–10

NOAA Office of Ocean Exploration and Research Benthic Deepwater Animal Identification Guide. Available from http://oceanexplorer.noaa.gov/okeanos/animal_guide/animal_guide.html. Accessed October 5, 2020

O'Hara TD, Rowden AA, Williams A (2008) Cold-water coral habitats on seamounts: Do they have a specialist fauna? Divers Distrib 14:925–934. https://doi.org/10.1111/j.1472-4642.2008.00495.x

Obura D, Mangubhai S (2011) Coral mortality associated with thermal fluctuations in the Phoenix Islands, 2002–2005. Coral Reefs 30:607–619. https://doi.org/10.1007/s00338-011-0741-7

Obura D, Stone G, Mangubhai S, Bailey S, Yoshinaga H, Barrel R (2011) Baseline marine biological surveys of the Phoenix Islands, July 2000. Atoll Res Bull 589:1–62. https://doi.org/10.5479/si.00775630.589.1

Orejas C, Ferrier-Pagès C, Reynaud S et al (2011) Long-term growth rates of four Mediterranean cold-water coral species maintained in aquaria. Mar Ecol Prog Ser 429:57–65. https://doi.org/10.3354/meps09104

Pante E, France SC, Gey D, Cruaud C, Samadi S (2015) An inter-ocean comparison of coral endemism on seamounts: the case of Chrysogorgia. J Biogeogr 42:1907–1918. https://doi.org/10.1111/jbi.12564

Parrish FA (2006) Precious corals and subphotic fish assemblages. Atoll Res Bull

Parrish FA (2007) Density and habitat of three deep-sea corals in the lower Hawaiian chain. Bull Mar Sci 81:185–194

Parrish FA (2015) Settlement, colonization, and succession patterns of gold coral *Kulamanamana haumeaae* in Hawaiian deep coral assemblages. Mar Ecol Prog Ser 533:135–147. https://doi.org/10.3354/meps11390

Parrish FA, Baco AR (2007) State of deep coral ecosystems in the US Pacific Islands region: Hawaii and the US Pacific Territories. In: Lumsden SE, Hourigan TF, Bruckner AW, Dorr G (eds). National Oceanic and Atmospheric Administration, pp 115–194

Parrish FA, Oliver TA (2020) Comparative observations of current flow, tidal spectra, and scattering strength in and around Hawaiian Deep-Sea Coral Patches. Front Mar Sci 7:1–16. https://doi.org/10.3389/fmars.2020.00310

Parrish FA, Roark EB (2009) Growth validation of gold coral *Gerardia* sp. in the Hawaiian Archipelago. Mar Ecol Progr Ser 397:163–172

Parrish FA, Baco AR, Kelley C, Reiswig H (2015) State of deep-sea coral and sponge ecosystems of the U.S. Pacific Islands Region. In: Hourigan TF, Etnoyer PJ, Cairns SD (eds) The state of deep-sea coral and sponge ecosystems of the United States. NOAA Technical Memorandum NMFS-OHC-4, Silver Springs, MD, (Chapter 7), pp 1–38

Parrish FA, Baco-Taylor A, Kelley C, et al (2017) Deep-Sea Coral Taxa in the Hawaiian Archipelago and other US Pacific Islands: depth and geographical distribution.

Parrish FA, Hourigan TF, Baco-Taylor AR, Kelley CD, Cairns SD (2022) List of Deep-Sea Coral Taxa in the Hawaiian Archipelago and Johnston Atoll: Depth and Geographic Distribution (v. 2021). Online Supplement to The state of deep-sea coral and sponge ecosystems of the United States. https://doi.org/10.25923/wj31-g055

Petersen S, Krätschell A, Augustin N et al (2016) News from the seabed – geological characteristics and resource potential of deep-sea mineral resources. Mar Pol 70:175–187. https://doi.org/10.1016/j.marpol.2016.03.012

Petruncio E, Weinnig A, Peters C, et al (2020) Pacific Remote Islands Marine National Monument: exploring the Howland and Baker Island unit and Johnston Atoll unit. In: New frontiers in ocean exploration: the E/V nautilus, NOAA Ship Okeanos Explorer, and R/V Falkor 2019 Field season, Raineault N, Flanders J (eds) Oceanography (33)1:46–47. https://doi.org/10.5670/oceanog.2020.supplement.01

Pires DO, Silva JC, Bastos ND (2014) Reproduction of deep-sea reef-building corals from the southwestern Atlantic. Deep Res Part II Top Stud Oceanogr 99:51–63. https://doi.org/10.1016/j.dsr2.2013.07.008

Polovina JJ (2005) Climate variation, regime shifts, and implications for sustainable fisheries. Bull Mar Sci 76:233–2144

Polovina JJ, Haight WR (1999) Climate variation, ecosystem dynamics, and fisheries management in the NWHI. Ecosystem approaches for fisheries management. Alaska Sea Grant College Program AK-SG-99-01.

Polovina JJ, Howell E, Kobayashi DR et al (2001) The Transition Zone Chlorophyll Front, a dynamic global feature defining migration and forage habitat for marine resources. Prog Oceanogr 49(1–4):469–483

Pourtalès LD (1867) Contributions to the fauna of the Gulf Stream at great depths. Bull Mus Comp Zool 1(6):1e

Pourtalès LD (1878) Reports on the results of the dredgings by the "Blake". Crinoids and corals. Bull Harv Univ Mus Comp Zool 5:197–212

Putts MR, Parrish FA, Trusdell FA, Kahng SE (2019) Structure and development of Hawaiian deep-water coral communities on Mauna Loa lava flows. Mar Ecol Prog Ser 630:69–82. https://doi.org/10.3354/meps13106

Pyle RL, Boland R, Bolick H et al (2016) A comprehensive investigation of mesophotic coral ecosystems in the Hawaiian Archipelago. PeerJ 4. https://doi.org/10.7717/peerj.2475

Raddatz J, Titschack J, Frank N et al (2020) *Solenosmilia variabilis*-bearing cold-water coral mounds off Brazil. Coral Reefs 39(1):69–83

Reed JK (1981) In situ growth rates of the scleractinian coral *Oculina varicosa* occurring with zooxanthellae on 6-m reefs and without on 80-m banks. In Proceedings of the fourth international coral reef symposium, vol 2, pp 201–206

Reed JK (2002) Comparison of deep-water coral reefs and lithoherms off southeastern USA. Hydrobiologia 471(1–3):57–69

Reed JK (2004) Deep-water coral reefs of Florida, Georgia and South Carolina: a summary of the distribution, habitat, and associated fauna. Report prepared for the South Atlantic Fishery Management Council. Appendix A, Comprehensive ecosystem-based amendment 1

Reed JK, Weaver DC, Pomponi SA (2006) Habitat and fauna of deep-water *Lophelia pertusa* coral reefs off the southeastern US: Blake Plateau, Straits of Florida, and Gulf of Mexico. Bull Mar Sci 78(2):343–375

Rix L, De Goeij JM, Mueller CE et al (2016) Coral mucus fuels the sponge loop in warm-and cold-water coral reef ecosystems. Sci Rep 6:1–11. https://doi.org/10.1038/srep18715

Roark EB, Guilderson TP, Flood-Page S et al (2005) Radiocarbon-based ages and growth rates of bamboo corals from the Gulf of Alaska. Geophys Res Lett 32:L04606. https://doi.org/10.1029/2004GL021919

Roark EB, Guilderson TP, Dunbar RB, Ingram BL (2006) Radiocarbon-based ages and growth rates of Hawaiian deep-sea corals. Mar Ecol Ser 327:1–37. https://doi.org/10.3354/meps327001

Roark EB, Guilderson TP, Dunbar RB et al (2009) Extreme longevity in proteinaceous deep-sea corals. Proc Natl Acad Sci U S A 106:5204–5208. https://doi.org/10.1073/pnas.0810875106

Roberts S, Hirshfield M (2004) Deep-sea corals: out of sight, but no longer out of mind. Front Ecol Environ 2(3):123–130

Roberts JM, Wheeler AJ, Freiwald A (2006) Reefs of the deep: the biology and geology of cold-water coral ecosystems. Science 312(5773):543–547. https://doi.org/10.1126/science.1119861

Roberts JM, Wheeler A, Freiwald A, Cairns S (2009) Cold-water corals: the biology and geology of deep-sea coral habitats. Cambridge University Press

Roberts JM, Murray F, Anagnostou E et al (2016) Cold-water corals in an era of rapid global change: Are these the deep ocean's most vulnerable ecosystems? In: Goffredo S, Dubinsky Z (eds) The Cnidaria, past, present and future. Springer, Cham, pp 593–606

Rogers AD (1999) The biology of *Lophelia pertusa* (Linnaeus 1758) and other deep-water reef-forming corals and impacts from human activities. Int Rev Hydrobiol 84:315–406. https://doi.org/10.1002/iroh.199900032

Rogers AD (2018) The biology of seamounts: 25 years on. Adv Mar Biol 79:137–224. https://doi.org/10.1016/bs.amb.2018.06.001

Rogers AD, Baco A, Griffiths H, et al (2007) Corals on seamounts. In: Pitcher TJ, Morato T, Hart PJB, et al. (eds). Blackwell Publishing, Oxford, pp 141–169

Rooney J, Donham E, Montgomery A et al (2010) Mesophotic coral ecosystems in the Hawaiian archipelago. Coral Reefs 29:361–367. https://doi.org/10.1007/s00338-010-0596-3

Ross SW, Quattrini AM (2007) The fish fauna associated with deep coral banks off the southeastern United States. Deep Sea Res Part I Oceanogr Res Pap 54(6):975–1007. https://doi.org/10.1016/j.dsr.2007.03.010

Rotjan R, Jamieson R, Carr B et al (2014) Establishment, management, and maintenance of the phoenix islands protected area. Adv Mar

Biol 69:289–324. https://doi.org/10.1016/B978-0-12-800214-8.00008-6
Rowden AA, Anderson OF, Georgian SE et al (2017) High-resolution habitat suitability models for the conservation and management of vulnerable marine ecosystems on the Louisville Seamount Chain, South Pacific Ocean. Front Mar Sci 4. https://doi.org/10.3389/fmars.2017.00335
Rudnick DL, Boyd TJ, Brainard RE et al (2003) From tides to mixing along the Hawaiian Ridge. Science 301:355–357. https://doi.org/10.1126/science.1085837
Sars M (1865) Om de I Norge forekommende fossile dyrelevniger fra Quartaerperioden. Christiania: University program første halvaar 1864, pp 1–134
Schlacher TA, Baco AR, Rowden AA et al (2014) Seamount benthos in a cobalt-rich crust region of the central Pacific: conservation challenges for future seabed mining. Divers Distrib 20:491–502. https://doi.org/10.1111/ddi.12142
Schlining K, von Thun S, Kuhnz L et al (2013) Debris in the deep: using a 22-year video annotation database to survey marine litter in Monterey Canyon, central California, USA. Deep Sea Res Part I Oceanogr Res Pap 79:96–105. https://doi.org/10.1016/j.dsr.2013.05.006
Schmitz Jr WJ (1996) On the world ocean circulation: volume 2 The Pacific and Indian Oceans: A global update (No. WHOI-96-08). Woods Hole Oceanographic Institution MA
Schott FA, Mccreary JP, Johnson GC (2004) Shallow overturning circulations of the tropical-subtropical oceans. Earth climate: The ocean–atmosphere interaction. Geophys Monogr 147:261–304. https://doi.org/10.1029/147GM15
Schröder-Ritzrau A, Freiwald A, Mangini A (2005) U/Th-dating of deep-water corals from the eastern North Atlantic and the western Mediterranean Sea. In: Friewald A, Roberts JM (eds) Cold-water corals and ecosystems. Springer, Berlin, pp 157–172
Seki MP, Lumpkin R, Flament P (2002) Hawaii cyclonic eddies and blue marlin catches: the case study of the 1995 Hawaiian International Billfish tournament. J Oceanogr 58:739–745. https://doi.org/10.1023/A:1022854609312
Spearman J, Taylor J, Crossouard N et al (2020) Measurement and modelling of deep sea sediment plumes and implications for deep sea mining. Sci Rep 10(1):1–14
Squires DF (1964) Fossil coral thickets in Wairarapa, New Zealand. J Paleontol 38(5):904–915
Stanley GD (1979) Paleoecology, structure, and distribution of coral buildups in western North America. U Kansas Paleo Contrib 65:1–68
Stanley GD, Cairns SD (1988) Constuctional azooxanthellate coral communities: an overview with implications for the fossil record. Palaios 3:233–242
Sweetman AK, Thurber AR, Smith CR, et al (2017) Major impacts of climate change on deep-sea benthic ecosystems. Elem Sci Anthr 5: Art. No. 4. doi:https://doi.org/10.1525/elementa.203
Talley LD (1993) Distribution and Formation of North Pacific Intermediate Water. J Phys Oceanogr 23:517–537. https://doi.org/10.1175/1520-0485(1993)023<0517:DAFONP>2.0.CO;2
Tchernia P (1980) Descriptive regional oceanography. Pergamon Press
Teichert C (1958) Cold- and deep-water coral banks. Bul Am Petrol Geol 42:1064–1082
Thresher RE, Tilbrook BD, Fallon S et al (2011) Effects of chronic low carbonate saturation levels on the distribution, growth and skeletal chemistry of deep-sea corals and other seamount megabenthos. Mar Ecol Ser 442:87–99. https://doi.org/10.3354/meps09400
Thurman HV (1981) Introductory oceanography, 3rd edn. Charles E. Merrill Publishing Company
Tissot BN, Yoklavich MM, Love MS et al (2006) Benthic invertebrates that form habitat on deep banks off southern California, with special reference to deep sea coral. Fish Bull 104:167–181
Tittensor DP, Baco-Taylor AR, Brewin P, Clark MR, Consalvey M, Hall-Spencer J, Rowden AA, Schlacher T, Stocks K, Rogers AD (2009) Predicting global habitat suitability for stony corals on seamounts. J Biogeogr 36:1111–1128
Tittensor DP, Baco AR, Hall-Spencer JM et al (2010) Seamounts as refugia from ocean acidification for cold-water stony corals. Mar Ecol 31:212–225. https://doi.org/10.1111/j.1439-0485.2010.00393.x
Tracey D, Rowden A, Mackay K, Compton T (2011) Habitat-forming cold-water corals show affinity for seamounts in the New Zealand region. Mar Ecol Prog Ser 430:1–22. https://doi.org/10.3354/meps09164
UNGA (2006) Resolution 61/105 Sustainable fisheries, including through the 1995 Agreement for the Implementation of the Provisions of the United Nations Convention on the Law of the Sea of 10 December 1982 relating to the Conservation and Mangement of Straddling Fish Stocks and Highly Migratory Fish Stocks, and related instruments. UNGA A/RES/61/105. Available from: https://www.un.org/en/ga/search/view_doc.asp?symbol=A/RES/61/105&Lang=E
University of Hawai'i Undersea Research Laboratory (HURL) Deep-sea Animal Identification Guide. Available from http://www.soest.hawaii.edu/HURL/HURLarchive/guide.php. Accessed July 2020
Van Oevelen D, Duineveld G, Lavaleye M et al (2009) The cold-water coral community as a hot spot for carbon cycling on continental margins: a food-web analysis from rockall bank (northeast atlantic). Limnol Oceanogr 54:1829–1844. https://doi.org/10.4319/lo.2009.54.6.1829
Vargas-Ángel B, Looney EE, Vetter OJ, Coccagna EF (2011) Severe, widespread El Niño-associated coral bleaching in the US Phoenix Islands. Bull Mar Sci 87:623–638. https://doi.org/10.5343/bms.2010.1095
Vaughan TW (1907) Recent Madreporarians of the Hawaiian Islands and Laysan. Bull US Nat Mus 59:427
Venegas RM, Oliver T, Liu G et al (2019) The rarity of depth refugia from coral bleaching heat stress in the Western and Central Pacific Islands. Sci Rep 9:19710. https://doi.org/10.1038/s41598-019-56232-1
Victorero L, Watling L, Palomares MLD, Nouvian C (2018) Out of sight, but within reach: a global history of bottom-trawled deep-sea fisheries from > 400 m depth. Front Mar Sci 5:98. https://doi.org/10.3389/fmars.2018.00098
Waller RG (2005) Deep-water Scleractinia (Cnidaria: Anthozoa): current knowledge of reproductive processes. In: Friewald A, Roberts JM (eds) Cold-water corals and ecosystems. Springer, Berlin, pp 691–700
Waller RG, Tyler PA (2005) The reproductive biology of two deep-water, reef-building scleractinians from the NE Atlantic Ocean. Coral Reefs 24:514–522. https://doi.org/10.1007/s00338-005-0501-7
Waller RG, Tyler PA, Gage JD (2005) Sexual reproduction in three hermaphroditic deep-sea Caryophyllia species (Anthozoa: Scleractinia) from the NE Atlantic Ocean. Coral Reefs 24:594–602. https://doi.org/10.1007/s00338-005-0031-3
Watling L, France SC, Pante E, Simpson A (2011) Biology of deep-water octocorals. Adv Mar Biol. 60:41–122. https://doi.org/10.1016/B978-0-12-385529-9.00002-0
Watling L, Guinotte J, Clark MR, Smith CR (2013) A proposed biogeography of the deep ocean floor. Prog Oceanogr 111:91–112. https://doi.org/10.1016/j.pocean.2012.11.003
Wells JW (1954) Recent corals of the Marshall Islands: Bikini and nearby atolls, Part 2, Oceanography. US Geol Surv Prof Pap 260 (I):382–486
Wishner K, Levin L, Gowing M, Mullineaux L (1990) Involvement of the oxygen minimum in benthic zonation on a deep seamount. Nature 346:57–59

Yesson C, Taylor ML, Tittensor DP et al (2012) Global habitat suitability of cold-water octocorals. J Biogeogr 39:1278–1292. https://doi.org/10.1111/j.1365-2699.2011.02681.x

Yesson C, Bedford F, Rogers AD, Taylor ML (2017) The global distribution of deep-water Antipatharia habitat. Deep Sea Res Part II Top Stud Oceanogr 145:79–86. https://doi.org/10.1016/j.dsr2.2015.12.004

Yesson C, Letessier TB, Nimmo-Smith A, Hosegood P, Brierley AS, Hardouin M, Proud R (2021) Improved bathymetry leads to >4000 new seamount predictions in the global ocean – but beware of phantom seamounts! UCL Open: Environment 2021(4):03. Available from: https://doi.org/10.14324/111.444/ucloe.000030

Zeng C, Rowden AA, Clark MR, Gardner JPA (2017) Population genetic structure and connectivity of deep-sea stony corals (Order Scleractinia) in the New Zealand region: implications for the conservation and management of vulnerable marine ecosystems. Evol Appl 10:1040–1054. https://doi.org/10.1111/eva.12509

Zhang R-H, Levitus S (1997) Structure and cycle of decadal variability of upper-ocean temperature in the North Pacific. J Clim 10:710–727. https://doi.org/10.1175/1520-0442(1997)010<0710:SACODV>2.0.CO;2

Zhang R-H, Rothstein LM, Busalacchi AJ (1998) Origin of upper-ocean warming and El Niño change on decadal scales in the tropical Pacific Ocean. Nature 391:879–883. https://doi.org/10.1038/36081

GPSR Compliance

The European Union's (EU) General Product Safety Regulation (GPSR) is a set of rules that requires consumer products to be safe and our obligations to ensure this.

If you have any concerns about our products, you can contact us on ProductSafety@springernature.com

In case Publisher is established outside the EU, the EU authorized representative is:

Springer Nature Customer Service Center GmbH
Europaplatz 3
69115 Heidelberg, Germany

Batch number: 10776259

Printed by Printforce, the Netherlands